About Island Press

Island Press is the only nonprofit organization in the United States whose principal purpose is the publication of books on environmental issues and natural resource management. We provide solutions-oriented information to professionals, public officials, business and community leaders, and concerned citizens who are shaping responses to environmental problems.

In 2005, Island Press celebrates its twenty-first anniversary as the leading provider of timely and practical books that take a multidisciplinary approach to critical environmental concerns. Our growing list of titles reflects our commitment to bringing the best of an expanding body of literature to the environmental community throughout North America and the world.

Support for Island Press is provided by the Agua Fund, Brainerd Foundation, Geraldine R. Dodge Foundation, Doris Duke Charitable Foundation, Educational Foundation of America, The Ford Foundation, The George Gund Foundation, The William and Flora Hewlett Foundation, Henry Luce Foundation, The John D. and Catherine T. MacArthur Foundation, The Andrew W. Mellon Foundation, The Curtis and Edith Munson Foundation, National Environmental Trust, The New-Land Foundation, Oak Foundation, The Overbrook Foundation, The David and Lucile Packard Foundation, The Pew Charitable Trusts, The Rockefeller Foundation, The Winslow Foundation, and other generous donors.

The opinions expressed in this book are those of the author(s) and do not necessarily reflect the views of these foundations.

About World Wildlife Fund

Known worldwide by its panda logo, World Wildlife Fund is dedicated to protecting the world's wildlife and the rich biological diversity that we all need to survive. The leading privately supported international conservation organization in the world, WWF has more than 30 national affiliates, programs in more than 100 countries, and a global membership of more than 5 million people, including 1.2 million in the United States.

WWF's mission is to stop the degradation of the planet's natural environment and to build a future in which humans live in harmony with nature by conserving the world's biological diversity, ensuring that the use of renewable natural resources is sustainable, and promoting the reduction of pollution and wasteful consumption. From working to save the giant panda, tiger, and rhino to helping establish and manage parks and reserves worldwide, WWF has been a conservation leader for 40 years. To learn more about WWF, visit www.worldwildlife.org.

About the African Wildlife Foundation

Founded in 1961, the African Wildlife Foundation (AWF) is the leading conservation organization focused solely on the African continent. AWF designs and implements conservation strategies that are based on science and that are compatible with human benefit. Since its inception AWF has ensured the continued existence of some of Africa's most rare species. AWF has also invested in training hundreds of African individuals who have gone on to play critical roles in conservation. AWF has pioneered the use of community conservation to demonstrate that wildlife can be conserved while people's well-being is also improved. To learn more about AWF, visit www.awf.org.

Freshwater Ecoregions
of Africa and Madagascar

Freshwater Ecoregions of Africa and Madagascar

A Conservation Assessment

Michele L. Thieme
Robin Abell
Melanie L. J. Stiassny
Paul Skelton
Bernhard Lehner
Guy G. Teugels
Eric Dinerstein
Andre Kamdem Toham
Neil Burgess
David Olson

World Wildlife Fund
United States

Washington • Covelo • London

Copyright © 2005 World Wildlife Fund

All rights reserved under International and Pan-American Copyright Conventions. No part of this book may be reproduced in any form or by any means without permission in writing from the publisher: Island Press, 1718 Connecticut Ave. NW, Suite 300, Washington, DC 20009.

ISLAND PRESS is a trademark of The Center for Resource Economics.

Library of Congress Cataloging-in-Publication data.

Freshwater ecoregions of Africa and Madagascar : a conservation assessment / Michele L. Thieme . . . [et al.].
    p. cm.
  Includes bibliographical references and index.
  ISBN 1-55963-365-4 (cloth : alk. paper)
  1. Biological diversity conservation—Africa.
2. Ecological regions—Africa. 3. Freshwater ecology—Africa. 4. Biological diversity conservation—Madagascar.
5. Ecological regions—Madagascar. 6. Freshwater ecology—Madagascar. I. Thieme, Michele L.
QH77.A35F74 2005
333.95'2816'096—dc22   2004026713

British Cataloguing-in-Publication data available.

Printed on recycled, acid-free paper

Design by Bea Hartman, BookMatters

Manufactured in the United States of America
10 9 8 7 6 5 4 3 2 1

This book is dedicated to all the individuals and institutions who work to conserve the freshwater ecosystems of Africa.

In memory of Guy Teugels, ichthyologist at the Africa Museum (Tervuren, Belgium), who was a leader in expanding our knowledge of African freshwater fish and who championed the building of a network of professional African freshwater conservationists.

Contents

List of Special Essays / ix

List of Figures / xi

List of Tables / xiii

List of Boxes / xv

Acronyms and Abbreviations / xvii

Preface / xix

Acknowledgments / xxi

1. Introduction / 1
   The Evolution and Diversity of Africa's Freshwater Systems / 2
   Threats / 5
   Structure of the Book / 5

2. Approach / 22
   Geographic Scope of the Study / 22
   Ecoregions and Bioregions / 22
   Freshwater Habitat Types / 23
   Conceptual Foundations / 27
   Elements of Analysis: Biological Distinctiveness and Conservation Status Indexes / 29
   Integrating Biological Distinctiveness and Conservation Status / 33

3. Biological Distinctiveness of African Ecoregions / 35
   Africa and Madagascar's Fresh Waters: Teeming with Life / 35
   Species Biological Values / 35
   Combining Species Richness and Endemism: Preliminary Biological Distinctiveness / 41
   Nonspecies Biological Values / 41
   Synthesis of Biological Distinctiveness Data / 44
   Conclusions / 46

4. Reversing the Flow: Conservation Status of Africa's Freshwater Ecoregions / 71
   Snapshot Conservation Status / 71
   Future Threat Assessment / 75
   Final Conservation Status / 76

5. Setting Priorities for Biodiversity Conservation among Africa's Freshwater Ecoregions / 90
   The Priority-Setting Matrix / 90
   Integrating Biological Importance and Conservation Status Indexes: Where to Act First / 91
   Priority Ecoregions / 94
   Comparison with Other Priority-Setting Exercises / 99
   Summary / 101

6. Africa's Freshwater Systems and Their Future / 103
   Who's Watching the Water? / 103
   Conservation Planning in Ecoregions and River Basins / 103
   Visionary Work in Freshwater Ecoregions and River Basins / 112
   Challenges to Freshwater Conservation and Sustainable Development in Africa and Madagascar / 112

Appendixes
- A. Methods for Assessing the Biological Distinctiveness of Freshwater Ecoregions / 139
- B. Methods for Assessing the Conservation Status of Freshwater Ecoregions / 145
- C. Data on the Species Richness and Endemism of Ecoregions / 149
- D. Data on the Biological Distinctiveness, Conservation Status, and Priority Class of Ecoregions / 157
- E. Statistical Analyses of Biological Distinctiveness and Conservation Status Data / 165
- F. Ecoregion Descriptions / 173

Glossary / 361

Literature Cited / 373

List of Authors / 417

Index / 425

List of Special Essays

1.1 Contribution of Inland Fisheries to Rural Livelihoods and Food Security in Africa: An Overview / 6
Christophe Béné

1.2 The Economic Value of Africa's Wetlands / 11
Lucy Emerton

1.3 Dragonflies: Sensitive Indicators of Freshwater Health / 19
Michael J. Samways

3.1 The Cichlid Fish Radiations of East Africa: A Model for Understanding Origin and Loss of Biodiversity / 48
Ole Seehausen

3.2 Freshwater Fish Biodiversity in the Congo Basin / 51
Guy G. Teugels and Michele L. Thieme

3.3 Baseline Fish Biodiversity Surveys: African Wildlife Foundation Experiences from the Zambezi River, Southern Africa / 53
Jimmiel Mandima and Henry Mwima

3.4 Cichlid Species Flocks in Small Cameroonian Lakes / 58
Uli Schliewen

3.5 The Niger Delta in Nigeria: A Hotspot and a Crossroad for Freshwater Fish Biodiversity / 60
Guy G. Teugels and C. Bruce Powell

3.6 Madagascar's Freshwater Fishes: An Imperiled Treasure / 62
John S. Sparks and Melanie L. J. Stiassny

4.1 Invasive Alien Species in African Freshwater Ecosystems / 78
Musonda Mumba and Geoffrey Howard

4.2 Impacts of Fishing on Inland Fish Populations in Africa / 81
Robin Welcomme

4.3 Climate Change, Human Water Use, and Freshwater Ecosystems in Africa: Looking toward the Future / 86
Bernhard Lehner

5.1 Safeguarding Human Health, Safeguarding Aquatic Snail Diversity: Conflicting Goals? / 101
David S. Brown

6.1 The Convention on Wetlands (Ramsar): An International Treaty Sparks Freshwater Conservation / 116
Aboubacar Awaiss and Denis Landenbergue

6.2 Lake Naivasha, Kenya: Community Management of a Ramsar Site / 119
IUCN Eastern Africa Regional Programme, Nairobi, Kenya

6.3 African Wildlife Foundation Experience in the Management of Fishery Resources in Two Southern African Landscapes / 123
Henry Mwima and Jimmiel Mandima

6.4 Fish out of Water? Competing for Water for Africa's Freshwater Ecosystems / 128
Patrick Dugan

6.5 Freshwater Ornamental Fishes: A Rural Livelihood Option for Africa? / 132
Randall E. Brummett

6.6 Some, for All, Forever: A New South African Water Law / 136
Carolyn (Tally) G. Palmer

List of Figures

1.1　Freshwater Ecoregions of Africa and Madagascar with Landcover / 3

1.2　Elevation Zones and Freshwater Realms of Africa and Madagascar / 4

1.3　Economic Values of Wetland Ecosystems / 12

2.1　Freshwater Ecoregions of Africa and Madagascar / 24

2.2　Freshwater Bioregions and Major Habitat Types / 28

2.3　Integration Matrix for Priority Setting / 34

3.1　Total Species Richness, Number of Endemic Species, and Percentage of Endemic Species / 36

3.2　Average Total Richness by Major Habitat Type / 36

3.3　Fish Richness, Number of Endemic Fish Species, and Percentage of Endemic Fish Species / 37

3.4　Aquatic Mollusk Richness, Number of Endemic Mollusk Species, and Percentage of Endemic Mollusk Species / 38

3.5　Herpetofauna Richness, Number of Endemic Herpetofauna Species, and Percentage of Endemic Herpetofauna Species / 39

3.6　Data Quality of Freshwater Ecoregions / 40

3.7　Preliminary Biological Distinctiveness Index and Rare Ecological or Evolutionary Phenomena / 42

3.8　Final Biological Distinctiveness Index / 43

3.9　The Niger Delta in Nigeria / 61

3.10　Species-Area Relationships for Madagascar and Other Oceanic and Continental Landmasses / 63

3.11　Plot of the Number of Native and Endemic Freshwater Fish Species Recorded from Madagascar in Recent Surveys / 68

3.12　Plot of the Total Number of Fish Species by Ecoregion for Madagascar / 68

3.13　Five Major Hydrographic Ecoregions Recognized on the Basis of Ichthyofaunal Diversity and Hydrographic Structure, Comprising the Eastern Highlands, Eastern Lowlands, Western Basins, Rivers and Lakes of the Northwest, and Southern Basins / 69

3.14　Some of the Highly Stenotopic Endemic Freshwater Fishes of Madagascar / 69

4.1　Estimated Land-Based Threats, Aquatic Habitat Threats, Biota Threats, and Future Threat Level / 72

4.2　Population Density and Major Dams / 73

4.3　Snapshot Conservation Status / 75

4.4　Final Conservation Status / 77

4.5　Number of Species as a Function of Lake Basin Area for 31 African Lakes and Number of Species as a Function of River Basin Area for 24 African Rivers / 82

4.6　Size Structure of Fish Assemblages in West Rivers / 82

4.7　Relative Percentages of Ichthyophages and Non-Ichthyophages by Maximum Length Class / 83

4.8 Illustration of the Fishing-Down Process Using the Historical Condition of the Oueme River Fishery (West Africa) and Representative Species / 83

4.9 Changes in Basic Characteristics of Fish Assemblages in Response to Increasing Exploitation Pressure / 84

4.10 Results of WaterGAP Calculations for 2025 / 88

4.11 Percentage of Ecoregion Area with Discharge Increases or Decreases of More Than 10 Percent by 2025 / 89

5.1 Priority Classes of Freshwater Ecoregions Using Final Conservation Status / 92

5.2 Populated Integration Matrix / 93

5.3 Percentages of North American and African Ecoregions in Each Priority Class / 93

5.4 Distribution of Data Quality Levels by Priority Class / 93

5.5 Number of Ecoregions by Major Habitat Type in Each Priority Class / 95

5.6 Overlap of Freshwater and Terrestrial Biological Distinctiveness Index and Overlap of Freshwater and Terrestrial Priority Classes / 100

6.1 Areas of Biological Importance in the Niger River Basin / 106

6.2 Freshwater Priority Areas in the Guinean-Congolian Freshwater Region / 107

6.3 Final Priority Areas for Biodiversity Conservation in Lake Malawi/Niassa/Nyasa / 109

6.4 African Wildlife Foundation Heartlands / 124

6.5 The Four Corners Transboundary Natural Resource Management Area / 125

6.6 The Zambezi Heartland / 126

6.7 In the Interaction Among Social, Economic, and Environmental Issues the Goal is to Increase the Intersecting Domain of Sustainability / 137

6.8 Options for Goods and Services Offered by Water Resources / 137

6.9 Summary Diagram of Flexible Water Resource Protection and Use / 138

A.1 Steps for Evaluating Biological Distinctiveness of Ecoregions / 140

A.2 Example of Method Used to Determine Natural Breaks for Number of Species or Number of Endemics in a Major Habitat Type / 142

E.1 Relationship between Ecoregion Area and Species Richness in the 93 Ecoregions / 169

List of Tables

1.1 Contribution of Fisheries of the Major River Basins and Lakes in West and Central to Employment and Income / 8

1.2 Contribution of Fisheries to Households' Cash Income in Different Parts of the Zambezi Basin / 8

1.3 Fish Consumption in African Countries with Moderate or Significant Inland Fisheries / 9

1.4 Some Threatened Dragonflies of Africa and Neighboring Islands / 20

2.1 Area of Major Habitat Types / 28

2.2 Categories of Threats Used to Assess the Integrity of Each Ecoregion and Examples for Each Category / 31

3.1 Evolutionary Phenomena and Rare Habitat Types in Freshwater Ecoregions / 44

3.2 Inland Sites with Globally or Continentally Outstanding Congregations of Wetland Birds in Africa / 46

3.3 Phylogenetic Diversity, Species Diversity, and Speciation Rates in Cichlid Radiations in East African Great Lakes / 49

3.4 Checklist of Native Malagasy Freshwater Fishes and Their Regions of Occurrence / 64

3.5 Occurrence of Native and Exotic Species at Selected Localities in Madagascar, 1989–1994 / 70

4.1 Distribution of Ecoregions by MHT for Snapshot and Final Conservation Status / 75

4.2 Examples of Infested Waterbodies in Some Freshwater Ecoregions of Africa / 79

4.3 Economic Costs and Control Expenditures Associated with Water Hyacinth / 79

6.1 Examples of Basin Projects with an Ecosystem Focus / 104

A.1 Data Sources for Biological Distinctiveness Analysis / 141

E.1 One-Way ANOVAs Testing for Differences between Major Habitat Types / 165

E.2 Post Hoc Mean Pairwise Comparisons for Fish Richness, Endemism, and Percentage Endemism / 166

E.3 Post Hoc Mean Pairwise Comparisons for Aquatic Mollusk Richness, Endemism, and Percentage Endemism / 166

E.4 Post Hoc Mean Pairwise Comparisons for Herpetofauna Richness, Endemism, and Percentage Endemism / 168

E.5 Pearson Correlations of Richness and Endemism for Each Taxon / 168

E.6 Area and Latitude of Ecoregions / 170

E.7 Results of Pearson Correlation Analyses of Log (Ecoregion Area) and Total Richness among Major Habitat Types / 171

List of Boxes

1.1 The Role of Wetlands in a Rural Economy: Pallisa District, Uganda / 13

1.2 The Economic Value of Wetland Services for an Urban Population: Nakivubo Swamp, Uganda / 14

1.3 Counting the Economic Costs of Freshwater Ecosystem Degradation: The Tana River Hydroelectric Schemes, Kenya / 15

1.4 Modeling Alternative Wetland Conservation and Development Scenarios for the Barotse Floodplain, Zambia / 16

1.5 Providing an Economic Justification for Freshwater Ecosystem Restoration: The Waza Logone Floodplain, Cameroon / 17

1.6 Identifying the Needs for Financial and Economic Incentives for Wetland Conservation: Lake Mburo National Park, Uganda / 18

2.1 Descriptions of the Freshwater Major Habitat Types of Africa and Madagascar / 26

3.1 Ecological Phenomena: The Case of Waterbirds / 45

3.2 Inland Deltas: Globally Rare Habitat Types? / 47

4.1 The Cost of Inaction: Nigeria / 80

6.1 Examples of Biodiversity Visions in Three African Freshwater Ecoregions / 105

6.2 Big Water: The Challenge of Conserving Biodiversity in Africa's Large Rivers and Lakes / 110

Acronyms and Abbreviations

AMCEN	African Ministerial Conference on the Environment	FAO	Food and Agriculture Organization
ANOVA	analysis of variance	GCM	general circulation model
ARWG	Aquatic Resources Working Group	GEF	Global Environment Facility
AWF	African Wildlife Foundation	GIS	geographic information system
BDI	Biological Distinctiveness Index	GISP	Global Invasive Species Program
CAR	Central African Republic	HCP	Heartland Conservation Process
CAS	catch assessment survey	IBA	Important Bird Area
CAW	Centre for African Wetlands	IBT	interbasin transfer
CEGEN	Centre for Environmental Management of Mount Nimba	ICLARM	The International Center for Living Aquatic Resources Management
CIA	Central Intelligence Agency	ICOLD	International Commission on Large Dams
CIESIN	Center for International Earth Science Information Network	ILEC	International Lake Environment Committee
		IPCC	Intergovernmental Panel on Climate Change
CLOFFA	Check-List of the Freshwater Fishes of Africa	IRBM	integrated river basin management
CSI	Conservation Status Index	ITCZ	Intertropical Convergence Zone
DANIDA	Danish Development and Aid Agency	IUCN	The World Conservation Union
DGIS	Directorate-General for International Cooperation, the Netherlands	LCBC	Lake Chad Basin Commission
		LHWP	Lesotho Highlands Water Program
DRC	Democratic Republic of the Congo	MHT	major habitat type
ECOFAC	Conservation and Rational Use of Forest Ecosystems in Central Africa	NASA	National Aeronautics and Space Administration
EPA	Environmental Protection Agency	NBA	Niger Basin Authority
EU	European Union	NBI	Niger Basin Initiative

NBI	Nile Basin Initiative
NCF	Nigerian Conservation Foundation
NEMA	National Environment Management Authority
NEPAD	New Partnership for Africa's Development
NGO	nongovernment organization
NPV	net present value
NWA	National Water Act of South Africa
ROC	Republic of the Congo
SADC	Southern African Development Community
SAIAB	South African Institute of Aquatic Biodiversity
SDC	Swiss Development Cooperation
SSC	Species Survival Commission
TBNRM	Transboundary Natural Resource Management
TDS	total dissolved solids
UNDP	United Nations Development Programme
UNEP	United Nations Environment Programme
UNESCO	United Nations Educational, Scientific and Cultural Organization
UNICEF	United Nations Children's Fund
UNPD	United Nations Population Division
USAID	United States Agency for International Development
WCD	World Commission on Dams
WCMC	World Conservation Monitoring Centre
WDPA	World Database on Protected Areas
WHO	World Health Organization
WI	Wetlands International
WRI	World Resources Institute
WWF	World Wide Fund for Nature
WWF-US	World Wildlife Fund (USA)

Preface

For millions of Africans, especially rural communities, rivers, lakes, and wetlands are critical for survival and livelihood, providing vital supplies of fish and materials as well as ecological services. Covering just 1 percent of the continent's surface, Africa's wetlands also support a wealth of biodiversity, including spectacular arrays of resident and migratory waterbirds, exceptional fish diversity, and flagship species.

Unfortunately, despite their economic, social, and ecological importance, Africa's freshwater systems are being degraded at an alarming rate, as evidenced by the declining status of freshwater biodiversity. Seventy-nine freshwater species in sub-Saharan Africa (including islands) are classified as critically endangered, 116 are endangered, and 103 are vulnerable (IUCN 2002). The status of freshwater plants is less well known, but two species are classified as critically endangered, one as endangered, and one as vulnerable. Of the known species, 30 percent of amphibian and 50 percent of freshwater mollusk species in Africa are at risk.

The Environment Initiative of the New Partnership for Africa's Development (NEPAD) identifies the maintenance of healthy wetland ecosystems as critical for securing sustainable development in Africa and sets out an ambitious program for conservation and management of Africa's wetlands. The objective of this program is to promote and attain a healthy and productive environment in which African countries and their people have wetlands and watersheds that can support fundamental human needs such as clean water, appropriate sanitation, food security, and economies. The adoption of the NEPAD Environment Initiative and its wetland component in June 2003, together with the growing number of African signatories to the Ramsar Convention (thirty-eight in 2003, compared with twenty in 1994), are indications of strong political will on the part of African decision-makers for conservation and wise use of Africa's wetlands.

For years WWF has played a key role in the conservation of Africa's natural resources, working with partners from local community groups to government policymakers, combining regional and global policy, research, and advocacy with on-site conservation action in many countries.

This study provides an in-depth analysis of the state of freshwater biodiversity across Africa, Madagascar, and the islands of the region. To produce this volume, WWF collaborated with hundreds of African and international scientists with expertise on the species, habitats, and ecological processes of the region. The ecoregional framework and data presented here provide a first overview of how freshwater biodiversity is distributed across the region, which systems are most threatened, and where we must act first if we are to retain the region's biological values.

This analysis of the freshwater ecoregions of Africa and Madagascar is the fifth installment of a series using ecoregions to identify biological and conservation priority areas. This study has built on the volume assessing North America's fresh waters (Abell et al. 2000) and three volumes assessing the terrestrial biodiversity of North America, the Indo-Pacific, and Africa and Madagascar (Ricketts et al. 1999a; Wikramanayake et al. 2002; Burgess et al. 2004). Together these assessments form the backbone of WWF's global strategies for conserving terrestrial, freshwater, and marine biodiversity in the Global 200 ecoregions. For a continent where data on biodiversity are inadequate and scattered, the two volumes on Africa constitute a most valuable and welcome contribution.

Dr. Yaa Ntiamoa-Baidu
Director, Africa and Madagascar Programme
WWF International

Acknowledgments

Numerous individuals and institutions gave generously of their time and knowledge to make this work possible. A few warrant special mention: Thomas Kristensen of the Danish Bilharziasis Laboratory for providing aquatic mollusk species lists by ecoregion for all of Africa; Christian Lévêque of the Centre National de la Recherche Scientifique for reviewing the ecoregion descriptions and fish species lists for West Africa; Uli Schliewen of the Zoologische Staatssammlung München for reviewing many of the Congo Basin ecoregion descriptions and species lists; Lincoln Fishpool of BirdLife International, who provided data on wetland bird congregations; and the Zoological Museum of the University of Copenhagen (Jon Fjeldså, Carsten Rahbek, Louis Hansen, and Steffan Galster), who helped to assemble a database of amphibians and aquatic mammal species by ecoregion.

Several collaborators authored multiple ecoregion descriptions: Ashley Brown and Emily Peck, formerly of the Conservation Science Program at WWF-US; Belinda Day and Helen Dallas of the Freshwater Research Institute at University of Cape Town; Liz Day of the Freshwater Consulting Group; Abebe Getahun of Addis Ababa University; Victor Mamonekene of Université Marien Ngouabi-Brazzaville; Dalmas Oyugi, formerly of Kenya National Museum; John Sparks of the Department of Ichthyology at the American Museum of Natural History; and Lucy Scott and Denis Tweddle of the South African Institute for Aquatic Biodiversity. Andrew Balmford of the Conservation Biology Group, Cambridge University, provided analytical advice. We also thank the World Resources Institute for providing human population data projected to the year 2025. Geographic information system and mapping software was kindly donated by the Environmental Systems Research Institute, Inc.

Within the WWF network, we have received much input and support. From WWF-US: Laurianne Cayet, Jennifer D'Amico, Tucker Gilman, Tim Green, Ken Kassem, John Lamoreaux, Colby Loucks, Miranda Mockrin, John Morrison, Kate Newman, the late Henry Nsanjama, Sue Palminteri, Taylor Ricketts, Emily Rowan, Kate Stanford, Meseret Taye, Emma Underwood, and Wes Wettengel all contributed. From the wider WWF-Africa and Madagascar Programme: Aboubacar Awaiss, Jonas Chafota, Monica Chundama, Papa Samba Diouf, Steve Gartlan, Sarah Humphrey, Lucy Kashaija, Marc Languy, David Lindley, Samuel Matagi, and Yaa Ntiamoa-Baidu. From the WWF Living Waters Programme: Denis Landenbergue, Jamie Pittock, and Chris Williams.

We would like to give special thanks to WWF South Africa for hosting the initial WWF African ecoregions workshop in Cape Town (August 1998). In addition to the many experts present in South Africa, we would like to acknowledge the input of experts at the Congo Basin ecoregion workshop (March 2000) and all others who have worked with the WWF-US Conservation Science Program since then.

Several organizations generously supported the publication of this book. This publication was made possible through support provided by the SC Johnson Fund of Racine, Wisconsin; the Office of Environment, Bureau for Economic Growth, Agriculture, and Trade, U.S. Agency for International Development, under the terms of Award No. LAG-A-00-99-0048-00; the Directorate-General for International Cooperation (DGIS), the Netherlands, under the terms of Activity No. WW204107; and the Swiss Agency for the Environment, Forests and Landscape (SAEFL).

The opinions expressed herein are those of the authors and do not necessarily reflect the views of the SC Johnson Fund; the U.S. Agency for International Development; the Directorate-General for International Cooperation, the Netherlands; or the Swiss Agency for the Environment, Forests and Landscape.

CHAPTER 1

Introduction

From the muddy waters of the Congo River spilling into equatorial swamp forests to the vast floodplains of the Inner Niger Delta, from the deep and shallow lakes of the Rift Valley in eastern Africa to the ephemeral streams of the Namib Desert, the fresh waters of Africa and Madagascar are incredibly diverse. Matching this array of systems is a wondrous diversity of life forms and strategies for survival. Below the surface of nearly every body of water, even the most ephemeral pools, freshwater organisms survive. More than 4,300 described species of freshwater vertebrates and invertebrates (fish, aquatic-dependent mollusks, amphibians and reptiles, and mammals) that live in the continent's waters were included in this conservation assessment. An abundance of new species and communities are yet to be described (Lundberg et al. 2000; Stiassny 2002a).

For hundreds of millions of African people, the health of these freshwater systems and of the diverse communities of organisms that depend on them often is inextricably linked to their own health and survival (Revenga et al. 2000; Jackson et al. 2001; Johnson et al. 2001; Everard and Harper 2002; Directorate of Fisheries–Ghana 2003). As in human communities everywhere, clean freshwater is a daily necessity, and healthy ecosystems are needed for its provision. For example, about one-third of twenty-five large cities in Africa draw their drinking water from forested protected areas (Dudley and Stolton 2003). Fish and other aquatic and amphibious animals are also critical sources of protein for many people in the region. The vital importance of Africa's wetlands in providing food security, tradable products, and cultural and aesthetic values for local communities is clear (Brouwer 2003; Denny 2001; essay 1.1). As recently recognized by the targets set at the World Summit on Sustainable Development, sustainable development in the region will not occur without clean and adequate sources of freshwater (United Nations 2002). The economic value of the ecosystem services that freshwater systems provide is extraordinary; one recent estimate places a global value of US$6.6 trillion annually for the goods, services, biodiversity, and cultural contribution provided by all inland waters and wetlands (Postel and Carpenter 1997; Costanza et al. 1997; Balmford et al. 2002; essay 1.2).

Yet freshwater systems across much of Madagascar and the African continent are under increasing pressures from the combined onslaught of introduced species, pollution from burgeoning populations and industries, dams and water withdrawals, and overall land use change (Cohen et al. 1996; Lévêque 1997; Davies and Day 1998; Chapman and Chapman 2003).

This multitude of threats requires us to establish priorities and to set goals and targets for the conservation of aquatic systems and their rich biodiversity. We need a framework that allows us to take conservation actions based on knowledge of what is important at global, regional, and local scales. This book aims to present an objective plan for large-scale biodiversity conservation in Africa and Madagascar, focusing on the global biodiversity values of the region.

To set priorities for future work, we need first to look backward to examine the evolution and natural diversity of Africa's freshwater systems. In this chapter, we introduce Africa's fresh waters (figure 1.1) and the forces considered responsible for the distribution and concentration of Africa and Madagascar's freshwater biodiversity. We then move to the modern era to review contemporary pressures and threats to African freshwater systems that will dramatically and permanently change the distribution of their biota unless addressed in the coming years.

The Evolution and Diversity of Africa's Freshwater Systems

Africa's river and lake basins are among the oldest in the world, having a much longer history than those in the temperate zone. The breakup of Gondwanan landmasses in the mid- to late Mesozoic (165–65 m.y.a.) led to the separation of Madagascar from Africa and India; Madagascar may have been isolated from mainland Africa for as long as 160 m.y. (Rabinowitz et al. 1983; Lourenço 1996). It is unclear whether the freshwater fauna of the island was present at that time or originated later via marine dispersal (see essay 3.6; Krause et al. 1997). The major catchments of mainland Africa, with the exception of the drainage systems of eastern Africa, have existed since before the Miocene (23.8–5.3 m.y.a.), although the details of their configuration have changed over time.

The late Miocene marked the end of a long period of tectonic stability across the continent (Beadle 1981). At this time many of the continent's basins, including the Niger, Chad, and Congo, were endorheic (i.e., forming large inland lakes without external drainage) (Roberts 1975). The relief of the continent is also thought to have been low, such that divides between basins were low and faunal barriers between systems were permeable, resulting in a widespread, uniform freshwater fauna (Lévêque 1997). Post-Miocene earth movements marked the beginning of the rifting and uplifting that caused greater separation of basins and promoted speciation. Because of these tectonic activities over the past 20 million years, African rivers generally have many more rapids and waterfalls than other rivers in the world, and none of the rivers have unimpeded access into the interior of the continent (Lévêque 1997). The end of the Miocene also marked the beginning of the rifting and formation of the Rift Valley (Beadle 1981). The Rift Valley divides into the western and eastern portions, with most of the East African lakes lying in the trenches of the rift; Lake Victoria is an exception, lying in a depression between the two rifts (Lévêque 1997).

Climatic changes have also influenced the connections and extent of waterbodies on the continent and on Madagascar. During the Quaternary, humid periods have alternated with droughts, drastically affecting discharge and interbasin connections. A detailed review of these changes is beyond the scope of this book; we refer readers to Lévêque (1997), who details climatic changes and their effects on freshwater systems from the Paleocene through present. At least five climatic zones currently exist across the Afro-Malagasy region, including equatorial (hot and humid with two rainy seasons), tropical (hot with summer rain), subtropical (hot and arid), mediterranean (arid summers and winter rains, with rare frost), and mountain (Denny 1993). These climatic conditions significantly affect discharge patterns and vegetation communities across the continent. The general precipitation pattern is one of highest rainfall near the equator in the Congo Basin and along the Gulf of Guinea, with progressively less at higher latitudes. However, the deserts in the north and south interrupt this general pattern, and the amount of rain and its distribution can vary greatly within Africa. Because of the high rainfall and low evaporation in the Congo Basin, this river and its tributaries, which account for only about 13 percent of the surface area of the continent, carry about 30 percent of the Africa's surface flow (FAO 1995). Arid, semi-arid, and dry subhumid areas (lands with a ratio of precipitation to potential evaporation of 0.05 to 0.65) cover about 43 percent of Africa's surface area (Bjørke 2002), and about 80 percent of the continent's surface waters are estimated to be lost to evaporation (Gleick 1993). Most of Africa's rivers (more than 90 percent) are less than 9 km long, with many flowing only seasonally (Lundberg et al. 2000).

The major basins in Africa are the Niger, Nile, Congo, Orange, Limpopo, and Zambezi rivers and lakes Chad, Malawi, Tanganyika, and Victoria. Africa is divided into High Africa (elevations mainly above 1,000 m) and Low Africa (150–600 m, with land above 1,000 m confined to a few peaks and plateaus). The dividing line runs from south of the Cuanza Basin in Angola eastward, then northward along the western boundary of the Congo Basin, then between the Ethiopian Highlands and the lowland portions of the Nile Basin toward the Red Sea (Roberts 1975) (figure 1.2). Low Africa includes the sedimentary basins of the Nile, Chad, Niger, and Congo, all of which contain large, shallow depressions and are separated from one another by low-lying divides. There are only a few mountain ranges in Low Africa; these include the Guinean and Cameroonian Highlands and several mountains in the southern Sahara. The Ethiopian Highlands, the Albertine Highlands, the Eastern Arc, and the Drakensberg and Maloti Highlands all occur in the east in

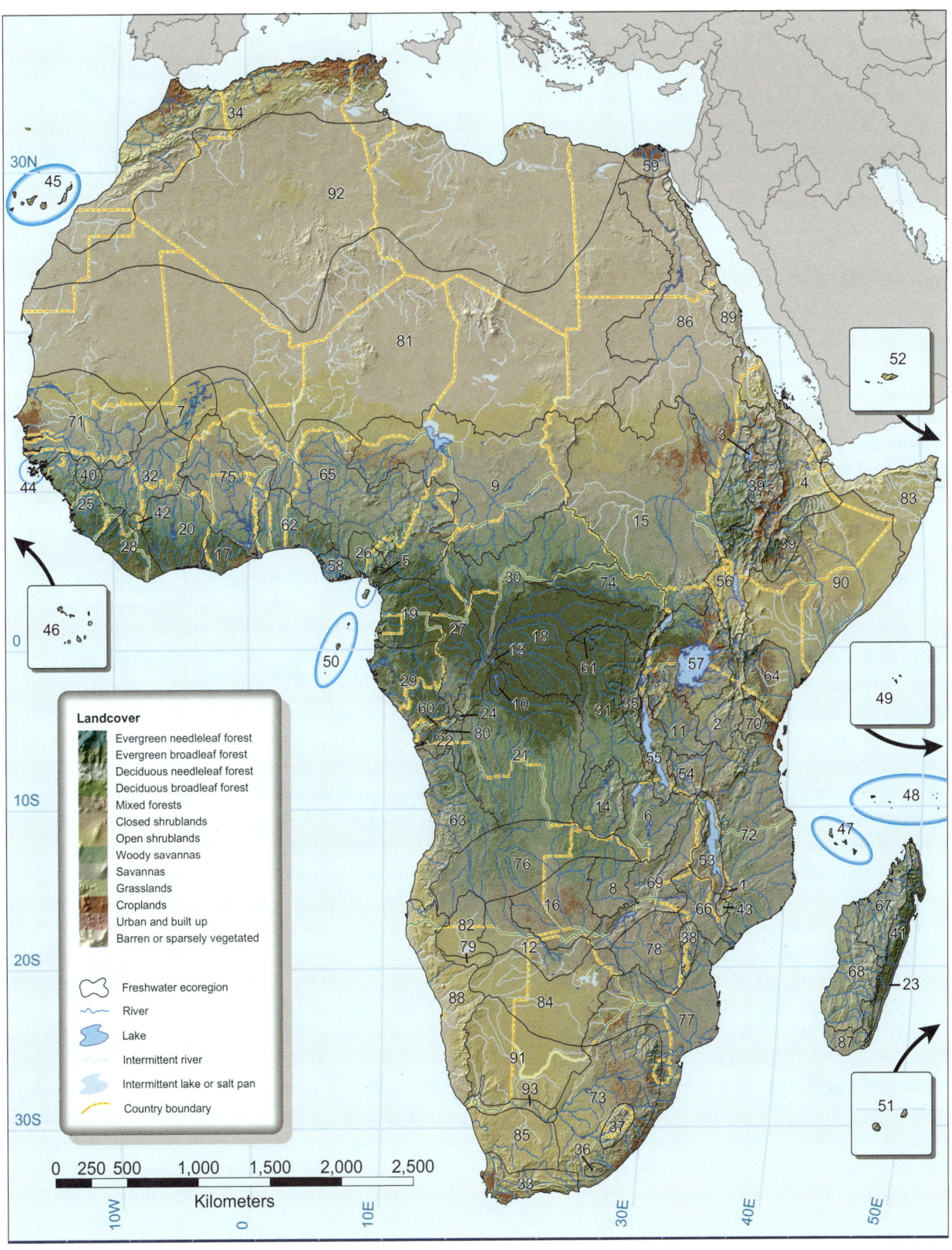

FIGURE 1.1. Freshwater ecoregions of Africa and Madagascar with landcover (ecoregion names provided in Figure 2.1). Landcover data from the MODIS global landcover product. (Hansen et al. 2000; Boston University 2001).

High Africa, creating a system of highland plains and mountains from southern Africa up to the Horn of Africa. The main feature of High Africa is the Central African Plateau, which dominates the interior of southern and eastern Africa with elevations of more than 1,000 m, except at the continent's edge (Kingdon 1989). Associated with the uplifted regions of High Africa is the Great Rift Valley, a geological fault system that extends about 3,000 km from the Red Sea to central Mozambique in eastern Africa. Many small lakes and several large lakes, including lakes Malawi and Turkana, occupy the eastern branch of the Rift Valley, and lakes Tanganyika, Kivu, Edward, and Albert occupy the western branch. Lake Victoria lies not in the Great Rift Valley but in the sunken plain between the two troughs of its eastern and western branches. The two largest river basins in High Africa are the Zambezi, which flows across a plateau at about 1,200 m before descending to its delta at the Indo-Pacific Ocean, and the Orange, with elevations of about 1,800 m in the west and 600 m in the east.

Accompanying this geographic and topographic variety is a diversity of freshwater species assemblages. Tropical communities in general are characterized by greater numbers of species for many taxonomic groups and more complex interactions compared with those in the temperate zone, and Africa's freshwater systems are no exception (Lowe-McConnell 1987). For example, Africa's freshwater fish diversity, at more than 3,000 species, rivals that of Asia (>3,500 species) and South America (>5,000 species) (Kottelat and Whitten 1996; Lundberg et al. 2000). However, Africa's evolutionary phenomena—its diverse species flocks (groups of two or more sister species that are endemic to a lake or river basin; Turner et al. 2001) and relictual "living fossils"—are what make its freshwater fauna particularly distinctive (Brown 1994; Lundberg et al. 2000). Africa has more archaic and phylogenetically isolated freshwater fishes than any other continent and outstanding species radiations among a variety of taxa in both rivers and lakes (Roberts 1975; Lowe-McConnell 1987; Brown 1994). Diverse riverine fish faunas (Upper and Lower Guinean, Nilo-Sudanian, and Congolian), including almost all of the archaic and phylogenetically isolated groups, occur in Low Africa. In contrast, the rivers of High Africa are species-poor for fish. However, this pattern does not hold for amphibians, odonates, and mollusks, with several ecoregions in coastal and highland eastern and southern Africa displaying high richness and endemism (Fjeldså et al. ongoing; LeBerre 1989; Brown 1994; Schiøtz 1999; Clausnitzer 2001) (essay 1.3). Also, High Africa holds the large Rift Valley lakes, which are re-

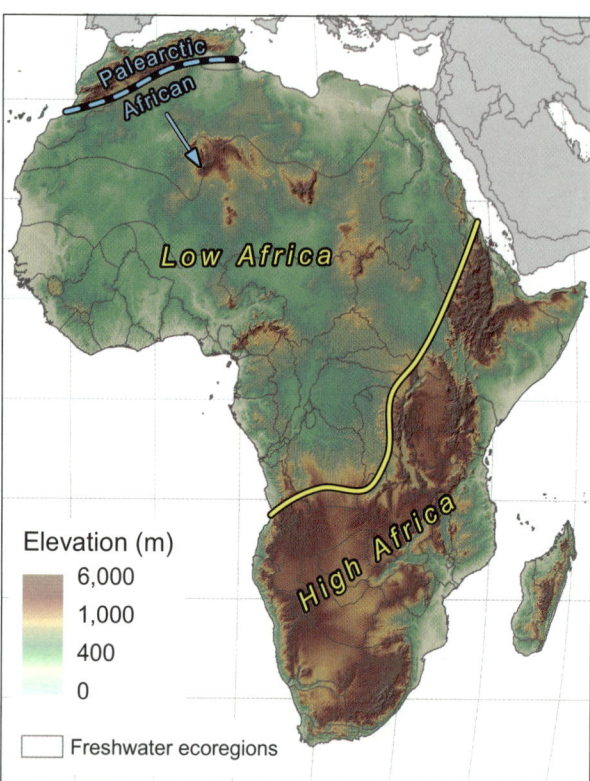

FIGURE 1.2. Elevation zones and freshwater realms of Africa and Madagascar.

nowned for their radiations of hundreds of cichlid fishes (Roberts 1975; Stiassny and Meyer 1999).

Based on freshwater fish species assemblages, most of continental Africa falls within the African realm. However, the fauna of the Atlas Mountains in the northwest of the continent has strong affinities with the fauna of the Palearctic zone and therefore is included in the Palearctic realm (figure 1.2) (Roberts 1975; Doadrio 1994; Banarescu 1995). Madagascar sometimes is considered to be a subregion of the African realm, but its fish fauna is not distinctly African (see essay 3.6).

Many of Africa's fresh waters are yet to be explored, and, despite recent advances in our knowledge of their biota, much taxonomic work remains to be completed (Lundberg et al. 2000; Stiassny 2002a). Future studies undoubtedly will reveal new species and relationships between taxa. We stress that this volume provides a synthesis of current knowledge of Africa's freshwater species diversity and distributions, and its results should be considered preliminary, given the lack of data for many waterbodies across the continent and Madagascar.

Threats

Increasingly, land and forest degradation, water withdrawals, dams, and pollution threaten Africa and Madagascar's distinctive freshwater systems. About 85 percent of Africa's total water withdrawals are estimated to be directed toward agriculture (FAO 1995), and as of the year 2000, about one-third of its surface area was estimated to be under agricultural land use (FAO 2001a). Runoff of sediments, nutrients, pesticides, and herbicides enters rivers and lakes from these agricultural lands. Pollution from domestic sewage and industrial facilities is also a large problem in many parts of Africa, where sanitation infrastructure often is inadequate and industrial point and non-point source pollution are underregulated (Institute of Marine Sciences et al. 1998; UNEP 1999).

Increasing fishing pressure and introduced aquatic species also have negative effects on the biota of many river and lake systems. Inland fish capture in Africa increased by an average of about 37,000 tons per year, or 2 percent, between 1984 and 1997 (FAO Inland Water Resources and Aquaculture Service: Fishery Resources Division 1999). Africa is now second only to the Asia-Pacific region in total catch of inland fisheries. Most systems are considered to be fished at levels where damage to the assemblages has already occurred (see essay 4.2). The well-known case of the introductions of Nile perch (*Lates niloticus*) (Ogutu-Ohwayo 1990) and Nile tilapia (*Oreochromis niloticus*) into Lake Victoria and the subsequent loss of many fish taxa is a stark example of the potential effects of introduced species, particularly when introductions occur in concert with habitat disturbance (Harrison and Stiassny 1999). Introduced aquatic plants, such as water hyacinth (*Eichhornia crassipes*) and Kariba weed (*Salvinia molesta*), are also problems in many river and lake systems, resulting in mounting economic, social, and ecological costs (see essay 4.1).

Not only are freshwater systems across Africa and Madagascar currently facing these threats, but human population growth points toward an increasing intensity of threat in the future. Projections show the population of Africa tripling over the period 1995 to 2050 (from 0.7 billion people in 1995 to 2.0 billion in 2050) (United Nations Population Division 2001). As inhabitants of the poorest continent in the world, many Africans are heavily reliant on natural resources (Economic Commission for Africa 2001). Increasingly, humans and freshwater species are forced into a destructive competition that both probably will lose without intervention. Healthy freshwater systems are vitally important to the survival of Africa's rich biodiversity and to African human communities.

Given the imperative to conserve healthy freshwater systems and the limited resources and time available for doing so, WWF has embarked on a project to identify the freshwater systems of highest conservation value. These are places that are highly distinctive for the numbers and types of species that inhabit them and that are under threat. The results of this prioritization are intended for both WWF and conservation partners throughout Africa and Madagascar.

Structure of the Book

This book is one in a series of continental conservation assessments of freshwater and terrestrial biodiversity and follows a structure similar to those of the previous assessments (Dinerstein et al. 1995; Olson et al. 1998; Ricketts et al. 1999a; Abell et al. 2000; Wikramanayake et al. 2002; Burgess et al. 2004). In chapter 2 we summarize our methods for ecoregion delineation and assessment. The results of the biological distinctiveness and conservation status analyses are presented in chapters 3 and 4, respectively. In chapter 5 we integrate the results of these two indexes to make recommendations for where freshwater conservation work should proceed first in the Afro-Malagasy region. The identification of priorities is not an end in itself, and in chapter 6 we provide examples of visionary efforts that are under way in important freshwater ecoregions across the continent. Throughout the text, each ecoregion is indicated by its name followed by the map code in brackets (e.g., Inner Niger Delta [7]).

The book can be read at two levels: the main text provides an outline of the methods and highlights the major results, and the appendixes include more detailed methods, results, and information. To make the analyses as transparent and repeatable as possible, we provide details of the methods, including the index values and statistical analyses, in appendixes A–E. We also provide descriptions of all the ecoregions in appendix F. These descriptions include information on the biogeography, biological distinctiveness, and conservation status of each ecoregion.

ESSAY 1.1

Contribution of Inland Fisheries to Rural Livelihoods and Food Security in Africa: An Overview

Christophe Béné

For millions of people in Africa, Asia, and Latin America, the fisheries of inland lakes, rivers, and other freshwater ecosystems provide an important source of food, employment, and income. In sub-Saharan Africa, for example, the fringing floodplains of Lake Chad and its associated rivers and the main course of the Zambezi River and its tributaries yield more than 200,000 tons of fish every year, which are worth more than US$70 million (Neiland and Béné 2004; Turpie et al. 1999). Several hundreds of thousands of households throughout Africa depend directly on fisheries and related activities such as fish processing and trade. When barriers to entry are low and other local economic activities have been eroded by macroeconomic forces, fisheries can also provide a safety net. Inland fisheries are particularly important in this context. Widely dispersed and easily accessible to marginal or isolated communities in most of Africa, these fisheries provide an alternative source of income and food when other livelihoods are deficient. Finally, for many poor in rural and even urban or periurban areas, fish often is the sole accessible and affordable source of animal protein.

Despite these crucial roles in rural development, poverty alleviation, and food security, government institutions and decision-makers so far have largely neglected river fisheries in developing countries. The fishery departments in many countries contend with limited budgets and limited or inadequate human resources. Other development projects (such as hydropower or irrigation dams) can also have enormous impacts on the sustainability of inland fisheries. The discrepancy often is large between the benefits of the fishery sector to the local and national economies and the resources allocated by governments to manage these fisheries.

Of the many factors explaining this neglect, one of the most important is the failure to fully recognize and account for the real contributions that freshwater fisheries, and more generally inland aquatic resources, play in providing income and nutrition to resource-poor households. This failure results largely from the fact that in many countries most officials and policymakers working in national capitals have little opportunity to engage with inland fishers. As a result, fisheries often are synonymous with the more readily visible marine or coastal fisheries, where a large part of the catch is traded in the capital or other large urban areas. Conversely, inland fisheries usually are perceived as a marginal activity undertaken by only a few people, mostly living in remote and rural areas, with consequently little economic or social importance.

In Africa, however, and in many other parts of the developing world, the reality is quite different: inland fisheries are widespread and involve several millions of rural dwellers. Along main rivers or smaller tributaries, on the edges of oxbows and other lakes, in the vicinities of small permanent or seasonal ponds, on the margin of marshes or floodplains, fishing is carried out in almost every location where humans have settled close to freshwater.

The second misperception shared by many decision-makers and officials is that this activity is undertaken only by full-time, professional fishers belonging to specialized fishing communities. Here again the reality is far more complex. Fishing on large inland waterbodies such as the Great Lakes in East Africa is indeed a full-time occupation for several thousands of specialized fishers, but fishing is also undertaken by many other groups as an opportunistic or seasonal activity in the rural or even periurban areas.

In this essay I use specific case studies and a general review to illustrate more concretely these points. The objective of this essay is to highlight the importance of inland fisheries to the livelihoods and food security of the rural African populations.

Fish as a Source of Income and Employment

Direct Income and Employment

Estimating the exact contribution of African inland fisheries to national economies is a challenge, and efforts to obtain or generate simple statistics such as the number of fishers or the economic wealth generated by the activity face a number of difficulties. Reliable information and statistics do not always exist or, when available, often do not make the distinction between inland and marine fisheries. Using various secondary data and FAO statistics, Neiland and Béné (2004) reconstructed information on employment and income for seven major river basins

Fishers with cast net fishing on the Dzanga River, Central African Republic. Photo credit: ©WWF-Canon/Martin Harvey

in West and Central Africa: Senegal-Gambia, Volta, Niger-Benue, Logone-Chari, Congo-Zaire, Atlantic coastal basins, and three major lakes in these basins (Lake Kanji, Lake Volta, and Lake Chad).

The results of this analysis (table 1.1) indicate that the main rivers and floodplains in West and Central Africa provide a livelihood to 227,000 full-time fishers[1] and yield an annual catch of about 570,000 tons, valued at US$295 million (first sale value, i.e., gross financial value). Table 1.1 also indicates that the potential total annual fishery production[2] for this region (about 1.34 million tons, with a potential annual value of US$750 million) is more than twice the actual estimated production (Neiland and Béné 2004).

Multiplier Employment Effect

In addition to those working in fish capture, many other people are involved full-time in upstream or downstream related activities (processing, packaging, transportation, and provision of inputs). Assuming a conservative 1:1 multiplier ratio[3] implies that 450,000–500,000 people are directly involved in full-time activities related to inland fisheries in West and Central Africa.

Despite the fact that these production and employment estimates are limited to the West and Central Africa regions, they clearly demonstrate that inland fisheries in Africa are not a marginal activity supporting only a few people, as some may imagine, but an important economic activity carried out by several hundreds of thousands of people. If adequately supported by development policies, inland fisheries could contribute significantly to the rural development of many African countries. In some areas fisheries are already a major element of the household economy. For example, in the Zambezi Basin the value of inland fisheries (estimated through their share of the household's cash income) is a major contributor to the economy of the local populations, systematically generating more cash than cattle and sometimes more than crops (Turpie et al. 1999) (table 1.2). However, this reality is often unknown to decision-makers.

All Fishers Are Not Necessarily Full-Time Professionals

Fishing activities include a wide range of degrees of labor involvement (e.g., occasional, opportunistic, seasonal, full time), and it would be misleading to reduce the contribution of inland fisheries to the number of full-time fishers. In some places the high level of investment necessary to enter the fishery limits entrants to fishing as a full-time profession; however, in other

TABLE 1.1. Contribution of Fisheries of the Major River Basins and Lakes in West and Central Africa to Employment and Income.

	Employment (fishers)	Fishery Production (t/yr)	Value of Production (million $/yr)	Potential Production (t/yr)	Potential Production (million $/yr)
River basins					
Senegal-Gambia	25,500	30,500	16.78	112,000	61.60
Volta (rivers)	7,000	13,700	7.12	16,000	8.32
Niger-Benue	64,700	236,500	94.60	205,610	82.24
Logone-Chari	6,800	32,200	17.71	130,250	71.64
Congo-Zaire	62,000	119,500	47.80	520,000	208.00
Atlantic coastal	6,000	30,700	46.66	118,000	179.30
Major lakes					
Volta	20,000	40,000	28.40	62,000	44.02
Chad	15,000	60,000	33.00	165,000	90.75
Kainji	20,000	6,000	3.30	6,000	3.30
Total	227,000	569,100	295.37	1,334,860	749.17

Source: Neiland and Béné (2004).

TABLE 1.2. Contribution of Fisheries to Households' Cash Income (US$/household/year) in Different Parts of the Zambezi Basin, Compared with Other Activities.

	Barotse Floodplain	Caprivi-Chobe Wetlands	Lower Shire Wetlands	Zambezi Delta
Cattle	120	422	31	0
Crops	91	219	298	121
Fish	180 (43%) 1st	324 (28%) 2nd	56 (13%) 2nd	100 (39%) 2nd
Wild animals	6	49	1	0.4
Wild plants	24	121	48	29
Wild foods	0	11	7	4
Clay	2	0	8	0.1

Source: Turpie et al. (1999).

places the number of people who receive a regular income from fishing is only a tiny proportion of the overall number involved in fishing. For example, in his socioeconomic profiling of the fishing community of the Mutshindudi catchment in northeast South Africa, van der Waal (2000a) found that less than 1 percent of the fishers in the area are deriving a regular income from fishing. In this area a major portion of the fishers consist of scholars (for the younger) and unemployed (for the older) who fish on a part-time basis. For these people, the low catch per unit effort precludes generating a significant income. However, even if they catch only few fish, the survey indicates that these represent a substantial part of the household's needs in animal protein and micronutrients, thereby contributing significantly to the food security and dietary health of the households.

Fish and Food Security

From the time of the first hunter-gatherers along the continent's rivers to the modern era, fish have always played an important role in food security. During periods of famine, fish have often provided a crucial product that was bartered for other staple foods, thus preventing the population from starving or from being forced to migrate. For instance, Couty and Duran (1968) report that during the 1902 famine in the Lake Chad area, local Massa populations were able to survive by exchanging dried fish for sorghum with Peul merchants. In normal years these Massa populations consume an average of about 40 kg of fish per capita.

Today, fish continues to be a central element of local diets in African countries, and inland fishery catch can represent a major portion of the total animal protein consumption. The FAO has recently estimated that fish provides 19 percent of the protein intake of developing countries (FAO 2000c). However, this share can exceed 25 percent in the poorest countries and is up to 90 percent in isolated parts of coastal or inland areas, where river-, floodplain-, or lake-related fishing activities play a crucial role in food supply. For example, fish from Lake Malawi are reported to provide up to 75 percent of the animal protein consumed in the eastern part of Malawi (Ribbink 1994). Fish also provide oils and essential micronutrients such as calcium, iodine, and certain vitamins (FAO 2003). The rate of fish consumption depends on factors such as the availability of the resource, its relative price with respect to other sources of protein, and the dietary habits of human populations (table 1.3).

At the international conference on the Sustainable Contribution of Fisheries to Food Supply (Kyoto, Japan) in 1995, the ninety-five participating states approved a declaration and a plan of action to enhance the contribution of fisheries to human food supply by conserving and better managing fish resources. The need to ensure food security was further emphasized at the World Food Summits in Rome in 1996 and 2002, which stressed the need for future sustainable management of natural resources. The predicted rises in global population and corresponding increases in demand for food, including fish, mean that many of the food security problems of today are likely to persist. The effects of the imbalance between supply and demand are not likely to be evenly felt across the world. Although many countries and regions have made progress in reducing food energy deficiencies, many others (notably in sub-Saharan Africa) have either experienced a worsening of food security or have managed to achieve improvements only through a greater reliance on food imports from developed countries. Under these conditions the role of inland fisheries in food security is even more crucial (Delgado et al. 2003).

TABLE 1.3. **Fish Consumption in African Countries with Moderate or Significant Inland Fisheries.**

Country	Fish Consumption (kg/person/yr)
Burkina Faso	1.4
Burundi	3.2
Cameroon	9.3
Central African Republic	4.2
Chad	6.5
Congo (Brazzaville)	20.7
Kenya	6.1
Malawi	5.6
Mali	8.5
Mozambique	6.0
Niger	0.5
Nigeria	5.8
Rwanda	1.6
Tanzania	8.5
Uganda	12.1
Zambia	5.8
Zimbabwe	1.8

Source: FAO (2002).

Fisheries as a Central Element of Household Livelihoods

Fisheries and Rural Development

The contribution of inland fisheries to local, rural, or national development cannot be reduced to income, food, and employment. Economists and socioanthropologists are starting to better understand the role of fishing in household livelihoods as part of a diversified and complex economic support system at both the household and community levels. In floodplain areas, for instance, fishing fits in a flexible matrix of various activities that constitute the basis of a diversified livelihood strategy on which households rely to spread risks between various economic activities in an uncertain environment and create synergy between the inputs and outputs of these activities and thereby enhance capital accumulation and income opportunities. For example, research in the Lake Chad Basin illustrates how fishing can constitute a powerful engine for capital accumulation and a central element in livelihood generation. Fish provide a source of cash to be reinvested in various fishing or nonfishing activities (Neiland et al. 2000; Béné et al. 2003b). In particular, better-off households in these regions use a large part of the income generated by fish catches to purchase farming inputs (e.g., fertilizers, seeds) and also to hire farming labor. The ability to hire extra labor is a critical advantage in the Sahelian region, where it is not so much land but rather labor that is the major constraint to farming production. The results also show how additional investments in fishing inputs (through new fishing gear or more labor allocated to this activity) can generate instantaneous income surplus, in contrast to farming activities, where several months (until harvest time) pass before any benefit is obtained from the investment. Given the high environmental and political uncertainty that characterizes the Sahelian regions, the capacity of fishing activities to generate instantaneous gains represents an enormous advantage over farming.

Fisheries and the Poor

Although fishing can generate significant wealth and therefore act as a catalyst for rural development, fishing also is an activity of last resort for the rural poor. The central place of fisheries in the livelihood strategies of the extremely poor is explained by the common or sometimes open access nature of the resource. This semi–open-access situation allows people to enter the sector when their access to other activities or resources is economically or institutionally limited or impeded (Béné et al. 2003a). The last resort dimension of fisheries therefore is of great importance to the poorest households because generally they have the most limited access to land or other capital.

The reliance on fisheries to provide income for the poorest also applies to processing and trading activities. This aspect adds an important gender dimension to the discussion, given that women usually are the main actors in that part of the sector. For example, Gordon (2003) describes the case of fish trading associated with the chisense fishery on Lake Mweru (at the border between Zambia and Congo) in the mid-1980s. The majority of the women involved in the trading activity were poor, generally lacking the financial support of their husbands. They "had to look for other activities to meet their daily needs" in a context where the traditional female activity—cassava farming—was becoming "increasingly difficult [due to] land scarcity and unprofitable prices" (Gordon 2003: 173). Under these circumstances, fish trading provided the activity of last resort for these poor women.

Finally, fisheries can also provide a critical safety net for vulnerable households when they face a sudden decline in their income. This can happen when the head of a household loses his or her job, when farm crops fail, or, on a larger scale, when the local or even national economy deteriorates. Recurrent civil wars or military conflicts, population displacement, and natural disasters, all common in Africa, also create circumstances in which those affected turn to fisheries as additional or alternative sources of income, food, or employment. Twice over the last 30 years, the Lake Kariba fishery has provided such a safety net for the population of the southern region. First, in the mid-1970s several thousand miners working in the copperbelt in Zambia lost their jobs, migrated to the lake region, and undertook fishing as their livelihood. Second, a few years later, during the Zimbabwean Independence War, hundreds of families moved to the lake region for security reasons and entered the fishery to ensure minimum revenues until the political situation in their region of origin had improved (Jul Larsen 2003).

Conclusions

As emphasized throughout this book, the rivers, lakes, and streams of Africa harbor an astonishing array of fish and other aquatic species that constitute an important part of the global aquatic biodiversity heritage. Unfortunately, these freshwater species and their habitats face a series of unprecedented threats imposed by increasing human uses through industrial and domestic pollution, water demand and consumption, and the effects of hydropower and irrigation dams on natural river flows.

In reaction to these threats, it is tempting to advocate a conservation approach promoting complete protection and isolation of the environment from human impacts (e.g., through the creation of strict protected areas or natural reserves). As this essay has shown, however, rivers, lakes, wetlands, and floodplains do not simply sustain aquatic resources. They also support millions of people, the majority of whom live in impoverished communities in rural and sometimes remote areas and who depend on these resources for the goods and services necessary to sustain their livelihoods. Promoting a conservation approach that denies access by these populations to aquatic resources would ignore the realities of economic development and, more fundamentally, would overlook a crucial dimension of the problem: nature and people are part of the same world and are bound to live together. Furthermore, it is essential to realize that in Africa—as in a large number of other developing areas where poverty, vulnerability, and food insecurity are rising issues—freshwater resources play a crucial and often irreplaceable role in supporting the groups that depend the most heavily on these natural resources, namely the rural poor (Horemans 1998; Welcomme 2001; Neiland and Béné 2004).

Acknowledgments

The author is grateful to Dr. Patrick Dugan (WorldFish Center) and Michele Thieme (WWF-US) for their useful comments on previous versions of this text.

Notes

1. Although the exact degree of labor involvement (full time or part time) was not initially indicated in the data, it is reasonable to assume that the figures from which this compilation was derived are accounting only for full-time (professional) fishers. The estimate given here (227,000 full-time fishers) is also likely to be the lower limit of what is thought to be the exact number, given that only recorded or observed fishers had been included in the data.

2. Potential productions are calculated using standard coefficients relating the surface of the waterbodies (river or floodplain) with potential production rates (t/ha/yr) as typically observed for these waterbodies and classically used in the literature (e.g., Welcomme 2001).

3. For instance, Horemans (1998) uses a multiplier of 3:1 for coastal artisanal fisheries.

ESSAY 1.2

The Economic Value of Africa's Wetlands

Lucy Emerton

From an economic viewpoint, freshwater ecosystems are among Africa's most undervalued resources. Decision-makers, developers, and land use planners have long perceived little economic benefit to conserving wetlands and few economic costs attached to their degradation and loss. Given this tendency to undervaluation, it is hardly surprising that wetlands all over the continent have been modified, converted, overexploited, and degraded in the interests of seemingly more productive land and resource management options that appear to yield much higher and more immediate profits. Wetland undervaluation has also been a problem in conservation planning and practice. Efforts have rarely been made to justify wetland conservation in development terms or to set in place mechanisms to ensure that the resulting activities are economically viable and financially sustainable.

In this essay we look at the advances made in wetland economic valuation over recent years and describe how these tools are starting to be used to address real-world challenges in wetland conservation and development planning in Africa.

Broadening the Definition of Wetland Economic Value

Wetland economic value is poorly understood and rarely articulated, and as a result is often omitted from decision-making. Traditional concepts of economic value are based on a very narrow definition of benefits and costs and usually exclude goods and services that do not appear in formal markets. As is the case for most environmental resources, however, wetlands generate economic benefits far greater than those of just physical or marketed products. Excluding these nonmarketed, or less tangible, benefits runs the risk of ignoring some of the most valuable wetland goods and services.

Fortunately, economic definitions and methods have moved forward and are now much better able to cope with wetland goods and services. The concept of total economic value has become one of the most widely used frameworks for identifying and categorizing environmental benefits. Total economic value encompasses the subsistence and nonmarket values, ecological functions, and nonuse benefits associated with wetlands. It involves considering the full range of characteristics of a wetland as an integrated system: its resource stocks or assets, flows of environmental services, and the attributes of the ecosystem as a whole (Barbier et al. 1997). Broadly defined, the total economic value of wetlands includes the following (figure 1.3):

- Direct use values: wetland raw materials and physical products that are used directly for production, consumption, and sale, such as those providing energy, shelter, foods, agricultural production, water supply, transport, and recreation.

FIGURE 1.3. The economic values of wetland ecosystems.

- Indirect use values: the ecological functions that maintain and protect natural and human systems through services, such as maintenance of water quality and flow, flood control and storm protection, nutrient retention and microclimate stabilization, and the production and consumption activities that the functions support.
- Option values: the premium placed on maintaining a pool of wetland species and genetic resources for future possible uses, some of which may not be known now, such as leisure, commercial, industrial, agricultural, and pharmaceutical applications and water-based developments.
- Existence values: the intrinsic value of wetland ecosystems and their component parts, regardless of their current or future use possibilities, such as cultural, aesthetic, heritage, and bequest significance.

Using Valuation for Wetland Management in Africa

Alongside the advances made in defining and conceptualizing economic value, techniques for quantifying environmental benefits and expressing the benefits in monetary terms have also moved forward over the last decade. Today, a wide range of market and nonmarket methods are available and used for valuing wetland goods and services (see Gren and Söderqvist 1994; Barbier et al. 1997; Emerton 1998). These have yielded impressive estimates of the total economic value of freshwater systems in Africa. It is now possible to quantify the economic benefits associated with specific wetland ecosystems; for example, four key components of southern Africa's Zambezi Basin (the Barotse Floodplain in Zambia, the Chobe-Caprivi Wetlands in Namibia, the Lower Shire Wetlands in Malawi, and the Zambezi Delta in Mozambique, together spanning more than 22,000 km^2) have been estimated to be worth almost $60 million a year in terms of direct use alone, and they are estimated to generate ecosystem service net present values of $182 million (Turpie et al. 1999). Data that express the importance of wetlands in national economies are also being generated. For example, an assessment of the value of biodiversity in Uganda found that freshwater resources make by far the biggest contribution (almost half) and that lakes, rivers, wetlands, and floodplains generate goods and services for the country valued at more than $334 million a year (NEMA 1999).

Yet although calculating the economic value of a wetland ecosystem can be an interesting academic exercise, valuation

> **BOX 1.1. The Role of Wetlands in a Rural Economy: Pallisa District, Uganda.**
>
> Pallisa District lies in eastern Uganda, containing a population of almost half a million people and covering an area of just under 2,000 km². More than 99 percent of the people are rural and depend on mixed farming as the basis for their subsistence and income. More than a third of the district, or 71,100 ha, is occupied by wetlands, which form an extremely important part of local livelihood systems. Although wetland goods and services generate extremely high economic benefits, little is known about these values. As a result, wetlands often are seen by district and national planners as wastelands rather than as valuable stocks of natural capital that, if managed sustainably and used wisely, can yield a flow of economic benefits for current and future generations.
>
> Most of the inundated and flood recession areas in and around Pallisa's wetlands are used for subsistence agriculture, mainly rice growing, and for grazing. Wetlands provide a wide range of other benefits to local communities, including handicraft and building materials, food resources such as fish and wild vegetables, medicine for various ailments, and transport. Wetlands also provide services such as flood control, water purification, and maintenance of the water table and are important habitats for fauna and flora.
>
> In total, wetland goods and services have been calculated to be worth more than $34 million a year to the Pallisa District economy, or almost $500/ha, including both direct and indirect use values as well as value added through processing and marketing of wetland products. The bulk of this value accrues from wetland agriculture, harvesting of wetland tree products, and the use of water for irrigation and domestic consumption. The majority of value accrues at the household subsistence level, although wetland resource use and marketing also generate appreciable local income and revenues for the district government. Wetland resources have a particularly high value, both in absolute terms and relative to other sources of livelihood, for poorer and more vulnerable sectors of the population and for women. They also provide a vital source of security during drought and dry seasons and when other sources of production (such as crops) fail. Yet because little is known about these values at present, wetlands tend to be downplayed in estimates of district output and income, in socioeconomic development planning, and in efforts to set in place mechanisms to alleviate poverty and support sustainable livelihoods in Pallisa District.
>
> *Source: Karanja et al. (2001)*

is not an end in itself. Rather, it is a means of providing information that can be used to make better and more informed choices about how resources are managed, used, and allocated. In particular, wetland valuation allows decision-makers to consider the relative gains from different land use and investment alternatives in a more inclusive and comprehensive way, including activities that contribute to wetland conservation and sustainable use as well as those that may lead to wetland loss. Valuation enables wetlands to be factored into development decisions and also helps to make sure that economic concerns are reflected in wetland conservation activities.

The Local Economic Value of Wetlands

One of the implications of failing to consider nonmarket values is that the economic importance of wetlands at the household or subsistence level has often been underrepresented. Wetland resources are a central component of day-to-day income and subsistence in many rural areas, and, as is the case in Pallisa (box 1.1), often are of particular significance for poorer groups or at times of uncertainty or stress. Wetland functions also often play an important role in filling the gap between the level of basic services that human populations need to survive and that which the government is able to provide, as in Nakivubo (box 1.2). Being able to express these values offers a means of ensuring that the needs of poor rural and urban communities are represented in decision-making and shows that wetland conservation and sustainable use can make a tangible contribution to meeting basic development needs.

Integrating Wetland Values into Development Planning

Environmental valuation is increasingly being used to question the economic and development wisdom of making land use and investment decisions that degrade or destroy wetlands. Failing to account for wetland values has often resulted in decisions that have given rise to high and often untenable economic costs. Valuation tools can be used to illustrate the hidden ecological and economic costs of developments that interfere with ecosystem integrity and present them in a format that can be incorporated into more conventional economic appraisal and analysis frameworks, as in the case of the Tana River hydroelectric scheme (box 1.3). Valuation also provides the basic data

> **BOX 1.2. The Economic Value of Wetland Services for an Urban Population: Nakivubo Swamp, Uganda.**
>
> Nearly half of Uganda's urban dwellers live in Kampala, where population is estimated to be increasing at a rate of nearly 5 percent a year, almost twice the national average. To cope with this rapidly rising population, settlement is expanding, construction is taking place, and urban infrastructure is being improved throughout the city. Many of these developments have involved draining and reclaiming wetlands. Settlement and industry have severely encroached on one wetland area in particular: Nakivubo, one of the largest wetlands in the city. There is a danger that Nakivubo may soon be modified and completely converted, at immense social and economic costs to some of the poorest and most vulnerable sectors of the population.
>
> One of the most important values associated with Nakivubo wetland is the role it plays in ensuring urban water quality in Kampala. Both the outflow of the only sewage treatment plant in the city, at Bugolobi, and the main drainage channel for the city, enter the top end of the wetland. The main drainage channel can carry highly polluted flows, and more than 90 percent of Kampala's people have no access to a piped sewage supply. Nakivubo functions as a buffer through which most of the city's industrial and urban wastewater passes before entering Murchison Bay, and the wetland physically, chemically, and biologically removes nutrients and pollution from these wastewaters. These functions are extremely important because the purified water flowing out of the wetland enters Murchison Bay only about 3 km from the intake to Ggaba Water Works, which supplies all of the city's piped water supplies.
>
> These economic benefits—and by implication the economic costs associated with their loss—must be balanced against the potential profits accruing from wetland conversion and development. The infrastructure needed to achieve a similar level of wastewater treatment would incur costs of $1–2 million a year in terms of extending sewage treatment facilities or improving water treatment. Meanwhile, Nakivubo plays an important role in filling the gap between the level of basic services that the urban population needs and those that the government is able to provide. There is a real danger that the ongoing conversion and reclamation of wetlands in Kampala, such as Nakivubo, will undermine the very aims of urban development: the better provision of basic services, the generation of income and employment, and the alleviation of urban poverty.
>
> *Source: Emerton et al. (1999)*

for modeling and predicting the likely impacts of future conservation or development actions, such as for different management scenarios in the Barotse Floodplain (box 1.4). It can also be an important means of justifying measures to mitigate or reverse the effects of wetland degradation or restoring freshwater ecosystems, as illustrated in the case of Waza Logone (box 1.5).

Generating Funding and Incentives for Wetland Conservation

Economic valuation provides important insights for conservation and development planning. Wetland conservation is not a free exercise, and there is a need to adequately cover both direct and indirect costs. In many cases these costs are substantial and often are incurred by the groups who are least able to bear them, most typically the local communities that must forgo certain land and resource uses and government agencies that must pay for wetland management. As the case of Lake Mburo illustrates (box 1.6), unless these costs can be gauged and sufficient financial resources and economic incentives made available to offset them, wetland conservation stands little chance of success over the long term.

Conclusion: Future Challenges in Valuing Africa's Wetlands

Economic valuation can provide a powerful tool for placing wetlands on the agenda of conservation and development decision-makers. By expressing people's preferences in monetary terms, valuation aims to make wetland goods and services directly comparable with other sectors of the economy when investments are appraised, activities are planned, policies are formulated, or land and resource use decisions are made. Although a better understanding of the economic value of wetlands does not necessarily favor their conservation and sustainable use, and valuation cannot by itself overcome the omission of wetland concerns from decision-making, it at least permits them to be considered as economically productive systems alongside other possible uses of land, resources, and funds. Economic arguments and data have a powerful influence on decision-making, and as long as wetland values are omitted, decisions will tend

> **BOX 1.3. Counting the Economic Costs of Freshwater Ecosystem Degradation: The Tana River Hydroelectric Schemes, Kenya.**
>
> The Tana River is one of Kenya's most important river systems. With a total length of some 1,000 km and a catchment area greater than 100,000 km^2, the Tana is the only permanent river in an extremely dry region and is associated with a range of highly productive natural ecosystems containing unique and endemic biodiversity. It is also heavily used for hydropower. To date, five major reservoirs have been built on the Tana, which together provide nearly three-quarters of the country's electricity. However, dam construction has had a major influence on the Tana's downstream flow and physical characteristics, most notably by regulating waterflow and decreasing the frequency and magnitude of flooding. In the past, the river flooded its banks, usually twice a year. These biannual floods inundated the floodplain and delta area up to a depth of 3 m, supporting grasslands, lakes, seasonal streams, riverine forests, and mangroves. Since 1989, when the last dam was commissioned, flooding has decreased dramatically in volume and frequency.
>
> A new hydropower scheme, the Mutonga–Grand Falls Dam, has recently been proposed for construction on the Tana River downstream of the existing schemes. With a projected reservoir area of 100–250 km^2, an impounding time between 9 months and 2.5 years, and rated power output of 60–180 megawatts, all of the options for dam construction that are being considered will compound the hydrological effects already resulting from existing dams. The proposed scheme would be the last stage in complete control of the Tana's waters because after construction there would be no appreciable addition to the river's flow except in extreme events. This would end the biannual flood pattern and significantly lower the local water table. Existing changes in downstream ecosystems would be hastened and exacerbated, including reduction in the area and composition of floodplain grasslands; lowering of surface water and groundwater levels; loss of fertile riverbank sediment deposits; reduction in swamps, oxbow lakes, and seasonal water bodies; senescence of riverine forest; and mangrove degradation caused by inadequate freshwater flows.
>
> More than a million people directly depend on the Tana's flooding for their livelihoods, and four times this number rely on it for their water supplies. Valuing dam-related changes in freshwater-dependent ecosystems shows that the net present cost of existing dams has been more than $26 million and that the construction of an additional dam could nearly double this figure. Economic appraisal of the dams has not considered the substantial costs of ecosystem degradation or related them to the need to invest in avoiding or mitigating hydrological, ecological, and socioeconomic impacts. This omission has potentially devastating consequences for the natural ecosystems and human populations that depend on the Tana's flooding regime.
>
> *Source: Acropolis Kenya (1994)*

to deemphasize their benefits and to marginalize the natural ecosystems that provide them.

To a large extent, valuation of wetlands in Africa has yet to reach its full potential. Freshwater ecosystems remain poorly represented in environmental valuation studies, particularly tropical systems; much more work has been carried out in temperate wetlands (Gren and Söderqvist 1994). Despite important advances in calculating and expressing the value of wetland goods and services in Africa, a major challenge remains: ensuring that the results of these studies, and the figures they generate, are actually fed into decision-making processes and used to influence conservation and development agendas. When properly measured, the total economic value of wetland ecological functions, services, and resources often exceeds the economic gains from activities that are based on ecosystem conversion or degradation (Barbier et al. 1997). For the most part, decision-makers remain unaware of these values, at immense cost to Africa's freshwater ecosystems and to the economic activities and users that depend on them.

BOX 1.4. **Modeling Alternative Wetland Conservation and Development Scenarios for the Barotse Floodplain, Zambia.**

The Barotse Floodplain and its associated wetlands cover more than 1.2 million ha in western Zambia, making it one of the largest wetland complexes in the Zambezi Basin. Almost a quarter of a million people live on the floodplain and depend on its natural resources for their day-to-day subsistence and income. In total, it is estimated that the wetland has a gross economic direct use value of some $12.25 million a year, yielding net financial benefits of more than $400 per household per year from fishing, livestock keeping, cropping, and plant and animal harvesting. At the same time it generates a wide range of services that enable and protect off-site production and consumption, including downstream flood attenuation (calculated to have a net present value [NPV] of $0.4 million), groundwater recharge ($5.2 million), nutrient cycling ($11.3 million), and carbon sequestration ($27 million NPV).

These environmental values have been largely excluded when land and water use decisions have been made in the region. Factoring in the economic benefits of wetland goods and services can substantially change the indicators of profitability and economic desirability of development decisions. For the case of the Barotse Floodplain, a dynamic ecological-economic model that simulated the effects of human activity on the wetland system over a 50-year period was used to show the economic and financial implications of different land management scenarios. These included various combinations of a do-nothing scenario of continuing resource use and human population growth, a wise use scenario based on sustainable wetland use and management, a protected area scenario that required some kinds of extractive resource use to be reduced or curtailed completely, and an agricultural development scenario that assumed the gradual transformation of the floodplain to large-scale irrigated rice.

This dynamic modeling indicated clearly that the most economically valuable future management option for the Barotse Floodplain was wise use and conservation of the wetland area. This yielded a NPV of almost $90 million, compared with just over $80 million under a do-nothing scenario, less than $70 million for strict protection, and less than $80 million for large-scale agricultural schemes. Whereas a highly protective management regime was found to incur high opportunity costs in terms of sustainable resource use forgone, both local and national economic benefits and financial profits generated by land conversion to agriculture were far outweighed by the economic costs of wetland goods and services lost. Interestingly, the economic and financial values derived from managing the Barotse Floodplain sustainably were most pronounced at the local level.

Source: Turpie et al. (1999)

BOX 1.5. **Providing an Economic Justification for Freshwater Ecosystem Restoration: The Waza Logone Floodplain, Cameroon.**

Covering an area of some 8,000 km^2 in northern Cameroon, the Waza Logone Floodplain represents a critical area of biodiversity and high productivity in a dry area, where rainfall is uncertain and livelihoods are extremely insecure. The floodplain's natural goods and services provide basic income and subsistence for more than 85 percent of the region's rural population, or 125,000 people. The biodiversity and high productivity of the floodplain depend to a large extent on the annual inundation of the Logone River. In 1979, however, the construction of a large irrigated rice scheme reduced flooding by almost 1,000 km^2. This loss of flooding has had devastating effects on the ecology, biodiversity, and human populations of the Waza Logone region.

The hydrological and ecological rehabilitation of the Waza Logone Floodplain, through reinundation, is an important element of the Projet de Conservation et de Développement de la Région de Waza-Logone. To date the project has already accomplished two pilot flood releases, which have led to demonstrable recoveries in floodplain flora and fauna and have been welcomed by local populations. Further restoration of the previously inundated area will involve constructing engineering works that allow flooding to take place. To make the case to government and donors for investment in reinundation, the Waza Logone Project recently carried out a study to value the environmental and socioeconomic benefits of flood release and the costs of flood loss to date.

This study found that the socioeconomic effects of flood loss have been devastating, incurring livelihood costs of almost $50 million over the 20 or so years since the scheme was constructed. Up to 8,000 households have suffered direct economic losses of more than $2 million a year through reduction in dry season grazing, fishing, natural resource harvesting, and surface water supplies. The affected population, mainly pastoralists, fishers, and dryland farmers, represent some of the poorest and most vulnerable groups in the region. Reinundation measures have the potential to restore up to 90 percent of the floodplain area, at a capital cost of approximately $10 million. The economic value of floodplain restoration will be immense. Adding more than $2.5 million a year to the regional economy, or $3,000/km^2 of flooded area, the benefits of reinundation will have covered initial investment costs in less than 5 years. Ecological and hydrological restoration will also have significant impacts on local poverty alleviation, food security, and economic well-being. Flood release will rehabilitate vital pasture, fisheries, and farmland areas used by nearly a third of the population, to a value of almost $250 per capita.

Source: IUCN (2001a)

BOX 1.6. **Identifying the Needs for Financial and Economic Incentives for Wetland Conservation: Lake Mburo National Park, Uganda.**

Lake Mburo National Park covers an area of some 260 km^2 of savanna, woodland, wetland, and lake habitats in southwestern Uganda. Traditionally an important rangeland for the Banyankole pastoralists, over the last 70 years the area has been subject to a series of progressively more protective management regimes. In the 1930s it was designated a controlled hunting area, in the mid-1960s it was established as a game reserve, and in 1982 it was gazetted as a strict national park. At the latter time, people who were living in the park area were evicted, and restrictions on land and resource uses in the national park were set in place.

For the last decade, the Uganda Wildlife Authority has followed a community-based approach to managing Lake Mburo National Park. This has involved conservation awareness and education activities, human-wildlife conflict resolution, and the allocation of a fixed proportion of park revenues to community development activities in the buffer zone. Yet a major question remains as to whether there are sufficient economic incentives for surrounding communities to be willing and able to support conservation of the national park.

Valuation of Lake Mburo under its current management regime shows that substantial economic benefits are generated for both local communities and the Uganda Wildlife Authority. Resource use, both legal and illegal, yields up to $125,000 a year for the 7,500 households who live around the national park, and just under $20,000 has been provided through community revenue-sharing projects. Meanwhile, the park managing authorities earn gross revenues of some $275,000 a year, mainly from donor grants, fishing licenses, and tourism revenues. When these benefits are compared with the economic costs of managing Lake Mburo as a conservation area, clear imbalances become apparent. Whereas the annual physical expenditures associated with protecting the park run to more than $275,000, local communities are estimated to incur costs of almost $500,000 a year in lost grazing land and resource use opportunities and crop and livestock damage caused by wildlife in the park. Serious issues arise as to whether, under current management conditions, it is economically feasible or financially sustainable to protect Lake Mburo. The national economy is incurring a net economic loss, the managing authority generates insufficient revenues to finance park operations, and local communities are effectively subsidizing conservation of Lake Mburo.

Source: Emerton (1999)

ESSAY 1.3

Dragonflies: Sensitive Indicators of Freshwater Health

Michael J. Samways

Indicator Qualities

Dragonflies (Odonata) have long been recognized as important indicators of habitat type and change. Both the larva and the adult must have the right water and riparian conditions for continued survival at a particular location. Landscape disturbance leads to a change in the dragonfly assemblage, with certain species leaving or no longer breeding and others entering temporarily or permanently into the modified area. Assemblage composition therefore is determined by the ecoregional pool of available species and by the extent, severity, and length of the local disturbance.

Dragonfly assemblages are rarely permanent, even in the most natural conditions. Water levels and vegetation canopies change with time, and this stimulates turnover in dragonfly species. Some species are even specialists of temporary pools, and others come and go with the vagaries of El Niño climatic cycles. Other species are more permanent and may be localized endemics and inhabitants of clear montane streams that have existed for millennia. These narrow endemic species not only need protection in their own right but also are highly sensitive indicators of changing environmental conditions. As such, they are valuable in conservation planning.

The pan-African value of dragonflies is that they are an invertebrate group that is well known taxonomically and biogeographically. Also they are conspicuous and easy to sample by simply using close-focus binoculars.

Species Richness of the Africa Fauna

About 720 species of Odonata have been recorded in Africa and neighboring islands, including Madagascar. The rate of description of new species from the continent is one species about every 2 years. The total is likely to be 750–800, although we may experience some extinctions in which species disappear through anthropogenic impact before they are scientifically described.

The richest area by far is equatorial Africa, with more than half the continent's species occurring there. East Africa is also rich, with a third of the species. Madagascar has about 150 species, and the Seychelles 18.

Arguably, there are three main centers of particularly high endemism for African Odonata species. These are Madagascar, especially the moist forests; the Cameroon highlands, including Mt. Cameroon; and the Western Cape. This endemism arises principally from montane stream fauna. These streams are at middle elevations, especially 500–1,200 m above sea level.

Threats to Dragonflies

By far the greatest threat to Odonata species is habitat loss. This is especially true for middle-elevation stream species. Landscape changes such as invasion of riparian grasslands by alien trees, removal of indigenous riparian trees in forested areas, trampling of banks by cattle, and overextraction of water for agriculture are the most serious disturbances. These impacts are also synergistic, with one threat compounding another.

There is no evidence yet that dams are detrimental to Odonata. Indeed, small farm dams with an abundance of grasses and reeds at the margins are distinctly beneficial and increase the area of occupancy. Most of the species that benefit from this landscape modification are abundant and widespread species, although a few highly localized species are given the opportunity to become much more common than they would when only in natural habitats. Dam pools that are rich in species are those with constant water levels and an abundance of marginal vegetation. Fluctuating levels, leading to loss of marginal vegetation, are highly impoverishing.

Although invasive alien trees and bushes often are detrimental to indigenous fauna, alien water weeds have mostly benefited dragonflies in both still and running water. Water weeds such as *Pistia stratiotes, Eichhornia crassipes,* and *Elodea* spp. generally increase oxygen levels in the water and provide shelter for the larvae and perches for the adults.

Invasive alien fishes, especially trout in South Africa, appear to have affected certain endemic Odonata species. The highly localized cape endemic dragonfly, *Ecchlorolestes peringueyi,* is

TABLE 1.4. **Some Threatened Dragonflies of Africa and Neighboring Islands That Have Recently Been Assessed.**

Species and Ecoregion	Threat and Current Status
Chlorolestes apricans Basking malachite Amatolo-Winterberg Highlands [36]	Highly threatened, especially by invasive alien trees, particularly *Acacia mearnsii*, and by the trampling of riverbanks by cattle.
Amanipodagrion gilliesi Pangani [70]	Highly threatened through loss of its forest habitat.
Metacnemis augusta Ceres stream damsel Cape Fold [33]	This species has not been rediscovered since 1920, despite intensive searches.
Metacnemis valida Kubusi stream damsel Amatolo-Winterberg Highlands [36]	Highly threatened by invasive alien trees, especially *Acacia mearnsii*, shading out its habitat.
Pseudagrion inopinatum Balinsky's sprite Southern Temperate Highveld [73]	On the verge of extinction; it is not entirely clear why it is so threatened.
Pseudagrion umsingaziense Umsingazi sprite Zambezi Lowveld [77]	Highly threatened by urbanization degrading its marshland habitat.
Africallagma polychromaticum Cape bluet Cape Fold [33]	Highly threatened by invasive alien trees and conversion of its stream habitat to agriculture. Has not been seen for more than 40 years.
Ceratogomphus triceraticus Cape thorntail Cape Fold [33]	Threatened by invasive alien trees along its riverine habitat. Also appears to be intrinsically rare.
Syncordulia gracilis Yello presba Cape Fold [33]	Highly threatened by invasive alien trees along its riverine habitat.
Syncordulia yenator Mahogany presba Cape Fold [33]	Highly threatened by invasive alien trees along its riverine habitat.
Orthetrum rubens Waxy-winged skimmer Cape Fold [33]	Apparently highly threatened, especially by invasive alien trees along its riverine habitat, but also appears to be intrinsically rare.
Sympectrum dilatatum St. Helena darter St. Helena Island (no ecoregion number)	Last seen in 1977, this species is thought to have been extirpated by a combination of factors, including impacts from an alien frog.

Note: These species are a selection that has been reassessed recently. Many others are data deficient but are suspected to be highly threatened. Some others have been reassessed and found not to be as threatened as previously thought.

Chlorolestes apricans. On the brink of extinction, this Amatola-Winterberg Highland species is precariously holding on in a few streams running through farmland. Photo credit: Michael Samways

now known only above waterfalls and in small streams out of reach of trout.

Threatened Species

Some of the species threatened in Africa are listed in table 1.4. All are island species or narrow endemics and are mostly subject to multiple impacts. There are likely to be more threatened species than this, especially in places such as West Africa where we have no recent assessments.

Conservation Action

Far more baseline data on the distribution of Odonata species across Africa are needed. The Congo Basin and the Albertine Rift areas are particularly in need of further inventories. Species known to be rare or threatened, some of which may already be extinct, are also in need of further searches. This applies especially to islands, the Cape, and Cameroon faunas.

Management action is needed in some areas. Protecting the Cape endemics entails alien tree removal, which is currently being undertaken through the huge Working for Water Programme. Trees in question include *Acacia mearnsii, A. longifolia,* and *Pinus* spp., which shade out habitats. Above all, however, it is critical to bring dragonflies and other focal aquatic invertebrates into the freshwater planning process.

Metacnemis valida. Only just avoiding extinction, this Amatola-Winterberg Highland species is severely threatened by the synergistic effects of invasive alien trees, cattle trampling of streambanks, trout, and detergent pollution. Photo credit: Michael Samways

CHAPTER 2

Approach

The political atlas of Africa displays the boundaries of fifty-four countries, but a map representing the biological and ecological units of the continent would be starkly different. In this chapter we present a preliminary map of the freshwater systems of Africa that reflects the distribution of selected aquatic animals across the continent. To be effective, conservation planning for freshwater systems requires units that reflect the ecological connectivity and distribution of aquatic communities across the landscape (Abell et al. 2000, 2002). In constructing a map of the freshwater ecoregions of Africa, we have made a first attempt at providing these freshwater planning units.

This chapter summarizes our current understanding of the freshwater biological framework of the Afro-Malagasy region. In the first part of the chapter we define freshwater ecoregions and discuss the rationale for their use as conservation units. We then outline a broad classification system of freshwater habitat types, which we apply to ecoregions to allow comparative analyses. Next, we summarize the general approach for analyzing data on the biological distinctiveness and conservation status of each ecoregion. We end with a process for integrating biological distinctiveness and conservation status indexes to determine conservation priorities.

Geographic Scope of the Study

Africa, associated islands (Canaries, Cape Verde, Bijagos, São Tomé, Príncipe, Annobón, Bioko, Socotra, the Seychelles, Comoros, and Mascarenes), and Madagascar encompass the study area (the Afro-Malagasy region). The freshwater species of North Africa have greater affinities with Palearctic taxa than with those of the Afrotropics (sub-Sahara), although the Nile River and parts of the Ethiopian Highlands are considered zones of overlap, containing both Afrotropical and Palearctic freshwater species for some taxa (Doadrio 1994; Brown 1994; Wishart et al. 2000). Although North Africa probably would be more accurately included in an assessment of Palearctic freshwater systems, we have opted to provide an overview of the entire continent by including North Africa in our assessment. Madagascar's freshwater fauna is distinct from that of continental Africa, and its origins are still under investigation (Benstead et al. 2000; see essay 3.6). Surrounding islands, particularly those furthest from the mainland, have limited freshwater biotas that often include only marine-derived species. We have included Madagascar and these islands in the assessment to make it comprehensive of the entire region.

Ecoregions and Bioregions

We define an ecoregion as a large area of land or water containing a distinct assemblage of natural communities and species, whose boundaries approximate the original extent of natural communities before major land use change. These communities share most of their species, dynamics, and environmental conditions and function together effectively as a conservation unit (Dinerstein et al. 1995). A biogeographic province or bioregion is a complex of ecoregions that share a similar biogeographic history and thus often have strong affinities at higher taxonomic levels (e.g., genera, families) (Abell et al. 2000).

Thirteen freshwater experts (ichthyologists, freshwater conservation biologists, and aquatic entomologists) attended a workshop in August 1998 at which they delineated a preliminary set of ecoregions and bioregions (figure 2.1). Roberts's (1975) ichthyogeographic provinces of Africa served as the basis for the bioregion delineation, with modifications based on more recent data (Greenwood 1983; Hugueny and Lévêque 1994; Doadrio 1994; Lévêque 1997) (figure 2.2a). Ecoregions are nested within these bioregions and are likewise based primarily on fish distributional data. A final set of eleven bioregions and ninety-three freshwater ecoregions for Africa and Madagascar was delineated based on further consultation with these and other experts (figure 2.1; see List of Contributors and table A.1). Ecoregion lines usually follow the boundaries of drainage basins because these basins often serve as biogeographic barriers. Where basin divides do not circumscribe species distributions or where basins contain internal barriers to dispersal, ecoregions straddle or divide basins. Specific information on the biogeographic history and delineation of boundaries for each ecoregion is given in individual descriptions in appendix F.

Freshwater Habitat Types

Whereas historical connections of rivers and lakes, past climatic conditions, and geological events largely shape the configurations of ecoregions and bioregions, the current structure and function of freshwater systems define the habitat types used in this assessment. Habitat types are not geographically defined units; rather, the freshwater major habitat types (MHTs) reflect groupings of ecoregions with similar biological, chemical, and physical characteristics. The MHTs refer to the dynamics of ecological systems and the broad habitat structures that define them, and these groupings provide a structured framework for examining and comparing the diversity of life in freshwater systems. We followed a simple system of freshwater habitat types to group ecoregions across the Afro-Malagasy region. Each ecoregion was assigned to its dominant habitat type (box 2.1 and figure 2.2b). However, because of the large scale of ecoregions, all contain patches of other habitat types. Ecoregions in the large lakes habitat type can contain swamps, floodplains, and grassy savannas in addition to the dominant lake habitat.

Aerial view of islands and waterways of central Okavango wilderness, Botswana. Photo credit: ©WWF-Canon/Martin Harvey

These smaller habitats cannot be mapped at the scale of this assessment; however, such habitat diversity contributes to species and ecosystem process diversity within ecoregions.

The major habitat types for the Afro-Malagasy region are as follows:

- closed basins and small lakes
- floodplains, swamps, and lakes
- moist forest rivers and streams
- mediterranean systems
- highland and mountain systems
- island rivers and lakes
- large lakes
- large river deltas
- large river rapids

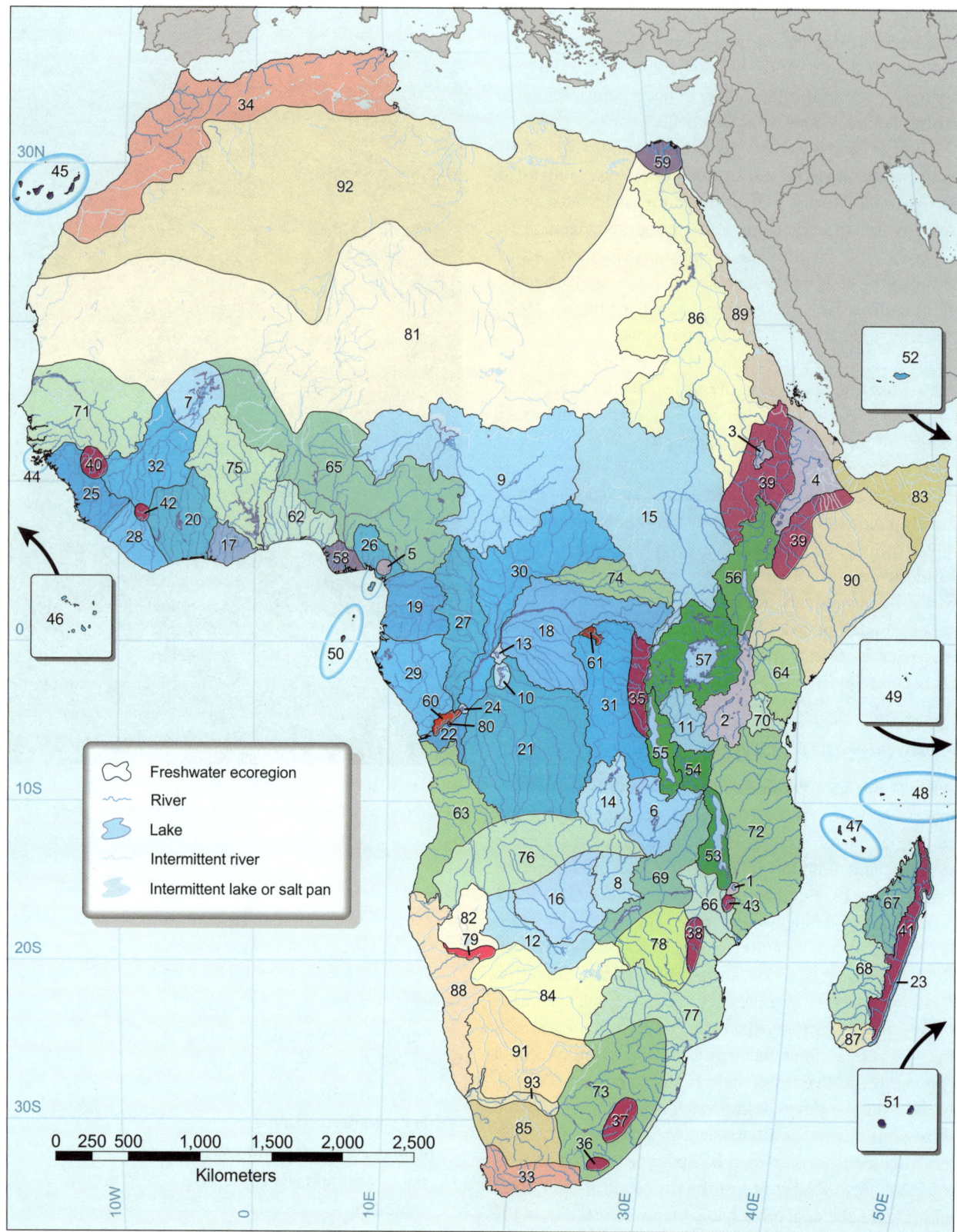

FIGURE 2.1. Freshwater ecoregions of Africa and Madagascar.

Closed Basins and Small Lakes
- 1 - Lakes Chilwa and Chiuta
- 2 - Southern Eastern Rift
- 3 - Lake Tana
- 4 - Northern Eastern Rift
- 5 - Western Equatorial Crater Lakes

Floodplains, Swamps, and Lakes
- 6 - Bangweulu-Mweru
- 7 - Inner Niger Delta
- 8 - Kafue
- 9 - Lake Chad Catchment
- 10 - Mai Ndombe
- 11 - Malagarasi-Moyowosi
- 12 - Okavango Floodplains
- 13 - Tumba
- 14 - Upper Lualaba
- 15 - Upper Nile
- 16 - Upper Zambezi Floodplains

Moist Forest Rivers
- 17 - Ashanti
- 18 - Cuvette Centrale
- 19 - Central West Coastal Equatorial
- 20 - Eburneo
- 21 - Kasai
- 22 - Lower Congo
- 23 - Madagascar Eastern Lowlands
- 24 - Malebo Pool
- 25 - Northern Upper Guinea
- 26 - Northern West Coastal Equatorial
- 27 - Sangha
- 28 - Southern Upper Guinea
- 29 - Southern West Coastal Equatorial
- 30 - Sudanic Congo (Oubangi)
- 31 - Upper Congo
- 32 - Upper Niger

Mediterranean Systems
- 33 - Cape Fold
- 34 - Permanent Maghreb

Highland and Mountain Systems
- 35 - Albertine Highlands
- 36 - Amatolo-Winterberg Highlands
- 37 - Drakensberg-Maloti Highlands
- 38 - Eastern Zimbabwe Highlands
- 39 - Ethiopian Highlands
- 40 - Fouta-Djalon
- 41 - Madagascar Eastern Highlands
- 42 - Mount Nimba
- 43 - Mulanje

Island Rivers and Lakes
- 44 - Bijagos
- 45 - Canary Islands
- 46 - Cape Verde
- 47 - Comoros
- 48 - Coralline Seychelles
- 49 - Granitic Seychelles
- 50 - São Tomé, Príncipe, and Annobon
- 51 - Mascarenes
- 52 - Socotra

Large Lakes
- 53 - Lake Malawi
- 54 - Lake Rukwa
- 55 - Lake Tanganyika
- 56 - Lake Turkana
- 57 - Lakes Kivu, Edward, George, & Victoria

Large River Deltas
- 58 - Niger Delta
- 59 - Nile Delta

Large River Rapids
- 60 - Lower Congo Rapids
- 61 - Upper Congo Rapids

Savanna–Dry Forest Rivers
- 62 - Bight Coastal
- 63 - Cuanza
- 64 - Kenyan Coastal Rivers
- 65 - Lower Niger-Benue
- 66 - Lower Zambezi
- 67 - Madagascar Northwestern Basins
- 68 - Madagascar Western Basins
- 69 - Middle Zambezi Luangwa
- 70 - Pangani
- 71 - Senegal-Gambia Catchments
- 72 - Eastern Coastal Basins
- 73 - Southern Temperate Highveld
- 74 - Uele
- 75 - Volta
- 76 - Zambezian Headwaters
- 77 - Zambezian Lowveld
- 78 - Zambezian (Plateau) Highveld

Subterranean and Spring Systems
- 79 - Karstveld Sink Holes
- 80 - Thysville Caves

Xeric Systems
- 81 - Dry Sahel
- 82 - Etosha
- 83 - Horn
- 84 - Kalahari
- 85 - Karoo
- 86 - Lower Nile
- 87 - Madagascar Southern Basins
- 88 - Namib Coastal
- 89 - Red Sea Coastal
- 90 - Shebelle-Juba Catchments
- 91 - Southern Kalahari
- 92 - Temporary Maghreb
- 93 - Western Orange

BOX 2.1. **Descriptions of the Freshwater Major Habitat Types of Africa and Madagascar.**

Closed Basins and Small Lakes. Small lakes, many of which are in closed basins, are young compared with most large lakes, and the smallest of them are vulnerable to drying out. The more ephemeral nature of small lakes translates into fewer opportunities for speciation. Also, as a result of their smaller size they tend to support fewer species. Small lakes, then, usually have less diverse biotas than large lakes and often support cosmopolitan species that tolerate a wide range of conditions (Moss 1998). In Africa, small lakes often have been formed via rifting and volcanic activity. Several lakes are the result of lava flows damming the course of a river. Crater lakes form in the craters of ancient volcanoes. There are also many small saline lakes in Africa. These lakes generally occur in semi-arid or subhumid climates where evaporation exceeds precipitation (Hammer 1986).

Floodplains, Swamps, and Lakes. Rivers and streams flow into large, depressed inland deltas or riverine lakes in these ecoregions. A variety of habitats are present, including swamps, flooded grasslands, river channels, and open water. Vegetation adapted to low-oxygen environments includes the typical *Cyperus papyrus* and *Typha domingensis* in swamps and *Oryza longistaminnis* in flooded grasslands. Certain fish in these ecoregions have also evolved adaptations for low-oxygen environments. Flooded grasslands and open water zones harbor more diverse fish species assemblages. These areas often are prime habitat for waterbirds and provide watering holes for many terrestrial species, particularly during the dry season.

Moist Forest Rivers and Streams. Gallery forests border the rivers and streams of this habitat type throughout much of their length. Generally these rivers and streams are well shaded in their upper reaches, becoming wider and more open to sunlight in the lower reaches. These rivers, located in humid areas, maintain continuous flow throughout the year. In their lower reaches, most of these rivers have both lotic and lentic components, with large backwaters and flooded swamp forests along their course. These rivers receive floods during the rainy season, with unimodal or bimodal flooding occurring, depending on the location of the ecoregion.

Mediterranean Systems. In these semi-arid ecoregions, subfreezing temperatures are infrequent (except at high elevations) and rainfall is highly seasonal. A large percentage of the annual rainfall occurs within a few months, although rainfall is highly variable between years and extended droughts are common. The freshwater fauna of these ecoregions is adapted to live under such potentially harsh conditions. Terrestrial vegetation ranges from mixed coniferous and broadleaf woodland to lower vegetation composed of dwarf shrubs with an understory of herbs and grasslike plants, often with a large diversity of flowering bulbs.

Highland and Mountain Systems. Clear, cold, and swift streams, moving quickly across well-weathered mountain rocks at high elevations, characterize this habitat type. Because of their swift nature and turbulent flow, these waters are also well oxygenated. Some of the higher-elevation ecoregions are temperate islands in the midst of tropical Africa. Certain ecoregions of this habitat type consist of the headwaters of multiple river basins that share similar faunal assemblages in their upper reaches.

Island Rivers and Lakes. These ecoregions are confined to the islands (excluding Madagascar) adjacent to Africa and Madagascar. Their freshwater fauna is depauperate because of the islands' isolation and small size. Often, species are of marine origin but permanently reside in fresh waters (vicarious species, sensu Myers 1949).

- savanna–dry forest rivers
- subterranean and spring systems
- xeric systems

Reflecting the dry nature of the region overall, the xeric systems and savanna–dry forest rivers habitat types together cover a majority (41 and 21 percent, respectively) of the land surface. Moist forest rivers and floodplains, swamps, and lakes (covering 14 and 12 percent, respectively, of the total area) are the next most extensive habitat types. All other habitat types cover less than 4 percent of the total area of the Afro-Malagasy region (table 2.1).

Freshwater habitat types play an important role in conservation planning at the regional scale. By organizing ecoregions within these habitat types, we can facilitate representation in regional conservation strategies. At the global

Large Lakes. Large lakes are distinguished by their size and generally old age, two features that encourage the evolution of a diverse biota. The large areal or vertical extent of these lakes provides for a diversity of habitats in the main pelagic, littoral, and profundal zones. For example, littoral areas may include rocky crevices, sandy shorelines, or swamp or marsh vegetation, each habitat with its specialized inhabitants. Smaller satellite lakes often border the main lake in these ecoregions and may support a subset of the same species. Large lakes tend to be stratified for at least part of the year because of their size and depth. The existence of a distinct epilimnion and anoxic hypolimnion creates a barrier that most organisms do not penetrate. The deeper lakes Malawi and Tanganyika have areas that are permanently unmixed and anoxic (Moss 1998).

Large River Deltas. As large rivers slow and deposit sediment over their floodplains before flowing into the ocean, they create vast expanses of silt deposition and a system of channels through an alluvial plain. Historically, the Niger and Nile rivers brought enormous amounts of nutrient-rich sediment from upstream areas to their deltas, creating highly productive systems. Deltaic habitats, including braided channels, coastal lakes, and mangroves, provide a variety of habitats for freshwater, brackish, and marine species. Large congregations of waterbirds also distinguish river deltas.

Large River Rapids. Rapids commonly result from a sudden steepening of the stream gradient, from the presence of a restricted channel, or from the unequal resistance of the rocks and boulders over which a stream flows. Specialized faunas often evolve to survive in the unique conditions present in these reaches of a river.

Savanna–Dry Forest Rivers. In these tropical and subtropical ecoregions, rivers flow through landscapes dominated by grasslands, sometimes interspersed with tall shrubs and trees in an open formation. The climate is generally hot, with high year-round temperatures, dry winters, and periods of heavy rainfall in the short summers. Input of allochthonous organic matter is low because of a lack of vegetative cover, and there is little cooling shade provided by riparian forests.

Subterranean and Spring Systems. Cave and spring systems often are overlooked in freshwater assessments because they generally are small landscape features. However, they tend to contain biotas that are quite distinct from those found in aboveground systems. These habitats are common in landscapes dominated by limestone and dolomite (karst). Karstic regions are particularly amenable to the formation of subterranean systems because of their high solubility (Hobbs 1992). Springs are created by discharge at the earth's surface from aquifers. Their magnitude of discharge and water temperature results from subterranean hydrologic and geologic factors. For example, thermal springs may be present in areas where volcanoes have recently been active (Hobbs 1992).

Xeric Systems. Located in xeric areas (generally receiving less than 250 mm rain per year), the aquatic systems in these ecoregions normally are ephemeral or become subterranean during the dry season. Pans, one of the habitats found in this habitat type, collect rainy season water that evaporates during the dry season, leaving a partially or completely dry depression. As the water evaporates from the pan, the salts in the remaining water become increasingly concentrated, in some cases turning a freshwater habitat into a saline one. Aquatic organisms that survive under these harsh conditions have unique adaptations, including the ability to move over land to another waterbody, to estivate in the mud until the rains come again, and to enter a state of cryptobiosis in which metabolism of the organism, seed, or egg is stopped.

---

scale, we can plan for representation within both habitat types and biogeographic realms (e.g., the African [or Ethiopian] realm).

Conceptual Foundations

We used freshwater ecoregions in this analysis because they provide appropriate units for representation of distinct species assemblages, habitats, and processes at the continental scale. The freshwater ecoregions that we offer in this assessment take into account aquatic species distributions and drainage basin divides. They are designed to serve as conservation planning units, as biogeographically meaningful units for subdividing a larger river or lake basin or for aggregating several basins sharing similar species assemblages and ecological processes into a single unit.

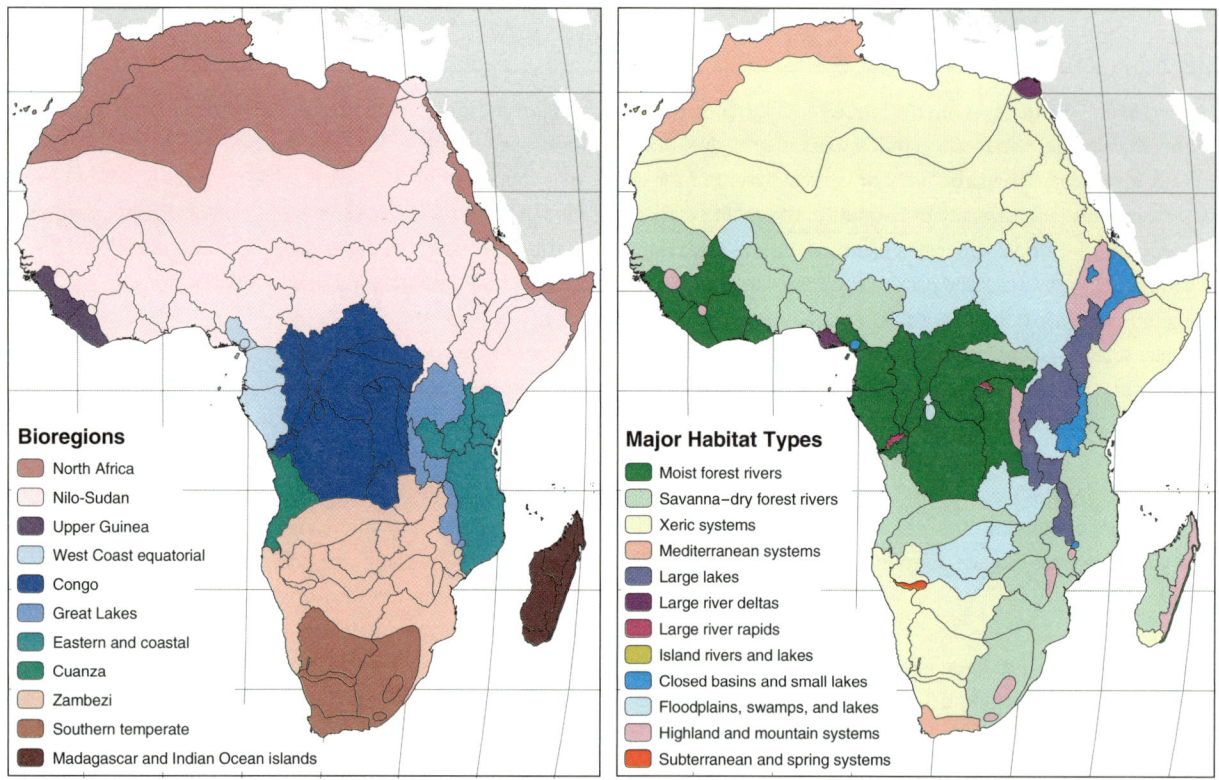

FIGURE 2.2. (a) Freshwater bioregions. (b) Freshwater major habitat types.

Four fundamental goals of biodiversity conservation provide the foundation for this work:

- representation of all natural communities
- maintenance of viable populations of species
- maintenance of ecological and evolutionary processes that create and sustain biodiversity
- design and management of the system to be responsive to short-term and long-term environmental change (Noss 1992)

To meet the goal of representation of all distinct natural communities, we ensured that representative ecoregions from each MHT and bioregion were included in the final set of priorities. We addressed the viability of populations, at a large scale, by integrating an analysis of each ecoregion's conservation status into our priority setting. Ecological and evolutionary processes often are ignored in priority-setting exercises; however, we also included these important elements in our analysis of each ecoregion's biological distinctiveness. Finally, the ecoregion provides an appropriate unit for evaluating and managing for the effects of short- and long-term environmental change.

TABLE 2.1. **Area of Major Habitat Types.**

Major Habitat Type	Area (km^2)	% Area
Closed basins and small lakes	385,385	1
Floodplains, swamps, and lakes	3,619,864	12
Moist forest rivers	4,213,424	14
Mediterranean systems	1,082,867	4
Highland and mountain systems	806,608	3
Island rivers and lakes	25,361	0.1
Large lakes	937,924	3
Large river deltas	106,569	0.4
Large river rapids	28,742	0.1
Savanna–dry forest rivers	6,356,973	21
Subterranean and spring systems	26,162	0.1
Xeric systems	12,362,292	41

Water at Sossusvlei, a rare 1-in-10-year event, Sossusvlei, Namib-Naukluft National Park, Namibia. Photo credit: ©WWF-Canon/Martin Harvey

Elements of Analysis: Biological Distinctiveness and Conservation Status Indexes

In this assessment we used two indexes—biological distinctiveness and conservation status—to assign priority levels to ecoregions. WWF first developed these indexes for a conservation assessment of terrestrial ecoregions in Latin America and the Caribbean (Dinerstein et al. 1995) and has used them in subsequent analyses (Olson et al. 1998; Abell et al. 2000; Wikramanayake et al. 2002; Burgess et al. 2004). Each index was evaluated using several criteria that are outlined in this chapter and described in detail in appendixes A and B. These appendixes also include full lists of data sources used to evaluate the criteria.

Biological Distinctiveness Index

Biodiversity assessments typically focus at the species level. Our use of the term *biological distinctiveness* invokes a broader definition of biodiversity: in addition to species, we incorporated ecosystem diversity and the ecological processes that sustain biodiversity. The Biological Distinctiveness Index (BDI) is a measure of the degree to which the biodiversity of an ecoregion is distinctive at a range of biogeographic scales. Four main criteria were used to evaluate biological distinctiveness:

- species endemism (number of endemics and percent endemism)
- species richness
- rare ecological and evolutionary phenomena (e.g., species radiations, mass migrations)
- rarity of habitat type

The relative biological distinctiveness of each ecoregion was evaluated within each major habitat type because the index is based largely on values of species richness and endemism, which vary widely between different habitats. For example, a direct comparison of species richness in mediterranean and moist forest river habitats would yield little useful information.

SPECIES BIOLOGICAL VALUES (ENDEMISM AND RICHNESS)

We assessed species endemism and richness for each ecoregion after systematically comparing our ecoregion map with the estimated ranges of more than 4,300 African freshwater species representing five taxonomic groups: fishes, aquatic mammals, frogs that depend on aquatic habitats, freshwater mollusks, and aquatic and semi-aquatic reptiles. These taxonomic groups were chosen because they are important elements of freshwater biotas and a reasonable amount of distribution data is available for them. Only organisms that spend a majority of their time in freshwater for at least one life stage were included in the assessment. We did not incorporate waterbird species in our species-level analysis because the vagility of most waterbirds makes their distributions correspond poorly with ecoregions (Lévêque 1997; Fishpool and Evans 2001). We assigned presence or absence to each species in each ecoregion, based on the best available information (see appendix A and table A.1 for a full list of sources and methods).

Within each major habitat type, we awarded points for both species endemism and species richness (0–6 and 1–3 points, respectively). Both number of endemic species and percentage endemism (based on totals from all taxa) were evaluated for each ecoregion, and the one that earned a higher number of points was used in the analysis. Endemism was awarded twice as many points as richness to account for the greater contribution of endemism to biological distinctiveness and the unique conservation challenges presented by endemic species. Before assigning point values, we adjusted species richness by ecoregion area to address the correlation between species numbers and ecoregion size (see appendix E). We summed endemism and species richness points and assigned a preliminary biological distinctiveness category to each ecoregion. The four categories reflect the distinctiveness of the ecoregion's biodiversity at different biogeographic scales: globally outstanding (9 points), continentally (e.g., biogeographic realm or regionally) outstanding (7–8 points), bioregionally (e.g., Nilo-Sudan) outstanding (5–6 points), or nationally important (4 points or less) (see appendix A for a full discussion of methods).

NONSPECIES BIOLOGICAL VALUES (PHENOMENA AND RARE HABITATS)

In addition to species endemism and richness, we included two nonspecies biological attributes in the BDI: ecological or evolutionary phenomena and rare habitat types. The biological distinctiveness of an ecoregion was raised to globally or continentally outstanding if it was judged to contain globally or continentally outstanding ecological or evolutionary phenomena or rare habitat types.

Ecological or evolutionary phenomena include globally outstanding centers of evolutionary radiation, endemism at higher taxonomic levels, and exceptional waterbird congregations (see box 3.1 and table 3.2). Incorporation of ecological or evolutionary phenomena is designed to capture ecoregion characteristics that numeric species measures miss. Our evaluation of evolutionary radiations is based primarily on expert assessment. Analysis of the species databases for each taxon provided information on higher-level taxonomic endemism, and BirdLife International provided data on waterbird congregations (see appendix A for a full description of data sources and decision rules).

We also evaluated the occurrence of rare habitat types, important for their contribution to global and continental biodiversity. These habitat types occupy a lower hierarchical level than MHTs and occur within ecoregions. Rarity can be either natural or human-induced. Thus, a habitat type that was once widespread but has been disrupted in the majority of its original locations was treated, in its remaining locales, as equal to a habitat type that was originally rare. An ecoregion was considered globally outstanding if fewer than eight ecoregions worldwide contain the particular habitat type and continentally outstanding if fewer than three contain it in Africa. We consulted with numerous experts and literature sources to evaluate the distribution of rare habitat types. This measure represents the number of opportunities to conserve this habitat type worldwide and the corresponding importance of an African ecoregion that contains it.

DATA QUALITY

At the initial workshop where the ecoregions were delineated, experts also evaluated the quality of data available for each ecoregion as high, medium, or low. They based the evaluations on the availability of recent biological data for each ecoregion and taxonomic revisions for species occurring there. Large portions of Africa have been poorly sampled, and the taxonomies of several freshwater taxa remain unresolved. We correlated the results of the data quality assessment with the results of the biological distinctiveness analysis to explore the effect of data quality on the distinctiveness categories assigned to each ecoregion. The data quality analysis also can inform future research priorities.

Conservation Status Index

The second major discriminator, conservation status, was designed to estimate the current and future potential of an ecoregion to meet three fundamental goals of biodiversity conservation: maintaining viable species populations and

TABLE 2.2. **Categories of Threats Used to Assess the Integrity of Each Ecoregion and Examples for Each Category.**

Land-Based Threats (land degradation)	Aquatic Habitat Threats	Biota Threats (exploitation and exotics)
Intensive logging and associated road building	Degraded water quality (e.g., point or non–point source pollution; changes in temperature, pH, other physical parameters; sedimentation or siltation)	Unsustainable fishing or hunting
Intensive grazing, particularly in riparian zone		Unsustainable extraction of wildlife or plants as commercial products
Widespread mining or other resource extraction	Habitat fragmentation from dams or other barriers to dispersal, migration, and general movement	Competition, predation, hybridization, or infection by established exotic species
Agricultural expansion and clearing for development	Excessive recreational impacts	
Urbanization and associated changes in runoff	Altered hydrographic integrity (flow regimes, water levels) resulting from dams, surface or groundwater withdrawals, channelization, interbasin water transfers, etc.	
Loss or conversion of riparian and floodplain vegetation		
Reduced organic matter input, including woody debris	Loss of aquatic habitat, such as from flooding by reservoirs or from desiccation	

communities, sustaining ecological processes, and responding effectively to short- and long-term environmental change. The snapshot conservation status is based on three criteria: degree of land-based threats (land degradation), aquatic habitat threats, and biota threats (effects of introduced species and species exploitation). We then assessed future threats over the next 20 years to arrive at the threat-modified conservation status, or final conservation status. A complete description of the conservation status index (CSI), its components and design, is found in appendix B.

CATEGORIES OF THREATS

As the conservation community has become aware of the perilous state of freshwater biodiversity (Ricciardi and Rasmussen 1999), scientists have documented the causes behind the decline of many species (O'Keeffe et al. 1987; Williams et al. 1989; Allan and Flecker 1993; Davies and Day 1998; Shumway 1999). Allan and Flecker (1993) describe six categories of threats: habitat loss and degradation, introduced species, overexploitation, secondary extinctions (extinction of a species caused by the extinction of another species), chemical and organic pollution, and climatic change. Shumway (1999) adds policy deficiencies and inadequate planning, and Stiassny (2002a) adds a knowledge impediment to the list. Noss and Cooperrider (1994) recategorize the threats into three overarching categories: resource misuse, pollution, and exotics. No matter how the threats are classified, most scientists agree that habitat degradation and species invasions are the top causes of loss of freshwater biodiversity, although any given system generally faces multiple threats (Allan and Flecker 1993; Harrison and Stiassny 1999).

For the purposes of this analysis we followed O'Keeffe et al. (1987), who categorized threats as related to the river, catchment, or biota. Given the lack of comprehensive, continental datasets on threats to freshwater systems in Africa, we classified threats into extent of land-based threats within the ecoregion (e.g., landcover change), direct threats to aquatic habitats (those that occur within the stream channel, lake, or other waterbody), and biota threats (effects of introduced species and overharvesting) (table 2.2). We did not include secondary extinctions or political shortcomings in our assessment of threats. Secondary extinctions are difficult to quantify and tend to result in local population losses rather than global extinctions. Policy deficiencies and inadequate planning generally manifest themselves as one or more of the threats identified in table 2.2. Unlike some other conservation assessments, we did not use individual species imperilment as a criterion because imperilment is a consequence of the threats that we evaluated. Additionally, information on imperilment for most African freshwater species is lacking (IUCN 2002).

We assessed future threats separately from current threats. Some freshwater systems may face impending threats (i.e., within the next 20 years) that, if they occurred, would change an ecoregion's conservation status ranking.

Many future threats are continuations of current activities or trends. Future threats could include anticipated high population growth and resultant effects, planned interbasin water transfers or dams, landcover and climate change and related effects on discharge, and planned mining or logging operations.

We used a combination of expert assessment and geographic information system (GIS) analysis of existing datasets to assess each of the three current threat categories and future threats. Each current threat category was awarded 1–3 points, based on the severity of threat (low threat = 1, medium = 2, and high = 3). To assess land-based threats we combined landcover data with population data to estimate the percentage of degraded land in each ecoregion. We evaluated aquatic habitat threats based largely on expert opinion and literature review, assigning points based on clear decision rules (see appendix B). However, we also conducted an analysis of the FAO database on African dams (FAO 2000a), elevating to the highest level of aquatic habitat threat the ecoregions that had more than fifty dams or one or more dams with a capacity of at least 150 billion m^3. We assessed biota threats largely using expert opinion and literature review.

The range of potential future threats and compounding effects among them make an exhaustive analysis of future trajectories and impacts impossible. For this analysis, we attempted to paint a general picture of possible trends to help assess which ecoregions are most likely to change and where proactive conservation efforts are needed most urgently. We used data on planned infrastructure or resource extractions, climate change, and projected population densities to categorize ecoregions into three levels of future threat: high, medium, and low. The full details on the methods, thresholds, and data sources used for evaluating each of the threat categories are provided in appendix B.

SNAPSHOT CONSERVATION STATUS

We assessed the conservation status of ecoregions in the tradition of the IUCN Red Data Book categories for threatened and endangered species (critical, endangered, and vulnerable), but instead of focusing on individual species we estimated the state of whole faunas, ecological processes, and ecosystems. We summed the points awarded for the three categories of current threats into a single index, from which five categories of conservation status were derived: critical, endangered, vulnerable, relatively stable, and relatively intact. These categories follow those of Dinerstein et al. (1995).

In this section we provide a generalized description of each conservation status category. Not all conditions listed for a given category need to occur to warrant that classification. The descriptions reflect how, with increasing habitat loss, degradation, and fragmentation, ecological processes cease to function naturally, populations no longer occur within their natural ranges of variation, and major components of biodiversity are eroded (adapted from Dinerstein et al. 1995 and Abell et al. 2000).

Critical: The remaining intact habitat is limited to isolated areas or stream segments where populations have low probabilities of persistence over the next 5–10 years without immediate or continuing habitat protection and restoration. Remaining habitat does not meet the minimum needs for maintaining viable populations of many species and ecological processes. Surrounding land use practices are incompatible with maintaining aquatic habitat structure and function. Hydrographic integrity has been severely modified by permanent, large-scale structures. Established exotic species seriously threaten native species populations. Consistently poor water quality excludes all but the hardiest species from large portions of remaining habitat. Many species are already extirpated or extinct.

Endangered: Remaining intact habitat is restricted to isolated areas or segments of varying size and length (a few larger intact areas or reaches may be present) where populations have medium to low probabilities of persistence over the next 10–15 years without immediate or continuing habitat protection or restoration. Remaining habitat does not meet minimum needs for many species populations and large-scale ecological processes. Surrounding land use practices are largely incompatible with maintaining aquatic habitat structure and function. Hydrographic integrity has been modified by structures of varying size and permanence. The spread of exotic species poses a potentially serious threat to native species populations. Poor water quality excludes many species from remaining habitat. Some species are already extirpated or extinct.

Vulnerable: Remaining intact habitat occurs in blocks or segments ranging from large to small; in many intact areas, populations probably will persist over the next 10–20 years, especially if these areas are given adequate protection and moderate restoration. Some remaining habitat meets minimum needs for species populations and large-scale ecological processes. Surrounding land use practices sometimes are compatible with maintaining aquatic habitat structure and function. Hydrographic integrity may be restored in some areas by the implementation of moderate changes. Established exotic species may be controllable. Some species may already be extirpated or extinct.

Relatively stable: Natural communities have been altered in certain areas, causing local declines in some populations and disruption of ecosystem processes. These disturbed areas can be extensive, but they are still patchily distributed relative to the area of intact habitats. Ecological links between intact habitats are still largely functional. Sensitive species are still present but at diminished densities. Hydrographic integrity, if altered, could be restored with the implementation of minor changes. Surrounding land use practices do not impair aquatic habitats or could be modified easily to minimize impacts. Exotic species pose little or no threat to natives. A nearly full complement of native species still exists.

Relatively intact: Native communities in an ecoregion are largely intact, with species, populations, and ecosystem processes occurring within their natural ranges of variation. Populations of sensitive species are not diminished. Species move and disperse naturally within the ecoregion. Ecological processes fluctuate naturally throughout largely contiguous natural habitats. Hydrographic integrity is unmodified, and surrounding land use does not impair aquatic habitat. Maintenance of current conditions will conserve native species over both the short and long term.

FINAL (THREAT-MODIFIED) CONSERVATION STATUS

We modified the snapshot conservation status of each ecoregion according to the degree of expected future threat. An ecoregion with high threat was elevated by one conservation status level to arrive at its modified conservation status. (For example, an endangered ecoregion with high threat was promoted to critical.) Conservation status for ecoregions with moderate or low threat remained unchanged (see appendix B).

Integrating Biological Distinctiveness and Conservation Status

Biological distinctiveness and conservation status are two essential discriminators for biodiversity conservation planning at large scales. They combine an evaluation of the relative biological importance with a measure of anthropogenic impacts, both current and projected, facing each ecoregion. Considered together, the two indexes provide a powerful tool for setting regional priorities when limited resources necessitate careful and strategic planning.

Some freshwater ecoregions might be so degraded that their biodiversity will decline even further without intensive restoration efforts and protection of the few remaining sites containing native biota. Alternatively, some ecoregions may offer opportunities to maintain the ecological integrity of large intact basins over the long term. Integrating the re-

Waterfall, Montagne d'Ambre National Park, Madagascar. Photo credit: ©WWF-Canon/Olivier Langrand

sults from the BDI and CSI helps to categorize ecoregions into these and other conservation trajectories.

To integrate the two discriminators, we used a matrix developed by Dinerstein et al. (1995). Biological distinctiveness categories lie along the vertical axis, and conservation status categories are along the horizontal axis (figure 2.3). Based on their classifications for both indexes, ecoregions fall into one of the twenty cells of the matrix. The entire assessment process, including this integration step, is carried out separately for each MHT to ensure representation of each MHT in the final prioritization. The twenty cells are organized into five classes, which reflect the nature and extent of the management activities likely to be required for effective biodiversity conservation:

Class I: Globally outstanding ecoregions that are highly threatened. These ecoregions contain elements

Biological Distinctiveness	Final Conservation Status				
	Critical	Endangered	Vulnerable	Relatively Stable	Relatively Intact
Globally Outstanding	I	I	I	III	III
Continentally Outstanding	II	II	II	III	III
Bioregionally Outstanding	IV	IV	V	V	V
Nationally Important	IV	IV	V	V	V

FIGURE 2.3. Integration matrix for priority-setting.

of biodiversity that are of extraordinary global value or rarity. Conservation actions in these ecoregions must be swift and immediate to protect the remaining source pools of native species and communities.

Class II: Continentally outstanding ecoregions that are highly threatened. Conservation actions must be immediate to protect the remaining native species and habitats, but the overall biological value is lower than that of the Class I ecoregions. Immediate protection of remaining habitat and potential restoration for altered areas are recommended.

Class III: Ecoregions with globally or continentally outstanding biodiversity values that are not highly threatened at present. These ecoregions represent some of the last places where intact basins, large expanses of intact habitat, and associated species assemblages might be conserved.

Class IV: Bioregionally outstanding and nationally important ecoregions that are highly threatened. These are lower-priority ecoregions that are under stress. Conservation actions include protection of remaining representative species assemblages. Proper management of water resources, conservation of remaining intact subbasins, and monitoring of ecological integrity are also needed.

Class V: Bioregionally outstanding and nationally important ecoregions that are not highly threatened at present. These ecoregions are some of the last places where large areas of intact habitat and associated species assemblages might be conserved, but they are less important biologically than the Class III ecoregions. The protection of representative intact basins and proper management of water resources are recommended.

This priority-setting should not discount the fact that biodiversity conservation is important in every ecoregion because naturally functioning ecosystems provide many services to animal, plant, and human communities (van Wilgen et al. 1996; Costanza et al. 1997; Jackson et al. 2001; Brismar 2002; Tockner and Stanford 2002; Dudley and Stolton 2003) (see essays 1.1 and 1.2). Conservation of natural areas in all ecoregions ensures preservation of distinct species and communities as well as of genetic and functional diversity of populations across species ranges (Hughes et al. 1997).

However, there are ecoregions around the world that warrant more immediate attention from conservationists because they are of global importance biologically, and they are at immediate risk from anthropogenic threats (Olson and Dinerstein 1998). With limited resources and time available for conservation, it is important to be strategic in the allocation and timing of conservation effort and funds. This framework is designed to assist in this process and to highlight the extraordinary nature of African freshwater biodiversity. We recognize that others might choose a different priority-setting scheme, placing greater emphasis on other cells than we have highlighted, and for this reason we include all data generated in the assessment so that others might set their own priorities.

CHAPTER 3

Biological Distinctiveness of African Ecoregions

Africa and Madagascar's Fresh Waters: Teeming with Life

The inland waters of Africa and Madagascar are extremely diverse both in form and in the life that they support. At the heart of Africa, the Guinean-Congolian forested rivers are some of the biologically richest waters on the continent, with species adapted to rapids, swamp forests, large and small rivers, and lateral lakes (Kamdem Toham et al. 2003). Equally rich are East Africa's great lakes, with their extraordinary species flocks of cichlid fishes (Turner et al. 2001). These large lakes, whose own evolutionary history is still being deciphered, also represent great repositories of information for understanding the mechanisms of evolutionary processes (Kaufman et al. 1997; Seehausen 2000; essay 3.1). At the tips of the continent in northern and southern Africa, endemic species and relicts of ancient lineages survive in mediterranean climates (Brown 1994; Doadrio 1994; van Nieuwenhuizen and Day 2000a). The crater lakes of the Cameroonian Highlands, the forested rivers of Upper Guinea, and the swamp forests of the Niger Delta in West Africa also support endemic families and genera of aquatic species (Lévêque et al. 1989; Stiassny et al. 1992; essay 3.5). Many of the largely xeric regions of the continent are punctuated by vast floodplains, such as the Sudd and the Okavango. Additionally, areas with saline lakes, such as the southern Eastern Rift Valley, support large congregations of wetland birds (Fishpool and Evans 2001). Madagascar's rivers and lakes are also home to a distinctive freshwater fauna, including endemic taxa of crayfish, aquatic insects, amphibians, and fish (Goodman and Benstead 2003).

The biological wealth of Africa and Madagascar's freshwater systems is indisputable. Yet biodiversity varies across the landscape, with some areas more distinctive or species-rich than others. As described in chapter 2, we used a biological distinctiveness index (BDI), which includes measures of species endemism, species richness, ecological and evolutionary phenomena, and habitat rarity, to compare the ecoregions across the Afro-Malagasy region. In this chapter we present the results of the biological distinctiveness assessment. Anthropogenic features are assessed within a conservation status index (CSI) presented in chapter 4.

Species Biological Values

Richness

The highest species numbers in Central Africa lie within the Congo Basin and the Lower Guinean ecoregions, in parts of West Africa, and in the Rift Valley lakes (figure 3.1a and appendix C). High species richness is typical for certain MHTs (see appendix E for a full statistical discussion; figure 3.2). The moist forest rivers and large lakes tend to have more species than the other habitat types. MHTs with characteristically low species richness at the ecoregion scale are xeric systems, subterranean and spring systems, and island rivers and lakes.

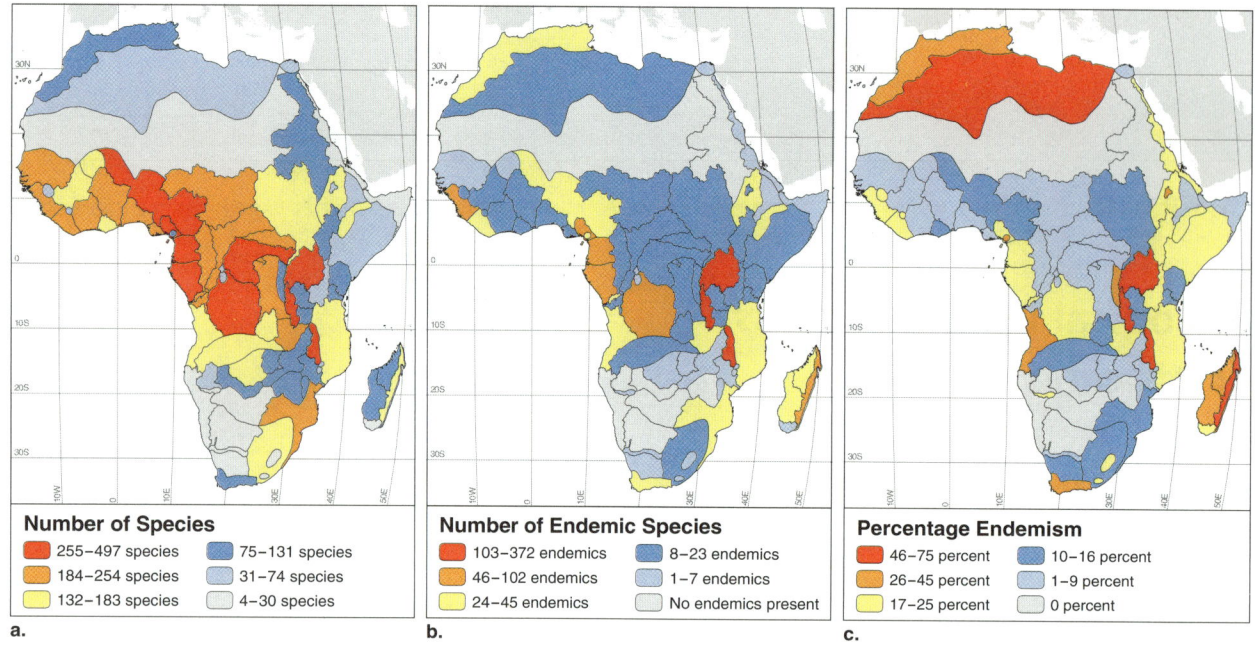

FIGURE 3.1. (a) Total species richness. (b) Total number of endemic species. (c) Percentage of endemic species. (Classification method: Natural Breaks [Jenk's optimization]. This method identifies breakpoints between classes using a statistical formula that identifies groupings and patterns inherent in the data.)

Within some MHTs, certain ecoregions stand out in terms of high total species richness (the sum of all taxa richness values) (appendix C). In the large lakes MHT, for instance, Lake Malawi [53], Lake Tanganyika [55], and Lake Victoria [57] have the highest total species richness. Among moist forest rivers ecoregions, Central West Coastal Equatorial [19], Kasai [21], Cuvette Centrale [18], and Northern West Coastal Equatorial [26] have the highest overall richness values. Even after accounting for area effects, richness in some ecoregions dwarfs that in other ecoregions of the same MHT (appendix E).

FISH

Moist forest rivers and large lakes ecoregions support significantly more fish species than ecoregions of all other MHTs except large river deltas and large river rapids (table E.2). Ecoregions with the highest fish richness in Africa are all in the Great Lakes bioregion: Lake Malawi [53], Lake Tanganyika [55], and Lake Victoria [57] (figure 3.3a). Each of these ecoregions has more than 280 described fish species, most of which are cichlids. The numbers of described species for these lakes are significant underestimates of the numbers expected to be described, although the total fish richness for each lake does reflect the expected and relative richness levels (Turner et al. 2001). Several ecoregions in the

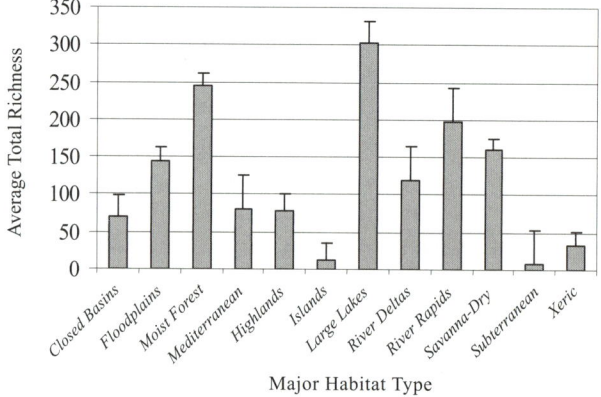

FIGURE 3.2. Average total richness by major habitat type with standard deviation bars.

West Coastal Equatorial and Congo River bioregions also support high fish richness (essay 3.2). The Central West Coastal Equatorial [19], Southern West Coastal Equatorial [29], Cuvette Centrale [18], Malebo Pool [24], Lower Congo [22], and Kasai [21] ecoregions all support more than 200 fish species. These numbers are expected to increase with further research because many waters of these ecoregions are undersampled. The Lower Niger-Benue [65] ecoregion, found in the Nilo-Sudan bioregion, also has more than 200

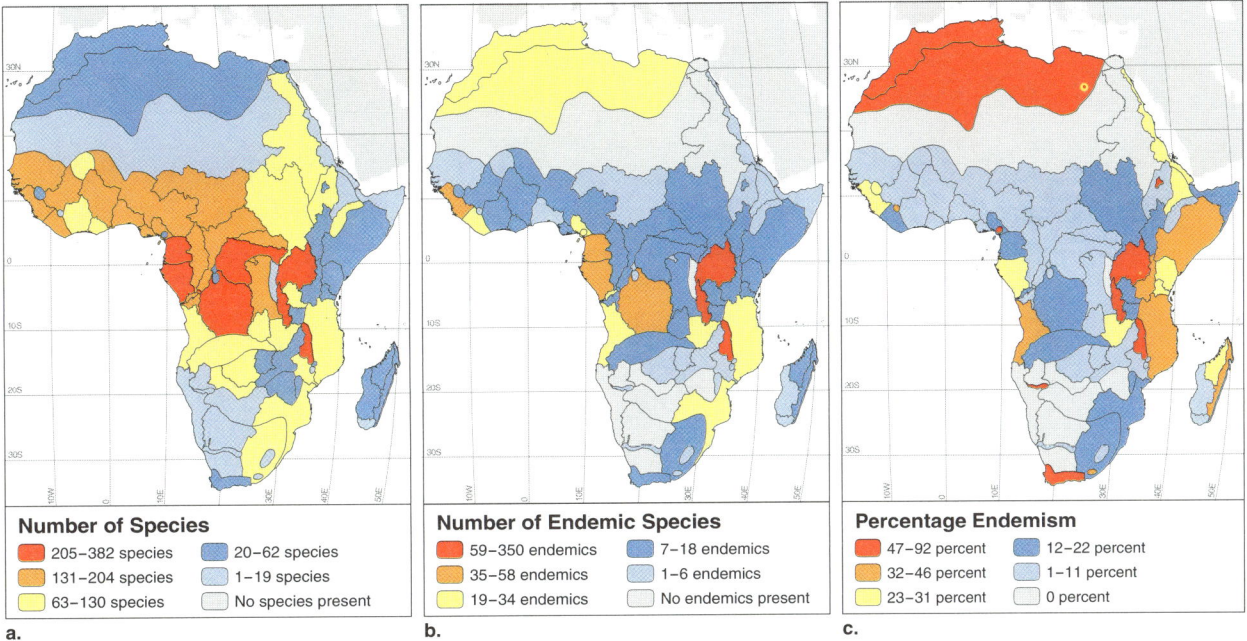

FIGURE 3.3. (a) Fish richness. (b) Number of endemic fish species. (c) Percentage of endemic fish species. (Classification method: Natural Breaks [Jenk's optimization]. This method identifies breakpoints between classes using a statistical formula that identifies groupings and patterns inherent in the data.)

fish species. An additional ten ecoregions have 150–200 species, and all of these are located in West and Central Africa. Fish richness tends to decrease outward from the equator, and xeric, island, and subterranean ecoregions have low species richness values. Unsurprisingly, the most species-poor ecoregions are all either small islands, localized caves, or desert. These depauperate ecoregions, all with zero to three freshwater fish species, are São Tomé, Príncipe, and Annobon [50]; Thysville Caves [80]; Southern Kalahari [91]; Bijagos [44]; Canary Islands [45]; Cape Verde [46]; Coraline Seychelles [48]; and Socotra [52].

AQUATIC MOLLUSKS

Aquatic mollusk richness generally shows similar patterns to that of fish, though with far fewer species (figure 3.4a). Large lake and savanna–dry forest river ecoregions contain significantly more species, on average, than all other MHTs (table E.3). Lake Malawi [53], Lake Tanganyika [55], Lake Victoria [57], Bangweulu-Mweru [6], Upper Congo [31], Lower Niger-Benue [65], Eastern Coastal Basins [72], Bight Coastal [62], and Eburneo [20] all support more than thirty species. As with fish, mollusk richness tends to decrease outward from the equator, and xeric, island, and subterranean ecoregions have few, if any, aquatic mollusks. Many of the highland and mountain ecoregions and other ecoregions with

An undescribed species of *Campylomormyrus* of the snoutfish family (Mormyridae) from the Congo. Photo credit: Carl Hopkins, Cornell University

Lanistes carinatus, a prosobranch snail. Photo credit: Henry Madsen, Danish Bilharziasis Laboratory

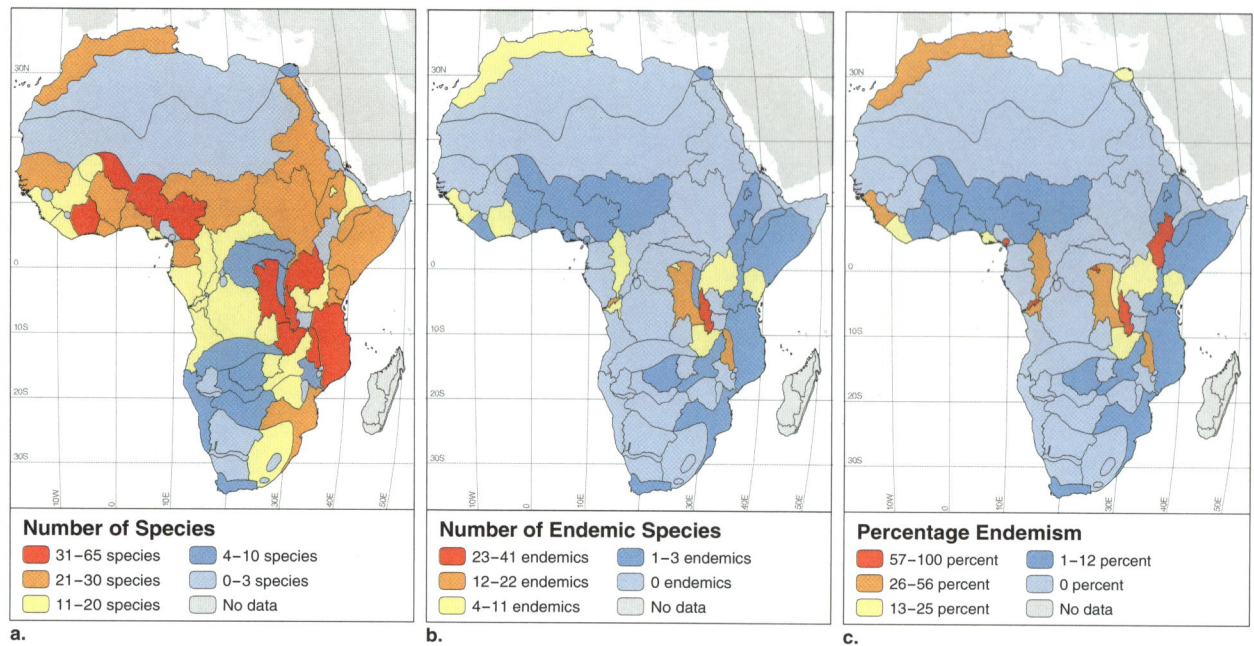

FIGURE 3.4. (a) Aquatic mollusk richness. (b) Number of endemic mollusk species. (c) Percentage of endemic mollusk species. (Classification method: Natural Breaks [Jenk's optimization]. This method identifies breakpoints between classes using a statistical formula that identifies groupings and patterns inherent in the data.)

a small areal extent are also depauperate. Data for aquatic mollusks were unavailable for Madagascar; therefore, we excluded mollusk counts from the ecoregional species totals of all MHTs represented on Madagascar (i.e., moist forest rivers and streams, highland and mountain systems, savanna–dry forest rivers, and xeric systems).

AQUATIC HERPETOFAUNA

Freshwater herpetofauna richness reaches its highest levels in ecoregions in Central and West Africa and in the Madagascar Eastern Highlands [41]. This Madagascan ecoregion, along with the Lower Niger-Benue [65], Northern West Coastal Equatorial [26], and Central West Coastal Equatorial [19], all support more than 100 species of aquatic-dependent reptiles and amphibians (figure 3.5a). Aquatic frogs make up most of these numbers. In many parts of Africa, new species are routinely being described, and these herpetofaunal numbers probably represent substantial underestimates of the true number of species.

AQUATIC MAMMALS

Only fourteen strictly aquatic mammals—those that spend a majority of their life in the water—occur in Africa and were considered in the analysis. Richness of this limited group was highest in Lake Victoria [57], Southern Upper Guinea

Arum lily (reed) frog (*Hyperolius horstocki*), Cape, Republic of South Africa. Photo credit: ©WWF-Canon/Martin Harvey

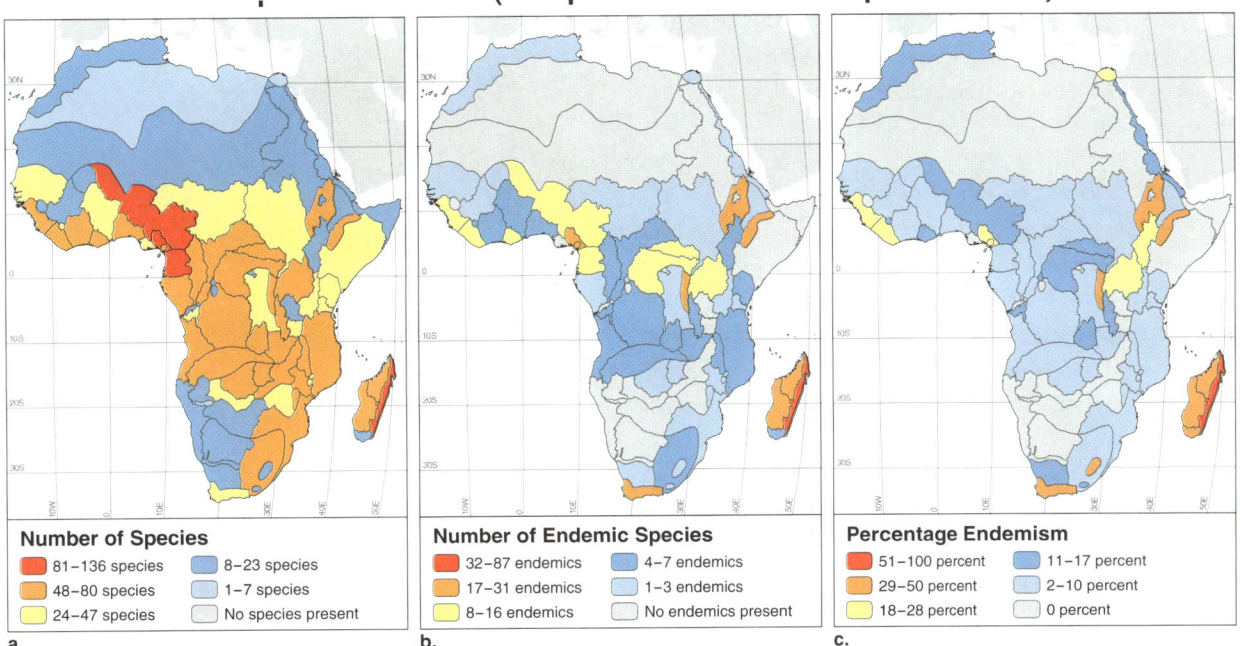

FIGURE 3.5. (a) Herpetofauna richness. (b) Number of endemic herpetofauna species. (c) Percentage of endemic herpetofauna species. (Classification method: Natural Breaks [Jenk's optimization]. This method identifies breakpoints between classes using a statistical formula that identifies groupings and patterns inherent in the data.)

[28], Upper Congo [31], and Northern West Coastal Equatorial [19], all of which had eight species (appendix C).

Endemism

Ecoregions supporting high numbers of endemic species are nearly all a subset of the ecoregions with the greatest total species richness (figure 3.1b). The three great lakes in East Africa rank highest for total number of endemics, followed by Madagascar Eastern Highlands [41], Kasai [21], Northern Upper Guinea [25], and several ecoregions in the West Coastal Equatorial bioregion. Within their respective MHTs, Western Equatorial Crater Lakes [5], Bangweulu-Mweru [6], Cape Fold [33], Permanent Maghreb [34], Lower Congo Rapids [60], Cuanza [63], Madagascar Northwestern Basins [67], Eastern Coastal Basins [72], and Temporary Maghreb [92] also rank high for total number of endemics.

FISH

Fish species endemism is by far the highest in the Great Lakes bioregion: Lake Victoria [57] (N = 224), Lake Malawi [53] (N = 350), and Lake Tanganyika [55] (N = 231) all have more than 200 endemics. More than 75 percent of the fish endemics for each of these lakes are members of the family Cichlidae, with most of them belonging to a small number of

Hippopotamus (*Hippopotamus amphibious*) wading, sub-Saharan Africa. Photo credit: ©WWF-Canon/Martin Harvey

closely related lineages (Turner et al. 2001). Fish endemism is much less, but still high, in parts of the Upper Guinea, West Coast Equatorial, and Congo bioregions (figure 3.3b). The Central West [19] and Southern West [29] Coastal Equatorial ecoregions, Kasai [21], and Northern Upper Guinea [25] contain forty to sixty endemic fish species.

AQUATIC MOLLUSKS

Lake Victoria [57] (N = 11), Lake Malawi [53] (N = 18), and Lake Tanganyika [55] (N = 42) are among the five richest ecoregions for endemic mollusks (figure 3.4b). The other two are the Upper Congo [31] and Lower Congo Rapids [60] ecoregions, with twenty-two and sixteen endemic aquatic mollusks, respectively.

AQUATIC HERPETOFAUNA AND MAMMALS

Highland regions tend to support the largest numbers of endemic herpetofauna and mammals. All of the Madagascan ecoregions, both lowland and highland, and the Albertine [35] and Ethiopian [39] Highlands, each support more than twenty endemic aquatic reptiles and amphibians (figure 3.5b). One ecoregion in the West Coast Equatorial bioregion, the Northern West Coastal Equatorial [26], also has twenty-eight endemic species. The Madagascar Eastern Highlands [41] has the highest number of endemic herpetofauna, with eighty-seven endemic frog species described. Endemic aquatic mammals, although few in number, are also found in highland ecoregions. Mount Nimba [42], Albertine Highlands [35], and Madagascar Eastern Highlands [41] support, respectively, the Mount Nimba otter shrew (*Micropotamogale lamottei*), Ruwenzori otter shrew (*Micropotamogale ruwenzorii*), and web-footed tenrec (*Limnogale mergulus*).

Percentage Endemism

Although our approach emphasizes the importance of conserving the largest number of endemic species possible, evaluation of the absolute number of endemics can obscure patterns of endemism in species-poor ecoregions. When we examine the percentage of each ecoregion's total number of species that are endemic, several ecoregions that were not highlighted for absolute numbers of endemics rank near the top (figure 3.1c). For example, Lake Tana [3], Northern Upper Guinea [25], Madagascar Eastern Lowlands [23], Socotra [52], and São Tomé, Príncipe, and Annobon [50] garnered the largest number of points possible for percentage endemism but not for total number of endemics. Several of these ecoregions are islands or cover a small areal extent. Many ecoregions showed congruence between number of endemics and percentage endemism, with thirteen receiving the highest number of points for both indexes.

Data Quality

Taxonomic data quality is considered lowest in the Congo Basin, Cuanza, the Horn of Africa, and several island ecoregions, in parts of southern and eastern coastal Africa, and on the west coast of Madagascar (figure 3.6). However,

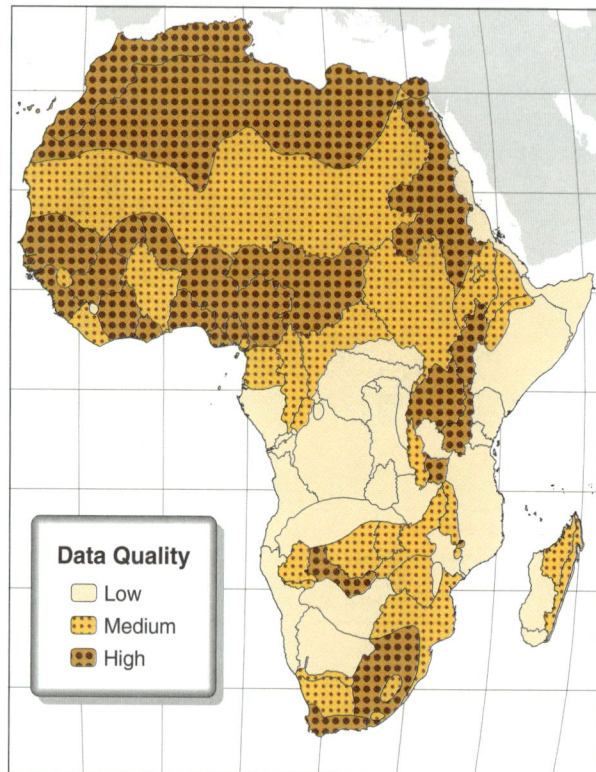

FIGURE 3.6. Data quality of freshwater ecoregions as assessed by experts.

availability of data is changing rapidly in many parts of the region as ongoing studies are completed and published (e.g., essay 3.3). For example, data quality of the Ethiopian Highlands [39] was considered low at the beginning of this study, but the level was elevated to medium after the publication of two recent studies (Getahun and Stiassny 1998; Golubtsov et al. 2002). Another example of an ongoing effort to fill data gaps is IUCN's work to collate data on the distribution and threatened status of freshwater mollusks, fishes, aquatic plants, decapod crabs, and Odonata (dragonflies) from Malawi, Tanzania, Burundi, Kenya, and Uganda (IUCN/SSC Freshwater Biodiversity Assessment Program 2003). The evaluation of data quality that we provide here indicates where the largest gaps in information remain and suggests potential areas for future sampling and cataloging of existing collections.

The data quality map does not mirror the biological distinctiveness map. A chi-square analysis showed that data quality and preliminary biological distinctiveness are not significantly related (χ^2 [df = 8, N = 93] = 4.0, p = .67). In other words, the most thoroughly sampled areas do not necessarily fall within the highest distinctiveness category.

Lesser flamingo (*Phoenicopterus minor*) in breeding plumage, Lake Bogoria, Kenya. Photo credit: ©WWF-Canon/Martin Harvey

Combining Species Richness and Endemism: Preliminary Biological Distinctiveness

The broad patterns of richness and endemism across Africa provide important insights for identifying conservation priorities, but we seek to go beyond the simple illustration of patterns. Our method (chapter 2) aims to synthesize the distinctiveness values of ecoregions based on species richness and endemism and to set species-based priorities within major habitat types. Furthermore, we seek to combine species values with nonspecies values to develop a single index that captures the overall biological importance of each ecoregion.

We combined the species richness and species endemism point values of each ecoregion as described in chapter 2 and placed each ecoregion into the category that reflected its species-based distinctiveness. This preliminary evaluation of biological distinctiveness was undertaken separately within each of the twelve MHTs to ensure representation from each MHT in our final priorities. When we summed the endemism and richness points for each ecoregion and assigned preliminary biological distinctiveness categories, twelve ecoregions were considered globally outstanding, nineteen were continentally outstanding, thirty-one were bioregionally outstanding, and thirty-one were nationally important (appendix D; figure 3.7).

Nonspecies Biological Values

Rare Ecological or Evolutionary Phenomena

Unlike richness or endemism, rare ecological and evolutionary phenomena are not easily counted, ranked, or comprehensively mapped. However, these phenomena represent essential components of biodiversity that we seek to conserve. Ten ecoregions were considered globally outstanding and two ecoregions were judged continentally outstanding based on species radiations or high-level taxonomic endemism (table 3.1, figure 3.7). Globally outstanding species radiations occur mostly in ecoregions from the large lakes and closed basins and small lakes MHTs. High-level taxonomic endemism often occurs in conjunction with these species radiations and is also

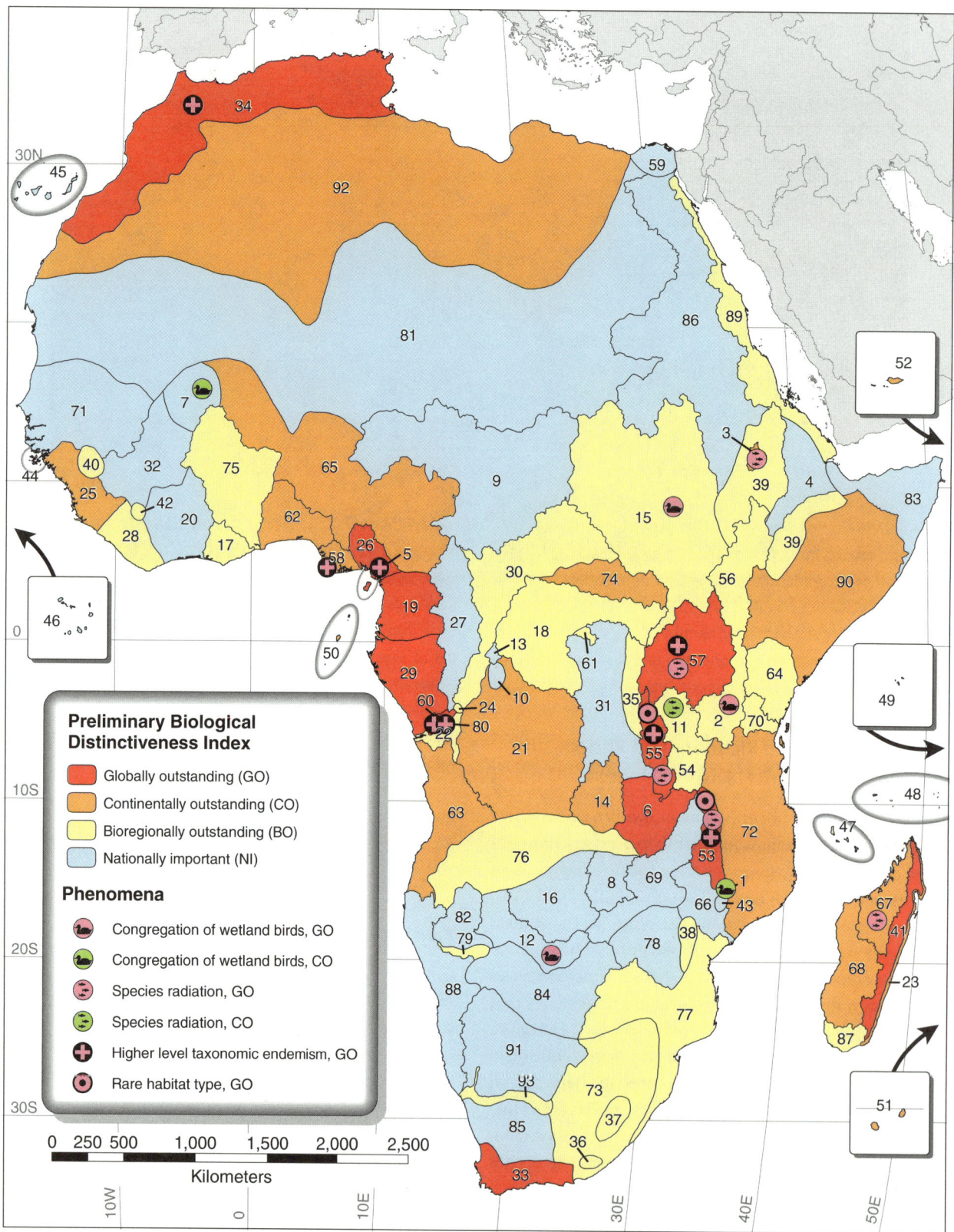

FIGURE 3.7. Preliminary Biological Distinctiveness Index and rare ecological or evolutionary phenomena.

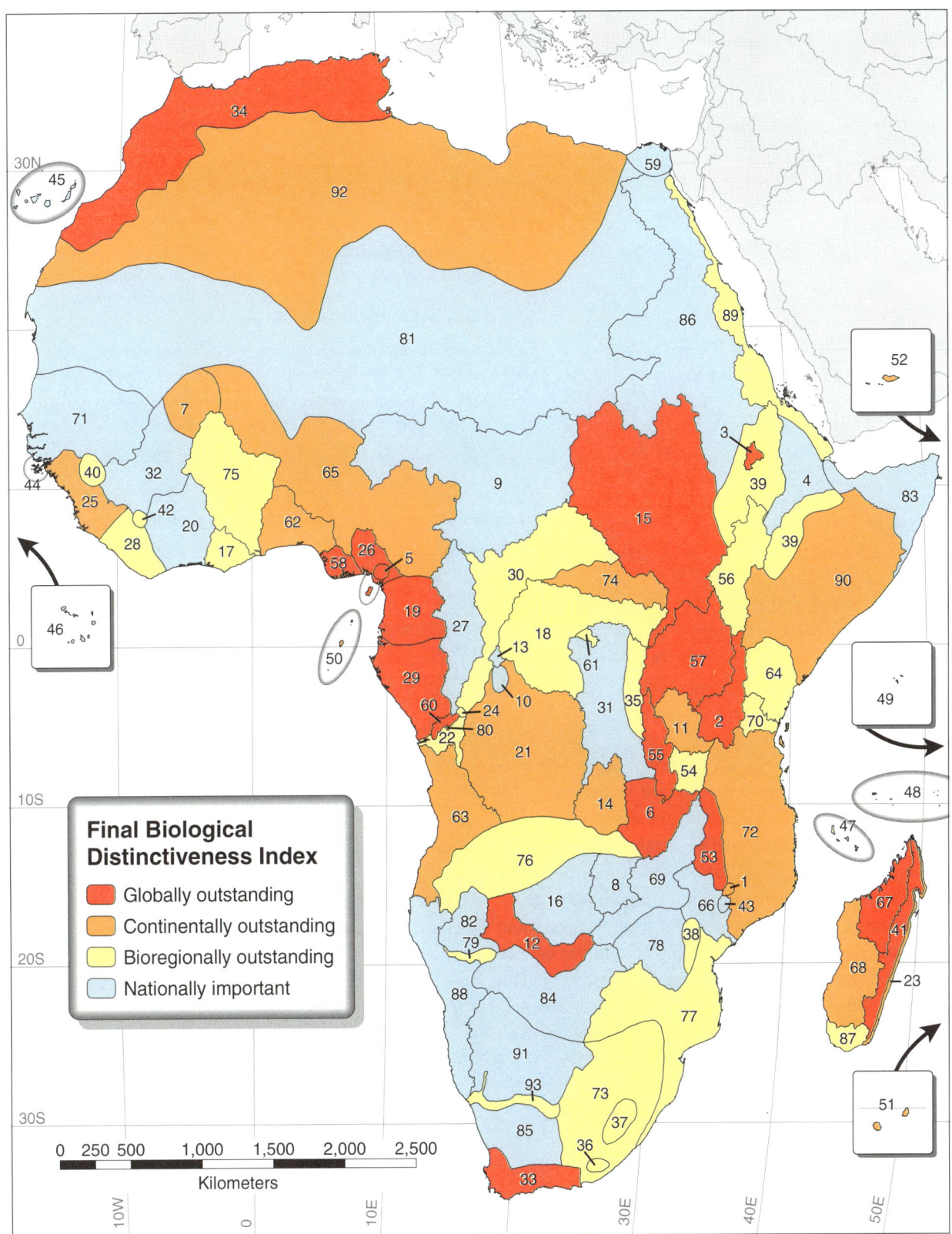

FIGURE 3.8. Final Biological Distinctiveness Index.

TABLE 3.1. Evolutionary Phenomena and Rare Habitat Types in Freshwater Ecoregions.

Ecoregion	Classification of Phenomena	Phenomena
Western Equatorial Crater Lakes [5]	GO	Higher-level taxonomic endemism: 4 endemic fish genera (*Konia, Myaka, Pungu, Stomatepia*).
Lakes Kivu, Edward, George, and Victoria [57]	GO	Cichlid species radiation with 22 endemic cichlid genera and 3 additional endemic genera (*Xenoclarias* [Clariidae], *Cynopanchax, Laciris* [Poeciliidae]).
Lake Tanganyika [55]	GO	Cichlid species radiation with 49 endemic cichlid genera and 8 additional endemic genera (*Tanganikaallabes* [Clariidae], *Lophiobagrus, Phyllonemus, Bathybagrus* [Claroteidae], *Stolothrissa* [Clupeidae], *Acapoeta, Xenobarbus* [Cyprinidae], *Lamprichthys* [Poecillidae]. 18 endemic aquatic mollusk genera. Rare habitat type of deep, rift valley lake.
Lake Malawi [53]	GO	Cichlid species radiation with 50 endemic cichlid genera and 1 endemic cyprinid genus (*Engraulicypris*). Rare habitat type of deep, rift valley lake.
Lake Tana [3]	GO	Cyprinid species radiation with 15 large barbs (1 of 2 cyprinid flocks in the world, the other being in the Philippines' Lake Lanao).
Niger Delta [58]	GO	Higher taxonomic endemism: endemic fish family (Denticipidae).
Thysville Caves [80]	GO	Higher taxonomic endemism: 1 of 2 fish genera (*Caecobarbus*) is endemic.
Permanent Maghreb [34]	GO	Higher taxonomic endemism: 8 endemic aquatic mollusk genera.
Lower Congo Rapids [60]	GO	Higher taxonomic endemism: 4 endemic aquatic mollusk genera and a fish species radiation including 6 *Caecomastacembelus* species (Mastacembelidae).
Madagascar Northwestern Basins [67]	GO	A small endemic radiation of etropline cichlids (*Paretroplus*). The number of species in this genus probably will not exceed 15.
Malagarasi-Moyowosi [11]	CO	Cichlid species radiation: an *Orthochromis* species flock of 7 species.

Note: Data on the ecological phenomenon of bird congregations are in box 3.1.
CO, continentally outstanding; GO, globally outstanding.

present in several of the smaller ecoregions that have deltas, rapids, or subterranean habitats. An additional three ecoregions support globally outstanding congregations of wetland birds (Okavango Floodplains [12], Upper Nile [15], and Southern Eastern Rift [2]), and two support continentally outstanding congregations (Lakes Chilwa and Chiuta [1] and Inner Niger Delta [7]). These ecoregions all contain either saline or soda lakes or extensive seasonal floodplain habitats in an arid or semi-arid environment (box 3.1).

Rare Habitat Type

The only rare habitat type identified was the deep (more than 500 m) rift valley lakes in Lake Malawi [53] and Lake Tanganyika [55]. These were considered globally rare because this habitat type is represented by fewer than eight examples worldwide. No other habitat types were considered globally or continentally rare (see box 3.2 for details of the methods as applied to inland deltas). Freshwater habitats that are not particularly rare but are of important biological value are described in the ecoregion descriptions (appendix F).

Synthesis of Biological Distinctiveness Data

After evaluating phenomena and subsequently elevating ecoregions with continentally or globally outstanding phenomena, the resulting tally was nineteen globally outstanding, nineteen continentally outstanding, twenty-six bioregionally outstanding, and twenty-nine nationally important ecoregions (figure 3.8). The most important ecoregions for freshwater biodiversity at a global or continental scale represent more than 40 percent of all ecoregions in the Afro-Malagasy region.

BOX 3.1. **Ecological Phenomena: The Case of Waterbirds.**

Africa's rivers, lakes, and associated wetlands provide important habitat for waterbirds. These include areas that support endemic or vulnerable species; those that serve as critical stopover areas for migrating birds; and those that provide habitat for enormous aggregations of both breeding and nonbreeding birds, comprised of a few to many dozens of species. Some of these habitats are permanent, such as several of the Rift Valley's alkaline lakes, which may attract up to a million or more greater (*Phoenicopterus ruber*) and lesser flamingos (*Phoenicopterus minor*). Many others are temporary, and their importance varies widely from year to year, depending on water levels; examples include flooded savannas, pans, and many smaller lakes. Some of the most important sites for waterbirds, such as mangroves, beaches, lagoons, and mudflats, are more saline than fresh, occupying locations on the coast. These may receive freshwater inputs, or they may be distinct from freshwater systems altogether, in which case they have not been included in this analysis. In general, in this continent characterized by large arid expanses, water sources are likely to be magnets for waterbirds (and other fauna) where they occur.

Many of the enormous congregations of waterbirds qualify as ecological phenomena at global or continental scales. Some places, such as the Inner Niger Delta and certain lakes of the Eastern Rift, support such extraordinary assemblages and numbers of birds that they easily rank as globally outstanding. However, it can be difficult to evaluate the relative importance of aquatic and wetland habitats that are less obvious standouts, particularly because bird congregations can vary substantially from year to year.

BirdLife International has spent a number of years identifying important bird areas (IBAs) around the world. BirdLife's IBAs of global significance include critical areas for globally threatened and restricted-range species, for biome-restricted assemblages, and for congregations (Stattersfield et al. 1998; Fishpool and Evans 2001). Taken together, these four criteria for IBAs allow for the identification of a broad range of sites. For example, in Egypt, thirty-four sites meet one or more of the criteria when all bird species are considered, and twenty-six of these support waterbirds (Baha El Din 1999). Using these criteria, one would expect most ecoregions to contain multiple IBAs. Therefore, the presence of IBAs in an ecoregion is insufficient to identify ecological phenomena at the continental or global scale.

Using data from Fishpool and Evans (2001), we identified sites that support phenomenal congregations, those documented as supporting 1 million or more waterbirds (table 3.2). Of this subset, those ecoregions with congregations greater than 1.5 million are considered globally outstanding, and those with fewer than 1.5 million are considered continentally outstanding. Many of the largest wetland bird congregations occur in arid regions that experience seasonal flooding and expansion of shallow wetland habitats. Because these habitats are ephemeral, they rarely coincide with high levels of biodiversity among strictly aquatic taxa. Therefore, we separated the phenomena of wetland bird congregations from the analysis of other aquatic phenomena. We represent globally outstanding congregations (those with more than 1.5 million birds) with a bird symbol on the map of biological distinctiveness (see figure 3.7). This approach allows us to highlight areas that are particularly important for wetland bird congregations without obscuring or diluting the importance of strictly aquatic biodiversity.

TABLE 3.2. **Inland Sites with Globally or Continentally Outstanding Congregations of Wetland Birds in Africa (islands and coastal sites not included).**

Sites	Ecoregion	Countries	Habitat Type	Level	Maximum Number of Wetland Birds Documented at Site
Lac Débo–Lac Oualado Débo	Inner Niger Delta [7]	Mali	Inland delta swamps, network of river channels and low floodplains	CO	1,011,086
Lake Chilwa and floodplain	Lakes Chilwa and Chiuta [1]	Malawi	Shallow lake surrounded by swamps and seasonally flooded grassland	CO	1,011,180
Lake Ngami	Okavango Floodplains [12]	Botswana	Seasonally flooded inland delta lake	GO	1,569,400
Lake Nakuru, Lake Bogoria, Lake Manyara, and Lake Natron and Engaruka basin	Southern Eastern Rift [2]	Kenya, Tanzania	Soda (alkaline) lake; shallow, brackish lake, with marshes and swamps	GO	1,526,580 (Nakuru) 1,092,240 (Bogoria) 2,294,349 (Manyara) 1,500,000 (Natron and Engaruka basin)
Sudd (Bahr-el-Jebel system)	Upper Nile [15]	Sudan	Seasonally inundated floodplains, permanent swamps, flowing waters, and lakes	GO	2,959,491

CO, continentally outstanding; GO, globally outstanding.

Source: Data provided by L. Fishpool (BirdLife International) from Fishpool and Evans (2001).

Globally Distinctive Ecoregions

Across Africa and Madagascar, several ecoregions stand out for their spectacular freshwater biodiversity. Unquestionably, the great lakes (ecoregions [53], [55], and [57]) of the Rift Valleys in eastern Africa are at the top of this list, with their incredible species radiations and high levels of endemism. However, several small highland lakes also contain fish species radiations and higher-level endemism; these are Lake Tana [3] in the Ethiopian Highlands and the Western Equatorial Crater Lakes [5] in Cameroon (essay 3.4). The moist forest rivers of the entire West Coastal Equatorial bioregion (ecoregions [29], [19], [26], and [5]) along the western coast of Central Africa are globally outstanding, with an extremely rich freshwater fauna. In the Congo Basin, only the Lower Congo Rapids [60] ecoregion is elevated to the highest level of distinctiveness, although as this basin is further explored, many other ecoregions are expected to be elevated; the Kasai [21] and Upper Lualaba [14] from the Congo Basin both were rated as continentally outstanding. At the northern and southern edges of the continent, the two mediterranean systems, Permanent Maghreb [34] and Cape Fold [33], both support large numbers of endemic fish.

The Cape Fold also supports an endemic herpetofauna, and the Permanent Maghreb has eight endemic aquatic mollusk genera. In West Africa, the Niger Delta [58] is globally outstanding and supports an endemic fish family, in addition to having representation of four monotypic fish families in the world (essay 3.5). Madagascar also provides habitat for an endemic and rich freshwater fauna, with Madagascar Eastern Highlands [41] and Madagascar Northwestern Basins [67] elevated to globally outstanding (essay 3.6). Thysville Caves [80] ecoregion, although depauperate, supports higher-level endemism, with one of two fish genera being endemic. Bangweulu-Mweru [6] supports a rich molluskan fauna, herpetofauna, and ichthyofauna, in addition to many endemic fish species. Globally outstanding congregations of wetland birds occur in the Okavango Floodplains [12], the Sudd floodplains of the Upper Nile [15], and the saline lakes of the Southern Eastern Rift [2] (figure 3.8).

Conclusions

The combination of species endemism, species richness, and nonspecies biological features presented here provides the first overall index of the biological importance of freshwater

BOX 3.2. **Inland Deltas: Globally Rare Habitat Types?**

Places such as the Okavango Floodplains [12] and the Inner Niger Delta [7] are well known landscape features in Africa, commonly called *inland deltas*. But what exactly is an inland delta, and is it a rare habitat type at the continental or global scale?

To our knowledge, there is no established definition of an inland delta. The Ramsar Bureau (see essay 6.1) lists "permanent inland delta" in its wetland classification, yet it has not developed its own definition. The Wetlands International staff has been unable to locate a definition in the literature (S. Frazier, pers. comm., 2000).

To begin to build a definition, we can look at the pieces of the term. *Inland* refers to land that is away from the sea. Most definitions for *delta* refer to a plain of alluvial sediments deposited when a river flows into a body of standing water and its velocity and transporting power are suddenly reduced.

If we put these two pieces together, we might derive a definition that reads, "an assemblage of sediments accumulated where a stream flows into an inland body of standing water and its velocity and transporting power are suddenly reduced." The river might terminate at the delta, or it might continue its course beyond the standing body of water. There are numerous examples around the world of places fitting this definition, such as the Peace-Athabasca Delta, Russia's Selenga Delta, Australia's Gwydir River Delta, and the Rheindelta Bodensee along the border of Austria and Switzerland. Interestingly, it is difficult to find a clear-cut African example fitting this definition, perhaps because natural lakes are rare outside East Africa.

The cases of the Okavango and the Inner Niger, as well as the Sudd (Upper Nile [15] ecoregion), represent a somewhat different definition of *inland delta*. Here, the term can be taken to mean an area where a river slows and may divide as it enters a flat area (an alluvial plain), and flooding of the river (often seasonal) creates a broad area of floodplain wetlands. Again, the river may terminate at the wetlands, or it may continue its course beyond them. Examples on other continents include the Inner Danube Delta, the Pantanal in Brazil, Paraguay, and Bolivia, and Argentina's Gran Chaco. This definition could apply to any place where a river spreads out to form a system of floodplain wetlands, but it generally has been used to refer to large areas.

These two definitions are not mutually exclusive. In fact, most examples fitting the first definition also fit the second. Extensive areas of wetlands providing important habitat for freshwater species, including waterbirds, tend to characterize inland deltas of both types.

We can combine the two definitions to obtain a more general one: "an area where a river slows and may divide as it enters a flat area or a standing body of water, and a broad area of seasonal or permanent wetlands is created." We have chosen to use this hybrid definition because it includes all common usages of the term and highlights what we believe are the most important features from a biodiversity perspective.

Using this hybrid definition, are inland deltas a rare habitat type? We define a globally rare habitat type as one that occurs in fewer than eight locations worldwide and a continentally rare habitat type as one that occurs in fewer than three locations in Africa (see appendix A for details). Using these criteria, inland deltas are neither globally nor continentally rare. However, many African floodplain wetlands support outstanding populations of waterbirds and for this reason have been highlighted (see figure 3.7 and box 3.1).

units in the Afro-Malagasy region. These results form one of the two major indexes used to develop a set of priorities for conservation investment in the region. As with biodiversity, threats vary widely from ecoregion to ecoregion across the Afro-Malagasy region. Some ecoregions have been almost completely degraded, whereas others are largely untouched. Chapter 4 develops our second index, that of measuring threats and opportunities for conservation, so that we can determine the current status of Africa's freshwater systems.

ESSAY 3.1

The Cichlid Fish Radiations of East Africa: A Model for Understanding Origin and Loss of Biodiversity

Ole Seehausen

More than 15 percent of the world's freshwater fish species live in a handful of large lakes in eastern central Africa. Lakes Malawi and Tanganyika are trough-shaped, narrow, and very deep, and together with the smaller lakes Kivu, Edward, and Albert lie at the bottom of the Western Rift Valley. Lake Turkana, of similar shape but more extreme hydrochemistry, lies in the semiarid Eastern Rift Valley. Lake Victoria, the biggest of all, is saucer-shaped and very shallow. Together with Lake Kyoga and many smaller satellites it fills up a large part of the plateau between the rifts. Each of the three largest lakes contains as many or more species of freshwater fish than all European rivers and lakes together. More than 1,500 species are currently known, but the real figure could be almost twice that. Even more striking than the numbers, 90 percent of the species belong to a single family, the perchlike cichlids (Cichlidae, Perciformes).

Cichlid Diversity in East African Lakes

The cichlid family is distributed over most of Africa, the Middle East, India, and tropical America, but nowhere else has this family undergone evolutionary radiations comparable with those in the East African lakes. The family is about 100 million years old and consists of about seven major lineages. More than 75 percent of its species have been produced by the youngest of these lineages, often called the East African lineage, the H-line, or, more broadly, haplochromines. This lineage originated about 5–10 million years ago in East Africa, possibly in Lake Tanganyika (Streelman et al. 1998). Lake Tanganyika is the oldest of the Great Lakes (about 10–12 million years old) and the only place in Africa where a phylogenetically broad range of representatives of this lineage has survived until today. Based on morphological, allozymic, and molecular genetic information, the H-lineage can be broken down to eight extant branches: Lamprologini (by some authors seen as the sister group to the H-lineage), Ectodini, Cyprichromini, Limnochromini, Perissodini, Eretmodini, Tropheini, and Haplochromini (table 3.3). Therefore, Lake Tanganyika can be viewed as a reservoir of ancient lineages in the East African cichlid radiation, which is of great interest for evolutionary biologists and conservationists. However, the lake's species richness is only moderate, with current estimates at about 200 (table 3.3).

Although much younger than Lake Tanganyika, Lake Malawi (about 2 million years old) and Lake Victoria (less than 750,000 years old) are more species rich, each having more than 500 extant species. However, these larger assemblages are phylogenetically much less diverse. Both are essentially composed of representatives of only one of the eight Tanganyikan branches, the Haplochromini, or haplochromines in the strict sense. Surprisingly, in Lake Tanganyika this branch has very few species. Although knowledge on the phylogeny of East African cichlids is incomplete, the current picture suggests that the radiations of lakes Malawi and Victoria have evolved independently from different haplochromine ancestors (Meyer 1993). Haplochromine cichlids closely related to the species flocks in these lakes are widely distributed in Africa but are not noticeably species rich elsewhere. Either early haplochromines left Lake Tanganyika and spread across the continent, or haplochromines evolved outside Lake Tanganyika from non-haplochromine ancestors, which may or may not have had their roots in Lake Tanganyika, and invaded Lake Tanganyika secondarily from rivers. The current phylogenetic knowledge does not allow us to discriminate between these possibilities, but it leaves no doubt that lakes Victoria and Malawi must have been colonized from rivers by haplochromines. Possibly just one or a few ancestral species were involved in each radiation, resulting in radiations in each of these two lakes of monophyletic species flocks. The cichlid assemblage of Lake Tanganyika usually is called a species flock, independent of monophyletic considerations. A comprehensive review of what is known about phylogenetic relationships between East African (mostly Lake Tanganyika) cichlids is provided by Nishida (1997).

Cichlid Species Flocks Can Help Us Understand Evolutionary Radiations

An evolutionary radiation is the origin of species diversity within a rapidly multiplying lineage (see Schluter 2000 for an in-depth treatment of the phenomenon). The Great Lakes of eastern Africa provide at least three independently evolved large-scale evolutionary radiations. Strikingly similar life forms have evolved in each of these lakes independently. Noticeable examples of this parallel evolution are mackerel-shaped pursuit

TABLE 3.3. **Phylogenetic Diversity, Species Diversity, and Speciation Rates in Cichlid Radiations in East African Great Lakes.**

Phylogenetic Lineage	Age (m.y.)	Number of Extant Species	Number of Founders	Speciation Interval*	Reference (A + B if one refers to age, the other to species numbers)
Lake Tanganyika					
H-lineage					
Lamprologines	4.1–10	~85	1	640–1,500	Kocher et al. (1995), Nishida (1997) + Stiassny (1997)
Ectodines	3.7–5.6	~30	1	754–1,141	Sturmbauer and Meyer (1992), Kocher et al. (1995)
Cyprichromini	2–6	10	1	602–1,806	Takahashi et al. (1998) + Konings (1998)
Limnochromini	2–6	8	1?	667–2,000	Takahashi et al. (1998) + Konings (1998)
Perissodini	2–6	8	1	667–2,000	Takahashi et al. (1998) + Konings (1998)
Eretmodini	2–6	5	1	861–2,584	Verheyen et al. (1996); Konings (1998)
Tropheini	2.8–6	39	1	530–1,135	Kocher et al. (1995), Nishida (1997) + Konings (1998)
Others					
Oreochromis	8–10	3	1–3	5,047–>10,000	Kocher et al. (1995)
Bathybatini	8–13	8	1	2,667–4,333	Kocher et al. (1995), Nishida (1997) + Konings (1998)
Trematocarini	8–12	9	1	2,524–3,786	Kocher et al. (1995), Nishida (1997) + Konings (1998)
Boulengerochromis	8–12	1	1	>8,000–12,000	Kocher et al. (1995), Nishida (1997)
Lake Malawi					
Haplochromini	0.6–2	600–1000	1	60–217	Meyer et al. (1990) + Konings (1995), Turner (1999)
Oreochromis	1–2	3	1	631–1,262	Sodsuk et al. (1995)
Lake Victoria					
Haplochromini	0.012–0.2	500–1000	1	1.2–2.2	Johnson et al. (1996) + Seehausen (1996)
Oreochromis	0.2	2	2	>200	
Lake Edward Haplochromini	0.2	>40	1–2	38–46	Greenwood (1980) + L. Kaufman (pers. comm., 2001)
Lake Kivu Haplochromini	0.01–0.05	16	1–2	2.5–16	Snoeks (1994); Lippitsch (1997)
Lake Turkana Haplochromini	0.012–0.5	4	1	6–250	Greenwood (1974)
Lake Albert Haplochromini	0.5	5–10	1?	86–151	Greenwood (1980)

Note the discordance between phylogenetic diversity and evolutionary potential in terms of speciation rate.

*In thousands of years; calculated as mean doubling time of species number, L (Turner 1999): $L = t \times \frac{\log_e(2)}{\log_e N_t - \log_e N_0}$, where N_t = number of extant species, N_0 = number of founding species, and t = age of radiation.

hunters of the open waters (*Bathybates* in Tanganyika, *Rhamphochromis* in Malawi, *Prognathochromis macrognathus* in Victoria), snail eaters that crunch their prey between unusually formed oral jaws (*Chilotilapia* in Malawi, *Macropleurodus* in Victoria), blunt-snouted algae scrapers with highly specialized dentition (*Tropheus* in Tanganyika, *Pseudotropheus* in Malawi, *Neochromis* in Victoria), and dwarfs that live and breed in empty snail shells (*Lamprologus* in Tanganyika, *Pseudotropheus* in Malawi), to mention just a few. Whereas ecological and morphological diversity of the younger flocks at least superficially resembles that of the Tanganyikan flock, behavioral diversity is much higher in the older Tanganyika flock. This is particularly apparent in terms of parental care and mating systems. The Lake Tanganyika flock is made up of monogamous, harem-forming, and polygynous cichlids; substrate, cave, and mouthbrooders; species in which both mother and father and even older sisters and brothers guard the fry; and others in which parental care is exclusively maternal (Kuwamura 1997). In contrast, all 1,000+ cichlid species in lakes Victoria and Malawi are mouthbreeders with exclusively maternal care. A similar discrepancy between the flocks can be observed on the level of molecular variation. Based on variation in mitochondrial DNA, the Tanganyikan species flock is five to ten times more diverse than the Malawian flock and fifty times more diverse than the Victorian flock. In fact, the Lake Victoria cichlids and parts of the Lake Malawi radiation are so genetically similar that molecular markers provide very little, if any, phylogenetic resolution (Meyer 1993). Therefore, rates of evolution differ widely between characters that differ in their exposure to natural selection. This can teach us about the roles of mutation, drift, selection, and constraint in the evolution of species diversity.

The striking difference between the East African Great Lakes and other lakes and rivers in Africa in terms of cichlid species diversity has often been explained by ecological opportunity and lack of competition from other fish taxa. Until recently, many scientists believed that the eastern Great Lakes were biogeographically isolated, with depauperate fish faunas before the radiation of cichlids. However, several reviews have pointed out that there is no reason to believe that these faunas were depauperate at all (Fryer 1996; Seehausen 2000). Therefore, there is no evidence that competition prevented the diversification of cichlids elsewhere. Although the environment in large lakes probably offers more ecological niches than the environment in smaller lakes or rivers, alone this fails to explain cichlid diversity because it does not account for the enormous variation in cichlid species richness among the three East African Great Lakes and between these and other big lakes such as lakes Turkana and Chad (table 3.3). Instead, an interaction between mating system (strength of sexual selection), genetic composition of the cichlid founder stock, and environmental conditions that affect sexual selection must be inferred to explain the variation. Research on Lake Victoria cichlids indicates that this complex interaction of factors may determine the rate of speciation by sexual selection. At the same time, the ecological flexibility and anatomical adaptability of cichlids allows larger numbers of different species to share ecological resources. Consequently, large numbers of species accumulate where speciation is an abundant process (Seehausen 2000).

There are many theories about the particular mechanisms by which lake cichlids have speciated. The role of lake-level fluctuations in breaking up formerly continuous habitats into geographically isolated patches between which migration is lacking (separate lake basins) or infrequent (habitat patches within a lake), thereby enabling allopatric speciation, is well established. However, some scientists point to a lack of evidence that such diverged populations were reproductively isolated at the time when changing water levels brought them together again. The explanatory power of this mechanism appears to be particularly limited in places where many species evolved in single lakes during periods of insignificant water level fluctuations. The most extreme such case is Lake Victoria, in which more than 500 species seem to have evolved from just a handful of founders within the last 12,500 years.

All the diverse groups of Lake Victoria cichlids, as well as the most species-rich groups of Lake Malawi cichlids, share at least one of two genetic polymorphisms that have been linked to speciation by disruptive sexual selection with or without geographic barriers to gene flow (Seehausen 2000). One is a polymorphism in chromosomally linked sex-determining and color genes and genes for mating preferences; the other polymorphism is in genes for male nuptial color and female mating preference. These examples constitute a specific case of an often restated hypothesis: that speciation in lake cichlids is a common byproduct of mating preferences evolving rapidly under sexual selection (Dominey/Lande/West-Eberhardt hypothesis). This mechanism can operate only in environments where the water is transparent and clear enough to allow broad-spectrum color vision. However, even under these conditions this seems to be a common process only in groups that possess the polymorphisms. A possibility that remains to be tested is that these polymorphisms explain why only some lineages have undergone the most rapid speciation known from vertebrates, and only in lakes with clear waters. Next to the rapidly radiating taxa in lakes Victoria and Malawi are lineages of haplochromines that are at least as old as the flocks but have not undergone a single speciation event.

Understanding Evolution Can Help Conservation

The radiations of some haplochromine cichlid lineages illustrate how understanding evolutionary processes can be relevant to conservation. Lake Victoria cichlids are famous not just for their radiation but also for their recent mass extinction. The shockingly rapid loss of species diversity was at first almost entirely attributed to effects of the predatory, introduced Nile perch (*Lates niloticus*), but it is now apparent that several factors are at work. One of them is loss of reproductive isolation among sympatric species caused by relaxation of vision-based mate selection, which is in turn a consequence of increased water turbidity and decreased population densities. In murky waters usually only one species per genus occurs at any one place, whereas up to six coexist where the water is still clear. This amounts to loss not only in species numbers but also in ecological diversity because many of the reproductively isolated species had evolved to use different foods or different parts of the habitat. This case has significant bearing on conservation priorities. Whereas conservation efforts traditionally have targeted the protection of genetic diversity at species or other taxonomic levels, today we recognize the importance of targeting evolutionary processes and the taxa that display particularly strong responses to such processes and therefore have the greatest potential to generate new replacement faunas. The so-called taxic/genic (product-oriented) approach and the so-called evolutionary front (process-oriented) approach to conservation can result in diametrically opposed recommendations with regard to conservation priorities. Whereas the first would give Tanganyikan cichlids (a reservoir of old, genetically very distinct lineages) priority over Malawian and Victorian cichlids, the latter would do the reverse because Victorian and Malawian cichlids are evolutionarily more dynamic. Clearly both concepts ought to be considered when conservation policies are made.

Suggested Readings on the Three Big Cichlid Fish Radiations

Fryer, G. and T. D. Iles. 1972. *The Cichlid Fishes of the Great Lakes of Africa: their biology and evolution.* Oliver and Boyd, London, UK. (predominantly on Lake Malawi)

Kwanabe, H., M. Hori, and N. Makoto, editors. 1997. *Fish communities in Lake Tanganyika.* Kyoto University Press, Kyoto, Japan.

Seehausen, O. 1996. *Lake Victoria rock cichlids: taxonomy, ecology and distribution.* Verduijn Cichlids, Zevenhuizen, the Netherlands. (not very broad, but no broader reader is available)

ESSAY 3.2

Freshwater Fish Biodiversity in the Congo Basin

Guy G. Teugels† and Michele L. Thieme

The Congo Basin has the highest species richness of any river system on the African continent and is second only to the Amazon Basin at a global level. Its enormous surface area and high diversity of habitats have facilitated the evolution of a highly diverse freshwater fauna. Geological events that have increased the number of barriers to faunal exchange have also provided opportunities for species evolution. Extraordinary species in the Congo Basin include spiny eels (Mastacembelidae) with reduced eyes for life in the lower Congo rapids, characids (Distichodontidae) that feed on the fins of other fish, and snout fishes (Mormyridae) with long rostrums that allow them to feed between rocks or in the mud and sand.

Surface estimations for the Congo Basin vary between 3,457,000 km^2 and 4,100,000 km^2, an area almost equal to the surface of the Indian subcontinent. Despite its name, it is an international basin, with waters flowing in Angola, Burundi, Cameroon, Central African Republic, Republic of the Congo, Democratic Republic of the Congo, Rwanda, Tanzania, and Zambia. Mean annual discharge of the Congo River is 40,487 m^3/second, and minima and maxima are not far apart, which is typical for equatorial rivers. Tributaries flowing from the north flood from August to November, and those from the south flood from May to June, together creating a bimodal peak in the

---

†Deceased, July 22, 2003.

Congo River's lower reaches (Lowe-McConnell 1987; Lévêque 1997).

The Congo River system supports a high diversity of habitats, including rapids, swamps, lakes, main rivers, marginal waters, and inundation zones. Habitat differences along the main river suggest subdividing it into several sections: the upper Congo, central Congo, Malebo Pool, and lower Congo. The entire Congo Basin is composed of fourteen freshwater ecoregions.

The upper Congo River, or Lualaba, arises in Zambia and flows northward to the town of Kisangani. Falls and rapids (formerly known as Stanley Falls) are found over the last 150 km of the Lualaba to immediately upstream from Kisangani. In this section the Luvua is one of the major tributaries.

The central Congo River (or Cuvette Centrale), located largely on the equator, flows westward in an arc from Kisangani to Kinshasa. Along its 2,000-km course the river drops only about 100 m and is thus wide and slow flowing, with widths often between 3 and 15 km. The lower reaches of the central Congo's two main tributaries, the Oubangui and the Kasai, are comparable with the central Congo River in size.

In the late Miocene and early Pliocene, the Congo was an endorheic basin that drained to a large lake in what is today the Cuvette Centrale. Lakes Mai-Ndombe and Tumba are thought to be remnants of this great lake. Today, floodwaters inundate large tracts of swamp and equatorial rain forests in the Cuvette Centrale for months at a time.

At Kinshasa, the Congo River expands into Malebo Pool (formerly known as Stanley Pool), a 500-km^2 enlargement of the river with an archipelago of islands and sandbanks. For the next 350 km down to the town of Matadi, thirty-two cataracts (Congo rapids) are interspersed with stretches of slow-flowing water. Along these 350 km, the river drops 275 m; therefore, this region probably contains one of the highest concentrations of potential hydroelectric power in the world. The lower Congo is defined as the area from Boma to its mouth at the Atlantic Ocean. This stretch of river is a mangrove-lined estuary (Roberts and Stewart 1976; Beadle 1981).

The number of species known from the basin is still increasing. In 1963, Poll and Gosse reported more than 408 primary freshwater fish arranged in twenty-four families. Hardly thirty years later, and based on specimens available in natural history museums, Teugels and Guégan (1994) reported 686 primary freshwater species belonging to twenty-six families. This number is far from complete because many parts of the basin are still unexplored, mainly because of accessibility problems and political instability, particularly over the last decade. Nevertheless, the known species number is far above the mathematically predicted number of 462 (Hugueny 1989).

Lower Congo Rapids, Democratic Republic of the Congo. Photo credit: Caroly Shumway, New England Aquarium

Malebo Pool [24] has the highest species diversity of any of the Congo Basin ecoregions, with 205 species, followed by the Kasai [21] (203 species) and the Cuvette Centrale [18] (excluding Lakes Tumba and Mai-Ndombe) (200 species). Siluriformes, or catfish, form almost one-fourth of the species composition in the Congo (161 species). They are followed by cypriniforms (111 species), mormyrids (110 species), and cichlids (89 species) (Teugels and Guégan 1994).

Endemicity for the basin is estimated at levels up to 80 percent (Lowe-McConnell 1987), but it should be noted that many Congolese species apparently also occur in rivers in Cameroon and Gabon. Moreover, since the 1901 publication of Boulenger's book *Les Poissons du Bassin du Congo,* there has been no overall revision of the fish fauna of the Congo Basin. A faunal guide based on recent revisionary studies is absolutely necessary because it will provide the basis for all other fundamental and applied research. Except for basic work in and around Malebo Pool (Poll 1959), around Kisangani (Gosse 1963), in and around Lake Tumba (Matthes 1964), in the Lower Congo rapids (Roberts and Stewart 1976), and in Upemba National Park (Poll 1976), little is known about the ecology of this very complex fauna. As with the Amazon Basin, there is still an enormous amount of investigative work to be undertaken on the Congo Basin to obtain even a moderately complete understanding of its biodiversity.

The conservation status of freshwater habitats and species is largely unknown because armed conflict has stymied field work in the Congo Basin in recent years. However, remotely sensed data on land uses in the region can provide a general idea of current and future threats to the river and its biota. The health of a freshwater system is intimately connected with land-based activities in its drainage basin, although activities vary in the type and degree of harm they cause (Allan and Flecker 1993; Davies and Day 1998).

Large expanses of the Congo Basin are still sparsely populated, and the Democratic Republic of the Congo supports some of the largest tracts of intact forest in Africa (Sayer et al. 1992). War and instability in the region have limited large-scale international operations in logging, grazing, mining, and agriculture (East 1999). Selective logging, pastoral activities in the north of the basin, artisanal mining, and small-scale farming may be having short-term effects on riverine biota at the local scale (D. Wilkie, pers. comm., 2002). Probably of more concern are the growing urban centers (Bangui, Brazzaville, Kisangani, Mbandaka, and Kikwit) with their associated increases in untreated sewage and sedimentation from dirt roads. Artisanal fishing is widespread in the basin, with many ethnic groups dependent on fish as their primary source of protein. Fish are traded and sold along the river and commonly eaten by rural and urban populations alike. The impact of harvest on fish populations is unknown.

Judging from land use activities in the region, local degradation of rivers and streams may be occurring throughout the basin, but there are probably many rivers and streams that still retain intact species assemblages. The lack of recent surveys in the region precludes an accurate evaluation of the current sta-

Women holding *Chrysichthys* sp. and *Malapterus* sp. (left to right) at the Bandundu market, Democratic Republic of the Congo. Photo credit: Caroly Shumway, New England Aquarium

tus of the basin. Of great concern for the future integrity of the fresh waters in the Congo Basin are the probable construction of dams, introductions of exotic species, and increases in industrial-scale extraction activities that will come with long-overdue political stability in the region.

ESSAY 3.3

Baseline Fish Biodiversity Surveys: African Wildlife Foundation Experiences from the Zambezi River, Southern Africa

Jimmiel Mandima and Henry Mwima

The Zambezi River and its tributaries provide important habitats for an abundance of freshwater fish including socially and economically significant species such as tigerfish (*Hydrocynus vittatus*), lungfish (*Protopterus annectens brieni*), and endemic cichlid (tilapias) and cyprinid species. Of the 239 species recorded in the Zambezi Basin (excluding Lake Malawi), 122 species occur in the mainstem of the Zambezi River (Skelton 2001). Significant work has been carried out to understand the biodiversity of fish in the Zambezi River, but there still remains a paucity of knowledge on fish diversity in specialized habitats, especially the upper reaches of the system. Anthropogenic activities and fishing pressure are known to adversely affect fish resources within the basin.

In the Four Corners Transboundary Natural Resources Management (TBNRM) area, the African Wildlife Foundation (AWF) and members of the Aquatic Resources Working Group (ARWG) undertook field studies to provide information on fish biodiversity in the region and developed standardized ecological monitoring methods that will be presented for adoption by all parties involved in the management of shared fishery resources in the region. AWF and partners have also conducted fish biodiversity surveys in the Middle Zambezi (see essay 6.3 and figure 6.4). A database of fish from these inventories forms a benchmark for conservation monitoring in the Four Corners TBNRM area and Zambezi Heartland over time.

Fish Biodiversity Surveys in Upper and Middle Zambezi

With leadership from the South African Institute for Aquatic Biodiversity (SAIAB), AWF conducted three field investigations of fish biodiversity in Zambia's Upper Zambezi. The Upper Zambezi surveys spanned a longitudinal transect from Zambezi River headwaters downstream to Victoria Falls, with latitudinal transects from midstream to the outer extremity of the river. Specifically, surveys occurred at the following sites:

- the headwaters of the Zambezi: West Lunga River and Kabompo River
- pre-floodplain mainstream rivers: Kabompo River at Kabompo, Zambezi River at Zambezi town
- Barotse floodplain in Mongu area
- lower floodplain, Senanga area
- the Zambezi River above Victoria Falls (Kazungula to Livingstone and Victoria Falls)

Surveys were timed to coincide with the low water, rising flood, peak flood, and declining flood periods to cover the range of flow conditions experienced at each site. Surveys began in the low water period of 2002 (August–September) and continued through the low-water period of 2003.

In the Middle Zambezi, researchers conducted field surveys in the Zambezi Heartland at the confluence of the Zambezi and Luangwa rivers, where local people make extensive use of the fishery. Sampling occurred four times during the declining and low water periods. Surveys targeted both nonfishing and fishing sites to generate information on the effects of restricted access areas on fish populations. Because the confluence site includes communities from three countries, it offers a laboratory to investigate the influence of different management regimes on transboundary, shared fishery resources. In addition to this site, surveys were conducted in Cahora Bassa Reservoir in Mozambique, where an active inshore fishery exists in the Zumbo Basin of the lake.

Collection Methods for Fish Biodiversity Surveys

To maximize the habitat types sampled and thus the number of species collected, a variety of gears, including seine nets, gillnets, fyke nets, hand-held D-nets, electrofishing, and rod and line, were used to collect fish. Representative (voucher) specimens are preserved and housed at the SAIAB and will also be housed with each respective country's natural history museum. Tissues for biochemical and genetic analysis have been taken from as many species as possible, preserved in 100 percent ethanol and lodged in the SAIAB tissue bank. A digital database is being prepared through AWF and will be used with the southern African freshwater fish geographic information system (GIS) atlas to plot species distribution maps.

Highlights from Biodiversity Investigations

Ninety-eight species were caught in the three surveys in the Upper Zambezi River. These surveys also added about twenty new species to the regional inventory mainly from the headwaters in northwestern Zambia (D. Tweddle, pers. comm., 2003). Research to accurately document these species is ongoing. Additional highlights include the following:

- Surveys were conducted in diverse habitats of the river system that included the river main channel, side channels, floodplain lagoons, and backwaters.
- Understanding was improved of the distribution, habitat preferences, and responses to flooding cycle of species found in the Upper Zambezi.
- Samples of specimens were acquired for taxonomic description of new, undocumented species that will be published as part of the project.
- Taxonomic issues were identified that must be addressed with follow-up research.

Further study of rare and threatened species is needed to characterize threats to their existence. Particularly lacking is information on the conservation status of species known only in upper reaches of tributaries. Intensified genetic studies are necessary to explore the differences in fish populations for both fisheries and conservation importance. Although we undertook a biological survey, there is a need for more socioeconomic data to complement biological data and to support local communities in the management and conservation of their own resources.

In the Middle Zambezi River, the four surveys yielded a total of thirty species, which is approximately 50 percent of the total species composition recorded for the Middle Zambezi to date (Skelton 2001). Key observations include the following:

- Catches from experimental gillnetting were dominated by five species from five different families, indicating a rich familial diversity.
- Fish caught in nonfished breeding sites were significantly larger than those from fished areas, providing evidence of the utility of closing some areas as a management strategy.

Juvenile stages of most species were present in the catches, suggesting that a viable breeding species assemblage occurs here.

Standardization of Aquatic Resource Ecological Monitoring Methods

This project was initiated to establish a fishery resource monitoring team for the Four Corners TBNRM area. This team would develop a suite of standardized ecological monitoring methods, to become an integral component of the management approach of the Southern African Development Community (SADC). The specific objective was to formulate and test standardized methods for monitoring fishery resources in the Four Corners TBNRM area, with the long-term aim of establishing a system of joint fishery management among the four nations of the Four Corners TBNRM area.

To satisfy these objectives, AWF, through the ARWG, conducted a workshop to design the methods and undertook field testing of the preliminary methods. These activities are described in more detail in this section.

Workshop to Design Standard Ecological Monitoring Methods

AWF convened a workshop with seven fish biologists and one socioeconomist in November 2002 at Victoria Falls, Zimbabwe. Experts at the workshop agreed that fish sampling should occur in a range of habitats with multimesh gillnets as the primary gear, supplemented by electrofishing, seine netting, and trap nets (fyke nets) (details on methods are available from AWF reports in preparation).

AWF also established a GIS subcommittee of the ARWG with the goal of building the capacity of the four countries to develop standardized data collection, documentation, and analytical procedures for processing results from the field surveys. The subcommittee is developing a shared GIS database to archive survey data and analytical products.

Field Testing of the Standardized Methods

The ARWG conducted two field expeditions at Senanga, a site on the lower reaches of the Barotse Floodplain in the Upper Zambezi, Zambia, where fishing is an important activity for local communities. The first expedition occurred in April and May 2003 and surveyed twelve sites representing floodplain microhabitats. Thirty fish species were caught. The butter catfish (*Schilbe intermedius*) dominated the catch, followed by the tigerfish (*Hydrocynus vittatus*). At the time of the survey, the flood levels were at a record high for the decade, restricting the survey to a limited range of fast-flowing, main river channel habitats.

The biodiversity survey was complemented by catch assessment surveys (CASs, which assess catches by fishers at landing sites, recording the fish caught, gear type and mesh size, and length and weight measurements) and frame surveys (detailed documentation of socioeconomic parameters of fishing communities including number of fishers, fishing gear, craft used, fishing times, and income). The effort included testing of forms for collecting socioeconomic data on the floodplain fishing industry and for recording catches from fishers.

The second survey was conducted during the low water period in September and October 2003. An electrofisher and seine net were used as additional gears for fishing in shallow lagoons and backwaters. This survey recorded a total of forty-one fish species. The species dominance pattern remained the same, with *Schilbe intermedius* contributing 55 percent of the total catch.

It is intended that lessons learned during the standardization of ecological monitoring methods will provide impetus for SADC's objective (under the Fisheries Protocol) of harmonizing fishery legislation in the region.

General Observations on the Fisheries in the Upper and Middle Zambezi Sites

In addition to the biological surveys, teams also made observations on general fishing activity and noted perceptions of local stakeholders about the conservation status of the fishery while in the field. This section briefly describes the socioeconomic aspects of the fishery that are relevant to conservation.

Upper Zambezi River

At all floodplain survey sites in Senanga and Mongu, fishing pressure was intense. All fish species, including the smallest species, are being harvested. On the floodplains, teams observed exceptionally intensive fishing. Drifting gillnets and large open water seines were in widespread use, leaving few sanctuaries for fish. The drifting gillnets were used close to the bank, with fish sheltering in riverbank vegetation being driven out of cover by beating of the vegetation in advance of the floating net. These methods place considerable pressure on both adults and juveniles of the larger species, which are vulnerable to gillnet capture.

Without exception, all people with whom we spoke had observed a decline in catch size in the last two decades, attributed to increased use of small-meshed nets. This was reported also in northern areas, such as Zambezi and Mwinilunga towns, in addition to the more recognized fishing areas such as the Barotse floodplains.

At meetings with leaders of traditional groups in the region—the *ngambela* and the *kuta* at Limulunga—the use of fishing with small-meshed gears was discussed. The Royal Establishment of the Lozi people of western Zambia have drafted proposals for fishery regulation, and it is recommended that the Fisheries Department, local administrative authorities, and Royal Establishment combine efforts to develop a management plan at the local level.

Middle Zambezi River

Most fishing activities take place along the Luangwa River upstream from the confluence with the Zambezi River. Fishers are mainly Mozambican and Zambian and use drifting gillnets, mosquito netting, traps, and poisons. Drifting gillnets are common, and these are often set throughout the day, with nets checked continuously during peak fishing seasons. All twenty members of a Trans-Boundary Local Area Committee, a committee set up by IUCN as part of the Zimbabwe, Mozambique, and Zambia TBNRM initiative to represent the local communities, believe that the area is overfished, and they speculate that drifting gillnets contribute significantly to excessive fishing pressure. Observations during field surveys confirmed widespread use of drifting gillnets throughout the day. Catches assessed from four fishing groups in Mozambique just below the confluence of the Zambezi and Luangwa Rivers were dominated by small, immature fish, which is an indicator of poor recruitment caused by overfishing.

The gender allocation of fishing activities is also of note in the Middle Zambezi. In contrast to the Upper Zambezi River, where women and children participate in the harvesting of fish in the shallow floodplain and lagoons using baskets, only men participate in active fishing in the Middle Zambezi. This is presumably because most of the fishing grounds are fast-flowing main river channels, and fishing involves following drifting nets in hand-paddled dugout canoes (*mokoros*), generally labor-intensive work. However, women are heavily involved with fish processing at the landing sites. In most cases, these women are fish traders from as far away as Lusaka and the Copperbelt in Zambia. Consequently, most fish from the confluence of the Luangwa and Zambezi Rivers and from the Cahora Bassa reservoir in Mozambique are destined for markets in Zambia. Women from Zambia spend 2–4 months in fishing camps, using sun and fire to dry the fish, and leave only after accumulating large quantities for sale in urban centers.

There is a close business relationship between the mostly Mozambican fishers and largely female and Zambian fish traders. The traders barter gillnets from Lusaka in exchange for fish from the fishers. Traders are required to purchase a license from Mozambique to buy fish, and they are not allowed to fish directly. Nonetheless, cases of such traders employing Mozambican nationals to work as their fishers are rampant.

Women and children using baskets for fishing in shallow backwaters, Upper Zambezi. Photo credit: W. Mhlanga, ARWG, Four Corners TBNRM area

Pile of dried fish on the shores of Cahora Bassa, Mozambique, ready for dispatch to the market in Zambia. Photo credit: J. Mandima, AWF, Zambezi Heartland

There is minimal fishing on the Zimbabwean side of the Zambezi Heartland because most of this stretch of the river is under the Parks Authority and fishing is generally prohibited on these lands. Limited fishing recently began near the confluence of the Luangwa and Zambezi rivers, where a local community group was granted a permit to fish using gillnets; however, they are not satisfied with the catches from the fishing grounds allocated. Gillnets with mesh less than 3.5 inches wide are prohibited, but policing is poor and catches are low, suggesting a likelihood of fishery depletion caused by excessive illegal fishing in waters under the parks estate, thereby threatening sites set aside for regeneration and breeding.

It is clear from these field observations of fishing activities that there is a need for an improved understanding of fish stocks and intersite fishery impacts to guide the sustainable use of fish resources. It is important to recommend suitable gear that support sustainable fishing for fisheries of commercial value.

Other Field Activities

In addition to the testing of ecological monitoring methods in the field and the fish biodiversity surveys, AWF and its partners (see essay 6.3) carried out a survey of aquatic plants and limnology of the system. The plant survey provides baseline information for fish habitat characterization that will be useful in correlating fish species distribution, diversity, and abundance to habitat type. The limnological assessments provide information on the status of the aquatic habitat.

From the aquatic plant survey in the Upper Zambezi River, sixty-seven aquatic plants were collected and identified from the floodplain at Senanga.

Results from the limnological assessments in the Middle Zambezi are still being processed, but preliminary results show that the total phosphorus (TP) levels ranged from 18.1 µg/L in the Zambezi River below the confluence with the Luangwa to 205 µg/L in the Luangwa River. A TP level of 47 µg/L or less suggests an oligotrophic system, between 47 µg/L and 115 µg/L is considered mesotrophic, and TP levels above 115 µg/L indicate eutrophy (van Grinkel 2002 in Magadza 2003). Our Luangwa River measurement of TP level (205 µg/L) is within the eutrophic range, suggesting potential problems related to nutrient loading. Human settlements and agriculture along the Luangwa River probably are the cause of the high phosphorus levels, although further investigation is warranted.

Conclusions

The activities carried out by AWF in southern Africa's largest freshwater system are aimed at improving the management of shared water and aquatic resources in the SADC. The Zambezi River traverses five countries that have a variety of water and fishery resource policies and regulations. Yet the activities of individual countries have direct or indirect impacts on all others in the catchment. It is evident from observations made in the Zambezi Heartland that fish are a common good being used by citizens of the transboundary region regardless of their nationalities. It is also clear that current fishing practices cannot allow the sustainable use of the resource, hence the need for interventions that span biological, socioeconomic, and legal issues.

The documentation of fish species distribution and abundance in parts of the Middle and Upper Zambezi will assist in resource allocations between different user communities that depend on fishing for their livelihoods. This baseline information allows informed decision-making by both resource managers and users and will enable more equitable and sustainable use of fish resources. This approach captures the key tenets of the ecosystem approach, defined by IUCN as "a strategy for management of land, water and living resources that promotes conservation and sustainable use in an equitable way" (Smith and Maltby 2003: 9).

Our experience in southern Africa clearly demonstrates that working with local partners with the requisite knowledge of the area, including local communities—who are often considered to be both threats to and beneficiaries of the resource—is a critical component of transboundary, landscape-level conservation. It also assists in building local and regional capacity that will promote the sustainability of such initiatives once an external organization leaves.

Acknowledgments

U.S. Agency for International Development Regional Center for Southern Africa (Four Corners TBNRM area component), the Netherlands Directorate-General for International Cooperation, and the Ford Foundation (Zambezi component) are acknowledged for their financial support to the activities described in this essay. We also would like to acknowledge the AWF partners who worked tirelessly to generate reports and data that provided key information for this essay; Professor Paul Skelton, Ronelle Kemp (our botanist), Wilson Mhlanga, Denis Tweddle, and all members of the ARWG and the SAIAB team are specially thanked. Finally, we are indebted to the following AWF colleagues for their invaluable comments on the draft: Helen Gichohi, Harry van der Linde, Joanna Elliott, David Williams, and Elodie Sampéré.

ESSAY 3.4

Cichlid Species Flocks in Small Cameroonian Lakes

Uli Schliewen

Like an archipelago of blue islands in the green sea of rainforests and grasslands, dozens of tiny lakes are scattered throughout the western provinces of Cameroon. Almost all of them are volcanic crater lakes and are located in a region that is called the Cameroon line, a mountainous area of high volcanic activity that has been continuously producing calderas over the past 25 million years (Fitton and Dunlop 1985). Rain filled some of these calderas with water and created many "empty" lakes devoid of fishes and shrimps. Because most of the approximately thirty-six crater lakes in Cameroon have no or extremely steep outflows (Kling 1987), they remained isolated from nearby river systems, which could have supplied colonizing fish and shrimp species. However, some fish species managed to enter a few of the tiny, isolated basins. In three of these lakes (Barombi Mbo, Bermin, and Ejagham), the evolutionary forces that acted on the founder fish populations quickly led to levels of endemism per lake area that are almost unrivaled on Earth.

Fish species from different families had entered the lake basins, but fishes of the family Cichlidae proved to be the most versatile competitors and speciated extensively. Cichlids now exploit almost all available resources in these lakes. Several lakes do not harbor endemic cichlids but support endemic fishes from other groups. For example, Lake Dissoni in the Rumpi Hills harbors one endemic poeciliid (*Procatopus lacustris*) and probably one endemic *Barbus* and *Clarias* (Trewavas 1962, 1974; Schliewen 1996b).

Barombi Mbo is the best known of the high-endemism lakes. The roughly circular lake basin has a diameter of 2.15 km and a maximum depth of 111 m. The area used by fishes is limited to the upper layers of the permanently stratified lake because below 40 m no oxygen is detectable. The most recent lava flow into the lake was almost exactly 1 m.y.a., dating the lake to be at least that old (Cornen et al. 1992). At present, fifteen fish species have been found in the lake, twelve of which are endemic. Except for the clariid catfish (*Clarias maclareni*), all endemics are tilapiine cichlid fishes. Four of the five tilapiine genera are endemic: *Konia* (two species), *Stomatepia* (three species), *Pungu* (one species), and *Myaka* (one species). The only nonendemic cichlid genus is *Sarotherodon* (four species) (Trewavas 1962; Trewavas et al. 1972; Schliewen 1996a). In addition to the endemic fishes, Barombi Mbo harbors at least one endemic sponge (*Corvospongilla thysi*) and one endemic but undescribed caridinid shrimp (*Caridina* sp.) (Trewavas et al 1972; E. Roth, pers. comm., 1996).

Some Barombi cichlids exhibit unique ecological and morphological specializations. For example, *Pungu maclareni* is a sponge-feeder that uses its very strong jaw musculature and specialized teeth to crush sponge spicules (Dominey 1987). *Konia dikume*, another fish that exhibits unusual behavior, temporarily enters deep waters with extremely low oxygen concentrations to feed on *Chaoborus* larvae. This unique fish can spend short amounts of time in deoxygenated water because a high hemoglobin concentration allows storage of large amounts of oxygen in its blood (Green and Corbet 1973).

Lake Bermin, with an approximate diameter of only 700 m and a maximum depth of approximately 16 m, is much smaller than Barombi Mbo but still is home to nine endemic cichlid species. All of these species belong to the tilapiine subgenus *Coptodon*, a taxon that is only distantly related to the *Sarotherodon* of Barombi Mbo. Except for one species, *Tilapia bemini*, all other species were described in 1992 (Thys van den Audenaerde 1972; Stiassny et al. 1992). No exact determination of the crater's age exists, although the degree of erosion of the crater rim suggests an age of much less than a million years (G. Kling, pers. comm., 1991). This finding is corroborated by the lesser degree of morphological specialization in comparison with the Barombi cichlids. However, several species exhibit striking features, including *Tilapia* (*Coptodon*) *spongotroktis*, which feeds predominantly on whole chunks of the massive freshwater sponge growth in Lake Bermin. Another species, *T. snyderae*, is approximately 5.5 cm long, making it the smallest known tilapiine cichlid fish. This species occurs in three different color morphs (Stiassny et al. 1992).

Lake Ejagham is a special case because it is not a volcanic crater lake but is ecologically and geomorphologically very similar to the other crater lakes. The lake's outlet is isolated from the nearby Munaya River by a waterfall that is insurmountable for cichlid fishes. Its oval-shaped lake basin (approximately

Lake Oku, a sacred crater lake in the Kilum Mountain Forest, Mount Kilum, Cameroon. Photo credit: ©WWF-Canon/Meg Gawler

1,050 × 700 m) has a maximum depth of 18 m. In contrast to Barombi Mbo and Bermin, this lake was colonized both by *Coptodon* and *Sarotherodon* tilapias. The *Coptodon* gave rise to at least five different species, whereas the *Sarotherodon* split into two. Except for *Tilapia deckerti,* all of them are still undescribed (Thys van den Audenaerde 1968; Schliewen et al. 2001).

In addition to their unrivaled levels of endemism, the cichlid fishes of these three lakes became famous among evolutionary biologists because their study helped to solve a long-standing question in evolutionary biology. This question was whether the origin of a new species necessitated the spatial separation of founder species (allopatric speciation) or whether new species could arise in the presence of the founder species (sympatric speciation). The molecular genetic analysis of the relationships of the cichlid fishes in the three lakes showed that cichlid species assemblages in Barombi and Bermin (Schliewen et al. 1994) and some species pairs in Lake Ejagham (Schliewen et al. 2001) are more closely related to each other than to any other riverine cichlid species or to cichlids from other lakes in the region. Because the lakes are isolated from nearby river systems, conical in shape, and geomorphologically homogenous, the results implied that speciation had taken place entirely within the limits of these lakes and therefore in full sympatry.

At present, the cichlid species flocks of the crater lakes of Cameroon remain the only widely accepted example of sympatric speciation in nature. Although there exist descriptive data on the ecology and behavior of some of the Barombi (Trewavas et al. 1972; Peters and Berns 1982; Dominey 1987; Dominey and Snyder 1988) and Bermin cichlids (Stiassny et al. 1992), the enormous scientific potential to study pattern and process of sympatric speciation in nature has not been fully explored except in Lake Ejagham (Schliewen et al. 2001).

Despite the high levels of species endemism and the unique behaviors of the fish that live in these lakes, effective conservation efforts have not been undertaken to protect the lakes' ecosystems (apart from designating the Lake Barombi Mbo Forest Reserve and Ejagham Forest Reserve). The small size of the lakes renders them extremely vulnerable to even minor disturbances. Conservation measures are long overdue. In particular, the most diverse and famous lake, Barombi Mbo, is under immediate threat. Partial deforestation of the interior crater rim has already taken place because of increased demand for agricultural land by the local Barombi people and by people from nearby Kumba town. This is likely to cause increased erosion and consequently increased sediment input into the oligotrophic lake system. Water extraction has also temporarily caused lake level alterations, which have changed breeding habitat needs for some of the endemic cichlid species, especially *Sarotherodon linnellii* (Schwoiser, pers. comm., cited in Dominey 1987). The use of modern gillnets supposedly has decreased populations of target fish species, although in contrast to previous reports (Reid 1990), all fish species are still present in the lake (pers. obs., 2002). Water pollution from insecticide use in small farms within the crater rim and from increased wastewater inflow from the small Barombi village is also likely to affect the lake's ecosystem. Last but not least, the introduction of exotic fish species most likely would have disastrous results. Although no direct action has been planned, the mere chance that either molluscivorous fish species for bilharzia control or nonindigenous tilapias for increased fishery revenue would be introduced is a serious threat (Reid 1990; Schliewen 1996a). These threats to Barombi Mbo also largely apply to the other lakes.

Preliminary suggestions for the protection of Barombi have been made (Reid 1990; Schliewen 1996a). These recommendations include undertaking a detailed analysis of threats and evaluating conservation potential.

ESSAY 3.5

The Niger Delta in Nigeria: A Hotspot and a Crossroad for Freshwater Fish Biodiversity

Guy G. Teugels† and C. Bruce Powell††

The Niger River, with its main course of about 4,200 km (Rzóska 1985), is the major river system of the West African subregion. With its upper reaches located in Guinea, it flows successively through Mali, where it has an important central inland delta, Niger, and Nigeria, where it receives large tributaries such as the Sokoto and the Benue. Its main affluents drain parts of Benin and Cameroon and more than half of Nigeria. Via a well-developed delta, the Niger enters the Atlantic Ocean in the Gulf of Guinea.

The fish faunas of the upper and central Niger are well known. The only available taxonomic work on the fishes of the Niger Delta in Nigeria is an annotated list published in 1963, including fifty-one species, of which only ten are primary freshwater species (Boeseman 1963). The delta, which covers an area of more than 10,000 km^2, is mostly freshwater. It contains the least known of Nigeria's four main blocks of closed forest (Lowe 1992), mangrove swamps are abundant, and the beach ridges contain large areas of freshwater swamp forest.

Between 1987 and 1997, one of us (C.B.P.) made extensive collections in the coastal drainages flowing from eastern Nigeria to the delta (lower Niger River and its immediate floodplain, Orashi River, Sombreiro River, and New Calabar River). Study of these collections revealed the presence of several species new to science (Coenen and Teugels 1989; Norris and Teugels 1990; Teugels and Roberts 1990; Teugels and Thys van den Audenaerde 1990) and increased the list of species from this area.

A remarkably high number of primary freshwater species (165) has been identified from the lower Niger; this number excludes permanent freshwater representatives of marine families such as Denticipitidae (denticle herrings), Clupeidae (herrings), and Eleotridae (sleepers). For the complete Niger Basin, covering 1,125,000 km^2 (Hugueny 1989), 225 primary freshwater species are known (Lévêque et al. 1991). In the delta, covering only about 1 percent of this surface area, some 74 percent of the freshwater fish species of the Niger are present. When the secondary freshwater species are included, 80 percent of the total species number of the Niger occurs in the delta.

The fish fauna of the Niger Delta is considered Nilo-Sudanian, meaning that it shares affinities with the fauna occurring in the major drainage basins of the Nile, Chad, Niger, Volta, and Senegal. However, a number of species are typical Lower Guinea elements (i.e., species present in coastal rivers from Cameroon to the mouth of the Congo). The fish fauna also contains disjunct populations of a few Upper Guinea species (species normally occurring in coastal basins from Guinea to Liberia). Thus, the delta can be considered a zoogeographic crossroads. The Nilo-Sudan area is mostly savanna, whereas the Lower and Upper Guinea areas are generally forested regions. The delta also has a small cluster of endemic species that are locally abundant.

A detailed study of the distribution patterns of fish species in the Niger Delta showed that their occurrence is not uniform. Instead, a dichotomy is apparent between acidic clear black-

---

†Deceased July 22, 2003.
††Deceased June 24, 1998.

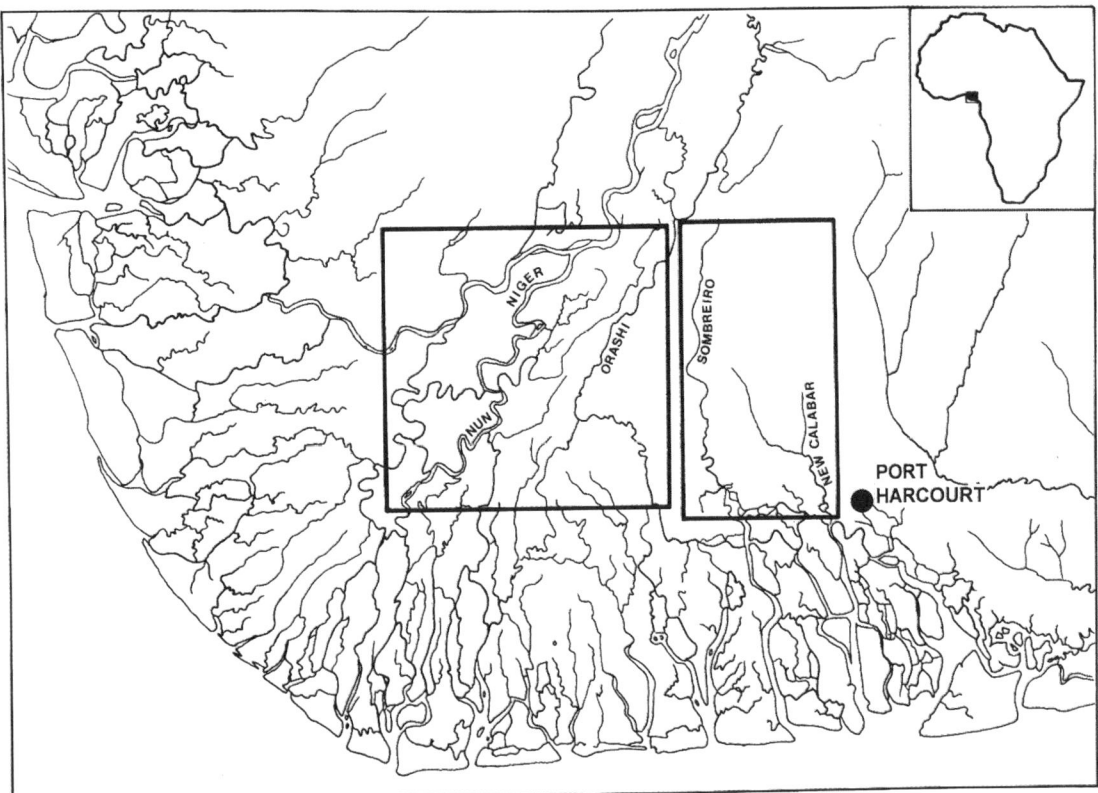

FIGURE 3.9. The Niger Delta in Nigeria. Based on distribution patterns of fish species, a dichotomy is apparent between blackwater (Sombreiro and New Calabar rivers) and whitewater systems (Niger, its immediate floodplain, and the Orashi River).

water systems (Sombreiro and New Calabar Rivers) and hard whitewater systems (the Niger River and its immediate floodplain and the Orashi River) (figure 3.9). Blackwater rivers contain up to 65 percent forest species and only 15 percent savanna species. Whitewater rivers have a species composition dominated by savanna species (46.5 percent), although they also contain an important number of forest species. The former are generally acidic, have a low conductivity, and are very transparent. They show little or no seasonal change in water level and have perennial densely vegetated banks, covered with mats of aquatic grasses and macrophytes, behind which swamp forest grows. Bottom vegetation is common, and the bottom consists of fine sand, beds of dead leaves, and anoxic organic mud. The whitewater systems are less acidic, with a higher conductivity and a low transparency. They show important changes in water level and support seasonal floodbank plain grasses; bottom vegetation is generally absent, and the bottom consists of sand, clay, and mud.

The observed dichotomy clearly is not related to physical barriers because the Orashi and the Sombreiro systems are connected by floodwaters each year. The Niger Delta has extended seaward over the last 6,000 years across the remains of an earlier delta into which it is embayed (Grove 1985). This historical deltaic process might contribute to explain the observed dichotomy: the Niger and its floodplain system together with the Orashi River are considered to be active channels, whereas the Sombreiro and New Calabar rivers can be seen as relict channels, containing relict populations of species that disappeared from other parts of the delta. The current conditions in the two types of rivers also provide habitats suitable to different types of species.

In addition to the little-studied impacts of upstream developments (e.g., irrigation and hydroelectric projects), the delta is subject to other ecological and hydrological alterations, linked mainly to activities of the oil industry. Swamp forest is destroyed rapidly by timbering and farming wherever access roads are made to new oil-drilling locations. Large tracts of swamp forest have also been marked for drainage for large-scale oil palm plantations, and more clearing is projected for rice plantations. Over the past decades there have been canalization projects for local boat transport; further alterations are expected along with several newly commissioned road projects, all of which will have unplanned effects on flow patterns and floodplain ecology. The delta itself lacks protected areas, although two have been advocated: a mangrove reserve and the Taylor Creek area in the forest zone (Stuart et al. 1990; Lowe 1992; IUCN 1993).

Our work in the eastern part of the delta is definitely not exhaustive. Species diversity in the floodplain lakes and beach-ridge swamps also must be investigated. More detailed information is also needed for the western part of the delta. A better knowledge of the freshwater fish assemblages in the Niger Delta will allow more efficient fishery management and wetland conservation strategies in an area under increasing pressure from intensive exploitation.

ESSAY 3.6

Madagascar's Freshwater Fishes: An Imperiled Treasure

John S. Sparks and Melanie L. J. Stiassny

The island of Madagascar is a global hotspot harboring exceptional concentrations of terrestrial biodiversity, and experiencing exceptional rates of habitat loss (Myers 1988; Myers et al. 2000). Although a global aquatic hotspot exercise remains to be done, there is little doubt that the importance of the Malagasy aquatic ecosystems matches, if not exceeds, that of their terrestrial counterparts. Although infrequently discussed, many of Madagascar's endemic freshwater fishes represent some of the most, if not the most, endangered vertebrates on the island (Stiassny and Raminosoa 1994; Wright 1997). Recent attention has been focused on the island's imperiled terrestrial biotas, and here we aim to direct some attention to the "quiet crisis" occurring beneath the water line and to bring the plight of Madagascar's extraordinary freshwater denizens to the forefront of conservation efforts on the island. (Our focus in this essay is exclusively on freshwater fishes, but a brief summary of other components of aquatic communities in Madagascar can be found in Benstead et al. 2000.)

In this essay we summarize new data from taxonomic revisions and ichthyofaunal surveys made during the past decade by us and others. A synthesis of these and earlier data reveals a new picture of Madagascan ichthyofaunal diversity and highlights new areas of importance for conservation efforts.

Madagascar's freshwater ichthyofauna traditionally has been viewed as being markedly depauperate, numerically falling well below the number predicted by the island's area (Kiener and Richard-Vindard 1972; Jenkins 1987; Stiassny and Raminosoa 1994; Riseng 1997). Our data show that this view is no longer tenable. In fact, the picture that emerges is of a native fauna that is numerically fully in line with that of other landmasses of similar size (figure 3.10) but also is globally outstanding in terms of local endemism and regional diversity (table 3.4, figures 3.11–3.13).

Of a total of 135 native fish species recorded from freshwater habitats (table 3.4), 90 are endemic to the broader Mascarene region and 84 species are endemic to the island itself. This number of Malagasy endemics represents a remarkable 45–71 percent increase over that reported by the most recent prior ichthyofaunal inventories. For example, Kiener and Richard-Vindard (1972) record 32, Stiassny and Raminosoa (1994) record 42, and de Rham (1996) records 49 endemics.[1] Most recently, Benstead et al. (2000) list 64 endemics, 58 of which have been verified in the context of the current study. Much of the increase in species numbers reported here is accounted for by the discovery of "new," that is, scientifically undescribed, species. Listed in table 3.4 are 36 of these taxa that are currently awaiting or have recently undergone formal scientific description (Stiassny and Rodriguez 2001; Stiassny et al. 2001; Sparks and Reinthal 2001; Stiassny 2002b; Sparks 2002). The existence of such a wealth of undescribed vertebrate diversity underscores the importance of continued survey efforts in freshwater habitats and suggests the magnitude of what may already have been lost in the degraded areas of the island. At higher taxonomic levels the degree of endemism is muted, with just two endemic families (9.5 percent) and thirteen endemic genera (24 percent) recognized (table 3.4).

Another notable aspect of the Madagascan ichthyofauna is its apparently relict nature (Pellegrin 1933, 1934; Kiener and Richard-Vindard 1972). Of the groups whose phylogenetic relationships have been investigated, many are found to represent the sister group (closest relatives) to large assemblages of very similar fishes that are more closely related to each other than to the Malagasy lineages (i.e., the Malagasy groups are phylogenetically plesiomorphic, or less derived) (Harrison and Howes 1991; Stiassny and de Pinna 1994). This is probably a result of the long-term isolation of Madagascar from other Gondwanan landmasses (Rabinowitz et al. 1983; Storey 1995; Storey et al. 1995). This repeated pattern is particularly intriguing, regardless of whether marine dispersal or vicariance is the process responsible for the origins of Madagascar's extant freshwater fishes. Whatever the explanation for this striking concentration of phylogenetically primitive taxa in Madagascan freshwater, the result is a major resource for evolutionary studies and an additional argument for the importance of incorporating aquatic systems in regional conservation programs on the island (Stiassny and Raminosoa 1994).

Despite the many interesting features of Madagascar's ichthyofauna, in some respects even more interesting is what is not found on the island. Many major groups of freshwater fishes found in but not necessarily restricted to Africa, including cyprinids, characins, polypterids, mormyrids, and osteoglossomorphs (Kiener and Richard-Vindard 1972; Stiassny and Raminosoa 1994), are absent. But perhaps even more puzzling, given our understanding of the sequence of events fragmenting Gondwanan landmasses (Rabinowitz et al. 1983; Storey 1995; Storey et al. 1995; Hay et al. 1999), is the fact that Madagascar lacks representatives of a whole series of families present in both Africa and India. These include anabantids, bagrids, channids, clariids, mastacembelids, notopterids, and schilbeids. To account for such a striking series of absences one is forced to postulate a series of major extinctions on Madagascar for which there is currently no evidence.

For students of historical biogeography, the origins of Madagascar's extant freshwater fish fauna remain a mystery. Representatives of all extant Malagasy freshwater fish groups are conspicuously absent from Cretaceous (145–65 m.y.a.) deposits on the island (Gottfried and Krause 1998). This lack of fossil evidence has led paleontologists to conclude that the present-day fish fauna did not colonize the island until the Cenozoic (less than 65 m.y.a.) (Krause et al. 1997; Gottfried et al. 1998), well after the breakup of Gondwanaland. Nevertheless, congruent biogeographic patterns for many of the extant taxa that have been investigated phylogenetically suggest a similar process (i.e., vicariance resulting from Gondwanan fragmentation) to explain their current distributions. In other words, contrary to the conclusions of paleontological research, phylogenetic analyses of extant taxa indicate that the breakup of Gondwana in the mid- to late Mesozoic (165–65 m.y.a.) may have isolated many components of Madagascar's present-day fish fauna on the island.

There are ecological anomalies as well. For example, the is-

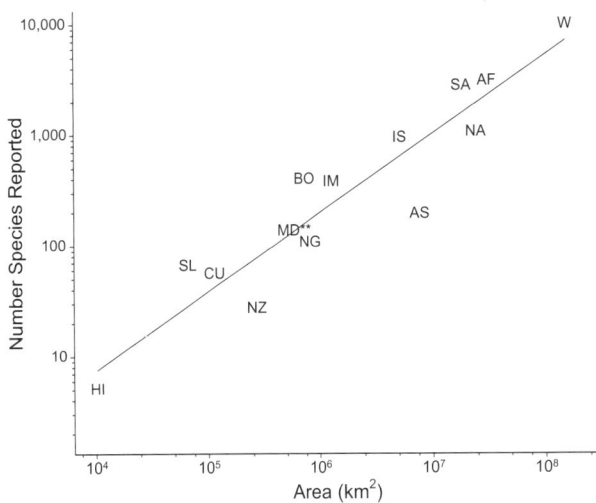

FIGURE 3.10. Species-area relationships for Madagascar and other oceanic and continental landmasses. The total for Madagascar is shown corresponding to the number of species reported in this study (MD**). Modified after Riseng (1997). AF, Africa (Daget et al. 1984, 1986a, 1986b; Skelton 1990a; Lundberg et al. 2000); AS, Australia (Allen 1989; Pollard et al. 1990; McDowall 1996); BO, Borneo (M. Kottelat [pers. comm., 2001] records 394 species from Borneo [shown] and estimates there are now approximately 450 described species [not shown]; Inger and Kong [1962] report 290 species from western Borneo [not shown]); CU, Cuba (55 [shown] recorded by Alayo [1973] and 36 [not shown] by Vergara [1980]); HI, Hawaii (Fitzsimons and Nishimoto 1990); IM, Indo-Malay archipelago (Zakaria-Ismail 1994 [minimum estimate]; Kottelat [1998] states that approximately 1,000 spp. are known from western Indonesia [not shown]); IS, Indian subcontinent (Talwar and Jhingran 1992); NA, North America (Williams and Miller 1990; Lundberg et al. 2000); NG, New Guinea [Munro 1967; Allen [1991] records 320–350 spp. including some estuarine forms [not shown]); NZ, New Zealand (McDowall 1990, 1996); SA, South America (Moyle and Cech 1996; Vari and Malabarba 1998); SL, Sri Lanka (Pethiyagoda 1991); W, world (Stiassny 1996; Lundberg et al. 2000). Geographic areas obtained from *Rand McNally Universal World Atlas* (1987).

land lacks any large, predatory fish except eels (*Anguilla* spp.). Equally noteworthy is the absence of primary freshwater families (those confined exclusively to freshwater) (Myers 1938) and the presence of only three secondary freshwater families (those tolerant of brackish water but normally found inland): the Aplocheiliidae, Cichlidae, and Poeciliidae. Despite the absence of a large primary freshwater fauna, there are small endemic radiations of bedotiids and cichlids, and the island is also home to a surprisingly speciose gobioid fauna, including a number of blind cave-dwelling species of the gobiid genus *Glossogobius*

TABLE 3.4. **Checklist of Native Malagasy Freshwater Fishes and Their Regions of Occurrence.**

Family	Genus	Species	Species Endemic to Madagascar	Southern Basins	Western Basins	Northwestern Basins	Eastern Highlands	Eastern Lowlands
Anguillidae	*Anguilla*	bicolor		X	X	X	X	X
Anguillidae	*Anguilla*	marmorata		X	X	X	X	X
Anguillidae	*Anguilla*	mossambica		X	X	X	X	X
Clupeidae	*Pellona*	ditchela			X	X		X
Clupeidae	***Sauvagella***	madagascariensis	M	X			X	X7
Clupeidae	***Sauvagella***	n. sp. "ambomboa"	M			X		
Clupeidae	*Spratellomorpha*	bianalis			X	X		
Ariidae	*Arius*	africanus			X	X		X
Ariidae	*Arius*	dussumieri			X	X		
Ariidae	*Arius*	madagascariensis	M			X		X
Ariidae	*Arius*	n. sp. "ankofia"	M			X		
Ariidae	*Arius*	n. sp. "sofia"	M			X		
Anchariidae	***Ancharius***	brevibarbus	M				X	X
Anchariidae	***Ancharius***	fuscus	M				X	X
Anchariidae	***Ancharius***	n. sp. "southwest"	M		X			
Atherinidae	*Atherinomorous*	cf. duodecimalis		X			X	X
Atherinidae	***Teramulus***	kieneri	M				X	
Atherinidae	***Teramulus***	waterloti	M			X		
Bedotiidae	***Bedotia***	geayi	M					X
Bedotiidae	***Bedotia***	madagascariensis	M					X
Bedotiidae	***Bedotia***	longianalis	M					X
Bedotiidae	***Bedotia***	tricolor	M					X
Bedotiidae	***Bedotia***	marojejy	M				X	
Bedotiidae	***Bedotia***	n. sp. "masoala"	M				X	
Bedotiidae	***Bedotia***	n. sp. "leucopeplos"	M				X	
Bedotiidae	***Bedotia***	n. sp. "nosivolo"	M				X	
Bedotiidae	***Bedotia***	n. sp. "ranomafana"	M				X	
Bedotiidae	***Bedotia***	n. sp. "vondrozo"	M				X	
Bedotiidae	***Bedotia***	n. sp. "garassa"	M					X
Bedotiidae	***Bedotia***	n. sp. "betampona"	M				X	
Bedotiidae	***Bedotia***	n. sp. "sambava"	M					X
Bedotiidae	***Bedotia***	n. sp. "bemarivo"	M					X

(continued)

Note: The five major hydrographic ecoregions listed correspond to those discussed in the text. Families, genera, or species in bold type are endemic to Madagascar.

X, presence of a species in a particular ecoregion; M, species endemic to Madagascar.

TABLE 3.4. *(continued)*

Family	Genus	Species	Species Endemic to Madagascar	Southern Basins	Western Basins	Northwestern Basins	Eastern Highlands	Eastern Lowlands
Bedotiidae	*Bedotia*	n. sp. "karikary"	M					X
Bedotiidae	*Bedotia*	n. sp. "manombo"	M					X
Bedotiidae	*Rheocles*	alaotrensis	M				X	
Bedotiidae	*Rheocles*	lateralis	M				X	
Bedotiidae	*Rheocles*	pellegrini	M				X	
Bedotiidae	*Rheocles*	sikorae	M				X	
Bedotiidae	*Rheocles*	wrightae	M				X	
Bedotiidae	*Rheocles*	n. sp. "derhami"	M			X		
Bedotiidae	*Rheocles*	n. sp. "large blackfin"	M				X	
Bedotiidae	*Rheocles*	n. sp. "andapa"	M				X	
Bedotiidae	*Rheocles*	n. sp. "ambatovy"	M				X	
Aplocheilidae	Pachypanchax	sakaramyi	M			X		
Aplocheilidae	Pachypanchax	omalonotus	M			X		
Aplocheilidae	Pachypanchax	n. sp. "large green"	M		X			
Aplocheilidae	Pachypanchax	n. sp. "ankofia"	M			X		
Aplocheilidae	Pachypanchax	n. sp. "sofia"	M			X		
Poeciliidae	*Pantanodon*	madagascariensis	M					X
Poeciliidae	*Pantanodon*	n. sp. "manombo"	M					X
Syngnathidae	Coelonotus	leiaspis				X	X	X
Syngnathidae	Hippichthys	cyanospilus			X	X		
Syngnathidae	Microphis	brachyurus					X	X
Syngnathidae	Microphis	fluviatilis			X	X		
Ambassidae	*Ambassis*	fontoynonti	M					X
Ambassidae	Ambassis	natalensis			X	X	X	X
Ambassidae	Ambassis	productus			X	X	X	X
Terapontidae	Terapon	jarbua			X	X		X
Terapontidae	*Mesopristes*	elongatus	M				X	X
Kuhliidae	Kuhlia	rupestris			X	X	X	X
Monodactylidae	Monodactylus	argenteus			X	X		X
Scatophagidae	Scatophagus	tetracanthus			X	X		X
Carangidae	Caranx	sexfasciatus			X	X		X
Chanidae	Chanos	chanos			X	X		
Cichlidae	*Paratilapia*	polleni	M	X	X	X	X	X
Cichlidae	*Paratilapia*	bleekeri	M		X	X		
Cichlidae	*Paratilapia*	n. sp. "garaka"	M			X		
Cichlidae	*Paratilapia*	n. sp. "all black"	M				X	

(continued)

TABLE 3.4. *(continued)*

Family	Genus	Species	Species Endemic to Madagascar	Southern Basins	Western Basins	Northwestern Basins	Eastern Highlands	Eastern Lowlands
Cichlidae	*Ptychochromis*	oligacanthus	M	X	X	X		
Cichlidae	*Ptychochromis*	n. sp. "black saroy"	M					X
Cichlidae	*Ptychochromis*	n. sp. "green saroy"	M				X	X
Cichlidae	*Ptychochromis*	n. sp. "inornatus"	M			X		
Cichlidae	*Ptychochromis*	n. sp. "kotro/onilahy"	M		X			
Cichlidae	*Ptychochromis*	nossibeensis	M			X		
Cichlidae	*Ptychochromoides*	betsileanus	M		X			
Cichlidae	*Ptychochromoides*	n. sp. "itasy"	M		X			
Cichlidae	*Ptychochromoides*	katria	M				X	
Cichlidae	*Ptychochromoides*	n. sp. "vondrozo"	M				X	
Cichlidae	*Oxylapia*	polli	M				X	
Cichlidae	*Paretroplus*	damii	M			X		
Cichlidae	*Paretroplus*	n. sp. "dridrimena"	M			X		
Cichlidae	*Paretroplus*	kieneri	M		X	X		
Cichlidae	*Paretroplus*	maculatus	M			X		
Cichlidae	*Paretroplus*	petiti	M			X		
Cichlidae	*Paretroplus*	polyactis	M	X			X	X
Cichlidae	*Paretroplus*	maromandia	M			X		
Cichlidae	*Paretroplus*	nourissati	M			X		
Cichlidae	*Paretroplus*	menarambo	M			X		
Cichlidae	*Paretroplus*	n. sp. "tsimoly"	M			X		
Cichlidae	*Paretroplus*	n. sp. "sofia"	M			X		
Cichlidae	*Paretroplus*	n. sp. "ventitry"	M					X
Mugilidae	*Agonostomus*	telfairii					X	X
Mugilidae	*Liza*	macrolepis			X	X		X
Mugilidae	*Liza*	alata			X	X		
Mugilidae	*Mugil*	cephalus			X	X		X
Mugilidae	*Valamugil*	buchanani			X	X		
Mugilidae	*Valamugil*	robustus			X	X		X
Gobiidae	*Acentrogobius*	audax			X	X		
Gobiidae	*Acentrogobius*	therezieni	M		X			
Gobiidae	*Awaous*	aeneofuscus			X	X	X	X
Gobiidae	*Chonophorus*	macrorhynchus	M			X		
Gobiidae	*Glossogobius*	biocellatus			X	X		X

TABLE 3.4. *(continued)*

Family	Genus	Species	Species Endemic to Madagascar	Southern Basins	Western Basins	Northwestern Basins	Eastern Highlands	Eastern Lowlands
Gobiidae	*Glossogobius*	callidus			X	X		X
Gobiidae	*Glossogobius*	giuris			X	X	X	X
Gobiidae	*Glossogobius*	**ankaranensis**	M			X		
Gobiidae	*Gobius*	hypselosoma				X		
Gobiidae	*Bathygobius*	**sambiranoensis**	M			X		
Gobiidae	*Bathygobius*	fuscus					X	X
Gobiidae	*Istigobius*	ornatus			X	X		
Gobiidae	*Oligolepis*	acutipennis						X
Gobiidae	*Oxyurichthys*	tentacularis			X	X		X
Gobiidae	*Papillogobius*	reichei			X	X		X
Gobiidae	*Redigobius*	balteatops						X
Gobiidae	*Redigobius*	bikolanus						X
Gobiidae	*Sicyopterus*	fasciatus					X	X
Gobiidae	*Sicyopterus*	laticeps					X	X
Gobiidae	*Sicyopterus*	**franouxi**	M				X	X
Gobiidae	*Sicyopterus*	**n. sp. "masoala"**	M				X	X
Gobiidae	*Stenogobius*	genivittatus			X	X		X
Gobiidae	*Taenioides*	gracilis						X
Gobiidae	*Yongeichthys*	nebulosus			X	X		
Eleotridae	*Butis*	butis			X	X	X	X
Eleotridae	*Eleotris*	acanthopoma						X
Eleotridae	*Eleotris*	fusca			X	X	X	X
Eleotridae	*Eleotris*	melanosoma			X			X
Eleotridae	*Eleotris*	**pellegrini**	M			X		X
Eleotridae	*Eleotris*	**vomerodentata**	M					X
Eleotridae	*Hypseleotris*	**tohizonae**	M			X		X
Eleotridae	*Ophiocara*	porocephala			X	X		X
Eleotridae	*Ophiocara*	macrolepidota					X	X
Eleotridae	**Ratsirakia**	**legendrei**	M				X	
Eleotridae	**Typhleotris**	**madagascariensis**	M	X				
Eleotridae	**Typhleotris**	**pauliani**	M		X			
Eleotridae	**Typhleotris**	**n. sp. "anomaly"**	M	X				
Megalopidae	*Megalops*	cyprinoides			X	X		
Total		135 species	84	10	48	69	48	66

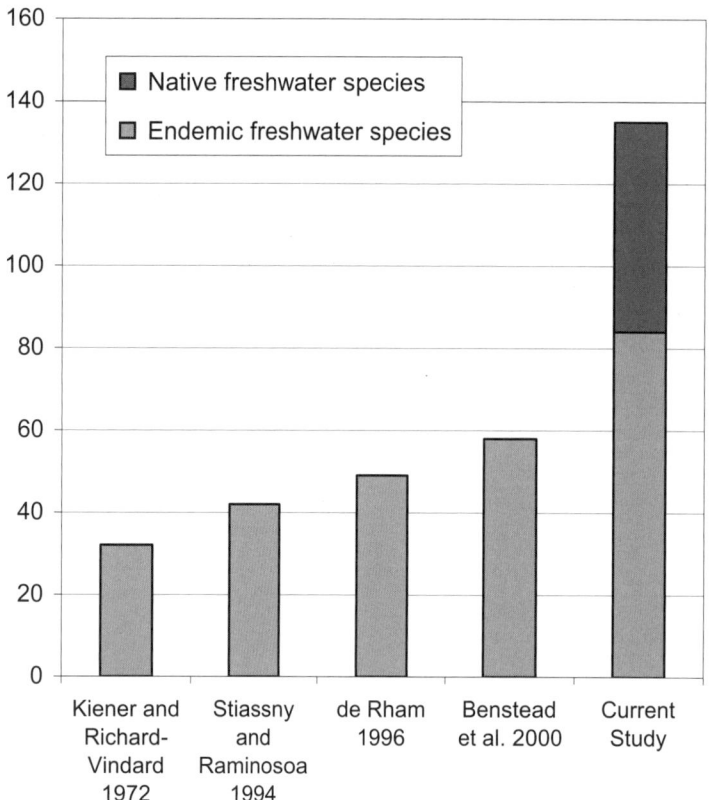

FIGURE 3.11. Plot of the number of native and endemic freshwater fish species recorded from Madagascar in recent surveys. A total of 135 native freshwater species are recorded in this study. The current survey results in an increase of 45–71 percent in the number of endemic species over previous summaries.

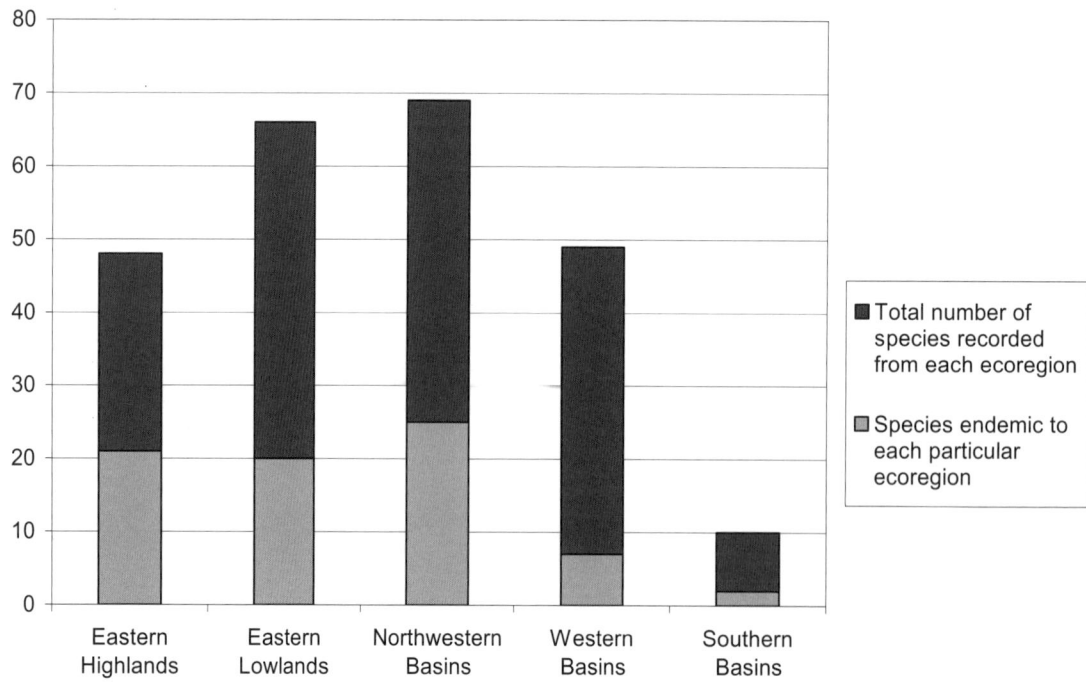

FIGURE 3.12. Plot of the total number of fish species by ecoregion for Madagascar (i.e., the number of native freshwater fish species recorded from each ecoregion), indicating the number of species that are endemic to each particular ecoregion.

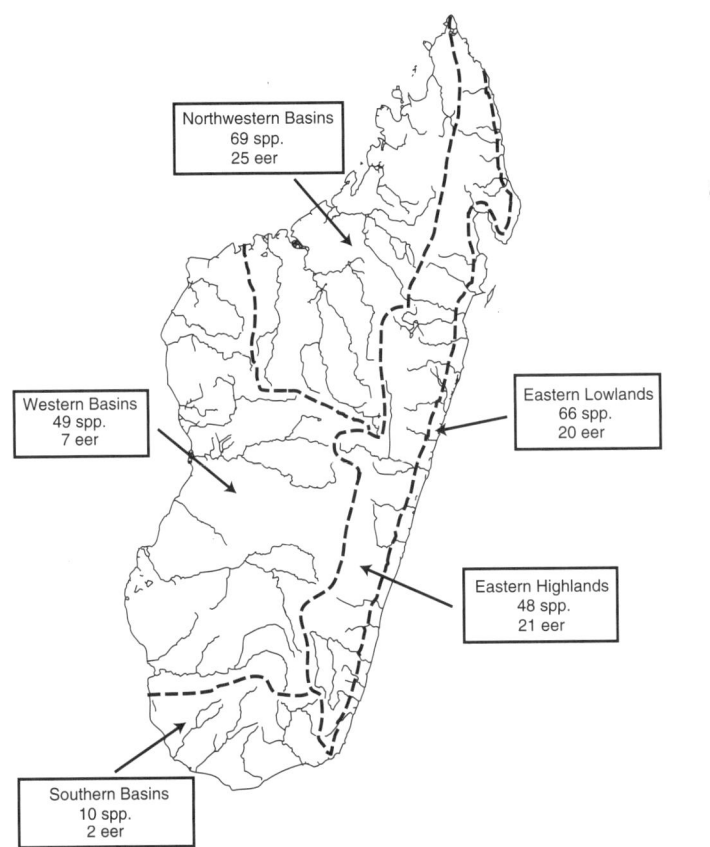

FIGURE 3.13. Five major ecoregions recognized on the basis of ichthyofaunal diversity and hydrographic structure, comprising the eastern highlands, eastern lowlands, western basins, rivers and lakes of the northwest, and southern basins. The values in boxes indicate the number of native freshwater fish species (spp.) recorded from the particular ecoregion and the number of those species endemic to that particular ecoregion (eer).

FIGURE 3.14. Some of the highly stenotopic endemic freshwater fishes of Madagascar. (a) *Oxylapia polli*, which is known only from a short stretch of the Nosivolo River, Mangoro drainage, near Marolambo in the eastern highlands. (b) *Teramulus waterloti*. Only a single surviving population is known from small forested streams in Bora Special Reserve, Anjingo drainage, northwestern Madagascar. (c) *Sauvagella* n. sp. "*ambomboa*," known only from the Ambomboa and Mangarahara rivers near Mandritsara, western drainages in northeastern Madagascar. (d) *Pantanodon* n. sp. "*manombo*," the only surviving member of the genus and known from only a single extremely small *Pandanus* swamp in Manombo Special Reserve, southeastern Madagascar. (e) *Typhleotris* n. sp., known from a single limestone cave near Lake Tsimanampetsotsa in southwestern Madagascar.

and the eleotrid genus *Typhleotris* (Arnoult 1959; Banister 1994).

From the perspective of ichthyological diversity, the island's freshwater systems often have been divided into two main regions: the eastern coastal drainages, containing the highest diversity of freshwater fishes on the island, and the much drier and supposedly species-poor western basins. East coast drainages are characteristically clear watered, with steep, short runs that terminate on a narrow coastal plain. The rivers of western Madagascar typically are much longer and generally slow-flowing, often forming shallow and turbid oligotrophic floodplain lakes (see Aldegheri 1972 for a detailed discussion of the hydrography of Madagascar). Our new distributional data suggest a somewhat more complex picture, and we recognize a third major ecoregion: the rivers and floodplain lakes of northwestern Madagascar (table 3.4 and figures 3.12 and 3.13). A total of sixty-nine native freshwater species are recorded from these northwestern rivers and lakes, whereas forty-five species have distributions restricted to western drainages in general. We have divided the eastern coastal drainages based on an elevational gradient and find that sixty-six species are recorded from the lowland eastern drainages, forty-eight species occur in highland eastern drainages, and fifty-three native species have distributions restricted to eastern drainages in general (table 3.4, figure 3.14). Compared with the dry western and southern basins, the eastern (highland and lowland) and northwestern basins are subject to more annual rainfall (Donque 1972), are substantially more diverse in habitat, and are not as influenced by vast seasonal fluctuations in flow rate that could easily wipe out existing fish populations during periods of complete desiccation.

Benstead et al. (2000) consider three main factors accounting for the current dire state of the Madagascan ichthyofauna: the degradation of aquatic habitats after deforestation, impacts of overfishing, and the interaction of native species with exotics. Sadly, this human-mediated encroachment will play an increasingly significant role in the attrition of native fishes because Madagascar has one of the highest population growth rates in the world at 3.2 percent (Kremen et al. 1999). To this list of factors should be added yet another, which is the extremely restricted geographic distribution of many of the island's species. Many are known only from a type locality, a much localized area, or a stretch of water. Examples of such highly stenotopic species (able to exist only in a narrow range of habitats) include *Oxylapia polli, Pantanodon* n. sp. *"manombo," Sauvagella* n. sp. *"ambomboa," Teramulus waterloti,* and a new species of blind *Typhleotris* restricted to limestone caves in southwest coastal Madagascar (figure 3.14). Such extremely localized distributions point to the vulnerability of the Malagasy ichthyofauna as a whole. Many endemic and native species have already been completely replaced by exotics, particularly on the central plateau and throughout much of western Madagascar (table

TABLE 3.5. Occurrence of Native and Exotic Fish Species at Selected Localities in Madagascar, 1989–1994.

Region	Number of Localities with		
	Natives	Natives and Exotics	Exotics
East coast	10	26	8
Central plateau	0	0	9
West coast	2	20	21
Total	12	46	38

Source: Benstead et al. (2000).

3.5). Unless measures are taken to protect this fragile fauna, as has been the case for many of Madagascar's terrestrial organisms, these fishes will be lost forever. Although it may already be too late to save all but a remnant of Madagascar's unique freshwater ichthyofaunas, it is our hope that the situation in Madagascar will serve as a warning and underscore the importance of early protection of vulnerable island freshwater ecosystems.

Note

1. These numbers are approximate because each study varies slightly in its definition of a freshwater fish species. For the purpose of this study we consider a taxon to represent a freshwater species if it is generally collected in, is commonly found well inland in, or is limited completely to freshwater habitats. Following this definition, we also include species whose life histories are poorly known but that appear to spend a significant portion of their lives in freshwater as opposed to entering freshwater for the sole purpose of reproduction. Marine taxa that enter the lower reaches of tidally influenced rivers are excluded.

CHAPTER 4

Reversing the Flow: Conservation Status of Africa's Freshwater Ecoregions

The health and resilience of aquatic systems in Africa, as elsewhere around the globe, are under increasing pressure related to growing human populations (Cohen et al. 1996; Lévêque 1997; Chapman and Chapman 2003). The threats are numerous, from dams and reservoirs to interbasin water transfers, regional overexploitation, climate change, pollution, the translocation of native species, and the introduction of invasive exotics (Malmqvist and Rundle 2002; Junk 2002; Tockner and Stanford 2002). However, a large proportion of Africa's freshwater resources remain relatively intact, and many of its rivers and lakes still support exceptional biodiversity. In this chapter we assess the severity of threat to each ecoregion to provide a basis for classifying actions necessary to protect their freshwater biodiversity.

We present broad patterns detected in our two-part conservation status analysis. The first part presents a snapshot conservation status—the current situation as of 2002—and the second part presents the final assessment as modified by a future threat analysis (projected threats over the next 20 years). More detailed information on specific threats to each ecoregion is available in appendix F.

Snapshot Conservation Status

Land-Based Threats (Land Degradation)

This category covers threats to freshwater organisms that originate in the terrestrial landscape. Land degradation is strongly correlated with human population density in areas of urban growth but may also be extensive in agricultural areas where population is low. Land degradation is medium or high throughout most bioregions and major habitat types (MHTs) in Africa. The only MHT that has generally low values (signifying low levels of degradation) for this indicator is xeric systems, where both human populations and agriculture tend to be low (figure 4.1a). However, in these areas there may be high densities of grazing animals, causing changes to the vegetation and soil degradation.

Land-based threats were assessed to be high in most of West Africa, parts of southern and East Africa, much of Madagascar and nearby islands, several Atlantic coast islands, and much of the Rift Valley and Ethiopian highlands. All of the ecoregions in the closed basin and small lakes MHT, particularly vulnerable to changes in land use, were rated high for land degradation. Human population densities are high bordering lakes Victoria and Malawi and parts of Lake Tanganyika, in parts of coastal East Africa, in much of West Africa, in the Nile and Niger deltas, along the northern and southern coasts of the continent, in parts of Madagascar, and in the Ethiopian highlands and northern Rift Valley (figure 4.2). Agricultural lands have a slightly different distribution. Cropland covers much of West Africa and the Rift Valley in East Africa, the Ethiopian highlands, the lower Senegal River, coastal Kenya and Tanzania, and parts of the Zambezian bioregion. Cropland is also highly concentrated in parts of the Cape Fold [33] and Permanent

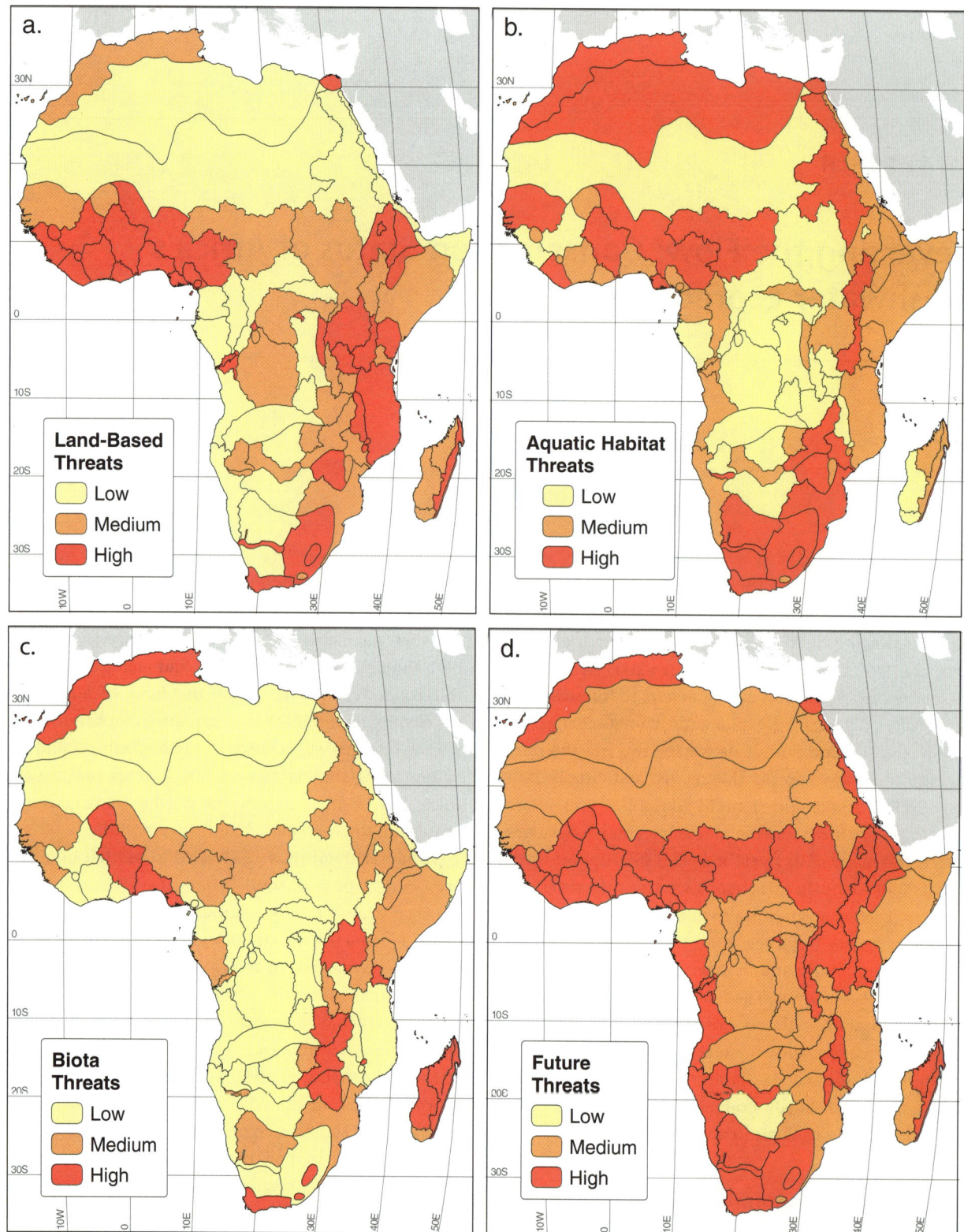

FIGURE 4.1. (a) Land-based threats (estimated). (b) Aquatic habitat threats (estimated). (c) Biota threats (estimated). (d) Future threat level (estimated).

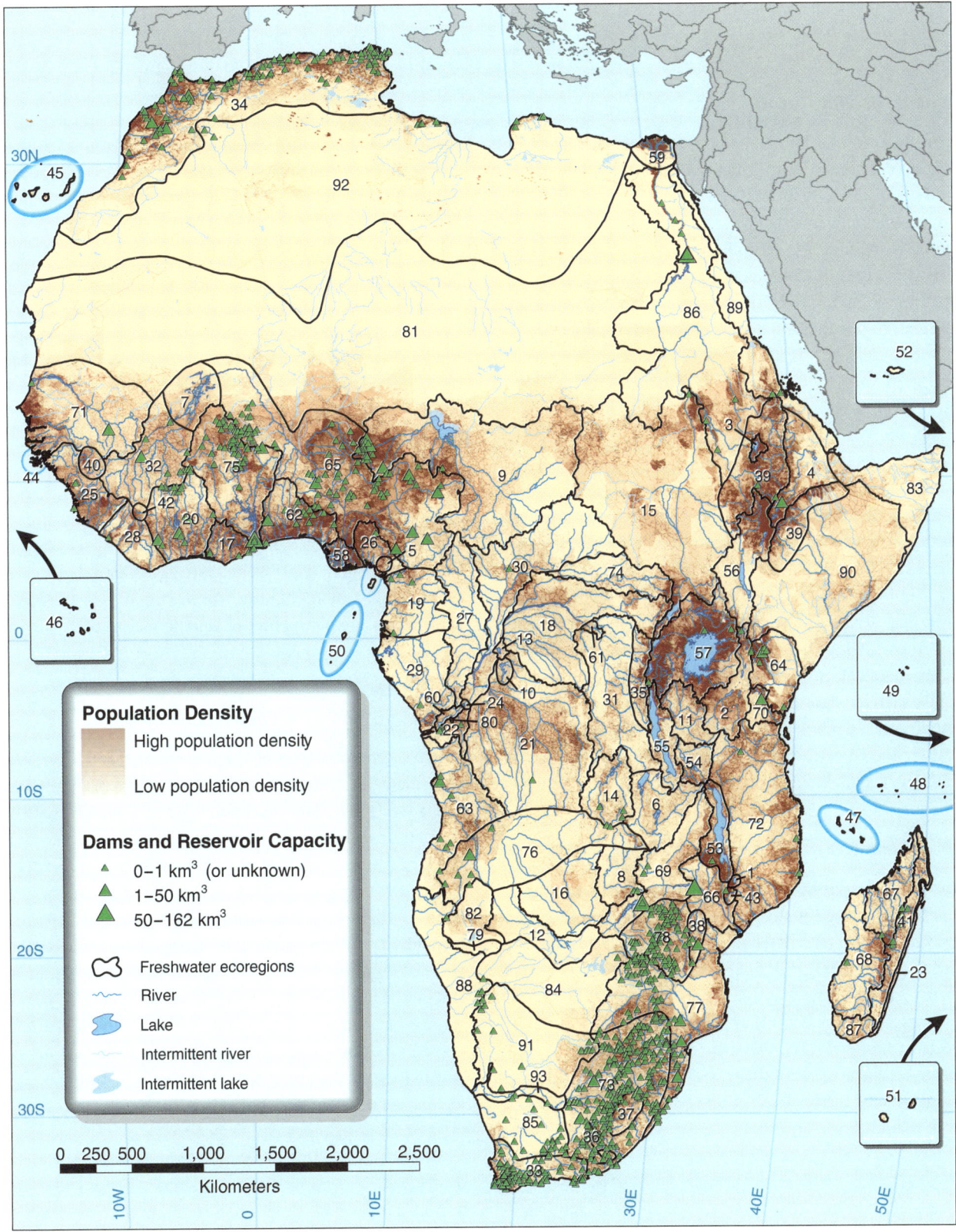

FIGURE 4.2. Population density from LandScan Global Population 2001 database (Dobson et al. 2000; Oak Ridge National Laboratory 2002) and major dams from Food and Agriculture Organization database on African dams (FAO 2000a).

Maghreb [34] and along the Lower Nile [86] and in the Nile Delta [59] (see figure 1.1).

Aquatic Habitat Threats

Threats to aquatic habitats that occur directly in rivers and lakes include such disturbances as dams, water diversions, and point source pollution. These alterations and inputs disrupt species movements; change water quality, temperature, and flow regimes; and generally degrade habitat (Davies and Day 1998; World Commission on Dams 2000; Bunn and Arthington 2002). Habitat threats were assessed to be high over large portions of the continent, including parts of North and West Africa and a large block of ecoregions in southern Africa (figure 4.1b). Among those with a high ranking, seven ecoregions—the Cape Fold [33], Permanent Maghreb [34], Lower Niger-Benue [65], Southern Temperate Highveld [73], Volta [75], Zambezian Lowveld [77], and Zambezian (Plateau) Highveld [78]—had more than fifty large dams (FAO 2000a) (see figure 4.2). The two ecoregions with the largest reservoir capacities—the Lower Nile [86], with Aswan High Dam (about 180 billion m^3), and the Middle Zambezi Luangwa [69], with Kariba Dam (about 160 billion m^3)—were also assessed as having high aquatic habitat threats. Only the floodplains, swamps, and lakes MHT had generally low values for this indicator, in large part because such habitats are less likely to be dammed. Moist forest rivers in parts of the continent where there were low land-based threats, such as parts of the Congo Basin, also had low aquatic habitat threats.

Biota Threats (Exploitation and Exotics)

Introduced species and overexploitation directly affect populations of native aquatic species. Potentially devastating declines of native fisheries, and the economic losses that can accompany introductions of exotic species, are concerns in many parts of Africa (essay 4.1) (Pimentel et al. 2001). Lake Victoria [57] provides a stark example of the potential effects of biotic threats; there, many cichlid species disappeared from the lake or their stocks suffered drastic reductions after the introduction of Nile perch (*Lates niloticus*) and four tilapiine species, although other ecological changes to the lake have also been implicated in the declines (Witte et al. 1992). Lake Victoria also provides hope for ecosystem rebound if biotic disturbances are mitigated; some cichlid species recently reemerged after a decline in the Nile perch stock (Witte et al. 2000; Balirwa et al. 2003). Nonnative species, including vertebrates, invertebrates, and plants, have also negatively affected the ecology of much of Madagascar, the Permanent Maghreb [34], the Amatolo-Winterberg Highlands [36], the Drakensberg-Maloti Highlands [37], the Canary Islands [45], parts of the Zambezi Basin (ecoregions [69] and [78]), and Pangani [70] (figure 4.1c). Interbasin transfers (IBTs) can be a mechanism for the introduction of alien species, and the increasing prevalence of IBTs is threatening the native biota in many parts of the region (Snaddon et al. 1998; Davies et al. 2000b).

Concerns about overexploitation of aquatic species touch nearly all ecoregions, although limited data exist on the status of inland fish and other species populations by ecosystem across the continent (essay 4.2). In a majority of sub-Saharan African countries, fish protein constitutes more than 20 percent of the animal protein intake of human populations, and in several countries this figure is more than 50 percent (FAO Fisheries Department 2002). Given the importance of fish to many of the poorest Africans and Malagasy, conserving the ecosystems that support these fisheries gains heightened significance (see essay 1.1). Lakes Chilwa and Chiuta [1], Bangweulu-Mweru [6], Inner Niger Delta [7], Lake Victoria [47], Niger Delta [58], Bight Coastal [62], and the Volta [75] are highlighted for the intense hunting and fishing pressure placed on the aquatic species there (figure 4.1c).

Synthesis of Snapshot Conservation Criteria

Broad patterns of ecoregion-scale degradation and intactness are evident both across the continent (figure 4.3) and within MHTs (table 4.1). The continent-wide map shows that six ecoregions are in critical condition: the Madagascar Eastern Lowlands [23], Cape Fold [33], Drakensberg-Maloti Highlands [37], Niger Delta [58], Volta [75], and Zambezian (Plateau) Highveld [78]. Endangered ecoregions are scattered throughout the continent. They occur in parts of western and eastern Africa, in eastern Madagascar, on select islands, and on the northern edge of the continent (figure 4.3). Ecoregions assessed as relatively intact are found in the sparsely populated Congo Basin, upper parts of the Zambezi Basin, and the major deserts of the continent. Most of the ecoregions are considered either vulnerable or relatively stable (45 percent and 30 percent, respectively).

Examination of the snapshot assessment results by MHT shows that ecoregions of certain habitat types are far more degraded on average than others (table 4.1). All of Mediterranean and large river delta ecoregions were either critical or endangered. In contrast, floodplains, swamps, and lakes; large lakes; moist forest rivers; and xeric systems had at least

TABLE 4.1. Distribution of Ecoregions by MHT for the Snapshot and Final Conservation Status.

Major Habitat Type	Snapshot Conservation Status (Final Conservation Status)					
	Critical	Endangered	Vulnerable	Relatively Stable	Relatively Intact	Total
Closed basins and small lakes	0 (1)	1 (3)	3 (1)	1 (0)	0 (0)	5
Floodplains, swamps, and lakes	0 (0)	0 (2)	4 (4)	5 (3)	2 (2)	11
Moist forest rivers	1 (1)	0 (7)	7 (2)	6 (4)	2 (2)	16
Mediterranean systems	1 (2)	1 (0)	0 (0)	0 (0)	0 (0)	2
Highland and mountain systems	1 (2)	1 (5)	7 (2)	0 (0)	0 (0)	9
Island rivers and lakes	0 (1)	1 (3)	4 (2)	3 (2)	1 (1)	9
Large lakes	0 (1)	1 (1)	1 (2)	3 (1)	0 (0)	5
Large river deltas	1 (2)	1 (0)	0 (0)	0 (0)	0 (0)	2
Large river rapids	0 (0)	0 (1)	1 (1)	1 (0)	0 (0)	2
Savanna–dry forest rivers	2 (4)	3 (6)	9 (5)	2 (1)	1 (1)	17
Subterranean and spring systems	0 (0)	0 (2)	2 (0)	0 (0)	0 (0)	2
Xeric systems	0 (0)	0 (2)	4 (6)	7 (3)	2 (2)	13
Total	6 (14)	9 (32)	42 (25)	28 (14)	8 (8)	93

half of their ecoregions in the relatively intact or relatively stable categories.

Overall, the results of the snapshot assessment suggest that most ecoregions have not yet experienced extreme degradation. About three-quarters of the ecoregions are considered either relatively stable or vulnerable, and only six are considered to be in critical condition. Africa's freshwater systems are at an important juncture; despite increasing pressure, these freshwater habitats and many of their species may remain into the future if we secure the opportunity to protect them now.

Future Threat Assessment

Growing human populations and associated land and water degradation and overexploitation, climate change, planned IBTs or dams, and new or expanded agricultural, mining, or logging operations are among the largest future threats to the freshwater landscape of Africa. Climate change, large resource use operations, and population growth were evaluated for the future threat assessment. Future species introductions are also of great concern, but it is difficult to predict with any precision where they might occur. Therefore, we excluded species introductions from the evaluation of future threats.

Using the results of a global model of water availability

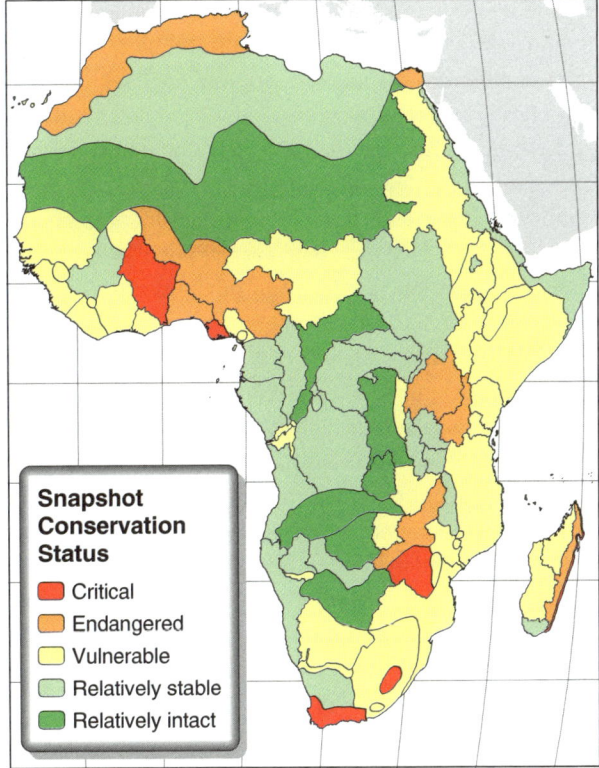

FIGURE 4.3. Snapshot conservation status.

and water use (Döll et al. 2003; Alcamo et al. 2003), we evaluated the effects of climate change and future water use on river discharge in 2025 (see appendix B and essay 4.3 for a full explanation of methods). Ten ecoregions, lying in a block that stretches along the Atlantic coast from the mouth of the Congo River to the southern tip of the continent and extending inland to the Okavango Floodplains [12], xeric Karoo [85], and Southern Kalahari [91], are estimated to experience a decrease of 10 percent or more in mean annual discharge by 2025. In contrast, the Upper Niger [32], Volta [75], and Bight Coastal [62] ecoregions are expected to experience an increase in discharge.

Proposed large-scale infrastructure and agricultural, mining, or logging operations that threaten freshwater organisms are planned or ongoing in eight ecoregions: the Permanent Maghreb [34], Drakensberg-Maloti Highlands [37], Lake Chad Catchment [9], Inner Niger Delta [7], Southern West Coastal Equatorial [29], Kenyan Coastal Rivers [64], Upper Nile [15], and Southern Temperate Highveld [73]. Consideration of small-scale activities, which are often widespread and undocumented, would add many more ecoregions to the list.

Increasing pressure on natural resources (including fisheries and forest cover), in addition to land and water degradation, are expected to be associated with increases in human populations. Population increases are likely to affect most heavily much of West Africa, the Ethiopian highlands, parts of East Africa and the Rift Valley (including the large lakes), much of Madagascar, and the northern edge of the continent. About half (forty-one) of the ninety-three ecoregions are predicted to have an average population of more than 50 people/km^2 in 2025, and nineteen ecoregions are predicted to have an average of more than 100 people/km^2.

Combining these analyses of climate change effects, resource use operations, and projected population density, future threats appear to be high over much of the continent, with large blocks of ecoregions expected to experience substantial degradation to their freshwater within the next 20 years. Many ecoregions in West Africa, the Rift Valley, and eastern and northern Madagascar and along the coast from Equatorial Guinea to South Africa face high future threats (table 4.1, figure 4.1d). All major habitat types are equally affected, with at least one ecoregion from every MHT having a high level of future threat.

Final Conservation Status

Nearly two-thirds (fifty-eight) of the ecoregions qualified for an elevation in conservation status level based on projected threats from climate change, planned developments, and human population growth. Blocks of ecoregions in West Africa, southern Africa, the Rift Valley, and northern and eastern Madagascar and along the northern edge of the continent are considered critical or endangered in the final (threat-modified) analysis. Much of the Congo Basin, the upper portions of the Zambezi Basin, Malagarasi-Moyowosi [11] and Lake Rukwa [54] in eastern Africa, several island ecoregions (Bijagos [44], Coralline Seychelles [48], and Socotra [52]), and several ecoregions in the xeric systems MHT (Dry Sahel [81], Horn [83], Kalahari [84], Madagascar Southern Basins [87], and Temporary Maghreb [92]) remain relatively intact or relatively stable (figure 4.4).

All MHTs have at least one ecoregion in critical or endangered condition in the final conservation status assessment (appendix D). The savanna–dry forest rivers, highland and mountain systems, and closed basin and small lakes MHTs appear to be the most threatened, with more than 60 percent of their ecoregions designated as critical or endangered. Large river deltas and Mediterranean systems, each with only two ecoregions, are also highly threatened, with all of their ecoregions in the top level of imperilment. Xeric systems and floodplains, swamps, and lakes have the largest numbers of relatively intact and relatively stable ecoregions. Xeric systems may be more threatened than revealed by our indicators because threats in xeric regions may result from activities (e.g., extraction of groundwater resources) that are not explicitly addressed by the indicators that rely heavily on population data and cropland land use. The least threatened ecoregions in the floodplains, swamps, and lakes MHT all occur in the Congo, the upper Zambezi, and the Malagarasi-Moyowosi basins.

After modification for future threat, fourteen ecoregions were assessed as critical, thirty-two were endangered, twenty-five were vulnerable, fourteen were relatively stable, and eight were relatively intact. In other words, about one-sixth of the ecoregions, containing about 500 endemic freshwater fish, aquatic mollusks, freshwater herpetofauna, and mammals, are expected to be critically degraded within the next 20 years if action is not taken to reverse trends immediately.

To prevent further loss and degradation of freshwater species and habitats in Africa, we must tackle some of the root causes of habitat loss and species endangerment and develop bold ways to enhance conservation efforts across the region in terms of both scope and impact. In the next two chapters we will see where we have the best opportunities for doing so and explore some of the approaches we have at our disposal to achieve this goal.

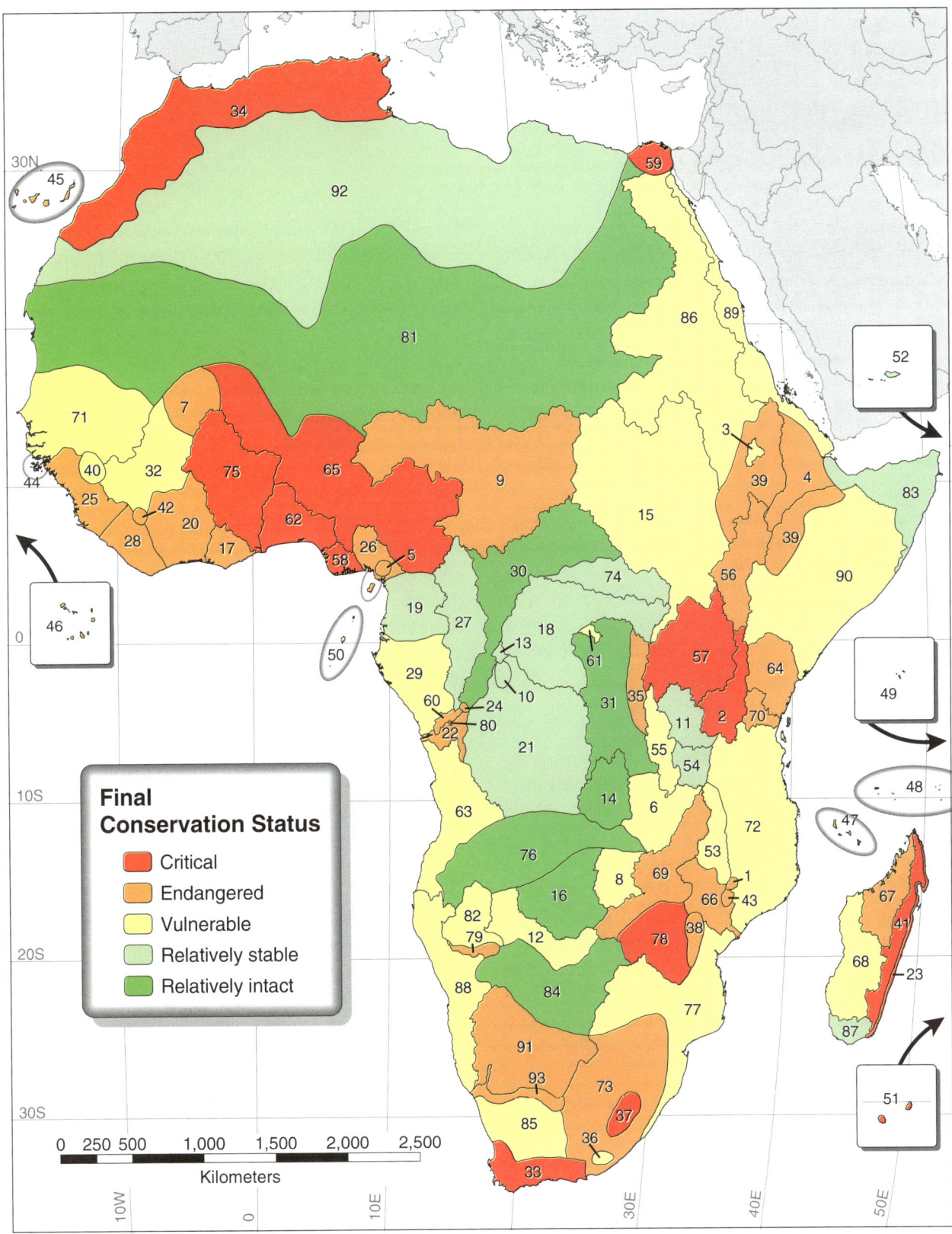

FIGURE 4.4. Final conservation status.

ESSAY 4.1

Invasive Alien Species in African Freshwater Ecosystems

Musonda Mumba and Geoffrey Howard

All aspects of life on the African continent are intimately interwoven with freshwater systems. The productivity and diversity of the freshwater ecosystems are extremely important to the livelihood, and often survival, of African peoples (Shumway 1999). An important issue for Africa has been to increase investment and cooperation among riparian countries in managing freshwater resources in an efficient and sustainable manner (UNEP 1996). Although the continent has abundant water resources, they are spatially and temporally distributed by the variability of climate, topography, and geology, making water very unevenly available across Africa. The continent has nine major river basins, nearly all of which are international, some traversing more than eight countries. Unfortunately, Africa's waters may be among the most vulnerable because of the large proportion of semi-arid and desert areas and the needs of a rapidly increasing human population (Stiassny 1996). Habitat degradation and loss are compounded by the effects of exotic species.

The Problem

The introduction of invasive alien species in freshwater ecosystems is believed to be the second greatest threat to global freshwater biodiversity after habitat loss (Shumway 1999). Increasing numbers of species introductions have occurred on the African continent in the last century. Fish introductions have occurred for many reasons, including fish culture, use of apparently empty niches, production of more or larger fish, and control of unwanted organisms (Lévêque et al. 1988). Seldom have the biological or economic implications of introductions been considered before species introductions.

Lake Victoria (table 4.2) has the dubious distinction of being the site of one of the world's most notorious vertebrate introductions in history (Shumway 1999). The introduction of Nile perch (*Lates niloticus*) and Nile tilapia (*Oreochromis niloticus*) to Lake Victoria in the 1950s—against the prevailing scientific advice of the day—has been linked to the elimination of many native fish species. Since the perch and several tilapiine species were introduced, the lake probably has lost 200 taxa of endemic cichlids, and the remaining 150 or more probably are endangered. Although there is evidence that habitat degradation may have been largely responsible for the declines, the introduction of the nonnative fish has certainly resulted in observable changes. In 1976, only 0.5 percent of the commercial catch was exotics; 7 years later exotics were 68 percent of the catch. Ramifications of the species introductions can also be seen in the surrounding landscape. The traditional fishery was based on small cichlids, which could be preserved through air drying. Drying the perch's oily flesh, on the other hand, requires firewood, and this has increased pressure on the area's limited forests. Cutting of forests has increased siltation and eutrophication, which in turn have further unbalanced the precarious lake ecosystem (Kaufman 1992; Kaufman, pers. comm., 2000).

Invasive plants have also caused havoc in most of the waterbodies on the African continent. The most problematic are water hyacinth (*Eichhornia crassipes*) and the kariba weed (*Salvinia molesta*). Both species colonize waterbodies and form dense floating mats of interlocking plants. Pollution, eutrophication, flow regulation (dams and reservoirs), and human-mediated transfer have all been implicated in the spread of invasive plants. The appearance of the water hyacinth in Africa has led to problems for many rural African communities. Fish landings have been completely blocked by weeds in some towns, such as in Epe, a Lagos community where fishing is one of few livelihoods (Odusile 2001).

Water Resources in Danger

Many inland waterbodies in Africa have been infested with invasive alien species. Data are lacking on the actual extent of distribution of most of these species, although there is some evidence that the invasive alien plant water hyacinth has spread throughout the entire continent over the last century (Phiri and Navarro 2000).

Most of the freshwater ecoregions in Africa have waterbodies that contain invasive alien species (table 4.2). Most African rivers and lakes and many wetlands are affected by some alien invasive species, be they fish, crustaceans, mollusks, submerged and emergent water plants, and semi-aquatic species that affect floodplains and the edges of larger waterbodies. But many often go unnoticed because they are out of sight beneath the water surface or because they are not recognized as alien and invasive until their impacts become severe. Some of the affected ecosys-

TABLE 4.2. **Examples of Infested Waterbodies in Some Freshwater Ecoregions of Africa.**

Ecoregion	Affected Protected Area	Countries	Some of the Invasive Species
Floodplains, swamps, and lakes			
Inner Niger Delta [7]	Inner Niger Delta (Ramsar site)	Mali	Water hyacinth (*Eichhornia crassipes*)
Kafue [8]	Lochinvar National Park (Ramsar site in Zambia)	Zambia	Water hyacinth, water lettuce (*Pistia stratiotes*), giant sensitive plant (*Mimosa pigra*)
Large lakes			
Lake Victoria [57]		Kenya, Uganda, and Tanzania	Water hyacinth, water lettuce, and Nile perch (*Lates niloticus*)
Large river deltas			
Nile Delta [59]		Egypt	Water hyacinth
Savanna–dry forest rivers			
Middle Zambezi Luangwa [69]		Zambia and Zimbabwe	Water hyacinth and water lettuce
Lower Zambezi [66]		Malawi and Mozambique	Water hyacinth

tems are in protected areas, such as the Kafue Flats floodplain system in Zambia and the Inner Niger Delta in Mali (table 4.2).

Are Invasives a Possible Threat to Food Security in Africa?

Freshwater fisheries form an essential part of the food security of many local peoples throughout the world. The rivers and lakes in Africa provide innumerable benefits in relation to food security. Africa's fisheries are economically important for both domestic and external markets. River basins (together with associated wetlands) are highly productive ecosystems, and an estimated 40 percent of fish in Africa comes from riverine and floodplain fisheries (Welcomme 1979). The total yield in 1995 of fish from African inland waters was nearly 1.8 million tons, larger than that from all North American waters (Shumway 1999). Unfortunately, many of the wild populations of fish in Africa are under serious threat from invasive alien species.

Economic Impact

Confronted with tough challenges to accelerate economic growth and reduce poverty, African countries are constantly under pressure to pursue short-term growth policies, which shift

TABLE 4.3. **Economic Costs and Control Expenditures Associated with Water Hyacinth.**

Country or Water Body	Cost (US$ thousand/yr)
Nigeria	50,000
Lake Victoria	9,660
Uganda	4,560
Egypt	7,000
Malawi	133
Zimbabwe	43
Total	71,396

Source: Joffe and Cooke (1997).

ecological and economic costs to the next generation (UNEP 2000b). The costs of environmental neglect in Africa can be substantial, arguing for proactive protection (box 4.1).

The control of water hyacinth has already proven to be costly. In one valuation analysis, the World Bank/GEF Lake Victoria Environment Management project estimated annual losses of $200,000 in local fisheries; $350,000 in beaches and water

> **BOX 4.1. The Cost of Inaction: Nigeria.**
>
> Traditionally, the government and private sectors have hesitated to implement new environmental protection measures because of assumed high costs. This narrow preoccupation has overshadowed the equally important consideration of the mounting economic, social, and ecological costs of not acting.
>
> A World Bank study of Nigeria provides a stark assessment of the risks and enormous costs if no remedial action is taken to address eight of the country's largest environmental problems. In the absence of near-term remedial and mitigation measures, the long-term losses to Nigeria from environmental degradation in these eight areas have been estimated.
>
> | | Annual Cost of Inaction (US$ million/year) |
> |---|---|
> | Soil degradation | 3,000 |
> | Water contamination | 1,000 |
> | Deforestation | 750 |
> | Coastal erosion | 150 |
> | Gully erosion | 100 |
> | Fishery losses | 50 |
> | Water hyacinth | 50 |
> | Wildlife losses | 10 |
> | Total | 5,110 |
>
> *Source:* FEPA (1991).

supply for domestic, livestock, and agricultural purposes; and $1.5 million in urban water supply lost to blocked intakes (Joffe and Cooke 1997). The total annual expenditures associated with the control of water hyacinth in five countries and Lake Victoria is estimated to be more than $70 million (table 4.3). These estimates represent the direct benefits that would have accrued from prevention (Kasulo 1999).

In another example, Alimi and Akinyemiju (1990) give the direct cost of manual, mechanical, and chemical control methods for some sites in Nigeria. Costs to clear 1 km^2 of water hyacinth were $9,500 for manual control, $8,000 for mechanical clearance, and $4,400 for chemical control. Although this analysis appeared to support chemical control, no mention was made of its environmental effects on nontarget species.

Information Gaps

There seems to be a general lack of awareness of the possible impacts of alien introductions and mechanisms for monitoring introductions and impacts across the continent. However, some efforts have been made by a few organizations such as the World Conservation Union (IUCN) and the Global Invasive Species Programme (GISP) to publish awareness materials (e.g., Howard and Matindi 2003; IUCN 2001b; Wittenberg and Cock 2001). Nevertheless, there is a dearth of readily available information for water and wetland managers that is specific to Africa, although efforts are beginning in some countries such as South Africa (Henderson and Cilliers 2002).

Policies directed at the control and prevention of invasive introductions will be developed and implemented only when research on species introductions has better quantified likely impacts. In many cases lack of data on the seriousness of an invasion prevents timely actions that might be able to avert the impact on peoples' livelihoods and on biodiversity. Although prevention and control technology is available elsewhere, it is still not readily available to those who need it in Africa, even when an invasion has been identified. Organizations responsible for quarantine, species identification, invasive species management, and ecosystem restoration are underfunded and lack the necessary capacity and often the legal instruments to act quickly and effectively.

Regional and Local Initiatives

The New Partnership for Africa's Development (NEPAD) has devoted an entire issue to this topic in its Environment Initiative (NEPAD and UNEP 2003). This initiative recognizes aquatic invasives as one of the most serious threats to both development and biodiversity in Africa. Within the NEPAD process, and supported by the GISP, a number of subregional initiatives on invasive species are associated with economic development organizations such as the Southern Africa Development Community and the East Africa Community. The GISP has developed global information (McNeely et al. 2001; Wittenberg and Cock 2001) on how to prevent and manage invasive species, which is currently being tailored to both aquatic systems and to Africa (see GISP Web site at http://globalecology.stanford.edu/DGE/Gisp/).

An example of a local initiative is the Working for Water program in South Africa, where serious aquatic invaders are addressed through manual labor and other forms of control (see the WFW Web site at http://www.dwaf.gov.za/wfw/).

Monitoring and Risk Assessment

Although some information is available about selected species and waterbodies, it is generally true that in continental Africa and its island states there is both a lack of knowledge about the potential harm of invasive species and an absence of experience on how to prevent aquatic invasive species. Even the invasion and subsequent control of water hyacinth in Lake Victoria remain largely undocumented or published. Monitoring freshwater systems for invasives and determining their impacts, in addition to monitoring for the presence of species with the potential for invasion, are effective ways to address this problem. Both approaches have been taken up by the Ramsar Convention (see resolution VIII.7) and resolutions adopted at the World Parks Congress. Ramsar requires that invasive species be listed in wetland site descriptions and that impacts of invasive species be monitored over time. The Durban Accord and recommendations to the Convention on Biological Diversity from the recent World Parks Congress state the importance of monitoring species in protected areas—for potential and actual invasiveness—as one way of improving management effectiveness (http://www.iucn.org/themes/wcpa/wpc2003/).

The intentional introduction of alien species often is necessary for extending crop production, agroforestry, production forestry, aquaculture, and livestock production, but it should not be done without consideration of the possible risks of these species becoming invasive in their new homes. This risk can be avoided through assessment of the particular species and waterbody of concern (Champion and Clayton 2001). A recent discussion about the reintroduction of genetically altered African fish for aquaculture addressed this issue and made recommendations for procedures to ensure the safety of such introductions to protect peoples' livelihoods and local biodiversity, in this case in reference to *Oreochromis niloticus* (ICLARM 2002). Together with FAO, the International Center for Living Aquatic Resources Management strives to increase global food production through aquaculture while protecting wild and managed ecosystems from invasion by species that could escape from fish farms.

Recommendations

The following recommendations may assist in the prevention and management of invasive alien species in Africa's wetlands:

- More information must be made available to decision-makers and water managers in Africa on the threats of invasive alien species (IASs).
- Local and regional efforts aimed at biodiversity and ecosystem management should take into account the current extent and possible future introductions and spread of aquatic IASs.
- There is a need for risk assessment before the intentional introduction of any exotic (alien) species and monitoring after introduction.
- Successes in aquatic IAS control in Africa should be published and monitored (FEPA 1991).

ESSAY 4.2

Impacts of Fishing on Inland Fish Populations in Africa

Robin Welcomme

Wetlands are common throughout most of Africa. Most of them are seasonal, associated with major rivers as enlarged floodplain areas such as the Central Delta of the Niger or the Yaeres of the Chari/Logone or as swamps associated with major lakes such as Lake Bangweulu or Lake Chad. Such areas have rich and varied fish faunas that depend on the wetlands for breeding, nurseries, feeding, and refuge. Fish populations in these environments reach very high densities and also undergo seasonally predictable migrations, both factors that favor fisheries. They also attract human population because of the richness of their soils, their water supply, and the opportunities for grazing cattle. Traditional local cultures, with customary ways of allocating and managing the various natural resources, have centered on the fishery and on other rural activities. However, population growth in recent decades, together with technological innovation in a range of fields, has introduced growing exploitive and environmental pressure, particularly on the fishery resources.

In 2000, the nominal catch of fisheries in the inland waters of Africa produced 2.2 million tons of freshwater and diadromous fish (figures from FAO at http://www.fao.org/fi/statist/statist.asp). This level of yield was reached after a long period of sustained growth, although trends over the last ten years indicate that further significant increases seem unlikely. The origin of the catch seems to be roughly equally divided between the major lake and reservoir fisheries and those of the major rivers.

Numbers of Species

In common with most tropical systems, African lake and river fish faunas are rich and complex. Species richness in African lakes correlates well with lake area in a sample of thirty-one lakes (figure 4.5a). The three great lakes of Africa—Victoria, Tanganyika, and Malawi—all have complex species flocks of several hundred species, and numbers of species remain to be distinguished and described, at least in lakes Victoria and Malawi. Similar relationships can be obtained for rivers (figure 4.5b). Some species are widely distributed throughout the continent or throughout certain basins. Elements of the Nilotic fauna, for example, extend over much of the Chad, Niger, and Senegal systems as well. However, there is a high degree of endemism in the rift valley lakes and in the coastal river systems of East and West Africa.

Lengths of Species in African Fish Assemblages

When species assemblages are analyzed for their length characteristics in terms of the maximum length to which each species grows, it is apparent that the majority of species are small. The length composition of the West African fauna is summarized in figure 4.6. This shows that 50 percent of species grow to less than 10 cm standard length, and 90 percent of species are less than 55 cm. This is similar to assemblage structures from other parts of the world (Welcomme 1999). Length is also linked to trophic habit. There is a close correlation between the percentage of fish-eating predators and length (figure 4.7), showing that a greater percentage of larger species are predatory in habit.

Effect of Fishing

Fish assemblages respond to most types of externally induced stresses, other than selective disturbances such as dam building and toxic discharges, by undergoing a series of changes centered on a general decline in mean size. This decline arises thorough the loss of larger individuals and species from the assemblage and their replacement by smaller ones. In Africa such shifts have been formally described from the Oueme River in

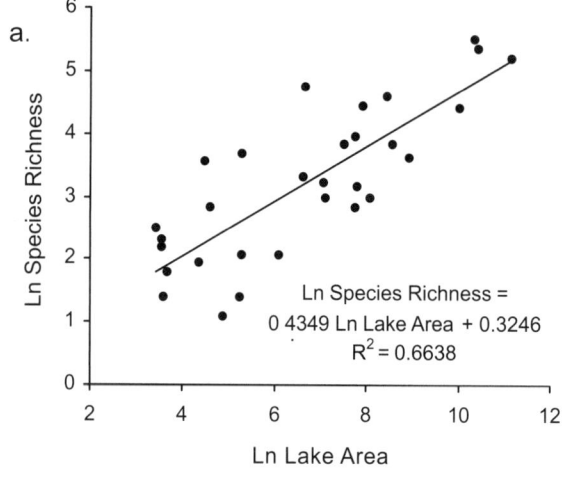

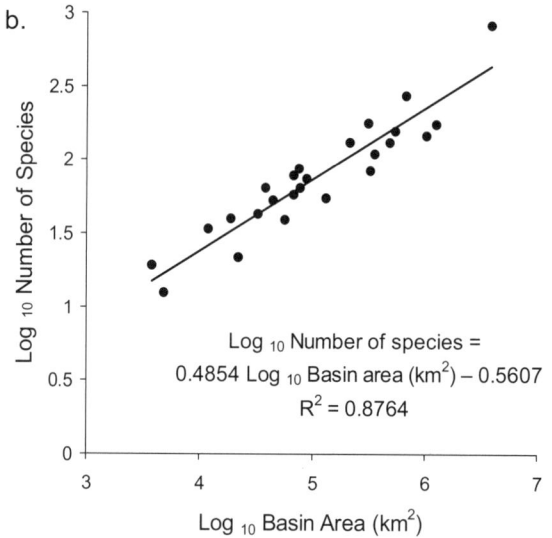

FIGURE 4.5. (a) Number of species as a function of lake basin area for 31 African lakes. (b) Number of species as a function of river basin area for 24 African rivers.

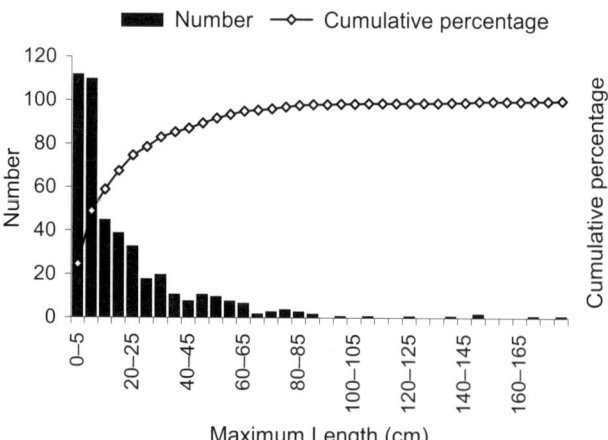

FIGURE 4.6. Size structure of fish assemblages in West African rivers.

Benin (Welcomme 1999), the Central Delta of the Niger in Mali (Laë 1994), the Shire River in Malawi, Lake Malawi (Turner 1981), and Lake Victoria (Balirwa et al. 2003). These declines probably are general throughout the continent; reduced sizes of fish caught have been noted less formally in many systems. Reduction in size is accompanied by changes to many of the fundamental parameters of the fish community. Mortality and growth rates increase, as does biological production. Biomass decreases, giving a net increase in the production:biomass (P:B) ratio. Fish stocks with high P:B ratios can withstand heavy fishing mortalities and thus sustain high yields, as in the pelagic *Limnothrissa* fisheries of lakes Kariba and Cahora Bassa.

The number of species in the fishery tends to increase as fishing-down proceeds. Large species are eliminated from the fishery but only rarely become locally extinct. The survival of species occurs through use of connected refugia. For example, in the Oueme, large species such as *Lates niloticus, Distichodus,* and *Citharinus* disappeared from the fishery and the lower reaches of the river in the 1960s but persisted in refugia at upstream sites (figure 4.8). Similarly, in Lake Victoria all large species had declined or disappeared from the fishery, and the larger species of the haplochromine stock were also declining in the trawl fisheries before the establishment of the Nile perch (*Lates niloticus*) (Balirwa et al. 2003). Many of these species survive in the satellite lakes located in the riparian wetlands of the lake. Other large species survive by reducing their mean length and size at first maturation and are thus able to persist in the fished area.

Indicators

Studies have identified a number of indicators that can be used to diagnose the state of a fishery or of aquatic biodiversity in general. Most of these indicators rely on historical knowledge of the original fish fauna as compared with present composition (figure 4.9).

Mean Length

The mean length of the assemblage or of the catch may be the simplest and most direct indicator of its condition. As a rule, if a significant portion of the fish are of moderate length (say, more than 60 cm), then the assemblage is still probably in good condition. Reduction of mean size to 20 cm and below would indicate severe damage caused by either overfishing or environmental degradation. As a correlate of mean length, the mesh size of gear used in the fishery is a rapid indicator of the state of the fishery. Where mesh sizes are universally small (2- to 3-cm bar), the mean length of the fish will be within the size indicating that the fishery may be experiencing difficulties. By the

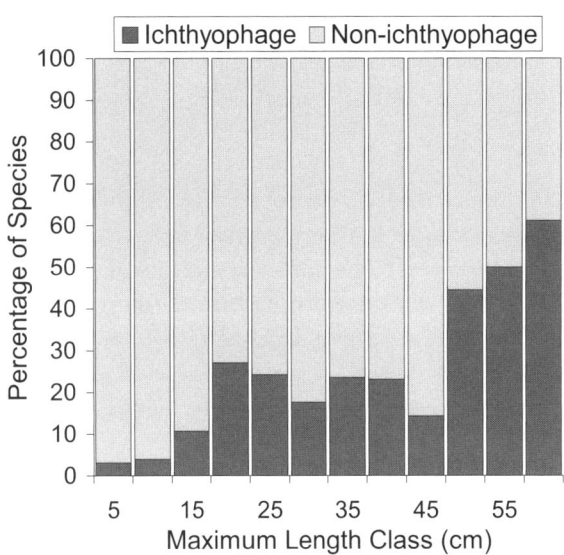

FIGURE 4.7. Relative percentages of ichthyophages and non-ichthyophages by maximum length class (data from West African rivers).

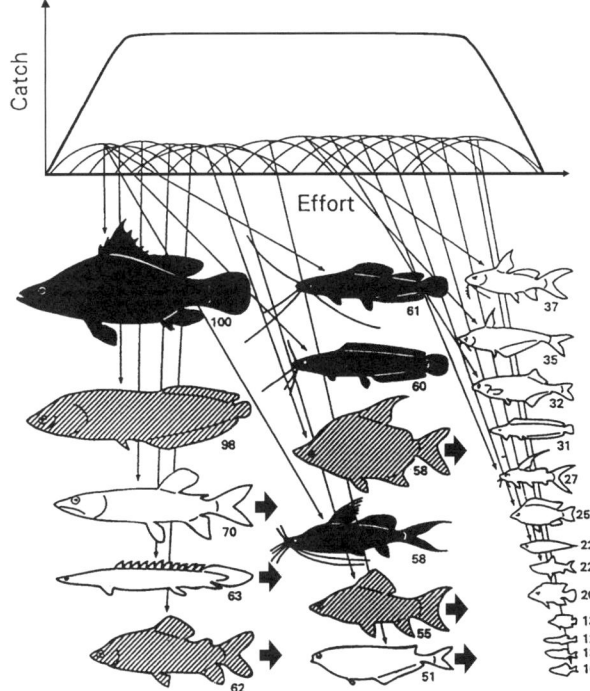

FIGURE 4.8. Illustration of the fishing-down process using the historical condition of the Oueme River fishery (West Africa) and representative species (from Welcomme 1999). *Black,* species that disappeared from the fishery before 1965; *hatched,* species numbers were seriously reduced; *black arrows,* species that reduced their breeding size. Peak yearly yields in the 1950–1960s were around 10,000 tons. Numbers next to fish indicate the maximum length (in cm) of each species.

same token, where larger mesh nets are in use, a greater range of sizes of fish in the catch is indicated.

Number of Species

One of the most immediate responses to environmental or fishery-induced stresses on a fish assemblage is the disappearance of certain elements of the fauna. Generally the larger species disappear first, and the fishery successively concentrates on smaller length classes as the assemblage is fished down. Because any fish assemblage consists of many more small species than large ones, this will result in an increase in the number of species forming the catch.

Type of Species

Major migratory species decline wherever overfishing or environmental stresses have occurred. A second widespread effect is the rise in abundance in invasive exotic species.

Response Time

In floodplain rivers and river-driven lakes, catches respond to the intensity of flooding in previous years. Here the lag in the response of the fishery to years of good or bad flooding may be taken to indicate the state of exploitation of the fishery. Fisheries that respond very rapidly (usually in the same year because their catch is based on age 0+ fish) are at risk, whereas longer response times (2–3 years) indicate much lower levels of exploitation relative to potential.

Predator-Prey Relationships

Because a greater percentage of large species are fish-eating predators, certain ratios between trophic groups are symptomatic of unstressed communities. Lack of large predators in a fish assemblage may be taken as a sign that disturbance is occurring.

Other Indicators

Other parameters—natural mortality rate, growth, P:B ratios, and changes in representation between K- and r-selected species—can be used to assess the status of the assemblage under study. However, these measures are more complicated and usually are direct correlates of mean length. This means that they may have value as confirmatory indices but do not supply information additional to that gained by simpler measures.

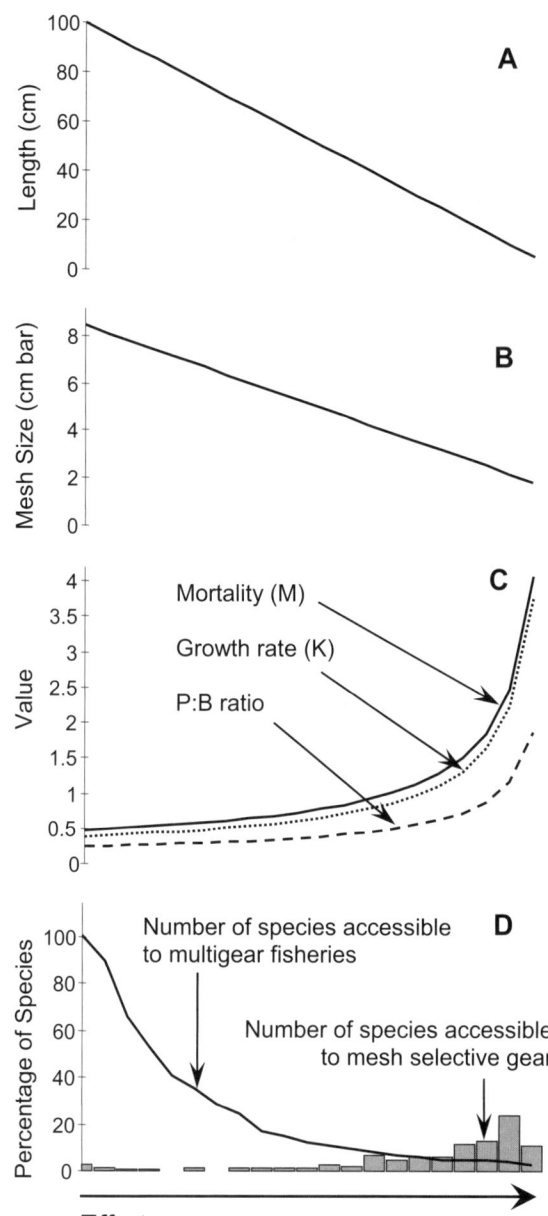

FIGURE 4.9. Changes in basic characteristics of fish assemblages in response to increasing exploitation pressure: (a) mean length of fish in catch, (b) mesh size of size selective gear, (c) changes in essential parameters, and (d) changes in relative number of species available to broad-spectrum multigear fisheries and size-selective fisheries.

Associated Problems

The nature of the inland fishery resource is influenced not only by fishing but also by a range of other pressures. On the whole, Africa has remained undamaged by the problems of water pollution and river modification that have taken place on other continents. Nevertheless, dam building has had an impact on the composition and abundance of fish populations in many rivers, leading to reductions in catch (World Commission on Dams 2000). Associated with dam building, there is an increasing tendency to use water for irrigated agriculture, and there are plans for massive diversions of water between basins across the continent. Lowered flows mean that the floodplains that are the basis of so many African river fisheries are inundated to a lesser degree and that dry season flows may be insufficient to support fish (Welcomme and Halls 2001). Such effects cannot be considered in isolation because their effects are cumulative and add to the overall pressure on the resource.

Resilience of Fish Assemblages

One of the questions that most often arises in connection with the fishing-down process is the degree to which it is reversible. Reversing ecological trends usually is a function of the resilience of the communities of organisms that are subjected to the stress. Indications in African waters are that populations are resilient and may recover when the stress is lessened or removed. In Lake Victoria, for instance, some of the haplochromine communities and many of the larger noncichlid species have recovered in abundance as the fishery has concentrated on the Nile perch and this invasive species itself has become fished down (Balirwa et al. 2003). The reestablishment of normal catches in rivers of the Niger Basin after the Sahelian drought (see, for example, Welcomme and Halls 2004) and a subsequent increase in catches associated with restored floods indicate resilience on the part of these fish populations. Indeed, the close relationship between flood levels and fish catches in many rivers around the world (see Welcomme 1985; Welcomme and Halls 2004) argues that rivers can recover rapidly from low flow conditions in one year through rapid increases in population size in subsequent years. In lakes that receive much of their water supply from incoming rivers (e.g., Lake Chad [Durand 1983]), the reestablishment of fish populations has followed the restoration of normal flood regimes after catastrophic events such as the recurring Sahelian droughts.

This resilience should not be taken as a license to abuse inland waters and the living aquatic resources that depend on them. Responses of fish populations to sustained human interventions such as dam building, water extraction, channelization, and pollution have been shown to be more drastic elsewhere in the world and to challenge any natural resilience fish populations may have (World Commission on Dams 2000; Welcomme 2001).

Conservation Measures

Anecdotal evidence from reports on many fisheries in Africa indicates that they are fished at levels where damage to the assemblages has occurred. In some fisheries, such as the Oueme and the Central Delta of the Niger, about 90 percent of the catch comes from young of the year. Clearly this situation is undesirable for conservation and probably unsustainable for fish production. The situation arises for two main reasons. First, there is an excess of effort in most fisheries. Second, top-down, centralized management has been unable to formulate and implement the type of flexible fishing policies needed.

Modern trends in management seek to resolve these problems by limiting access to the fishery and by devolving much of the policy formulation and enforcement to all stakeholders in the resource. Many African fisheries have a history of traditional management. Traditional systems fell into disrepute over much of the continent during the nation-building of the 1960s to 1990s, but now such approaches are looked to as the basis for co-management. Traditional and co-management systems seek not to maximize the size and quantities of fish caught but rather to address problems of equity, resource sharing, and conflict reduction. Other measures, such as the establishment of conservation and harvest reserves and systematic conservation-based stocking programs to support endangered species, are only just entering into the thinking of many managers.

Experiences such as those on the Niger River (Welcomme and Halls 2004) and Lake Victoria (Balirwa et al. 2003) show that population pressures are leading to increasing numbers of fishers on those fisheries, and this trend probably extends to fisheries across the continent. Inevitably such pressures accelerate the fishing-down process, robbing systems of resilience and possibly leading to the eventual collapse of fish stocks. It appears that the only solution is to reduce fishing pressure by limiting access to the fishery. At the same time, environmental issues, such as the withdrawal of water for agriculture, must be urgently addressed so that the natural habitat of the fish is protected.

ESSAY 4.3

Climate Change, Human Water Use, and Freshwater Ecosystems in Africa: Looking toward the Future

Bernhard Lehner

Africa is one of the most vulnerable regions in the world to climate change. This vulnerability and the limitations of poor countries to adapt to climate change were highlighted in *Climate Change 2001,* the third assessment report of the Intergovernmental Panel on Climate Change (IPCC 2001). The report establishes how human activity is modifying the global climate, with temperature rises and precipitation changes projected for the next 100 years that could affect human welfare and the environment (Desanker 2002). Historical climate records show that during the last century Africa experienced a warming of approximately 0.5°C over most of the continent, accompanied by changes in precipitation patterns. Climate change simulations indicate a significant acceleration of these trends into the future. Moderate scenarios project a persistent warming of 0.5°C over the next 25 years, and other scenarios warn of even more severe changes (Hulme et al. 2001).

The most simplified explanation of climate change is that an increase in greenhouse gas emissions causes the global average temperature to rise. A warmer atmosphere holds more energy, which increases the potential of evaporation and, as a consequence, precipitation. At a global level, the increased energy intensifies the magnitude, frequency, and persistence of atmospheric circulation patterns. Finally, amplified rainfall events of the future may cause severe flooding, dry spells may grow into severe droughts, and today's storms may turn into tomorrow's cyclones.

The consequences of climate change in Africa are anticipated to be varied and far reaching, and include land degradation, desertification, sea level rise, loss of biodiversity, reduction of water resources, spread of diseases, and impacts on food security (Desanker 2002). Yet climate change is only one aspect of the ongoing global change, which includes all human-induced changes to our environment. For example, future demographic and socioeconomic trends and human-related landcover change are also expected to seriously affect the state of the global environment in the future.

Climate and global change threaten the freshwater systems of Africa in several ways. In general, changes in precipitation and air temperature will affect water temperature, water quality, and water quantity. The consequences for aquatic systems can include alterations in the composition of riparian vegetation, changes in river morphology and dynamics, reduced oxygen concentrations and changes in chemical load caused by higher temperatures, changes in species distributions, and loss of habitats such as spawning areas, refuges, and nurseries (Carpenter et al. 1992). Besides direct climatic effects, the physical characteristics of a river basin also have a strong influence on its water balance. For example, changes in landcover or land use can lead to alterations in soil conditions, infiltration rates, erosion, evaporation, groundwater recharge, and runoff generation. In many cases the connection between cause and effect is spatially or temporally discrete and thus less obvious; for example, deforestation in sensitive headwater areas can increase flood risk in areas far downstream, or irrigated agriculture can deplete groundwater levels and lead to the drying out of rivers in subsequent years. In addition to changes in mean annual flows, changes in the seasonal hydrological regime are important. For example, a shift in timing and duration of flooding or low flow periods can negatively affect many aquatic species whose life cycles are adapted to the natural hydrological regime and limit water availability for human populations.

Limitations in Simulating the Future

Numerous physical processes, parameters, and complex feedback mechanisms must be considered in potential climate change scenarios. Computer simulation models assist in these efforts. General circulation models (GCMs) have been developed to simulate future climate characteristics and to study the outcome of different scenario assumptions (e.g., varying concentrations of greenhouse gases). However, these models are inherently uncertain, and despite ever-increasing computer power the results are still coarse on both spatial and temporal scales.

Additionally, interpreting model outcomes is difficult. For the analysis of extreme weather or discharge events, time series are needed that are long enough to derive statistically significant results. However, actual observations often exist for only the most recent decade or two. Thus, data may be unavailable or too inaccurate to distinguish between normal fluctuations and extremes. As a result, the projections of future climatic trends are blurred by both uncertainties in the GCM calculations and the incompleteness of available reference data.

Long-term average values (e.g., mean global temperatures or mean annual discharges over long time periods) are the most reliable outputs of climate and large-scale hydrological models (Hulme et al. 2001). However, the conclusions that can be drawn from these values are limited. Rising average water levels could have either positive or negative effects on the availability of water resources in a river or lake basin. If excess discharges mainly reflect more frequent and stronger flood events, then local characteristics (e.g., levees, natural inundation areas) and the magnitude of change will determine whether the effects are positive (water storage in reservoirs for later use, flooding as an essential part of aquatic life cycles) or negative (hazards and destruction). A decrease in flows, on the other hand, generally leads to a lower availability of water resources. However, seasonal variability is still of high importance. Are water levels lessened equally throughout the year, or do prolonged dry spells occur, broken by singular flash floods? Even stable long-term annual discharges, suggesting no future change, can mask strong seasonal changes, such as increases in both summer droughts and winter floods for the same area (Lehner et al. 2001).

All in all, the incomplete understanding of physical processes, the sketchy assumptions of driving forces and anticipated trends, the uncertainty of the calculations, and the difficulty in interpreting the results of the models limit the reliability of climate and global change studies. Therefore, simulations of the future should not be interpreted as forecasts or predictions but as likely paths or scenarios of future development.

Despite all these limitations, when based on plausible sets of assumptions, scenarios can give valuable insight about possible future changes in complex systems and highlight critical issues of concern (Henrichs et al. 2002). Scenarios are powerful tools to assess and better understand causes, effects, and the dimension of climate change, so they should be used to explore the sensitivity of a range of African environmental and social systems (Hulme et al. 2001).

Case Study: Effects of Climate and Global Change on African Freshwater Ecoregions

Several assessments have examined the possible effects of climate and global change on Africa, its environment, and socioeconomic issues (e.g., Desanker 2002; Hulme et al. 2001; IPCC 2001). In this case study we focus on the effects of climate and global change on water supplies for human populations and the health of freshwater ecosystems. For that, we first need to evaluate the extent to which changes in temperature, precipitation, and evaporation will lead to an overall increase or decrease in runoff, groundwater recharge, and river flows in the freshwater ecoregions of Africa. Because of their capacity to store, infiltrate, or intercept water, soils and vegetation play an important role in the translation of climate change to changes in the hydrological regime. But human activities can also strongly influence the water balance of a river basin, for example through agricultural land use changes, deforestation, increase of impervious areas in settlements, construction of dams, or water extraction for irrigation, industry, and household use.

In our case study for the African continent, we therefore analyze the results of an integrated global discharge model that includes both water availability and human water use calculations. For future scenarios, the applied WaterGAP 2 model (Water Global Assessment and Prognosis, version 2) is driven by GCM outputs for climate change and a set of scenario assumptions for changes in human water use (Alcamo et al. 2003; Döll et al. 2003). The model calculates water balances at a spatial resolution of 0.5° (approximately 50 km × 50 km at the equator), routes the results along a global river network system taking lakes and reservoirs into account, and finally computes monthly discharge values for every grid cell. These discharges can be interpreted as the total amount of water that drains from each cell. In other words, the values represent the maximum renewable water resources available in each cell after human water use has been taken into consideration.

The applied assumptions and driving forces for future scenarios are largely consistent with the no-climate-policy IS92a scenario estimates of the IPCC (1992) and the intermediate Baseline-A scenario developed by the Dutch National Institute of Public Health and Environment (Alcamo et al. 1998). They represent intermediate assumptions about population growth, economic growth, and economic activity. For example, it is assumed that for Africa the population will grow from about 640 million in 1990 to 2.2 billion in 2050 and that the gross domestic product will rise from about $650 to $1,950 per capita in the same period. As for climate change, an average annual increase of global carbon dioxide emissions of about 1 percent per year is anticipated in the simulations. Yet even with similar driving forces, climate projections can vary widely between GCMs (IPCC 2001). To allow for critical comparisons, WaterGAP applies climate projections calculated by two different state-of-the-art GCMs, the HadCM3 model (Gordon et al. 2000) and the ECHAM4 model (Röckner et al. 1996).

Discharge calculations of WaterGAP 2 were analyzed in

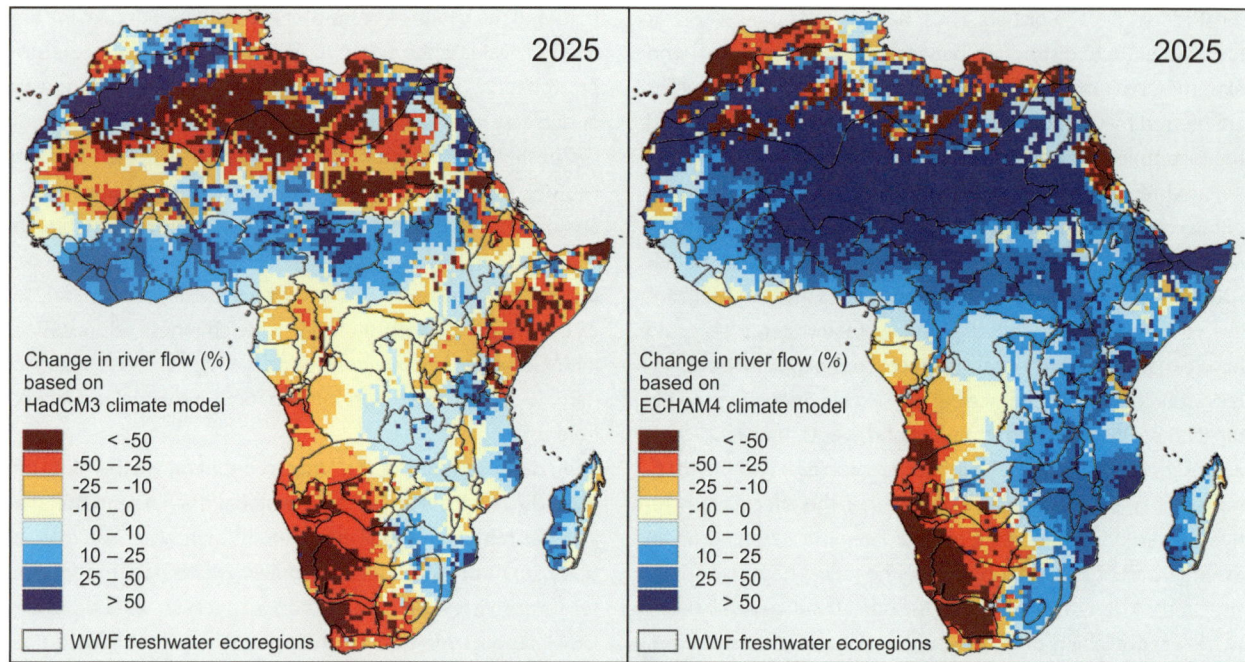

FIGURE 4.10. Results of WaterGAP calculations for 2025. The left map is based on climate change outputs of the HadCM3 general circulation model (GCM), and the right map is based on the ECHAM4 GCM; for both maps the same changes in water use apply. The colors represent changes in river discharge between today and 2025, with red colors indicating lower long-term average discharges and blue colors indicating higher long-term average discharges in 2025. Changes are illustrated in percentages because the same absolute change is assumed to have a different impact in water-rich as opposed to water-poor areas. Caution is advisable when looking at arid areas. For the Sahara Desert, for example, current discharge values are negligible, so a slight absolute change equals a strong relative change. Illustrated changes for extremely arid areas therefore may be strongly influenced by the uncertainties of the model calculations.

terms of their projected change between the present (long-term means of 1961 to 1990) and the year 2025 (long-term means of 2011 to 2040). The results suggest a largely unstable trend and strong climate change effects on the future discharges of African rivers. The main results of the model simulations can be summarized as follows (figure 4.10):

- Based on the climate change projections of two different GCMs for Africa, the derived changes in river discharges for the year 2025 agree in some areas but contradict in others.
- Areas where both model results agree are the Mediterranean coast of North Africa, with decreasing discharges; a band including parts of the Sahel and areas south of it, showing increasing water resources; large parts of southwestern Africa, highlighting the strongest decreases in renewable water resources; and Madagascar, with discharge increases in the west and decreases in the east.
- Decreases in water resources can be explained by a shift toward a drier climate or a significant increase in human water use.
- Independent of the direction of change, the magnitudes of changes across Africa are generally high. More than

half of the African land surface shows a change in the long-term average discharge of plus or minus 10 percent or more, with frequent changes of more than 50 percent.

To determine potential future effects on Africa's freshwater ecosystems, we further analyzed the discharge changes using the freshwater ecoregion framework. Following a more conservative approach, we tried to identify areas in which the simulations with both GCMs agree that large parts of ecoregions are strongly affected in the same direction. The most consistent increases in river discharge are found for freshwater ecoregions of tropical West Africa, whereas the ecoregions of southwestern Africa are threatened by the strongest reductions in water resources (figure 4.11). In the most affected ecoregions it can be expected that the strong changes in long-term average discharges are concurrent with increases in extreme events (floods and droughts). However, additional studies are necessary to verify this assumption.

In conclusion, this case study yields strong indications that, following moderate climate and global change scenario assumptions, severe future threats are expected for both fresh-

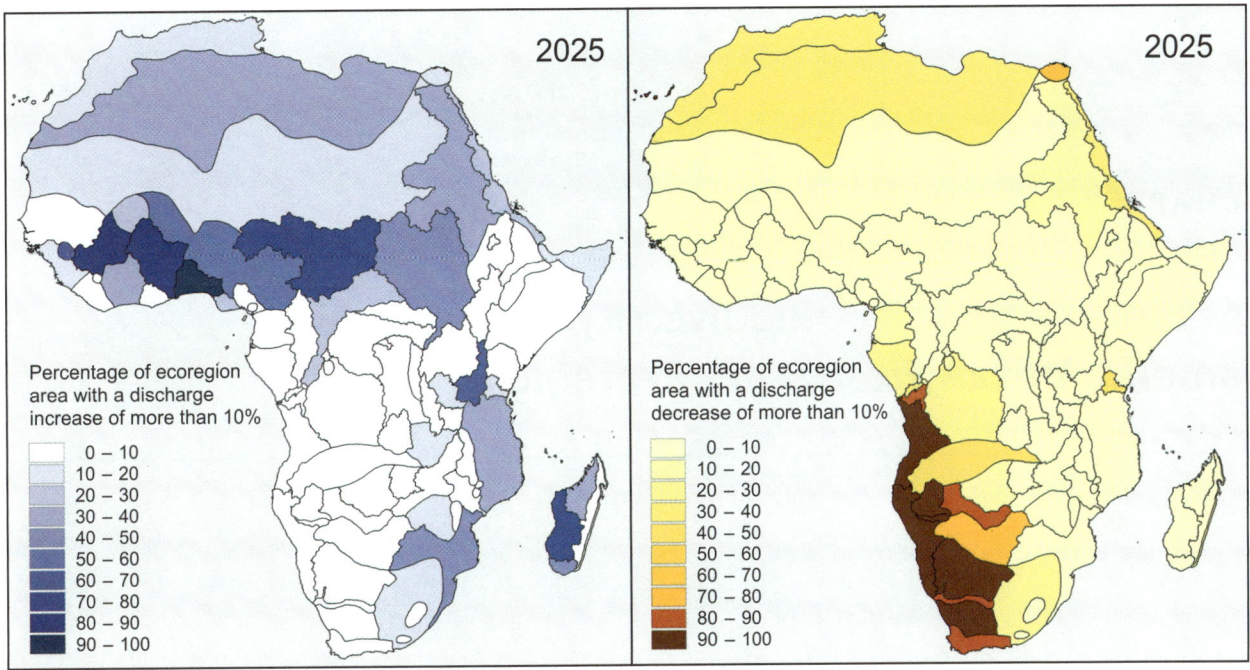

FIGURE 4.11. Percentage of ecoregion area with discharge increases or decreases of more than 10 percent by 2025. The left map shows the percentage of ecoregion area where, according to both applied general circulation model realizations, a discharge increase of at least 10 percent occurs; the right map presents results of discharge decreases in the same manner.

water ecosystem functioning and water availability for human populations. In parts of Africa, changes in the amount of renewable water resources and in the hydrological regime—that is, frequencies and magnitudes of floods and droughts—would put additional pressure on often already disturbed and destabilized aquatic ecosystems and could surpass the ability of the system to adapt and recover.

The results presented here are considered a starting point for more detailed discussions about anticipated climate and global change effects on African water resources and freshwater ecosystems. Additional studies are necessary to validate the results. From an ecological perspective, for example, it is important to identify quantitative thresholds that mark the limits of alteration to which an ecosystem can adapt. However, it can generally be assumed that any change, regardless of direction or magnitude, tends to destabilize ecosystems. Too little water threatens aquatic species, but too much water may pose just as large a threat because increasing flow velocities or flood intensities may change important habitat characteristics or result in direct destruction. It is also likely that the stronger and faster a change, the more severe the impact.

Finally, caution is advised in interpreting the findings of this case study because model computations are uncertain and scenario assumptions are incomplete. Still, the results of this study strongly suggest that strategies to mitigate the impacts of climate and global change on freshwater ecosystems and water resources in Africa should be developed. Or, as Desanker (2002: 6) summarizes, "an integrated approach to environmental management is needed to ensure sustainable benefits for Africa."

CHAPTER 5

Setting Priorities for Biodiversity Conservation among Africa's Freshwater Ecoregions

Aquatic systems around the globe harbor a vast array of organisms and complex interactions that match those of the far better known terrestrial realm. In the previous chapters, we have highlighted the most distinctive of these systems in Africa, Madagascar, and offshore islands, a region that has long been overlooked for its freshwater biodiversity. In this chapter we use the biological distinctiveness and conservation status indexes to set priorities among freshwater ecoregions, identifying those that we believe should be targeted for immediate conservation action. Our priorities are based on a combined assessment of biological distinctiveness and level of threat.

Given that many of the people in the Afro-Malagasy region are among the poorest in the world, one might reasonably question the value of focusing on freshwater biodiversity and setting priorities for its conservation. However, it is exactly for this reason that it is critically important to focus on freshwater conservation now. Food security is tied to freshwater resources and biodiversity across much of the region, and over the long term, the health of human populations will be tied to the health of the region's fresh waters (Denny 2001; Johnson et al. 2001) (essay 5.1). Raising the standard of living across the entire region in a sustainable manner is inextricably linked to protecting the region's freshwater systems (United Nations 2002; World Water Council 2003).

Because of the roles that each of the region's freshwater systems plays in ecosystem functioning, food and water supply, groundwater recharge, and other services, all ecoregions are worthy of conservation interventions (van Wilgen et al. 1996; Baron et al. 2002; Dudley and Stolton 2003). However, the ecoregions that support outstanding levels of biological diversity, contain globally unique species and assemblages, or harbor the last viable examples of important ecological processes clearly vie for priority attention. In this chapter we introduce a decision-making matrix that provides guidance as to where, given limited resources, biodiversity conservation efforts are likely to have the greatest effect or are most urgently needed.

Our aims in setting priorities among Africa's ninety-three freshwater ecoregions are as follows:

- to identify ecoregions that harbor unique biological values and high level of threat
- to highlight ecoregions of outstanding diversity that contain largely intact basins and that face moderate to low threats, where cost-effective investments can be made now to safeguard these freshwater systems and their biodiversity over the long term
- to highlight ecoregions that are research priorities because data quality is low but the ecoregions are suspected to be highly biodiverse
- to ensure that there is a high level of representation by major habitat type (MHT) and bioregion in the final set of ecoregion priorities

The Priority-Setting Matrix

We integrated the biological distinctiveness and conservation status indexes by building on matrixes developed by Dinerstein et al. (1995) and Abell et al. (2000). The integration matrix assigns each ecoregion to one of five priority classes based on the combination of their biological distinctiveness and conservation status categories (see figure 2.3). The matrix we use gives highest priority to globally outstanding ecoregions in critical, endangered, or vulnerable condition. Secondary priority is given to globally or continentally outstanding ecoregions that are the most intact (those in the relatively stable and relatively intact categories).

Abell et al. (2000) and Olson et al. (1998) used a similar matrix for freshwater ecoregions but downgraded ecoregions in critical condition by one level. Because we believe that it is still necessary and possible to save the remaining remnants of the critically threatened Afro-Madagascan freshwater ecoregions, the matrix we use reflects this perspective. Given the importance of all of the region's freshwater systems, efforts are needed to expand the funding base to conserve and maintain these systems as much as possible. Other conservation philosophies might assign different priority levels to each combination of biological distinctiveness and conservation status, and we provide all component data to allow other prioritizations (appendixes C and D).

The five classes of conservation priority in the matrix are described here. These categories provide direction for conservation investment in the future.

> Class I contains globally outstanding ecoregions that are highly threatened.
>
> Class II contains continentally outstanding ecoregions that are highly threatened.
>
> Class III contains globally or continentally outstanding ecoregions with relatively intact aquatic systems.
>
> Class IV contains bioregionally outstanding and nationally important ecoregions that are highly threatened.
>
> Class V contains ecoregions bioregionally outstanding and nationally important ecoregions with relatively intact aquatic systems.

For full descriptions of each class, see chapter 2.

Integrating Biological Importance and Conservation Status Indexes: Where to Act First

The majority of the nineteen ecoregions in most urgent need of conservation attention (Class I) are in the Great Rift Valley, Madagascar, and the Guinean-Congolian region (the West Coastal Equatorial and Congo bioregions), with seven Class I ecoregions (Cape Fold [33], Permanent Maghreb [34], Lake Tana [3], Niger Delta [58], Okavango Floodplains [12], Upper Nile [15], and Northern Upper Guinea [25]) occurring elsewhere (figure 5.1). Class III ecoregions, with high conservation value and relatively intact condition, include parts of the Guinean-Congolian region (Upper Lualaba [14], Central West Coastal Equatorial [19], Kasai [21], and Uele [74]), Malagarasi-Moyowosi [11], Socotra [52], and Temporary Maghreb [92]). Continentally distinctive ecoregions under high threat (Class II) include several island ecoregions (including two ecoregions on Madagascar), several coastal ecoregions (Bight Coastal [62], Shebelle-Juba Catchments [90], and Eastern Coastal Basins [72]), two ecoregions in the Niger Basin (Inner Niger Delta [7] and Lower Niger Benue [65]), Lakes Chilwa and Chiuta [1], and Cuanza [63] (figure 5.2, appendix D).

Although conservation is important in every ecoregion, throughout the rest of the chapter we give highest prominence to the Class I and Class III ecoregions because they represent the highest priorities for restoration and protection based on their biodiversity values and level of threat. Of the ninety-three ecoregions of the Afro-Madagascar region, nineteen are Class I (20 percent) and seven are Class III (8 percent), about one-third of the total. Class I ecoregions include about 1,450 of the species considered in this analysis, and Class III ecoregions include about 200 species.

Comparison with North America

We compared the results of the Afro-Malagasy assessment with those for North America (Abell et al. 2000) after applying the African integration matrix to the North America data. Predictably, nearly twice as many ecoregions (in percentage) in North America are in critical condition. Africa and North America have nearly equal percentages of ecoregions in priority Class I, whereas North America has more than three times the percentage of ecoregions in Class II, reflecting the larger percentage of ecoregions classified as continentally outstanding for North America (figure 5.3).

Data Quality and Priority-Setting

In providing available data on freshwater biodiversity for each of the ninety-three ecoregions, experts also estimated the quality of that data. Thus, we were able to analyze the distribution of ecoregions across priority classes by data quality. This analysis reveals that each priority class contains ecoregions from each data quality level (figure 5.4) and that

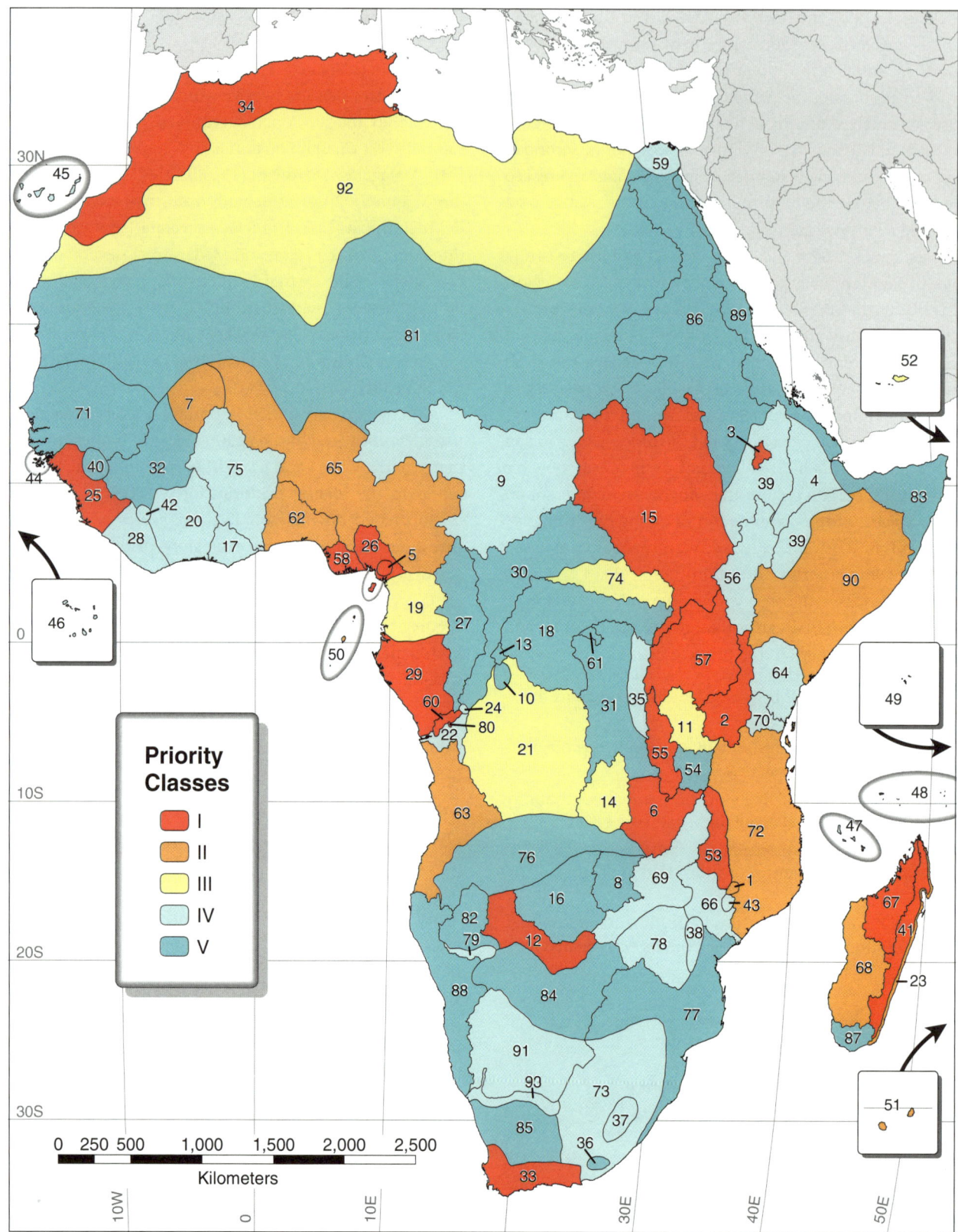

FIGURE 5.1. Priority classes of freshwater ecoregions using final conservation status.

	Final Conservation Status				
Biological Distinctiveness	Critical	Endangered	Vulnerable	Relatively Stable	Relatively Intact
Globally Outstanding	I (6)	I (5)	I (7)	III (1)	III (0)
Continentally Outstanding	II (4)	II (3)	II (6)	III (5)	III (1)
Bioregionally Outstanding	IV (2)	IV (14)	V (5)	V (3)	V (2)
Nationally Important	IV (2)	IV (10)	V (7)	V (5)	V (5)

FIGURE 5.2. Populated integration matrix.

priority level and data quality appear to be independent χ^2 (df = 8, N = 93) = 12.23, p = .27, although 20 percent of the cells have low sample sizes. Several ecoregions are Class I and III priorities despite their low data quality (Bangweulu-Mweru [6], Malagarasi-Moyowosi [11], Upper Lualaba [14], Kasai [21], Southern West Coastal Equatorial [29], Socotra [52], Lower Congo Rapids [60], Uele [74], and Thysville Caves [80]), suggesting that the highest-priority classes are not biased to include only the best-known ecoregions.

It is highly probable that a number of ecoregions will rise in priority as more data become available. For example, Cuanza [63], considered continentally outstanding in this analysis, is extremely poorly sampled but is suspected to harbor a highly diverse and distinctive freshwater fauna and should be a research priority for the future. Recent studies in the West Coast Equatorial rivers provide an example of how much we have yet to discover in many of the Afro-Malagasy region's fresh waters; these studies have described numerous new mormyrid and cichlid fishes (Lamboj and Snoeks 2000; Lamboj 2002, 2003; Sullivan et al. 2002; Lamboj and Stiassny 2003; Schliewen and Stiassny 2003). Indeed, all ecoregions that are of low data quality should be research priorities. These ecoregions are located throughout much of the Congo Basin, Cuanza, the Horn, and several islands, in parts of southern and eastern coastal Africa, and on the west coast of Madagascar (see figure 3.6). Those that are currently considered continentally outstanding are of particular interest because new data may reveal that they contain globally outstanding species assemblages (appendix D).

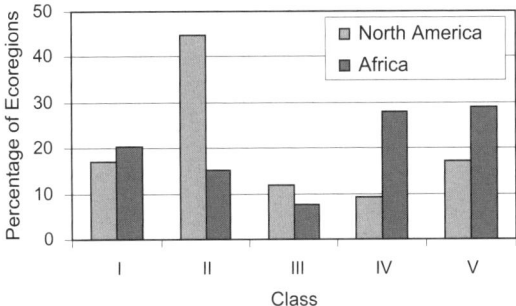

FIGURE 5.3. Percentages of North American and African ecoregions in each priority class.

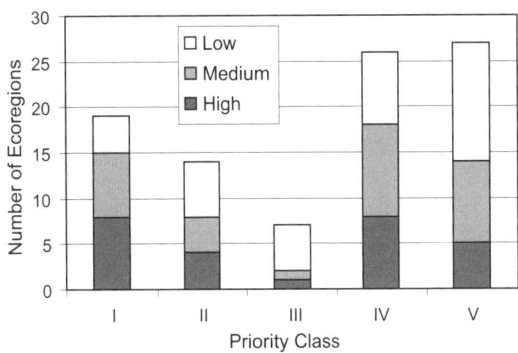

FIGURE 5.4. Distribution of data quality levels by priority class.

Representation

We examined the distribution of ecoregions among each MHT and bioregion to ensure representation. Most MHTs with more than eight ecoregions are represented in each of the five priority classes, although the island rivers and lakes MHT and xeric systems have no Class I ecoregions, and highland and mountain systems lacks Class II and III (figure 5.5). All MHTs, including those with fewer than eight ecoregions, have at least one ecoregion as either priority Class I or Class III.

In addition to examining representation of major habitat types, we examine representation across bioregions. All bioregions except Cuanza and Upper Guinea have at least one ecoregion as either priority Class I or Class III. The Cuanza bioregion encompasses only the Cuanza [63] ecoregion, which has been identified as a research priority. The Upper Guinea bioregion includes Bijagos [44] and the Northern [25] and Southern [28] Upper Guinea ecoregions, which are identified as nationally important, continentally outstanding, and bioregionally outstanding, respectively. The Northern Upper Guinea [25] ecoregion, assessed as endangered, originally was in Class II. To better achieve bioregion representation, we have elevated Northern Upper Guinea [25] from Class II to Class I.

Priority Ecoregions

A brief description of each Class I and Class III ecoregion is given in this section. Full descriptions of each and all references are presented in appendix F.

Class I

CLOSED BASINS AND SMALL LAKES

Southern Eastern Rift [2]. Shallow lakes, rivers, and streams; hot and cold water springs; and marshes, swamps, and salt pans dot the landscape of the eastern portion of Africa's Great Rift Valley. This rift valley system, which is also known as the Gregory Rift Valley, stretches for more than 700 km from central Kenya to northern Tanzania and covers an area of approximately 3,800 km². Flocks of flamingos are larger here than anywhere else in the world; half of the world's lesser flamingo (*Phoenicopterus minor*) population depends on the lakes of this ecoregion, feeding on the abundant blue-green algae *Spirulina platencins* and diatoms. Globally outstanding congregations of waterbirds are found at lakes Nakuru, Bogoria, Manyara, and Natron (with maximum total bird counts from 1 to 1.3 million). There is also a small monophyletic species flock of cichlids in Lake Natron. The aquatic fauna of this ecoregion is under threat from a variety of factors including pollution, overfishing, introduced species, agricultural runoff, deforestation, and water extraction. The principal drivers of future change in the Southern Eastern Rift [2] ecoregion are human population pressure and climate change.

Lake Tana [3]. This highland lake in Ethiopia supports the only intact cyprinid species flock in the world. About 70 percent of the fish species in Lake Tana are endemic to it, with eighteen endemic cyprinids identified thus far. Fishing is the largest threat to the ecoregion, although diversion of water for irrigation and pressures associated with human population growth also affect the supply and quality of water entering the lake. Observers have suggested that a fishery management plan administered and accepted by local communities might be the best solution to mitigate fishing pressures. Educational programs about the possible negative effects of introduced species on the lake ecosystem are also necessary.

Western Equatorial Crater Lakes [5]. Located in Cameroon, this ecoregion's lakes display an extremely high level of endemism, with cichlid species flocks occurring in several of the lakes (lakes Bermin, Barombi Mbo, and Ejagham). More than two-thirds of the thirty-eight fish species and approximately one-third of the aquatic insects found in the crater lakes are endemic to the ecoregion. Threats vary substantially from lake to lake, but introduced species, water abstraction, and pollution (chemicals, sediment, and nutrient inputs) are predominant.

FLOODPLAINS, SWAMPS, AND LAKES

Bangweulu-Mweru [6]. The swampy lakes of the Bangweulu-Mweru ecoregion host a diverse aquatic fauna. One-third of the currently known 111 fish species are endemic to this ecoregion, which is a crossroads for the Congolian and Zambezian freshwater faunas. The region has not been well surveyed in recent years and probably harbors much more undescribed diversity. The region also has a rich molluscan fauna, with thirty-seven species, seven of which are endemic; four endemic aquatic-dependent frogs; and two near-endemic dragonflies that are of conservation concern. The encroachment of people and cattle and overfishing threaten these lakes, but further investigation is needed to generate a comprehensive understanding of current and future threats.

Okavango Floodplains [12]. The Okavango Floodplains ecoregion contains the largest expanse of wetlands in south-

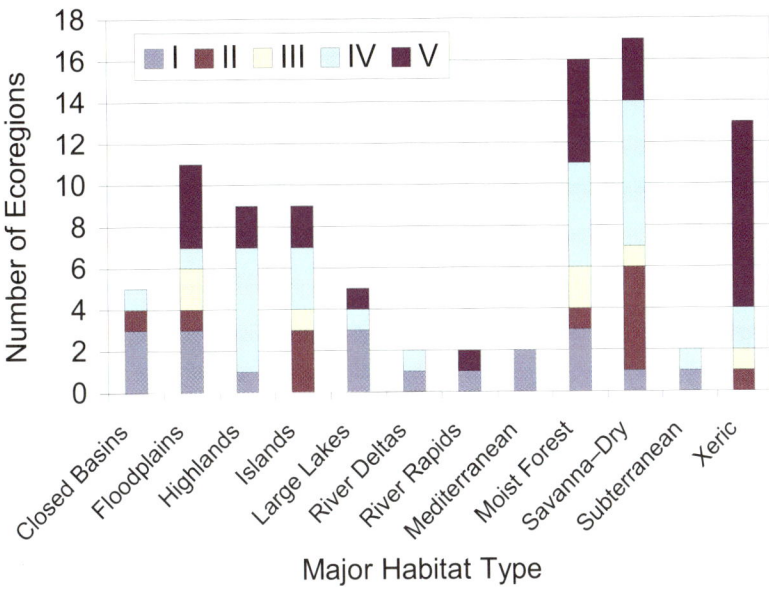

FIGURE 5.5. Number of ecoregions by major habitat type in each priority class.

ern Africa, with a wide range of habitats that support a rich bird fauna. A total of 450 bird species have been recorded in the delta, and large mixed roosts of herons, egrets, storks, and ibis occur in selected localities. Lake Ngami has attracted a globally outstanding congregation of more than 1.5 million wetland birds. Several threatened birds and fish occur in the ecoregion. The largest future threat to the Okavango Delta comes from possible large-scale water extraction from the Okavango and its tributaries in Botswana, Namibia, and Angola. The peripheries of the delta are also of some concern because of the rapid encroachment of people and cattle into the panhandle and drainage areas.

Upper Nile [15]. Situated at the heart of the Upper Nile ecoregion, the vast swamps and floodplains of the Sudd (Bahr-el-Jebel) are among the biologically most important wetlands in Africa. More than 115 fish species have been documented, including 16 endemics. Migratory birds stop over and wetland birds inhabit the extensive Sudd floodplains, and large populations of mammals follow the changing water levels and vegetation. For example, a million white-eared kob (*Kobus kob*) have been documented undertaking a migration of more than 1,500 km as they track the availability of floodplain grasses. In the Sudd system, a globally outstanding congregation of nearly 3 million wetland birds has been documented. Civil war and conflict in Sudan pose the greatest threat to human populations and wildlife conservation in the Upper Nile ecoregion: as of September 2004 an estimated 50,000 people had been killed and an additional 1.2 million displaced (CBS News 2004). Diversion of water through the proposed Jonglei Canal would cause the Sudd swamps and associated floodplains to shrink dramatically, threatening the fauna and flora that depend on them for their survival. Future climate change as a result of global warming will also threaten the Sudd through altered rainfall patterns and water regimes.

MOIST FOREST RIVERS

Northern Upper Guinea [25]. The forested coastal streams and rivers of Northern Upper Guinea support a rich and endemic aquatic fauna in Guinea-Bissau, Guinea, and Sierra Leone. About 28 percent of the 160 fish species are endemic. Ten endemic frogs, four endemic freshwater crabs, at least two endemic dragonflies, and five endemic mollusks also live in these waters. Deforestation and pollution from mining, agriculture, settlements, and industry threaten the rivers, small lakes, and wetlands of Northern Upper Guinea. Civil strife in this region has also affected the environment and complicated conservation efforts.

Northern West Coastal Equatorial [26]. This ecoregion encompasses the coastal rivers and streams that feed the Gulf of Guinea, from the Cross River in Nigeria to the Bay of Cameroon in Cameroon. Rivers in this ecoregion are particularly rich, with more freshwater fish species than other rivers with similar catchment areas in West Africa. More than 200 fish species and about 130 frog species depend on the waters of this ecoregion. About one-fifth of the freshwater fish and about one-quarter of the aquatic frogs are estimated to be endemic. Runoff of sediments and chemicals from logging and agriculture are the major future threats to the freshwater systems of the ecoregion. The burgeoning needs of a growing human population will also put pressure on freshwater resources into the future.

Southern West Coastal Equatorial [29]. The waters of this rainforest ecoregion are exceptionally rich in freshwater species. The major rivers, the Ogooué, Kouilou-Niari, and Nyanga, descend slowly to the coast from interior plateaus and rolling hills. Diversity is high, with more than 230 fish, more than 50 reptiles and amphibians, and about 15 mollusks. Among the fish, cyprinids, cyprinodonts, and mormyrids show particularly high species diversity. About one-quarter of the approximately 230 fish species are endemic to this ecoregion. Although the rivers, streams, and headwater forests are pristine, many areas are coming under increasing pressure from deforestation and subsequent erosion. Oil exploitation and logging are the main activities that pollute the freshwater systems of this ecoregion.

MEDITERRANEAN SYSTEMS

Cape Fold [33]. The Cape Fold in South Africa is a small ecoregion that encompasses a phenomenal diversity of landscape types and correspondingly high levels of biodiversity. The exceptional terrestrial floral diversity of the Cape Floral Kingdom is world-renowned. Although the aquatic fauna of this ecoregion is incompletely documented, richness and endemism of freshwater invertebrates, amphibians, and fish are remarkably high. About half of the thirty freshwater fish and the thirty-eight described aquatic frogs are considered endemic to the Cape Fold. The dark, acid, peat-stained waters of the southwestern cape contain a particularly distinctive fauna. The main threats to freshwater biodiversity include habitat destruction or degradation, caused by agricultural and urban development, and predation by or competition with alien invasive fish.

Permanent Maghreb [34]. This ecoregion, which extends along the northernmost portion of Africa, contains a freshwater fauna with many European elements and low species richness but high endemism. Rivers and streams drain the Atlas Mountains and flow into the Mediterranean Sea in the northeast and into the Atlantic Ocean in the west. Productive *chotts* (shallow, irregularly flooded depressions) also provide important habitat for waterbirds. The highly fragile freshwater systems of this mediterranean ecoregion face numerous threats, including runoff from industrial and agricultural sources, fragmentation from dams, possible oil and gas spills, depletion of underground water sources, pressures from tourism, desertification, and salinization.

HIGHLAND AND MOUNTAIN SYSTEMS

Madagascar Eastern Highlands [41]. This ecoregion has much higher richness and endemism among its aquatic species than other highland ecoregions in Africa and Madagascar. Endemism is particularly high among aquatic frogs, and preliminary studies indicate that richness and endemism of aquatic insects are high as well. More than fifty species (some awaiting description) of freshwater fish, about half of which are endemic, are known to inhabit the rivers and streams of the Madagascar Eastern Highlands. Two fish families that are endemic to Madagascar, Bedotiidae (rainbowfish) and Anchariidae (catfish), also occur in the ecoregion. Extensive deforestation and the spread of exotic species are major threats to aquatic systems in Madagascar's eastern highlands.

LARGE LAKES

Lake Malawi [53]. Lake Malawi boasts one of the richest lacustrine fish faunas in the world, hosting several species flocks of endemic cichlids and an endemic clariid flock (Lowe-McConnell 1987). Endemism in the ecoregion is remarkably high. Among fish, 99 percent of the more than 800 cichlid species and more than 70 percent of the seventeen clariids are endemic (Ribbink 2001). Equally, the lacustrine invertebrates appear to have high levels of endemism, although exact numbers are unknown for these largely unstudied groups (Fryer 1959). Richness of other taxa is also exceedingly high in the ecoregion, with about 200 mammals, 650 birds, more than 30 freshwater mollusks, and more than 5,500 plant species (Brown 1994). This ecoregion encompasses the entire lake basin, including inflowing tributaries and a portion of the outflowing Shire River. The main threats to the lake are deforestation, agricultural practices, burning, runoff of excessive sediments and nutrients, and overexploitation of fish resources.

Lake Tanganyika [55]. Lake Tanganyika is exceptional not only for its high level of species richness (animals and plants estimated at more than 1,400 species) but also for high levels of endemism in several taxa (fish, copepods, ostracods, shrimp, crabs, and mollusks). At 1,470 m deep, Tanganyika is the second deepest lake in the world, after Lake Baikal. The lake lies in a deep graben in the Great Rift Valley in eastern Africa, and its basin covers parts of Tanzania, Democratic Republic of the Congo (DRC), Zambia, and Burundi. The major threats are sedimentation from deforestation, water pollution near urban areas, and overfishing; these factors are changing species compositions and disrupting community interactions.

Lakes Kivu, Edward, George, and Victoria [57]. Lake Victoria is the second largest freshwater lake in the world and the largest tropical lake. In addition, it harbors one of the world's most important examples of rapid species radiations. About 500 species of haplochromine cichlids are known from Lake Victoria alone, although estimates of species numbers vary widely. The total number of cichlid species in Lakes Edward and George is nearly 100, most of which are endemic. Species introductions (Nile perch [*Lates niloticus*], several exotic tilapiine species [*Oreochromis miloticus, O. leucostictus, Tilapia zillii,* and *T. rendalli*], and water hyacinth [*Eichhornia crassipes*]), overfishing, and eutrophication have caused a catastrophic loss of biodiversity in the lakes of this ecoregion. In only a decade, the populations of more than 200 fish species were thought to have been lost from Lake Victoria or to have been drastically reduced (Seehausen et al. 1997b). However, in the last several years the Nile perch fishery has showed signs of overfishing, and, subsequently, some of the indigenous fish have showed signs of reemergence in lakes Kyoga, Victoria, and Nabugabo (Chapman et al. 2003; Balirwa et al. 2003). Deforestation, rapidly expanding human populations, and improper disposal of agricultural and industrial waste also threaten the biological integrity of the lakes.

LARGE RIVER DELTAS

Niger Delta [58]. Both black and whitewater rivers flow into the highly productive Niger Delta, which supports an extremely rich freshwater fauna and five monotypic fish families, the highest concentration worldwide. Nutrient-rich silt-laden whitewater rivers (e.g., the Niger and Orashi rivers) and nutrient-poor blackwater rivers (e.g., the Sombreiro and New Calabar rivers) converge at the delta. The Niger Delta is under intense pressure from urbanization and industrialization, oil exploration and exploitation, upstream impoundment, logging, agriculture, hunting, population increase, invasive species, and overfishing. Growing human populations combined with a growing oil industry are the biggest future threats to this ecoregion.

LARGE RIVER RAPIDS

Lower Congo Rapids [60]. This stretch of the Congo River encompasses thirty-two falls and rapids that constitute the Cristal Mounts Rapids. An endemic aquatic fish and snail fauna is adapted to the swift waters of the rapids. About one-quarter of the fish species are endemic, as are sixteen of eighteen aquatic snails known from the ecoregion. Within the snail fauna are four endemic monotypic genera (*Congodoma, Liminitesta, Septariellina,* and *Valvatorbis*). Most of the fishes have morphological and behavioral adaptations to fast-running water, such as a reduction of eye size or modified body form. The world's only know blind cichlid, *Lamprologus lethops,* is endemic to these rapids. Runoff of untreated sewage, sediments, and industrial chemicals from Brazzaville and Kinshasa flows downstream through the rapids. Future threats include plans to develop the Grand Inga plant, with a projected capacity of 39,000 megawatts. Constructing this plant would involve blocking the river's flow at the rapids and creating a large reservoir.

SAVANNA—DRY FOREST RIVERS

Madagascar Northwestern Basins [67]. Freshwater habitats in this ecoregion include *tsingy* (karst) formations of the Ankarana reserve in the far north; steep, rocky clearwater streams draining the Tsaratanana Massif and Montagne d'Ambre; floodplain lakes; large perennial rivers; and numerous crater lakes on the island of Nosy Be. A number of rivers in this ecoregion, most notably the Mahavavy du Nord, Sambirano, Ankofia-Anjingo, and Mangarahara-Amboaboa (Sofia tributary), contain rich and highly endemic freshwater fish faunas. A recent review found that seventy-one native freshwater fish and twenty-six endemics have been recorded from northwestern rivers and lakes. It is likely that fish diversity is still substantially underestimated for this ecoregion, given that a majority of the basins draining the Tsaratanana Massif, as well as headwaters and upper-middle reaches of most of the ecoregion's major rivers, remain poorly surveyed. Major threats to freshwater biodiversity stem from deforestation, overfishing, and the spread of exotic species.

SUBTERRANEAN AND SPRING SYSTEMS

Thysville Caves [80]. A depauperate but distinctive aquatic fauna lives in the subterranean streams of Thysville Caves. The caves are located just south of the mainstem Congo and downstream of Pool Malebo. The Thysville Caves ecoregion hosts two fish species, one of which is subterranean, and several bird, aquatic mammal, and mollusk species. The most famous inhabitant is the Congo blind barb (*Caecobarbus geertsii*), which belongs to an endemic monotypic fish genus. The region is heavily populated, and the caves are located on the edge of a degraded forest. Because the habitat in the caves depends on groundwater levels, the primary threats to the Thysville Caves include changes in the linked groundwater–surface water hydrology of the region, as well as increasing human population and associated deforestation in the area.

Class III

FLOODPLAINS, SWAMPS, AND LAKES

Malagarasi-Moyowosi [11]. The highly productive wetlands of the Malagarasi-Moyowosi ecoregion in Tanzania host a rich aquatic fauna with numerous wetland birds and a high diversity of fish and freshwater mollusks. A recent survey of the Malagarasi drainage revealed the presence of at least 108 fish species belonging to 20 families and 48 genera, without taking into account numerous lacustrine Tanganyika species that enter the lagoons of the river delta. A radiation of the riverine mouthbreeding goby cichlids (genus *Orthochromis*) includes at least eight species that inhabit the eastern portion of the ecoregion. Influxes of refugees, agriculture, and overgrazing pose the largest threats to the wetlands of this ecoregion. In 2000, the government of Tanzania designated the wetland as its first Ramsar site.

Upper Lualaba [14]. The waters of the upper Lualaba River flow through the swampy valley of the Kamolondo Depression. In this depression, a continuous swamp belt fringes the river, and a series of lakes, including Lake Upemba, are connected to the river through narrow channels. The Upper Lualaba ecoregion possesses a rich fish fauna and abundant waterbirds and is suspected to possess an endemic odonate assemblage. The ecoregion lies in the DRC, in the southeastern portion of its Shaba Province. The fish fauna is incompletely known for this ecoregion, although it is well sampled compared with other Congo Basin ecoregions (Banister and Bailey 1979; Banister 1986). Rapids tend to contain specialized species, and the rapids of the Lufira River are known to host several endemic fishes. Continued civil strife, deforestation, agriculture, mining, and overfishing are among the largest future threats to the ecoregion.

MOIST FOREST RIVERS

Central West Coastal Equatorial [19]. The forested rivers and streams of this ecoregion support a rich and endemic freshwater fauna. Coastal rivers in Gabon, Equatorial Guinea, and Cameroon flow through rainforests on their way to the Gulf of Guinea and the Atlantic Ocean. The main rivers are the lower Sanaga, Nyong, Ntem, Benito, and northern tributaries of the Ogooué (Abanga, Okano, and Ivindo). High numbers of fish, aquatic mollusks, and aquatic amphibians and reptiles inhabit these freshwater systems. More than 270 fish species are known from this ecoregion, 58 of which are endemic, but ongoing investigations indicate that much higher numbers of species and endemics are present in this ecoregion. Erosion from logging, alterations caused by dam construction, and the collection of aquarium fishes are some of the threats to the freshwater species of this ecoregion.

Kasai [21]. The sometimes torrential rivers of the Kasai basin possess a rich aquatic fauna with high fish endemism. More than 200 fish species are known from this ecoregion, and about one-quarter of these are endemic. However, few biological studies have been completed in recent years, and one would expect further surveys to reveal new species and endemics. The Kasai River originates on a plateau in Angola and flows through channels characterized by falls and rapids before discharging into the Congo River. The ecoregion largely follows the boundaries of the Kasai River catchment, which encompasses a vast area (900,000 km^2) in southwestern DRC and the northeast corner of Angola. Future increases in human population threaten the Kasai's freshwater systems, along with continued civil unrest, diamond and coltan mining, and deforestation. Continuing or worsening political instability in the Kinshasa area would almost certainly accelerate unsustainable use of the ecoregion's natural resources.

ISLAND RIVERS AND LAKES

Socotra [52]. This island ecoregion possesses few permanent watercourses, but its temporary streams support an aquatic fauna that includes endemic crabs and mollusks. Pollution is a problem in the ecoregion because there is no well-developed infrastructure to deal with sewage and other waste disposal. The main future threats to the island's environment are improperly planned infrastructure development, breakdown of traditional land management practices, introduction of exotic plants and animals, and climatic changes.

SAVANNA–DRY FOREST RIVERS

Uele [74]. The rivers and streams in the Uele ecoregion drain from a high plateau along the northeastern border of the Congo Basin in the DRC. The basin of the Uele River forms the boundary of the ecoregion. The ecoregion supports a rich aquatic fauna for its size and MHT. More than 130 fish and about 70 frogs have been described from the ecoregion; however, few biological studies have been completed in recent years, and one would expect further surveys to reveal new species and endemics. Civil unrest and movements of refugees are the largest threats to the natural resources of this region. Population growth and increased mining and agricultural activity also threaten the future integrity of these freshwater systems.

XERIC SYSTEMS

Temporary Maghreb [92]. One perennial river, several temporary rivers, isolated oases, and saline lakes are among the few waterbodies in the xeric Temporary Maghreb. Less than 100 mm of rain falls each year in much of this desert-covered ecoregion, which extends from the northwestern coast of Egypt, across Libya and Algeria, and through the northern portions of Mali and Mauritania and ends at the Atlantic Coast in Western Sahara (Morocco). Overall, richness of the aquatic fauna of the Temporary Maghreb is low, although fish richness and endemism are both high considering the scarcity of permanent waterbodies. There are about forty fish species known from this ecoregion, about twenty of which are endemic. The largest threat to this ecoregion is the exploitation of subterranean water for human use. Agriculture and urban growth are also putting pressures on the wetlands of the Temporary Maghreb.

Comparison with Other Priority-Setting Exercises

Excepting a global analysis that covers the Afro-Malagasy region (Groombridge and Jenkins 1998), we know of no other recent region-wide identification of freshwater biodiversity priorities, although several excellent syntheses of Africa and Madagascar's freshwater biodiversity information have been completed (e.g., Hughes and Hughes 1992; Lévêque 1997; Shumway 1999). In their description of a blueprint for conservation in Africa, Brooks et. al (2001) suggest that data on freshwater and marine priorities for the continent are needed urgently and should be integrated with terrestrial conservation priorities for a comprehensive conservation strategy. We suggest that the freshwater priorities presented here could provide the foundation for such an integration of a freshwater, marine, and terrestrial conservation strategy for the Africa and Madagascar region.

The globally outstanding and highest-priority freshwater ecoregions (Classes I and III) overlap extensively with priority terrestrial ecoregions identified by Burgess et al. (2004) in their companion assessment of Africa's terrestrial biodiversity (figure 5.6). We describe these areas of overlap in this section.

Biological Distinctiveness Overlap

Some clear patterns emerge from this analysis. First, the eastern part of Madagascar is of high importance for both terrestrial and freshwater biodiversity, with freshwater biological values declining in the drier west and south. The mediterranean climate habitats of North Africa and South Africa also are of high biological importance for both terrestrial and freshwater taxa, with both containing ancient relict elements from lineages that in some cases can be traced back to Gondwanaland. In the tropics, the western margins of the Guinean-Congolian region are of the highest importance for both sets of priority ecoregions. The forests and lakes of the Rift Valley of eastern Africa, from the Sudd swamps in the north to Lake Malawi in the south, are also considered globally outstanding in both assessments. Perhaps surprisingly, the main montane regions of tropical Africa (Albertine Rift, Ethiopian highlands, and Eastern Arc Mountains) are poor in terms of aquatic fauna, whereas they are of global importance for terrestrial biodiversity. Similarly, the lowland coastal forests of eastern Africa possess low aquatic values while being of global importance for terrestrial species. Such differences may reflect a lack of study of their fish faunas—which is a primary driver of aquatic value in this assessment—or may reflect a genuine mismatch between terrestrial and aquatic priorities. Or perhaps the aquatic biodiversity of these forests is more highly comprised of aquatic elements that we did not analyze. At least for the Eastern Arc and the eastern African coastal forests, a notable Odonata fauna is known, confined to the forest patches and often extremely rare (Clausnitzer 2001). Further research is needed to assess this possibility.

Priority Class Overlap

The overlap of priority classes demonstrates further agreement between terrestrial and freshwater priorities. Areas that feature in the overlap of the Class I (high biodiversity and high threat) and III (high biodiversity and low threat) priorities for the two sets of ecoregions include the Cameroon highlands, mediterranean habitats of North and South Africa, the Rift Valley region, parts of the Guinean-Congolian region, and eastern Madagascar. The northern portion of the Upper Guinea forests is also added to this list of priorities, as are Lake Tana in the Ethiopian highlands and a portion of the Okavango ecoregion in southern Africa. These are all areas where conservationists should focus their attention to prevent extinction because both ter-

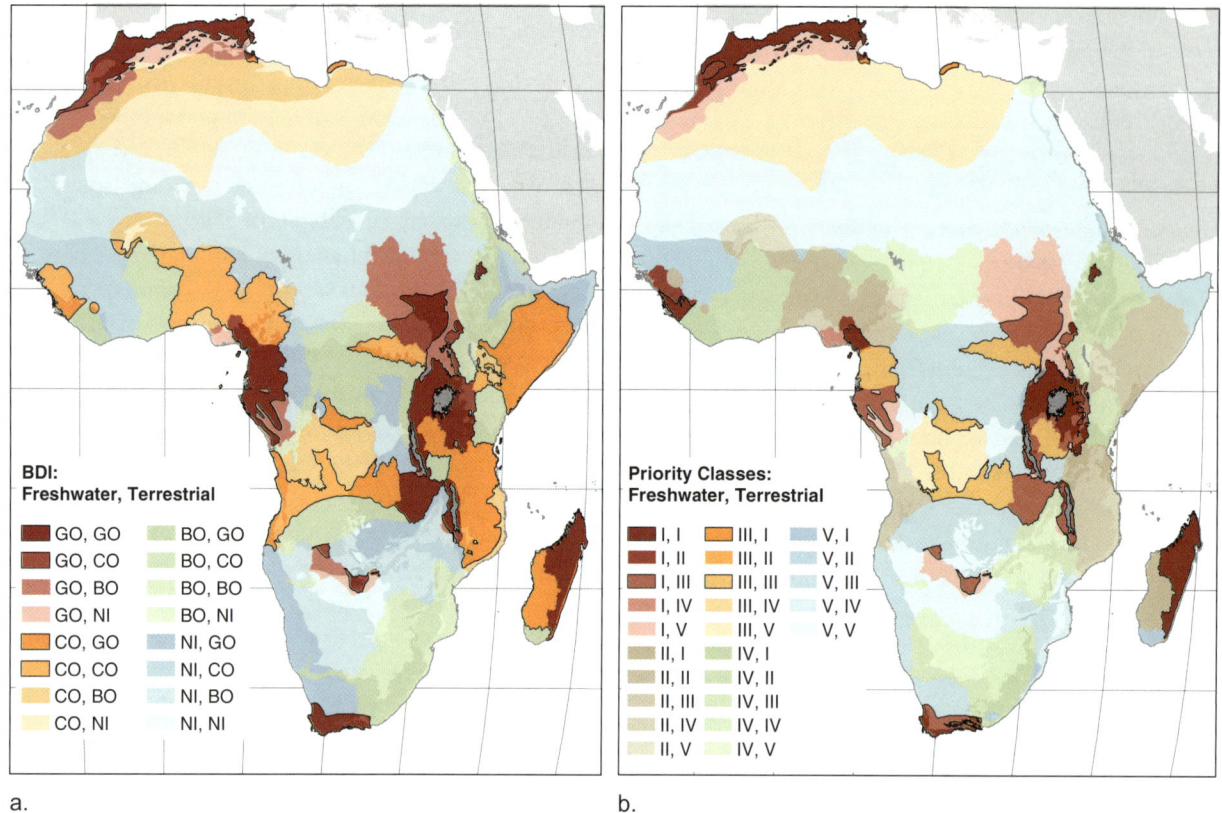

FIGURE 5.6. (a) Overlap of freshwater and terrestrial Biological Distinctiveness Index. (b) Overlap of freshwater and terrestrial priority classes. BO, bioregionally outstanding; CO, continentally outstanding; GO, globally outstanding; NI, nationally important.

restrial and freshwater conservation targets may be achieved at the same time. Once again, the major tropical mountain regions of Africa do not feature in this set of overlaps, even though they were of the highest importance for conservation investment to conserve terrestrial biodiversity. A few areas of highest importance for freshwater biodiversity are poorly captured by the terrestrial priorities, including the Kasai [21], Niger Delta [58], and Lower Congo Rapids [60] (figure 5.6b).

Like that of Burgess et al. (2004), other terrestrial priority-setting schemes omit several of the most important areas identified for freshwater biodiversity in Africa. For example, Conservation International's hotspots omit the entire Congo Basin and Lower Guinea (West Coastal Equatorial) region, all of the Rift Valley lakes, and several other distinctive freshwater systems (Myers et al. 2000). BirdLife's endemic bird areas miss most of the Congo Basin, the Permanent Maghreb, and large parts of the basins of the Rift Valley lakes, among other areas (Stattersfield et al. 1998). Clearly, these and other sets of priorities focus mostly on terrestrial species. However, because much conservation effort and funding are directed to these priority areas (Conservation International 2003), it is important to recognize that these priorities fail to cover a comprehensive set of representative freshwater systems.

One important exception to these terrestrially focused assessments is a preliminary global analysis of areas important for freshwater biodiversity completed by the World Conservation Monitoring Centre (WCMC; Groombridge and Jenkins 1998). The priorities identified in the WCMC assessment match fairly well with our Class I priorities. However, WCMC identified Lake Turkana and the Volta Basin as priorities, whereas these ecoregions were both considered Class IV in our analysis. These differences result from the lower weighting that we gave to richness in our analysis. We also elevated three ecoregions for their globally outstanding wetland bird congregations: the Southern Eastern Rift [2], Okavango Floodplains [12], and Upper Nile [15]; these were not recognized as of priority in the WCMC assessment. The Permanent Maghreb [34] ecoregion, highlighted for its high levels of fish endemism, is also not highlighted in Groombridge and Jenkins (1998).

Summary

This chapter has identified the most important freshwater systems for conservation investment across the Afro-Malagasy region. The next challenge is to increase our conservation efforts to fully conserve the biological diversity of this region and more fully link conservation approaches to economic and social development. Conservation in these priority systems will be achieved only by integrating freshwater efforts with ongoing and planned terrestrial and marine approaches, with an emphasis on working toward the sustainable use of water and land resources. The final chapter of this book shows how WWF envisages operationalizing large-scale conservation initiatives in Africa and highlights some of the particular challenges facing freshwater conservation.

ESSAY 5.1

Safeguarding Human Health, Safeguarding Aquatic Snail Diversity: Conflicting Goals?

David S. Brown

The precise number of snail species living in the fresh and brackish waters of Africa is unknown, but it is probably about 400 (Brown 1994). Most of the species occur nowhere else in the world; many have evolved in Africa's ancient lakes and river systems. Shells can be bizarrely shaped and sometimes color-patterned. Specialized species have adapted to diverse conditions, ranging from ephemeral rainpools to torrents in rocky gorges and soft sediments in the still depths of lakes. Africa's freshwater snails are highly worthy of conservation on three counts: the beautiful variety of bodily form displayed by different species, the major contribution these organisms make to the ecology and biodiversity of African inland waters, and the story they tell of the evolution of aquatic life in Africa.

Happily, Africa's freshwater snails have not suffered, as far as is known, any disaster comparable to the extinction of the endemic cichlid fishes of Lake Victoria (see essay 4.1). Yet snails are certainly threatened, not only by water extraction, dam-building, and pollution. All snail species face an additional threat because a few are intermediate hosts of parasites that infect people and domestic livestock. As a result, some people 2mlieve that the only good snail is a dead snail.

Snail-borne diseases are caused by parasitic worms that pass the early part of their life cycle in water snails and develop to maturity after transfer to a mammalian host. Schistosomiasis (*Bilharzia* disease), caused by blood flukes of the genus *Schistosoma*, is the most important snail-borne infection of people in tropical Africa, and other kinds of schistosomes infect cattle. Schistosome eggs are excreted by infected people, and if passed into water they hatch to produce tiny swimming larvae that seek a suitable snail species in the pulmonate genera *Biomphalaria* or *Bulinus*. After entering the snail's body the young parasites multiply, and after a few weeks hundreds of a new type of swimming larva emerge from the snail and are attracted to human skin. Because young children often play in the water, this age group has a high prevalence of infection. Having penetrated the host's skin, the young schistosome travels through the blood vessels to take up residence near the bladder or the lower intestine, depending on the species. The worms reach sexual maturity in about 2 months, when the females lay spined eggs that work their way through the host's tissues to be excreted. Schistosomiasis is aggravated greatly by increases in human population and the lack of sanitation in rural areas.

The medical importance of schistosomiasis in Africa has led many national and international health authorities to allocate resources to its control. Chemotherapy, sanitation, and education are used in varying combinations with snail control. Much effort has been devoted to breaking the parasite's life cycle by removing its snail hosts. Attempts to do this have been of three types: biological control, habitat modification, and the destruction of snails by molluscicidal chemicals. Mollusciciding is the only method that has been used extensively. If the results had met expectations and the chemicals had been used on a larger scale, much damage probably would have already been done to the aquatic ecosystems of Africa. In practice, molluscicides have had only limited success because of high cost and the difficulty, especially in large waterbodies, of killing a high enough percentage of snails. However, mollusciciding is still a recommended strategy against local foci of schistosomiasis transmitted in small waterbodies. Despite efforts to develop controlled release formulations for delivering molluscicide selectively to target snail species (Thomas 1998), the molluscicides available for practical use are toxic to a wide range of aquatic organisms, including apparently all mollusks. This is highly unfortunate because only a small proportion of the aquatic snail fauna, probably less than 5 percent of the species, is known to transmit schistosomiasis.

About 200 species, or half of the freshwater snail species in Africa, are imperiled to some degree by one or more threats (Baillie and Groombridge 1996; Brown and Kristensen 1998; Kristensen and Brown 1999). Most of the species considered to be at risk are operculate (prosobranch) snails. Of these, the family Thiaridae has the greatest number of imperiled members because many thiarids are locally restricted to parts of rivers and streams in West and Central Africa, and more than thirty species are unique to Lake Tanganyika. The Ampullariidae and Bithyniidae are also of high conservation interest because many species are known only from a few small waterbodies or from a single lake. The list of threatened species probably will change with advances in knowledge of species taxonomy and geographic distribution. In some parts of the continent, such as South Africa, the freshwater fauna is fairly well known, but other areas have been scarcely visited by mollusk investigators.

Some species may be removed from the threatened list if new populations are discovered, but the protection of range-restricted species that are adapted to specialized conditions in lakes and rivers will always be a high priority. About 150, or 70 percent, of the threatened species are concentrated in only four areas: the large eastern lakes, including lakes Tanganyika and Malawi; the Congo and Lualaba river systems; the western area including Cameroon and extending west into Sierra Leone; and the southern coastal region. To what extent are aquatic snails in these areas at risk from operations to control schistosomiasis?

First we can eliminate the southern coastal region, where the endangered species live mostly in areas of temperate climate lying outside the southern limit for transmission of schistosomiasis. In the western area, too, the endemic snail species appear to be at little or no risk because they live in highland areas that are not known to be foci of the disease. More worrying is the lower basin of the Congo River, where the specialized snail fauna living in the rocky rapids of the main river could be endangered by inflow of molluscicide from snail control operations in tributary streams, where schistosomiasis is transmitted (Mandahl-Barth et al. 1974).

It is in the larger lakes in the eastern rift valleys that acute conflicts between biodiversity conservation and schistosomiasis control seem most likely to arise. Lakes Malawi and Tanganyika are outstanding for their many endemic operculate snail species. Unfortunately, schistosomiasis transmission also takes place on the lake shores, though usually in patches of marsh rather than adjacent to the open lake. Increasing human populations around these lakes and developments in agriculture, industry, and tourism are likely to provoke demands for schistosomiasis control by all means available, including mollusciciding.

It may be that health authorities will be deterred from applying molluscicides to the open shores of large lakes such as Malawi and Tanganyika by the high cost and the practical difficulties encountered in other lakes. For example, no more than a short-term reduction in infection was achieved by intensive mollusciciding in Lake Volta, Ghana (Chu et al. 1981). However, there may be calls for chemical treatment of small sites along lake shores, especially near tourist access points, if only to satisfy the concerns of tour operators. It would be difficult to ensure that treatment of marshes and tributary streams would not harm the nontarget snail species living nearby in the open lake.

It is preferable that molluscicides should not be used at all in or near lakes and other habitats where there are snails of conservation interest. In such areas, schistosomiasis could be controlled by chemotherapy and by providing people with sanitation and a safe water supply. To identify areas where the use of molluscicides should be avoided, there should be a liaison between public health and conservation authorities. It should always be possible to devise a strategy that minimizes conflict between the objectives of disease control and conservation. It is important that staff with responsibility for applying chemicals to the environment, whether molluscicides or insecticides, are well informed about conservation objectives in their area.

It is not putting conservation values before public health to constrain measures against the snail hosts of schistosomiasis so as to minimize damage to nontarget snails and to the overall biodiversity of African inland waters. Draining wetlands and channelizing watercourses might destroy snail habitats, but these practices would also accelerate soil erosion, which already severely reduces the productivity of land in many parts of Africa. Killing snails is not necessary to rid communities of infection. More effective and long-lasting results would be obtained through a combination of chemotherapy, provision of safe water supplies, and education. The people of Africa need healthy aquatic ecosystems, in which there is high biodiversity with minimal erosion and pollution, as reliable sources of good-quality water.

CHAPTER 6

Africa's Freshwater Systems and Their Future

A vision without action is just a dream;
an action without vision just passes time;
a vision with an action changes the world.
BARKER ET AL. (1990)

Who's Watching the Water?

Whereas terrestrial ecosystems and their inhabitants are clearly visible, freshwater biodiversity is largely hidden from view and therefore often neglected. Large-scale projects with a focus on conservation of terrestrial biodiversity are numerous in Africa (see Burgess et al. 2004 for a comprehensive review), especially compared with freshwater biodiversity–focused initiatives. Nevertheless, there are several ongoing large-scale freshwater projects in the Afro-Malagasy region, some of which have components that address biodiversity conservation. These include basin-wide initiatives for several of the Rift Valley lakes and for the Niger River and Lake Chad Basin (table 6.1).

We face the challenge of turning the set of large-scale continental priorities presented in this assessment into conservation achievements at finer geographic scales. This process involves developing spatially explicit and measurable plans defining what is needed to sustain biodiversity over the long term. For WWF, these plans are developed in the form of a biodiversity vision (Dinerstein et al. 2000; Abell et al. 2002). Biodiversity visions are based on a fundamental but rarely asked question: from the perspective of conservation biologists, what would successful conservation for an ecoregion look like 30–50 years hence? Answering this question requires an understanding of what biological features must be saved now and what diminished features must be restored over the coming decades. The biodiversity vision also asks, if we cannot conserve everything everywhere, then what should we conserve? For African and Madagascan ecoregions, the biodiversity vision must also balance conservation with the needs of a growing human population and its reliance on natural resources. Meeting the needs of people and those of aquatic species will remain a major concern and challenge into the future.

Typically, a biodiversity vision identifies a set of priority sites or larger landscapes in an ecoregion that, if effectively conserved, would maintain the key biological values of the larger ecoregion. The vision may also identify parallel sets of actions in the policy and legal arenas that are needed to mitigate or remove threats to critical landscapes and ensure their long-term viability. In most freshwater ecoregions, many of the most important interventions are policy related; for example, maintaining water flow into a wetland may entail policies related to upstream dams and water withdrawals.

This chapter outlines the WWF approach to developing freshwater biodiversity visions. We conclude with what we perceive as the greatest challenges to freshwater conservation in the Afro-Malagasy region.

Conservation Planning in Ecoregions and River Basins

It is essential to plan and implement conservation actions at scales appropriate to the physical and biological processes that shape biodiversity features. Historically, inappropriate

TABLE 6.1. **Examples of Basin Projects with an Ecosystem Focus.**

Basin Projects*	Funding and Implementing Organizations	Sources
Large-scale basin projects in the Nile, Niger, Okavango, Senegal, Lake Chad, Lake Tanganyika, Lake Victoria, and Volta	Global Environmental Facility, International Waters	Overview at http://www.iwlearn.net/region/africa.php Specific sites: http://www.nilebasin.org, http://www.ltbp.org, http://www.lvemp.org
Various projects in the Niger, Lake Chad, Zambezi (especially River Kafue and Lake Malawi/Niassa/Nyasa), Great Ruaha River, River Mara and Lake Victoria, and Lake Bogoria	WWF	http://www.panda.org/about_wwf/what_we_do/freshwater/index.cfm
Cape Action for People and the Environment	CAPE	http://www.capeaction.org.za/ and van Nieuwenhuizen and Day (2000a, 2000b)
Various projects in the Senegal, Niger, Volta, Kamadugu Yobe, Tanganyika, Victoria, Naivasha, Zambezi, Rufiji, Pangani, and Limpopo basins	IUCN Water and Nature Initiative and IUCN African Regional Offices	http://www.waterandnature.org/projects.html and http://www.iucn.org/themes/wetlands/project.html

*Not a comprehensive list.

planning units or inattention to biological targets and biophysical processes have hampered freshwater conservation efforts (Abell et al. 2002; Groves 2003). Trying to conserve a floodplain system and its species without considering the source of the floodwaters has obvious limitations, yet this has been the dominant paradigm in many river conservation and restoration schemes (Wissmar and Beschta 1998; Frissell and Ralph 1998). Similarly, efforts that have focused on a single management site have done little to protect species that use a range of habitats over their life cycles (Saunders et al. 2002; Robinson et al. 2002). For freshwater systems, river and lake basins provide the obvious organizing unit for conservation planning, and basin-scale planning promises to provide a significant advance over traditional single-site approaches. Yet basins are scale-independent and in themselves may not be the best representations of aquatic biogeographic patterns.

Current distributions of aquatic species are the result of multiple factors, including historical evolutionary processes that may have occurred in different hydrographic basins than those that exist today (Poff 1997). As a result of past connections, two river basins or portions of river basins may share many or most species assemblages, and biologists developing a plan to conserve those assemblages might group the two areas into a single unit. Alternatively, barriers to dispersal within a basin can facilitate the evolution of multiple distinct biotas; examples include a high waterfall isolating upstream species or a large river preventing dispersal between tributaries. From a biogeographic perspective, it is clear that a given river or lake basin may fail to capture the biogeographic patterns that should underpin biodiversity conservation planning. The freshwater ecoregions that we offer in this assessment, which take into account both aquatic species distributions and drainage basin divides, are designed to serve as these biogeographic conservation planning units.

The fact that a biogeographically derived planning unit may not correspond to a single basin requires us to ask how to best reconcile the differences between these two planning units. The answer lies in the distinct uses of ecoregions and drainage basins in conservation. A freshwater ecoregion provides the template for a biological assessment. Through that assessment, we can identify the river reaches, floodplains, lakes, swamps, or other freshwater systems that support particularly important biodiversity features. The biogeographic unit therefore defines the geographic universe of that assessment, allowing evaluation of key features across the entire area where they occur.

However, identifying important biological features is not equivalent to identifying management strategies for conserving them. Developing strategies entails assessing threats, and a threat assessment immediately takes us from the aquatic realm into the terrestrial, forcing consideration of the entire area of influence, upstream and even downstream. Even if a biogeographic unit consists of a portion of a basin, we inevitably must enlarge the scope of the analysis to include the entire basin. It is at this point that basin-wide and biogeographic conservation planning approaches merge.

(text continues on page 112)

BOX 6.1. **Examples of Biodiversity Visions in Three African Freshwater Ecoregions.**

Niger Basin Initiative. The Niger River flows through five ecoregions in West Africa, and its basin lies in eleven countries. Among these ecoregions are the globally outstanding Niger Delta [58], with one endemic freshwater fish family; the continentally outstanding Inner Niger Delta [7], with congregations of more than 1 million waterbirds; and the continentally outstanding Lower Niger-Benue [65], with high fish, mollusk, and herpetofaunal richness. The river is considered the cultural, economic, and ecological backbone of West Africa because human populations have depended on its vital freshwater supply in this essentially xeric region for thousands of years. Unfortunately, with burgeoning human populations and ever-increasing sources of pollution and degradation, the river is under severe threat.

The Niger Basin Initiative (NBI) is an environmental partnership between WWF, Wetlands International (WI), and the Nigerian Conservation Foundation created in May 2001 to address biodiversity conservation of the Niger River on a basin-wide scale. A cooperation agreement was then signed in October 2003 between the international organizations working on nature conservation in the basin (WWF, WI, BirdLife, and IUCN) and the Niger Basin Authority. The partner organizations are working together to ensure that biodiversity conservation and the sustainable use of natural wetland and forest resources are built into development plans for the basin.

In April 2002, the NBI hosted a workshop at which participants identified and mapped the most important areas in the basin for biodiversity, identified the most pressing socioeconomic issues, planned developments in the basin, and overlaid this information to prioritize conservation actions. Areas of biological importance were selected based on biodiversity data for fish, birds, and other vertebrates. Additionally, subbasins within the basin were evaluated for their ability to contribute to maintaining ecological and hydrological processes. About forty biologists, ecologists, and hydrologists from seven countries in the Niger River basin selected nineteen priority areas for the long-term conservation of freshwater biodiversity (figure 6.1). —*Aboubacar Awaiss*

Guinean-Congolian Forest and Freshwater Region. The Guinean-Congolian Region, including twenty-two freshwater ecoregions, harbors some of the world's greatest tropical forests and rivers. Its enormous surface area and its high diversity of habitats, as well as climatic and environmental stability over a long period, have facilitated the evolution of a highly diverse freshwater fauna.

Although biological data and knowledge are incomplete for these forests and rivers, the rapidly changing political situation and emerging conservation opportunities demand the development of strategies based on the best available data. In March 2000, WWF convened a workshop to determine biological priorities in the Guinean-Congolian Forest and Freshwater Region (Kamdem Toham et al. 2003). The freshwater group focused on the hydrographic Congo Basin (excluding Lake Tanganyika [55] but including the Bangweulu-Mweru [6] ecoregion) and included the freshwater ecoregions of the West Coastal Equatorial bioregion [ecoregions 19, 26, and 29] and the Western Equatorial Crater Lakes [5] and Niger Delta [58] freshwater ecoregions.

The experts at the workshop produced the following vision statement for the freshwater systems of the region:

> Our vision for the Guinean-Congolian Freshwater Region is to conserve, to the fullest possible extent, its globally outstanding richness, diversity, and uniqueness in terms of the habitats, fishes, and other aquatic taxa. There must be clear and achievable conservation goals and programs must be properly planned, knowledge-based, and incorporate sound science—including fundamental data on species identity, distributions, and life cycles. Specially designed aquatic reserves, the protection of headwaters, and the minimization of the impacts of commercial aquaculture and damming are required. These aquatic conservation goals should be fully integrated into terrestrial conservation programs and vice versa. Conservation planning must be sensitive to the sustainable requirements of various stakeholders, such as fishing peoples and the agricultural community. Sound education, awareness, and training programs, and a regional network for the study and interpretation of aquatic biodiversity, will be crucial in securing the vision. The essentially unspoiled nature and vast scale of this aquatic ecosystem make it a singularly compelling conservation challenge. (Kamdem Toham et al. 2003: 42)

Freshwater experts first identified regions for which insufficient data were available to evaluate biodiversity importance and then identified areas of biodiversity priority within known portions of the region of analysis (figure 6.2). Experts identified known priorities to be the Ivindo River, the Kouilou-Niari, the Cuvette Centrale, Maï Ndombe, the middle Congo River mainstem, rapids upstream from Kisangani, Thysville Caves, Upemba, Kalengwe Rapids, Lac Fwa, and several areas in the larger Cameroonian highlands and Niger Delta regions. The experts expect further studies to reveal other biological and conservation priorities. Because data gaps largely overlap with areas of insecurity and conflict, many of the priority research

(continued)

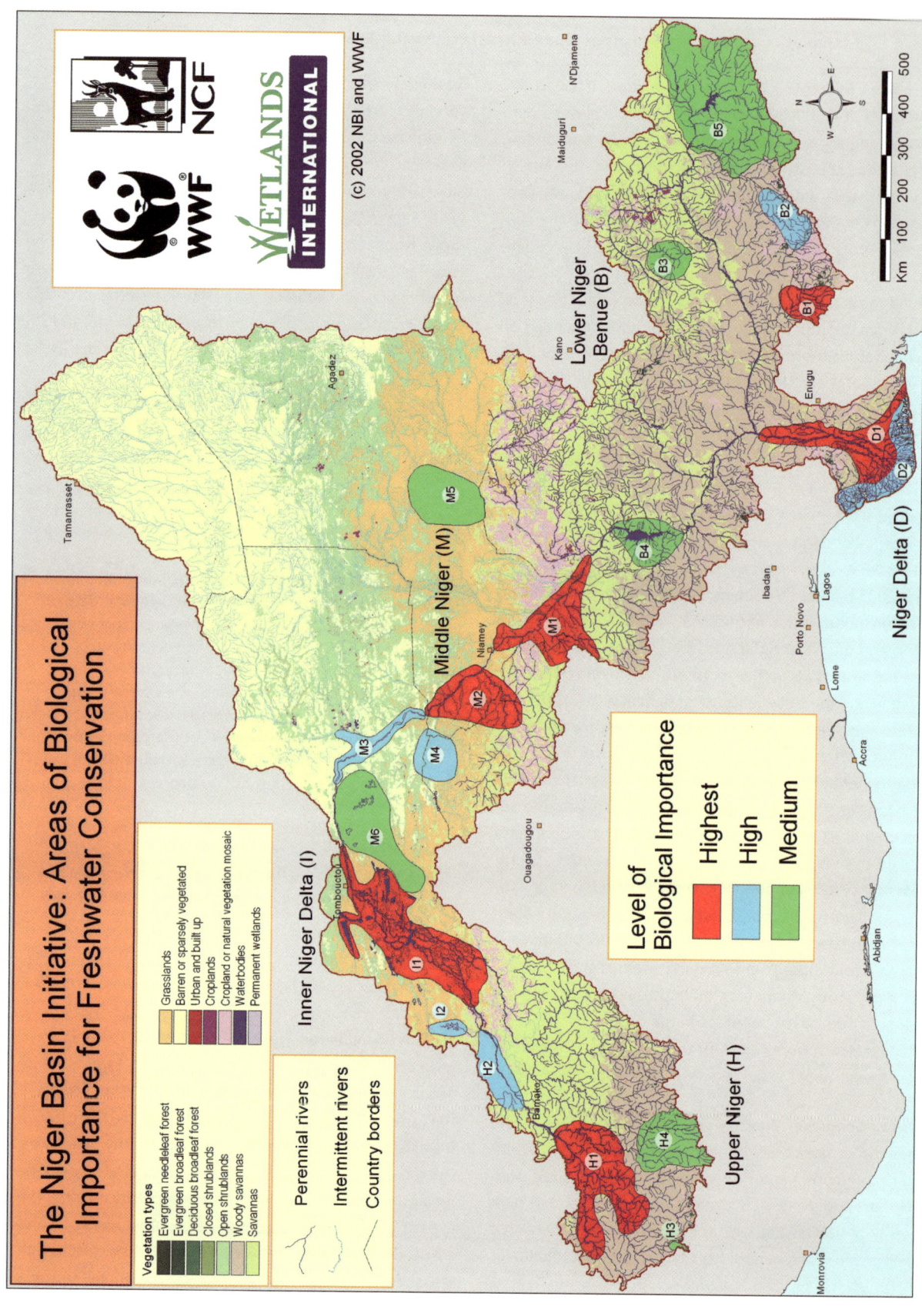

FIGURE 6.1. Areas of biological importance in the Niger River Basin.

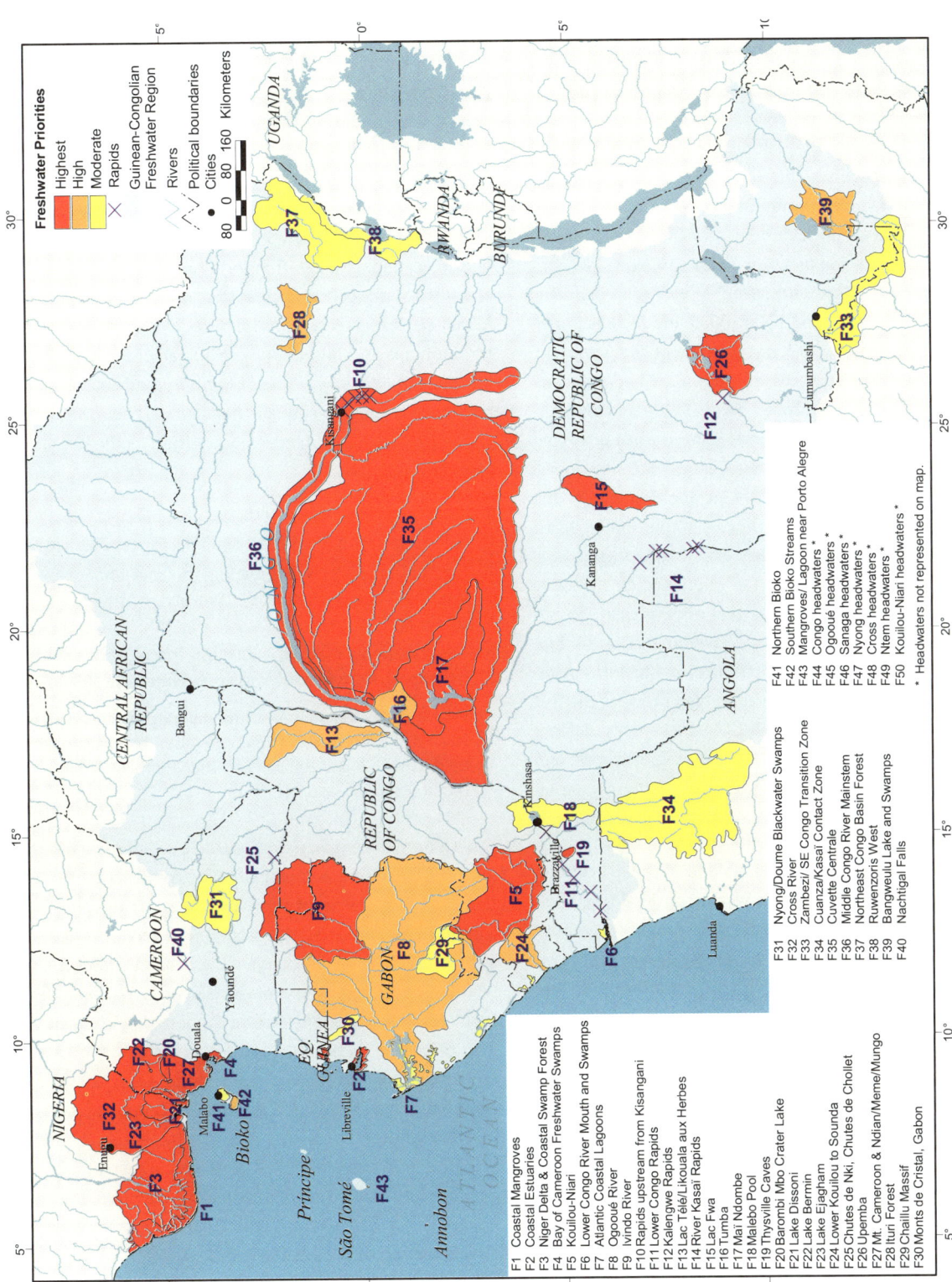

FIGURE 6.2. Freshwater priority areas in the Guinean-Congolian freshwater region.

BOX 6.1. *(continued)*

areas are located in regions where conservation action is unfeasible. In addition to the areas shown in figure 6.2, headwater regions were also considered of priority because of their role in sustaining the flow regime.

After the workshop, the biological vision for the Guinean-Congolian terrestrial systems was adopted by country signatories of the Yaoundé Declaration as the blueprint for conservation in the region. Other initiatives are also adopting the results of the vision as a foundation for large-scale conservation projects. The Brazzaville Priority Action Plan outlines targets for the period 2002–2005 and focuses implementation on transborder forest areas identified within the biodiversity vision. Gabon has recently established a network of thirteen national parks based on the vision and on IUCN critical sites. Cameroon has refined its network of protected areas based on the vision, and the Central African Republic, Republic of the Congo, and Democratic Republic of the Congo will undergo the same process in the coming years. Although these efforts are largely terrestrial in focus, some work has also begun in freshwater priority areas, especially in those that overlap with terrestrial priority areas. To our knowledge, the endorsement by governments of the biological vision and the multinational effort to establish and implement a region-wide conservation plan provides the first operational model in tropical Africa of the benefits of working to implement conservation at the ecoregional scale, and we encourage the further integration of the freshwater vision into these plans. — *Andre Kamdem Toham*

Lake Malawi. Lake Malawi boasts one of the richest lake fish faunas in the world, hosting several species flocks of endemic cichlids and an endemic clariid flock (Lowe-McConnell 1987). The Lake Malawi/Niassa/Nyasa ecoregion encompasses the lake basin and Lake Malombe to the south. The lake and its basin cover about 130,000 km² and include much of Malawi, the southwestern corner of Tanzania, and the northwestern corner of Mozambique. Each of the three riparian states manages its own portion of the lake according to national policies and strategies. At a workshop in October 2001, experts from the three bordering countries identified twenty priority conservation areas (figure 6.3) in the lake basin using overlays of geospatial data on biodiversity and threats (Chafota et al. 2002). In addition to areas in the lake itself, the experts identified seven river catchments (the Songwe, Lufirio, Ruhuhu, Dwangwa, Bua, Kaombe, and Linthipe basins) as priority areas for conservation activities, in recognition of the need to lessen impacts from land use in the basin. Workshop participants concluded that runoff from agricultural practices in southern Malawi and parts of Tanzania is the largest threat to the freshwater biodiversity of the ecoregion. In terms of fisheries, the most exploited section of the lake is at the southern and shallower end, where a combination of fishing pressure, eutrophication, and fish translocations is causing concern about the status of fish populations. Workshop participants identified conservation opportunities along the shores of Mozambique and Tanzania where human population levels are low and where large areas of undisturbed coastal habitats and less exploited aquatic habitats still occur.

The most important conservation targets for the lake were identified as follows:

- Reduce eutrophication of the lake's waters to levels that will not decrease biodiversity and fishery productivity.
- Maintain stocks of river-breeding fish species above levels that could decrease biodiversity and fishery productivity.
- Maintain populations of cichlid fish species above levels that could decrease biodiversity and fishery productivity.
- Establish and maintain institutional mechanisms that facilitate collaboration and cooperation between stakeholders who affect the use and management of lake-based resources and habitats.

Experts from the region suggested the establishment of a trilateral mechanism for coordinating basin management activities. Since the workshop, Malawi, Mozambique, and Tanzania have announced a partnership with WWF, the Ramsar Secretariat, and the Swiss Agency for the Environment to work on joint conservation of the lake on a transboundary basis. The three countries will be supported in designating their respective parts of the lake as Ramsar sites. With assistance from the Danish Development and Aid Agency, Tanzania has already begun designation of its portion of the lake as a Ramsar site.

A significant investment of time and resources will be needed to achieve the recommendations outlined here. This vision is larger than what any one organization could achieve alone. Therefore, partnerships with relevant government, nonprofit, and business organizations will be crucial.

— *Jonas Chafota*

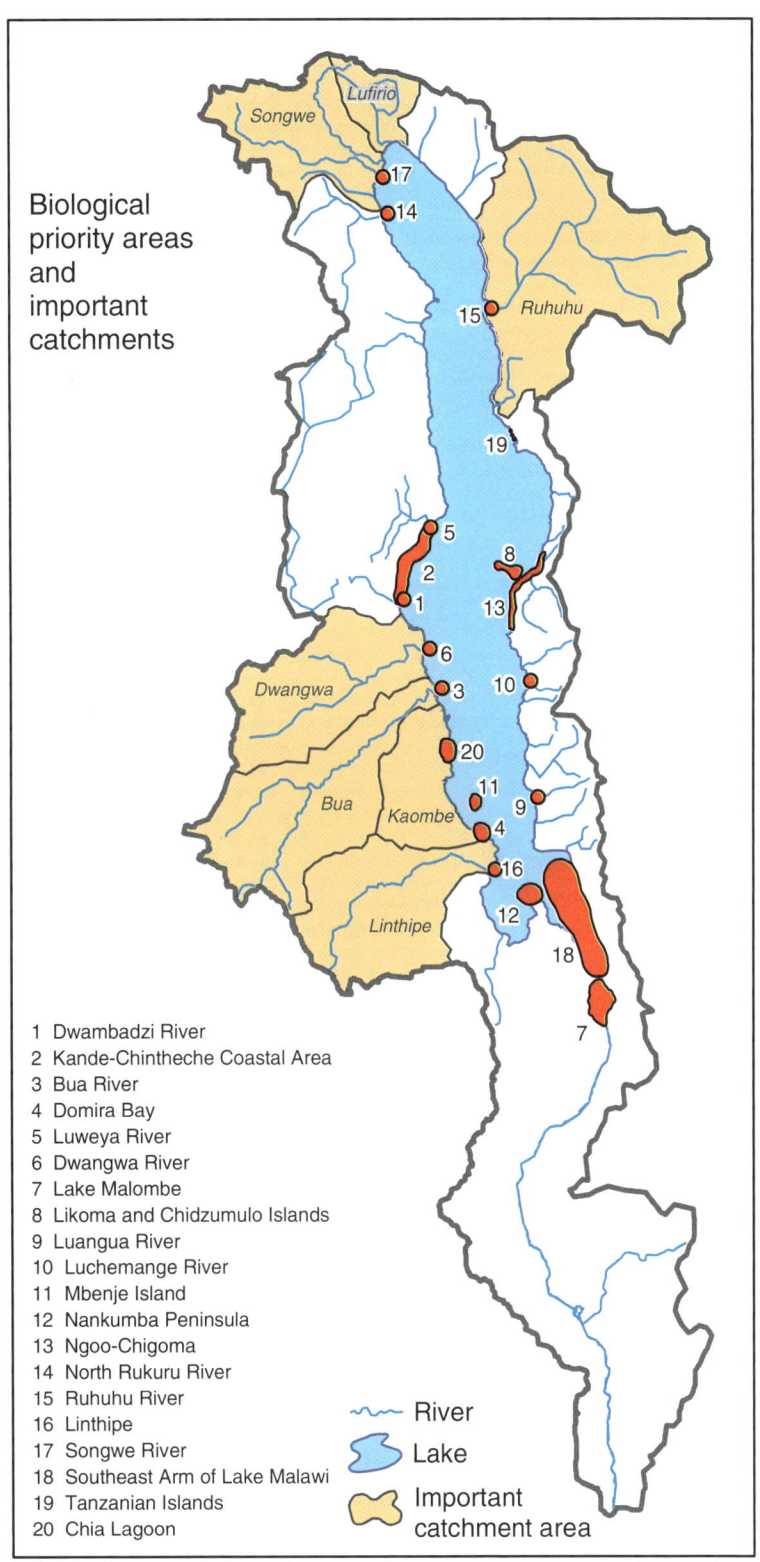

FIGURE 6.3. Final priority areas for biodiversity conservation in Lake Malawi/Niassa/Nyasa.

BOX 6.2. Big Water: The Challenge of Conserving Biodiversity in Africa's Large Rivers and Lakes.

The complexities of conserving biodiversity in the ever-moving environment of freshwater are detailed elsewhere in this volume. Those complexities are compounded by the huge size of some African river and lake basins and the challenges of managing large freshwater systems within a patchwork of jurisdictions and interests. Consider the Zambezi. This junior partner in the fraternity of great African rivers drains eight countries and supports a population of 65 million people (Zambezi River Authority 2003). Within each country is a dizzying array of local governments and pastoral, agricultural, urban, and industrial interests that vie for their share of the freshwater resource, and their activities, such as logging, irrigation, and diversion, have profound impacts. This situation exists on almost all of Africa's great rivers and lakes.

The solution lies in sewing the patchwork together into an integrated management framework within which governments can work together, stakeholders can be adequately represented, and decisions can be made at a basin-wide scale. Concerns about biodiversity and ecological integrity should underpin such a framework. This is the central tenet of integrated river basin management (IRBM).

IRBM is defined as a "process of coordinating conservation, management, and development of water, land, and related resources across sectors within a given river [or lake] basin, in order to maximize the economic and social benefits derived from water resources in an equitable manner while preserving and, where necessary, restoring freshwater ecosystems" (Jones et al. 2003: 2). Recognizing a central challenge of managing the freshwater resource — the fact that it moves — IRBM provides a tool for developing management solutions from headwaters to mouth and a strategic framework for action at basin-wide or smaller scales. Within an IRBM planning initiative, managers can bring the whole array of conservation tools to bear: protected areas, ecological restoration, improved forestry and agricultural practices, innovations in dam operations and ship design, and new energy and sanitation technologies. On the scale of a large African river, none of these interventions would be effective in isolation, but as parts of an integrated strategy, they could be tremendously effective.

This is easier said than done. In fact, very few IRBM schemes are in place and in full operation around the world. IRBM is a new idea, and it is an immensely complicated undertaking. The basins that are closest to full implementation — the Danube, the Everglades, and the Murray Darling — still have a long way to go before they are functioning optimally. However, planning and implementation are ongoing in dozens of river and lake basins, and governments are building the international frameworks necessary for IRBM on transboundary rivers and lakes (Gilman et al. 2004).

In Africa, this approach is being pursued on the largest scale imaginable: the River Nile. The Nile Basin Initiative (NBI) is a joint effort by ten of the eleven Nile basin countries to facilitate the common pursuit of sustainable development and management of Nile waters. The NBI, supported in large part by the World Bank, United Nations Development Programme, and Canadian International Development Agency, is developing a basin-wide strategic action program with projects to address energy, agriculture, water resource planning, stakeholder involvement, and the environment. All these projects are intended to be mutually supporting, creating basin-wide engagement, capacity, common strategic and analytical frameworks, and demonstration projects. The Nile Transboundary Environmental Action Project will support basin-wide action to address transboundary environmental issues including pollution, forest management, wetland conservation, and protected area establishment and management. Projects at the community level that fit within the overall strategy are also an important piece of the program (Nile Basin Initiative 2003).

Transboundary programs such as these are at some stage of development on many of Africa's rivers and lakes, through groups such as the Niger Basin Authority, the Organization for the Development of the Senegal River, the Lake Chad Basin Commission, and the Zambezi River Authority (Jones et al. 2003). Many such authorities are more talk and paper than action at present, but that is changing rapidly as planning advances and international donors take greater interest. Results are starting to appear.

A good example can be found in Zambia on the Kafue River, a major tributary and subbasin of the Zambezi where WWF has been working with local conservation partners since 1998. The floodplain of the river, known as the Kafue Flats, is home to zebra, buffalo, cheetah, wild dog, and an endemic ungulate, the Kafue lechwe or marsh antelope (*Kobus leche kafuensis*). The floodplain and wetlands are an important stopover point for migratory birds. More than 450 species of birds occur there, including the vulnerable wattled crane (*Bugeranus carunculatus*).

The construction of two large dams at either end of the Kafue Flats disrupted the natural hydrology of the floodplain, resulting in serious habitat loss and precipitous declines of many plant and animal species. For example, the Kafue lechwe population has fallen more than 50 percent in recent

BOX 6.2. *(continued)*

years. In addition, local communities blame the decline in fish yields and forage in riparian grazing areas on flow alterations produced by the dams.

The dams have important beneficiaries, however, including the sugar industry on the flats, which produces most of Zambia's crop for domestic consumption and export, and the private companies and government ministries concerned with energy production. The power station at the Kafue Gorge dam produces more than half of Zambia's electricity and a large amount that is sold to neighboring Zimbabwe and South Africa.

An integrated river basin management approach in the subbasin, bringing the key stakeholders of the flats together, has yielded a plan that addresses the needs of all major stakeholders and could restore the declining biodiversity of the flats. The sugar industry, the Zambian Electrical Supply Company, and the Ministry of Energy and Water Development have agreed to a set of protocols to change the operations of the dams to more closely mimic natural flows through the flats and plan to begin testing them in 2004. In addition, the sugar growers are exploring the use of biofilters such as wetlands and reedbeds to pretreat effluent from their farms. Local communities are participating in awareness-raising activities and exploring ecotourism opportunities. To date, through a partnership approach, WWF has also formally established the Mwanachingwala Conservation Area, through a combination of donations of private (mainly from the sugar growers) and traditional lands. The reintroduction of a select number of species into the conservation area is due to commence soon, and 500 km^2 of the flats are to be restored in the newly created conservation area (Jones et al. 2003). In terms of ecotourism, the Blue Lagoon National Park has reopened for visitation after receiving support for development of its infrastructure.

The benefits to freshwater biodiversity from such integrated management approaches could be enormous. Successfully implemented plans that address the hydrology, water quality, and connectivity of river systems from headwaters to estuary would be a boon for freshwater ecosystems, even if biodiversity conservation were not the stated goal of such an effort. Indeed, integrated strategies that benefit both human populations and biodiversity are the gold standard of conservation action and are entirely possible in the context of freshwater. This is particularly true in Africa and other parts of the developing world, where many freshwater ecosystems remain intact compared with those in much of the developed world (Richter et al. 2003).

The definition of IRBM mentioned earlier is not universally accepted, and the term often is used interchangeably with others such as "integrated water resource management," "watershed management," and "catchment management." Whatever the label, it is unfortunate that many such planning efforts lack emphasis on functioning ecosystems as the source of freshwater and a prerequisite to successful basin management. The challenge for scientists and conservationists is to ensure that these strategies are informed by sound conservation biology and that stakeholders, governments, and river basin managers are committed to preserving the ecological integrity of the freshwater resources on which we all depend. Only then can the potential of IRBM as a tool for biodiversity conservation be fully realized and the animals and plants of Africa's big water conserved.

—*Christopher E. Williams*

Kafue lechwe (*Kobus leche kafuensis*) with cattle egrets (*Bubulcus ibis*), Kafue Flats, Southern Province, Zambia. Photo credit: ©WWF-Cannon/Sarah Black.

Visionary Work in Freshwater Ecoregions and River Basins

Ecoregion planning and the creation of a biodiversity vision provide an overview of the actions necessary at local, national, and international scales for biodiversity conservation in an ecoregion or river basin. Looking first at an ecoregion or river basin in its entirety allows us to understand what tradeoffs are associated with different land use and development decisions and how projects can and should relate to each other. Actions that may be ongoing or nascent gain a new importance and synergy as they are implemented within a more coherent framework (Abell et al. 2002).

When undertaking ecoregion planning, we ask ourselves whether our vision for the future is ambitious enough to conserve an ecoregion's distinct biodiversity features over the long term. Creating and acting on a vision for the conservation of an ecoregion's biodiversity can be a powerful exercise. Within a year of completion, the freshwater visions outlined in box 6.1 assisted in catalyzing the formation of a transboundary management commission, the creation of new protected areas, and agreements for implementing international conventions and management recommendations. Numerous types of mechanisms, at the landscape or site scale and above, exist for implementing biodiversity visions, including community-based fishery or wetland management, Ramsar site designation and management, and the activities of river basin authorities (see essays 6.1, 6.2, and 6.3). Policies related to landscape development are critical components of freshwater conservation because of the highly interconnected nature of freshwater systems with upstream, upland, downstream, and groundwater areas. Effective implementation of a vision will incorporate many of these mechanisms, among others (box 6.2).

Working at a large scale has its challenges, including the need for long-term commitments and sustained financial resources. The realities of financial support, governance, poverty, political stability, peace, and human capacity are critical factors that can prevent or foster successful implementation. Recent priority-setting exercises have attempted to integrate socioeconomic factors into determining where conservation success might best be achieved (Balmford et al. 2002; O'Connor et al. 2003; Smith et al. 2003). Rivers and lakes also often cross national boundaries, necessitating international agreements and cooperation for their management; political obstacles to transboundary work can often seem insurmountable (van der Linde et al. 2001). Yet the conservation achievements produced by first planning at a large scale and then implementing at finer scales have the potential to be far more substantial than those produced by small projects conceived separately from one another (Abell et al. 2002; Groves 2003).

Challenges to Freshwater Conservation and Sustainable Development in Africa and Madagascar

In box 6.1 we provide examples of visioning exercises undertaken in several of the freshwater ecoregions of the Afro-Malagasy region. In this section we outline what we see as the greatest challenges to achieving sustainable use and conservation of the region's freshwater systems.

Political Instability and Civil Unrest

Across parts of Africa, civil unrest is a daily reality, affecting the lives of millions of Africans. Shambaugh et al. (2001) recommend that basic support for conservation continue through times of war and civil unrest because past experience has shown that this provides a solid foundation for postconflict conservation efforts. However, environmental degradation often occurs during times of civil unrest and war (Hart and Mwinyihali 2001; Blom and Yamindou 2001). In terms of freshwater systems, pollution of lakes and rivers, hunting of aquatic mammals, and exploitation of fish stocks are wartime activities that appear to have the most severe impact (e.g., see ecoregion descriptions for Northern Upper Guinea [25], Upper Nile [15], Cuvette Centrale [18], and Kasai [21]). Human populations rely more heavily on bushmeat and fisheries as protein sources are limited by decreased agricultural production and the suspension of trade. For example, in the Democratic Republic of the Congo (DRC) and Sudan, populations of large terrestrial and aquatic vertebrates have been affected by bushmeat hunting during the long years of civil conflict (Bakarr et al. 2001; Morjan et al. 2001). Movements and large settlements of refugees also cause land degradation, which in turn degrades freshwater systems. The relative effect of conflict on freshwater and terrestrial biodiversity in the region should be evaluated as studies resume in areas that have suffered from long-term unrest (e.g., Sudan, Angola, DRC).

Environmental Governance

In nations at war or at peace, environmental governance is critically important. Mugabe and Tumushabe (1999: 15) define *environmental governance* as "a body of values and norms that guide or regulate state-civil society relationships in the use, control, and management of the natural environment . . . providing a conceptual framework within which public and private behavior is regulated in support of sound ecological stewardship." The need for effective policies, laws, and regulations that support conservation and sustainable use of freshwater systems is critical (essay 6.4). In many parts of Africa, laws are weak or mechanisms for enforcing them are not operational. Junk (2002) identifies policy deficiencies, deficient planning concepts, limited information and awareness, and institutional weakness as the main administrative limitations for sustainable wetland management in the tropics.

In the last several decades many tropical countries have experienced a resurgence of community-based management of natural resources that often integrates new governance structures with traditional management practices (see essays 6.2 and 6.5). These community-based initiatives are particularly relevant to the successful management of wetlands in Africa because of the high dependence of local communities on wetland resources for livelihoods and food security. An extensive review of these projects is beyond the scope of this book; we refer readers to a few texts with examples (e.g., IUCN 1995–2003; Acreman and Hollis 1996; Palfreman 2001; Gawler 2002).

Resource Extraction in Natural Resource–Based Economies (Logging, Mining, Agriculture, Oil)

In addition to pollution from municipal and industrial sewage, many freshwater systems receive sediments and associated chemicals from agricultural, mining, or forestry operations. About 85 percent of Africa's total water withdrawals are estimated to be directed toward agriculture (FAO 1995), and as of the year 2000, about one-third of its surface area was estimated to be under agricultural land use (FAO 2001a). Although the total land area of Africa under irrigation is small and rarely exceeds 1 percent of an individual basin's total area, it can represent a large proportion of the cultivated land in basins in the semi-arid and arid regions of Africa. As a percentage of total internal renewable water resources, water withdrawals vary widely from one region to another. For example, agricultural irrigation consumes only about 0.01 percent of runoff in the Congo Basin, but 20 percent of runoff in the Limpopo and 70 percent in the Nile are used in irrigation (FAO 2001a). In North Africa, about 95 percent of renewable water resources are withdrawn (Shiklomanov 1999 in Junk 2002). Some of the wetlands that have been degraded by large-scale irrigation schemes include the Nile Delta and floodplains along the lower Nile River, the Logone floodplain and the floodplain of the Benue River, the Hadejia-Nguru wetlands, the Phongolo floodplain, and the Senegal Delta (Junk 2002). By 2025, total water abstraction in Africa is expected to rise by 54 percent to 337 km^3/year, with agricultural use accounting for 53 percent of this amount (IUCN 2000).

Forestry operations can cause erosion and contribute sediments to freshwater systems, sometimes causing significant disturbance. Mining is similar to forestry in that it often causes pulses of sediment input, with the addition of heavy metals. Local populations may contribute significantly to loss of tree cover through cutting for building materials, firewood, or charcoal. Loss of forest cover can significantly affect the hydrology of an area, leading to changes in annual, dry season, and peak flows and significantly altering biogeochemical processes and in-stream habitat.

Land Use Change and Habitat Loss

Africa still contains large areas of natural or near-natural vegetation and stream flow. Current economic patterns indicate that most of the next generation of Africans will continue to live subsistence lifestyles, farming for food and deriving their fuelwood, protein, medicines, and building materials from natural resources. Given that future estimates show the population of Africa tripling between 1995 and 2050 (increasing from 0.7 billion people in 1995 to 2.0 billion in 2050) (United Nations Population Division 2001), Africa is predicted to undergo tremendous agricultural expansion. At a global level, an additional 30 percent of the remaining forests and natural woodlands are set to disappear in the next 50 years (Tilman et al. 2001), and much of this will occur in Africa. Rising water demand, a growing risk of pollution and physical destruction of habitats, and general changes in land use practices and traditions, eventually culminating in large-scale vegetation loss or desertification, are likely consequences. These changes obviously have enormous implications for the integrity of freshwater systems and their biotas.

Invasive Species

About fifty fish species have been introduced into or translocated within the inland waters of Africa, twenty-three of which are from outside Africa (Welcomme 1988). The well-known case of the introductions of Nile perch (*Lates niloticus*), four tilapiines, and water hyacinth (*Eichhornia crassipes*) into Lake Victoria and the subsequent decline and loss of populations of many fish taxa is a stark example of how species introductions, in tandem with other environmental changes, can negatively affect a system's ecology (Ogutu-Ohwayo 1990; Seehausen et al. 1997a). We suggest that many well-intentioned aquaculture projects may also be introducing invasive species into continental African waters. For example, the widespread translocation and introduction of tilapias (*Oreochromis* and *Tilapia* spp.) throughout the continent and into Madagascar have been associated with the decline of native species in several systems (e.g., lakes Victoria, Itasy, and Alaotra) (Reinthal and Stiassny 1991; Lévêque 1997). Although these projects may be productive in the short term, they may lead to longer-term ecological problems that will affect native freshwater species and human communities that rely on functioning systems.

Introduced aquatic plants, such as water hyacinth (*Eichhornia crassipes*) and giant salvinia (*Salvinia molesta*), are problems in many river and lake systems, resulting in mounting economic, social, and ecological costs (see essay 4.1). For example, in South Africa, about 150 species of more than 8,000 introduced plants are considered invasive (Department of Water Affairs and Forestry 2003). In the dry south, alien species generally consume more water than indigenous vegetation, in addition to threatening biodiversity and constituting a significant fire hazard (Davies and Day 1998; Dye et al. 2001; Binns et al. 2001). The Working for Water Programme provides an innovative approach to combating these problems. The program aims to increase water supply in select catchments by removal of invasive plants, with the work undertaken by trained people from the most marginalized sectors of South African society (Binns et al. 2001; Department of Water Affairs and Forestry 2003).

Water Management: Dams and Interbasin Water Transfers

The many and varied water management schemes that exist across the Afro-Malagasy region have a wide range of effects on freshwater ecosystems. The World Commission on Dams outlines many of the negative effects that large dams (more than 15 m high) have had on aquatic systems and biodiversity, floodplain ecosystems, and fisheries in Africa. Recent studies also indicate that reservoirs may make a significant contribution to global greenhouse gas emissions (St. Louis et al. 2000). According to FAO data, current dams are concentrated in South Africa, Zimbabwe, Burkina Faso, and Nigeria (see figure 4.2). Three of the four largest rivers, the Nile, Niger, and the Zambezi, have all been significantly affected by the construction of large dams along their course. The two dams with the largest capacities, Aswan High Dam (about 180 billion m^3) and Kariba Dam (about 160 billion m^3), are sited along the Nile and Zambezi Rivers, respectively. Both of these dams have changed downstream ecology, largely through loss of seasonal high and low flows and sediments trapped behind the dam wall. In the delta floodplain of the Zambezi River, lowered shrimp catches, declines in the productivity of artisanal fisheries, floodplains invaded by upland vegetation, dying mangroves, and decreased wildlife populations are some of the results of the altered hydrology (Soils Incorporated [Pty] Ltd. and Chalo Environmental and Sustainable Development Consultants 2000). The Nile Delta is subsiding and eroding because of the lack of sediment input, and seawater intrusion threatens to increase salinity of the coastal lakes and rivers (Stanley and Warne 1998; Baha El Din 1999). Additionally, diminished nutrient levels in the flow of the Nile to the Mediterranean are considered largely responsible for the declines in sardine (*Sardinella aurita* and *Sardinella maderensis*) landings (96 percent and 36 percent, respectively) since the closure of Aswan Dam (Lévêque 1997). The biggest river in Africa, the Congo, is largely undammed; however, the government of the DRC recently announced plans to move forward quickly with development of the Grand Inga plant, which, unlike the current Inga Dam, would extend across the entire river (Société Nationale d'Electricité (SNEL), République Démocratique de Congo 2003).

Davies et al. (2000b) outline the implications of interbasin water transfers for river conservation. In addition to changing flows within basins, water transfers break down biogeographic barriers and introduce species across basins. For example, as a result of the Orange–Sundays–Great Fish River water transfer, several fish species have been introduced to the Great Fish River, despite devices installed to prevent their transfer (Skelton 1980; Laurenson and Hocutt 1984; Davies et al. 2000b). Large-scale interbasin transfers that are currently under negotiation or construction include the Lesotho Highlands Water Project (LHWP) and the Eastern National Water Carrier project in Namibia.

The Lesotho Highlands Water Project is the largest interbasin transfer in Africa and will transfer 2.2 billion m^3 of water per year from the headwaters of the Orange River to the tributaries of the Vaal River in South Africa (Snaddon et al. 1998; Lesotho Highlands Water Project 2002). Many questions remain about the long-term environmental effects of this controversial project, despite numerous environmental studies and mitigation projects; altered flow regimes, water quality changes, introductions of alien species to both donor and recipient rivers, and declines in the critically endangered Maloti minnow (*Pseudobarbus quathlambae*) are among the many expected impacts (Davies and Day 1998).

Namibia is negotiating the extraction of an estimated 20 million m^3 of water annually from the Okavango River via the Eastern National Water Carrier. Although this is less than 10 percent of the river's annual flow, it would be a significant amount during the dry season and could deleteriously affect functioning of the delta ecosystem, transfer species from the Okavango to more southerly drainages, and affect groundwater levels in the region (Davies and Day 1998).

Among several proposed future projects, there have been recent discussions of transferring water from the Oubangui Basin to the Lake Chad Basin. Such an action could introduce Congolian freshwater species into the Nilo-Sudanian bioregion and vice versa, with likely negative consequences for one or both native faunas.

Inland Fisheries and Overexploitation

According to a recent FAO study, the single most important issue for the future of inland fisheries is not increasing fishing pressure but the degradation of the environment and subsequent loss of freshwater habitats (FAO Inland Water Resources and Aquaculture Service, Fishery Resources Division 1999). Indeed, habitat loss is of highest concern. However, overexploitation of certain fisheries has been documented in select waterbodies of Africa and Madagascar, and sustainable fishery management will be increasingly important for the long-term viability of freshwater fish populations. Inland fish capture in Africa increased by an average of about 37,000 tons per year, or 2 percent, between 1984 and 1997, making Africa second only to the Asia-Pacific region in total catch of inland fisheries (FAO Inland Water Resources and Aquaculture Service, Fishery Resources Division 1999). Four of the top ten countries (Uganda, Tanzania, Egypt, and Kenya) for inland capture fishery production occur in Africa (FAO Fisheries Department 2002). Lévêque (1997) distinguishes between impacts on individual species, primarily large-bodied species of low reproductive capacity, and impacts on fish communities from heavy exploitation (Durand 1980; Daget et al. 1988; Coulter 1991; Witte et al. 1992; Kolding 1992; Turner 1994). Both types of effects have been documented in the Afro-Malagasy region, and Welcomme (see essay 4.2) considers many of the region's fisheries to have been fished at levels where damage to the assemblages has already occurred.

The region's fisheries are also of vital importance to human communities, providing livelihoods and a source of protein (see essay 1.1 and essay 6.4). On average, Africans get more than 20 percent of their animal protein from fish, and populations in several countries in the region get more than 50 percent of their animal protein from fish (FAO 2003). For example, in Ghana the fishery sector supports the livelihoods of 1.5 million people, and fish is the most important source of animal protein, supplying 60 percent of the daily intake of the average Ghanaian (Aggrey-Fynn 2001; Directorate of Fisheries–Ghana 2003).

Climate Change

Future climate changes are expected to severely affect the hydrological cycle of freshwater systems in many parts of Africa (see essay 4.3). Despite the inherent uncertainties in model calculations, some general trends indicate decreases in the available water resources for the Mediterranean coast of North Africa and particularly for large parts of southwestern Africa. Increasing tendencies are projected for West Africa south of the Sahara (including the Niger basin) and for parts of southeastern Africa and western Madagascar. Many other areas, such as the Congo Basin, show different trends depending on the applied climate change models and scenario assumptions. In addition to expected changes in long-term trends, increases in the seasonal variability of regional climate patterns are projected. These climatic changes may cause significant shifts in the magnitude, timing, and duration of river flows. As a consequence, not only alterations in long-term average flows or reductions in water resources are expected but also increasing intensities and frequencies of extreme events such as floods and droughts (IPCC 2001; Hulme et al. 2001; Desanker 2002). Although climate change is a global problem, the particular vulnerability of Africa has been highlighted by various authors and is largely attributed to the limited adaptive capacities of African countries to cope with the effects of climate change (IPCC 2001).

Data Gaps and Research Needs

Significant gaps in our knowledge of the Afro-Malagasy region's freshwater ecosystems remain, including a lack of basic species inventories for many areas, accurate taxonomic classifications and naming of many organisms, and a fundamental knowledge of ecosystem functioning in many parts of the Afrotropics. Also of critical importance to the sustainable management of freshwater systems are hydrologic data. Unfortunately, the collection of this essential information is declining in the region. The number of monitoring stations for water flow and water quality in Africa declined by 90 percent between 1990 and 2000 (Vörösmarty et al. 2001). Obviously, these knowledge gaps are a significant hindrance to freshwater conservation activities across the region. Denny (2001) outlines a two-pronged approach to address some of the outstanding questions. He suggests that inventories, assessment, and monitoring of freshwater systems are of immediate urgency, and research into processes, structure, and functioning of the systems will be important in the long term. The generation and application of both biological and hydrological data ultimately will depend on the availability of an active network of freshwater scientists across the Afro-Malagasy region. The Environment Plan of the New Partnership for Africa's Development provides a recent opportunity for addressing many of the gaps and research needs. Under "Programme Area 2: Conserving Africa's Wetlands," it includes plans for the development of national wetland policies, the development of subregional networks of wetland scientists, long-term inventory and monitoring of Africa's wetlands, wetland restoration, and capacity building, along with subregional targets for meeting each of the goals (NEPAD 2003). One example of a recent regional initiative to build research capacity is the creation of the Centre for African Wetlands in West Africa (CAW 2003). As many others have suggested previously, the continued expansion and strengthening of a wetland network to achieve these goals in the global south should receive significant financial and technical investment from the north and information and experience exchange with other parts of the global south (Wishart and Davies 1998; Denny 2001; Tiéga 2001; Junk 2002; NEPAD 2003).

Despite the many challenges outlined in this chapter, there are several points of hope for the future of the freshwater systems in Africa and Madagascar. Compared with wetlands in much of the north and in some other parts of the tropics, the wet tropics of Central Africa and other areas with low population densities retain areas with functioning freshwater systems. In these areas, there is an opportunity to limit or prevent the widespread degradation that has occurred in much of the north. The action plan of the Environment Initiative of NEPAD, recently endorsed by the African Ministerial Conference on the Environment, provides a new framework for wetland conservation in Africa with its detailed goals and targets (AMCEN 2002). The continued and active expansion of the Ramsar network in Africa is also a positive development, with thirty-eight countries now contracting parties to the Ramsar Convention and 130 sites designated as of March 2004 (Wetlands International 2004). The recently revised and enacted progressive water law in South Africa also warrants mention; this piece of legislation is based on the principles of sustainability of use and equity of distribution. By explicitly recognizing the intrinsic value of freshwater systems, this new law marks a shift in public consciousness, recognizing the need to maintain healthy ecosystems to sustainably provide freshwater to human and aquatic communities over the long term (essay 6.6; Palmer 1999; Palmer et al. 2002). The value of the ecosystem services that freshwater systems provide is remarkable, and sustainable use is essential to maintaining the healthy systems that will continue to provide these services (Postel and Carpenter 1997; Costanza et al. 1997; essay 1.2). As nations across the region continue to grow and develop, it is our hope that they will follow a path that maintains these natural values, in the best of Afro-Malagasy traditions. The world will be greatly diminished if we fail to recognize the importance of the Afro-Malagasy region's freshwater habitats, species, and biological processes before it is too late.

ESSAY 6.1

The Convention on Wetlands (Ramsar): An International Treaty Sparks Freshwater Conservation

Aboubacar Awaiss and Denis Landenbergue

Freshwater ecosystems have increasingly become a focus of conservation attention in recent years, and this has been greatly assisted by the Ramsar Convention. Also known as the Convention on Wetlands (Ramsar, Iran, 1971), this intergovernment treaty had a total of 138 contracting parties as of December 31, 2003, of which 38 were from Africa (out of a total of 53 countries). Several additional African countries have been progressing toward becoming parties to the convention, some of them with support from WWF's Living Waters Programme.

The vision of the convention, as adopted by the Conference of Contracting Parties during its seventh session (May 1999), is "to develop and maintain a network of wetlands that are of international importance due to their ecological and hydrological functions, for the conservation of worldwide biological diversity and the endurance of human life" (Ramsar 1999).

In Africa, the seemingly inherent conflict between human livelihoods and wetland conservation presents a challenge to the successful implementation of the Convention on Wetlands mission. It is imperative to devise incentives for wetland conservation that maintain the standard of living of human populations in addition to the ecological functioning of the wetland (Tiéga 1998). Often, this mission is hampered by a lack of government

Local fisher with fish traps in the flooded savanna on the northern side of Lake Chad on the edge of the Sahara Desert, Chad. Photo credit: ©WWF-Canon/Jens-Uwe Heins

guidelines and country policies and programs that acknowledge the importance of wetland biodiversity conservation.

One of the pillars of the convention is the "List of Wetlands of International Importance." As of December 31, 2003, a total of 1,328 Wetlands of International Importance (or Ramsar sites) had been registered by contracting parties, covering 111,900,000 ha worldwide. Out of this, Africa had designated 29,300,000 ha in 131 Ramsar sites.

Although the total area of wetlands designated in the Africa-Madagascar region represents a valuable contribution, there is still much more to achieve, especially in terms of conservation of existing sites and representation of wetland types in new sites. WWF's Living Waters Programme has set an objective of designating at least 250 million ha of new freshwater protected areas worldwide by 2010. This objective was endorsed as a Ramsar Convention target by the 8th Conference of the Parties held in November 2002 in Valencia, Spain.

Freshwater Ecosystems in Africa: Conserving Their Biological Diversity and Maintaining Their Crucial Role in Livelihoods

An increasing number of Ramsar contracting parties have adopted policies to limit wetland degradation and loss and to foster sustainable management. As of the Ramsar Convention in November 2002, ten African countries had established national wetlands policies, and another eight had taken initial steps toward doing so.

In Africa, wetlands also are crucial contributors to economic and sociocultural activities, and Africans generally have an acute awareness of the importance of access to fresh water (Awaiss and Saadou 1998). This is well illustrated in the framework of the "Strategy and Action Plan for the Integrated Management of Africa's Wetlands," a component of the Environment Initiative of the New Partnership for Africa's Development (NEPAD 2002).

Indeed, for many African communities residing near wetlands of local, regional, or international importance, ensured access to a diversity of biological resources is an essential condition for survival. The huge dependence of Africa's populations on these resources makes the continent particularly vulnerable to ecological deterioration (Awaiss and Seyni 1998).

In the event of reduced productivity caused by ecological deterioration, there are few available alternatives for development, and the financial resources needed to restore the environment are limited. This is why it is imperative to highlight the important link between sustainable development and preservation of biological diversity and to foster this understanding within society as a whole as well as within the political arena. Mechanisms by which sustainable development and conservation can jointly be achieved, such as the implementation of Ramsar site management plans to maintain the ecological character of each site, must be more widely adopted. Unfortunately, too few of the African sites registered on the list currently have a management plan for the aquatic resources of designated wetlands as stipulated in Article 3.1 of the convention.

The Ramsar List: A Critical Component of Integrated River or Lake Basin Management in Africa

Ramsar sites in Africa and Madagascar can limit the rate of regional biodiversity loss if innovative combinations of traditional and modern conservation methods are implemented both inside and outside sites. In this regard, WWF and Ramsar have promoted integrated river or lake basin management in Africa since 2000, with the goal of conserving and sustainably managing freshwater ecosystems and their natural resources while also contributing to livelihoods and poverty reduction.

Emphasizing Ramsar sites as crucial tools for the implementation of integrated river or lake basin management, this approach was first developed in pilot basins or ecoregions such as the Lake Chad Basin (jointly with the Lake Chad Basin Commission, its five member states, and the Global Environment Facility [GEF]), the Niger River Basin (with the Niger Basin Authority, its nine member states, and the GEF), and the Lake Malawi/Niassa/Nyasa Basin (with its three riparian countries and cooperation agencies such as the Swiss Development Cooperation). A key aspect of river or lake basin management in Africa has been the promotion of basin-wide networks of Ramsar sites and the development of networks of wetland managers based on the successful MedWet model. The MedWet Initiative, developed in 1991 for the Mediterranean Basin, is a long-term collaborative effort toward the conservation and wise use of Mediterranean wetlands. MedWet brings together all the governments of the region, several international agencies and conventions, nongovernment organizations, and wetland centers to work for wetland conservation (MedWet 2004). It is Ramsar's first regional initiative and considered to be a model for future regional initiatives elsewhere.

Whatever approach is adopted, the conservation of African Ramsar sites can succeed only with the cooperation of local populations. The convention has always emphasized that use by people, under sustainable conditions, is compatible with registration on the Ramsar List and with the preservation of wetlands in general (Ramsar 1999).

Peuhl woman washing up at sunset, Youwarou, Lake Walado-Débo, Inner Niger Delta, Mali. Photo credit: ©WWF-Canon/Meg Gawler

Conclusions

Registration of a wetland on the Ramsar List can help to maintain biological diversity in Africa if the issues of African values, priorities, and practices are taken into account in conservation plans. Until the recent past, international values, rather than national or local values, tended to dominate biological diversity conservation programs. For economic and cultural reasons, Africa's dependence on biological resources has not always received the attention it deserves, but there are signs that this is changing quickly. For example, in November 2005, the 9th Ramsar Conference of the Parties will for the first time ever take place in Africa (Kampala, Uganda) and will focus on wetlands and poverty reduction.

ESSAY 6.2

Lake Naivasha, Kenya: Community Management of a Ramsar Site

IUCN Eastern Africa Regional Programme, Nairobi, Kenya

Lake Naivasha is a rare example of a Ramsar site where the mandate to manage the site resides with the local community. Community members led the effort to develop a management plan and seek designation for the site. A voluntary association of concerned residents and commercial, agricultural, administrative, and municipal interests continues to meet regularly to discuss progress and matters arising in the implementation of the management plan. Lake Naivasha is a successful case of community-based natural resource management, and it may serve as a model for community participation in the management of other wetlands of international importance.

Lake Naivasha covers about 145 km^2 and lies on the floor of the eastern branch of the Rift Valley in East Africa, at 0°45'S and 36°21'E. At 1,887 m a.s.l., it is the highest of Kenya's Rift Valley lakes. Its freshwater is unusual among the Eastern Rift Valley lakes, most of which are sodic.

Lake Naivasha drains a basin of some 3,400 km^2, and it has two main influent rivers: the Gilgil and the Malewa. Together, these account for around 90 percent of the surface water entering the lake. Rainfall at Naivasha is around 650 mm a year, but evapotranspiration is an estimated 2,141 mm a year, and evaporation from the lake is 1,529 mm (Abiya 1996). The contribution of the lake's subsurface influents to the reduction of this deficit is significant. Intriguingly, Naivasha has no surface outlet, so its ability to flush out excess salts probably results from substantial water seepage into surrounding sediments.

Over the past 10,000 years, Lake Naivasha's water level has fluctuated widely, from drying out completely on several occasions to reaching 100 m higher than its current level. Water level fluctuations continue to affect the lake, although fluctuations of more than 5 m rarely occur. The lake is shallow, with a maximum depth of around 8 m.

Biodiversity

Lake Naivasha has a noteworthy avian fauna. An estimated 495 bird species either reside in or pass through the Naivasha area; this estimate represents one of the highest counts in Kenya. Some of Kenya's largest congregations of waterfowl also occur at the lake: between 1991 and 1997, an average of 22,000 waterbirds gathered there (Bennun and Njoroge 1999). Additionally, the lake has held 1 percent or more of three bird species' biogeographic populations: red-knobbed coot (*Fulica cristata*), African spoonbill (*Platalea alba*), and little grebe (*Tachybaptus ruficollis capensis*).

The lake's aquatic fauna is spartan. Its fishery is dominated by introduced species, including three fish species (*Oreochromis leucostictus*, *Tilapia zillii*, and *Micropterus salmoides*) and a crayfish (*Procambarus clarkii*). The lake is home to large numbers of hippopotamus (*Hippopotamus amphibious*) and an estimated fifty-five different mammal species occur in the Lake Naivasha area.

Forests and Other Vegetation

There are four main forest blocks in the lake's basin, all of which play a vital role in the basin's hydrology. Riparian vegetation includes *Acacia* woodland (*Acacia xanthophea*) and open grassland. Historically, most of the lakeshore was fringed with papyrus (*Cyperus papyrus*); currently, however, only about 12 km^2 of papyrus fringes the lake (Lopez 2002). Like its aquatic fauna, Lake Naivasha's vegetation became increasingly dominated by exotic species, particularly the free-floating aquatic fern (*Salvinia molesta*) and the notorious water hyacinth (*Eichhornia crassipes*). These formed dense and extensive floating mats on the lake. Efforts to control these weeds with introduced, host-specific weevils have been largely successful, and the plants are no longer the problem that they used to be.

Lake Naivasha's Economy

Most of the lake's riparian land is under private stewardship, in many cases by large-scale horticultural and floricultural farms. Farms cover an estimated 50 km^2 of land and yield net returns of some $63 million a year. These farms employ an estimated 30,000 people, who are attracted to the area from all over Kenya (Sayeed 2001).

Almost all of these farms are irrigated from the lake. An estimated 63.7 million m^3 of water is abstracted annually for human use. Irrigated agriculture continues to increase in the larger Naivasha Basin, as do water transfers out of the basin. Nakuru, a major town lying outside the basin to Naivasha's northwest,

Pelicans (*Pelecanus onocrotalus*) along the shores of Lake Naivasha. Photo credit: Sarah Higgins, Lake Naivasha Riparian Association

extracts 17,500 m^3 of water daily from the Turasha, a major tributary of the Malewa River.

An additional and important economic activity in the basin is geothermal power generation. The first plant at Ol Karia, close to Lake Naivasha, is Africa's first geothermal power station, producing 45 megawatts of electricity. The second plant at Ol Karia, completed recently, has a capacity of 70 megawatts, for a total capacity of 155 megawatts, or about 12.5 percent of the power flowing through Kenya's national grid.

Finally, tourism plays an important role in Naivasha, attracting day-trippers from Nairobi and foreign tourists keen to enjoy the lake's spectacular setting, visit the parks in the basin, or birdwatch around the lakeshores.

Key Threats

Almost all of the problems that threaten Lake Naivasha relate to unplanned development in the basin and the introduction of exotic species. These threats may be summarized as follows:

- *Water abstractions.* Abstractions from Lake Naivasha's tributaries and the lake itself are cause for concern, given the continued high rate of development in the irrigated agricultural sector. Also of concern are abstractions to supply the needs of growing numbers of migrant workers around the lake, of general population growth in the basin, and of an expanding industry.

- *Pollution.* The main sources of pollution are agricultural runoff and organic waste from growing migrant worker settlements. Pollution is a serious threat, given the loss of much of the lake's papyrus, which has been cleared to make room for agriculture and settlement. Previously, papyrus played a key role in filtering water entering the lake.

- *Catchment degradation.* Most catchment degradation originates from poor farming practices, including intensive land use and farming on riverbanks and steep slopes. Additional degradation arises from mining riverbanks for sand and from forest clearance for agriculture, timber, and charcoal manufacture.

- *Ecological changes caused by species introductions and overfishing.* The ecology of the lake continues to change as a result of species introductions and intense fishing pressure. For example, intense fishing of exotic black bass (*Micropterus salmoides*) has caused populations of exotic crayfish, a prey item of bass, to explode. Crayfish foraging, in turn, has decimated submerged macrophyte vegetation beds, the main food of the red-knobbed coot (*Fulica cristata*), such that this species now rarely occurs in large congregations. The red-knobbed coot features highly in the diet of Naivasha's fish eagles (*Haliaeetus vocifer*), which used to occur here in some of the highest densities on the African continent. With declining coot and fish populations, the undernourished fish eagles have failed to breed successfully on the lake for several years.

The Management of Lake Naivasha

Since 1929, the Lake Naivasha Riparian Association (LNRA) has managed the riparian land lying between Lake Naivasha's low and high water marks. The lake's management has traditionally fallen within the jurisdiction of multiple government agencies, with no clear leader among them and no clear management strategy for the lake.

Aware of the increasingly acute environmental problems facing Lake Naivasha, the LNRA has sought to address them via a two-pronged approach. First, the LNRA designed a management plan for the lake and its basin. Completed in 1996, the plan is being implemented by the Lake Naivasha Management Implementation Committee (LNMIC). On October 1, 2004, the Lake Naivasha Management Committee was officially gazetted by the Kenyan Minister for Environment and Natural Resources and charged with implementing the Lake Naivasha Management Plan. The committee includes representatives from the LNRA, government ministries and departments, the Naivasha Municipality, the local district authority, the geothermal power generation industry, the lake's fishers, and IUCN. There are also ten subcommittees, each representing a sector of managerial concern. Among the largest of these is the Lake Naivasha Growers' Group, which represents the lake's horticultural and other farming interests. The other subcommittees are composed of those representing the livestock industry, biodiversity conservation, the Naivasha Municipality, power production, fisheries, and tourism, among others.

The Lake Naivasha Management Plan provides an initial framework describing the lake's main threats and management objectives, and outlining the broad parameters within which management will operate. The plan's main operational guidelines are contained in a series of codes of conduct, developed by each of the LNMIC's subcommittees, which are responsible for implementing them and for dealing with any violators. In the event that a subcommittee faces challenges too large or complex for it to solve alone, the difficulty can be referred to the LNMIC for action.

The LNMIC encourages each subcommittee to solve its problems independently and to exercise as much autonomy as possible. The LNMIC has no legal policing powers, and much of the success of the management plan has been obtained through voluntary compliance.

The second part of the LNRA's strategy has involved gain-

Community members' sign showing their support for the Lake Naivasha Ramsar site. Photo credit: Sarah Higgins, Lake Naivasha Riparian Association

ing alternative political support for implementing its eventual management plan and counterbalancing possible resistance from the state to community-based natural resource management. In 1993, as the Kenyan government sought to increase the number of Kenyan Ramsar sites, it occurred to the LNRA that some of the political strength it needed might be obtained via Ramsar. With technical help from IUCN and other agencies, the LNRA persuaded the Kenyan government to nominate Lake Naivasha as Kenya's second Ramsar site in 1995. Doing so placed Lake Naivasha's problems and the efforts made to deal with them on an international stage, so that any local resistance to the LNRA's activities would be scrutinized globally. In this way, Ramsar helped to overcome local administrative and bureaucratic objections to the LNRA's activities. At the same time, the Ramsar designation could yield important commercial dividends: flowers and other horticultural produce from the area could be advertised as having been grown on a Ramsar site, augmenting the environmental friendliness of the product. Finally, the designation was a source of pride to many of the lakeshore's inhabitants and helped them to maintain their energy and momentum in managing the lake.

The Ramsar designation has not solved all problems facing the lake's management. However, recent changes in Kenyan legislation should improve LNMIC's ability to deal with problems that still persist. Section 42 of the new Environmental Management and Co-ordination Act deals with lakes and rivers and provides the Minister for the Environment with a series of sweeping powers that include enacting any measure deemed necessary for the adequate management of lake and river resources. Through this act, the LNRA hopes to be recognized as a management authority for Lake Naivasha.

To date, much of the success of the LNRA and LNMIC is derived from a series of key components in the management process:

- *Consensus and dialogue.* These lie at the heart of the management process and provide stakeholders with ample opportunities to vent their concerns, contribute to the process, and together find solutions to the challenges they face.
- *High transaction costs.* The costs related to arriving at decisions are high, particularly in terms of time, discussion, and commitment. However, the dividends are significant. Once unanimous agreement has been reached, cohesion among stakeholders is maintained and conflict minimized.
- *Adequate management forums.* The LNMIC meets every 6 weeks and can call special general meetings to discuss matters of importance if necessary. Consensus and dialogue cannot occur in the absence of adequate forums designed for these purposes.
- *Adaptive or dynamic management.* The Lake Naivasha management plan is dynamic in that it is periodically reviewed and updated to meet challenges as they arise. This is particularly the case with the plan's sectoral codes of conduct.
- *Representation and equity.* The management process seeks to ensure that all stakeholders are represented and are given the opportunity to contribute meaningfully. Each member of the LNRA has one vote, regardless of size or wealth of the member's organization or company. In addition, subscriptions are kept as low as possible to ensure that impoverished stakeholders are not excluded from the process.
- *Awareness and wealth.* The generally high standards of living of riparian residents and their awareness of environmental issues have contributed substantially to the success of the LNRA and the LNMIC.
- *Leadership and commitment.* The management process benefits from strong and inspired leadership and a very high degree of commitment and initiative among committee members and a majority of riparian residents.

Among the most serious problems that the LNMIC has faced are the deep-rooted suspicions held by commercial horticultural interests of efforts designed to conserve the environment and promote its wise use. When the LNRA was seeking Ramsar designation for the lake in 1994, many growers erroneously thought that the designation would pave the way for the government to declare the lake a protected area and prevent all water abstraction from it. Others thought it would mean that the Ramsar Bureau, based in Switzerland, would dictate what stakeholders were allowed to do in the basin. Although the majority of voters were in favor of Ramsar designation, consensus was necessary. Detractors were given 3 months to investigate Ramsar to their satisfaction. In the meantime, a radio report alleging extensive environmental damage to Lake Naivasha by the flower industry caused several flower export orders to be canceled. Seen in this light, the Ramsar designation suddenly seemed attractive because it implied that agricultural production in the basin could be environmentally friendly. Consensus was subsequently obtained. Similarly acute suspicions have plagued the development of the management plan. In every case, through a process of extraordinary patience, tact, and persuasion, consensus has been achieved and the management process carried forward.

An additional problem faced by the LNMIC is limited funding. Most funding comes from the judicious investment of contributions made by LNRA membership and small grants from nongovernment organizations and other international organizations. In 2002, for example, one of Naivasha's most important British flower purchasers, a major supermarket chain, partially funded an aerial survey of the lake and its riparian areas.

The outcome of the LNMIC's style of management has been impressive. Many of the lake's large-scale farmers have introduced more efficient irrigation systems, installed water meters so that abstractions can be monitored, submitted themselves to environmental audits, and implemented pesticide and fertilizer application practices that exceed international standards. The LNMIC's subcommittees actively seek to protect basin forests and the lake's fishery, provide environmental education and awareness, reduce the amount of sewage entering the lake, monitor all new developments on the lake's shores, and ensure that they are subjected to an environmental impact assessment. These achievements have all been the result of voluntary actions. In recognition of its successes, the LNRA was awarded the Ramsar Wetland Conservation Award in the nongovernment organization category in 1999 at the 7th Conference of Parties in Costa Rica. The LNRA chair gave a keynote address on the Lake Naivasha experience to this international audience.

The challenges facing the LNMIC remain substantial, particularly as it seeks to expand its activities into the Naivasha Basin. However, Lake Naivasha today is an outstanding wetland, rich in biodiversity and lying in the heart of intense economic activity. It is part of one of Africa's most successful community-based natural resource management systems.

ESSAY 6.3

African Wildlife Foundation Experience in the Management of Fishery Resources in Two Southern African Landscapes

Henry Mwima and Jimmiel Mandima

The African Wildlife Foundation (AWF) is the leading international conservation organization focused solely on the African continent. For more than 40 years, AWF has concentrated its efforts on building the capacity of Africa's people and institutions to manage natural resources and to protect the unique and rich biodiversity of the African continent. From the day AWF was founded in 1961, it has recognized that Africa's wildlife resources and ecosystems are key to the future prosperity of Africa and its people. Over the past 5 years AWF has established and built the African Heartlands Program, an integrated approach to conservation and development in selected large, wildlife-rich landscapes, or Heartlands, that offer both ecological and economic viability for the long term.

Heartlands are cohesive conservation landscapes that are biologically important and cover areas large enough to maintain healthy populations of wild species and natural processes well into the future. AWF currently works in eight Heartlands covering parts of eleven countries in central, eastern, and southern Africa (Botswana, Democratic Republic of the Congo, Kenya, Mozambique, Namibia, Rwanda, South Africa, Tanzania, Uganda, Zambia, and Zimbabwe; figure 6.4). Each Heartland forms a sizable economic unit in which tourism and other natural resource–based activities can contribute significantly to local livelihoods. Most Heartlands include a combination of government lands (such as national parks), community-owned lands, and private property. In these vast conservation landscapes, which often cross national boundaries, AWF works with a broad range of local partners to improve natural resource management and conservation practices and to mitigate threats to valuable resources. The number of Heartlands is expected to increase with time and resources to encompass other geopolitical areas and ecosystems of Africa (Muruthi 2004).

As the Heartland Program has developed, aquatic resources—rivers and wetlands and the species that depend on them (primarily fish)—have taken center stage as priority conservation targets[1] in several of our Heartlands. The Four Corners Transboundary Natural Resource Management (TBNRM) area[2] and the Zambezi Heartland are particularly rich in water-related conservation targets as they span large sections of the Zambezi River Basin. Waters of the Zambezi, Chobe, Kwando-Linyati system, Kafue, Okavango Delta, and Luangwa support a thriving tourism industry; commercial, subsistence, and recreational fisheries; and irrigation of commercial crops. All these activities have direct and indirect effects on the health of aquatic resources in the area. On the basis of extensive consultations with stakeholders, AWF and partners are working to identify and mitigate threats to water resources in the Zambezi Basin and to develop best practices that will enable local people to sustainably use and benefit from these resources.

This essay describes how AWF has applied its Heartland Conservation Process (HCP)[3] in identifying conservation targets and related threats in these two landscapes and discusses experience gained in the management of fishery resources. A brief description of the two landscapes follows.

Four Corners

The Four Corners TBNRM area covers approximately 220,000 km². including the eastern Caprivi Strip in Namibia, Ngamiland in Botswana, Hwange District in Zimbabwe, and parts of Southern and Western Provinces in Zambia (figure 6.5). The Zambezi River is the major drainage system and forms the core of the Four Corners TBNRM area ecosystem. The area extends along the Zambezi River from about 50 km below Victoria Falls upstream to the Chobe-Linyanti floodplains and parts of tributaries such as the Kwando and Machili. National parks and wildlife reserves in the area include Chobe and Moremi in Botswana; Mamili, Mudumo, and Bwabwata in Namibia; Mosi-Oa-Tunya and Sioma Ngwezi in Zambia; and Hwange and Zambezi in Zimbabwe. National parks and other protected areas (safari areas, game management areas, forest reserves, conservancies, and Moremi Wildlife Reserve) constitute about 50 percent of the total area. The Four Corners TBNRM area is a prime wildlife and tourism area and contains some of the most important terrestrial and freshwater ecosystems in Africa (such as Okavango Delta and the Victoria Falls; see figure 6.5).

Zambezi Heartland

The Zambezi Heartland extends from Lake Kariba to Cahora Bassa Reservoir and covers an area of approximately 39,000 km², consisting of 6,500 km² of national parkland, 4,900 km²

FIGURE 6.4. African Wildlife Foundation Heartlands (as of January 2004). Map produced by AWF Spatial Analysis Laboratory.

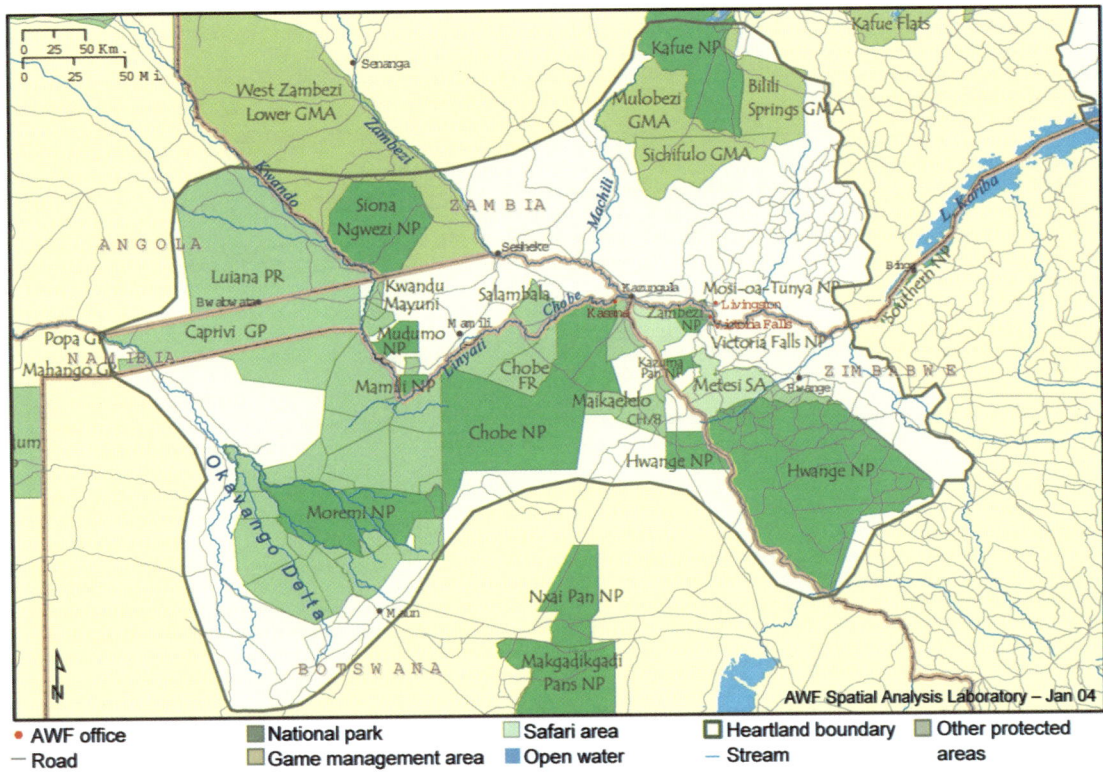

FIGURE 6.5. The Four Corners Transboundary Natural Resource Management Area.

in game management areas, 11,000 km² in safari areas, and the remainder in open communal areas (figure 6.6). In addition to the high diversity of terrestrial wildlife and plants in this Heartland, the Zambezi River and its tributaries also provide habitat for freshwater fish, including many species important for commercial and subsistence fisheries such as the tigerfish (*Hydrocynus vittatus*) and a wide variety of cichlid (tilapias) and cyprinid species.

Heartland Conservation Planning

In April 2000 and June 2001, AWF held planning meetings for the Zambezi Heartland and Four Corners TBNRM area, respectively. These meetings brought together local people, wildlife agencies, representatives of the private sector, and nongovernment organizations who identified conservation targets, threats to these targets, and related threat abatement strategies. AWF considers this a crucial step for successful conservation work in every Heartland (AWF 2003a, 2003b).

Conservation Targets

AWF and partners identified the following conservation targets for both the Four Corners TBNRM area and the Zambezi Heartland: river systems, wetlands, wildlife corridors, habitat complexes, native fishes, endangered and threatened species, and species assemblages. The aquatic conservation targets are described in this section.

River Systems

The key conservation target is the Zambezi River and its tributaries. The Zambezi is one of Africa's great rivers, with a catchment area of 1.42 million km² (Griffin et al. 1999). Along its entire stretch, the river is naturally divided into three sections: the Upper Zambezi [16, 76] from the source to Victoria Falls, the Middle Zambezi downstream from Victoria Falls to Cahora Bassa in Mozambique [69], and the Lower Zambezi from Cahora Bassa to the Delta at the Indian Ocean in Mozambique [66]. The Zambezi River's natural flood regime is the driving ecological process supporting pervasive wetland and riparian habitats throughout the Four Corners TBNRM area.

The Upper Zambezi is separated from the lower sections by a natural barrier, Victoria Falls, so fish species composition is different from that in the lower sections of the river. The Middle Zambezi has been largely modified by the creation of artificial reservoirs at Kariba, Kafue, and Cahora Bassa gorges. The dams created permanent artificial barriers, and the resultant river fragmentation prohibits migration of fishes along the river channel (World Resources Institute 2003). These dams also alter the flood

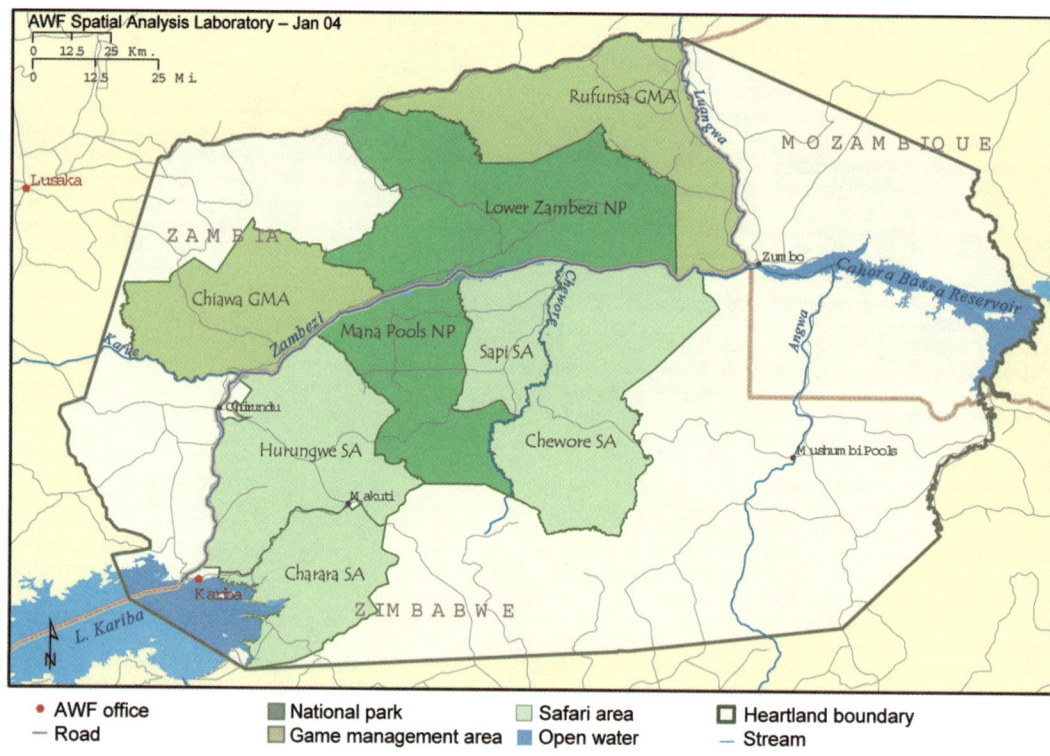

FIGURE 6.6. The Zambezi Heartland.

regime downstream in the Lower Zambezi, thereby influencing the viability of aquatic fauna and flora that use the fresh-brackish-seawater habitat continuum. AWF and partners identified the following threats for river systems: altered flow regime, habitat loss, siltation, eutrophication, and point source pollution.

Wetlands

Chobe/Kwando/Linyanti/Liambezi, Zambezi Floodplain (East Caprivi), Okavango Delta, Kazungula floodplains, Kazuma depressions, and dambos and pans constitute important habitats for aquatic and terrestrial biodiversity. These systems are critical for aquatic biodiversity, providing specialized fish breeding and feeding grounds, and as watering points for wildlife in the entire landscape. Some of these vleis, pans, and swamps are the only habitats for aquatic species such as the lungfish (*Protopterus annectens brieni*) that have limited distributions in southern Africa and are remnants of an ancient lineage of fishes related to the coelacanths (Skelton 2001).

AWF and partners identified a falling water table and habitat degradation and destruction as threats for wetlands. These threats are caused by a combination of factors such as dam operations, incompatible human settlements, deforestation, uneven elephant populations, incompatible tourism developments, and incompatible grazing practices.

Native Fishes

Fish are important in the region both for their contribution to biological diversity and for their role in commercial, sport, and subsistence fishing. A large portion of the freshwater fish catch for the basin's riparian countries comes from the Zambezi River and its tributaries, and fish provide a significant percentage of protein to the Zambezi Basin states. Zambezi fisheries also contribute substantially to the national economies of Botswana, Malawi, Mozambique, Zambia, and Zimbabwe. For example, the Barotse Floodplain fisheries in the Upper Zambezi support about 300,000 people (Chenje 2000). The commercial offshore pelagic fishery for the freshwater sardine (*Limnothrissa miodon*, locally called *kapenta*) on Lake Kariba yielded 30,000 tons of fish worth US$55 million in 1993 (Chenje 2000). Demand for fish has continued to rise as a result of human population growth, contributing to declines in some fish populations and altered species composition and structure.

There is subsistence fishing along the entire Zambezi River, from the Upper Zambezi to the Delta. Most fishers are from riverbank communities that undertake fishing as a livelihood in addition to traditional subsistence agriculture. In the Middle Zambezi, subsistence fishing is prevalent along the Kafue, Luangwa, and Zambezi rivers on the Zambian and Mozambican sides and to a lesser degree on the Zimbabwean side. Recreational fishing is increasing along the river, with a concurrent increase in tourist lodges and related facilities.

The Zambezi River system is endowed with a rich fish fauna, although the fauna is incompletely known. A basin-wide inventory of fishes of the Zambezi River and its tributaries (excluding Lake Malawi) totals 239 species (Skelton 2001). Recent surveys conducted by AWF and partners (the Aquatic Resources Working Group [ARWG] and the South African Institute of Aquatic Biodiversity [SAIAB]) suggest that there are about twenty additional species to be described, mainly from the headwater streams in northwestern Zambia. Research on these species is ongoing (D. Tweddle, pers. comm., 2004).

Threats to native fishes include river regulation and water withdrawals, poor land management, and water pollution. Globally, 20 percent of all freshwater fish species are threatened or endangered because of dams and water withdrawals that have destroyed or degraded free-flowing river ecosystems (Ricciardi and Rasmussen 1999). The mighty Zambezi River is no exception, with two of southern Africa's largest hydrological schemes: Kariba and Cahora Bassa (figure 6.6). Further threats to fishery resources include poor land management that causes erosion and deposition of silt in rivers and streams and can destroy breeding grounds. Water pollution from agricultural activities and urban settlements causes eutrophication, which often leads to the proliferation of invasive weeds and deoxygenation.

Intervention Strategies

Based on the conservation targets and threats identified in this essay, AWF and partners decided to focus its first intervention strategies on improving the management and condition of the native fisheries in the Four Corners and Zambezi Heartland (AWF 2000, 2001). Two intervention strategies have proven beneficial to collaborative management of fishery resources: establishment of the ARWG in the Four Corners TBNRM area and the creation of multi-institutional partnerships in both Heartlands. These interventions are discussed in this section.

Aquatic Resources Working Group

As part of the Four Corners TBNRM Initiative, AWF undertook a consultative process through which we identified partners and mechanisms for supporting projects aimed at joint and improved management of fisheries and strengthened collaboration between freshwater fish scientists, government officials, and conservation organizations. This culminated in the establishment of the ARWG. Subsequently, the Southern African Development Community (SADC) formally recognized the ARWG as the working group to address management of shared fishery resources in the Four Corners TBNRM area.

ARWG membership consists of representatives of fishery authorities from Botswana, Namibia, Zambia, and Zimbabwe and a representative of the SADC Inland Fisheries Sector Technical Coordinating Unit and the SAIAB. The group members also sit on the SADC Inland Fisheries Technical Committee, which plans and implements strategies for regional cooperation in the management of shared aquatic resources. The ARWG is composed of seasoned and experienced fishery biologists and ecologists, who have more than 150 years of combined experience in fishery biology and ecological monitoring.

ARWG's primary goal is to promote collaboration in the management of shared fishery resources and information exchange in the region. To this effect, AWF and its partners are working toward standardization of monitoring systems, surveys of fish biodiversity and related socioeconomic conditions, and establishment of a geographic information system database. When fully under way the ARWG will contribute significantly to a more efficient planning of conservation interventions in the Four Corners Heartland (see essay 3.3).

Multi-Institutional Partnerships

In recognition of the pool of expertise available in the subregion, AWF sought to work with institutions that would extend and complement its conservation activities. AWF has established multi-institutional technical teams to implement its work on monitoring water resources in the Zambezi Heartland. Key partners include the following:

- Lake Kariba Research Station (ULKRS), University of Zimbabwe
- Lake Kariba Fisheries Research Institute (LKFRI), Zimbabwe Parks and Wildlife Management Authority
- Zambia Department of Fisheries (DOF)
- Environmental Council of Zambia (ECZ)
- Zambezi River Authority (ZRA)
- Tchuma Tchato Community-Based Natural Resource Management (CBNRM) Program Staff in Mozambique

AWF has been coordinating the field activities of these institutions to enhance transboundary management of aquatic resources in the Zambezi Heartland. This strategy facilitated collaboration between institutions whose previous management activities were driven by national agendas. Because the impacts of many management decisions in the Zambezi basin transcend national boundaries, the multi-institutional management of these shared waters offers the best opportunity to improve the viability of its resources for the benefit of wildlife and the people of the region. This strategy has proven to be useful in bringing people with diverse expertise from different countries together to conserve shared fishery resources. Furthermore, smaller groups have emerged and are interacting in generating information necessary for formulating joint freshwater resource management plans. For example, fishery ecologists from ULKRS and DOF teamed up with AWF in August 2003, to do the following:

- Assess the diversity and relative abundance of fish species in different habitats in the vicinity of the confluence of the Zambezi and Luangwa rivers
- Measure key water quality parameters to assess the nutrient status of the aquatic systems in the vicinity of the confluence of the Zambezi and Luangwa rivers

These surveys improve the understanding of the distribution and abundance of species in the Zambezi ecosystem and contribute to more targeted management plans, as will water quality results when they become available. The pooled resources and expertise provided by the international collaboration extended the scope of the surveys, and the sharing of results will extend their impact across borders. Such international collaboration represents a vital step toward transboundary conservation management.

Acknowledgments

The U.S. Agency for International Development Regional Center for Southern Africa, the Netherlands, Directorate-General for International Cooperation, and the Ford Foundation are acknowledged for their financial support to the activities addressed in this essay. The work described in this essay would not have been possible without the efforts of our partners from ARWG, SAIAB, ULKRS, LKFRI, DOF, ECZ, ZRA, and Tchuma Tchato CBNRM Program. We would also like to thank our AWF colleagues, Joanna Elliot, Helen Gichohi, Harry van der Linde, Philip Muruthi, Elodie Sampéré, and David Williams, for their insights and invaluable remarks in the preparation of this essay.

Notes

1. Conservation targets are elements of biodiversity at a site and the natural processes that maintain them. These are the focus of Heartland planning around which strategies are developed. The intent of target identification is to develop a short, effective list of species, communities, or large-scale ecological systems whose protection will capture all the biodiversity at the site (AWF 2003b).

2. The area is called Four Corners TBNRM area because AWF is currently implementing a regional transboundary program defined largely by USAID. Revision of the Heartland boundary is anticipated considering ecological (including key species) and administrative factors.

3. HCP is a customized science-driven participatory conservation planning, implementation, and monitoring process developed by AWF with help from The Nature Conservancy.

ESSAY 6.4

Fish out of Water? Competing for Water for Africa's Freshwater Ecosystems

Patrick Dugan

The present volume highlights the multiple values of Africa's aquatic ecosystems and the benefits they bring at local, national, and international levels. Béné (essay 1.1), in particular, has highlighted the central role of inland fisheries in supporting livelihoods, ranging from those who catch the fish to fish processors and traders. Given their contribution to African livelihoods and their dependence on the quality of the continent's aquatic ecosystems, Africa's inland fisheries provide an important barometer of the state of these ecosystems and the lives of the people who depend on them.

Other essays in this volume also highlight the threats facing Africa's aquatic ecosystems, including physical encroachment and loss, pollution, and overharvesting of resources (e.g., essays 4.1 and 4.2). However, by far the largest threat comes from the changes in land and water management that alter the hydrological dynamics that have driven the seasonal productivity of these ecosystems for millennia. As we look to the future and consider what is needed to sustain Africa's aquatic ecosystems, the single greatest challenge is to ensure that the quality and quantity of water needed to sustain this productivity are maintained. This essay considers some of the policy and management issues that must be addressed if we are to meet this challenge successfully and the importance of science as a basis for informed debate.

Sustaining Fisheries and Ecosystems in the Face of Growing Demand for Water

A first step toward providing water for aquatic ecosystems and the people who depend on them lies in recognizing the scale of the growing competition for water. Although there are already many examples of African rivers, such as the Senegal, Niger, and Zambezi, where the flow regime has been altered significantly by dams and irrigation schemes and the associated riverine floodplains and other wetlands have been degraded, the coming decade is likely to see a substantially increased investment to harness the continent's water resources for agricultural, industrial, and urban purposes. For example, Rosegrant et al. (2002) report that even under an optimistic scenario of "sustainable" water use, total mean water withdrawal in sub-Saharan Africa will rise from 128 km^3 in 1995 to 173 km^3 in 2025. Under a crisis scenario withdrawal is projected to be 247 km^3. In addition, there is continuing demand from national governments and the private sector for greater investment in the hydropower potential of Africa's rivers. How much of this potential will be realized, and how much will remain untapped, will be an issue of intense debate in coming years.

In the face of this growing demand for greater investment to harness the multiple benefits of Africa's freshwater, there is an urgent and growing need for the conservation movement to embrace a much more ambitious engagement in water management issues. This engagement should be rooted in recognition of Africa's development needs and in a willingness to join the growing policy debate over how best to use the continent's water resources for the long-term benefit of national and local economies. In this section I review the three major issues that will be central to a constructive debate: governance, valuation of goods and services, and water needs of aquatic ecosystems.

Governance and Institutions

A central premise of the conservation case for aquatic ecosystems is that many hundreds of thousands of people will benefit from investments in water management that sustain ecosystems and the livelihoods that depend on them. However, to translate this premise into policies and management practices that sustain water flow, effective systems of governance at the local and basin level are essential. These systems of governance should foster more effective engagement of all stakeholders and equitable sharing of the benefits from the water resources and the aquatic ecosystems under their jurisdiction. In simple terms this means that the people who are most closely dependent on these resources need to be better engaged in land and water use decisions at all scales.

At present, however, effective governance over aquatic resources is the exception rather than the norm in most developing countries, and all too often the majority of society is excluded from any involvement in policy-making. As a result, policy decisions often favor certain powerful sectors rather than the wider society and, in particular, the poor. This is especially true when the poor are located far from urban centers, as is the case for many of Africa's rural communities dependent on aquatic resources. If access by the poor to aquatic resources is to be improved and management of these resources is to be sustainable, then major reform of aquatic resource governance, policies, and institutions is needed. Such efforts to improve policies and systems of governance and to strengthen institutions must be grounded in a better understanding of how these policy-making processes function, how responsibilities for managing aquatic resources can be shared between government and community organizations, how different stakeholder groups in society affect policy-making and implementation, and how improved information can result in decisions that benefit the poor (Béné and Neiland 2004).

In conjunction with these changes in governance, institutions, and policies, information systems that support effective governance of aquatic ecosystems (and other natural resources) must be developed. The conventional view of policy-making and implementation assumes that policymakers will use new information and better understanding to improve policies for the benefit of society. In this situation, researchers provide information for policymakers, who make policy decisions and then hand these decisions down to administrators (managers) for implementation through various management arrangements. However, in many developing countries, policy-making and implementation systems do not function in this way. Instead, many decisions are made to favor certain powerful sectors of society rather than society as a whole. This problem is compounded by the fact that much current information about poor people's livelihoods and natural resource management issues tends to be disseminated within limited networks. At present, technical information gathering and dissemination is mainly in print, often in English, and usually packaged for presentation to a fairly well defined audience. In contrast, most poor people tend to share knowledge through local language text and oral and visual communication systems. As a result, natural resource users often are prevented from participating in technical information networks. More flexible, decentralized systems of information exchange are needed (Dugan et al. 2002).

Valuation of Ecosystem Goods and Services

The major investments needed to develop and implement improved policies, institutions, and governance systems for aquatic ecosystems and fisheries will occur only when stakeholders are better aware of the value of aquatic ecosystems and their resources. However, only rarely is the information needed to build such awareness available and used to influence policy change. Rather, across much of Africa existing data are fragmentary, dispersed, and dated. Even for fisheries, which are generally the best-documented aquatic resource, there is widespread skepticism about the accuracy and relevance of current statistics. Most of these statistics are collected from a small number of monitored landing sites, an approach of limited value in assessing the importance of river fisheries in the tropics (Van Zalinge et al. 2000). Actual catches of many freshwater fisheries are believed to be at least twice the reported figures (FAO 1999; Welcomme 2001). Therefore, there is an urgent need to improve the quality of information available on aquatic resource use by various communities and social groups, the economic and social values of these resources, and their contribution to sustaining or enhancing livelihoods, reducing poverty, and improving food security, in addition to the potential cost to society of the loss or degradation of these systems.

The value of aquatic ecosystems varies widely between ecosystem types, depending on the biological characteristics of individual sites and the ways people use them (Neiland et al 2004). Although it can often be helpful to draw on information from several different river systems when trying to illustrate the importance of Africa's aquatic ecosystems and their resources, the information needed to improve policy and management in individual rivers must be drawn from the river system under consideration. Therefore, there is an urgent need for much greater investment in efforts to assess the value of Africa's aquatic ecosystems, wherever such information will assist in improving governance and the quality of decision-making about aquatic ecosystems and water use. At the same time, greater capacity to complete such analyses must be developed.

Managing Water for Aquatic Ecosystems

Although much effort is needed to build awareness of the value of Africa's aquatic ecosystems, these efforts can build on the growing international profile that aquatic ecosystems and their resources have received over the past few years. In particular, processes such as the World Water Vision, the Global Water Dialogue, and the World Commission on Dams (WCD) have increased awareness of the need for new approaches to managing water at the basin level so that benefits from natural ecosystems can be sustained. For example, Guidelines 15 and 16 of the WCD call for "Environmental Flow Assessments" and "Maintaining Productive Fisheries" and specify the need to assess water needs for fish populations. However, if the international awareness and policy frameworks generated by these and other initiatives are to bring sustained benefits to poor communities dependent on Africa's aquatic ecosystems, then they must lead to water management decisions at local, national, and regional levels that take account of the needs of aquatic ecosystems. This effort will include detailed information on the value of specific ecosystems and on the volumes and distribution of water needed to sustain these ecosystems and different levels of ecosystem benefits.

In practice, few aquatic ecosystems exist under natural hydrological conditions, and many are subject to highly modified flow regimes. Many systems continue to provide a range of goods and services to society, whereas the character of others has been so altered that previous uses are no longer sustained and serious health and other effects have been incurred. Thus, in the face of increasing competition for water, there is a critical need to be able to assess how ecosystems respond to changes in quantity, distribution, and quality of water and the relationship between changes in the flow regime and the level of benefits that they yield. Once the relationships between river flow and benefits of an aquatic ecosystem are established, the impacts of various management strategies can be assessed. Such information can then be used at a local or national level to inform decisions on the allocation of water from an ecosystem in such a manner as to optimize the overall benefit to society (Dugan et al. 2002).

As argued earlier, inland fisheries with their central role in sustaining food security and the livelihoods of millions of poor people across the developing world are one of the most important and visible benefits of natural aquatic ecosystems. Yet for most rivers little information is available on the water management regime needed to sustain the fishery and its benefits in the face of increasing demand and competition for water. Therefore, there is a particularly urgent need to develop methods to assess the impact of changes in flow regime on fish populations, fishery productivity, and fishing communities; to use these methods to provide such information for selected rivers; and to strengthen the capacity of local, national, and regional institutions to use these tools in making water allocation and river basin management decisions that improve food security and livelihoods of fishing-dependent communities.

To help address this need, a recent study (Arthington et al. 2004) reviewed existing environmental flow assessment methods and recent advances in modeling fish production in river fisheries. The results highlight the advantages of the Downstream Response to Imposed Flow Transformations method (Brown and King 2000; Tharme 2000), which relates the flow ecosystem response with the economic and social values of the aquatic resources provided. However, this method has been applied only in small river systems with limited fisheries in southern Africa and Australia. It remains to be tested and expanded for use in larger and more complex floodplain river systems and adjusted to incorporate recent advances in modeling of fish population dynamics and their responses to changes in river flow (Arthington et al. 2004). Channels for delivering the results of flow assessments to poor stakeholders through decentralized institutional arrangements also must be identified and developed. The relevance of the information to stakeholders, and the efficacy with which it is transferred from the research arena to local communities and authorities, will dictate its ultimate influence on and value to the development agenda.

Building Partnerships

A major investment is needed if the institutions responsible for sustaining Africa's aquatic ecosystems are to emerge and be supported with the information needed for effective decision-making. In addressing this challenge it is important to build effective partnerships with other stakeholders concerned with the management of Africa's water resources. Although historically there has been little effective dialogue between the conservation community and those concerned with water management for agricultural, industrial, and urban uses, the scale of the growing water crisis provides a climate in which new approaches and partnerships are emerging. As a recent study by the International Food Policy Research Institute has emphasized, "business as usual" will lead to steady growth in environmental degradation, food insecurity, and a long-term water crisis (Rosegrant et al. 2002). New approaches to water and land management that can resolve these problems must be developed.

In this context, those concerned with the conservation and use of aquatic ecosystems have much in common with many other water users. For example, more equitable and effective use of irrigation water for agriculture entails a reassessment of water rights and devolution of water management roles, responsibilities, and resources to the local level. The debate over where and how to achieve this reassessment of water rights in agriculture has much in common with that concerning the development of governance structures for rivers and aquatic ecosystems. Conservationists will gain much by engaging in partnerships with those who are seeking a policy environment in which the water rights of poor farmers, poor fishers, and others who use aquatic ecosystems can be upheld.

Similarly, there is growing recognition that agriculture and water productivity can be improved by greater investment in rain-fed agriculture rather than the conventional approach of expanding or intensifying irrigation. The potential benefits of rain-fed agriculture include increased on-farm productivity, reduced encroachment on marginal lands, reduced erosion, and greater sustainability of hydrological flows.

At present, many of these benefits remain theoretical, and achieving them will necessitate a sustained investment. However, it is clear that there is much to be gained from investing in dialogue between the agricultural and environmental communities in the field of water management. The Dialogue in Water, Food, and Environment provides an international forum for fostering such partnerships, but this type of dialogue must be replicated and implemented at the local level. The conservation community and the agriculture sector need to seize this opportunity.

Conclusions

Among the challenges being faced by Africa's aquatic ecosystems, competition for the water that sustains them is the most critical. As argued here, a major investment is needed to address this challenge, focusing on the development of governance systems that take into account the needs of people whose livelihoods depend on these ecosystems. These governance systems must be informed by high-quality, locally relevant information on the value and water needs of aquatic ecosystems.

In addressing these challenges there is much room for partnerships with other interest groups concerned with water management, particularly those engaged in water management for agriculture. The case for conservation should be based on the best possible data on aquatic ecosystems, particularly on their values, uses, and flow needs. As demand for water grows and the need for hard decisions increases, the case for conservation of Africa's aquatic ecosystems must be rooted in the best possible science.

ESSAY 6.5

Freshwater Ornamental Fishes: A Rural Livelihood Option for Africa?

Randall E. Brummett

The impressive aquatic biodiversity of Africa documented in this volume is at risk of extirpation, if not extinction, in large parts of its current distribution. Unlike in the marine realm, where overfishing poses a dominant threat, the largest threats to freshwater fishes in most parts of Africa are competition for water, habitat loss, and attendant changes in hydrographic regimes. Freshwater habitats continue to be degraded as forests are cleared, swamps drained, and streams and lakes silted by unsustainable agriculture or polluted by rapid urbanization and unregulated industry. Focal areas of concern are those where fish biodiversity is the highest: the rainforest rivers of Central and West Africa (Upper and Lower Guinean Ichthyological Provinces and the Congo Basin) and the East African lakes Albert, Edward, Kivu, Malawi, Rukwa, Tanganyika, Turkana, and Victoria (including a number of small satellites). Without substantial changes to local strategies for natural resource management, these biodiversity assets will continue to be lost.

It is hypothesized that the fish fauna of West and Central African rivers is derived from an older and more widely distributed fauna that inhabited the continent at least since the Miocene (25 m.y.a.) and possibly much earlier (Reid 1996). In these rivers live an estimated 1,000 species, between 50 and 80 percent of which are thought to be endemic (Roberts 1975; Lowe-McConnell 1987). Major families are the Mormyridae, Cyprinidae, Alestiidae/Citharinidae (old Characidae), Aplocheilidae (old Cyprinodontidae), and the Siluridae and Mochokidae (Lowe-McConnell 1987; Lévêque and Paugy 1999). The lakes of East Africa are young, the oldest being Lake Tanganyika at 6–10 million years (Lévêque 1997). In these lakes live at least 1,500 fish species, of which the Cichlidae represent some 90 percent, often in the form of endemic, sympatric species flocks of closely related forms (Goldschmidt 1996).

Overexploitation, particularly with the use of chemical poisons, has been increasing in recent years, but West and Central African freshwater ecosystems are most threatened by deforestation. Deforestation results in sedimentation and serious water quality changes (Kamdem Toham and Teugels 1998) and disrupts important trophic relationships between the forest and the rivers that sustain it (Reid 1996; Chapman and Chapman 2003). Estimates of deforestation in Central Africa are in the range of 11,000 km^2/year (Revenga et al. 1998; Somé et al. 2001).

Overfishing is a much more serious problem in the lakes of East Africa than in the rivers to the west. Beach seining with very fine mesh nets, even in protected areas, has had significant impacts on fish populations in Lake Malawi (pers. obs.). Introduction of alien species has seriously eroded the biodiversity of Lake Victoria. In addition, watersheds in much drier eastern and southern Africa have been deforested by 40 to 80 percent through a combination of firewood harvesting and slash-and-burn agriculture, resulting in the transfer from croplands of an average of some 20 tons/ha/year of silt onto fish feeding and spawning grounds (Revenga et al. 1998; Lévêque and Paugy 1999; Jamu et al. 2003).

In Africa, as elsewhere, conflict between commerce and conservation generally results in loss of biodiversity. However, it is not only the wildlife that suffers; local human communities also suffer the consequences of environmental degradation. Communities that have relied for generations on forest and aquatic resources for their livelihoods may find themselves dispossessed of their natural inheritance, often by vested commercial interests based outside of the area (Jansen 1997; Godoy et al. 2000; Somé et al. 2001).

Community-Based Natural Resource Management

To justify conservation of aquatic biodiversity from the point of view of local communities, the value of resources must be substantial and accrue locally. There must also be a system of governance that empowers local communities and authorities to set and enforce exploitation methods and quotas. Adaptive co-management, in which communities undertake to sustainably manage their own resources, is an emerging concept that has been used in a number of places. By transferring management and enforcement to local communities, this structure aims to increase control over natural resources while reducing central government expenditures. Adaptive co-management of freshwater capture fisheries is being tested in a number of African countries (Khan et al. 2004). To date, the track record of community management and conservation interventions is mixed, but new knowledge about how such efforts might be improved and what time frames to expect is encouraging (Hulme and Murphree 2001).

A key aspect of adaptive co-management is the valuation of resources from the point of view of indigenous people (Sheil and Wunder 2002). For example, in the case of forests, timber may not be the largest potential source of income (Peters et al. 1989), but because timber companies have already made substantial investments in equipment, infrastructure, and market development, there is a comparative advantage of large-scale tree exploitation in terms of short-term realizable profits. Also, profits accrue at a level and in such a way as to be more accessible to policymakers. A similar logic applies to large- versus small-scale capture fisheries, hence the continued presence of foreign fishing fleets off the coast of Africa at a time when local fishing communities are suffering extreme poverty and declining catches. In contrast to these large-scale operations, the value of most nontimber forest products and artisanal fisheries accrues locally and in a dispersed manner that makes accounting and taxation impossible.

However, it has been shown that such small businesses can produce wider economic growth. Delgado et al. (1998: 4) reviewed results from Burkina Faso, Niger, Senegal, and Zambia and found that "even small increments to rural incomes that are widely distributed can make large net additions to growth and improve food security." Winkelmann (1998: 10) identified interventions that lead to improved incomes at the level of the rural resource manager as "having a larger impact on countrywide income than increases in any other sector."

From the point of view of rural communities, directly confronting the timber and large fishing companies over ownership of resources is an uphill task. For artisanal fishers who are being required to increase mesh sizes and respect closed seasons, watching even small trawlers take several tons of fish in a single haul seriously undermines the credibility of regulatory bodies, whether local or national, whether or not the fish stocks are related. In fact, rather than struggling to protect remaining resources, local fishing communities confronted with expropriation have often joined in the ravaging of their own resources to capture whatever profit they can before the big companies arrive (pers. obs.).

Rather than competing with large companies, small operators might be able to create value for heretofore underexploited resources that offer a competitive advantage. By targeting new species, new products, or new (local or international) marketing channels or using culture-based and satellite exploitation systems, smaller investors could have an advantage over larger companies that lack the same economies of scale.

Children play a major role in the ornamental fish industry in many less developed tropical countries. Here, a young fisher from the Ntem River valley in southern Cameroon shows his catch of killifish (*Epiplatys* spp.) and freshwater prawns (*Macrobrachium vollenhovenii*) destined for aquaria in Yaoundé. Photo credit: R. E. Brummett, WorldFish, Cameroon

The Ornamental Fish Trade

Exploitation or culture of ornamental fishes for the aquarium trade is one such small-scale operation that could potentially accrue large benefits for both local communities and ecosystems. Ornamental fishes generally are not amenable to mass capture techniques and are seldom available in such large quantities that wholesale extrapolation is economically feasible for investors with high overheads. Typical ornamental fishes tend to be small, prefer habitats with large amounts of submerged structure, and are most often found singly or in small groups. Ornamentals are most commonly exploited by individual fishers, often children, who wade in the water with hand nets or set small basket traps.

Sustainable exploitation of this aquatic resource for local markets and export to the international ornamental fish trade may offer lucrative livelihood alternatives to local communities. A successful example of a locally managed ornamental fishery occurs in the Brazilian state of Amazonas, where some sixty to seventy small businesspeople manage the activities of hundreds of fish collectors in the communities of Barcelos and Santa Isabel do Rio Negro. On trade of 30–50 million ornamental fishes (of which more than 80 percent are cardinal tetras, *Paracheirodon axelrodi*), local communities earn some $250,000 per year, approximately 60 percent of total income for the region. Ornamental exports from the state of Amazonas as a whole generate some $3 million/year (Chao and Prang 2002).

Capture and sale of ornamental fishes are widespread in tropical countries. FAO estimates the export value of ornamental fishes at $200 million/year, of which more than 60 percent accrues to developing countries (FAO Forest Resource Assessment Programme 1999). Major importers of ornamental fishes are the United States, Japan, Germany, France, the United Kingdom, the Netherlands, Belgium, Italy, Singapore, and Spain, each purchasing at least $6 million worth of fish per year. Overall, ninety-eight countries imported ornamental fishes in 1998 (Yap 2002). Only 5–10 percent of total trade in ornamental fishes is of captured specimens; most are cultured under controlled conditions. The dominant exporting countries include Singapore, Haiti, the United States, the Czech Republic, Hong Kong, Malaysia, Sri Lanka, Japan, Israel, and the Philippines, each with more than $4.5 million in overseas sales. Impressive as they are, these figures represent only a fraction of the total market because some of the major producers are also major consumers (United States, Singapore, Japan). Overall trade, domestic and international, in ornamentals in 1998 was $4.5 billion (Yap 2002).

More than 1,500 ornamental fish species are regularly traded, and 95 percent of these are destined for home aquaria managed by nonspecialists (Chapman et al. 1997; Olivier 2001). A fish suitable as an ornamental is fairly hardy, takes artificial food, is not aggressive, is brightly colored, does not eat aquatic plants, and is not afraid of light. The major groups traded are as follows:

- Minnows (Cyprinidae): *Barbus* from Asia and Africa; *Capoeta* from West Asia; *Puntius, Brachydanio, Danio, Rasbora, Epalzeorhynchus,* and *Labeo* from South and Southeast Asia, and *Tanichthys, Carassius* (goldfish), and *Cyprinus* (koi) originally from China but now produced widely
- Tetras (Alestiidae): *Astyanax, Colossoma, Gymnocorymbus, Hemigrammus, Hyphessobrycon,* and *Paracheirodon,* primarily from South America
- Catfishes (Callichthyidae and Loricariidae): *Corydoras, Callichthys, Hoplosternum, Ancistrus, Hypostomus,* and *Pterygoplichthys* from South America
- Rainbow fishes (Melanotaeniidae, Pseudomugilidae, Telmatherinidae): *Bedotia* from Madagascar; *Glossolepis* and *Melanotaenia* from Australia and New Guinea; and *Telmatherina* from Indonesia
- Livebearers (Poeciliidae): *Molienesia, Poecilia,* and *Xiphophorus* from Latin America but now widely bred and introduced
- Cichlids (Cichlidae): *Pterophyllum, Symphysodon, Geophagus, Apistogramma, Gymnogeophagus, Astronotus, Cichla,* and *Cichlasoma* from Latin America; and *Haplochromis, Hemichromis, Pelvicachromis, Melanochromis, Pseudotropheus, Tropheus, Julidochromis,* and *Neolamprologus* from Africa
- Killifish (Aplocheilidae): *Aphyosemion, Nothobranchius,* and *Epiplatys* from Africa; and *Rivulus* from South America
- Anabantoids (Belontiidae, Helostomatidae): *Betta, Colisa, Trichogaster, Macropodus, Pseudosphromenus,* and *Helostoma* from South and Southeast Asia

Typically, the chain of distribution from the wild to market for ornamental fishes is as follows: fish breeders or exporters sell to wholesalers, wholesalers then sell to retailers, and finally the retailers sell to aquariophiles. With each transfer, substantial losses of fish are incurred due to stress, and aquarium shops try to recoup the lost revenue by increasing retail prices. There is some debate as to the exact magnitude of losses, with importers claiming much higher mortality rates than exporters. Nevertheless, whereas the retail prices of the major species start at $10 per fish, the wholesale price often is between 10 and 50 cents (Olivier 2001).

Lessons for Africa

Although several of the most valuable and widely traded genera are African, none of the major ornamental fish importing or exporting countries are in Africa. However, a number of exporters based in Malawi, Nigeria, Tanzania, and Democratic Republic of the Congo make regular shipments to Europe and Asia of mostly wild-caught fish. From the rivers of West and Central Africa come representatives of the Mormyridae (*Campylomormyrus, Gnathonemus*), Mochochidae (*Synodontis, Auchenoglanis*), and Cichlidae (*Pelvicachromis, Nanochromis, Teleogramma,* known in the aquarium trade as dwarf cichlids). From the rift lakes of eastern Africa come the popular small cichlids known as *mbuna*.

Unfortunately for local exporters, the bulk of the profit on sales of these easy-to-spawn fish accrues to the large wholesale breeders in Asia and the United States (S. Grant, pers. comm., 1995). South Africa has a number of breeders, but they focus primarily on the more common alien species such as *Molienesia, Poecilia, Xiphophorus, Pterophyllum, Trichogaster, Apistogramma,* and *Tanichthys* (Hoffman et al. 2000). Reports from Europe indicate that a large percentage of the indigenous (wild capture) fishes exported from Africa arrive in very poor condition, eroding confidence and harming the standing of reputable traders (C. Eon, pers. comm., 2003).

There are four major markets for ornamental fishes that African producers might access. The largest (95 percent of total trade) and most regular is that for home aquaria and includes the species and countries mentioned earlier. There is also a specialty market, dominated by Germany, for wild-caught fish, particularly the dwarf cichlids and *mbuna*. A third market is for new broodstock for commercial breeders in Florida and Singapore. Finally, museums and public aquaria pay high prices for rare species or particularly large individuals.

Like any global business, the international ornamental fish trade is highly competitive. To make a successful venture based on the culture or capture of ornamentals, African traders would do well to consider some of the lessons learned by trial and error in other parts of the world:

- Fishers should focus on a few species for which capture, handling, feeding, and markets can be standardized.
- Exporters need to maintain a wider range of species to overcome problems of seasonal supply and increase market opportunities. For example, wholesalers in Europe prefer minimum shipments of forty to fifty boxes with about 100 fish per box, depending on individual size.
- To maximize profits at the local level, the number of intermediaries should be kept to a minimum. Cooperative management has been successful in some parts of the world, and such a management structure might be helpful in cases of limited capital for investment (Bérubé 1992).
- Local markets may offer opportunities to sell excess stock or species of less interest to importers and can help improve overall cash flow in times of low overseas demand.
- Fish should be handled carefully and kept in good condition. Bad handling leads to the majority of losses and a bad reputation with clients.
- Use of antibiotics and other quarantine treatments can increase survival, but regulations on use of these substances are becoming increasingly strict.
- Fish should not be colored or otherwise altered. Importers and animal rights groups are watching the fish trade closely for abuses.
- Fishers should work with local communities and government to obtain exclusive access to fishing grounds. Too many operators in an area depresses prices (Brichard 1980).
- The fishery should be well managed and wisely used. Overexploitation is the fastest route out of business.
- Producers should concentrate on species that cannot be reproduced easily outside their native habitat because of their large adult size or peculiar reproductive ecology. Monopoly, even if short-lived, provides important market protection for new investors.
- Investors should consider branching out into aquatic crustaceans and plants to diversify marketing options and spread risk.
- Rather than relying on capture, producers interested in export markets should attempt to reproduce and culture as many species as possible. Reliance on seasonal capture complicates marketing, and tank-reared individuals handle and ship with fewer mortalities.

These conditions for success are not specific to Africa. However, Africa is by far the continent with the least development of its capacity to capture or culture ornamental fishes for profitable exportation. Consequently, new investors on the continent will come into the international market at a great disadvantage in terms of production technology and market knowledge. To be profitable in the long term, careful management of the natural fish populations on which the exploitation is based is essential.

On the other hand, Africa has a strong comparative advantage in having a large number of species that are unknown to the majority of aquariophiles, and many of them are difficult or impossible to reproduce outside their native habitat. Whereas Asian, American, and European producers rely predominantly on alien species and often have difficulty obtaining new broodstock when existing fish become either inbred or contaminated by antibiotic-resistant diseases (Olivier 2001), African breeders working with naturally spawning species can obtain new broodstock easily and thus maintain the high quality that will be increasingly demanded by the international ornamental fish industry.

ESSAY 6.6

Some, for All, Forever: A New South African Water Law

Carolyn (Tally) G. Palmer

Water for people and people for water.
PALMER ET AL. (2002: 1)

Everyone depends on water for life, well-being, and economic prosperity. In homes water is used for drinking, cooking, and washing. In workplaces water is used for agriculture and industry. Water also provides for recreation and meets aesthetic and spiritual needs. Water is so important and is used in so many ways that if it is overused, we risk damaging our very life source. Potentially damaging overuse of water comes mainly from taking too much water out of aquatic ecosystems and putting in too much waste.

South Africa is a dry country. We share an average rainfall of 500 mm/year with Australia and Canada, but in Canada 67 percent of rainfall becomes runoff, whereas in Australia the percentage is 9.8 and in South Africa only 8.6 because of high levels of evapotranspiration. Additionally, uneven rainfall distribution makes it difficult to distribute water to all users. In the past decade, some of the most advanced water law and policy in the world has come from South Africa. The originality of the approach lies in clearly setting two primary objectives—equity and sustainability—while accepting the complexity of water resources.

The National Water Act (NWA) (No. 36 of 1998) recognizes that water resources are part of the integrated water cycle made up of freshwater ecosystems such as rivers, wetlands, lakes, estuaries, and groundwater and the processes of precipitation, transpiration, infiltration, and evaporation (South African Government 1998). Closely connected to the water cycle is human use of water resources. The NWA promotes protection of water resources precisely so that people can use water both now and into the future. Water is at the heart of a better life for all.

Water and Democracy

In 1994, after South Africa's first democratic elections, water law was recognized as an important area for legal reform. There were two main reasons for this. Before 1994 there was enough water for people, but previous governments had not provided the pipes, pumps, and purification works to allow access to safe, clean water for all of the population. The other reason for legal reform was more fundamental. The 1956 Water Act was based on a principle called riparianity. This meant that the right of people to use water was linked to land ownership. Land ownership was discriminatory, so a privileged few had much more access to water. The NWA abolished riparianity and requires the South African government to be the public trustee of water resources and therefore to be responsible for water resource protection in order to ensure long-term, sustainable water resource use.

The main aims of the NWA are encapsulated in a Department of Water Affairs and Forestry slogan: "Some, for all, forever." *Some* acknowledges that water is a limited resource, *for all* emphasizes fairness (all people must be able to use the resource), and *forever* reminds us to use water and water-linked ecosystems wisely to protect them for the future.

The NWA provides for only two rights to water: water for basic human needs (washing, cooking, and drinking) and water to sustain aquatic ecosystems, called the Reserve. These rights ensure adequacy of supply, not delivery. The Water Services Act ensures that water is equitably distributed to people. The Reserve has priority before water is allocated to users in the broader domestic, industrial, and agricultural sectors (where water use includes both abstraction and waste disposal). Given that water is a scarce, essential resource, the NWA recognizes that aquatic ecosystems need protection so that they can continue to provide goods and services. A flexible approach to resource protection is needed, which promotes simultaneous social and economic benefits. South African corporate governance requires management and auditing against a triple bottom line: economic, social, and environmental (figure 6.7).

Resource Protection and Use

Integrated water resource management is a balance between water resource protection and water resource use. It is in the interest of all to protect water resources and use them efficiently. Both overprotection and underprotection are inefficient and expensive. The NWA provides two mechanisms to ensure water

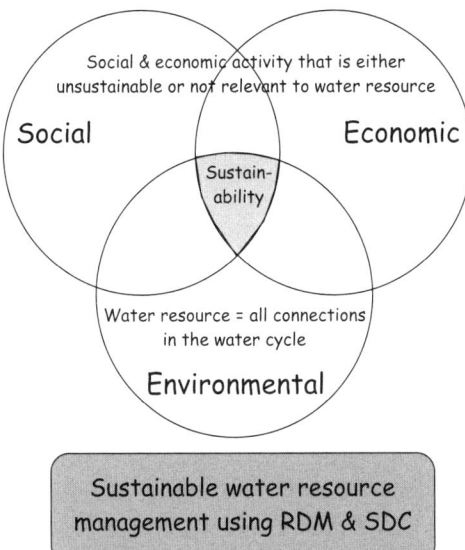

FIGURE 6.7. In the interaction among social, economic, and environmental issues the goal is to increase the intersecting domain of sustainability. RDM (resource-directed measures) and SDC (source-directed controls) are National Water Act approaches to sustainable water resource management.

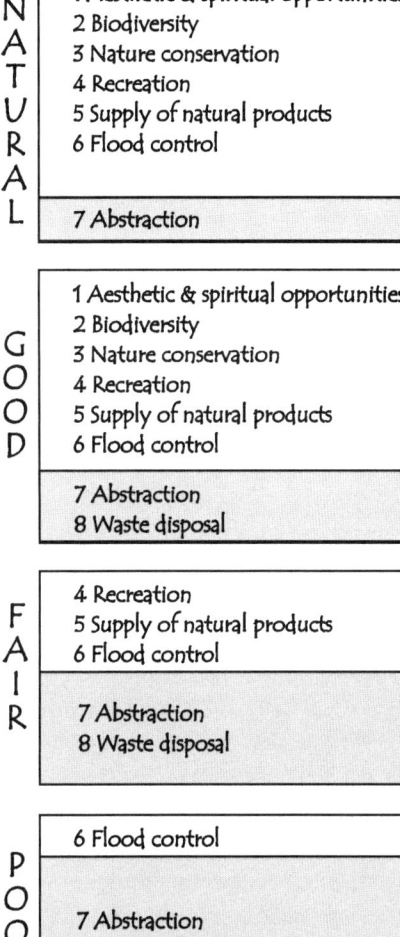

resource protection at an appropriate level: resource-directed measures and source-directed controls. Resource-directed measures provide descriptive and quantitative goals for the state of the resource, and source-directed controls specify the criteria for controlling impacts in authorizations such as waste discharge and abstraction licenses.

The national strategy for implementing the NWA recognizes two sources of ecosystem goals: "ecospecs" are quantitative and descriptive objectives specifying ecosystem conditions, and "userspecs" describe user requirements. Ecospecs ensure the maintenance of ecosystem processes at a designated level of ecosystem health; userspecs allow water users with more sensitive needs than provided for by ecospecs to promote a stricter objective. Userspecs that will not impair ecosystem function are combined with ecospecs into the formal resource quality objectives that will guide water resource management. (A decision support system for providing quantitative and qualitative objectives for different water levels of ecosystem health may be found at http://www.ru.ac.za/iwr; click on link to "Hydrological Models & Software").

Different levels of ecosystem health are described by a classification system ranging from "Natural" (unaffected), to "Good" (slightly to moderately degraded), to "Fair" (heavily degraded), to "Poor" (unacceptably heavily degraded) (figure 6.8).

FIGURE 6.8. Decisions about selecting a class for a water resource are informed by an understanding of the different goods and services offered by each ecosystem health class. Of the goods and services listed, water abstraction and waste disposal offer the greatest economic benefits of water use, whereas the others offer mainly social and environmental benefits. Resources in the Poor class offer few benefits because of increasing uncertainty of supply, degraded water quality, and high costs of purification. In South Africa, systems may be classified as Natural, Good, Fair, or Poor but may be managed only for Natural, Good or Fair condition. Poor systems must be rehabilitated.

Africa's Freshwater Systems and Their Future 137

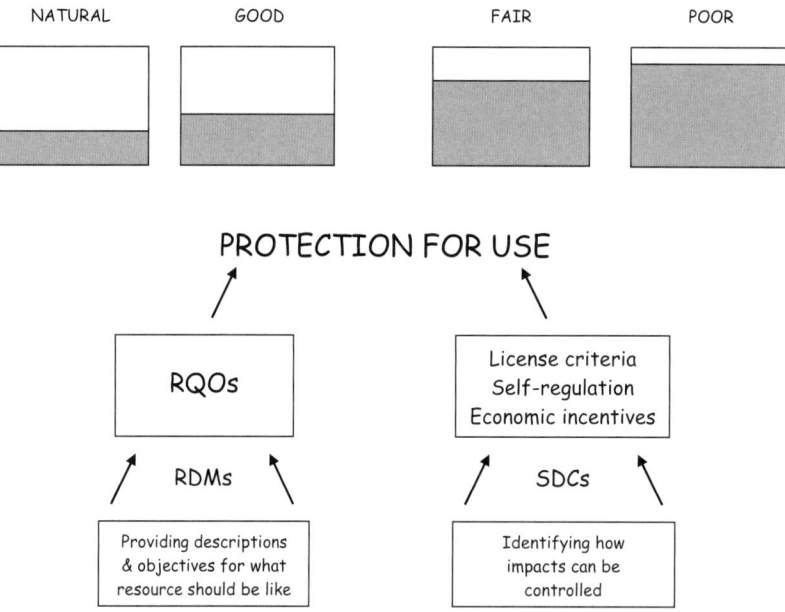

FIGURE 6.9. Summary diagram of flexible water resource protection and use. Monitoring is needed to document progress toward the achievement of resource quality objectives (RQOs) and to meet license criteria. RDMs, resource-directed measures; SDCs, source directed controls.

This classification is a key step in resource protection. Each resource class has its own specific ecospecs and therefore can be protected and used to a different degree. Specialists, stakeholders, and resource managers identify the appropriate class for a particular water resource, and the government, as the public trustee, decides on the class and the particular range of resource quality objectives that will guide management decisions. Managers then define the appropriate license criteria to control water use. This flexible approach provides for both resource protection and resource use (figure 6.9).

Sustainability and Governance

It is difficult to evaluate the relative values of all the goods and services offered by aquatic ecosystems, although environmental economists have attempted to quantify the value of natural systems and social impacts. The equity and sustainability aims of the NWA will be met only if social, economic, and environmental outcomes are taken into consideration in both the short term and the long term. Government and institutions need an integrated approach so that their actions reflect real environmental and social values as well as economic values. It is a huge challenge to actually achieve the lofty aims of protecting water resources for long-term benefits.

Nevertheless, South Africa has taken significant steps in this direction. The first step has been to recognize that a catchment or river basin is the natural unit for water resource management and to put into place the necessary management structures. The NWA provides for catchment management agencies to manage groups of catchments in large water management areas. The decisions of these catchment management agencies will affect the quality of life for both this generation and generations to come.

An integrated approach to water policy, law, and implementation means recognizing links and understanding how abiotic ecosystem components (e.g., water quality, flow, and physical shape) influence the responses of living organisms and ecosystem processes. The combined biophysical processes link to social and economic processes through the human use of water resources.

Traditionally, understanding natural systems and their biophysical characteristics has been separated from social science approaches to understanding the needs and aspirations of people and from economic approaches to financial and governance issues. Sustainable resource management demands that these three areas be integrated. The key linking concept is that only functioning ecosystems can provide people with valuable goods and services.

APPENDIX A

Methods for Assessing the Biological Distinctiveness of Freshwater Ecoregions

General explanations and the rationale for the four categories of biological distinctiveness are presented in chapter 2. The design of the index is summarized in figure A.1, and in this appendix we describe the evaluation methods in detail. The information sources used to evaluate biological distinctiveness are listed in table A.1.

Species Distribution and Endemism Data

We collected published and unpublished range and distribution data for African species in five taxonomic groups, comprising more than 4,000 species. The groups were fish, aquatic mollusks, aquatic mammals, amphibians dependent on aquatic habitats, and aquatic and semi-aquatic reptiles. Amphibians and reptiles have been grouped together as herpetofauna in this study. Biogeographic data on aquatic plants and other invertebrates were unsynthesized, difficult to locate, or of poor quality. Species that are considered brackish but are commonly found in coastal rivers and streams were counted in this analysis but were not considered endemic to any freshwater ecoregion.

For fish, data were in the form of written distribution accounts and, for parts of Africa, in the form of range maps. The primary data source was the CLOFFA (Check-List of the Freshwater Fishes of Africa) series, although more recent regional references were used where available (table A.1). We used the written distribution information or range map for each species to record the species as present or absent in each ecoregion and to indicate whether the species was endemic to the ecoregion. When range maps comprised occurrence data, species were counted as present in an ecoregion only if an occurrence was located within the ecoregion boundary. Where range maps were composed of range rather than point data, any range overlap with an ecoregion was recorded as species presence in that ecoregion. Ichthyologists then reviewed species lists for each ecoregion and updated them with the most current available data (table A.1). All data were compiled in a Microsoft Access database containing a record for each species and the ecoregions in which they occurred.

Data for aquatic frogs and aquatic mammals for sub-Saharan Africa were provided by the Zoological Museum of the University of Copenhagen from their species distribution database. Field guides and additional references supplied data and range maps for these taxa for North Africa and Madagascar (table A.1).

For aquatic mollusks, a comprehensive source of range data was unavailable. Aquatic mollusk data were supplied by Dr. Thomas Kristensen of the Danish Bilharziasis Laboratory. These data did not include the distributions of aquatic mollusks in Madagascar. The lack of aquatic mollusk data for Madagascar required that mollusks be excluded from the analysis of richness (and endemism) of moist forest rivers, highland and mountain systems, savanna–dry forest rivers, and xeric major habitat types (MHTs).

New freshwater species are continually being discovered in Africa, and many await formal description before the scientific community accepts them as valid species. Many taxa that are not yet formally named may have fairly well-known distributions. These unnamed species were not included in our biodiversity assessment. Additionally, only full species (as opposed to subspecies) were counted in the assessment, although many subspecies are considered at risk. These decisions were made to standardize the assessment method.

In the strictest sense, an endemic species would be found in a single ecoregion and nowhere else. We relaxed this definition of endemism in two ways. First, we regarded strictly and nearly endemic (more than 75 percent range in an ecoregion) species as endemic. Second, we considered as endemic all species with highly restricted ranges (less than 50,000 km^2) regardless of whether they occurred in multiple ecoregions. We applied this interpretation to concentrate attention on where near-endemics occur and to ensure that we captured restricted-range species

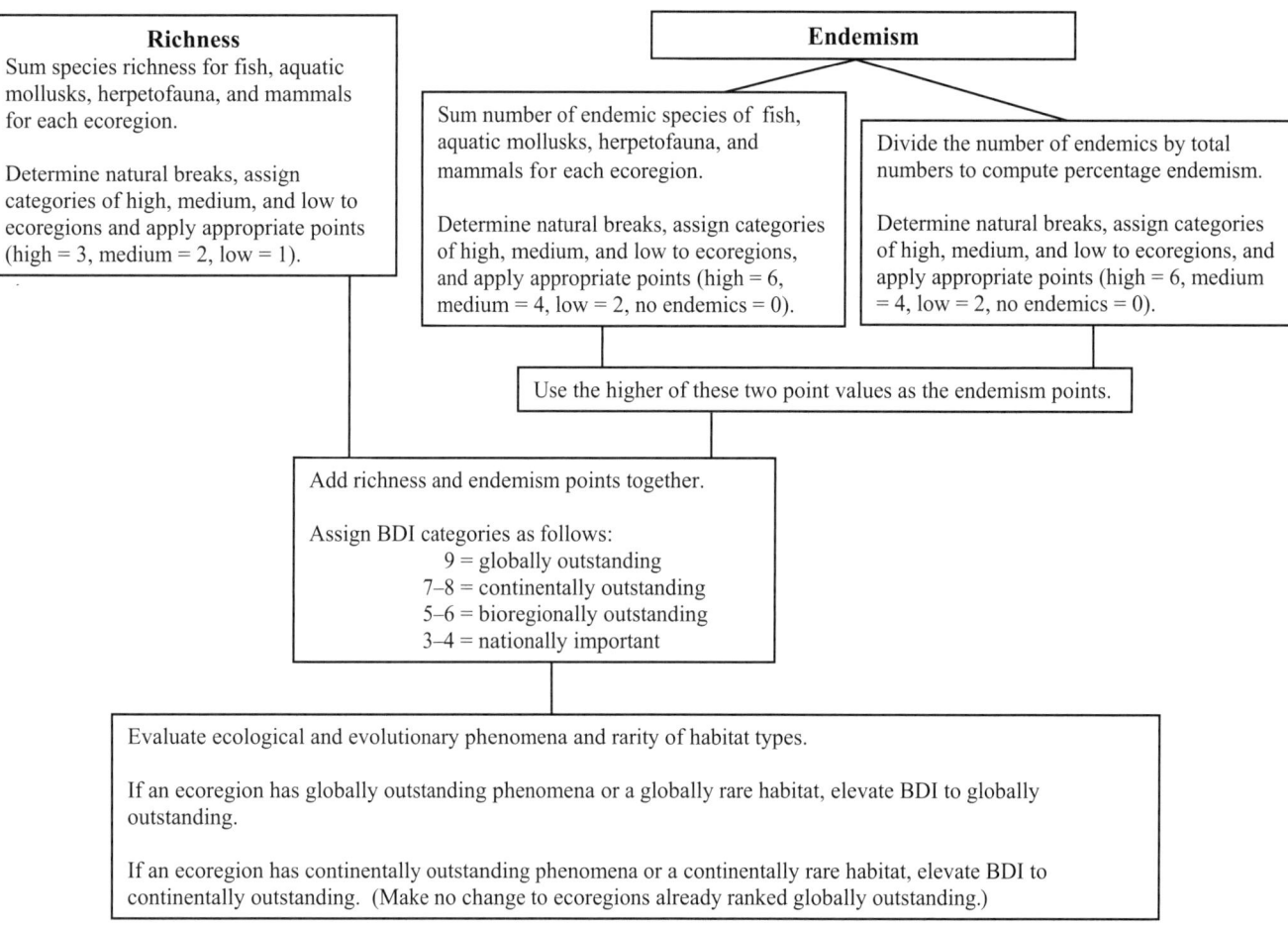

FIGURE A.1. Steps for evaluating biological distinctiveness of ecoregions. BDI, Biological Distinctiveness Index; MHT, major habitat type.

that are often at high risk of extinction from habitat loss. This modified definition of endemism is codified in the following set of decision rules:

1. Total species range more than 50,000 km²:
   a. A single ecoregion contains 75–100 percent of the species' range: species recorded as endemic to that ecoregion, recorded as present in all others containing it.
   b. No single ecoregion contains more than 75 percent of the species' range: species recorded as present in all ecoregions where it occurs.
2. Species range less than 50,000 km²:
   a. Species occurs in five or fewer ecoregions: species is recorded as endemic in all ecoregions where it occurs.
   b. Species occurs in more than five ecoregions (many small, disjunct areas): species is recorded as present in all ecoregions where it occurs.

We chose the 50,000-km² threshold for narrow-range endemics following BirdLife International's classification for endemic bird species (Bibby et al. 1992). The 75 percent threshold for the proportion of a range in a single ecoregion is an arbitrary threshold but represents most of the total range. The point here is to highlight ecoregions that represent the only practical opportunity to conserve a certain species in the wild. An ecoregion containing more than 75 percent of a species' range fits that description much more than one containing less than 25 percent. A species whose range extends outside of the Afro-Malagasy region was not treated as endemic regardless of its distribution in the region.

Richness

We compared species richness between ecoregions of the same MHT by summing richness values for each of the four taxa analyzed and using the combined value. These taxa are assumed to serve as a proxy for richness of the entire freshwater biota

TABLE A.1. Data Sources for Biological Distinctiveness Analysis.

Indicator	Data Source
Species richness and endemism	
Fishes	1–22
Aquatic mollusks	23, 24
Aquatic mammals	25, 26
Aquatic frogs	25, 27–30
Aquatic reptiles	27, 28, 30–33
Rare ecological or evolutionary phenomena	Expert assessment, 34
Higher-level taxonomic diversity	Sources same as for richness and endemism
Rare habitat type	Expert assessment and literature review

Sources:

1. Daget et al. (1984).
2. Daget et al. (1986a).
3. Daget et al. (1986b).
4. Daget et al. (1991).
5. Lévêque et al. (1990).
6. Lévêque et al. (1992).
7. Skelton (1993).
8. Dr. C. Lévêque, CNRS-Programme Environment, France (ecoregions 7, 9, 17, 20, 25, 28, 32, 40, 42, 62, 65, 71, 75, 81).
9. Dr. G. G. Teugels, Africa Museum, Belgium (ecoregions 10, 13, 14, 18, 19, 21, 22, 24, 27, 29–31, 35, 58, 60, 61, 74, 80).
10. Dr. M. L. J. Stiassny, American Museum of Natural History, USA (ecoregions 1, 4, 15, 23, 41, 59, 67, 68, 83, 86, 87, 89, 90).
11. Dr. A. Getahun, Addis Ababa University, Ethiopia (ecoregions 3, 39, 59, 83, 86, 89, 90).
12. Dr. P. Skelton, South African Institute for Aquatic Biodiversity, South Africa (ecoregions 6, 8, 12, 16, 33, 36–38, 43, 63, 66, 69, 73, 76–79, 82, 84, 85, 88, 91, 93).
13. Dr. J. Sparks, American Museum of Natural History, New York (ecoregions 23, 41, 67, 68, 87).
14. Dr. U. Schliewen, Zoologische Staatssammlung München, Germany (ecoregions 5, 26).
15. Dr. L. De Vos, National Museum of Kenya, Kenya (ecoregions 2, 11, 54, 64).
16. Dr. I. Doadrio, Spanish National Museum of Natural Sciences, Spain (ecoregions 34, 92).
17. Dr. A. Konings, Cichlid Press, USA (ecoregion 55).
18. Dr. M. K. Oliver, USA (ecoregion 53).
19. Dr. A. Ribbink, South African Institute for Aquatic Biodiversity, South Africa (ecoregion 53).
20. Dr. L. Seegers, Germany (ecoregion 56).
21. Dr. O. Seehausen, University of Hull, UK (ecoregion 57).
22. Dr. D. Thys van den Audenaerde (ecoregion 50).
23. Dr. T. Kristensen, Danish Bilharziasis Laboratory, Denmark (all ecoregions except Madagascar).
24. Brown (1994).
25. Fjeldså et al. (ongoing).
26. Kingdon (1997).
27. Henkel and Schmidt (2000).
28. Glaw and Vences (1994).
29. Schiøtz (1999).
30. LeBerre (1989).
31. Ross (1998).
32. Ernst and Barbour (1989).
33. Branch (1998).
34. Fishpool and Evans (2001).

(Ricketts et al. 1999b; Heino 2002; Kretschmar 2003). This method gives all species equal value; in other words, we considered one frog species as important as one fish.

Total richness values were adjusted to compensate for ecoregion area before being used to define biodiversity richness categories of high, medium, and low (see appendix E for method). For each MHT, the total adjusted richness values were plotted in increasing order, and cut points for each level were set where a sharp increase in slope between points occurred (figure A.2). After each ecoregion was assigned a richness level, corresponding point values were applied: high = 3 points, medium = 2 points, and low = 1 point.

Endemism

We collected endemism data in the same manner and from the same sources as richness. Conserving areas with the highest absolute levels of endemism is desirable, but many species-poor ecoregions have a high number of endemics relative to their small number of total species (Abell et al. 2000). To capture both

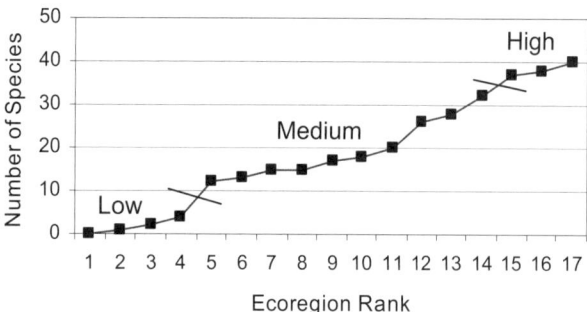

FIGURE A.2. Example of method used to determine natural breaks for number of species or number of endemics in a major habitat type.

absolute and relative endemism, we used two endemism measures. The first is simply the sum of all endemic species across the four selected faunal groups. The second is a calculation of the percentage endemism, which is the proportion of an ecoregion's richness that is endemic to that ecoregion. We totaled endemism and richness over all taxa before calculating percentage endemism. We did this to avoid inflated levels of percentage endemism in ecoregions that contain few species in each taxon (e.g., an ecoregion that has only two mollusks, of which one is endemic).

For both endemism and percentage endemism, we followed the same method as for species richness: ecoregions were separated by MHT, and then the values for all faunal groups were summed, cut points were set, and levels of high, medium, and low were assigned. Because endemism is a highly significant factor in determining an ecoregion's distinctive biodiversity value, we gave it twice as much weight in the index as richness. The point values for each level were high = 6, medium = 4, low = 2, no endemics or endemism = 0. Next, for each ecoregion we compared the endemism and percentage endemism categories and took the higher point value of the two.

Species Distinctiveness Categorization

The point totals for the component criteria (richness and endemism or percentage endemism) were added to yield the total for the Biological Distinctiveness Index (BDI). Based on these totals, ecoregions were placed into one of four categories:

Distinctiveness Category	Points
Globally outstanding	9
Continentally outstanding	7 or 8
Bioregionally outstanding	5 or 6
Nationally important	4 or less

Biological (Species and Nonspecies) Distinctiveness Categorization

To determine the final distinctiveness ranking for each ecoregion, we examined the presence of ecological and evolutionary phenomena and rare habitat types in each ecoregion. If a globally or continentally outstanding evolutionary phenomenon or rare habitat type was present in the ecoregion, then that ecoregion was elevated to globally or continentally outstanding, respectively.

Rare Ecological or Evolutionary Phenomena

Ecological and evolutionary phenomena also contribute to the distinctiveness of an ecoregion's biodiversity (Lovejoy 1997). These biological values are included to capture important biodiversity features that are otherwise difficult to measure using quantitative methods. We considered the evolutionary phenomena of distinctive centers of evolutionary radiation and higher-level (e.g., genus, family) taxonomic diversity in our analysis. The only ecological phenomenon that we evaluated was extraordinary concentrations of wetland birds.

Species radiations were identified by experts at the assessment workshop, where they listed any appropriate phenomena occurring in each ecoregion. They also were asked to describe the level of rarity of each phenomenon, emphasizing those that are globally or continentally outstanding in terms of rarity. Species-level data gathered for the evaluation of richness and endemism were used to identify ecoregions containing higher-level taxonomic endemism. We followed decision rules used by Olson and Dinerstein (1998) in a global analysis of freshwater biodiversity in determining the distinctiveness of higher-level taxonomic endemism. If an ecoregion harbored one or more endemic families of fish, aquatic reptiles, amphibians, aquatic mammals, or aquatic mollusks, or if 30 percent of all genera of any of these groups was considered endemic to the ecoregion, then the ecoregion was considered globally outstanding for higher-level taxonomic endemism. For wetland bird concentrations, we relied on data provided by BirdLife International from Fishpool and Evans (2001), using an arbitrary cutoff of 1 million wetland birds to designate ecoregions as containing continentally outstanding concentrations and a cutoff of 1.5 million for globally outstanding concentrations.

Ecoregions that were judged to contain globally outstanding ecological or evolutionary phenomena were categorized as globally outstanding in the overall BDI. Therefore, ecoregions without extraordinarily high richness or endemism but that nevertheless contained globally important and rare phenomena could be categorized as globally outstanding. Ecoregions that were categorized as continentally outstanding were automatically categorized as continentally outstanding in the overall BDI. The ecoregions elevated for evolutionary phenomena are noted in table 3.1.

Rare Habitat Type

The rarity of habitat types was evaluated to identify the broad habitat types that offer few opportunities for conservation. The rarity of habitat types contained in ecoregions was evaluated based on expert assessment and a recently completed global analysis (Olson and Dinerstein 1998). Ecoregions were placed in one of three categories: globally rare, continentally rare, and not applicable, based on the following decision rules:

Fewer than eight ecoregions occur globally that contain the habitat type in question: globally rare.

Fewer than three ecoregions occur in Africa that contain the habitat type in question: continentally rare.

Otherwise: not applicable.

Similar to the rare ecological and evolutionary phenomena criterion, an ecoregion that was judged to contain globally rare or continentally rare habitat types was then categorized as globally or continentally outstanding in the overall BDI.

APPENDIX B

Methods for Assessing the Conservation Status of Freshwater Ecoregions

In assessing conservation status, we evaluated threats from four broad categories: land-based threats (land degradation), aquatic habitat threats, biota threats (exploitation and exotics), and future threats. Although several continental-scale datasets exist for Africa, none is comprehensive enough to cover the full suite of threats included in each category. Thus, we combined analyses of spatial datasets with expert consultation and literature review. Prior conservation assessments for both terrestrial and freshwater ecosystems have used expert opinion to evaluate conservation status over large areas (O'Keeffe et al. 1987; Dinerstein et al. 1995; Olson and Dinerstein 1998; Ricketts et al. 1999a; Abell et al. 2000). This approach, informed by continental datasets, has proved useful as a qualitative tool that circumscribes strict information requirements.

Determination of Snapshot Conservation Status

The snapshot conservation status estimates the current status of each ecoregion, based on evaluations of land degradation, aquatic habitat threats, and threats to the aquatic biota in each ecoregion. Points were assigned for each indicator on a scale from 0 (least degraded) to 3 (most degraded).

Land-Based Threats (Land Degradation)

The first category covers threats to freshwater organisms that originate in the terrestrial landscape. We combined landcover data (MODIS global landcover product) with population data to determine the percentage of degraded land in each ecoregion. The University of Maryland (UMD) global landcover data classes "urban" and "cropland" were used to define areas of converted landcover (Hansen et al. 2000; Boston University 2001). This information was then combined with population density, derived from the LandScan Global Population 2001 database (Dobson et al. 2000; Oak Ridge National Laboratory 2002). We followed Sanderson et al. (2002) and used a population density of 10 or more people per square kilometer to determine where population pressures start to affect the environment negatively, although figures between 5 and 20 people per square kilometer have been used elsewhere (Mittermeier et al. 1998; Hoare and Du Toit 1999; Eswaran et al. 2001). LandScan models human populations along all roads, including those in protected areas. Under the presumption that human populations are low in protected areas, we removed the population information in all protected areas of IUCN categories I–IV. These categories of protected areas include strict nature reserves, wilderness areas, national parks, national monuments, and habitat and species management areas, as defined by IUCN (1994). We combined the two maps (UMD and LandScan) using geographic information system (GIS) tools and measured the extent of land degradation.

Unfortunately, data from the human footprint analysis of the Wildlife Conservation Society (Sanderson et al. 2002) were unavailable when we were completing our analysis of land degradation. However, we were subsequently able to compare the results of their analysis with ours. The human footprint is a "quantitative evaluation of human influence on the land surface based on geographic data describing human population density, land transformation, access, and electrical power infrastructure" (Sanderson et al. 2002: 896). Sanderson et al. (2002) used nine datasets to determine the extent of degradation on a scale of 0–100 in each 1-km^2 grid cell. We compared the percentage of each ecoregion considered degraded based on the human footprint (value of more than 20) and the percentage degraded based on our analysis. The two measures of level of degradation were highly correlated ($r = .80$, $p < .0001$).

We used the following guidelines in designating the level of land-based threats in each ecoregion:

Percentage of Ecoregion Area Degraded	Points
>50%	3
25–50%	2
5–25%	1
<5%	0

We recognize that our GIS analysis of land degradation measures only a subset of the activities on land that may be adversely affecting freshwater systems. For that reason, we used expert opinion and literature review to validate the results of the map-based analyses. In the few cases where there was disagreement between expert opinion and GIS analyses, expert opinion took precedence, and the more appropriate level was assigned. Seven ecoregions were considered to be at a higher level of degradation than the GIS analysis showed: Okavango Floodplains [12], Cape Fold [33], Nile Delta [59], Southern Temperate Highveld [73], Kartsveld Sink Holes [79], Etosha [82], and Western Orange [93].

Aquatic Habitat Threats

The aquatic habitat threat category addresses the threats that occur directly in the aquatic habitat. These could include degraded water quality (e.g., point or non–point source pollution; changes in temperature, pH, dissolved oxygen, or other physical parameters; sedimentation or siltation); reduced organic matter input, including woody debris; habitat fragmentation from dams or other barriers to dispersal, migration, and general movement; excessive recreational impacts; altered hydrographic integrity (flow regimes, water levels) resulting from dams, surface or groundwater withdrawals, channelization, interbasin water transfers; and other losses of aquatic habitat, such as from inundation by reservoirs or from desiccation. We recognize that many of these threats may overlap with or be caused by land-based threats. We evaluated aquatic habitat threats based largely on expert opinion and literature review, using the following decision rules:

Description of Aquatic Habitat Threats	Points
Low habitat quality for sensitive species. Abandonment and disruption of migratory or breeding movements caused by physical barriers and habitat destruction. Pollutants or linked effects widespread (e.g., recorded in several trophic levels).	3
Dynamic physical processes modified. Some abandonment and disruption of migratory or breeding movements caused by physical barriers or habitat destruction. Pollutants or linked effects commonly found in target species or assemblages.	2
Dynamic physical processes somewhat modified. Some abandonment and disruption of migratory or breeding movements. Pollutants or linked effects may be found in target species or assemblages.	1
No significant aquatic threats known for the ecoregion.	0

In addition, we analyzed the FAO database on African dams (FAO 2000a). This dataset is based primarily on the International Commission on Large Dams' (ICOLD) register of large dams (those more than 15 m in height) with updates from several African countries (ICOLD 1989). Ecoregions that had more than fifty dams were assigned to the highest level of threat for this indicator. We also elevated ecoregions with fewer dams but an extremely large reservoir capacity (more than 150 billion m^3), indicating the presence of large dams on major rivers.

Biota Threats (Exploitation and Exotics)

Examples of biota threats include unsustainable fishing or hunting, unsustainable extraction of wildlife or plants as commercial products, and competition, predation, hybridization, and infection by established exotic species. Several global databases exist on invasive species and were used as references in designating the level of threat (FAO 1998; Invasive Species Specialist Group 2000). However, data for Africa have been collected at the country level, making direct evaluations of number of invasives per ecoregion difficult.

Similarly, there is no comprehensive, continental dataset that evaluates the level of overexploitation of aquatic organisms. Biota threats therefore were assessed largely using expert opinion and literature review (see appendix F for descriptions of biota threats specific to each ecoregion and literature sources).

Description of Biota Threats	Points
High intensity of exploitation or disturbance by exotics	3
Moderate levels of exploitation or disturbance by exotics	2
Low exploitation or disturbance by exotics	1
Little or no exploitation or disturbance by exotics	0

Calculation of the Value of Snapshot Conservation Status

The summed snapshot conservation status index has a point range of 0–9, with higher values denoting a higher level of endangerment. The point thresholds for the categories of conservation status are as follows:

Snapshot Conservation Status	Points
Critical	9
Endangered	8
Vulnerable	6–7
Relatively stable	4–5
Relatively intact	1–3

Likelihood of Future Threats

Planned Infrastructure

We obtained information on planned infrastructure projects and resource extractions from the literature and experts (e.g., ICOLD 1989; U.S. Geological Survey 2001b; FAO 2001c, 2001d). We classified the level of threat to each ecoregion as high, medium, or low based on the presence, extent, and magnitude of the activity.

Climate Change

To examine the effects of climate change on river discharge, we analyzed discharge estimations for 2025 calculated by the global integrated water model WaterGAP (Alcamo et al. 2003; Döll et al. 2003). WaterGAP simulates the effects of climate change on water availability and incorporates the effects of projected demographic, socioeconomic, and technological change on water use. We analyzed WaterGAP results from two different model runs, based on climate change calculations of two independent, widely accepted general circulation models (GCMs): the ECHAM4 model (Röckner et al. 1999) and the HadCM3 model (Gordon et al. 2000). Because there remains much uncertainty in modeling the effects of climate change, we assessed the results only where both model runs were in agreement. We calculated the area of each ecoregion over which the model results showed a change in mean annual discharge of more than 10 percent (i.e., an increase or a decrease of more than 10 percent). Areas (grid cells of the WaterGAP model) where both model runs delivered discharge changes over this threshold and in the same direction were marked as areas of strong discharge change. If more than 80 percent of an ecoregion's area was marked in this way, then the future threat to the ecoregion was considered to be high. We assume in this approach that a change of 10 percent or more of mean annual discharge would substantially change the ecological dynamics of the system and consequently affect the freshwater biota.

Population Increase

Finally, we examined projected population density to help identify areas where anthropogenic pressures are most likely to be greatest. Projected population density modeled to the year 2025 was calculated by the World Resources Institute, based on data from CIESEN et al. (2000) and UNPD (1999). Ecoregions with an estimated mean population density of more than 50 people/km^2 in 2025 across the ecoregion were assigned a high future threat level.

Final Conservation Status, as Modified by Future Threat

If any of the three analyses (i.e., planned infrastructure, climate change, and population increase) produced a ranking of high threat, then the overall future threat level was designated as high. We raised the conservation status of the ecoregions with high future threats by one level to determine the final conservation status. Ecoregions with low or medium future threats were left unchanged.

APPENDIX C

Data on the Species Richness and Endemism of Ecoregions

Ecoregion	Major Habitat Type	fish			herpetofauna		
		richness	endemism	% endemism	richness	endemism	% endemism
Lakes Chilwa and Chiuta [1]	CBSL	23	1	4	46	2	4
Southern Eastern Rift [2]	CBSL	27	9	33	25	2	8
Lake Tana [3]	CBSL	26	18	69	15	0	0
Northern Eastern Rift [4]	CBSL	8	2	25	16	1	6
Western Equatorial Crater Lakes [5]	CBSL	38	27	71	60	17	28
Bangweulu-Mweru [6]	FSL	111	31	28	67	4	6
Inner Niger Delta [7]	FSL	111	2	2	12	1	8
Kafue [8]	FSL	62	2	3	53	1	2
Lake Chad Catchment [9]	FSL	140	4	3	44	3	7
Mai Ndombe [10]	FSL	30	3	10	16	0	0
Malagarasi-Moyowosi [11]	FSL	88	15	17	14	0	0
Okavango Floodplains [12]	FSL	80	0	0	43	0	0
Tumba [13]	FSL	48	2	4	19	2	11
Upper Lualaba [14]	FSL	101	11	11	55	7	13
Upper Nile [15]	FSL	118	16	14	25	1	4
Upper Zambezi Floodplains [16]	FSL	85	1	1	55	1	2
Ashanti [17]	MFR	105	11	10	63	8	13
Cuvette Centrale [18]	MFR	238	14	6	50	8	16
Central West Coastal Equatorial [19]	MFR	279	57	20	107	9	8
Eburneo [20]	MFR	130	10	8	59	4	7
Kasai [21]	MFR	224	49	22	67	5	7
Lower Congo [22]	MFR	200	11	6	34	4	12
Madagascar Eastern Lowlands [23]	MFR	54	10	19	72	31	43
Malebo Pool [24]	MFR	231	14	6	18	0	0
Northern Upper Guinea [25]*	MFR	160	44	28	57	10	18
Northern West Coastal Equatorial [26]	MFR	187	30	16	128	28	22
Sangha [27]	MFR	170	8	5	76	6	8
Southern Upper Guinea [28]	MFR	151	33	22	60	11	18
Southern West Coastal Equatorial [29]	MFR	236	58	25	55	3	5
Sudanic Congo (Oubangi) [30]	MFR	164	13	8	54	4	7
Upper Congo [31]	MFR	182	10	5	40	3	8
Upper Niger [32]	MFR	153	8	5	16	1	6
Cape Fold [33]	MS	28	16	57	38	18	47
Permanent Maghreb [34]	MS	40	21	53	13	2	15
Albertine Highlands [35]	HMS	16	0	0	56	22	39
Amatolo-Winterberg Highlands [36]	HMS	8	3	38	19	2	11
Drakensberg-Maloti Highlands [37]	HMS	10	1	10	9	3	33
Eastern Zimbabwe Highlands [38]	HMS	32	4	13	53	3	6

Major habitat types: CBSL, closed basins and small lakes; FSL, floodplains, swamps, and lakes; MFR, moist forest rivers; MS, mediterranean systems; HMS, highland and mountain systems; IRL, island rivers and lakes; LL, large lakes; LRD, large river deltas; LRR, large river rapids; SDFR, savanna–dry forest rivers; SSS, subterranean and spring systems; XS, xeric systems.

*Totals exclude mollusk richness and endemism for ecoregions in major habitat types that include Madagascan ecoregions (see appendix A for details on methods).

	mollusk			mammal		total		
richness	endemism	% endemism	richness	endemism	richness*	endemism*	% endemism
0	0	N/A	4	0	73	3	4.1
15	1	7	4	0	71	12	16.9
15	1	7	5	0	61	19	31.1
11	0	0	4	0	39	3	7.7
1	1	100	2	0	101	45	44.6
37	7	19	6	0	221	42	19.0
16	0	0	4	0	143	3	2.1
12	0	0	3	0	130	3	2.3
25	3	12	4	0	213	10	4.7
0	0	N/A	5	0	51	3	5.9
13	0	0	3	0	118	15	12.7
4	0	0	4	0	131	0	0.0
0	0	N/A	5	0	72	4	5.6
12	0	0	7	0	175	18	10.3
28	0	0	3	0	174	17	9.8
9	1	11	4	0	153	3	2.0
21	0	0	5	0	173	19	11.0
8	0	0	7	0	295	22	7.5
22	0	0	7	0	393	66	16.8
34	4	12	6	1	195	15	7.7
11	0	0	7	0	298	54	18.1
17	5	29	7	0	241	15	6.2
N/A	N/A	N/A	0	0	126	41	32.5
4	2	50	5	0	254	14	5.5
18	5	28	6	0	223	54	24.2
1	0	0	7	0	322	58	18.0
17	5	29	6	0	252	14	5.6
14	2	14	8	0	219	44	20.1
15	0	0	7	0	298	61	20.5
11	0	0	6	0	224	17	7.6
41	22	54	8	0	230	13	5.7
18	0	0	5	0	174	9	5.2
9	1	11	2	0	77	35	45.5
26	8	31	1	0	80	31	38.8
4	1	25	7	1	79	23	29.1
0	0	N/A	2	0	29	5	17.2
0	0	N/A	2	0	21	4	19.0
0	0	N/A	2	0	87	7	8.0

Ecoregion	Major Habitat Type	fish			herpetofauna		
		richness	endemism	% endemism	richness	endemism	% endemism
Ethiopian Highlands [39]	HMS	78	5	6	53	22	42
Fouta-Djalon [40]	HMS	58	14	24	12	0	0
Madagascar Eastern Highlands [41]	HMS	35	14	40	136	87	64
Mount Nimba [42]	HMS	17	6	35	42	7	17
Mulanje [43]	HMS	19	1	5	29	1	3
Bijagos [44]	IRL	1	0	0	9	2	22
Canary Islands [45]	IRL	1	0	0	2	0	0
Cape Verde [46]	IRL	1	0	0	0	0	N/A
Comoros [47]	IRL	10	0	0	1	1	100
Coralline Seychelles [48]	IRL	0	0	0	1	0	0
Granitic Seychelles [49]	IRL	4	1	25	2	1	50
São Tomé, Príncipe, and Annobon [50]	IRL	2	0	0	7	5	71
Mascarenes [51]	IRL	19	5	26	4	0	0
Socotra [52]	IRL	0	0	N/A	0	0	N/A
Lake Malawi [53]	LL	382	350	92	79	4	5
Lake Rukwa [54]	LL	53	11	21	45	0	0
Lake Tanganyika [55]	LL	288	231	80	53	6	11
Lake Turkana [56]	LL	56	11	20	23	5	22
Lakes Kivu, Edward, George, and Victoria [57]	LL	283	224	79	69	13	19
Niger Delta [58]	LRD	149	8	5	31	0	0
Nile Delta [59]	LRD	29	0	0	5	1	20
Lower Congo Rapids [60]	LRR	162	26	16	13	0	0
Upper Congo Rapids [61]	LRR	170	3	2	14	0	0
Bight Coastal [62]	SDFR	152	6	4	59	6	10
Cuanza [63]	SDFR	74	34	46	59	6	10
Kenyan Coastal Rivers [64]	SDFR	56	13	23	46	4	9
Lower Niger-Benue [65]	SDFR	204	16	8	101	16	16
Lower Zambezi [66]	SDFR	74	1	1	52	1	2
Madagascar Northwestern Basins [67]	SDFR	55	17	31	58	21	36
Madagascar Western Basins [68]	SDFR	37	3	8	57	25	44
Middle Zambezi Luangwa [69]	SDFR	53	1	2	56	0	0
Pangani [70]	SDFR	48	13	27	45	2	4
Senegal-Gambia Catchments [71]	SDFR	159	1	1	33	3	9
Eastern Coastal Basins [72]	SDFR	92	31	34	65	6	9
Southern Temperate Highveld [73]	SDFR	78	9	12	58	6	10
Uele [74]	SDFR	149	9	6	75	11	15
Volta [75]	SDFR	145	9	6	42	4	10
Zambezian Headwaters [76]	SDFR	106	13	12	72	5	7
Zambezian Lowveld [77]	SDFR	124	23	19	80	3	4
Zambezian (Plateau) Highveld [78]	SDFR	39	0	0	47	0	0
Karstveld Sink Holes [79]	SSS	3	2	67	6	0	0
Thysville Caves [80]	SSS	2	1	50	2	0	0
Dry Sahel [81]	XS	15	0	0	11	0	0
Etosha [82]	XS	19	0	0	18	0	0
Horn [83]	XS	8	1	13	14	0	0
Kalahari [84]	XS	4	0	0	22	0	0

	mollusk			mammal		total		
richness	endemism	% endemism	richness	endemism	richness*	endemism*	% endemism
24	2	8	4	0	135	27	20.0
1	0	0	4	0	74	14	18.9
N/A	N/A	N/A	1	1	172	102	59.3
0	0	N/A	4	1	63	14	22.2
0	0	N/A	3	0	51	2	3.9
2	0	0	4	0	16	2	12.5
1	0	0	0	0	4	0	0.0
6	0	0	0	0	7	0	0.0
12	0	0	0	0	23	1	4.3
5	0	0	0	0	6	0	0.0
6	1	17	0	0	12	3	25.0
3	0	0	0	0	12	5	41.7
10	2	20	0	0	33	7	21.2
4	2	50	0	0	4	2	50.0
32	18	56	4	0	497	372	74.8
2	0	0	4	0	104	11	10.6
65	41	63	5	0	411	278	67.6
1	1	100	4	0	84	17	20.2
53	11	21	8	0	413	248	60.0
16	2	13	6	0	202	10	5.0
5	1	20	0	0	39	2	5.1
18	16	89	4	0	197	42	21.3
9	6	67	4	0	197	9	4.6
30	3	10	5	0	216	12	5.6
20	0	0	7	0	140	40	28.6
29	6	21	3	0	105	17	16.2
32	1	3	6	0	311	32	10.3
8	0	0	4	0	130	2	1.5
N/A	N/A	N/A	0	0	113	38	33.6
N/A	N/A	N/A	0	0	94	28	29.8
15	1	7	4	0	113	1	0.9
21	2	10	3	0	96	15	15.6
26	0	0	5	0	197	4	2.0
37	2	5	3	0	160	37	23.1
20	0	0	4	0	140	15	10.7
7	0	0	7	0	231	20	8.7
25	1	4	5	0	192	13	6.8
6	0	0	5	0	183	18	9.8
27	1	4	4	0	208	26	12.5
18	0	0	3	0	89	0	0.0
0	0	N/A	0	0	9	2	22.2
2	0	0	1	0	7	1	14.3
1	0	0	0	0	26	0	0.0
0	0	N/A	1	0	38	0	0.0
1	0	0	0	0	22	1	4.5
4	0	0	1	0	27	0	0.0

Ecoregion	Major Habitat Type	fish			herpetofauna		
		richness	endemism	% endemism	richness	endemism	% endemism
Karoo [85]	XS	4	0	0	20	3	15
Madagascar Southern Basins [87]	XS	10	1	10	11	4	36
Namib Coastal [88]	XS	3	0	0	18	0	0
Red Sea Coastal [89]	XS	7	2	29	14	2	14
Shebelle-Juba Catchments [90]	XS	33	13	39	31	0	0
Southern Kalahari [91]	XS	2	0	0	14	0	0
Temporary Maghreb [92]	XS	40	21	53	4	0	0
Western Orange [93]	XS	13	1	8	15	0	0

mollusk			mammal		total		
richness	endemism	% endemism	richness	endemism	richness*	endemism*	% endemism
2	0	0	1	0	25	3	12.0
N/A	N/A	N/A	0	0	21	5	23.8
7	0	0	0	0	21	0	0.0
0	0	N/A	0	0	21	4	19.0
24	1	4	1	0	65	13	20.0
1	0	0	0	0	16	0	0.0
3	0	0	0	0	44	21	47.7
1	0	0	2	0	30	1	3.3

APPENDIX D

Data on the Biological Distinctiveness, Conservation Status, and Priority Class of Ecoregions

Ecoregion	Major Habitat Type	Data Quality	Richness Points (1–3)	Endemism Points (2–6)	Total Richness and Endemism Points (3–9)	Biological Distinctiveness	Phenomenon
Lakes Chilwa and Chiuta [1]	CBSL	H	3	2	5	BO	CO
Southern Eastern Rift [2]	CBSL	H	2	4	6	BO	GO
Lake Tana [3]	CBSL	M	2	6	8	CO	GO
Northern Eastern Rift [4]	CBSL	M	1	2	3	NI	
Western Equatorial Crater Lakes [5]	CBSL	M	3	6	9	GO	GO
Bangweulu-Mweru [6]	FSL	L	3	6	9	GO	
Inner Niger Delta [7]	FSL	H	2	2	4	NI	CO
Kafue [8]	FSL	M	2	2	4	NI	
Lake Chad Catchment [9]	FSL	H	2	2	4	NI	
Mai Ndombe [10]	FSL	L	1	2	3	NI	
Malagarasi-Moyowosi [11]	FSL	L	2	4	6	BO	CO
Okavango Floodplains [12]	FSL	H	2	0	2	NI	GO
Tumba [13]	FSL	L	2	2	4	NI	
Upper Lualaba [14]	FSL	L	3	4	7	CO	
Upper Nile [15]	FSL	M	1	4	5	BO	GO
Upper Zambezi Floodplains [16]	FSL	M	2	2	4	NI	
Ashanti [17]	MFR	H	2	4	6	BO	
Cuvette Centrale [18]	MFR	L	2	4	6	BO	
Central West Coastal Equatorial [19]	MFR	M	3	6	9	GO	
Eburneo [20]	MFR	H	2	2	4	NI	
Kasai [21]	MFR	L	2	6	8	CO	
Lower Congo [22]	MFR	L	3	2	5	BO	
Madagascar Eastern Lowlands [23]	MFR	M	2	6	8	CO	
Malebo Pool [24]	MFR	L	3	2	5	BO	
Northern Upper Guinea [25]*	MFR	H	2	6	8	CO	
Northern West Coastal Equatorial [26]	MFR	H	3	6	9	GO	
Sangha [27]	MFR	M	2	2	4	NI	
Southern Upper Guinea [28]	MFR	M	2	4	6	BO	
Southern West Coastal Equatorial [29]	MFR	L	3	6	9	GO	
Sudanic Congo (Oubangi) [30]	MFR	M	2	4	6	BO	
Upper Congo [31]	MFR	L	2	2	4	NI	
Upper Niger [32]	MFR	H	1	2	3	NI	
Cape Fold [33]	MS	H	3	6	9	GO	
Permanent Maghreb [34]	MS	H	3	6	9	GO	GO
Albertine Highlands [35]	HMS	L	2	4	6	BO	
Amatolo-Winterberg Highlands [36]	HMS	M	1	4	5	BO	
Drakensberg-Maloti Highlands [37]	HMS	M	1	4	5	BO	
Eastern Zimbabwe Highlands [38]	HMS	L	2	2	4	NI	
Ethiopian Highlands [39]	HMS	M	2	4	6	BO	

Major habitat types: CBSL, closed basins and small lakes; FSL, floodplains, swamps, and lakes; MFR, moist forest rivers; MS, mediterranean systems; HMS, highland and mountain systems; IRL, island rivers and lakes; LL, large lakes; LRD, large river deltas; LRR, large river rapids; SDFR, savanna–dry forest rivers; SSS, subterranean and spring systems; XS, xeric systems.

Data quality and likelihood of future threats: H, high; M, medium; L, low.

Biological distinctiveness, phenomenon, and final biological distinctiveness: GO, globally outstanding; CO, continentally outstanding; BO, bioregionally outstanding; NI, nationally important.

*Northern Upper Guinea [25] was elevated to Class I for representation of the Upper Guinea bioregion in the results of the prioritization.

Final Biological Distinctiveness	Land-Based Threat (1–3)	Aquatic Habitat Threat (1–3)	Biota Threat (1–3)	Snapshot Status Points (3–9)	Snapshot Conservation Status	Likelihood of Future Threats	Final Conservation Status, Modified by Future Threat	Priority Class (I–V)
CO	3	1	3	7	Vulnerable	H	Endangered	II
GO	3	3	2	8	Endangered	H	Critical	I
GO	3	1	1	5	Relatively stable	H	Vulnerable	I
NI	3	2	2	7	Vulnerable	H	Endangered	IV
GO	3	2	1	6	Vulnerable	H	Endangered	I
GO	2	1	3	6	Vulnerable	M	Vulnerable	I
CO	2	2	3	7	Vulnerable	H	Endangered	II
NI	2	2	2	6	Vulnerable	M	Vulnerable	V
NI	2	3	2	7	Vulnerable	H	Endangered	IV
NI	2	1	1	4	Relatively stable	M	Relatively stable	V
CO	3	1	1	5	Relatively stable	M	Relatively stable	III
GO	2	2	1	5	Relatively stable	H	Vulnerable	I
NI	3	1	1	5	Relatively stable	M	Relatively stable	V
CO	1	1	1	3	Relatively Intact	M	Relatively Intact	III
GO	2	1	1	4	Relatively stable	H	Vulnerable	I
NI	1	1	1	3	Relatively Intact	M	Relatively Intact	V
BO	3	2	1	6	Vulnerable	H	Endangered	IV
BO	2	1	1	4	Relatively stable	M	Relatively stable	V
GO	1	2	1	4	Relatively stable	L	Relatively stable	III
NI	3	2	1	6	Vulnerable	H	Endangered	IV
CO	2	1	1	4	Relatively stable	M	Relatively stable	III
BO	3	2	1	6	Vulnerable	H	Endangered	IV
CO	3	3	3	9	Critical	H	Critical	II
BO	3	1	2	6	Vulnerable	H	Endangered	IV
CO	3	1	2	6	Vulnerable	H	Endangered	I*
GO	3	2	1	6	Vulnerable	H	Endangered	I
NI	1	2	1	4	Relatively stable	M	Relatively stable	V
BO	3	3	1	7	Vulnerable	H	Endangered	IV
GO	1	1	2	4	Relatively stable	H	Vulnerable	I
BO	1	1	1	3	Relatively Intact	M	Relatively Intact	V
NI	1	1	1	3	Relatively Intact	M	Relatively Intact	V
NI	3	1	1	5	Relatively stable	H	Vulnerable	V
GO	3	3	3	9	Critical	H	Critical	I
GO	2	3	3	8	Endangered	H	Critical	I
BO	3	2	1	6	Vulnerable	H	Endangered	IV
BO	2	2	3	7	Vulnerable	M	Vulnerable	V
BO	3	3	3	9	Critical	H	Critical	IV
NI	2	2	2	6	Vulnerable	H	Endangered	IV
BO	3	2	2	7	Vulnerable	H	Endangered	IV

Ecoregion	Major Habitat Type	Data Quality	Richness Points (1–3)	Endemism Points (2–6)	Total Richness and Endemism Points (3–9)	Biological Distinctiveness	Phenomenon
Fouta-Djalon [40]	HMS	M	2	4	6	BO	
Madagascar Eastern Highlands [41]	HMS	M	3	6	9	GO	
Mount Nimba [42]	HMS	L	2	4	6	BO	
Mulanje [43]	HMS	M	2	2	4	NI	
Bijagos [44]	IRL	L	2	2	4	NI	
Canary Islands [45]	IRL	H	1	0	1	NI	
Cape Verde [46]	IRL	H	1	0	1	NI	
Comoros [47]	IRL	L	3	2	5	BO	
Coralline Seychelles [48]	IRL	M	2	0	2	NI	
Granitic Seychelles [49]	IRL	H	3	4	7	CO	
São Tomé, Príncipe, and Annobon [50]	IRL	M	2	6	8	CO	
Mascarenes [51]	IRL	L	3	4	7	CO	
Socotra [52]	IRL	L	1	6	7	CO	
Lake Malawi [53]	LL	M	3	6	9	GO	GO
Lake Rukwa [54]	LL	H	1	4	5	BO	
Lake Tanganyika [55]	LL	M	3	6	9	GO	GO
Lake Turkana [56]	LL	H	1	4	5	BO	
Lakes Kivu, Edward, George, and Victoria [57]	LL	H	3	6	9	GO	GO
Niger Delta [58]	LRD	H	3	4	7	CO	GO
Nile Delta [59]	LRD	H	1	2	3	NI	
Lower Congo Rapids [60]	LRR	L	3	6	9	GO	GO
Upper Congo Rapids [61]	LRR	L	3	2	5	BO	
Bight Coastal [62]	SDFR	H	3	4	7	CO	
Cuanza [63]	SDFR	L	1	6	7	CO	
Kenyan Coastal Rivers [64]	SDFR	L	1	4	5	BO	
Lower Niger-Benue [65]	SDFR	H	3	4	7	CO	
Lower Zambezi [66]	SDFR	L	2	2	4	NI	
Madagascar Northwestern Basins [67]	SDFR	M	1	6	7	CO	GO
Madagascar Western Basins [68]	SDFR	L	1	6	7	CO	
Middle Zambezi Luangwa [69]	SDFR	M	1	2	3	NI	
Pangani [70]	SDFR	L	2	4	6	BO	
Senegal-Gambia Catchments [71]	SDFR	H	2	2	4	NI	
Eastern Coastal Basins [72]	SDFR	L	1	6	7	CO	
Southern Temperate Highveld [73]	SDFR	H	1	4	5	BO	
Uele [74]	SDFR	L	3	4	7	CO	
Volta [75]	SDFR	M	2	4	6	BO	
Zambezian Headwaters [76]	SDFR	L	2	4	6	BO	
Zambezian Lowveld [77]	SDFR	M	2	4	6	BO	
Zambezian (Plateau) Highveld [78]	SDFR	M	1	0	1	NI	
Karstveld Sink Holes [79]	SSS	M	1	4	5	BO	
Thysville Caves [80]	SSS	L	2	4	6	BO	GO
Dry Sahel [81]	XS	M	1	0	1	NI	
Etosha [82]	XS	M	3	0	3	NI	
Horn [83]	XS	L	2	2	4	NI	
Kalahari [84]	XS	L	2	0	2	NI	
Karoo [85]	XS	M	2	2	4	NI	

Final Biological Distinctiveness	Land-Based Threat (1–3)	Aquatic Habitat Threat (1–3)	Biota Threat (1–3)	Snapshot Status Points (3–9)	Snapshot Conservation Status	Likelihood of Future Threats	Final Conservation Status, Modified by Future Threat	Priority Class (I–V)
BO	3	2	1	6	Vulnerable	M	Vulnerable	V
GO	3	2	3	8	Endangered	H	Critical	I
BO	3	2	1	6	Vulnerable	H	Endangered	IV
NI	3	2	1	6	Vulnerable	H	Endangered	IV
NI	2	1	1	4	Relatively stable	M	Relatively stable	V
NI	2	2	3	7	Vulnerable	H	Endangered	IV
NI	3	2	1	6	Vulnerable	H	Endangered	IV
BO	3	3	1	7	Vulnerable	H	Endangered	IV
NI	1	1	1	3	Relatively Intact	L	Relatively Intact	V
CO	2	3	2	7	Vulnerable	M	Vulnerable	II
CO	3	1	1	5	Relatively stable	H	Vulnerable	II
CO	3	3	2	8	Endangered	H	Critical	II
CO	2	1	1	4	Relatively stable	M	Relatively stable	III
GO	3	1	1	5	Relatively stable	H	Vulnerable	I
BO	2	1	2	5	Relatively stable	M	Relatively stable	V
GO	2	1	2	5	Relatively stable	H	Vulnerable	I
BO	2	3	1	6	Vulnerable	H	Endangered	IV
GO	3	2	3	8	Endangered	H	Critical	I
GO	3	3	3	9	Critical	H	Critical	I
NI	3	3	2	8	Endangered	H	Critical	IV
GO	3	2	2	7	Vulnerable	H	Endangered	I
BO	3	1	1	5	Relatively stable	H	Vulnerable	V
CO	3	2	3	8	Endangered	H	Critical	II
CO	1	2	1	4	Relatively stable	H	Vulnerable	II
BO	3	2	2	7	Vulnerable	H	Endangered	IV
CO	3	3	2	8	Endangered	H	Critical	II
NI	2	3	1	6	Vulnerable	H	Endangered	IV
GO	2	2	3	7	Vulnerable	H	Endangered	I
CO	2	1	3	6	Vulnerable	M	Vulnerable	II
NI	2	3	3	8	Endangered	M	Endangered	IV
BO	2	2	3	7	Vulnerable	H	Endangered	IV
NI	2	3	2	7	Vulnerable	M	Vulnerable	V
CO	3	2	1	6	Vulnerable	M	Vulnerable	II
BO	3	3	1	7	Vulnerable	H	Endangered	IV
CO	1	2	1	4	Relatively stable	M	Relatively stable	III
BO	3	3	3	9	Critical	H	Critical	IV
BO	1	1	1	3	Relatively Intact	M	Relatively Intact	V
BO	2	3	2	7	Vulnerable	M	Vulnerable	V
NI	3	3	3	9	Critical	M	Critical	IV
BO	2	3	2	7	Vulnerable	H	Endangered	IV
GO	3	2	1	6	Vulnerable	H	Endangered	I
NI	1	1	1	3	Relatively Intact	M	Relatively Intact	V
NI	2	1	1	4	Relatively stable	H	Vulnerable	V
NI	1	2	1	4	Relatively stable	M	Relatively stable	V
NI	1	1	1	3	Relatively Intact	L	Relatively Intact	V
NI	1	3	1	5	Relatively stable	H	Vulnerable	V

Ecoregion	Major Habitat Type	Data Quality	Richness Points (1–3)	Endemism Points (2–6)	Total Richness and Endemism Points (3–9)	Biological Distinctiveness	Phenomenon
Lower Nile [86]	XS	H	3	0	3	NI	
Madagascar Southern Basins [87]	XS	L	2	4	6	BO	
Namib Coastal [88]	XS	L	2	0	2	NI	
Red Sea Coastal [89]	XS	L	2	4	6	BO	
Shebelle-Juba Catchments [90]	XS	L	3	4	7	CO	
Southern Kalahari [91]	XS	L	1	0	1	NI	
Temporary Maghreb [92]	XS	H	2	6	8	CO	
Western Orange [93]	XS	M	3	2	5	BO	

Final Biological Distinctiveness	Land-Based Threat (1–3)	Aquatic Habitat Threat (1–3)	Biota Threat (1–3)	Snapshot Status Points (3–9)	Snapshot Conservation Status	Likelihood of Future Threats	Final Conservation Status, Modified by Future Threat	Priority Class (I–V)
NI	1	3	2	6	Vulnerable	M	Vulnerable	V
BO	2	1	2	5	Relatively stable	M	Relatively stable	V
NI	1	2	1	4	Relatively stable	H	Vulnerable	V
BO	1	2	1	4	Relatively stable	H	Vulnerable	V
CO	2	2	2	6	Vulnerable	M	Vulnerable	II
NI	1	3	2	6	Vulnerable	H	Endangered	IV
CO	1	3	1	5	Relatively stable	M	Relatively stable	III
BO	3	3	1	7	Vulnerable	H	Endangered	IV

APPENDIX E

Statistical Analyses of Biological Distinctiveness and Conservation Status Data

Patterns of Biodiversity by Major Habitat Type

We tested whether major habitat types (MHTs) differed in average levels of species richness and endemism, using one-way ANOVAs for fish, aquatic mollusk, and herpetofauna species richness, endemism, and percentage endemism, with MHT as the classification factor. All categories showed significant differences between MHTs, except for herpetofauna endemism (table E1). These results confirm that, for the Biological Distinctiveness Index (BDI) analysis, a direct comparison of all ecoregions, regardless of MHT, probably would undervalue ecoregions of depauperate MHTs (e.g., xeric and highland habitat types).

To explore the differences between MHTs further, we conducted post hoc mean pairwise comparisons using the conservative Tukey HSD test with a significance level of .05 (tables E.2–E.4). Significant differences between MHTs were found for fish species richness, endemism, and percentage endemism; aquatic mollusk richness, endemism, and percentage endemism; and herpetofauna species richness. These pairwise comparison tests highlight the most striking differences between MHTs. They show that, on average, large lakes and moist forest rivers are far more speciose for fish than are most other MHTs (table E.2). Large lakes also exhibit significantly higher rates of fish endemism than all other MHTs, as do subterranean and spring systems for fish percentage endemism. As with fish, mollusk richness and endemism are highest, on average, in the large lakes MHT; aquatic mollusk richness is also high in savanna–dry forest rivers (table E.3). Herpetofauna biodiversity appears to be more evenly distributed between MHTs, with herpetofauna richness being highest in savanna–dry forest rivers and moist forest rivers (table E.4).

TABLE E.1. One-Way ANOVAs Testing for Differences between Major Habitat Types.

Taxon	Richness	Endemism	Percentage Endemism
Fish	15.1***	8.9***	7.4***
Aquatic mollusks	3.9***	4.4***	5.3***
Herpetofauna	7.9***	1.8	2.4*

*$p < .05$, **$p < .01$, ***$p < .001$; no entry, $p > .05$; $N = 93$ for fish and herpetofauna, $N = 88$ for aquatic mollusks.

Relationship between Richness, Endemism, and Percentage Endemism

Correlations between richness, endemism, and percentage endemism suggest that the percentage endemism statistic (number of endemic species divided by total number of species) often gives different information than the number of endemic species taken alone. There was a strong positive relationship between fish richness and endemism ($r = .65$, $p < .0001$) but only a significant, weak correlation between fish richness and percentage endemism ($r = .20$, $p = .03$). Similarly, there was a strong correlation between herpetofauna richness and endemism ($r = .63$, $p < .0001$) and mollusk richness and endemism ($r = .65$, $p < .0001$) but no significant relationship between herpetofauna richness and percentage endemism ($r = .11$, $p = .12$) or between mollusk richness and percentage endemism ($r = .14$, $p = .07$) (table E.5). This finding reinforces the use of percentage endemism as an indicator for the BDI because high levels of endemism within species-poor ecoregions might have been overlooked otherwise.

TABLE E.2. Post Hoc Mean Pairwise Comparisons Using the Conservative Tukey HSD Test with a Significance Level of .05 for Fish Richness, Endemism, and Percentage Endemism between Major Habitat Types.

Major Habitat Type	LL			MFR			LRR			SDFR			LRD			FSL		
	R	E	%	R	E	%	R	E	%	R	E	%	R	E	%	R	E	%
LL					*	*	*			*	*	*		*	*	*	*	*
MFR	*	*	*							*						*		
LRR		*																
SDFR	*	*	*	*														
LRD		*	*															
FSL	*	*	*	*														
MS	*	*		*														*
HMS	*		*	*														
CBSL	*	*		*														*
XS	*	*	*	*						*								
IRL	*	*	*	*						*						*		
SSS	*	*		*		*						*		*			*	

Major habitat types: LL, large lakes; MFR, moist forest rivers; LRR, large river rapids; SDFR, savanna–dry forest rivers; LRD, large river deltas; FSL, floodplains, swamps, and lakes; MS, mediterranean systems; HMS, highland and mountain systems; CBSL, closed basins and small lakes; XS, xeric systems; IRL, island rivers and lakes; SSS, subterranean and spring systems.

R, richness; E, endemism; %, percentage endemism.

TABLE E.3. Post Hoc Mean Pairwise Comparisons Using the Conservative Tukey HSD Test with a Significance Level of .05 for Aquatic Mollusk Richness, Endemism, and Percentage Endemism.

Major Habitat Type	LL			MFR			LRR			SDFR			LRD			FSL		
	R	E	%	R	E	%	R	E	%	R	E	%	R	E	%	R	E	%
LL					*	*				*	*					*	*	
MFR		*	*					*										
LRR						*					*			*			*	
SDFR		*	*					*										
LRD								*										
FSL		*	*					*										
MS																		
HMS	*	*	*					*		*								
CBSL		*						*										
XS	*	*	*					*		*								
IRL	*	*	*					*		*								
SSS		*	*					*										

Major habitat types: LL, large lakes; MFR, moist forest rivers; LRR, large river rapids; SDFR, savanna–dry forest rivers; LRD, large river deltas; FSL, floodplains, swamps, and lakes; MS, mediterranean systems; HMS, highland and mountain systems; CBSL, closed basins and small lakes; XS, xeric systems; IRL, island rivers and lakes; SSS, subterranean and spring systems.

R, richness; E, endemism; %, percentage endemism.

Major Habitat Type

	MS			HMS			CBSL			XS			IRL			SSS		
	R	E	%	R	E	%	R	E	%	R	E	%	R	E	%	R	E	%
	*	*		*	*	*	*	*		*	*	*	*	*	*	*	*	
	*			*			*			*			*			*		*
										*			*					*
																		*
			*						*	*			*					*
												*			*			
												*			*			
			*						*									*
			*						*									*
												*			*			

Major Habitat Type

	MS			HMS			CBSL			XS			IRL			SSS		
	R	E	%	R	E	%	R	E	%	R	E	%	R	E	%	R	E	%
				*	*	*	*			*	*	*	*	*	*	*	*	
						*			*			*			*			*
				*						*			*					

TABLE E.4. Post Hoc Mean Pairwise Comparisons Using the Conservative Tukey HSD Test with a Significance Level of .05 for Herpetofauna Richness, Endemism, and Percentage Endemism.

Major Habitat Type	LL			MFR			LRR			SDFR			LRD			FSL		
	R	E	%	R	E	%	R	E	%	R	E	%	R	E	%	R	E	%
LL																		
MFR																		
LRR																		
SDFR																		
LRD																		
FSL																		
MS																		
HMS																		
CBSL																		
XS				*						*								
IRL	*			*						*						*		
SSS				*						*								

Major habitat types: LL, large lakes; MFR, moist forest rivers; LRR, large river rapids; SDFR, savanna–dry forest rivers; LRD, large river deltas; FSL, floodplains, swamps, and lakes; MS, mediterranean systems; HMS, highland and mountain systems; CBSL, closed basins and small lakes; XS, xeric systems; IRL, island rivers and lakes; SSS, subterranean and spring systems.

R, richness; E, endemism; %, percentage endemism.

Relationships of Diversity between Faunal Groups

Because of a lack of distributional data for other aquatic groups, we rely on fish, aquatic mollusks, and selected amphibians, mammals, and reptiles as the basis for our biodiversity measures. We are forced to assume that these groups serve as reasonable proxies for the distribution of other taxa, particularly those with entirely aquatic life cycles (Heino 2002; Kretschmar 2003). Although we cannot test the validity of this assumption directly (because we have no distributional data for unmapped taxa), we were able to test the correspondence between three of our mapped groups (we did not include aquatic mammals because of the small numbers of mammals in the dataset).

Pearson correlation correspondence analyses show strong associations between fish and mollusk richness ($r = .61, p < .0001$) and fish and herpetofauna richness ($r = .54, p < .0001$). A still significant yet weaker correlation exists between herpetofauna and mollusk richness ($r = .41, p < .0001$). Fish and mollusk endemism ($r = .67, p < .0001$) are highly correlated; however, fish and herpetofauna endemism ($r = .07, p = .48$) and mollusk and herpetofauna endemism ($r = .06, p = .56$) are not significantly correlated. This suggests that general habitat conditions favoring high fish and mollusk diversity may also favor amphibians and reptiles but that different mechanisms may have caused localized speciation in aquatic herpetofauna.

TABLE E.5. Pearson Correlations of Richness and Endemism for Each Taxon.

Taxon	Richness vs. Endemism	Richness vs. Percentage Endemism	Endemism vs. Percentage Endemism
Fish	.65***	.20	.65***
Mollusks	.65***	.14	.56***
Herpetofauna	.63***	.11	.55***

*$p < .05$, **$p < .01$, ***$p < .001$; no entry, $p > .05$.

Relationship of Ecoregion Area to Richness and Endemism

Ecoregions vary widely in area (table E.6). The mean ecoregion area is 322,066 km^2, with a minimum of 34 km^2 and a maximum of 4,539,226 km^2. We tested the assumption that ecoregion size was not significantly associated with the number of species. Broadly, the species–area relationship is nearly ubiquitous in ecology (Rosenzweig 1995). Numerous studies show a relationship between basin area and freshwater species richness in Africa and other parts of the world (Eadie et al. 1986; Hugueny 1989).

Major Habitat Type

	MS			HMS			CBSL			XS			IRL			SSS		
	R	E	%	R	E	%	R	E	%	R	E	%	R	E	%	R	E	%
														*			*	
										*				*			*	
											*			*			*	
														*				
														*				
					*													

Richness in all taxa and overall richness were significantly related to area: fish, $R^2 = .11$, $p = .0013$; mollusks, $R^2 = .16$, $p < .0001$; herpetofauna, $R^2 = .15$, $p < .0001$; and total richness, $R^2 = .14$, $p = .0002$ (figure E.1). Several major habitat types with a sample size of at least nine ecoregions (floodplains, swamps, and lakes; xeric systems; and highland and mountain systems) also had a significant correlation between log of ecoregion area and total species richness (table E.7). To account for this relationship, richness values have been adjusted by ecoregion area, using the methods outlined in the next section. Although these results were not significant for some MHTs, this may result from small sample sizes, so we controlled for ecoregion area in all MHTs. However, total endemism and percentage endemism were not significantly related to area (endemism, $R^2 = .016$, $p = .22$; percentage endemism, $R^2 = .002$, $p = .70$), so we did not adjust endemism values by ecoregion area.

Adjusting Species Richness by Area

Theoretically, species richness adjusted by area could be estimated for each ecoregion in each MHT by sampling several plots of fixed area inside each ecoregion to determine a value for richness adjusted by area (A. Balmford, pers. comm., 2000). However, such data are not available. An alternative is to correct the

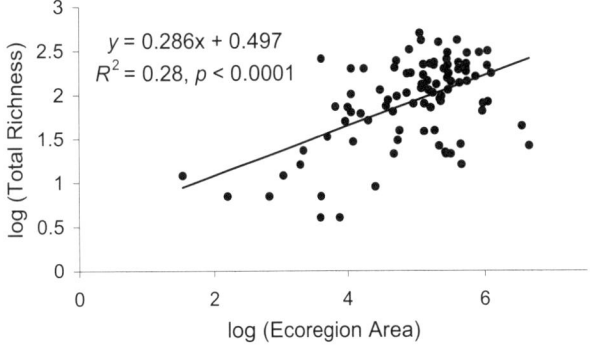

FIGURE E.1. Relationship between ecoregion area and species richness in the 93 freshwater ecoregions.

richness score according to the area of each ecoregion. We have taken this approach and used the following equation to estimate the corrected species richness:

$$\log(\text{species richness}) = 0.25 \, (\log[\text{area km}^2])$$

We then used the residuals of the fitted relationship to determine the relative richness of each ecoregion in an MHT, tak-

TABLE E.6. **Area and Latitude of Ecoregions.**

Ecoregion	Area (km²)	Latitude	Ecoregion	Area (km²)	Latitude
Lakes Chilwa and Chiuta [1]	6,559	−15.37	Coralline Seychelles [48]	34	−9.40
Southern Eastern Rift [2]	163,912	−2.82	Granitic Seychelles [49]	635	−4.67
Lake Tana [3]	15,665	11.87	São Tomé, Príncipe, and Annobon [50]	1,028	0.11
Northern Eastern Rift [4]	188,033	10.66	Mascarenes [51]	2,065	−21.12
Western Equatorial Crater Lakes [5]	11,216	4.70	Socotra [52]	3,818	12.51
Bangweulu-Mweru [6]	292,069	−10.89	Lake Malawi [53]	113,719	−12.22
Inner Niger Delta [7]	147,144	15.34	Lake Rukwa [54]	73,138	−7.85
Kafue [8]	121,041	−14.99	Lake Tanganyika [55]	119,554	−6.29
Lake Chad Catchment [9]	1,086,924	10.32	Lake Turkana [56]	231,600	4.96
Mai Ndombe [10]	282,679	−2.06	Lakes Kivu, Edward, George, and Victoria [57]	399,912	−0.24
Malagarasi-Moyowosi [11]	119,538	−4.79	Niger Delta [58]	48,808	5.33
Okavango Floodplains [12]	130,918	−18.16	Nile Delta [59]	57,760	30.52
Tumba [13]	20,214	−0.68	Lower Congo Rapids [60]	17,438	−5.02
Upper Lualaba [14]	10,018	−9.36	Upper Congo Rapids [61]	416,417	0.49
Upper Nile [15]	1,253,426	7.80	Bight Coastal [62]	182,807	7.96
Upper Zambezi Floodplains [16]	286,223	−15.96	Cuanza [63]	323,588	−11.24
Ashanti [17]	74,120	6.17	Kenyan Coastal Rivers [64]	176,731	−2.27
Cuvette Centrale [18]	11,303	0.45	Lower Niger-Benue [65]	1,105,219	11.04
Central West Coastal Equatorial [19]	213,055	2.28	Lower Zambezi [66]	129,823	−16.29
Eburneo [20]	255,296	7.90	Madagascar Northwestern Basins [67]	153,049	−15.74
Kasai [21]	554,077	−5.97	Madagascar Western Basins [68]	235,349	−20.04
Lower Congo [22]	51,681	−5.99	Middle Zambezi Luangwa [69]	293,623	−14.15
Madagascar Eastern Lowlands [23]	30,042	−18.97	Pangani [70]	52,719	−4.40
Malebo Pool [24]	4,946	−4.14	Senegal-Gambia Catchments [71]	532,138	14.54
Northern Upper Guinea [25]	161,957	9.65	Eastern Coastal Basins [72]	739,196	−11.71
Northern West Coastal Equatorial [26]	79,774	5.31	Southern Temperate Highveld [73]	554,780	−28.72
Sangha [27]	833,863	1.15	Uele [74]	539,910	3.80
Southern Upper Guinea [28]	128,315	6.60	Volta [75]	414,582	10.30
Southern West Coastal Equatorial [29]	288,091	−2.61	Zambezian Headwaters [76]	536,321	−13.79
Sudanic Congo (Oubangi) [30]	3,924	2.45	Zambezian Lowveld [77]	520,418	−24.64
Upper Congo [31]	183,926	−3.64	Zambezian (Plateau) Highveld [78]	222,705	−19.11
Upper Niger [32]	300,226	12.01	Karstveld Sink Holes [79]	25,526	−19.42
Cape Fold [33]	132,687	−33.17	Thysville Caves [80]	91,659	−5.11
Permanent Maghreb [34]	950,180	31.14	Dry Sahel [81]	4,539,226	21.81
Albertine Highlands [35]	84,355	−2.60	Etosha [82]	198,914	−17.23
Amatolo-Winterberg Highlands [36]	12,017	−32.67	Horn [83]	268,024	8.09
Drakensberg-Maloti Highlands [37]	47,494	−29.92	Kalahari [84]	445,216	−21.82
Eastern Zimbabwe Highlands [38]	39,221	−18.91	Karoo [85]	216,667	−30.78
Ethiopian Highlands [39]	430,775	9.72	Lower Nile [86]	1,124,184	19.69
Fouta-Djalon [40]	36,545	11.20	Madagascar Southern Basins [87]	46,501	−24.59
Madagascar Eastern Highlands [41]	128,424	−18.38	Namib Coastal [88]	321,078	−22.08
Mount Nimba [42]	11,255	8.10	Red Sea Coastal [89]	272,427	20.53
Mulanje [43]	9,218	−16.28	Shebelle-Juba Catchments [90]	923,233	3.65
Bijagos [44]	1,844	11.47	Southern Kalahari [91]	456,437	−25.28
Canary Islands [45]	7,557	28.40	Temporary Maghreb [92]	3,564,107	26.81
Cape Verde [46]	3,909	15.11	Western Orange [93]	54,273	−28.22
Comoros [47]	160	−12.18			

ing area into account. The coefficient of 0.25 represents a generalized value derived from numerous continental terrestrial and freshwater datasets of species and area (Welcomme 1979; Hugueny 1989; Rosenzweig 1995; Matthews and Robison 1998). Applying this equation provides an estimation of species richness in each ecoregion, with the effects of area largely removed, and permits a better assessment of the relative species richness values of the ecoregions in each MHT.

Relationship of Latitude to Biodiversity Index Scores

Biogeographic studies have established that species richness of aquatic vertebrates increases with decreasing latitude, although this may not be true for aquatic invertebrates (Allan and Flecker 1993; Ricketts et al. 1999b; Heino 2002). We investigated the relationship between latitude and species richness for our ecoregions. We calculated the latitude of each ecoregion's centroid with a geographic information system and correlated these latitudes with richness of fish, aquatic mollusk, and herpetofauna richness (mammal richness was not used because of the small number of species). The relationships between latitude and fish

TABLE E.7. **Results of Pearson Correlation Analyses of Log (Ecoregion Area) and Total Richness among Major Habitat Types.**

Major Habitat Type	r	N
Floodplains, swamps, and lakes	.69**	11
Moist forest rivers	.06	16
Highland and mountain systems	.53*	9
Island rivers and lakes	.02	9
Savanna–dry forest rivers	.35*	17
Xeric systems	.01	13

Note: $*p < .05$, $**p < .01$, $*** p < .001$; no entry, $p > .05$.

($r = .14$, $p = .18$), aquatic mollusk ($r = .12$, $p = .26$), and herpetofauna richness ($r = .18$, $p = .08$) were nonsignificant and positive. It is surprising that latitude does not appear to have a significant effect on the richness of Afro-Malagasy ecoregions, but this finding may result from confounding effects such as ecoregion area and MHT.

APPENDIX F

Ecoregion Descriptions

Ecoregion Number: **1**
Ecoregion Name: **Lakes Chilwa and Chiuta**
Major Habitat Type: **Closed Basins and Small Lakes**
Final Biological Distinctiveness: **Continentally Outstanding**
Final Conservation Status: **Endangered**
Priority Class: **II**
Author: **Denis Tweddle**
Reviewer: **Paul Skelton**

Location and General Description

Chilwa and Chiuta are shallow lakes at the headwaters of the Lugenda River, a tributary of the Ruvuma River, a large river that forms the border of Tanzania and Mozambique and flows to the Indian Ocean. Chilwa is a unique, closed system with extensive *Typha* swamps, saline waters, and highly variable lake extent and water quality. The lakes lie in the southern part of a northeast-southwest trending tectonic depression found at 622 m a.s.l. and located to the east of the main Rift Valley. Chiuta straddles the border between Mozambique and Malawi, whereas Chilwa is almost entirely in Malawi. Chiuta is intermittently linked with the Lugenda River, and Chilwa is separated from Chiuta by a sandbar some 25 m high and approximately 9,000 years old (Lancaster 1979). The ecoregion is defined by the drainage basins of these lakes and the unique assemblages of species they contain.

The climate in the ecoregion is strongly seasonal, resulting from the interaction of the southern African anticyclone and the Intertropical Convergence. This results in a hot wet season, usually from November or December to March or April, a cool dry season from May to September, and a hot, dry period in October and November. Mean monthly maximum temperatures range from 24°C in July to 34°C in October.

Lake Chilwa is shallow (generally 2–3 m deep) and saline and fluctuates greatly in size from year to year, depending on rainfall in the catchment. When full it covers almost 2,000 km². The lake has frequent major recessions and dries out occasionally, such as in 1960 and 1967. After the 1967 recession, the lake refilled to normal levels in one rainfall season. The water is alkaline, with a pH between 8 and 9. The lake has a high chloride concentration, unusual in inland saline waters, perhaps because of three underground hot springs (McLachlan 1979). In normal years, when the lake is about 3 m deep, conductivity ranges from 1,000 to 2,500 µS/cm. Surface temperature varies from 20°C to 39°C. The lake has no stable thermal stratification, and turbidity is high. These conditions have a major impact on the lake's flora and fauna.

Ten perennial streams and numerous seasonal streams flow into Chilwa and its marginal swamps and marshes (Lancaster 1979). These rivers and streams drain from isolated mountains, including the Mulanje and Chiradzulu mountains, the Zomba and Malosa plateaux, Chikala Hills, and a range of large hills east of Mecanhelas in Mozambique.

Lake Chiuta is slightly less shallow (3–4 m) and ranges in size from 25 to 130 km² according to season and rainfall. No permanent rivers enter the lake, but there are three major seasonal streams on the Malawi side (Tweddle 1983). Unlike Chilwa, Chiuta is a permanent lake, perhaps because of its greater depth and more reliable rainfall in the catchment. Conductivity measurements ranged from 120 to 200 µS/cm, and the average pH is 7.

The plains surrounding Lake Chilwa are heavily cultivated, and only isolated areas of the original *Brachystegia* savanna-woodland remains on the highest parts. Old beach terraces at altitudes of 627, 631, 637, and 652 m a.s.l. indicate large historical changes in lake level. On the more sandy, upper beach terraces, savanna vegetation, including *Terminalia sericea* trees and scattered *Combretum* shrubs, is interspersed with tracts of open grassland. Closer to the lake and on more poorly drained soils, a few huge individual trees of *Faidherbia albida, Combretum imberbe,* and *Kigelia africana* grow adjacent to riparian groves of *Acacia xanthophloea* and *Ficus* species. Trees disappear on the terrace about 4–5 m above the present lake. Baobabs (*Adansonia digitata*) cover the lower slopes of the volcanic (long extinct) Nchisi Island, which is about 3 km offshore in the west of the lake. Extensive swamps cover up to half of the lake's surface area, with swamps particularly common in the northern portion of the lake.

Outstanding or Distinctive Biodiversity Features

Lake Chilwa is important for the large congregations of waterbirds that occur on its floodplain (Fishpool and Evans 2001). Lakes Chilwa and Chiuta contain few, if any, endemic species, and aquatic species richness is low because of the high salinity of the water. Lake Chiuta's more diverse aquatic flora is attributable to its somewhat less saline water.

The existence in Chilwa of extensive beds of *Typha*, rather than *Cyperus papyrus*, can almost certainly be ascribed to the high conductivity and alkalinity of the lake muds, giving *Typha* a competitive advantage (Howard-Williams 1979). The growth of floating plant species in Chilwa is retarded by the high salinity of the water. For example, *Salvinia hastata,* common in the nearby Shire River, is absent. In Lake Chiuta, the open lake historically contained extensive beds of *Utricularia* species and *Potamogeton schweinfurthii* to a lesser extent (Tweddle 1983). The southern area was densely covered by *Vossia cuspidata*, with a 10- to 30-m wide *Typha* zone along the western shore (Tweddle 1983). An increase in open water seine netting, beginning in the 1980s, adversely affected the aquatic plant beds.

The extensive marshes, *Typha* swamp, floodplain grasslands, open water, rice paddies, and human-made lagoons provide habitat for many waterbirds (Dudley et al. 1979). More than 150 species of resident birds and 37 Palearctic birds have been observed at Chilwa; of these species, 22 are regular visitors from September to April (Dudley et al. 1979). BirdLife has identified Lake Chilwa as an important area for congregations of waterbirds. Significant populations of *Egretta ardesiaca, Plegadis falcinellus, Platalea alba, Dendrocygna bicolor, Amaurornis flavirostra, Gallinula angulata, Porphyrio alleni, P. porphyrio,* and *Rynchops flavirostris* have been documented on the floodplains of the lake (Dowsett-Lemaire et al. 2001). The seasonal and long-term changes in lake level have major impacts on floodplain inundation and consequently on waterbird populations. The site also holds populations of the vulnerable lesser kestrel (*Falco naumanni*), the locally rare pallid harrier (*Circus macrourus*), and great snipe (*Gallinago media*).

The ecoregion is rich in amphibian and reptile species, with about forty-five aquatic or wetland species reported. Thirteen amphibian species are known from the vicinity of Lake Chilwa, although none occur in Lake Chilwa itself because of the saline water (Dudley et al. 1979). However, the clawed frog *Xenopus muelleri* is common in the temporary pools in floodplain depressions. Other species around the periphery of the lake include *Phrynobatrachus acridoides, Ptychadena mascareniensis, Hyperolius parallelus albofasciatus,* and *Pyxicephalus adspersus.* The soft-shelled turtle (*Cycloderma frenatum*) inhabits the main lake, and the serrated terrapin (*Pelusios sinuatus*) occupies parts of the swamp and river marshes. The Nile crocodile (*Crocodylus niloticus*) is very rare in Lake Chilwa because of intensive hunting, whereas the Nile monitor (*Varanus niloticus*) is fairly common. *Python sebae* is found in the vicinity of streams, rivers, and lake edges. The aquatic mammal fauna includes *Hippopotamus amphibius* in very small numbers, along with Cape clawless otters (*Aonyx capensis*) and water mongoose (*Atilax paludinosus*).

In Lake Chiuta, little systematic sampling of organisms other than fish has taken place. Thirty-eight fish species have been recorded from Chiuta, and thirty-one of these same species have been recorded from the Chilwa catchment (Tweddle 1979). About half of this ecoregion's fish species also occur in the neighboring Lake Malawi [53] system. The fauna is typical of the East Coast bioregion, including *Barbus atkinsoni, B. zanzibaricus, Bagrus orientalis,* and *Pareutropius longifilis. Haplochromis tweddlei* is known only from Lakes Chilwa and Chiuta and may be endemic to the larger Ruvuma system (Roberts 1975; Tweddle 1983). The presence of the tank goby (*Glossogobius giuris*) in Chiuta suggests that there is (or historically was) no barrier to upstream movement of fishes in the Ruvuma and Lugenda River system. The absence in the Chilwa catchment of Chiuta species *Mormyrus longirostris, M.* cf. *brevianalis, Labeo* sp., *B. orientalis,* and *Synodontis* sp. probably is related to a lack of suitable habitat and historical access. The absence of *G. giuris* is less easy to explain because it can survive in saline conditions.

Changes in the algal community of Lake Chilwa have been described as the lake dried out and subsequently refilled in the late 1960s and early 1970s (Moss 1979). In the predrying phase, the water was dominated by dense growth of blue-green algae (*Oscillatoria planctonica* and *Anabaena torulosa*). In the drying phase, when the water was very alkaline and saline, it contained only the planktonic, filamentous blue-green algae of the genera *Arthrospira, Spirulina,* and *Anabaenopsis.* After many successional changes during and after refilling, the lake eventually displayed a much richer flora, in which *Oscillatoria* sp. was most abundant, with *Trachelomonas* spp., *Euglena spirogyra, Phacus* sp., *Cyclotella* sp., *Nitzschia* sp., *Anabaena* sp., *Scenedesmus quadricauda,* and *Peridinium* sp. all present. The zooplankton and their adaptations to changes in lake level were reviewed by Kalk (1979), who showed that in all phases of the lake, there were only three abundant species: the cladocerans *Diaphanosoma ex-*

cisum and *Daphnia barbata* and the calanoid copepod *Tropodiaptomus kraepelini.*

The invertebrate communities of the lake and swamp were described by Cantrell (1979), who distinguished four major habitats: swamp margins, *Typha* swamp, marsh and channels, and lagoons. Floating surfaces such as *Pistia*, rafts of dead *Typha,* and marsh grasses support many invertebrates, particularly midges and water beetles, whereas *Typha* swamp, although covering a large area, has a species-poor fauna. *Bulinus globosus,* the intermediate host of *Schistosoma haematobium*, is particularly common at the end of canoe channels but absent from open waters.

Status and Threats

Current Status

In 1997, Malawi designated Lake Chilwa a Ramsar site, and a Lake Chilwa Wetland and Catchment Management Project began in 1998. Twenty-two studies conducted in 1999 resulted in the Lake Chilwa Wetland State of the Environment Report in 2000 (Environmental Affairs 2000), and a Wetland Management Plan is in preparation. Although the lake is a closed basin, chemical pollution has not been a threat up to the present. The region is primarily agricultural with little industry, and farm inputs such as fertilizers and pesticides are minimal. However, the Chilwa Basin is heavily affected by human population pressure. Excessive fishing in rivers and river mouths affects reproduction of matemba, *Barbus paludinosus,* and fishing continues in the closed season with nets of very small mesh size. There is major catchment degradation from the cutting of trees, uncontrolled bush fires, and cultivation on marginal land, all of which are leading to increased levels of sedimentation.

High levels of trapping and hunting of birds near Lake Chilwa are causing the decline of populations of several species. It is estimated that more than a million waterfowl were snared between December and April 1998 and more than 70,000 birds were shot. The commercial exploitation of waterfowl started on a large scale in 1996 and has already caused several species, including two crane species and *Ephippiorhynchus sengalensis,* to be eliminated from the area. A Danish-funded project is attempting to set up refugia for breeding birds and to promote sustainable management of waterfowl resources by the local villages (Dowsett-Lemaire et al. 2001).

Excessive exploitation of Lake Chiuta's natural resources is also a problem. An influx of fishers using small-meshed open water seine nets (*nkacha*) from nearby Lake Malombe has resulted in extensive destruction of submerged vegetation, an important nursery habitat for tilapias and other species, and also in excessive exploitation of the juvenile fish (D. Tweddle, pers. obs., late 1980s). Conflicts over fishing rights and methods have arisen between local fishers and newcomers; attempts are under way to develop community structures to manage the fisheries of both lakes.

Future Threats

The human population continues to expand, so existing problems are likely to be exacerbated. The risk of pollution in the closed Chilwa Basin still exists, but with no immediate threat apparent.

Justification for Ecoregion Delineation

The Chilwa-Chiuta Basin dates back to the close of the Cretaceous period, 65–70 m.y.a., when a broad, arched highland area drained north along the axis of the present depression. Rifting in the Tertiary isolated the Shire River to the west, and uplifting along the Shire highlands' axis and relative downwarping of the Chilwa plain in the later Tertiary and Quaternary ponded back part of the former drainage to form a large lake and separated it from the Shire River drainage. The formation of the sandbar later separated the original lake into the two present lakes. The Chilwa-Chiuta system has an east coast fauna (Roberts 1975), isolated from Zambezian elements found in the Malawi and Shire systems immediately to the west. The ecoregion is defined by the drainage basins of lakes Chilwa and Chiuta.

Data Quality: High

Lake Chilwa was the subject of an intensive and extensive research project coordinated by the University of Malawi from 1966 to 1976, a period that encompassed a drying out of the lake and subsequent refilling and recovery of the ecosystem. The lake is thus a very well-known system (Kalk et al. 1979). A checklist of the plants and animals of Lake Chilwa is included in Kalk et al. (1979). Since 1999, the lake has been the subject of studies related to the development of the Wetland Management Plan (Environmental Affairs 2000). Lake Chiuta has been less well studied but nevertheless is reasonably well known (Tweddle 1983). Tweddle (1983) provided an identification key for the fishes of both lakes. Several of the fish species (e.g. *Synodontis* sp., *Mesobola* cf. *brevianalis*, *Labeo* sp., the Lake Chiuta form of *Labeo cylindricus,* and three small *Barbus*) warrant further taxonomic study.

Ecoregion Number: **2**
Ecoregion Name: **Southern Eastern Rift**
Major Habitat Type: **Closed Basins and Small Lakes**
Final Biological Distinctiveness: **Globally Outstanding**
Final Conservation Status: **Critical**
Priority Class: **I**
Authors: **Ashley Brown and Robin Abell**
Reviewers: **Samuel Matagi and Nathan Gichuki**

Location and General Description

The Southern Eastern Rift ecoregion, with its highlands, lowlands, freshwater and saline lakes, and wetlands, hosts outstanding congregations of flamingos and several endemic freshwater fish. This Rift Valley system, which is also known as the Gregory Rift Valley, stretches for more than 700 km from central Kenya to northern Tanzania and covers an area of approximately 3,800 km^2 (Hughes and Hughes 1992). The valley varies in width from 50 to 100 km and in elevation from 1,850 m a.s.l. in central Kenya to 600 m a.s.l. in northern Tanzania. The highlands on either side of the valley range in altitude from 2,000 m a.s.l. to 3,300 m a.s.l., and especially in Kenya they are characterized by forest ecosystems, woodlands, and open grasslands (often on old volcanoes). They receive annual precipitation of 1,200–2,000 mm, whereas the valley floor receives 600–900 mm per year. Volcanic activity manifests itself through steam vents and geysers, plugs, gorges, cliffs, calderas, and cones.

Shallow lakes, rivers, and streams, hot and cold water springs, marshes, swamps, and salt pans occur in the ecoregion. There are also several artificial wetlands, such as dams, fish ponds, sewage lagoons, and irrigated fields. The major freshwater lakes from north to south are Baringo (130 km^2) and Naivasha (156 km^2) in Kenya, and Babati, Burungi, and Kitangiri in northern Tanzania. The main saline or alkaline lakes are Bogoria (42 km^2), Nakuru (49 km^2), Elementaita (19 km^2), and Magadi (105 km^2) in Kenya and Natron (900 km^2), Manyara (470 km^2), Eyasi (1050 km^2), Barangida, and Singidani in Tanzania. There are also other small lakes (less than 20 km^2) on the valley floor, including lakes Solai, Kabongo, Kwenia, Sonachi, Oloidien, and Ol'bolossat in Kenya and Bdrangida, Lelu, and Momera in Tanzania. These lakes are salty, and some of them have high concentrations of fluoride and sodium carbonate salts. The lakes lie in the primarily evergreen bushland portion of the Eastern Rift Valley.

Saline and soda lakes are more abundant in the Eastern Rift than in any other African ecoregion. Streams that feed the lakes flow over highly alkaline volcanic rocks, bringing natron (a naturally occurring salt consisting of sodium carbonate and sodium bicarbonate) into the lakes' waters (Cole 1994). Many of these lakes are endorheic, and high ambient temperature in the Rift Valley increases the rate of evaporation, thereby increasing the water's alkalinity by raising the concentrations of Na$^+$, HCO$_3^-$, and CO$_3^{2-}$. Whereas the pH of freshwater lakes ranges from 6 to 8, that of the soda lakes ranges from 9 to 12. Many of the soda lakes fluctuate in size and change in water salinity with dry and wet periods. These soda lakes vary in shape from broad, shallow pans to narrow, deep depressions (Livingstone and Melack 1984).

In the Tanzanian Rift Valley, the largest saline and soda lakes are Eyasi, Manyara, and Natron. The bed of Lake Eyasi usually is dry but may occasionally flood up to a depth of 1 m (Hughes and Hughes 1992). The Sibiti River occasionally flows into the lake, but its waters usually evaporate before reaching the dry lakebed. Even though permanent springs lie along the lakeshores, their waters evaporate quickly. To the east of Lake Eyasi lies Lake Manyara. Manyara is also shallow, with a maximum depth of 3.7 m (Hughes and Hughes 1992). Lake Natron lies to the north on the border between Kenya and Tanzania. Seasonal streams drain into Lake Natron from the Ngorongoro highlands south of the lake, Mt. Lengai (2,942 m a.s.l.) in the southeast, and the Nkito Hills in the west (Hughes and Hughes 1992). The Ewaso Ngiro River, which is the principal affluent, flows through the Ngare Ngiro Swamp or Shompole Swamp before entering the lake. About twenty-eight springs, most of them saline, also feed into Lake Natron. Despite these inflows, most of the lake's water is derived from direct precipitation. Evaporation exceeds precipitation, and the maximum depth of Lake Natron is only 2 m. A large portion of the lake's bed is covered by a salt crust that dissolves during the rainy season (Wetlands International 2002).

The northern soda lakes, such as Magadi, Elementaita, Nakuru, and Bogoria, are smaller than the southern lakes and tend to fluctuate more in depth and area. For instance, most of Lake Magadi is a dry lakebed, containing water only after heavy rains when water reaches the northern portion of the lake via three wadis (temporary watercourses). The lakebed is composed of trona (solid sodium carbonate) and other associated salts. Shallow lagoons in the northern and southern ends of the lake are fed by hot springs all year, maintaining lake water temperatures at about 40°C (Hughes and Hughes 1992). The strongly alkaline Lake Nakuru typically covers between 0.35 km^2 and 0.49 km^2 in area but has dried out completely several times over the past 50 years, for unknown reasons (ILEC 2001). This endorheic lake, with a mean depth of 2.3 m, is small for its large drainage area of 1,760 km^2 (ILEC 2001). The three major rivers feeding the lake are Njoro, Makalia, and Nderit, and it also receives water from several alkaline springs. In 1998, heavy rains generated by El Niño in Kenya flooded the lake, lowering its salinity (R. Thampy, pers. comm., 2000). Lake Bogoria, with a mean area of 34 km^2 and a mean depth of 5.4 m, differs from these other lakes by having minimal water level fluctuations (Wetlands International 2002; ILEC 2001). The Wasenges-Sandai River system and numerous springs from the nearby escarpment feed Lake Bogoria, and Loboi marshes on the northwestern floodplain help maintain the lake's hydrology. The lake sits in a graben, which is situated in an area of active faulting, as evidenced by the presence of geysers, steam vents, and hot springs along the lake's shores (Wetlands International 2002).

There are also several freshwater lakes in the ecoregion. In the southwest of the ecoregion, Lake Kitangiri is fed by the Wembere River, which flows through an extensive floodplain (105 km long and up to 20 km wide) before it enters the lake from the south (Hughes and Hughes 1992). Lake Naivasha lies at 1,879 m a.s.l., and it is the highest of the major Rift Valley lakes. With a volume of 4.6 km^3, it is the largest freshwater reservoir in this ecoregion. There are no known surface outflows from the lake, but scientists believe that underground outflows must exist for the lake to maintain its freshwater condition (Wetlands International 2002). About 16 percent of the lake's water comes from subterranean sources, including seepage of surface flow and inflow from temporary watercourses that drain the slopes of the Ol Doinyo Epuru Ridge. The lake's water level, which averages 6.5 m, is known to fluctuate by several meters from year to year. Lake Baringo, a freshwater lake with an area of 166 km^2 and depths recorded between 3 and 8 m, lies in the north of the ecoregion and contains a large volcanic island in the center as well as several other small islands (Wetlands International 2002). Six rivers drain into the lake from the Mau Escarpment, the Lodiani Mountains, and the Aberdare Range. The main affluent of the lake, the Molo River, flows through extensive swamp belts. Hot springs on the islands and edges of the lake discharge saline water into the lake. There may be an outflow from the lake under the northern lava bed (Hughes and Hughes 1992).

The climate of this ecoregion is influenced by the seasonal hot and dry winds that blow from the northeast and southeast. In central Kenya, the Rift Valley Basin is located in the rainshadow of Aberdare Mountains, whereas the southern Kenya and northern Tanzania portions of the Rift Valley lie in the rainshadow of Mount Kilimanjaro. Temperatures and rainfall vary widely with altitude. The wet season in the southern part of the ecoregion occurs between January and May, with an average annual rainfall of about 600 mm. The daily average temperature ranges from a mean minimum of 23°C to a mean maximum of 35°C. The average rainfall decreases to about 411 mm per year around Lake Magadi, and its temporal distribution pattern is bimodal, with the wettest period in March–May and a smaller wet season in November and December. In the northern part of this ecoregion, especially around lakes Nakuru and Elementaita, annual rainfall is about 850 mm, with the rainy season extending from April to August (Ojany and Ogendo 1973). Because of higher land elevation, ambient temperatures are also lower in the region of the northern lakes. For example, Lake Nakuru, which is located at an average elevation of 1,890 m a.s.l., has a mean annual temperature of 18°C, a mean minimum of 10°C, and a mean maximum of 29°C (Hughes and Hughes 1992; Ndede et al. 2000; ILEC 2001).

Wetland vegetation in this ecoregion is highly variable and can be grouped into two categories: flora of saline lakes and wetlands and flora of freshwater lakes and river basins. The immediate margins of saline lakes are bare, but open waters support a rich community of phytoplankton, which is dominated by filamentous blue-green algae (*Spirulina platensis*) (Vareschi and Jacob 1985). Algal blooms of this species can last for several years and maintain photosynthetic rates that are close to the highest measured for terrestrial tropical plant communities (Livingstone and Melack 1984). Next to the bare mudflats lie the soda flats with highly alkaline soils. The soda flats support tall grassland communities, which are characterized by *Sporobolus spicatus*, short grassland communities dominated by *Diplachne fusca*, and sedge formations dominated by *Cyperus laevigatus* in permanently wet spots. Next to the grassland community lies a belt of *Acacia-Commiphora* woodland, which may contain *Acacia xanthophloea*, *A. tortilis*, *Tamarix nilotica*, and *Combretum* spp. (Sarunday 1999). Effluent rivers have gallery woodland and floodplains dominated by grasses and tall trees of *Acacia tortilis*, *A. seyal*, and *A. xanthophloea*, especially along the river valleys and in areas where underground water is close to the surface of the floodplain. Tall grasses and sedges dominate freshwater swamps, especially *Cyperus papyrus* in permanent water (Mwalyosi 1991).

There are three distinct vegetation zones in clear succession from the water toward land among vegetative communities of freshwater lakes and swamps in the Eastern Rift Valley (Njuguna 1984). The edges of the lake contain diverse emergent, submerged, and floating macrophytes. Swamp vegetation dominates the lake edges and the mouths of affluent rivers. Tall sedges such as *Cyperus papyrus*, *Phragmites mauritianus*, *Typha domingensis*, and *T. capensis* and tallgrasses such as *Panicum ripens*, *Echinochloa pyramidalis*, and *E. scabra* dominate permanent and seasonal swamps and floodplains (Wetlands International 2002). In the open water, characteristic submerged plants are *Nymphaea* spp., and floating plants are *Pistia stratiotes*, *Salvinia molesta*, and *Eichhornia crassipes*. The terrestrial areas immediately surrounding the freshwater lakes are covered by a mosaic of bushland, acacia woodland, and grassland. As the elevation increases, dense acacia forest occurs, containing such species as *A. albida*, *A. polycanta*, and *Albizia versicolor* (Mwalyosi 1991). The streams feeding the freshwater lakes have sparse gallery forest with tree species such as *Tamarindus indica*, *Ficus sycomorus*, and *Phoenix reclinata* (Loth and Prins 1986; Mutanga et al. 2000).

Outstanding or Distinctive Biodiversity Features

Large numbers of flamingos inhabit the lakes of this ecoregion, and a surprisingly endemic aquatic fauna is adapted to the seemingly inhospitable soda lakes. The flocks of flamingos are larger than anywhere else in the world; half of the world's lesser flamingo (*Phoenicopterus minor*) population resides in this ecoregion, feeding by filtering the microscopic blue-green algae, especially the abundant *Spirulina platencins* and diatoms (Hammer 1986). Many other aquatic species, including several endemic fish species, are capable of withstanding the thermal and saline waters of many of the lakes.

The lesser flamingo populations of this ecoregion are exceptional by any measure. Some estimates put lesser flamingo populations on Lake Nakuru at 1.4 million birds, and up to 2 million flamingos feed at Lake Bogoria (Tuite 1979; Livingstone and Melack 1984; Hughes and Hughes 1992). In some soda lakes, lesser flamingos consume as much as 93 percent of the daily primary algae production (Livingstone and Melack 1984). The salt flats of Lake Natron are the only known regular breeding site of lesser flamingos in East Africa (Hammer 1986). The flamingos nest in dense colonies on open mud and are known to migrate from lake to lake depending on food availability (Githaiga 1997; Nasirwa 2000). The lesser flamingo is confined to soda lakes and in this part of the Rift Valley occurs from Lake Manyara in the south to Lake Bogoria in the north. Hundreds of thousands of these birds breed regularly at Lake Natron and occasionally in Lake Magadi. There is also some evidence that lesser flamingos move between breeding areas in East Africa and southern Africa (Namibia and Botswana) (Simmons 2000). Greater flamingo (*Phoenicopterus ruber*) is also present, occasionally in the thousands (Wetlands International 2002), and feeds primarily on invertebrates from the bottom mud.

The saline and freshwater lakes of this ecoregion also provide breeding and feeding grounds for numerous other bird species. For example, recorded bird species include 385 species from Lake Bogoria, about 418 from Lake Naivasha, more than 400 in the Lake Manyara Basin, 458 from Lake Nakuru, and 473 from Lake Baringo (Gichuki 2003). In addition to flamingos, important bird populations include coot (*Fulica cristata*) in Lake Naivasha and white pelican (*Pelecanus onocrotalus*), pink-backed pelican (*P. rufescens*), little grebe (*Tachybaptus ruficollis*), greater cormorant (*Phalacrocorax carbo*), cattle egret (*Bubulcus ibis*), yellow-billed stork (*Mycteria ibis*), marabou (*Leptoptilos crumeniferus*), and pochard (*Aythya ferina, A. fuligula,* and *A. nyroca*) in Lake Nakuru (Hammer 1986; Hughes and Hughes 1992; Zimmerman et al. 1996; Baker and Baker 2002; Wetlands International 2002). Internationally important numbers of lesser flamingo, Abdim's stork (*Ciconia abdimii*), glossy ibis (*Plegadis falcinellus*), chestnut-banded plover (*Charadrius pallidus*), and little stint (*Calidris minuta*) occur in Lake Natron (Baker and Baker 2002).

About one-third of the ecoregion's nearly thirty freshwater fish are endemic. There is a small species flock from Lake Natron, composed of *Oreochromis alcalicus, O. ndalalani,* and *O. latilabris*. The endemic fish species are capable of surviving in waters of high temperatures and salinity. The three species endemic to Lake Natron and *O. grahami*, endemic to Lake Magadi, have all been documented to live in waters up to 40°C. Lakes Kitangiri, Eyasi, Singida, and Manyara host the endemic *Oreochromis amphimelas* in hot springs along their lake margins (FishBase 2001).

Two large, widespread aquatic vertebrates inhabit the freshwater lakes. Small numbers of hippopotamus (*Hippopotamus amphibius*) are found in several lakes of this ecoregion, including lakes Naivasha and Nakuru, and the Nile crocodile (*Crocodylus niloticus*) lives in lakes Baringo, Naivasha, and Nakuru. The lake margins also support populations of large mammals, particularly around lakes Natron and Manyara in Tanzania (Baker and Baker 2002).

Status and Threats

Current Status

The aquatic fauna of this ecoregion is under severe threat from a variety of factors including pollution, overfishing, introduced species, agricultural runoff, deforestation, and water abstraction. Climate changes pose a more general but real threat to the ecoregion. For instance, El Niño effects recently increased water volume in Lake Nakuru, causing lower salinity that in turn resulted in lower production of *Spirulina* (R. Thampy, pers. comm., 2000).

Overfishing and introduced species threaten the native fauna of some of the lakes. There is a large fishery on Lake Baringo, the yields of which declined from 670 tons/year in the 1960s to less than 500 tons/year in the early 1980s because of overfishing (Hughes and Hughes 1992). Baringo has also been invaded by Nile cabbage (*Pistia stratiotes*) at the river mouths, which is apparently responding to nutrient inputs from surrounding agricultural lands (Wetlands International 2002). The North American red swamp crayfish (*Procambarus clarkii*), introduced into Lake Naivasha, has caused the decline of the native *Aplocheilichthys antinorii* by decimating the invertebrate communities on which this fish feeds (Davies and Day 1998). Also in Naivasha, introduced water hyacinth (*Eichhornia crassipes*) and *Salvinia molesta* cover up to 25 percent of the surface area of the lake, usually with floating mats persisting for 2–3 years after a period of lake level rise (Harper et al. 1994; Wetlands International 2002). *Oreochromis esculentus* and *Tilapia rendalli* were introduced into Lake Kitangiri from Lake Victoria.

Lake Nakuru provides an example of the cascading changes that can result from the introduction of species into this ecoregion's lakes. Lake Nakuru has no native fish species, but in 1953, 1959, and 1962 Lake Magadi's *Oreochromis alcalicus* was introduced to feed on mosquitos. By the 1970s, the species had become one of the main consumers of algae, which had historically been eaten primarily by lesser flamingos. Populations of other birds, such as great white pelican, increased in response to the presence of the fish. For reasons that remain unknown, the algae *Spirulina platensis* disappeared from the lake in 1974, ultimately resulting in a substantial reduction in algal biomass, primary productivity, and resident numbers of flamingos (ILEC 2001). In 1991, a massive die-off of the fish occurred, again for unknown reasons (Githaiga 1997; Wetlands International 2002).

Agriculture and grazing in the lakes' watersheds increase the flow of sediments and chemicals into the waters. For example, about 25 percent of the watershed of Lake Baringo is cultivated (Gichuki 2003), and increasing erosion has led to increased siltation and turbidity of the lake and affluent rivers (Hughes and Hughes 1992). Overgrazing is also a serious problem around Lake Baringo, especially along parts of the shoreline (Wetlands International 2002). The Lake Naivasha basin contains numerous flower farms, which are treated with fertilizers, pesticides, and fungicides.

Other land-based threats to the biodiversity of Rift Valley lakes include urban pollution and deforestation. Raw sewage from Nakuru town has been discharged into Lake Nakuru, causing eutrophication. The pollution of the lake water has been aggravated by nutrient runoff from the surrounding agricultural lands. About 35 percent of Lake Baringo's basin has been deforested through agricultural conversion and the cutting of wood for fuel, building materials, and other uses. Many lakes also suffer from destruction of wetland vegetation along their shores. This vegetation is used as livestock fodder and thatch material for huts and sheds (Wetlands International 2002).

In this arid region, the lakes and rivers are sources of water. Lake Baringo has already undergone reductions in water level as a result of water abstractions from the Molo and Pekerra rivers and from the damming of Endau River (Wetlands International 2002).

Flamingo populations are adversely affected by pollution, soda mining, and other human disturbances (Nasirwa 2000). Soda mining deleteriously affects flamingos by changing water quality and microflora. This has occurred in Lake Magadi, with half a million metric tons of sodium carbonate being taken from the lake every year for industrial use (Smith 1988). Recent die-offs of flamingos have been documented in Bogoria and Nakuru, although the causes are unknown. One hypothesis is that although birds are able to accumulate individual toxins without lethal effects, high levels of multiple toxins might lead to death or make the birds more susceptible to other stresses (R. Thampy, pers. comm., 2000).

The ecoregion contains many protected areas that buffer and include the lakes. Lake Manyara National Park protects the northwest end of the Manyara lakeshore and the northern two-thirds of the lake. Tarangire National Park covers a large section of the Tarangire swamps, and the Ngorongoro Crater Conservation Area protects the swamps and lakes in the crater (Hughes and Hughes 1992). Serengeti National Park borders Lake Eyasi and covers the northern portion of the lake's catchment (Sayer et al. 1992). Lakes Baringo, Bogoria, Nakuru, Naivasha, and Natron are all Ramsar sites, although this designation does not necessarily confer protection. For example, Tanzania's Lake Natron, which is a Ramsar site, is unprotected, except for the regulation of hunting (Baker and Baker 2002). Kenya's Lake Bogoria is managed as a National Reserve and Nakuru as a National Park, whereas Lakes Naivasha and Baringo are managed by local communities (Wetlands International 2002).

Future Threats

Future threats to the ecoregion include increased water abstraction, municipal development, catchment deforestation, hydropower development, and soda ash mining. Continued deforestation of Lake Natron's Mau catchment could increase siltation and dilution of the lake through increased freshwater inputs. Proposed soda ash mining in Lake Natron would entail the construction of dikes in the lake, a pumping area, and new infrastructure. Wastewater returned to the lake would be heated and potentially polluted (Wetlands International 2002).

A large dam has been proposed for construction on River Malewa in Nyandarua highlands, with the intention of supplying water to the municipality of Nakuru. Increased wastewater input could raise water levels in Lake Nakuru by as much as 0.7 m, with corresponding reductions in salinity. Naivasha is threatened by removal of water for irrigation and by the planned impoundment on the Turasha River for rural water supply (Harper et al. 1994).

In the Lake Natron Basin, there is a plan to build a hydropower reservoir of approximately 50 km^2 area on the Ewaso Ngiro River in Kenya. This development would change the river's hydrology, thereby changing the ecology and natural productivity of Ewaso Ngiro floodplain. Damming of the river and increased water abstraction from other affluent streams for soda ash mining industry are likely to reduce water supply into Lake Natron, especially during the dry period. This situation has serious implications for the survival and reproductive success of the largest concentration of flamingoes in eastern Africa.

Attempts to increase geothermal energy production, especially in the basins of lakes Naivsha, Bogoria, and Baringo in Kenya, may pose a serious threat to the lakes' water supply. To produce geothermal energy, water is abstracted from subterranean sources and is not normally replaced. A site north of Lake Baringo and the western part of Lake Bogoria basin have been investigated and approved for production of geothermal energy (Smith 1988).

The principal drivers of change in the Southern Eastern Rift ecoregion are climate change and human population pressure together with their associated impacts on economic production and natural ecosystems in Kenya and Tanzania (Ondieki 2000).

Justification for Ecoregion Delineation

The formation of the East African Rift led to the isolation of many of the ecoregion's lakes from other lakes and rivers. The rift formed when tectonic plates below Somalia and the rest of Africa began separating (Lévêque 1997). The plates are still separating, and the walls of the Rift Valley drift apart at a rate of 4 mm per year (Cromie 1982). Historic connections existed between some lakes; for example, about 6,000–13,000 years ago lakes Naivasha, Elementaita, and Nakuru were part of a larger lake that may have later contracted because of changes in climate (Hughes and Hughes 1992; ILEC 2001). Thus, this ecoregion is delineated to include the small lakes in the Gregory Rift Valley. The ecoregion is distinguished by its large congregations of lesser flamingos and the endemic fauna that inhabit its saline and freshwater lakes.

Data Quality: High

Ecoregion Number:	3
Ecoregion Name:	Lake Tana
Major Habitat Type:	Closed Basins and Small Lakes
Final Biological Distinctiveness:	Globally Outstanding
Final Conservation Status:	Vulnerable
Priority Class:	I
Authors:	Michele Thieme and Ashley Brown
Reviewer:	Leo Nagelkerke

Location and General Description

Lake Tana, a lake in the highlands of Ethiopia, lies in the north of Ethiopia and is the source of the Blue Nile. Lake Tana was formed by a volcanic blockage that reversed the previously north-flowing river system (Beadle 1981). Numerous seasonal streams and four perennial rivers feed the lake, but only one, the Blue Nile, leaves it (Nagelkerke 1997). Shortly after leaving Lake Tana, the Blue Nile descends Tissisat Falls (c. 40 m high), which effectively isolates the lake's freshwater fauna from the rest of the Nile. The total area of the Lake Tana basin is 16,500 km^2, and the lake itself covers about 3,150 km^2.

The lake is situated in the highlands of Ethiopia at about 1,800 m and experiences a tropical highland climate. Air temperatures range widely, between 7°C and 31°C, whereas water temperatures stay mild, normally between 18°C and 26°C (Nagelkerke 1997). The dry season lasts from October or November to May or June, with maximum monthly rainfall (up to 500 mm/month) in July. Annual rainfall in the vicinity of the lake averages 1,315 mm/year, but evaporation is higher (about 1,800 mm/year) (Burgis and Symoens 1987).

Because evaporation exceeds rainfall, the hydrology of this shallow lake depends largely on the local climate (Burgis and Symoens 1987). Lake level varies depending on seasonal rains. The average difference between the lowest lake level (May–June) and the highest (September–October) is 1.5 m (Nagelkerke 1997). The lake has a mean depth of 8 m and is well mixed by strong winds in the evenings (Nagelkerke 1997). The water of the lake is clear, and in places at the lake bottom volcanic peaks form reefs (Sibbing et al. 1998). *Cyperus papyrus* and other *Cyperus* spp. line the shores of the lake (Beadle 1981).

The isolation of the lake from all but inflowing rivers has led to a highly endemic freshwater biota. It is likely that the Lake Tana barbs evolved from one ancestral species that probably resembled *Barbus intermedius* (Nagelkerke 1997). Fish species in the lake are most closely related to those of the Nilo-Sudanian biogeographic region.

Outstanding or Distinctive Biodiversity Features

Lake Tana hosts the only extended cyprinid species flock in Africa. The only other known flock, in the Philippines' Lake Lanao, has been decimated by introduced species. Fifteen species of large barbs have been described from Lake Tana (Nagelkerke 1997; Nagelkerke and Sibbing 1998, 2000). The species flock is believed to be less advanced in its evolution than Lake Lanao's cyprinid flock (Mina et al. 1996). Eight of the large barbs are piscivorous, and *Barbus humilis* and the newly described small species *Barbus tanapelagius* are thought to be the major prey species (De Graaf et al. 2000).

About 70 percent of the fish species in this highland lake are endemic to it. Eighteen endemic cyprinids have been identified. The fifteen large barb species of the subgenus *Barbus* (*Labeobarbus*) and the three or four small barb species of the subgenus *Barbus* (*Enteromius*) are under revision (L. Nagelkerke, pers. comm., 2001). The tilapia (*Oreochromis niloticus*) of Lake Tana belongs to a widespread species but is described as an endemic subspecies, *Oreochromis niloticus tana* (Seyoum and Kornfield 1992). The only river loach (family Balitoridae) known from Africa, *Nemacheilus abyssinicus,* was described from Lake Tana in 1902 and rediscovered in 1992 in the lake and in the upper Omo River (Dgebuadze et al. 1994). The large catfish *Clarias gariepinus*, widespread throughout Africa, also lives in the lake and forms an important part of the fishery.

The invertebrate fauna is depauperate. Fifteen mollusk species, dominated by the Planorbidae family, have been described, including one endemic. An endemic freshwater sponge, *Makedia tanensis,* has recently been discovered in the lake. The sponge is small (specimens found were up to about 2 cm), white, and of an encrusting form belonging to a monotypic genus (Manconi et al. 1999).

A high diversity of wetland birds also lives by the lake, including the piscivorous little grebe (*Tachybaptus ruficollis*), great white pelican (*Pelecanus onocrotalus*), great and African cormorants (*Phalacrocorax carbo* and *P. africanus*), and darter (*Anhinga rufa*). Many Palearctic migrant waterbirds also depend on the lake as feeding and resting grounds.

Status and Threats

Current Status

This lake is fairly pristine, with no established exotic species and little pollution (Nagelkerke et al. 1995; Mina and Golubtsov 1995; Nagelkerke 1997). Fishing is the largest threat to the ecoregion, although diversion of water for irrigation and pressures associated with population growth also affect the supply and quality of water entering the lake.

As of 1992 most of the fishing in the lake was artisanal, with 90 percent of the catch consisting of *Barbus* and *Clarias* species (Hughes and Hughes 1992). Commercial fishing is increasing, however, particularly in the southern portion of the lake near the town of Bahar Dar. The use of motorized boats is allowing access to deeper parts of the lake that were traditionally unfished, and selective catching of larger adults eventually may decrease the fish populations of the lake (Wudneh et al. 1999). Increased artisanal fishing and use of poison derived from seeds of the berberra tree also threatens the stability of native fish populations by killing both spawning and younger fish (Nagelkerke and Sibbing 1996). Large numbers of barbs are caught at their spawning grounds in the streams that feed the lake, and recent data suggest that fish stocks are in decline because of overexploitation of spawning fish (Nagelkerke 1997). Observers have suggested that a fishery management plan administered and accepted by local communities may be the best solution to mitigating fishing pressures (Nagelkerke 1997).

The population in Ethiopia is more than 64 million and is growing at a rate of 2.76 percent/year (UNESCO 2001). One consequence of population pressure is the cutting of forests for fuelwood. Overall, forest cover in Ethiopia declined from 40 percent in 1950 to 2.4 percent in 1987 (Sayer et al. 1992). Changes to hydrology have resulted. Currently there are no known national parks or reserves that protect the lake and its surrounding vegetation.

Future Threats

Diversion of water for irrigation is increasing as Ethiopian farmers build small dams on tributaries that flow into the lake and other tributaries of the Blue Nile, restricting flow from Ethiopia to Egypt and Sudan by as much as 3 million m^3 annually (Inventory of Conflict and Environment 1997). These hill reservoirs could decrease flow to the lake, lower lake levels, and cause changes in the water quality of the lake.

The introduction of an exotic species could decimate the unique fish fauna of the lake, as has occurred in other lakes in Africa and in the Philippines' Lake Lanao. Comprehensive educational programs to prevent the introduction of exotic species to the lake are critically important. Increased deforestation and fishing also pose future threats, barring a shift in current trends.

Justification for Ecoregion Delineation

The ecoregion is isolated from the rest of the Nile by Tissisat Falls (c. 40 m high), and this isolation has played a role in the evolution of the endemic fauna of the lake. This ecoregion is distinguished by the only extended cyprinid fish species flock in Africa; at present fifteen species of large barbs have been described from this endemic flock.

Data Quality: Medium

Ecoregion Number:	**4**
Ecoregion Name:	**Northern Eastern Rift**
Major Habitat Type:	**Closed Basins and Small Lakes**
Final Biological Distinctiveness:	**Nationally Important**
Final Conservation Status:	**Endangered**
Priority Class:	**IV**
Author:	**Abebe Getahun**

Location and General Description

Numerous highly productive lakes lie in the Northern Eastern Rift Valley, which cleaves the eastern and western sections of the high-altitude Ethiopian dome and extends from the edge of the xeric Red Sea Coastal [89] ecoregion in the north to Lake Awassa in the south. Two closed basins occur in this section of the Rift Valley: in the south, Lake Awassa basin, consisting of Lake Awassa and the swampy Lake Shallo; and the Oromo Lakes (previously known as the Galla Lakes) basin in the north, composed of a series of four interconnected lakes (Abijata, Langano, Shala, and Zwai). There are also a number of hot springs adjacent to the lakes.

The mean annual air temperature of this ecoregion varies with altitude and ranges between 20°C and 2?°C (Tudorancea et al. 1999). There is a 4-month dry season from November to February and an 8-month rainy season from March to October. The main rains occur between July and September.

Five major lakes and several rivers lie in this ecoregion. Lake Zwai, the most northerly lake, is located about 160 km south of Addis Ababa at an altitude of 1,840 m. The lake covers 434 km^2, and its average depth is 2.5 m (Balarin 1986). Extensive marshes of papyrus (*Cyperus papyrus*) border the lake and produce large numbers of mosquitoes (*Anopheles pharoensis, A. mauritianus,* and *Taeniorhynchus uniformis*) (Omer-Cooper 1930). The Makki River flows into the lake from the northwest, and the River Kattar flows into it from the northeast. The lake's waters flow out through the River Sucsuci into Lake Abijata.

The next three lakes in the chain are Lake Langano, Lake Abijata, and Lake Shala. Lake Langano, located at an altitude of 1,582 m, covers 241 km² and has a maximum depth of 47.9 m and a mean depth of 17 m. Salinity is 1.88 g/L (Wood and Talling 1988). Lake Langano receives most of its water from small rivers that drain from the Arsi Mountains, which make up the eastern wall of the Rift Valley. The only outlet from the lake is the Hora Kelo River, which flows into Lake Abijata (Ethiopian Wildlife and Natural History Society 1996). Lake Abijata is found at an altitude of about 1,600 m and is an alkaline lake. The shores slope gradually and are muddy, with areas of *Juncus* vegetation. The depth is about 10 m, and the bottom of the lake is sandy. Three rivers feed Lake Abijata: Gogessa, Bulbula, and Hora Kelo. Water from Lake Zwai also flows into Lake Abijata through River Sucsuci. Lake Abijata has no outlet, and it loses its water by evaporation only. Lake Shala, located at 7°28'N 38°30'E, is at 266 m the deepest of the Ethiopian Rift lakes. It is 28 km long and 15 km wide and is surrounded primarily by Pleistocene volcanic rocks. The eastern and western shores are covered by lacustrine deposits and Holocene sands, occasionally blackened by obsidian detritus (Mohr 1961). The great depth of Lake Shala may be related to the origin of the basin by intense faulting. Two permanent creeks and several seasonal streams flow into its closed basin.

Lake Awassa has a smaller surface area than Shala and is completely enclosed by faulting. Located at an altitude of 1,680 m, the lake lies south of the other lakes in the ecoregion. It has a surface area of 88 km², a maximum depth of 22 m, and an average depth of 11 m. Lake Awassa is a polymictic lake. The water is murky and alkaline, with a pH between 8.75 and 9.05 (Tudorancea et al. 1988). Like lakes Abijata and Shala, it is a terminal lake without any visible outlet. Its main tributary is the Tikur Wuha River, which drains swampy Lake Shallo. An extensive belt of submergent and emergent rooted vegetation, which extends about 150 m offshore, covers the littoral zone.

This ecoregion also contains a number of crater lakes (Bishoftu, Aranguade, Hora, Kilotes, and Pawlo) located at the northwestern edges of the Rift Valley around the town of Debrezeit (Mohr 1961), at an altitude of about 1,900 m. These lakes lie in volcanic explosion craters produced about 7,000 years ago. The Awash River, which begins in the Ethiopian Plateau, flows north in the Rift Valley and terminates in Lake Abhe, a closed lake near the border between Ethiopia and Djibouti.

Outstanding or Distinctive Biodiversity Features

The lakes and streams of the Northern Rift ecoregion support a depauperate freshwater fauna with few endemic species. Only eight fish, thirteen aquatic frogs, three aquatic reptiles, twelve aquatic mollusks, and four aquatic mammals live in or near the freshwater lakes and streams. Three endemic fish, *Barbus ethiopicus, B. microterolepis,* and *Garra makiensis,* inhabit Lake Zwai and its adjacent rivers. Introduced species of *Tilapia zilli, Clarias gariepinus,* and carps also live in this lake. There are large numbers of pelicans, cormorants, ducks, snipes, stilts, egrets, grebes, ibis, herons, gulls, and darters around the lake. One endemic frog, *Bufo langanoensis,* lives along the shores of Lake Langano and its tributaries. Lakes Zwai and Langano also harbor hippopotamus (*Hippopotamus amphibius*). The fish fauna of Lake Awassa consists of two *Barbus* species (*B. intermedius* and *B.* cf. *amphigramma*), the catfish *Clarias gariepinus,* and *Oreochromis niloticus. Oreochromis niloticus* is abundant in the lakes and rivers of this ecoregion. Despite their low number of species, the lakes of this ecoregion support most of the fish production of Ethiopia.

Several hot, somewhat sulfurous springs are found around the shores of some of these lakes. Hot springs occur in the Lake Awassa basin at two locations: in a creek valley south of Shashemene and near lakes Shala and Langano.

In the 1970s and 1980s more than 400 bird species were recorded from the Abijata-Shala National Park. This park is positioned in one of the narrowest parts of the Great Rift Valley, which is a major flyway for both Palearctic and African migrants, particularly raptors, flamingos, and other waterbirds (Ethiopian Wildlife and Natural History Society 1996). Many of these birds stop over to rest and feed in the ecoregion. The shallow waters of Lake Abijatta are remarkably rich in insect life, with large swarms of *Corixa* in particular, although plankton diversity is low. The zooplankton fauna in the Ethiopian Rift lakes is dominated, in terms of biomass, by copepods and cladocerans (Tudorancea et al. 1999).

The flora varies between the different lakes. Three major groups of algae dominate: Chlorophyceae, Cyanophyceae, and Diatomophyceae (Tudorancea et al. 1999). The most common emergent plants are *Scirpus* spp., *Typha angustifolia, Paspalidium germinatum,* and *Phragmites* sp. *Nymphaea coerulea* and *Potamogeton* spp. are the dominant species of floating and submerged vegetation. Abijata and Shala lakes lack aquatic macrophytes.

The vegetation to the east and south of Lake Shala is *Acacia-Euphorbia* savanna. The most common trees are the woodland acacias (*A. etbaica, A. tortilis,* and *Euphorbia abyssinica*) and bushes of *Maytenus senegalensis*. There is also a rich grass and herb flora (Ethiopian Wildlife and Natural History Society 1996). Beds of bulrushes and sesbania occur where the hot springs and rivers enter the lake, but most of the shore has steep cliffs. No fish are recorded from this lake.

Status and Threats

The main threats to the freshwater systems of this ecoregion are deforestation, land conversion for agriculture, and overgrazing. These activities, which strip the ground of its protective cover, make the fine-textured alluvial soils susceptible to wind and water erosion. Extraction of water from Lake Zwai and its tributaries for irrigation purposes could be a future threat.

Although at present there appears to be no overexploitation

of fishes in the commercial fishery on lakes Awassa and Zwai, proper management measures should be taken before overfishing becomes a problem.

Introductions of nonnative fish in some of the rivers and lakes threaten the native aquatic biodiversity. For example, *Tilapia zilli, Clarias gariepinus,* and *Cyprinus carpio* have been introduced into Lake Zwai and are beginning to dominate the fish fauna. Studies are being conducted to determine the level of impact.

Lakes Abijata and Shala lie in the Abijata-Shala National Park, which was created to protect the high diversity of waterbirds and the scenic beauty of the area. Islands on Lake Shala are important breeding sites for birds, and Lake Abijata is their feeding grounds. Unfortunately, fish-eating birds have mostly abandoned the park since the numbers of fish in Lake Abijata have decreased because of overfishing.

Justification for Ecoregion Delineation

This ecoregion is defined by the northern lakes of the Ethiopian Rift Valley and distinguished by lakes with a distinctive fauna when compared with the more southern Rift Valley lakes of Chamo and Abaya. The fish fauna in the northern lakes appears to have been derived from Awash and associated rivers, whereas the fish fauna of the latter is Nilo-Sudanic. These faunal affinities may be explained by the tumultuous geologic past of the ecoregion, which was exposed to six volcanic events between the Oligocene and the present (Woldegabriel et al. 1990). Analyses of invertebrate and fish fossils found in the sediments of the lakes indicate that lakes Zwai, Abijata, Langano, and Shala were once united into a single freshwater lake draining northward into the Awash River (Grove et al. 1975; Gasse and Street 1978). The present-day lakes are the result of subsequent tectonic or volcanic activity (Tudorancea et al. 1999).

Data Quality: Medium

Although baseline information is still lacking, this is one of the most studied ecoregions in Ethiopia. Physical and chemical features of some of the lakes are available in Wood et al. (1978) and Wood and Talling (1988). Zooplankton and phytoplankton studies of Lake Awassa are found in Mengistou and Fernando (1991a, 1991b), Mengistou et al. (1991), and Kifle and Belay (1990). Tudorancea and Harrison (1988), Tudorancea and Zullini (1989), and Tudorancea et al. (1989) studied the benthic communities of some of these lakes. Teferra (1987, 1988, 1989), Tadesse (1988), Teferra and Fernando (1989), Admassu (1994), and Tudorancea et al. (1988) have studied the biology of *Oreochromis niloticus,* the dominant fish species in these lakes. Nevertheless, the invertebrate fauna, the phytoplankton, and the macrophytes of many of the lakes have not been identified or studied. Moreover, the biodiversity of the rivers and streams is largely unknown.

Ecoregion Number:	**5**
Ecoregion Name:	**Western Equatorial Crater Lakes**
Major Habitat Type:	**Closed Basins and Small Lakes**
Final Biological Distinctiveness:	**Globally Outstanding**
Final Conservation Status:	**Endangered**
Priority Class:	**I**
Authors:	Uli Schliewen, Emily Peck, and Neil Burgess

Location and General Description

Situated in southwestern Cameroon along the Cameroon Line, a volcanic ridge running southwest-northeast, this ecoregion's crater lakes host endemic species flocks of cichlid fishes and endemic insects and shrimps. The ecoregion lies adjacent to the Atlantic Ocean and extends inland along the northwestern side of Cameroon's interior plateau. This area has been continuously producing calderas over the past 25 million years or so (Fitton and Dunlop 1985), and some of these have become filled with water. Approximately thirty-six crater lakes are now known from Cameroon. Most have no outflows or extremely steep outflows (Kling 1987), isolating them from nearby river systems.

The Crater Lakes ecoregion lies in the Tropical Humid Sudanian climatic zone (Kling 1987) and experiences a single, well-defined rainy season (Stiassny et al. 1992). Rains fall from May through August, tapering off in September, and the dry season extends from October through April (Stiassny et al. 1992). The mean annual rainfall for the ecoregion is approximately 1,570 mm (Hughes and Hughes 1992).

Lakes of the ecoregion include Barombi Mbo, Bermin, Dissoni (Soden), Ejagham, Kotto, and Mboandong, among many others. The largest is Lake Barombi Mbo, which is situated at 300 m a.s.l. and has an open water area of about 5 km^2. It has a mean depth of 69 m and a maximum depth of 111 m (Hughes and Hughes 1992). The most recent lava flow into the lake was almost exactly 1 million years ago, dating the lake to be at least that old (Cornen et al. 1992). Just south of Lake Barombi Mbo is Lake Barombi Kotto. It lies 110 m a.s.l. and covers about 3 km^2. Lake Dissoni (Soden), with an area of 3.6 km^2, lies just north of Lake Barombi Mbo, and Lake Bermin is situated still further north, with an area of little more than 0.5 km^2 (Hughes and Hughes 1992; Stiassny et al. 1992). Lake Ejagham is located in western Cameroon, has a surface area of about 0.5 km^2, and is about 18 m deep. Unlike the other lakes of this ecoregion, Ejagham is of nonvolcanic origin; it is probably a solution basin produced by groundwater and probably was formed during the last Ice Age glaciation of these highlands (Schliewen et al. 2001).

Vegetation in the ecoregion consists of submontane forests between 900 and 1,800 m and at higher elevations a mixture of montane elements including distinct montane forests and patches of montane grasslands, bamboo forests, and subalpine communities. Five tree species characterize the forested montane zone: *Nuxia congesta, Podocarpus latifolius, Prunus africana, Rapanea melanophloeos,* and *Syzygium guineense bamendae.* These trees become increasingly covered with an epiphytic flora, especially orchids and mosses, at higher altitudes (Letouzey 1985).

Outstanding or Distinctive Biodiversity Features

The western equatorial crater lakes of Cameroon support a highly endemic aquatic fauna with more than 75 percent endemism in fish. In particular, lakes Barombi Mbo and Bermin have experienced extensive species radiations of cichlids (Stiassny et al. 1992), resulting in an index of endemic fish per area that is unrivaled on Earth. The ecoregion also hosts an endemic and species-rich amphibian fauna, with about one-third of the nearly sixty frog species endemic to the area, primarily the surrounding forests. Additionally, there is one endemic aquatic mollusk, *Bulinus camerunensis.*

Lake Barombi Mbo is the most studied of the crater lakes in this ecoregion. At present, fifteen fish species have been found in the lake, twelve of which are endemic. Except for the clariid catfish *Clarias maclareni,* all endemics are tilapiine cichlid fishes. Four of the five tilapiine genera are endemic: *Konia* (two species), *Stomatepia* (three species), *Pungu* (one species), and *Myaka* (one species). The only nonendemic cichlid genus is *Sarotherodon* (four species) (Trewavas 1962; Trewavas et al. 1972; Schliewen 1996a). In addition to the endemic fishes, Barombi Mbo harbors at least one endemic sponge (*Corvospongilla thysi*) and one endemic but undescribed caridinid shrimp (*Caridina* sp.) (Trewavas et al. 1972; E. Roth, pers. comm., 1996).

Lake Bermin has an endemic radiation of nine tilapiine cichlids, all belonging to the subgenus *Coptodon,* a taxon that is only distantly related to the *Sarotherodon* of Barombi Mbo. Except for one species, *Tilapia bemini,* all other species in this lake were described in 1992 (Thys van den Audenaerde 1972; Stiassny et al. 1992). Although most of the tilapiines feed on detrital material, at least two trophic specialists are currently recognized: *T. imbriferna* (a phytoplanktivore) and *T. spongotroktis* (a sponge eater). Lake Bermin's noncichlid ichthyofauna includes a small cyprinid of the *Barbus aboinensis* group and an aplocheilid of the *Fundulopanchax mirabilis* group (Stiassny et al. 1992). The invertebrate fauna of Lake Bermin includes various species of freshwater crab, one rotifer, one cladoceran, one copepod, and the fish-eating colubrid snake *Afronatrix anoscopus* (Stiassny et al. 1992).

Lake Ejagham is a special case because it is not a volcanic crater lake but is ecologically and geomorphologically very similar to those that are. The lake's outlet is isolated from the nearby Munaya River by a waterfall that is insurmountable for cichlid fishes. Its oval-shaped lake basin (approx. 1,050 by 700 m diameter) has a maximum depth of 18 m. In contrast to Barombi Mbo and Bermin, this lake was colonized by both *Coptodon* and *Sarotherodon. Coptodon* gave rise to at least five different species, whereas *Sarotherodon* split into two species. Except for *Tilapia deckerti,* all of these species are still undescribed (Thys van den Audenaerde 1968; Schliewen et al. 2001).

Lake Dissoni in the Rumpi Hills harbors one endemic poeciliid (*Procatopus lacustris*), probably an endemic *Barbus,* and an endemic *Clarias* (Trewavas 1962, 1974; Schliewen 1996b).

Status and Threats

Despite their high conservation value and the dependence of local communities on fish from some of these lakes, effective conservation efforts have not been undertaken to protect the lake ecosystems (apart from designation of the Lake Barombi Mbo Forest Reserve and Ejagham Forest Reserve). Deforestation threatens the health of many of these lakes, and forestry activities occur even in the designated reserves. Slash-and-burn agriculture techniques also contribute to deforestation, causing soil erosion and siltation in the basins of some of the lakes. The use of chemical pesticides and fertilizers often accompanies the clearing of forests for agricultural purposes and threatens the health of these small lake basins.

The small size of the lakes renders them extremely vulnerable to even minor disturbances. The most diverse and famous lake, Barombi Mbo, is under immediate threat. Partial deforestation of the interior crater rim has already taken place with increased demand for agricultural land by the local Barombi people and by people from nearby Kumba town. This is likely to cause increased erosion and consequently increased sediment input into the oligotrophic lake system. Water extraction has also temporarily caused lake-level alterations, which are reported to have changed breeding habitat needs of some of the endemic cichlid species, especially *Sarotherdon linnellii* (Schwoiser, pers. comm., cited in Dominey 1987). The use of modern gillnets supposedly has had negative effects on population size of target fish species, although in contrast to previous reports (Reid 1990), all fish species are still present in the lake (U. Schliewen, pers. obs., 2002). Water pollution from insecticide use in small farms within the crater rim and from increased wastewater inflow from the small Barombi Village is also likely to affect the lake ecosystem.

Justification for Ecoregion Delineation

This ecoregion is distinguished by its crater lakes that host endemic species flocks of cichlid fishes as well as endemic insects and shrimps. Most of the crater lakes known from this ecoregion have no outflows or extremely steep outflows such that they are isolated from nearby river systems. This isolation has led to the evolution of highly endemic aquatic faunas.

Data Quality: Medium

Ecoregion Number: **6**

Ecoregion Name: Bangweulu-Mweru

Major Habitat Type: Floodplains, Swamps, and Lakes

Final Biological Distinctiveness: Globally Outstanding

Final Conservation Status: Vulnerable

Priority Class: I

Author: Lucy Scott

Reviewers: Paul Skelton, Jean-Jacques Symoens, and Harry Chabwela

Location and General Description

This ecoregion is situated in the southeastern corner of the Democratic Republic of the Congo (DRC) and northeastern Zambia. The ecoregion is a component of the southern headwaters of the Congo River and is typified by an extremely rich and endemic aquatic fauna in the permanent swamps and shallow lakes of the Bangweulu-Mweru system.

Rainfall varies from about 900 mm per year near Lake Mweru to 1,300–1,500 mm per year near Lake Bangweulu (Hughes 1997). A cool, dry season extends from April to August, followed by a hot, dry season from September to October and a hot, wet season from November to April.

There are six major lakes in the Bangweulu complex: Bangweulu, Walilup, Chifunauli, Kampolombo, Kangwena, and Chale (Bailey 1986). None of the lakes are very deep (3–10 m), but all have extensive swamps associated with them that are important contributors to productivity. The Bangweulu swamps are fed by a series of rivers, the largest of which is the Chambeshi River, which also flows out of the swamps and joins with the Luapula River. The Luapula flows for about 480 km before reaching Mweru. This stretch of river contains the Mumbatuta and Johnson falls, each of which impedes fish movement in the dry season (Bell-Cross 1965). The Luvua River then leaves Lake Mweru and this ecoregion and eventually joins the Lualaba River. Lakes Bangweulu and Mweru, situated on a plateau at about 1,000 m elevation, are deflation lakes (floodplain lakes formed by erosive processes in which wind carries off alluvium) (Bowmaker et al. 1978; Burgis and Symoens 1987).

Lake Bangweulu (2,070 km^2) lies to the west of a large swamp (5,700 km^2) and floodplain (6,000 km^2). The Bangweulu Basin is an old cratonic platform that has been subsiding over the last 20 million years (Denny 1985). There are indications that the lake was much larger in its recent geological history. The present-day lake is shallow (10 m) and unproductive, probably for edaphic reasons. The lake offers a limited range of habitats, with sandy beaches to the west and predominantly *Papyrus*, *Eleocharis*, and *Nymphaea* vegetation bordering the remaining shoreline. The Bangweulu swamps are permanent swamps, characterized by shallow lakes and a series of lagoons generally covered with *Ceratophyllum*, *Utricularia*, *Nymphaea*, *Nyphoides* sp., *Potamogeton richardi*, *Trapa natans*, *Pistia stratiotes*, and *Cyperus papyrus*. These lagoons and lakes are connected by channels lined with vegetation such as *papyrus* and *Vossia cuspidata*. The shallow swamps (1 m deep), are vegetated by *Eleocharis* and *Nymphaea*, which gradually merge into grassy, annually flooded plains surrounding the swamps. During the dry season, decomposing vegetation generates low oxygen levels and high acidity on the margins of the swamps, confining fish to the interiors of the swamps and the main channels connecting them. With flooding, which peaks in May, input of freshwater alleviates the anoxic conditions, and fish disperse into the surrounding swamps and floodplains, where breeding occurs.

Lake Mweru is deeper (37 m) and larger (4,413 km^2) and has a higher plankton production than Lake Bangweulu. A *Vossia* swamp system marked by extensive stands of large ambatch trees occurs at the southern end of the lake at the entrance of the Luapula River. The rest of the shoreline is sandy and rimmed with the sedge *Eleocharis* in shallow waters, with occasional rocky outcrops (Bowmaker et al. 1978).

Granites underlie the Chambeshi River and much of the area around the Bangweulu swamps and extend to the Lower Luapula River. However, the entire region is dominated by recent alluvial deposits, shales, sandstones, quartzites, and conglomerates (Grimsdell and Bell 1975). The soils associated with granites are similar to those overlying rocks of the Katanga system, whereas those associated with the alluvial deposits have a peaty organic horizon. The vegetation of upper areas is predominantly miombo (*Brachystegia* and *Julbernardia*) woodland, and alluvial areas are characterized by grassland and wetlands (Stuart et al. 1990; Hughes 1997).

Outstanding or Distinctive Biological Features

The swampy lakes of the Bangweulu-Mweru ecoregion host a diverse aquatic fauna that includes many endemic species. One-third of the about 100 described fish species are endemic, including five species of endemic *Barbus*, three *Nothobranchius*, nine cichlids, three kneriids, five mochokid catfishes, and three mormyrids (Poll 1976; Balon and Stewart 1983). The region also has a rich molluscan fauna, with thirty-seven species and seven endemics found in Lake Mweru, some of which are also in the lower Luapula River (Brown 1994). Three near-endemic frogs (*Bufo fuliginatus*, *B. melanopleura*, and *Hyperolius kibarae*) and one strict endemic (*H. polystictus*) live along the rivers and streams that feed the lakes. Two dragonfly species (*Aciagrion rarum* and *Moardithemis flava*) are of conservation concern and are near-endemics.

Black lechwe (*Kobus leche smithemani*) is endemic to the Bangweulu floodplains and upper Chambeshi River, where its numbers have dropped substantially (Huntley 1978). The Bangweulu tsessebe antelope (*Damaliscus superstes*) is also endemic to the Bangweulu flats (Cotterill 2003). The vulnerable slaty egret (*Egretta vinaceigula*) has been recorded in the Bangweulu swamp, probably as a nonbreeding visitor. The Bangweulu swamp is one of three principal breeding areas of the vulnerable wattled crane (*Grus carunculatus*), and the vulnerable papyrus yellow warbler (*Chloropeta gracilirostris*) occurs in swamps along the Luapula River near the southern end of Lake Mweru. The swamp-dwelling weaver *Ploceus katangae* also inhabits the swamps of along Lake Bangweulu (Cotterill 2004). The slender-snouted crocodile (*Crocodylus cataphractus*) is present in Lake Mweru, but its life history and conservation needs are unknown (Stuart et al. 1990). In addition, the region provides important habitat for the shoebill stork (*Balaeniceps rex*) (Kabii 1997).

Status and Threats

Current Status

The area is an important subsistence fishing area for local peoples. Fishing, hunting, and cattle grazing are the main occupations of those who live near the wetlands. More than 50 percent of fish production for Zambia originates in the Bangweulu Basin, and Kafue Flats and Mweru-Luapula are also significant for fish production (Chabwela 1992). Cichlid stocks in the Bangweulu Basin have been greatly depleted by overfishing, and increased fishing pressure continues to be placed on the remaining populations (Subramaniam 1992).

A set of protected areas is concentrated in an area of swamps, floodplains, and miombo woodlands south of the lakes. These include the Kasanka (390 km^2), Lavushi-Manda (1,500 km^2), Isangano (840 km^2), and Mweru-Wantipa (3,134 km^2) national parks in Zambia (Stuart et al. 1990).

Future Threats

Major threats to the ecoregion are the rapidly growing population and encroachment, overfishing, overhunting, and deforestation of the surrounding areas (Chabwela 1994c). Nearly all national parks and game management areas are severely depleted of large mammals. However, policy changes are being pursued in promoting co-management of fisheries and in community-based natural resource management of wildlife and forest resources.

Justification for Ecoregion Delineation

This ecoregion is distinguished by its swampy floodplain lakes, which host a rich and endemic freshwater fauna including endemic fish (five species of endemic *Barbus*, three *Nothobranchius*, nine cichlids, three kneriids, five mochokid catfishes, and three mormyrids), mollusks (*Cleopatra johnstoni*, *C. mweruensis*, *Melanoides crawshayi*, *M. mweruensis*, *Bellamya crawshayi*, *B. mweruensis*, and *B. pagodiformis*), mammals (*Damaliscus superstes* and *Kobus leche smithemani*), and wetland birds (*Ploceus katangae*). This ecoregion's waters were historically part of the Zambezi system, until the Congo River captured the Luvua River near Mkuka at the southern end of the Bangweulu swamps during the early Tertiary period (Moore and Larkin 2001). The freshwater fauna remains mostly Zambezian in origin, and this ecoregion therefore is included in the Zambezi bioregion (Banister 1986; Cotterill 2004). The Chambeshi is now confluent with the Luapula, but the entry of Congo fauna into Lake Bangweulu is prevented by the Mumbatuta Falls (Jackson 1986). The fish family Clupeidae is an example of a Congolian group present in Lake Mweru but absent from Lake Bangweulu. Bangweulu has fewer species than Mweru but shares a larger proportion (49 percent versus 32 percent) of them with the Kafue.

Data Quality: Low

The level of biological investigation is reasonable, but more work is needed to quantify the current status of these systems. Investigation into the conservation status of some of the threatened species and the impacts of the fishery are suggested.

Ecoregion Number:	**7**
Ecoregion Name:	Inner Niger Delta
Major Habitat Type:	Floodplains, Swamps, and Lakes
Final Biological Distinctiveness:	Continentally Outstanding
Final Conservation Status:	Endangered
Priority Class:	II
Authors:	Ashley Brown and Miranda Mockrin
Reviewers:	Bakary Kone, Christian Lévêque, and Eddy Wymenga

Location and General Description

The Inner Niger Delta is located in central Mali in the semi-arid Sahelian zone, just south of the Sahara Desert, roughly situated between Djenné in the south and Tombouctou in the north. The dune ridges on the Sahara's edge funnel the waters of the Inner Delta north and east through Mali. Each year during the rainy season, floodwaters of the Niger and Bani rivers spill over their banks, and the Inner Niger Delta in Mali is inundated to an area of 30,000 km^2, on average. In contrast, the delta contracts to 3,900 km^2 or less during the dry season (Welcomme 1986b; Zwarts and Diallo 2002). However, the surface of the inundated zone is highly variable according to river discharge: from 1956 to 2002 the maximum flooded area varied between 9,500 km^2 (1984, severe drought) and 44,000 km^2 (1957, high floods) (Quensière 1994).

The flooded grassland, lakes, and channels of the delta provide vital habitat for Afrotropical and migratory Palearctic birds as well as for many fish species, hippos, and manatees. The delta is also an essential resource for nearly 1 million Malians, supporting livelihoods in fishing, farming, and pastoralism in an otherwise arid country.

The Inner Niger Delta lies in a depression that formed the bed of a large lake during the Quaternary period (Welcomme 1986b). The delta extends for 425 km, with an average width of 80 km, tapering into a braided river near Tombouctou where the Niger curves to the east. The floodplain drops only 8 m over its course (Hughes and Hughes 1992), and its topography is a complex mix of submerged lower areas and higher, unflooded areas known as *tougérés*. A vast network of river channels (*mayos*) with levees separated by low, clay-based floodplains forms networks across the delta. As waters flow through the delta, they pass over Pleistocene and recent alluvium overlying Paleozoic sandstone (Hughes and Hughes 1992). The upper margin of the delta is delimited by the 280-m contour and it is surrounded by sandstone massifs. To the north, huge dunes of the Erg Ouagadou block a former westward course of the river.

Located in the Sahel just south of the Sahara Desert, the Inner Niger Delta covers the transition between a humid Guinean climate in the south and a dry climate at the edge of the Sahara. In the south, the rainy season begins in July and lasts through October, with a mean annual precipitation of 750 mm. In the north, the rainy season begins in July and lasts through September, with a mean annual precipitation of 250 mm (Dumont 1987). Temperatures are strongly seasonal, with an average maximum in May at Tombouctou (43°C) and Mopti (40°C), and the coldest months are December and January, when temperatures drop to an average minimum of 3–6°C in the north of the delta.

Local rainfall has a negligible impact on the flood regime in the Inner Delta. Flows in the delta depend completely on river discharge from the Niger and Bani headwaters and a few smaller and temporal streams that flow down from the Dogonland Plateau. Rainfall occurs in the Niger's headwaters from May through September, with a clear peak in August, creating a surge that reaches the inland delta in October (Zwarts and Diallo 2002). The Niger River is the longest river in West Africa and the third longest in Africa. Rising in the Fouta Djalon highlands of Guinea, the river extends for 4,100 km before flowing into the Atlantic Ocean on the Nigerian Coast. The Bani River is 1,100 km in length with sources in Côte d'Ivoire and Burkina Faso. Dogonland streams provide an insignificant contribution to the delta but do fill some of the southwestern lakes (Hughes and Hughes 1992). The surge of water that reaches the delta from these two rivers dissipates as it continues downstream, with about half of the Niger's water volume lost to evaporation (Quensière 1994; Zwarts and Diallo 2002). As a result of lateral expansion, the flood slows in the delta. It takes 1 month for the flood to reach Mopti 160 km downstream and nearly another month to reach Lake Debo, where the maximum flood occurs in November and December (Dumont 1987; Laë 1997; Zwarts and Diallo 2002). The extensive swamps and vegetation of the delta filter silt and salt from the water so that the water leaving the delta is clear, low in dissolved salts, and silt-free (John et al. 1993). Dry, landlocked Mali is completely dependent on these rivers for its water resources.

Several natural and artificial lakes border the delta. Lake Debo is a shallow lake that expands and contracts as the river level rises and declines. Lake Horo is separated from the river system by a dam and a sluice gate, opened in November to allow floodwaters to enter (Wetlands International 2002). Several ephemeral lakes dot the landscape of the western and eastern periphery of the delta. Today the lakes situated on the east side of the delta are mostly dry, with the exception of Lake Aougoundou and Lake Korarou, the latter being fed by rainwaters from the Dogonland Plateau.

Vegetation defines the different habitat types of the delta. Three main plant associations have been identified: submerged and floating plants in shallow or stagnant water, partially submerged and marginal vegetation dominated by grasses, and plants that grow on seasonally exposed sandy soils (Hiernaux 1982; John et al. 1993). Along the rivers, a typical scrub of *Mimosa pigra* and *Salix chevalieri* is found, often together with a vegetation of *Vetiveria nigritana*. Partly floating, long-stemmed grasses (*Echinochloa stagnina, E. pyramidalis, Oryza barthii,* and *Andropogon gayanus*) dominate in the floodplains. Permanent pools are richer and host the submerged macrophytes *Ceratophyllum* spp. and *Utricularia* spp., in addition to floating *Nymphaea* spp. The lakes, particularly Lake Debo and Walado Debo in the central part of the delta, are surrounded by *Echinochloa* spp. and *Vossia cuspidata* (Dumont 1987). Flooded forests of *Acacia kirkii* are also characteristic but increasingly rare because of overharvesting. Although these forests are dominated by *A. kirkii, Ziziphus mauritiana* may also occur. Algal blooms are common on the lakes and can reduce the water transparency.

Grasses such as *Acroceras amplectens, Echinochloa pyramidalis, E. stagnina,* and *Eragrostis atroviriens* dominate the low-lying floodplain in the southern half of the delta. Large areas of the flooded delta are occupied by wild rice (*Oryza longistaminata*) and a characteristic vegetation of *E. stagnina,* known as bourgou fields. Bourgou is used as feed for domestic animals and therefore is often planted by local people. Other typical species of the flooded pastures that occur higher in the inundation zone are *Vetivera nigritiana* and *Vossia cuspidata.* Along the heavily grazed outer fringes, *Andropogon gayanus, Cynodon dactylon,* and *Hyparrhenia dissoluta* dominate. Forests are scarce in the delta, having been heavily exploited, overgrazed, and harvested for firewood. Trees such as *Acacia seyal, Diospyros* sp., and *Kigelia africana* grow on higher levees. The northern half of the delta is characterized by emergent sand ridges, with palms such as *Hyphaene thebaica* and *Borassus aethiopum* (Gallais 1967).

Outstanding or Distinctive Biodiversity Features

The Inner Niger Delta provides habitat for a rich aquatic fauna and wetland-associated avifauna. The migration, feeding, and breeding of fish, birds, and other wildlife are synchronized with seasonal flooding in the delta.

The fish fauna of the delta is composed of about 130 species, of which the dominant families are Mormyridae, Mochokidae, and Cyprinidae. Two species are limited to the delta and the Upper Niger: *Synodontis gobroni* and the rapids-dwelling *Gobiocichla wonderi* (Welcomme 1986a). Many of the fish have life cycles that take advantage of the habitats and resources associated with the floodplain, with species migrating upriver and downriver as well as laterally out to the floodplain as the water rises (Quensière 1994). Waters of the flooded delta are initially well oxygenated, providing favorable habitat for spawning fish, developing eggs, and larval fish. The initial flooding of the savanna enriches the water with nutrients from decomposition of vegetation and animal droppings, creating a surge of bacteria, algae, and zooplankton that provides a rich feeding ground for fish. When the waters recede, the fish move upriver or risk becoming trapped in small, isolated ponds. Some fish species can survive in these dwindling pools by aestivating or by breathing air (Lowe-McConnell 1985).

Fish migrations include both lateral movements onto floodplains and long-distance, longitudinal movements. There is anecdotal evidence of several fish moving as much as 640 km up the Niger River into the Inner Delta with the onset of floods (Welcomme 1986a). One of the African tetras, *Brycinus leuciscus,* has been observed moving 50 km from the river mainstream to the edge of the floodplain and may move 125–400 km upstream from the Inner Delta to the Markala dam as floods subside (Lowe-McConnell 1985).

The Inner Niger Delta provides essential habitat for huge numbers of wetland birds, including Afrotropical resident species and migrants that spend the Palearctic winter in Africa (Roux and Jarry 1984; Fishpool and Evans 2001; Wymenga et al. 2002). As the water recedes, after peak levels in October and November, birds concentrate in the central part of the delta (Lac Debo, Walado Debo and Lac Korientzé). More than 500,000 garganey (*Anas querquedula*) and up to 200,000 northern pintail (*Anas acuta*) stay here during the northern winter, along with large numbers of ferruginous duck (*Aythya nyroca*), white-winged tern (*Chlidonias leucopterus*), ruff (*Philomachus pugnax*), black-tailed godwit (*Limosa limosa*), and other waterbirds (van der Kamp and Diallo 1999; Wetlands International 2002; van der Kamp et al. 2001, 2002b). The total waterbird numbers in the delta depend heavily on water level and can reach more than 1 million in favorable years. In addition, more than 1 million wetland-related passerines, particularly sand martin (*Riparia riparia*) and yellow wagtail (*Motacilla flava*), pass through the delta during their autumn migration to and from their breeding grounds. Despite these impressive numbers, several species in the delta are under serious threat, particularly large Afrotropical species such as the locally rare black-crowned crane (*Balearica pavonina pavonina*), which has a population of only fifty birds (van der Kamp et al. 2002b).

The delta is also of international importance for several bird species because a high proportion of their populations occurs in the delta. More than twenty-seven species have 1 percent of their population occurring in the delta, including the African cormorant (*Phalacrocorax africanus*), purple heron (*Ardea purpurea*), glossy ibis (*Plegadis falcinellus*), collared pratincole (*Glareola pratincola*), gull-billed tern (*Gelochelidon nilotica*), and Caspian tern (*Sterna caspia*). For example, 20 percent of the population of the *Plegadis falcinellus,* breeding in Europe and the Black Sea region, spends the northern winter in this area (van der Kamp et al. 2002b). The northern lakes, including the Ramsar site Lac Horo, often hold a large proportion of the West African wintering population of *Aythya nyroca* (up to 50 percent), although high numbers of this species have not been recorded recently (Scott and Rose 1996; Girard and Thal 2001). The delta is also nationally important for breeding purple swamp-hens (*Porphyrio porphyrio*) (Wetlands International 2002).

The Inner Niger Delta is also known for its large waterfowl breeding colonies, with 80,000 breeding pairs of birds of fifteen cormorant, heron, spoonbill, and ibis species (Skinner et al. 1987). These colonies contain about 50,000–60,000 cattle egrets (*Bubulcus ibis*), 18,000–20,000 African cormorants (*Phalacrocorax africanus*), 1,500–1,800 pairs of great white egrets (*Egretta alba*), 250–300 pairs of African darters (*Anhinga rufa*), and several other species. Breeding occurs during high flood (September–November), and the colonies are situated in flooded forests of (mainly) *Acacia kirkii*. Of the seven mixed breeding colonies present in 1985–1986 (Skinner et al. 1987), only two remain at present (1998–2002). The presence of the remaining two results largely from conservation efforts by the IUCN in close collaboration with local communities (van der Kamp et al. 2002a).

Several mammal species are closely linked to the wetlands of the Inner Niger Delta. The delta harbors an important but dwindling population of the vulnerable West African manatee (*Trichechus senegalensis*), which has suffered from hunting and severe droughts. Hippos (*Hippopotamus amphibius*) are present in the central and southern delta, with an estimated population of forty to sixty individuals (Stuart et al. 1990; Wymenga et al. 2002). Antelope populations have been seriously reduced by droughts, the bushmeat trade, and conflicts with grazing livestock. Buffon's kob (*Kobus kob kob*) was once numerous in the Inner Niger Delta but is no longer present. This also appears to be the case for roan antelope (*Hippotragus equinus*), dorcas gazelle (*Gazella dorcas*), and dama gazelle (*Gazella dama*). A small population of the vulnerable red-fronted gazelle (*Gazella rufifrons*) is believed to remain, although little information is available (Wymenga et al. 2002). Species such as clawless otter (*Aonyx capensis*), spotted-neck otter (*Lutra maculicollis*), African civet (*Civettictis civetta*), caracal (*Felis caracal*), serval (*Felis ser-*

val), striped hyena (*Hyaena hyaena*), and spotted hyena (*Crocuta crocuta*), once recorded from these regions, appear to have been extirpated. During four years of field work and several aerial surveys, none of these species nor any other large native mammals, with the exception of hippopotamus, were seen in the delta (Girard and Thal 2001; Wymenga et al. 2002). The vast floodplains still provide habitat for Nile crocodile (*Crocodylus niloticus*), Nile monitor (*Varanus nilotica*), and *Python sebae*. The Nile crocodile is believed to be on the edge of extinction, and the Nile monitor and python are facing heavy human pressure (Wymenga et al. 2002).

Aquatic plants of the delta are highly adapted to the yearly floods. One of the most important of these species is bourgou (*Echinochloa stagnina*). Bourgou plays a role in maintaining fish diversity by providing breeding and feeding habitat (Roggeri 1995; Oyebande and Balogun 1996). For example, *Marcusenius abadii*, *Mormyrus macrophthalmus*, *Polypterus bichir*, and *P. endlicheri* prefer bourgou mats for breeding (Welcomme 1986a).

Status and Threats

Current Status

Over the last half century, drought combined with the impact of dams, increased fishing pressure, and increased agriculture have changed the landscape of the Inner Delta. Two severe droughts, one in 1973 and one in 1984 (with persistent low water levels until 1994), created severe stress on the environment and human population of the Inner Delta. The effects of dams and changes in resource use have only worsened the situation (Laë 1995).

Several dams have changed the flow regime in parts of the delta. A dam constructed in 1946 at Markala irrigates the floodplain for the production of rice and grain crops (Oyebande and Balogun 1996). The Selingue dam, constructed in 1984 on the Sankarani tributary, provides electric power, irrigation, and flood control (Oyebande and Balogun 1996). Dams may disrupt fish migrations, affecting the ability of fish to move upstream and onto the floodplains to spawn (Laë 1997). A decision not to release artificial floods from the Selingue Dam during drought in the early 1980s resulted in a severe reduction of bourgou (*Echinocloa stagnina*) growth, creating a loss of feeding and breeding grounds for fish (Oyebande and Balogun 1996). Overall, dams and drought have caused the loss of about 5,000 tons of fish from the fishery in the central delta (Laë 1997).

Overfishing and poaching are also of concern. Ninety percent of Mali's freshwater fish catch comes from the Inner Delta, yet the size of the catch has declined since 1969, as has the average size of landed fish (Quensière 1994; Hughes and Hughes 1992). Before 1960, traditional management determined fishing practices for the delta (Quensière 1994; Ticheler 2000). Two ethnic groups, the Bozos and the Somonos, were the primary fishers in the Inner Delta. The change from traditional management to government regulation in 1960 opened fishery access to all citizens (Moorehead 1991; see also Ticheler 2000). This resulted in an increase in fishers as other ethnic groups became users of the fishery. The number of fishers in the delta doubled from 1977 to 1997 (Laë 1997). This increase, combined with the use of more sophisticated fishing equipment (such as nylon nets), has led to a decline in catch of certain species, including the economically valuable *Polypterus senegalus* and *Gymnarchus niloticus* (Quensière 1994; Ticheler 2000). Poaching also poses a continued threat to manatee populations (Happold 1987). Wild bovids and hippos have declined in recent decades; crocodiles were once common in rivers and some of the lakes but have also declined, as have populations of aquatic turtles (*Trionyx*, *Pelusios*) (Wymenga et al. 2002).

Traditional management systems under the Dina-law once regulated pastoralists, resident as well as nomadic. The nomadic farmers (Peul) move their herds onto arid, rain-fed lands as waters rise, and then move them back onto the floodplain to graze as water recedes (Gallais 1967). Livestock herds in the delta are considered to have the highest density in Africa, with as many as 2 million cattle and 3 million sheep present. Excessive grazing combined with the severe droughts of the past two decades has resulted in land degradation, pasture loss, and deforestation (Heringa 1990). Habitat degradation is also caused by erosion and agricultural mismanagement in the Niger watershed. Erosion leads to increased deposits of sediments that then fill in small pools and streams, eradicating aquatic vegetation and mollusks (Ticheler 2000). Deforestation in the delta also decreases bird habitat by removing roost and breeding trees. Wymenga et al. (2002) documented the loss of flooded forests as a result of drought and deforestation, despite their essential role in providing breeding habitat for colonial breeding waterbirds and rearing areas for fish.

Three Ramsar sites were declared in 1987: Lac Horo, Lac Debo, and the Séri floodplain complex, together comprising 1,620 km^2. The Ramsar sites are owned by the state and are used by local residents and nomadic pastoralists for drinking water, fishing, seasonal agriculture, and livestock rearing. Lac Horo is separated from the Niger by a dam and a sluice gate. The remainder of this ecoregion is unprotected and heavily used for different human activities (Kone 2002).

Future Threats

Human population pressure, changing patterns of resource use, and dams coupled with prolonged periods of drought pose continuing threats to the Inner Niger Delta (Shumway 1999; Quensière 1994; Wymenga et al. 2002). The combined effects of drought and anthropogenic influences may reduce fish species richness and alter the habitat of the ecoregion (Denny 1991). Decreased flooding may also change the composition of the fish assemblage as species more tolerant to perturbation increase (Laë 1995).

Justification for Ecoregion Delineation

The Inner Niger Delta ecoregion is delineated based on the general extent of the floodplains of the Delta and associated grasslands.

Data Quality: High

Ecoregion Number:	**8**
Ecoregion Name:	**Kafue**
Major Habitat Type:	**Floodplains, Swamps, and Lakes**
Final Biological Distinctiveness:	**Nationally Important**
Final Conservation Status:	**Vulnerable**
Priority Class:	**V**
Author:	**Lucy Scott**
Reviewers:	**Musonda Mumba, Paul Skelton, and Monica Chundama**

Location and General Description

This ecoregion encompasses the Kafue River drainage basin from central Zambia south to the Kafue Gorge where the river enters the Middle Zambezi River. It contains the extensive and seasonally inundated floodplains of the Kafue Flats and the large Lukanga swamp. However, the headwaters of the Kafue, including parts of the Lufupa, Lunga, Luswishi, and upper Kafue rivers, are excluded from this ecoregion and contained in the Zambezian Headwaters [76] ecoregion, with which they have faunal affinities. The Kafue River is about 1,000 km long from its source to the confluence of the two rivers (Beadle 1981). The river is a major tributary of the Zambezi and is found entirely in Zambia. It is a source of potable water for about 40 percent of the Zambian population and is the major source of water for the capital city, Lusaka (Chabwela and Mumba 1998).

Rainfall in the ecoregion is highly seasonal, averaging 600–900 mm annually, with the majority falling in the summer months of November–February. The maximum temperature varies from 20°C to 45°C, depending on the season.

The high rainfall combined with a gentle gradient in the main river has produced extensive swamps and floodplains. The Kafue is called a reservoir river (sensu Jackson 1961, 1963), meaning that the floodplains regulate the flood, releasing it slowly back to the river so that river levels seldom exhibit large variations in height. Inundation of the floodplains occurs from January to June, after the rains (Williams 1971). The floodplains are inundated to an average depth of 3 m and have water on them for long periods. Marginal vegetation is abundant and provides cover for small and juvenile fish (Williams 1971; Marshall 2000a). The 250-km-long Kafue floodplain, stretching from Itezhi-Tezhi to the Kafue Gorge, is up to 40 km wide during seasonal floods (Hughes and Hughes 1992).

The terrestrial vegetation in the Kafue ecoregion is a diverse mosaic of miombo (*Brachystegia* and *Julbernardia*), *Acacia-Combretum* and mopane woodland, and grasslands dominated by rice grass (*Oryza barthii*), *Echinochloa pyramidalis*, *Vetivaria nigritana*, *Acroceras macrum*, and *Setaria avettae*. The aquatic vegetation on permanent water and on the seasonally inundated floodplains of the Kafue Flats and Lukanga swamp is characterized by *Vossia cuspidata*, *Polygonium* sp., *Cyperus papyrus*, *Potamogeton* sp., *Apanogeton* sp., *Typha* sp., and *Leersia hexandra*. Floating rafts of vegetation are known to break away from the banks and float out to open water (Williams 1971; Stuart et al. 1990).

Outstanding or Distinctive Biological Features

The rivers, floodplains, and swamps of the Kafue ecoregion support a moderately rich freshwater fauna, but these habitats support few fish endemics compared with other large river and floodplain ecoregions. Only one killifish (*Nothobranchius kafuensis*) and one cyprinid that is known only from its type locality (*Barbus altidorsalis*) are endemic. However, near-endemism is high for large riverine cichlids in this ecoregion, and about sixty fish species live in its waters (Bell-Cross 1972). The Kafue Flats supports the highest abundance of waterbirds in the Zambezi Basin (Mundy 2000), and the area is an important bird area (Fishpool and Evans 2001).

Seasonal flooding in the Kafue is the most important ecological process maintaining biodiversity in the region. During periods of inundation, fish migrate out onto the floodplains to spawn, taking advantage of increased habitat and protective vegetative cover. Fish known to use the rich floodplain habitats in this way include the largemouth breams and tilapias and many small barbs. Females of the endemic Kafue killifish (*Nothobranchius kafuensis*) lay their eggs on the sediments of the floodplain or in pans during the wet season, but hatching is delayed until the following year when water returns.

Large concentrations of waterbirds, including the most significant population of the vulnerable wattled crane (*Grus carunculatus*), amass in the Kafue Flats. The vulnerable slaty egret (*Egretta vinaceigula*) has also been recorded in the flats and probably also breeds here. Congregations of long-tailed cormorant (*Phalacrocorax africanus*), cattle egret (*Bubulcus ibis*), African openbill (*Anastomus lamelligerus*), fulvous whistling duck (*Dendrocygna bicolor*), comb duck (*Sarkidiornis melanotos*), collared pratincole (*Glareola pratincola*), Caspian plover (*Charadrius asiaticus*), and ruff (*Philomachus pugnax*) are also known (Fishpool and Evans 2001).

Several migratory mammals graze on the grasses that grow seasonally on the floodplains. About half of all of the 100,000 remaining lechwe (*Kobus leche leche* and *K. l. kafuensis*) in Africa occur in the Kafue floodplain (Cotterill 2000). The red lechwe (*K. l. leche*) occupies extensive areas of the dry floodplain in parts of Kafue National Park and probably benefits the fishery by increasing the floodplain fertility (Williams 1971). The Kafue lechwe (*K. l. kafuensis*) is endemic to the Kafue Flats and inhabits low-lying land in the vicinity of the Kafue Gorge and Itezhi-Tezhi Dam (Williams 1971; Stuart et al. 1990).

Status and Threats

Current Status

Pressures on the Kafue floodplain are great, mostly through hydrological changes resulting from dam construction but also from heavy fishing levels and inappropriate fishing practices, illegal hunting of antelope, deforestation of parts of the immediate catchment, commercial sugar plantations just outside to the south, agricultural pollution, and infestation by aquatic weeds. Severe competition for water from the Kafue River and the presence of a major urban center (Lusaka) close by exacerbate these threats. Human population pressures are high. However, there is substantial tourism potential.

Because of poor land use practices in the southern portion of the Kafue Flats, the wetlands of this ecoregion are under threat. Land surrounding the floodplains has experienced soil erosion, overgrazing, and overuse of chemicals, and the runoff of pollutants and excess sediment have decreased water quality (Chabwela 1992). There is also concern about pollution of Kafue waters from runoff from mining sites and other industrial point sources. Bioaccumulation of heavy metals and pesticides has been documented downstream of some sites (Norrgren et al. 2000).

Two hydroelectric dams, Kafue Gorge and Itezhi-Tezhi, have altered the flow regime of the river, and there is some evidence that the new hydrology has negatively affected the fish stocks that need access to the floodplains for breeding (Subramaniam 1992). The Kafue Gorge Dam also influences the Kafue Flats by flooding of the flats at the peak of storage in the dam (Marshall 2000a). The Itezhi-Tezhi Dam was designed to release artificial discharges of water to simulate flooding conditions on the Kafue floodplain. Unfortunately, the dam still has a regulating effect on the flood regime and has caused the decline of flood-dependent grasslands and consequently lowered the quality of food available for the Kafue lechwe on the Kafue Flats (Davies and Day 1998). The WWF Partners for Wetlands Project, initiated in 1999, is working on a variety of projects in the flats, including the establishment of a conservation area in a previously degraded section of the flats (Mwanachingwala Conservation Area), support for ecotourism in partnership with investors in tourism, and the development and implementation of an integrated water management strategy for the flats. In regard to the latter, the sugar industry, the Zambian Electrical Supply Company, and the Ministry of Energy and Water Development have agreed to a set of protocols to change the operations of the dams to more closely mimic natural flows through the flats and aim to begin testing them in 2004.

The Kafue floodplain fishery is one of the most productive in the Zambezi system, and pre-impoundment the fishery provided an annual average catch of about 8,000 tons (Jackson 1961). The floodplain extends for approximately 250 km from Itezhi-Tezhi to the Kafue Gorge. Gillnets are set on the floodplain in the wet season, and fish are caught in the mainstem with seine nets in the dry season (Jackson 1961; Everett 1971). Since closure of the Itezhi-Tezhi Dam, average annual catch on the flats has been reduced to about 7,000 tons, probably through a combination of natural weather cycles (drought) and impacts of the dam (Centre for Ecology and Hydrology 2001).

Important protected areas in the Kafue ecoregion are the Kafue National Park (22,400 km^2) and an associated Game Management Area, the Blue Lagoon National Park (450 km^2), and the Lochinvar National Park (410 km^2). The Blue Lagoon and Lochinvar are both part of the Kafue Flats Ramsar site. Lochinvar, situated 177 km upstream from the Kafue Gorge, is on the south bank of the river and covers part of the Nampongwe River (Williams 1971; Stuart et al. 1990).

Future Threats

A large proportion of Zambia's human population and most of its urban areas and industrial potential lie in the Kafue drainage (Marshall 2000a). Future threats to the integrity of the ecoregion include those associated with urban and industrial development, including pollution and habitat destruction. Floodplain river fish communities that are adapted to compensate for high natural removal rates appear to be less affected by heavy fishing than the equivalent lacustrine populations, but any overfishing remains a threat to the integrity of these fish populations (Jackson 1986).

In addition, alien species pose a threat to the native aquatic fauna. The nonnative *Oreochromis niloticus* has become established in the Kafue drainage and is likely to spread throughout the flats, negatively affecting the native *O. andersonii* and *O. macrochir* (Schwanck 1995; Marshall 2000a). The spread of water hyacinth (*Eichhornia crassipes*) up the Kafue River may be related to the high nutrient input of runoff from sugar plantations. Given the hyacinth's invasive nature, its presence is of great concern (Fishpool and Evans 2001). In Lochinvar National Park, *Mimosa pigra* is spreading and shows signs of becoming a weed that might need control. Unfortunately, this plant is encroaching on the favored feeding grounds of the lechwe and the nesting sites of aquatic birds (M. Mumba, pers. comm., 2002).

Justification for Ecoregion Delineation

This ecoregion encompasses the Kafue River Basin from central Zambia south to the Kafue Gorge where the river enters the Middle Zambezi River. The Kafue is part of the west Zambezian aquatic faunal arena and has ichthyofaunal affinities with the Okavango, Upper Zambezi, and Cunene rivers. The Kafue also has some species in common with the southern tributaries of the Congo River, especially the Chambeshi (Bell-Cross 1972; Skelton 1994). The Kafue River has long stretches of rapids in the upper two-thirds of its course, but these present no barrier to fish movement. However, a series of falls downstream in the 48-km-long Kafue Gorge create two major physical barriers to fish movement from the middle Zambezi (Bell-Cross 1965; Bell-Cross 1972). Observed biogeographic affinities with Zambezian ichthyofauna can be explained by a series of stream captures. Before the Pleistocene, the Kafue may have flowed westward to join the Okavango, Upper Zambezi, and Cunene, and later it may have been captured by a tributary of the middle Zambezi (Bell-Cross 1972; Beadle 1981). It is thought to have separated from the Upper Zambezi in the mid-Tertiary, but its headwaters retain ichthyofaunal affinities with the Upper Zambezi and are included in the Zambezian Headwaters [76] ecoregion. The Kafue was separated from the Zambian Congo drainage when the Chambeshi was captured by the Luapula, and as a consequence it was not invaded by the Congo River species that invaded the Upper Zambezi (Marshall 2000a).

Data Quality: Medium

The quality of ecological data is fair. More research is needed in all aspects pertaining to this ecoregion, especially into the effects of overfishing, other anthropogenic activities, and the functioning of the system as a whole.

Ecoregion Number:	**9**
Ecoregion Name:	**Lake Chad Catchment**
Major Habitat Type:	**Floodplains, Swamps, and Lakes**
Final Biological Distinctiveness:	**Nationally Important**
Final Conservation Status:	**Endangered**
Priority Class:	**IV**
Author:	Ashley Brown
Reviewers:	Christian Lévêque and Emmanuel Obot

Location and General Description

Located at the southern edge of the Sahara Desert, the Chad Basin is bounded in the north by the Aïr and Tibesti Mountains, in the east by the Ennedi and Jebel Marra, and in the west by the Jos Plateau. The lake falls within Niger, Chad, Cameroon, and Nigeria. Lake Chad swells with seasonal river floods and supports a rich fish fauna and large waterbird congregations in an otherwise xeric region. Because of Chad's flat lakebed, small changes in the water budget cause large variations in its area within and between years. For example, in 1965 the lake occupied 25,000 km^2, but in 1973 it was reduced to 6,000 km^2 (Carmouze et al. 1983a). By 2001 the lake was further reduced to less than 2,500 km^2 by the combined effects of increased water extraction upstream and climate desiccation (NASA 2001).

The climate of Lake Chad is semi-arid to arid (Olivry et al. 1996). Conditions are dry and hot from March to June and dry and cool from November to February. On the east coast of the lake, the minimum and maximum air temperatures are about 14°C and 31.4°C in January, 24.2°C and 38.5°C in April, and 24.2°C and 31°C in August (Olivry et al. 1996). Precipitation occurs from June to October, with the northward movement of an unstable maritime air mass. Average rainfall over the lake is 212 mm per year in the north and 288 mm per year in the east.

The lake's water comes primarily from precipitation on the Adama Plateau in the south via the Logone River and from the northern highlands of the Central African Republic via the Chari River. These two rivers make up 95 percent of the total lake inflow, with the Chari contributing 50 percent of the lake's total water input in October and November (Carmouze and Lemoalle 1983; Olivry et al. 1996; Evans 1996). The Yobe in the northwest and the seasonally inundated riverbed El Beid contribute the remaining inflow. Flooding of the Logone and Chari occurs from September to December, and the El Beid floods from November to January. During high floods, water may pass from the Logone floodplains to the Niger River through the Mayo Kebbi System in the south of Chad.

The water chemistry of the lake is tied closely to climatic conditions. The Harmattan winds and dry season aridity contribute to high evaporation that often equals or exceeds water influx and can reach rates of up to 2,300 mm/year (Carmouze et al. 1983a; Hammer 1986). Wind also contributes to mixing of the shallow, polymictic lake, so that waters are always turbid: measurements of transparency have fluctuated from a mean of 1 m in 1965, when the water level was high, to 0.1 m in 1973, at the time of Sahelian drought (Carmouze and Lemoalle 1983). Despite the lake's endorheic nature and arid environment, water remains fresh as a result of hydrochemical regulation mechanisms. However, water salinity increases from the Chari Delta to the north of the lake. The influx of freshwater from the south pushes the denser saline water to the north, where it leaves by seepage into a subterranean system (Dejoux 1983).

There are extensive floodplains on the Logone and Chari rivers. These include the 'Yaéré floodplain in Cameroon, composed of more than 5,000 km^2 of inundated land located between the Chari and Logone in the east and the Mandara Mountains in the west (Carmouze and Lemoalle 1983). Estimates of the total inundated area south of Lake Chad during the rainy season are as high as 90,000 km^2 (Lowe-McConnell 1985). On the northwest side, the Hadejia-Nguru floodplain (6,000 km²), an important wetland for resident waterbirds and Palearctic migrants, drains to the Yobe River (Hollis et al. 1993; Ezealor 2002).

During a Normal Chad period, the lake landscape is a mixture of open water (38 percent), archipelagos (23 percent), and reed belts (39 percent) (Dumont 1992). Separating the lake into north and south basins is the Grand Barrier, a ridge of land submerged when the lake is fully inundated. The south basin, with more inflow, usually is the larger basin. These basins and the southeastern archipelago zone, an area studded with sandy islands, are the three main sections of the lake (Carmouze and Lemoalle 1983).

Vegetation of the south basin is composed mainly of *Cyperus papyrus*, *Phragmites mauritianus*, *Vossia cuspidata*, and other associated wetland macrophytes. The more saline northern basin supports *Phragmites australis* and *Typha australis*. The shoreline is lined with the reed *Cyperus laevigatus* in the north and *Cyperus articulatus*, *Pycreus mundtii*, and *Leersia hexandra* in the south (Iltis and Lemoalle 1983). *Hyphaene thebaïca* (doum palm) marks the base of the slope of the dunes, which are colonized by *Balanites aegyptiaca*, *Leptadenia pyrotechnica*, *Calotropis procera*, and *Acacia* spp. The Sahara Desert, which borders the northernmost section of the lake, has little or no vegetative cover (Wanzie 1990). Submerged vegetation (such as *Ceratophyllum demersum*, *Potamogeton schweinfurthii*, and *Vallisneria spiralis*) grows in the lake. More than 1,000 algae species have also been described from the lake (Compére and Iltis 1983).

Outstanding or Distinctive Biodiversity Features

Large bird congregations and a rich fish fauna adapted to seasonal flooding distinguish the Lake Chad ecoregion (Ezealor 2002). Waterbirds flock to the productive waters of Lake Chad during the wet season. As the influent Chari and Logone rivers swell as a result of seasonal rains, fish migrate from the lake up the rivers and onto productive floodplains for feeding and breeding. Flooding brings high phytoplankton and zooplankton productivity to the floodplains and increases macrophytic growth, creating ideal feeding and spawning habitat (Carmouze et al. 1983b). For example, macrophytic vegetation of floodplains, such as the Yaéré floodplains, provides shelter, breeding, or feeding habitat for juveniles of many different species (Bénech et al. 1983; Dumont 1992). Migratory species that move to floodplains to spawn include *Alestes baremose*, *A. dentex*, *Distichodus rostratus*, *Brachysynodontis batensoda*, and *Marcusenius cyprinoides*. During the period of Normal Chad, species exhibiting a wide distribution in the lake and its tributaries have included *Lates niloticus*, *Synodontis schall*, *Labeo senegalensis*, *Distichodus rostratus*, *Hydrocynus forskalii*, and *Schilbe mystus*. Twenty-five aquatic mollusk species are reported from the ecoregion and three endemic species (*Gabbiella neothaumaeformis*, *G. tchadiensis*, and *Biomphalaria tchadiensis*); however, Brown (1994) suggests that further studies may reveal that these species are more widespread. The ecoregion is also home to thirty-six water-dependent frog species, with only one species strictly endemic to the ecoregion (*Astylosternus nganhanus*).

The lake's aquatic biota, particularly its fish fauna, are highly susceptible to changes in water levels and chemistry, as evidenced by the 1972–1975 drought (Bénech et al. 1983; Dumont 1992). During those years, decaying vegetation caused deoxygenation of waters and local fish mortality. The north basin, isolated from the rest of the lake, suffered large mortalities caused by fishing pressure and degraded environmental conditions in the waters that remained in the lake basin (Bénech et al. 1983). Low flows in rivers constrained the seasonal fish migrations (Dumont 1992). Lacustrine species, often migratory and more selective in spawning preference, suffered from high mortality and fewer accessible spawning sites (Bénech 1992). Local extinctions occurred for several species such as *Heterotis niloticus* and *Hydrocynus brevis*. Natural selection operating on the fish communities during this dry period favored the development of marshy species adapted to survive in harsh environments. Fish that became more dominant included lungfish (*Polypterus senegalus*), *Oreochromis niloticus*, *Oreochromis aureus*, *Sarotherodon galilaeus*, *Brienomyrus niger*, and *Clarias* spp. (Bénech et al. 1983; Dumont 1992). Lungfish cope with anoxic conditions through aerial breathing, and *Oreochromis* and *Sarotherodon* tolerate low oxygen conditions.

Lake Chad is located along a major route for migratory birds, serving as a resting stop for southbound migrants (Dejoux 1983). Wading birds frequent the mudflats, and more than 1 million wintering ducks congregate along the edges of Lake Chad (Dejoux 1983; Denny 1991). The Hadejia-Nguru wetlands host large populations of several overwintering ducks, including the white-faced whistling-duck (*Dendrocygna viduata*) (maximum count 47,879 in 1996) and fulvous whistling-duck (*D. bicolor*) (maximum count 4,080 in 1992) (Scott and Rose 1996; Ezealor 2002). The ducks often congregate around mats of submerged vegetation in the floodplain (Dejoux 1983). In a July 1997 count, 41,386 individual waterbirds from thirty-eight species were counted in the Hadejia-Nguru wetlands. This number increased to sixty-five species and 274,993 individuals with inclusion of overwintering migrants (Dodman et al. 1999). On the Logone floodplains large populations of *Anas querquedula* have been reported (maximum count 25,000) (Scott and Rose 1996).

Status and Threats

Current Status

The Lake Chad Basin formed during the Cretaceous period (between about 160 and 65 m.y.a.) and has changed little in physical form since that time (Carmouze et al. 1983b). However, the water level of the lake has changed drastically in response to climatic changes. The ancient drainage basin of Lake Chad covered 2.5 million km², but the present endorheic lake has varied in area from 25,000 km² to as little as 2,500 km² in recent times. Six thousand years ago the area experienced a wet climatic period, and the lake grew to 300,000–400,000 km² and has been called "Mega-Chad." However, the sand dunes on the north and east lakeshore, now partially submerged, also provide evidence of a recent dry period (Beadle 1981).

Lake Chad has undergone several severe size changes in the last century. The Lake Chad of the late nineteenth century ("Greater Chad") had an area ranging from 20,000 to 25,000 km² (John 1986). In the period 1964–1971, when the lake level was more than 280 m a.s.l., the so-called Normal Chad covered an area of about 10,000 km² in the north basin and 11,000 km² in the south basin, reaching up to 25,000 km² total in 1964. The lake was very shallow, with a mean depth of 4–6 m (Carmouze and Lemoalle 1983; Dumont 1992). Over the last four decades, water levels have fluctuated widely in response to climatic change. In 1972–1975, when drought drastically reduced the water flow into the lake, the lake decreased to a mere 2,500 km² ("Lesser Chad") (Carmouze et al. 1983a). In 1973 the lake divided into three barely connected areas of water, and the north basin dried completely in 1975 (Olivry et al. 1996). Vegetation grew rapidly over the exposed lakebed, building up the Grand Barrier such that it obstructed flow from the south to the north basins. The lake expanded after this period, and there was no break between the open waters and the southeastern archipelago, although the north basin was cut off by the Grand Barrier and its vegetation (Dejoux 1983). To date this barrier has not been significantly breached or submerged. Two more major droughts occurred in 1984–1985 and 1992–1994 (Moguedet et al. 1990; WWF Living Waters 2000). As of 1992, the lake was almost one-tenth of its size in the 1960s (Hutchinson et al. 1992). The decline in area has continued, and recent satellite imagery shows the remnant lake area in 2001, a decline of 95 percent from the area 35 years previously (NASA 2001).

Since 1971, flows into the lake have been reduced to 50 percent of the previous long-term mean annual flows of more than 40 km³, although rainfall is estimated to have fallen by only 25 percent (Evans 1996). Some scientists claim that the recent drier climate and high agricultural water demands are the primary reasons why Lake Chad is shrinking. Irrigation projects on the Chari and dams on the Jama'are and Hadejia rivers, along with natural changes in the environment, have reduced water flow into the lake (WWF Living Waters 2000). Most of the flow from the north basin's only tributary, the Komadougou-Yobe River, is diverted to agricultural lands and never reaches the lake (Hutchinson et al. 1992). In the west, water supply to the Hadejia-Nguru floodplains has dropped because of increased withdrawals for irrigation and reduced flows through dams such as the Tiga Dam. Irrigation demand increased fourfold between 1983 and 1994, accounting for about half of the decrease in the size of the lake. Irrigation accounted for only 5 percent of the decrease from 1966 to 1975 (Earth Observatory–NASA 2001).

The Southern Chad Irrigation Project involves the eventual irrigation of 670 km² and the resettlement of 55,000 farming families (Kolawole 1987). However, natural fluctuations in lake level make it difficult to assess the viability of this program. When the lake falls below 279.9 m a.s.l., irrigation cannot take place, and the system has operated for only 6 of its first 10 years. During these operational years, the project experienced low water efficiency (U.S. Geological Survey 2001a). The unlined canals of the project are highly susceptible to water seepage and the growth of emergent hydrophytes such as *Typha australis*, which now congests many of the canals (WWF Living Waters 2000).

An increasing human population in the basin, estimated at 20 million people as of 2001, also stresses the scant water and fish resources of this arid region (Jauro 1998). Excessive grazing and overtrampling have occurred on lands adjacent to water sources as wildlife and livestock migrate to these already strained wetlands. Erosion and desertification are possible outcomes (Wanzie 1990). Local peoples cut wood and burn the surrounding rangelands during drought to increase production of fodder.

In January 2002 the Lake Chad Basin Commission met and declared Lake Chad a transboundary Ramsar site (Ramsar 2002b). The Lake Chad Basin Commission, created in 1964, consists of five member countries whose aim is to regulate use of the basin's resources. Current protected areas in this ecoregion include the Chad Basin National Park, which has a number of sectors, a wildlife sanctuary, and five forest reserves in the Hadejia-Nguru wetlands; the Manda National Park on the west bank of the Chari in Chad; and the Mandelia Faunal Reserve, on the floodplain between the Chari and Logone in Chad (Hughes and Hughes 1992; Ezealor 2002). The Hadejia-Nguru Wetlands also contain a Ramsar site and have been the subject of an extensive conservation project started in 1985, jointly undertaken by IUCN, BirdLife International, and the Nigerian Conservation Foundation (Hollis et al. 1993).

Future Threats

Although fluctuations in size appear to be part of the lake's history, drought, irrigation, and population pressures are further limiting water supply to the lake. The return of drought is a constant threat to the lake and surrounding floodplains. Continued low water levels especially threaten migrating fish that exploit both the lake and the riverine-floodplain systems (Bénech 1992). Pressure from a growing population means increased fishing, irrigation, and grazing.

Several proposed projects pose risks for the ecoregion. Development of a 1,070-km-long Chad-Cameroon oil pipeline, part of a World Bank funded project, could pose an environmental threat to the basin in the form of potential oil spills and development of infrastructure needed for the pipeline (World Bank 2000). Pressures to use water from the lake and its tributaries are high in this arid Sahelian region, and a number of ambitious irrigation projects have also been planned. The South

Chad Irrigation Project was put on hold during the disastrous droughts, but the installations are being maintained, awaiting higher water levels in the lake (Evans 1996). Irrigation schemes for a 2,400-km canal to replenish Lake Chad with waters from the Congo Basin have been suggested but are unlikely (FAO 1997d).

Justification for Ecoregion Delineation

This ecoregion encompasses the Lake Chad Basin, a system with a large, floodplain lake in a xeric region on the edge of the Sahara. Like the Inner Niger Delta [7], the Lake Chad ecoregion is distinguished by the ecological role it plays for migrating wetland birds. The ecoregion is also characterized by a widespread, Sudanian freshwater fauna and is included in the Nilo-Sudan bioregion. During the Pleistocene, Lake Chad probably was connected to the Niger, Nile, and probably Congo River basins. At present, the Chad Basin occasionally connects with the Niger Basin during flooding of the Mayo Kebbi system in southern Chad (Olivry et al. 1996). However, the Gauthiot Falls between the Benue River (in the Niger Basin) and the Mayo-Kebi (Chad Basin) has prevented the dispersal of some Niger Basin aquatic species into the Lake Chad Basin. Fish species such as *Citharidium ansorgii, Arius gigas, Synodontis ocellifer,* and *Cromeria nilotica* inhabit the Benue River but do not occur in the Chad Basin (Lévêque 1997). The recent isolation of the lake basin and the large fluctuations in water level explain the absence of aquatic endemics in such a historically large lake (Dumont 1992; Lévêque 1997).

Data Quality: High

Ecoregion Number:	**10**
Ecoregion Name:	**Mai Ndombe**
Major Habitat Type:	**Floodplains, Swamps, and Lakes**
Final Biological Distinctiveness:	**Nationally Important**
Final Conservation Status:	**Relatively Stable**
Priority Class:	**V**
Author:	**Emily Peck**
Reviewer:	**Uli Schliewen**

Location and General Description

Mai Ndombe is a dynamic lacustrine system that occupies the lowest interior elevation of the biologically diverse Congo River Basin. Lake Mai Ndombe, formerly called Lake Inongo and Lake Léopold II, covers an area of approximately 2,300 km^2, although the lake fluctuates in size according to season and rainfall (Welcomme 1972). The irregularly shaped lake has a north-south orientation, a length of 135 km, a width ranging from 17 to 55 km, and an average depth of 3 m (Hughes and Hughes 1992). The ecoregion probably contains the largest block of shallow blackwater lake and flooded forest habitat in the eastern Congo Basin. Situated entirely in the DRC, the ecoregion extends from the southwest part of Salonga National Park in the north to its confluence with the Lukenie-Fimi River in the south.

Located just south of the equator, the ecoregion is subject to a hot and humid climate. The mean annual temperature is close to 25°C, and relative humidity remains high at about 80–90 percent throughout the year. Rainfall varies with latitude and season. A mean annual rainfall of 1,900 mm in the north decreases slightly to 1,700 mm in the south. The wet season occurs during October and November, with a maximum monthly rainfall of up to 225 mm documented in October. The dry season occurs from June through August, with a minimum monthly rainfall of 10–50 mm recorded in July (Hughes and Hughes 1992).

Extensive areas of permanent swamp forest occur in the northeast of the ecoregion and extend southeast, along the Lokoro River. Permanent swamp forests also line the banks of Mai Ndombe's major affluents. Dense *Raphia* swamps proliferate along the shores of the wettest of these forests. These permanently inundated swamp forests grade into less permanently inundated forests of *Oubanguia africana* and *Guibourtia demeusei,* which eventually grade into evergreen rainforest in unflooded areas (Hughes and Hughes 1992).

Seasonal flooding of tributaries to the lake, combined with seasonal rainfall, have a marked effect on the hydrological cycle and ecology of Mai Ndombe. Fourteen major affluents flow across swampy lands and into the lake; the largest of these, Lokoro, Loti, and Olongo-Lule, are located at the northern end of the lake. After the onset of heavy rains in October, floodplains along the tributaries and around the lake become inundated. In the north, floodwaters flow into permanent swamp forests, bringing oxygenated water and nutrients. In the south, flooding of the Lukenie River periodically inundates swamp forests along the southern portion of the lake (Hughes and Hughes 1992). Flooding increases the productivity of the lake's shallow waters, characterized by high acidity and humic content. During years of particularly heavy flooding, the swamp can connect with Lac Tumba to the north.

At the end of the wet season, floodwaters leave Lake Mai Ndombe via the Lukenie-Fimi River. The Lukenie River flows from the east to its confluence with the waters leaving Mai Ndombe; after this junction, the river is called the Fimi. Further downstream, the Kasai and Fimi rivers join together and become the Kwa, which eventually joins the Congo. Mai Ndombe's black waters do not mix readily with the clear waters of the Lukenie-Fimi and are distinct for many kilometers downstream of the confluence of the Fimi and Kasai rivers (Hughes and Hughes 1992; Remane 1997).

Outstanding or Distinctive Biodiversity Features

The few studies and historical collections of Mai Ndombe suggest that species richness may be much higher than currently recorded (Roberts 1973). Representatives from several families of fish are common in the lake, including Alestiidae, Amphiliidae, Cichlidae, Claroteidae, Clariidae, Mochokidae, and Mormyridae. More than thirty fish species are described from the lake, and three, *Amphilius opisthophthalmus, Hemichromis cerasogaster,* and *Nanochromis transvestitus,* are considered endemic.

Although Mai Ndombe is similar ecologically to Lake Tumba, the species composition of the two lakes differs in several aspects, indicating different biogeographic histories. However, the ichthyofauna of the Lukenie-Fimi River, which receives Mai Ndombe's outflow, is among the least-studied parts of the Congo Basin, so any theories about the origin of the Mai Ndombe ichthyofauna remain speculative.

Mai Ndombe provides habitat for the peculiar cichlid species *Nanochromis transvestitus* (Stewart and Roberts 1984). The females of this fish show a striking coloration, with bright hues and contrasts, whereas the males are completely dull. This form of sexual dimorphism is uncommon in cichlids and hints at an unusual breeding system. Historical collections include other cichlids with peculiarities (e.g., the thick-lipped endemic *Hemichromis cerasogaster*).

The permanent swamp forests surrounding Lake Mai Ndombe contain fish species with adaptations to survive in the anoxic, semistagnant water. *Clarias buthupogon* and *Clarias gabonensis* are two such species (Lowe-McConnell 1987; Hughes and Hughes 1992).

Reconnaissance surveys of the region suggest that it is rich in freshwater mammals and waterbirds. Freshwater mammal species include marsh mongoose (*Atilax paludinosus*), giant otter-shrew (*Potamogale velox*), Cameroon clawless otter (*Aonyx congicus*), spotted-necked otter (*Lutra maculicollis*), and African water rat (*Colomys goslingi*). The rare kingfisher (*Corythornis leucogaster*) has been recorded adjacent to the lake (Hughes and Hughes 1992). Ten species of aquatic frogs, three species of aquatic turtles, three species of crocodiles, and several semi-aquatic snakes have also been recorded.

Occupying the lowest point in the Congo River Basin, the lake receives large volumes of organic matter from the surrounding catchment during the rainy season. The influx of organic matter results in high levels of fermentation and decomposition, which reduce the amount of dissolved oxygen in the lake water. However, the influx of floodwaters also brings oxygenated water to the flooded areas (Roberts 1973; Remane 1997). During heavy rains in October and November, fish move into the seasonally inundated swamp forests to feed and breed. Because Mai Ndombe is a blackwater lake, most of the food available to fish comes from terrestrial sources or floating vegetation, both of which are more commonly encountered during the flood season. The swamp forest's warm, still, shallow water, its dense vegetation, and its rich organic mud provide not only an abundance of fish food and ideal conditions for breeding but also protection for juvenile fish from predators (Roberts 1973; Lowe-McConnell 1987).

Status and Threats

Throughout the northern reaches of the ecoregion, much of the permanent swamp forest is in pristine condition. Population density around the lake is low (2–4 people/km^2); therefore, there is minimal human impact on the land. Human populations along the northern edge of the lake practice small-scale agriculture along its shores but have minimal impacts on the ecology of the area.

The lake also supports several artisanal fishing communities, which have traditionally partitioned fishing rights around lake sectors. These fishing activities are considered to be sustainable, but an increase of fishing probably would lead to overexploitation of the purely lacustrine species of the low-nutrient lake system (Shodjay 1985).

Throughout the southwestern reaches of the ecoregion, rice cultivation has catalyzed the clearing of millions of hectares of seasonally inundated forest along the banks of the Lukenie and Fimi rivers. Continued clearing of this land poses a serious threat to the integrity of the ecoregion (Hughes and Hughes 1992). Additionally, the DRC national plan states that the principal economic activities in the Mai Ndombe region are traditional and industrial forestry, traditional agriculture, and agroindustrial activities, which would be expected to have effects on the downstream systems depending on their scale (République Démocratique du Congo 1998).

Justification for Ecoregion Delineation

This ecoregion is defined by the extent of Lake Mai Ndombe and surrounding flooded forest and is distinguished by the lacustrine faunal elements that occur in its waters. Mai Ndombe sits in the Cuvette Centrale, a saucer-shaped depression in the central Congo River Basin. During the late Miocene and early Pliocene, a large endorheic lake occupied the entire Cuvette Centrale. Lake Mai Ndombe is thought to be a remnant of this great interior lake, although geological details supporting this are lacking (see essay 3.2). As a large and ancient lacustrine area, Mai Ndombe may have played an important role in the evolution of the endemic fish fauna of the Congo Basin. It may also have served as a refuge for the ancient lacustrine faunal elements that inhabited the larger former inland lake system (Lowe-McConnell 1987; Shumway 1999). Lake Mai Ndombe is connected to the Kasai River system such that its fauna is more related to the Kasai than to the Congo mainstem.

Data Quality: Low

Because only a few historical collections exist for the lake and its tributaries, species richness and endemism probably are underestimated. A survey should include all affluent streams and the entire flooded forest, with priority given to fish and mollusks.

Ecoregion Number:	**11**
Ecoregion Name:	Malagarasi-Moyowosi
Major Habitat Type:	Floodplains, Swamps, and Lakes
Final Biological Distinctiveness:	Continentally Outstanding
Final Conservation Status:	Relatively Stable
Priority Class:	III
Authors:	Ashley Brown and Michele Thieme
Reviewer:	Luc De Vos

Location and General Description

Productive swamps cover large areas of the Malagarasi-Moyowosi ecoregion, which includes the Malagarasi River Basin above Uvinza and the headwaters of the Rugufu and Luiche rivers. These rivers and their tributaries flow onto a central plateau (1,000–2,000 m) and eventually drain into Lake Tanganyika [55]. The majority of the ecoregion falls in northwest Tanzania, and a small section extends into southeastern Burundi.

Three major rivers, the Malagarasi, Moyowosi, and Ugalla, flow onto the plateau and form the extensive swamps of this ecoregion. The Malagarasi, which flows onto the plateau from the northwest, has its headwaters in the hills (elevation about 1,800 m) adjacent to Lake Tanganyika on the Burundi-Tanzania border. The Moyowosi River and its tributaries flow from the north and east. The Ugalla River, the main tributary to the Malagarasi, originates in the south of the ecoregion. The Malagarasi and the Ugalla are each about 500 km long, and the Moyowosi is about 200 km long (De Vos and Seegers 1998).

The Malagarasi drainage includes a large variety of biotopes, including swampy areas; small and medium-sized river channels with different bottom types and water chemistries; a large, slowly flowing large river with a few moderate rapids; and a large delta divided into two major branches (De Vos and Seegers 1998). On the upper Malagarasi and its headwaters, at altitudes of about 1,200–1,250 m a.s.l., strips of permanent swamp intermittently line both banks. There is also a large swamp stretching from where the Nikonga and Kigosi rivers meet the Moyowosi, continuing to where the Moyowosi joins the Malagarasi. The gradient of the Moyowosi River and its tributaries decreases as they reach the plateau, causing them to overflow their banks and divide into multiple channels. This permanent swamp zone is about 160 km long and up to 36 km wide, with an area of about 3,200 km^2. During the wet season, about 2,500 km^2 of inundated floodplain occurs on two tributaries of the Moyowosi, the Kigosi and Gombe rivers. There are also about 1,800 km^2 of floodplains and permanent swamps on the Ugalla River and its tributaries, the Ziuwe and the Wala (Hughes and Hughes 1992).

Vegetation varies with biotope. Papyrus (*Cyperus papyrus*) dominates the swamp vegetation, with *Typha, Carex,* and other emergent vegetation also present. Grasses, such as *Oryza* spp. and *Echinochloa pyramidalis,* occur on the seasonal floodplains. On the surrounding land, *Brachystegia spiciformis–Julbernardia globiflora* miombo woodland grows. *Brachystegia mirophylla* is dominant on granitic outcrops present in the area (IRA 2002). Gallery forests with species such as *Acacia* spp., *Borassus aethiopum,* and *Phoenix reclinata* also occur along the headwater streams (Beadle 1981; Hughes and Hughes 1992). In the water column, species from the genera *Potamogeton, Ceratophyllum, Chara,* and *Utricularia* are common. Water lilies, water chestnut (*Trapa*), water fern (*Azolla*), and Nile cabbage (*Pistia*) are also present (Beadle 1981).

Rainfall patterns vary throughout this tropical ecoregion, and three wind systems affect the climate. Rains are brought to the ecoregion via the southwest monsoon from the Congo Basin to the west and via the southeast trade winds off of the Indian Ocean. Northeast trade winds bring dry air from Ethiopia and Somalia (Hughes and Hughes 1992). Rains generally occur between January and May, with an additional peak from October to November. Average annual precipitation in Tanzania is 937 mm (FAO Inland Water Resources and Aquaculture Service: Fishery Resources Division 1999). However, on the plateau in the north, rainfall is generally below 600 mm per year and at higher elevations in the southern portion of Burundi rainfall can vary between 1,300 and 1,600 mm per year. Flooding of the periodically inundated gallery forest usually occurs after rains between November and December and also between March and April (Hughes and Hughes 1992).

The Malagarasi River is thought to be the relict headwaters of the pre-rift Congo Basin (Lévêque 1997). The presence of several fish species in the Congo and Malagarasi basins, but not in the intervening Lake Tanganyika, suggests a connection between the Malagarasi and the Congo River before the formation of the lake (Beadle 1981).

Outstanding or Distinctive Biological Features

The highly productive wetlands of the Malagarasi-Moyowosi ecoregion host a rich aquatic fauna with numerous wetland birds and a large diversity of fish, frogs, and freshwater mollusks.

A recent survey of the Malagarasi drainage revealed the presence of at least 108 fish species belonging to 20 families and 48 genera, without taking into account numerous lacustrine Tanganyika species that enter the lagoons of the river deltas (De Vos et al. 2001a). Several of those species are new to science and await formal scientific description. The ichthyofauna of the ecoregion consists of a mixture of taxa with various origins. Several fishes from the system have a wide distribution on the African continent (e.g., *Mormyrops anguilloides, Hydrocynus vittatus, Schilbe intermedius, Auchenoglanis occidentalis, Bagrus docmac, Clarias gariepinus,* and *Heterobranchus longifilis*). Almost 15 percent of the fish fauna from the Malagarasi have a Congolese origin. Typical species of this group include *Polypterus ornatipinnis, P. endlicheri, Distichodus maculatus, Citharinus gibbosus, Tetraodon mbu, Alestes macrophthalmus, Labeo weeksii, Micralestes stormsi, Raiamas salmolucius,* and *Bryconaethiops microstoma.* Those species are most commonly found in the lower courses of the Malagarasi and its delta. A few Malagarasi fishes have a Nilotic or Lake Victoria origin (e.g., *Gnathonemus longibarbis, Synodontis victoriae, S. afrofisheri, Brycinus sadleri,* and *Pollimyrus nigricans*), evidence of recent contacts or older connections between the system and the Nile drainage. A small proportion of the ichthyofauna belongs to an old typical Central African group of fishes; these include *Amphilius jacksonii, Clarias liocephalus, Mormyrus longirostris,* and *Barbus apleurogramma.* Additionally, several species suggest old contacts with East African hydrographic systems (e.g., *Barbus macrolepis, B. innocens,* and *Mesobola spinifer*). It is also clear from the fish composition that there has been and may continue to be an important connection between the Malagarasi drainage and the Lake Rukwa system. Seegers (1996) reported that at least two-thirds of the fish species of the Rukwa Basin have been derived from Malagarasi stocks.

Despite these many connections, nearly 15 percent of the Malagarasi fishes are endemic to the system. Most endemics belong to the Cichlidae family, particularly to the riverine mouth-breeding goby cichlids of the genus *Orthochromis.* This genus has undergone a species radiation on the east side of the ecoregion, where it is represented by at least eight species (De Vos and Seegers 1998). *Orthochromis* species are also found in drainages to the west of Lake Tanganyika (Congo drainage), which again indicates old contacts between the two systems. Other endemic Malagarasi cichlids belong to the genera *Haplochromis, Oreochromis,* and *Serranochromis.* Further endemic species belong to the catfish genera *Clariallabes* and *Chiloglanis,* the killifishes *Aplocheilichthys* and *Nothobranchius,* the cyprinids *Opsaridium* and *Barbus,* and the spiny eels *Mastacembelus.* Additionally, two lungfish species (family Protopteridae), fishes that have accessory organs for surviving periods of low oxygen, occur in the swamps of this ecoregion.

In addition to a rich fish fauna, the Malagarasi and its tributaries host a diverse assemblage of aquatic mollusks, frogs, reptiles, mammals, and important waterbird populations. Thirteen mollusk species, including several members of the *Bulinus* genus, live in this ecoregion. There are twelve species of aquatic-dependent frogs, none of which is endemic. Nile crocodile (*Crocodylus niloticus*), slender nosed crocodile (*Crocodylus cataphractus*), hippopotamus (*Hippopotamus amphibius*), marsh mongoose (*Atilax paludinosus*), sitatunga (*Tragelaphus spekei*), and clawless otter (*Aonyx capensis*) also inhabit the waters and swamps of the ecoregion. The ecoregion is important for resident and migratory waterbirds, with several hundred species occurring (Wetlands International 2002). This is one of only four ecoregions where the shoe-billed stork (*Balaeniceps rex*) lives (Hughes and Hughes 1992), with populations of more than 2,000 individuals (Baker and Baker 2002). Populations of the vulnerable wattled crane (*Grus carunculatus*) also occur.

Status and Threats

Current Status

Influxes of refugees, agriculture, and overgrazing pose the largest threats to the wetlands of this ecoregion. Civil strife in Burundi has been ongoing since 1993 (East 1999). Refugee camps in the north of Tanzania support large numbers of refugees. As of the end of 2001, there were about 500,000 refugees and an additional 300,000–470,000 Burundians who resided in western Tanzania in refugee-like circumstances without official refugee status (U.S. Committee for Refugees 2002).

At present, fishing occurs on an artisanal level in the rivers and floodplains of the ecoregion, and often illegal fishing nets are used (i.e., less than 3-inch mesh size). However, efforts to encourage use of legal mesh size nets are proving beneficial in certain areas (e.g., in Lake Sagara). Fishing and hunting occur primarily in the main valley of the upper Malagarasi (Hughes and Hughes 1992). In the Moyowosi-Kigosi and Ugalla River game reserves, illegal activities such as poaching and harvesting of forest products occur (East 1999).

Intensive agriculture, timber harvesting, and grazing for livestock are occurring in different parts of this ecoregion. Land use activities have recently increased in the southern portion of the ecoregion, particularly along the Ugalla River. There is concern that these activities are increasing silt and nutrient inputs to the wetlands (Nkotagu, unpublished project proposal). The clearing of virgin land for tobacco farming and the associated cutting of timber for curing tobacco is a major force for land use change. Local Natural Resource Government officers stated that for each acre of tobacco farmed, 5 acres of woodland are cut for fuelwood. Because of decreasing soil fertility (and the cost of replacing nutrients with artificial inputs) and nematode infestation, tobacco is cultivated in an area for only 2 years.

Tanzania has recently (2000) designated the Malagarasi-Moyowosi wetlands as a Ramsar site. These wetlands encompass 32,500 km^2 of riverine floodplains and swamps in the Malagarasi Basin. The site contains lakes and open water as well as permanent papyrus swamps and floodplains. The majority of the site lies in game and forest reserves, such as the Moyowosi Game Reserve and the Tongwe East Forest Reserve (Ramsar 2000).

Future Threats

The largest future threat is an increasing population and clearing of vegetation. Continued civil strife and an influx of refugees also pose threats to the ecoregion.

Justification for Ecoregion Delineation

This ecoregion is defined by the Malagarasi-Moyowosi Basin. The entire Malagarasi catchment drains an area of 130,000 km^2 and includes the area from the mountains bordering Tanzania and Burundi through to the Malagarasi-Moyowosi swamps (Patterson and Makin 1998). The Malagarasi catchment composes about 30 percent of the Lake Tanganyika drainage basin, and the Malagarasi is Lake Tanganyika's largest affluent (De Vos and Seegers 1998). Two smaller affluents, the Luiche and Rugufu rivers, also drain into Lake Tanganyika in Tanzania. The Malagarasi is flanked by these rivers, with the Luiche entering Lake Tanganyika about 30 km to the north and the Rugufu entering about 14 km to the south. Several shared ichthyofaunal elements suggest that these rivers were connected formerly with the Malagarasi, forming the Proto-Malagarasi (De Vos and Seegers 1998). Because of their shared ichthyofauna with the Malagarasi, the headwaters of the Luiche and Rugufu rivers are included in this ecoregion.

Data Quality: Low

Ecoregion Number:	**12**
Ecoregion Name:	Okavango Floodplains
Major Habitat Type:	Floodplains, Swamps, and Lakes
Final Biological Distinctiveness:	Globally Outstanding
Final Conservation Status:	Vulnerable
Priority Class:	I
Author:	Lucy Scott
Reviewers:	Paul Skelton and Ketlhatlogile Mosepele

Location and General Description

The Okavango Delta, situated in northern Botswana, is one of few major inland endorheic deltas on the continent and one of the most important wetlands in southern Africa. It has high bird richness and moderate fish richness for a wetland complex and provides habitat for a number of threatened species (Stuart et al. 1990).

The climate of the ecoregion is subtropical, with summer rainfall and precipitation increasing slightly from west to east. An average annual rainfall of about 650 mm falls mostly from December to April (Jubb and Gaigher 1971). Because of the ecoregion's location on the fringe of the semi-arid Kalahari Desert, 96 percent of the inflow to the delta is lost through evapotranspiration (Ellery and McCarthy 1998).

The Okavango River system, which has an average altitude of 900–1,200 m, rises in the central highlands of Angola on the Benguela Plateau. It shares its watershed divide with the westward-flowing Cunene and Cuanza, and it flows south and east into the mainly arid Okavango Basin to diverge into the many arms of the Okavango swamps (Beadle 1981). The Cuito and Cubango rivers, the two major tributaries of the Okavango, merge at the southern border of Angola. The Okavango River enters Botswana as a single, broad river but forms an extensive riverine floodplain known as the panhandle before subdividing into a number of channels and forming the delta. The permanent Okavango swamp covers 6,000–8,000 km^2, but annual floods from the highlands of southern Angola inundate the floodplains, expanding the seasonally inundated area to 12,000–15,000 km^2. Near the tributaries, the swamps are intersected with numerous deep, perennial channels, shallow floodwater channels, and backwaters or lagoons. Stony rapids are absent from the swamps, and the water is clear throughout the year.

In years of high flows, as occurred in the early 1990s, water flows through the swamp and into the Thamalakane River, which flows in a southwesterly direction until about 16 km below Maun, where it divides into the Ngabe River, which flows west into Lake Ngami, and the Botleti River, which continues southeastward into Lake Dow. Lake Ngami, located in the southwestern corner of the delta, was an imposing sheet of water until the late 1800s, when it began to dry up. Channel avulsion caused by siltation of the channel and the obstruction by aquatic vegetation of the Thaoge, a major western channel of the Okavango, possibly caused the desiccation of this lake (Roberts 1975). The Thaoge is moribund, such that the Nqoga is now one of the main outflow rivers from the delta (Porter and Muzila 1989). In years of good (high) flows, the Matsebe-Xudum-Kunyere system feeds into Lake Ngami (SMEC 1986; Porter and Muzila 1989).

During seasons of exceptional rainfall in the highlands of Angola, a floodwater link exists between the Cuando and Linyanti and Upper Zambezi and Chobe rivers and the Okavango swamps via the Chobe (Linyanti) River and the Selinda (Makwegana) spillway (Jubb and Gaigher 1971; Curtis et al. 1998). Consequently, the delta's fish fauna is very similar to that of the Upper Zambezi (Beadle 1981).

The swamps in the delta vary in character mainly according to the perennial or intermittent occurrence of floodwaters. In the northwestern section of the delta, which is permanently flooded, the swamps are mostly covered with a dense growth of papyrus (*Cyperus papyrus*), stands of reeds (*Phragmites*), and bullrush (*Typha*). In the lower part of the delta, where the floodwater is more seasonal, the dry depressions are periodically filled with water and generally have only reeds and papyrus around the edges (Jubb and Gaigher 1971). A region of thorn scrub and mopane (*Colophospermum mopane*) savanna is located in the north. It includes stretches of dense dry *Baikiaea plurijuga* forest, mopane and *Pterocarpus* woodland savanna, thorn scrub (*Acacia* and *Sclerocarya*) savanna, and semi-arid grassland. Forest savanna and dry woodland predominate in the drier southern portion of the ecoregion (Stuart et al. 1990).

Outstanding or Distinctive Biological Features

The Okavango Floodplains ecoregion contains the largest expanse of wetlands in southern Africa and supports a range of habitats and a rich bird fauna. The Okavango Delta is one of the largest inland deltas in the world (Allanson et al. 1990). The extensive floodplains support a moderately rich aquatic fish, herptile, and mammal fauna when compared with other floodplain systems in Africa. About 200 semi-aquatic plants, 80 fish species, and 4 aquatic mammals have been recorded from the rivers and swamps of the Okavango.

The seasonal floods of the Okavango are the most important driver of ecological processes in the region, and the distribution of fishes is generally governed by the flood cycle. The migration of fish out of the main channels into adjacent flooded areas and swamps coincides with the first floods each year in November and December (Merron and Bruton 1995). Floodplain spawning offers the advantages to juvenile fish of abundant food and well-oxygenated conditions.

A total of 450 bird species have been recorded in the delta. Among these are *Pelecanus onocrotanus, P. rufescens,* eighteen heron species, and breeding *Rhynchops flavirostris*. Large mixed roosts of herons, egrets, storks, and ibis occur in selected localities in the delta. Among the many Anatidae, *Nettapus auritus* has the largest congregations, with up to 15,000 breeding pairs and 40,000 nonbreeding visitors recorded (Tyler and Bishop 2001).

The Okavango floodplains are important for several threatened species. They are the most important breeding site for the vulnerable slaty egret (*Egretta vinaceigula*). Breeding and non-breeding individuals of the vulnerable wattled crane (*Grus carunculatus*) also occur in the delta (Tyler and Bishop 2001). Three fish species with restricted distributions in scarce rocky habitats are the ocellated spiny-eel (*Aethiomastacembelus vanderwaali*), largemouth squeaker (*Synodontis macrostoma*), and broad-headed catfish (*Clariallabes platyprosopos*) (Skelton 1993).

Status and Threats

Current Threats

The Okavango Delta is in good condition overall. Despite allegations of overfishing in the delta (Haggett 1999), the most recent fish stock assessment has shown that there is no overfishing (Mosepele 2000; Mosepele and Kolding 2002) in the delta, although incidences of localized overfishing in lagoons cannot be discounted. Natural fish kills occur in some parts of the swamps, caused by the first seasonal floodwaters pushing deoxygenated water up from under papyrus mats into open waters. In the Okavango River in Namibia, the fish stocks are still in a moderately robust state, although numbers have declined since 1984 (Hay et al. 1996). The periphery of the delta is of some concern because of the rapid encroachment of people and cattle into the panhandle and drainage areas (R. Bills, pers. comm., 2001; Alonso and Nordin 2003).

Invasive plants threaten to affect the aquatic system of the Okavango. Water hyacinth (*Eichhornia crassipes*) has not yet reached the Okavango Basin, but the highly invasive Kariba weed (*Salvinia molesta*) is present in restricted sections of the delta (Alonso and Nordin 2003).

Insecticides have been used since the 1920s to control malaria and the tsetse fly (*Glossina morsitans*), and spraying continues over parts of the delta. The long-term effects of these chemicals may be harmful to other aquatic and terrestrial organisms in the food chain, including people (Merron and Bruton 1995).

The central part of the delta lies in formal reserves and game management concessions, which are leased out to tourists and hunting companies. The Chobe National Park (11,000 km^2) protects an important part of the ecoregion, conserving globally significant breeding populations of slaty egret; the Mahango Game Park (244 km^2) in western Caprivi also protects breeding habitat of that species. The Moremi Game Reserve (3,880 km^2) in the east of the delta protects a good representation of all habitat types except papyrus swamp. Kgalagadi Game Reserve (52,800 km^2) includes part of the southern portion of the ecoregion, and the Khaudom Game Park (3,841 km^2) covers some tributaries of the Okavango River (Stuart et al. 1990; Fishpool and Evans 2001).

Future Threats

The largest future threat to the Okavango Delta comes from possible water abstraction from the Okavango and its tributaries by Botswana, Namibia, and Angola. Namibia is negotiating the extraction of an estimated 20 million m^3 of water annually from the Okavango River via the Eastern National Water Carrier project, which is the largest interbasin water transfer in Namibia. Although this is less than 10 percent of the river's yearly flow, it would be a significant amount during the dry season (J. Day, pers. comm., 2001). When complete, the water carrier will transfer water from the Okavango River via pipelines and open canals about 700 km to Von Bach Dam, located in the Swakop River Basin in Namibia. Excess withdrawal from the Okavango River could impair functioning of the delta ecosystem and result in the transfer of species from the Okavango to more southerly drainage systems. Research into impacts of the transfer was begun only after construction had commenced and raises some concerns. The transfer of fish from the Okavango to other drainage systems will provide direct interspecific competition with existing populations in the Omatako Reservoir and the Swakop River. In addition, evaporative losses are increased by the exposed water surfaces in the canal and are estimated at 70 percent (Davies and Day 1998).

Another large threat comes from the competing land uses of cattle grazing or farming and conservation or tourism (Fishpool and Evans 2001). There are increasing numbers of cattle in the region, and local effects of overgrazing are evident (Stuart et al. 1990). Little or no forest remains along the Okavango River because of clear-cutting for agriculture and settlement. In Namibia, overgrazing, together with the removal of riparian vegetation, causes erosion and the consequent siltation of rivers (Hay et al. 1996). Veterinary fences, placed to control the movement of cattle and minimize the potential for spread of diseases between cattle and wild animals, have restricted the movements of large mammals in the delta region, although there have been some recent successes in reducing their impact on animal movements (Ross 2003).

The alien Mozambique tilapia (*Oreochromis mossambicus*), common carp (*Cyprinus carpio*), and largemouth bass (*Micropterus salmoides*) have all been introduced to pans in the Omatako Omuramba region. These species could enter the Okavango Delta during floods, when the drainage is linked, and have deleterious effects on the fauna of the Okavango. For example, *Oreochromis mossambicus* is likely to hybridize with the indigenous *Oreochromis macrochir* and *Oreochromis andersonii* should they come into sustained contact (DeMoor and Bruton 1988).

The current biggest threat facing the Okavango Delta is the proposed Popa Falls Hydroelectric Power Scheme in Namibia. Sediment loading into the delta will be reduced, resulting in erosion of river channels and reduced flooding of the delta's swamps and grasslands. Consequently, reduced sediment loading and water flows will alter the delta's hydrological regime, with potentially negative impacts on the entire system (Kolding et al. 2003).

Justification for Ecoregion Delineation

This ecoregion is defined by the boundaries of the Okavango Basin, excluding the upper reaches of the river, which were historically connected with the river systems of the Zambezian Headwaters [76] ecoregion. The biogeographic affinities of the aquatic fauna of the Okavango are Zambezian, although it also shares a number of species with the southern tributaries of the Congo River, especially the Kasai. The faunal affinities of the region can be best explained by the theory that present-day headwaters of the Congo and Cuanza were formerly part of the Okavango and Zambezi (Skelton 1994). As a result of tectonic faulting, the Okavango was impeded and deflected eastward to the Upper Zambezi and was ultimately captured by the middle Zambezi River (Beadle 1981). Skelton (1994) considers the Okavango to be a western sector subunit of the Zambezian Province and to be the heart of the western Zambezian evolutionary arena.

Data Quality: High

The state of ecological knowledge is reasonable, and the area is well known taxonomically. In June 2000, Conservation International completed a rapid ecological assessment of the delta (Alonso and Nordin 2003). In terms of biological investigation, fish assessments and investigations into how the system functions as a whole are needed (R. Bills, pers. comm., 2001).

Ecoregion Number: **13**
Ecoregion Name: **Tumba**
Major Habitat Type: **Floodplains, Swamps, and Lakes**
Final Biological Distinctiveness: **Nationally Important**
Final Conservation Status: **Relatively Stable**
Priority Class: **V**
Author: **Ashley Brown**
Reviewers: **Uli Schliewen and Michael Brown**

Location and General Description

Lake Tumba, a blackwater lake surrounded by seasonal and permanent swamps, connects with the Congo River near its confluence with the Oubangui (Lowe-McConnell 1987). The 765-km^2 lake lies near the equator in the DRC (Marlier 1973).

Located at a low point in the Congo Basin, Lake Tumba is flooded for most of the year (Marlier 1973). The lake is extremely shallow and receives much of its nutrient input from small blackwater forest streams that flow from the surrounding, inundated swamps into the lake (Marlier 1973; Lowe-McConnell 1987). The pH of the lake water is low, with values of 4.0–4.9 recorded (Lévêque 1997). The lake is dotted with several small islands, and small deltas have formed at the mouths of some inflowing streams (Hughes and Hughes 1992). The average depth of the lake is only 3–5 m, and the maximum is 8 m (Marlier 1973). Despite its shallow depth, dissolved oxygen is abundant in the entire water column throughout the year because the waters are churned frequently by strong winds. Lake waters flow into the Congo River via the Irebié Canal, although the flow direction through the canal sometimes reverses during high flood, and floodwaters from the Congo enter the lake (Hughes and Hughes 1992).

The climate is predominantly tropical and wet. Mean annual temperature is about 25°C, and the mean daily minima and maxima are about 21°C and 31°C (Hughes and Hughes 1992). Mean annual rainfall is 1,800 mm. The months with highest rainfall are October and November, with 200 to 220 mm/month, and February to April, with 170 to 200 mm/month. In contrast, only 70 mm falls in July (Hughes and Hughes 1992).

The watercourses of the central Congo Basin exhibit two minimum and maximum flows each year, causing the lake to undergo a small flood between May and June and a large flood between September and January (Lowe-McConnell 1987; Hughes and Hughes 1992). Lake level fluctuates both within (about 4 m) and between years.

Swamp and riverine forests are found in a mosaic with swamp grassland and islands of dryland forest in this ecoregion. Lake Tumba is surrounded by seasonally or permanently inundated forest (Hughes and Hughes 1992). In calm cove waters, mats of *Echinochloa pyramidalis* and *Panicum parviflorum* grow. *Jardinea congoensis* and *J. gabonensis* grow in beds along the banks and sometimes are interrupted by thickets of *Cyrtosperma senegalense* and *Rhynchospora corymbosa*. These thickets occasionally break off and drift freely. Swamp forests grow in areas along the shore that are only minimally exposed at periods of low water. These forests may be submerged up to 4 m in depth during high water (Hughes and Hughes 1992). *Irvingia smithii* dominates the swamp forests, with *Alchornea cordifolia* and *Cynometra schlechteri* also present in high abundance. This forest type merges into *Guibourtia* and *Uapaca* forest, which transitions to evergreen rainforest as the inundation zone ends (Hughes and Hughes 1992).

Outstanding or Distinctive Biodiversity Features

Matthes (1964) reports 107 fish species from Lake Tumba. This lake is similar ecologically to Lake Mai Ndombe, and inundated swamps connect these two lakes during the rainy season. However, probably only swamp-adapted fishes move between the lakes, not the purely lacustrine fishes, such as the Lake Tumba endemic *Clupeocharax schoutedeni* and the probably endemic *Tylochromis microdon*. Mormyrids, clariid and bagrid catfishes, characoids, clupeids, rivulins, and cichlids occur in Lake Tumba. The two most speciose families are Cichlidae and Clariidae. Three species of clupeids inhabit the lake, one of which, *Nannothrissa parva*, is known only from Tumba, the Oubangui, and the upper Congo rapids. Some fish species enter the lake from the Congo River during high floods but do not establish populations in the lake (Hughes and Hughes 1992). The lake has no known mollusk species, probably because of the low calcium content of the water (Beadle 1981).

The very low mineral content of Lake Tumba means that allochthonous particulate organic matter from the surrounding forests forms the base of the lake's food chain (Beadle 1981). Most fish species spawn between August and September, at the start of the main flood, often moving many tens of kilometers upstream and entering inundated forests to breed and feed (Lowe-McConnell 1987).

Shoaling fish (such as *Barbus* and *Microthrissa*) live in the open waters of the lake and feed on small plankton (Hughes and Hughes 1992). *Odaxothrissa losera* also inhabits the pelagic zone and feeds on small fish. Insects and detritus provide an abundant source of food for fish inhabiting near-shore areas of vegetation. *Phago boulengeri* eats the fins of other fishes (Hughes and Hughes 1992).

Aquatic frogs are abundant in this ecoregion, with tadpoles taking refuge in vegetation near the banks of the lake (Hughes and Hughes 1992). Twelve known frog species rely on the lake. Two of these are endemics (*Cryptothylax minutus* and *Phlyctimantis leonardi*), both members of the Hyperoliidae family.

Several large aquatic reptiles and mammals live in the ecoregion. *Hippopotamus amphibius* is present but rare. Two crocodile species, *Crocodylus cataphractus* and *C. niloticus*, occur (Hughes and Hughes 1992). Many piscivorous snakes live near the lake as well.

The bird species composition is similar to that of the lower Congo River region south of the lake, which has several hundred species. The African openbill (*Anastomus lamelligerus*), the pink-backed pelican (*Pelecanus rufescens*), and ducks (*Anas* spp.) inhabit the ecoregion (Hughes and Hughes 1992).

Status and Threats

Artisanal fishing provides an important protein source to populations that live along the lake and move into the ecoregion to fish (e.g., Ntomba, Ngombe) (Bahuchet 1992). However, use of illegal equipment (e.g., gillnets and mosquito nets with overly small mesh), fishing in spawning areas along riverbanks or swamp forest zones during reproductive periods, and use of chemicals in spawning areas are threats to the fish populations of this ecoregion (Shumway et al. 2003). The FAO (2001b) estimates annual fishery yields for the Mbandaka region as 100,000–120,000 tons, but the true yield is unknown. The DRC National Biodiversity Action Plan notes that *Citharinus* is becoming rare in Lake Tumba as a result of overfishing and non-selective fishing techniques (République Démocratique du Congo 1998). A recent survey at Mbandaka-Ngombe in March–April 2003 found that more than 74 percent of the catch of four species (*Chrysichthys* sp., *Heterotis niloticus*, *Distichodus lusosso*, and *Mormyrops anguilloides*) were juveniles, indicating that these stocks are overexploited (ERGS Research Group, University of Kinshasa 2003). Work is ongoing through the Congo River Environment and Development Project to assist with institutionalizing improved management practices in this region (M. Brown, pers. comm., 2003).

Although most of the human population is involved in fishing, some agriculture is found adjacent to the lake (Hughes and Hughes 1992). Slash-and-burn leads to erosion, which can alter nearby river water quality (Marlier 1973). Currently no part of the lake is protected. However, there are proposals to include the lake in a community-based reserve, and, related to this, a major participatory mapping exercise of about 12,000 km² has recently been completed in the ecoregion. All the villages in the perimeter of the lake were mapped, along with the forest on the road to Mbandaka and the swamp forest at the confluence of the Oubangi and Congo rivers (M. Brown, pers. comm., 2003).

Justification for Ecoregion Delineation

This ecoregion is defined by Lake Tumba and its adjacent swamp forests and is characterized by migratory species that enter the ecoregion to feed and breed and by swamp-adapted species. Lake Tumba was formed when a tributary river became blocked close to the point of discharge into the Congo (Hughes and Hughes 1992). Sediments deposited in the main river may have been responsible for blocking the tributary (Bailey 1986). Located in one of the lowest points in the basin, Lake Tumba probably is also a remnant of the large Pliocene lake that existed before the Congo River became connected to the Atlantic Ocean. Thus, it is not surprising that the known species composition of the lake is similar to that of much of the Cuvette Centrale and the larger Congo Basin (Shumway et al. 2003). However, the fauna differs somewhat from that of Lake Mai Ndombe, which is connected to the Kasai River system rather than to the Congo mainstem. The fauna of Lake Tumba has been little studied recently; a greater understanding of the biogeography of this ecoregion probably will come from further investigation of its fauna.

Data Quality: Low

Matthes (1964) conducted a survey of the fishes. A survey of the fish and other biodiversity took place in September and October 2002 (Shumway et al. 2003). More surveys are needed for fishes and other aquatic species.

Ecoregion Number:	**14**
Ecoregion Name:	**Upper Lualaba**
Major Habitat Type:	**Floodplains, Swamps, and Lakes**
Final Biological Distinctiveness:	**Continentally Outstanding**
Final Conservation Status:	**Relatively Intact**
Priority Class:	**III**
Authors:	**Ashley Brown and Robin Abell**
Reviewer:	**Luc De Vos**

Location and General Description

The waters of the upper Lualaba River, teeming with life, flow through the swampy valley of the Kamolondo Depression. In this depression, a continuous swamp belt fringes the river, and a series of lakes are connected to the river through narrow channels. In the south of the ecoregion the vegetation cover is primarily savanna, and it changes to taller and denser forest to the north (Hughes and Hughes 1992). The ecoregion lies in the DRC in the southeastern portion of its Shaba Province.

The Kamolondo Depression covers an area 100 km wide and 400 km long between the Hakansson Mountains to the west (up to 1,200 m a.s.l.) and the Manika Plateau and Mulumbe Mountains to the east (up to 1,889 m a.s.l.) (Bailey 1986). The depression itself is at an elevation of about 1,000 m a.s.l. at its southwestern end, declining steeply to about 600 m a.s.l. and flattening out to form a series of lakes (Hughes and Hughes 1992). The well-mixed lakes and marshes of Upper Lualaba begin in this flatter portion of the depression, where elevations stay at about 500–600 m a.s.l. and extend for a total of 225 km. From south to north, the lakes are Kabwe, Kabale, Mulenda, Upemba, Lukanga, Kisale, Kalombwe, Zimbambo, and Kabamba. Kabwe and Kisale are expansions of the main river channel, with the Lualaba flowing directly through them. All of the lakes range in depth from 1 to 7 m (Bailey 1986). By some estimates there are about 8,000 km² of lakes and wetlands in the depression and an additional 800 km² of wetlands adjacent to the Lululwe, Kilubi, and Lovoi rivers to the west. During the flood season, wetlands in the depression expand to cover about 11,840 km² (Welcomme 1979; Hughes and Hughes 1992).

Lake Upemba, with an area ranging between 500 and 800 km² and a 70-km-long basin, is the largest of the lakes (Bailey 1986). Like the other lakes, it is shallow, with maximum depth reported to be only 3.2 m (Hughes and Hughes 1992). The lake is regularly stirred by winds and becomes supersaturated with oxygen at its bottom during the day. It is highly eutrophic and the site of intense algal production (Beadle 1981). The northern half of the lake is covered almost entirely with plants such as *Typha domingensis* (Hughes and Hughes 1992).

The Lualaba River winds through the lake region and is joined below Lake Upemba by its largest tributary, the Lufira River. The Lufira derives much of its flow from calcareous subterranean streams and affluents that themselves receive input from warm saline springs. As the Lualaba leaves the depression, it is about 100 m wide (Bailey 1986). It then travels through a swampy belt that extends for 80 km and contains a cluster of small lakes, including lakes Kittongola and Towe (Hughes and Hughes 1992).

The climate of the ecoregion is tropical and moist, with a mean annual temperature of about 24°C. About 1,000 mm of rain falls each year, with rainfall increasing toward the equator (Hughes and Hughes 1992). The wet season lasts for approximately half the year, and December is the wettest month. The Lualaba and Lufira flood seasonally in response to the rains. Water level in the Lualaba is highest from February to April and lowest from August to October, with a total fluctuation of 2.8 m in the southern section of the ecoregion. Lake water levels also fluctuate; Lake Upemba is high from March to June and low from October to January (Hughes and Hughes 1992).

Tall herbaceous vegetation dominates the swamps along the rivers and the lakes, with *Cyperus papyrus* and *Typha domingensis* prevalent. The papyrus swamps of this ecoregion are extensive. During the dry season the Lualaba flows through a navigable (usually dredged) channel, but in the wet season the areas surrounding the river become inundated, and floating mats of papyrus can obstruct the channel (Bailey 1986). Common floating aquatic vegetation includes Nile lettuce (*Pistia stratiotes*), water chestnut (*Trapa natans*), *Nymphaea caerulea*, *N. lotus*, and *Nymphoides indica* (Marlier 1973). *Pycreus mundtii* and *Paspalidium geminatum* commonly grow on the swamp edges (Hughes and Hughes 1992). Swamp woodland patches occur on elevated sandy sites and are dominated by ambatch (*Aeschynomene elaphroxylon*) and *Hibiscus diversifolius* (Hughes and Hughes 1992). These sites often are inundated from 40 to 60 cm. Grasslands dominate the plateau, and *Uapaca*, *Brachystegia*, and *Isoberlinia* woodlands occur on lands adjacent to the wetlands (Demey and Louette 2001).

Geologically, the Kamolondo Depression is a graben, a valley formed by an extension of the earth's crust, probably during the Miocene-Pliocene (Hughes and Hughes 1992; Poll and Renson 1948). It is likely that an ancient lake once covered the depression (Bailey 1986). Sedimentation from the Lualaba, Luvua, Lufira, and Luvoi rivers would have infilled the lake to form the swampy region present today (Denny 1985).

Outstanding or Distinctive Biodiversity Features

The swamps, shallow lakes, and river channels of this ecoregion host a rich aquatic fauna and suspected high endemism of Odonata (Kamdem Toham et al. 2003). The crustacean *Thermobathynella adami* lives in a hot spring in Upemba National Park and probably is endemic (Schminke 1987). The ecoregion has twelve mollusk species, none of which is endemic, although there are reports of endemic mollusks at Kalengwe Rapids (Kamdem Toham et al. 2003). Burrowing bivalves are abundant (Bailey 1986).

The fish fauna is incompletely known, although it is well sampled compared with other Congo Basin ecoregions (Poll 1976; Banister and Bailey 1979; Banister 1986). Poll (1976) reports 182 fish species from the Upper Lualaba–Upemba system and its affluent river, the Lufira. Both systems have distinct fish communities, with at least fourteen endemic fish taxa each for the Lualaba-Upemba and the Lufira. Dominant fish families include Cyprinidae (more than thirty species in the Lualaba-Upemba), Mormyridae (nearly twenty species), and Alestidae, Mochokidae, and Cichlidae (each with ten or more species). Fish biomass is high, and many species are swamp and open water adapted. Fish of the productive inundated zones include *Hydrocynus*, *Clarias*, *Synodontis*, and *Schilbe* species (Lowe-McConnell 1987). The main Lualaba channel supports the small plankton-feeding *Microthrissa* and small *Barbus*, as well as piscivores such as *Hydrocynus vittatus* (Lowe-McConnell 1987). Many of the tributary streams of Upemba contain primarily small *Barbus* species, although it is assumed that they also contain larger, seasonally spawning *Barbus* and *Varicorhinus* species (Banister 1986). Rapids tend to contain specialized species, and the rapids of the Lufira River are known to host several endemic fishes including cichlids (e.g., *Lamprologus symoensi*), kneriids (e.g., *Kneria katangae*), and mochokids (e.g., *Chiloglanis lufirae*). Cichlids are numerous in the lakes and rivers of the Upemba area, and *Oreochromis upembae*, which is found in only three ecoregions, is an abundant species (Banister 1986).

This ecoregion hosts a rich aquatic herpetofauna compared with other ecoregions of the floodplains, swamps, and lakes habitat type. Of forty-seven known species of frogs, six are endemic. The slender-snouted crocodile (*Crocodylus cataphractus*) and the Nile crocodile (*Crocodylus niloticus*) are also present in the ecoregion.

The aquatic mammalian fauna, made up of hippopotamus (*Hippopotamus amphibius*), swamp otter (*Aonyx congicus*), giant otter shrew (*Potamogale velox*), and marsh mongoose (*Atilax paludinosus*), resembles that of other Congolian ecoregions.

Several wetland birds of global conservation concern occur in this ecoregion, including the data-deficient *Estrilda nigriloris*, the vulnerable *Grus carunculatus*, and the near-threatened *Balaeniceps rex* (Hughes and Hughes 1992; Demey and Louette 2001).

Status and Threats

Current Status

The ecoregion lies in the DRC's Shaba Province, which possesses rich mineral wealth, including deposits of copper, zinc, cobalt, cadmium, manganese, tin, silver, and gold (Bailey 1986). This ecoregion is moderately populated, with people concentrated in the mining belt south of Lake Upemba and along the rivers and streams. Malemba-Nkuluis is the largest town in the Kamolondo Depression. Deforestation occurs in and around mining towns (Malaisse and Kapinga 1986), and there are many roads transecting the ecoregion (Kamdem Toham et al. 2003). In addition to the pressures from mining and human settlements, slash-and-burn agriculture and the use of pesticides and fertilizers threaten the freshwater systems of this ecoregion (UNEP 1999). Civil conflict has plagued the region since 1996, halting many conservation and research efforts as well as international investments in mining and industry. The conflict has had devastating effects on the economy and on the welfare of people in the ecoregion.

The Kamolondo Depression is the chief fish-producing center for Shaba. Fishing is locally intense, and along the Lufira River and the lakes of the Upemba system there are many fishing villages. Commercial fishing centers are located at Maka,

Kialo, Nyonga, Kalombe, Kikondja, Masango, Malemba-Nkulu, Kabala, and Mulongo (Hughes and Hughes 1992). Lake Upemba supports a productive fishery (Bailey 1986). The use of toxic plant products (such as a substance from *Tephrosia vogelii*) or engineered pesticides (such as DDT and Gamulinz) to catch fish appears to be increasing (D. Wilkie, pers. comm., 2000). These toxins can destroy fish stocks, fingerlings, eggs, and other aquatic fauna (Mino-Kahozi and Mbantshi 1997).

Mining in Shaba historically consumed almost 50 percent of the DRC's total generated power (Coakley 2001a). Upstream of the ecoregion, dams built for the mining industry create the reservoirs of Lac del Commune and Marinel Lake (Bailey 1986). A large dam also exists on the Lufira River just outside the ecoregion, forming Lake Lufira (Tshanga Lele) (Bailey 1986).

Approximately half of the Lake Upemba Basin is covered by Upemba National Park. This area includes lakes Kabwe, Kabele, Sangwe, Malenda, Tungwe, Upemba, and Kisale (Hughes and Hughes 1992). Poaching occurs in Upemba National Park, and during the political disturbances in Shaba in the early 1990s, local and regional authorities authorized hunting and the cutting of firewood in the park (East 1999). The status of fishing in the park is not known.

Future Threats

Continued civil strife, deforestation, agriculture, mining, and overfishing are among the largest future threats to the ecoregion.

Justification for Ecoregion Delineation

This ecoregion is defined by the Kamolondo Depression and additionally by the wetlands, swamps, and lakes along the Lualaba River below the depression and along the adjacent Lululwe, Kilubi, and Lovoi rivers to the west. It is likely that an ancient lake once covered the depression (Bailey 1986). Sedimentation from the Lualaba, Luvua, Lufira, and Luvoi rivers would have infilled the lake to form the swampy region present today (Denny 1985), which is characterized by freshwater species adapted to swamps and open water.

Data Quality: Low

Recent surveys have not been conducted in this ecoregion, and little historic information on fish and aquatic invertebrates exists.

Ecoregion Number:	**15**
Ecoregion Name:	**Upper Nile**
Major Habitat Type:	**Floodplains, Swamps, and Lakes**
Final Biological distinctiveness:	**Globally Outstanding**
Final Conservation status:	**Vulnerable**
Priority Class:	**I**
Authors:	**Emily Peck and Michele Thieme**
Reviewers:	**Fiesta Warinwa and M. S. Farid**

Location and General Description

The vast swamps of the Sudd are the primary feature of the Upper Nile ecoregion, which is situated mainly in Sudan with smaller areas in the DRC, Uganda, and Ethiopia. In Arabic, *sudd* refers to a floating mass of vegetation that blocks navigation. The dynamic Sudd wetland, whose size varies substantially in response to seasonal and interannual changes in water input, contains a diversity of habitats and supports a rich aquatic and terrestrial fauna (Rzóska 1974). Water entering the Sudd swamps drains from the hills of the Nile-Congo watershed divide, the escarpment of the Uganda Plateau, the Imatong Mountains, and the Ethiopian Plateau.

The ecoregion encompasses the basin of the White Nile River; its major tributaries, the Sobat River and Bahr el Ghazal; Lake Albert; and Lake Albert's main influent, the Semliki River. The Bahr el Ghazal and its tributaries—the Sue, Jur, Pongo, Lol, and Bahr el Arab—drain the southwestern portion of the ecoregion and flow into the Sudd, which sits in the heart of the ecoregion. The Sobat River and its tributaries—the Kangen, Akobo, Baro, and Pibor—drain the southeastern portion of the ecoregion. Running north between the Bahr el Ghazal and the Sobat is the Albert Nile (named Bahr el Jabal in Sudan), which drains Lake Albert, flows into the Sudd, and merges downstream with the Bahr el Ghazal to form the White Nile (Bahr el Abyad, in Arabic). The point at which the White Nile joins with the Blue Nile marks the northernmost border of the ecoregion (Rzóska 1974; Dumont 1986; Hughes and Hughes 1992).

The climate of the Upper Nile ecoregion ranges from subequatorial in the south to xeric in the north. Rainfall at the Sudd swamps is highest between June and November, with little falling during the rest of the year (Rzóska 1974). Near Lake Albert, average rainfall is near 1,400 mm (Murakami 1995). In the Sudd region, rainfall ranges from about 900 mm per year in the south to 800 mm in the north (FAO 1997e). In nearly 50 years of observations from one reliable station (Shambe on the Bahr El Jabal), annual rainfall varied from 454 to 1,507 mm, showing the high level of variability from year to year. The mean annual temperature in the ecoregion ranges from 20°C to 35°C (Rzóska 1974).

The southern and highest upstream part of this ecoregion is

situated in the Rift Valley. This area is rugged, with mountain peaks over 4,600 m. The Semliki River flows through the Rwenzori Mountains from Lake Edward to Lake Albert, but a series of rapids prevents faunal exchange between the two lakes; for this reason, Lake Edward is placed outside this ecoregion. Most of the Semliki's flow is from Lake Edward, although tributaries entering it from the northern slopes of the Rwenzoris also contribute some water. Lake Albert sits at 615 m, with two escarpments up to 2,000 m high bordering it. The lake is about 150 km long and 35 km wide on average and has a maximum depth of 56 m. Most of the lake's inflow comes from the Semliki River, although the much larger Victoria Nile, draining lakes Kyoga and Victoria, flows into Lake Albert's far northern end, just before the Albert Nile outlet. The Victoria Nile has only a small effect on Lake Albert's water quality but maintains a fairly constant outflow from the lake (ILEC 2001).

After leaving Lake Albert, the Albert Nile (Bahr el Jabal) continues its northward course through rocky gorges (Murakami 1995). About 125 km into Sudan, the river descends below 500 m and enters the vast, shallow depression of the Sudd. The depression ranges in elevation between 420 and 380 m a.s.l. and stretches for about 600 km from end to end. Most of the depression is flat, with a gradient of 0.01 percent or less, and it is underlain by clay soils (FAO 1997e). The extent of the swamps is highly variable because they expand and contract with seasonal flooding and respond to annual changes in water input. For instance, the swamps expanded substantially in the early 1960s, following a near doubling of water inflows from an average of about 27 trillion m^3 per year before 1960 to about 50 trillion m^3 from 1961 to 1980 (Hughes and Hughes 1992). These higher inflows were attributed to high precipitation in the Equatorial Lake Plateau headwaters during that time. The swamps receive their flow from both the Bahr el Jabal and the Bahr el Gazal and from direct precipitation (FAO 1997e). Welcomme (1979) notes that the ratio of area at low water to area at peak flood is only 11 percent, based on numbers from Rzóska (1974).

High flows in inflowing rivers peak between September and November, with flow volumes lowest between April and June. Widespread flooding occurs from July to November from the combined effects of rains and spillover from the rising rivers (Rzóska 1974). More than half the water entering the swamps is lost through evaporation and evapotranspiration in the swamps (FAO 1997e). Because greater inflows appear to correlate with greater losses, the outflow from the swamps into the Bahr el Jabal is nearly constant. This outflow ultimately contributes about half of the White Nile's flow; the remainder comes from the Sobat River (Murakami 1995).

The floodplain ecosystem supports a variety of plant species, with a succession from those adapted to mesic environments to those adapted to more xeric environments. Moving from the interior of the swamps, the floral zones of the Sudd grade outward from open water and submerged vegetation of the river-lake, to floating fringe vegetation, to seasonally flooded grasslands, to rain-fed grasslands, and finally to floodplain woodlands (Hickley and Bailey 1987). *Cyperus papyrus* is dominant at the riverine fringe and in the wettest swamps. According to Howell et al. (1988), *Cyperus papyrus* forms a fringe along the Bahr el Jabal that is up to 30 km broad in the south, declines to 50 m in the north, and disappears completely east of Wath Wang Kech. Associated species are few, but those that are present are climbers and tend to be most common at channel margins where there is more light. Species include *Coccinia grandis, Cayratia ibuensis, Luffa cylindrical, Zehneria minutiflora, Vigna luteola,* and the fern *Cyclosorus interruptus. Phragmites* and *Typha domingensis* swamps are present behind the papyrus stands. They are most extensive in the central and northern parts of the Sudd away from the main river channels, but they have been little studied because of their inaccessibility. There is also an open water area of about 1,500 km^2 that has floating and submerged aquatic plants. Flooding may be too deep in these areas to allow successful colonization by *Typha* (Howell et al. 1988).

Seasonally river-flooded grassland covers about 16,000 km^2, and seasonal floodplains up to 25 km wide exist on both sides of the main swamps. Wild rice (*Oryza longistaminata*) and *Echinochloa pyramidalis* grasslands dominate these seasonal floodplains. *Oryza longistaminata* is perennial and provides high-quality grazing for much of the year, although in dry years the regrowth after burning is sparse. *Echinochloa pyramidalis*, on the other hand, produces some regrowth in the dry season and thus provides important year-round pastures. The river-flooded grasslands, called *toic* by the Dinka tribe, are important to them as dry season pasturelands.

Beyond the floodplain are rain-flooded grasslands that cover about 20,000 km^2 (Robertson 2001c). These are dominated by *Hyparrhenia rufa* and are the major local source of thatching material. Mixed woodlands of *Acacia seyal, Ziziphus mauritiana, Combretum fragrans,* and *Balanites aegypticaca* border the grasslands (Denny 1991). Of these species the most extensive are *Acacia seyal* (5,400 km^2) and *Balanites aegypticaca* (5,300 km^2) (Howell et al. 1988).

Outstanding or Distinctive Biodiversity Features

The swamps and floodplains of the Sudd are among the most important wetlands in Africa and support a rich biota. Aquatic animals abound, with more than 115 fish species documented in the entire ecoregion, including 16 endemics. The Sudd occurs on the major eastern flyway between Europe and Asia and Africa, it is one of the most important wintering grounds in Africa for Palearctic migrants, and it provides essential habitat for millions of intra-African migrants (Howell et al. 1988). In the past, large populations of mammals, including elephant, giraffe (*Giraffa camelopardalis*), tiang (*Damaliscus lunatus tiang*), reedbuck (*Redunca redunca*), and Mongalla gazelle (*Gazella thomsonii albontata*), followed the changing water levels and vegetation.

Abundant grasses, crustaceans, mollusks, and aquatic insects also contribute to the diversity and high productivity of this system.

Many fish species migrate from the rivers into the nutrient-rich floodplains during the seasonal floods to feed and breed (Welcomme 1979). Twenty-two families and 118 fish species are known to occur in the Upper Nile ecoregion. Cyprinidae is the most speciose family, with Alestiidae, Cichlidae, Mochokidae, Mormyridae, Poeciliidae, and Schilbeidae also represented by high numbers of species. Lake Albert has a Nilotic riverine ichthyofauna comprised of fifteen families and forty-six species. Of the sixteen endemic species in the ecoregion, Lake Albert provides habitat for seven. These are *Lates macrophthalmus, Haplochromis loati, Thoracochromis albertianus, T. avium, T. bullatus, T. mahagiensis,* and *Neobola bredoi* (Lowe-McConnell 1987).

The floodplains of the Sudd are globally outstanding for their congregations of waterbirds, with about sixty species known and a maximum of nearly 3 million birds recorded at one time (Robertson 2001c). Herons, storks, ibises, and other waterbirds are abundant and occur at the swamp margins. During the dry season these birds exploit the fish that become available as pools dry up. Palearctic populations of the endangered great white pelican (*Pelecanus onocratalus*) fly more than 2,000 km from Eastern Europe and Asia to overwinter in the Sudd wetlands (Shmueli et al. 2000). The floodplains of the Sudd also support the largest population of shoebill (*Balaeniceps rex*) in Africa, estimated at 5,000 individuals (Stuart et al. 1990). Populations of the near-threatened black-crowned crane (*Balearica pavonina*) also depend on the wetlands (Newton et al. 1996; Shmueli et al. 2000).

Many mammals follow the expansion and contraction of grasslands and changing water levels of the Sudd. For example, a million white-eared kob (*Kobus kob*) used to undertake a migration of more than 1,500 km to track the availability of floodplain grasses (Denny 1991). More than 800,000 individuals were estimated to inhabit Boma National Park in 1982–1983, with population densities up to 1,000/km^2 near food sources during the dry season (Fryxell and Sinclair 1988). The endemic and threatened Nile lechwe (*Kobus megaceros*) has its center of population in this ecoregion.

Status and Threats

Current Status

Much of the Sudd swamps remain as a vast near-wilderness area, but it appears that years of civil war in Sudan have left protected areas seriously compromised, with poaching of large mammals going unchecked (Stuart et al. 1990). Morjan et al. (2001) suggest that there has been a drastic decline in nearly all mammal populations in many parks in southern Sudan since estimates were made in 1988 (Fryxell and Sinclair 1988). For example, populations of the white-eared kob have declined precipitously, experiencing a 74 percent decline between 1980 and 2001 at the rate of approximately 25,000 individuals lost a year. The Nile lechwe's imperilment is also caused in part by hunting and by competition with cattle for resources (Kingdon 1997). The decline in most mammal populations is attributed to excessive hunting pressure, an increase in human populations, the breakdown of law and order, and the absence of conservation programs (Morjan et al. 2001).

The effective habitat block of the Sudd swamp is at least 30,000 km^2, and associated habitats may extend over a larger area. However, the incomplete, empty Jonglei Canal is acting as a game trap and possibly having significant effects on some populations of large mammals. Reports from Sudan also suggest that vehicle tracks through the floodplains increased because of war-related movements (Stuart et al. 1990) and that the availability and use of automatic weapons has facilitated poaching.

Before the onset of the recent conflict, southern Sudan had five national parks, fourteen reserves, and numerous game-controlled areas. The vast swamps of the Sudd contain all or part of the Zeraf Island (6,750 km^2) and Mongalla (75 km^2) game reserves and the Boma (22,800 km^2), Badingilo (16,500 km^2), and Shambe (620 km^2) national parks. However, the IUCN/SCC Antelope Specialist Group considered the level of protection and management in Boma and Badingilo National Parks to be nil by the mid-1990s (East 1999). Boma National Park is extremely remote, with access roads from Ethiopia and Kenya being impassable except during the dry season (January–April). As the government of Sudan and the Sudan Peoples' Liberation Army contest the area, the main routes have been mined and alternative roads are in need of maintenance and repair. The regional park headquarters for the Upper Nile is based in Boma, despite the park's inaccessibility. The park system lacks the necessary equipment and supplies for adequate management, and park staff are poorly trained.

Located in the southwestern corner of the Darfur Province, on the border of the Central African Republic, Radom National Park is also designated as a biosphere reserve. It covers 112,509 km^2 of woodland savanna, and its wet meadows provide water and fodder for wildlife during the dry season. The park maintains populations of waterbuck (*Kobus ellipsiprymnus*), white-eared kob, and hippopotamus (*Hippopotamus amphibious*), as well as numerous other large terrestrial mammals. However, poaching has seriously depleted wildlife numbers (East 1999).

The human population density in the Sudd is less than 20 people/km^2 and is concentrated along major rivers, lakes, and floodplains. In addition to grazing their cattle, the Dinka, Nuer, and Shiluk tribes fish to supplement their livelihoods. The effect of these activities on fish populations is unknown.

Future Threats

Resumption of civil war poses the greatest threat to human populations and wildlife conservation in the Upper Nile ecoregion. After Sudan's independence from Britain in 1956, political tension between the Arab north and African south led to two civil wars. The civil war disabled protected area management, and the increased use of automatic weapons and vehicles led to uncontrolled hunting and poaching in previously inaccessible areas. Furthermore, international support for Sudan's conservation programs stopped with the resumption of civil war in 1983 (East 1999). A fragile cease-fire was in effect from October 2002, but widespread civil conflict erupted again in February 2003 in western Sudan's Darfur region. As of September 2004, an estimated 50,000 people had been killed and an additional 1.2 million displaced (CBS News 2004).

Several major planned developments pose serious threats to the ecological integrity of the ecoregion. The Jonglei Canal is a controversial engineering project whose purpose is to divert floodwaters from the Sudd downstream for agricultural development. The project calls for the excavation of a canal approximately 360 km long, 65 m wide, and 5 m deep. Excavation of the canal began in 1978 but was halted in 1983 because of the outbreak of the second civil war. Diversion of water at the upstream end of the Sudd through the canal would prevent much of the evaporative loss of water that occurs in the Sudd but would also shrink the swamps and floodplains. It would also decrease fish habitat and result in a reduction in river-flooded grasslands, removing habitat for wild mammals and cattle herds. Wild mammals will face risks when trying to cross the canal. Movements of local people and their livestock will also be affected (Howell et al. 1988). The Jonglei Canal is also likely to have a substantial impact on climate, groundwater recharge, sediment distribution, and water quality.

Justification for Ecoregion Delineation

This ecoregion is defined mainly by the Sudd swamp and its basin. It encompasses the basin of the White Nile River; its major tributaries, the Sobat River and Bahr el Ghazal; Lake Albert; and Lake Albert's main influent, the Semliki River. However, the ecoregion excludes lakes Victoria, Kyoga, and Edward and the Victoria Nile, all of which feed directly or indirectly into Lake Albert. Lake Albert is included in this ecoregion because of its Nilotic riverine ichthyofauna, distinct from the lacustrine fauna of the Lakes Kivu, Edward, George, and Victoria [57] ecoregion.

Data Quality: Medium

Ecoregion Number:	**16**
Ecoregion Name:	**Upper Zambezi Floodplains**
Major Habitat Type:	**Floodplains, Swamps, and Lakes**
Final Biological Distinctiveness:	**Nationally Important**
Final Conservation Status:	**Relatively Intact**
Priority Class:	**V**
Author:	**Lucy Scott**
Reviewers:	**Paul Skelton and Brian Marshall**

Location and General Description

Regular flooding events inundate the extensive floodplains and swamps of this ecoregion and are essential to maintaining an abundant aquatic flora and fauna. The ecoregion begins at the confluence of the Lungwebungu, Zambezi, and Kabompo rivers. The Zambezi River then flows south through the Barotse floodplains and turns east at the Caprivi Strip before descending the 98-m-high Victoria Falls above the Batoka gorge. The falls form an impassable barrier between the middle and upper portions of the Zambezi River (Davies 1986). The ecoregion is located in southwestern Zambia, southeastern Angola, the Caprivi Strip of Namibia, and the northern edge of Botswana.

The Upper Zambezi Floodplains ecoregion has a highly seasonal rainfall pattern, with most rain falling in the austral summer months. The flow of the river reflects this pattern; flow is lowest in October, rises in January, peaks in February–April, then declines in May–June (Bell-Cross 1965; van der Waal and Skelton 1984; Davies 1986). Precipitation usually is 800–1,100 mm per year, with evapotranspiration levels averaging 1,000 mm per year. Many tributaries are intermittent because of the high evaporation and highly seasonal rainfall (Roberts 1975).

The Zambezi River drains a huge, shallow alluvial basin more than 1,220 m above sea level, and because of its gentle gradient and moderate rainfall it is lined by extensive swamps and floodplains. The floodplains are inundated for long periods, and low flows are of short duration. The Upper Zambezi is an important reservoir river, seldom exhibiting large variations in height because of its gentle slope and the fact that floodwaters are not confined to a channel but spread out across the vast floodplains. The average flood height of the river is 5.2 m (Jackson 1961, 1963; van der Waal and Skelton 1984).

The Upper Zambezi receives water from the Cuando (also known as Kwando or Mashi, with a catchment of 57,000 km^2) and Lungwebungu rivers and rarely from the Okavango via the Chobe River (1,120 km) during times of very high rainfall (Bell-Cross 1972). Complicating the hydrology is the tendency of the Chobe to reverse flow when the Upper Zambezi floods, filling the former as far back as Ngoma (Jackson 1986). When the flood recedes, the flow changes direction and the Zambezi drains the southern part of the eastern floodplain (van der Waal and Skelton 1984).

Along the Upper Zambezi, long stretches of low-gradient river with associated floodplains and swamps alternate with extensive rapids. Rapids are present in several areas between Nangweshi and Katima Mulilo (176 km) and from Mombava to the Victoria Falls (72 km) (Roberts 1975). The Gonye Falls, about 300 km upstream from Victoria Falls, has a maximum height of 21 m and is the only waterfall, other than Victoria Falls, on the main river that is high enough to constitute a barrier to the movement of fish in the dry season (Bell-Cross 1972).

Three major floodplains are of major importance for the ecoregion's freshwater biodiversity. Along the Zambezi the Barotse (or central Barotse) floodplain extends from Lukulu in the north to Nangweshi in the south and is approximately 240 km long and up to 35 km wide. In flood, it covers an area of about 7,500 km^2. The eastern Caprivi (or southern Barotse) floodplain is approximately 100 km long and is located between Sesheke and Maramba (Hughes and Hughes 1992). Marginal vegetation is abundant and provides cover for small fish and juveniles. The eastern Caprivi floodplain is contiguous with the eastern portion of the Chobe-Linyanti floodplain system, which begins at the point where the Cuando River enters Botswana (van der Waal and Skelton 1984). The Chobe flows in an east-northeast direction to join the Zambezi, at which point the Chobe floodplain system becomes part of the southern Barotse floodplain. The Linyanti swamp is about 300 km^2 in area, but its size varies according to the extent of flooding in the Upper Zambezi (Marshall 2000a). The floods in the Linyanti system are dampened by the moderating effect of the vast *Phragmites-Typha-Cyperus* swamps in the river valley, delaying the flood peak in the Linyanti until August. Lake Liambezi, which periodically dries up, lies at the end of the Linyanti swamp, has an open water surface of 100 km^2 when full, and is bordered by a swamp of 200 km^2 (van der Waal and Skelton 1984). Total estimates of the extent of the swamp and floodplain in the east Caprivi area vary from 250 km^2 to more than 1,000 km^2 and from 1,670 to 1,870 km^2, respectively (Timberlake 2000b). The water is nutrient poor and unproductive because the rivers flow over belts of Kalahari Sands and constantly change course because of erosion and deposition.

The terrestrial vegetation of the ecoregion is predominantly miombo (*Brachystegia* and *Julbernardia*) woodland, with extensive wetlands and floodplains around the Zambezi River and important grasslands on the Liuwa plain. The Barotse floodplain is clearly separated from the surrounding woodland, whereas in eastern Caprivi the floodplains form a mosaic with the woodlands (Timberlake 2000b). Extensive and stable seasonally flooded grasslands grow on the Barotse floodplains, along with some swamp vegetation. Floodplain grasses in the wetter areas include *Acroceras macrum, Brachiaria arrecta, Digitaria* sp., *Echinochloa pyramidalis*, and *Oryza longistaminata*, among many others. In deeper channels and lagoons, the reed *Phragmites mauritianus* is the most abundant species. Vegetation stands also include *Cyperus papyrus, Hibiscus diversifolius, Urena lobata, Persicaria senegalensis, Aeschynomne uniflora*, and the swamp grasses *Echinochloa pyramidalis, E. stagnina*, and *Vossia cuspidata*. The Cuando (Kwando or Mashi) River is bordered by floodplain and riparian woodland (Timberlake 2000b).

Outstanding or Distinctive Biological Features

The vast floodplains and swamps of the Upper Zambezi Floodplains ecoregion provide breeding and feeding grounds for a moderately rich fish fauna and herpetofauna. The ecoregion supports a near-endemic radiation of large riverine cichlids (Skelton 1994).

The floodplain fishes are adapted to migrate to and from the floodplain with the flood cycle beginning with the first floods each year, in November and December. Floodplain spawning is advantageous to juvenile fish because of the abundant food, well-oxygenated conditions, and cover from predators. Fish on the floodplains exhibit a wide variety of life histories and behavioral patterns (Bowmaker et al. 1978). Fish migration is species specific; some fish exhibit seasonal migratory behavior, using the floodplain temporarily in the wet season, whereas others are semipermanent residents, depending on the availability of water (Winemiller 1991; Winemiller and Kelso-Winemiller 1994, 1996; Merron and Bruton 1995; van der Waal 1996).

Cyprinids, cichlids, and mochokid catfishes dominate the fish fauna of this ecoregion. The Cichlidae include two radiations of large predatory forms and a number of small, more widely distributed omnivorous species. The former include two species flocks, one of six *Serranochromis* and one of five *Sargochromis* species (Marshall 2000a). The *Sargochromis* typically eat snails, freshwater mussels, and crustacea, whereas *Serranochromis* generally prey on other fish (Winemiller 1991; Skelton 1993). There is also a species radiation of mochokid catfishes of the genus *Synodontis* (seven species) (Marshall 2000a).

The floodplains of this ecoregion are rich in water-dependent herpetofauna compared with other parts of the Zambezi Basin. The Barotse floodplain, with eighty-nine species currently known, has the richest wetland herpetofauna in the basin, and at least one frog, *Ptychadena mapacha*, is endemic to the ecoregion (Broadley 2000).

The Liuwa Plain National Park, east Caprivi wetlands, and the Barotse floodplain host large congregations of waterbirds, some of which come to breed. The Liuwa is particularly important for populations of black-winged pratincoles (*Glareola nordmanii*), and the Barotse floodplain hosts more than 20,000 ruff (*Philomachus pugnax*), a nonbreeding Palearctic migrant, and more than 10,000 cattle egrets (*Bubulcus ibis*). It also supports significant populations of *Phalacrocorax africanus, Anastomus lamelligerus, Charadrius asiaticus*, and *Chlidonias hybridus* (Leonard 2001).

Several threatened species of waterbirds, reptiles, and fish depend on the freshwater habitats of this ecoregion. The Linyanti swamp and the Liuwa plain are extremely important breeding grounds for the vulnerable wattled crane (*Grus carunculatus*) and the vulnerable slaty egret (*Egretta vinaceigula*), and wattled cranes also occur on the Barotse floodplains and Chobe-Linyati swamps (Leonard 2001). The Nile crocodile (*Crocodylus niloticus*) occurs in the Upper Zambezi, where moderately sized populations appear to be vulnerable. Of the fish, *Clariallabes platyprosopos* has an extremely limited range, with two of its four known localities occurring in this ecoregion at Katima Mulilo and Impalila. The striped killifish (*Nothobranchius* sp.), which is endangered, is known from two small rainwater pans at Gunkwe and Bunkalo in the Caprivi Strip (Marshall 2000a).

Status and Threats

Current Status

Two large and many small fisheries occur in this ecoregion. The main fishery is on the Barotse floodplain between Lukulu and Senanga, and the other large fishery (about half the size of the former) is on the eastern Caprivi floodplain between Sesheke and Maramba (Jackson 1986). Because the water of these regions is nutrient poor and unproductive, the fishery yield from the southern Barotse floodplain is low. For example, it is only half of that of the Kafue [8] floodplain, even though they are of comparable size (Marshall 2000a). Earthen bunds (impervious embankments) known as *maalelo* weirs are commonly constructed on the central Barotse floodplain to channel and trap fish as they return to the main river after the floods. Traditional fishing methods such as thrust baskets and spears are still used, but gillnets and seine nets were introduced in the 1920s (Jackson 1986). Similar types of fisheries existed in the semipermanent Lake Liambezi on the Chobe before the lake dried up during the droughts of the 1980s (Bell-Cross 1965; Marshall 2000a). The fish communities of floodplain rivers appear to be less affected by heavy fishing than equivalent lacustrine populations because they are adapted to compensate for high natural mortality rates (Jackson 1986). Although heavy fishing intensity on the Barotse floodplains is likely to affect species composition by taking out the larger individuals and larger species, the fish stocks currently appear to be intact (Marshall 2000a).

Persistent pesticides such as dieldrin and DDT were formerly used to control tsetse fly (*Glossina* sp.) and malarial mosquitoes, and they are still used in places along the Upper Zambezi Floodplains. These chemicals may have entered the food chain, and monitoring should be established to determine whether they are having harmful effects on humans or other organisms (Leonard 2001; P. Skelton, pers. comm., 2002).

There are several important protected areas in the ecoregion, but poaching is a serious problem in all of them. The Linyanti swamp contains substantial populations of large predatory fish and is protected by the Mamili National Park (320 km^2). The Caprivi Game Reserve (1,750 km^2), Mudumu National Park, Liuwa Plain National Park (3,660 km^2), Sioma Ngwezi National Park (5,276 km^2), and the West Zambezi Game Management Area are important protected areas but need improved management and protection (Jackson 1986; Stuart et al. 1990).

Future Threats

The major future threats to the Upper Zambezi Floodplains include overfishing, the use of small-meshed nets in an unregulated fishery, overgrazing, the introduction of alien species, and the removal of riparian vegetation (J. Day, pers. comm., 2001). Future threats to the endangered *Nothobranchius* sp. include habitat destruction from road building (both pans in which the species is found are adjacent to gravel roads) and pollution from domestic laundry being washed in these pans. There is also a potential threat from collectors for the aquarium trade (Davies and Day 1998). Many fish species have been introduced into the Middle and Lower Zambezi, but there are no records yet of alien species in the Upper Zambezi Floodplains. Marshall (2000a) warns against the future introduction of new species, particularly *Oreochromis niloticus*, into the Upper Zambezi. Additionally, the introduced water hyacinth (*Eichhornia crassipes*) is a problem in much of the Lower and Middle Zambezi, and efforts should be made to prevent the spread of this nuisance species into the Upper Zambezi. The floodplains seem secure from a conservation point of view because the area lacks significant agricultural or mineral potential and the human population density is fairly low (Marshall 2000a).

Justification for Ecoregion Delineation

This ecoregion is defined by the floodplains of the Zambezi and Cuando rivers above Victoria Falls. The floodplain fauna of the Upper Zambezi Floodplains ecoregion has similarities with the floodplain faunas of the Okavango Floodplains [12] and Kafue [8] (Skelton 1993; Marshall 2000a). The biogeographic affinities of Zambezian fish could best be explained as a result of a previous separation of the Upper Zambezi Basin from the Middle and Lower Zambezi, with consequent different origins of much of their faunas. Before the Pleistocene, it is postulated that the Upper Zambezi may have constituted a common faunal arena that probably drained first into the Atlantic via the southwest and later into the Indian Ocean via the Limpopo valley. The Upper Zambezi was later captured by the headwaters of the Middle Zambezi and changed direction to flow northeast to reach the Indian Ocean via the Lower Zambezi system (Bell-Cross 1972; Beadle 1981; Skelton 1994; Marshall 2000a).

Data Quality: Medium

The quality of ecological data is moderate, but more research is needed in all aspects pertaining to the ecology of this ecoregion and the impacts of fishery activities. The report *Biodiversity of the Zambezi Basin Wetlands* provides a comprehensive summary of current knowledge of biodiversity in the basin (Timberlake 2000a).

Ecoregion Number: **17**
Ecoregion Name: **Ashanti**
Major Habitat Type: **Moist Forest Rivers**
Biological Distinctiveness: **Bioregionally Outstanding**
Conservation Status: **Endangered**
Priority Class: **IV**
Authors: **Ashley Brown and Michele Thieme**
Reviewer: **Christian Lévêque**

Location and General Description

This ecoregion is located in Ghana's previously forested southwestern corner and a small portion of Côte d'Ivoire's southeast. The short rivers of this ecoregion descend gradually from historically forested lowlands into coastal lagoons and swamps. Moving from north to south, semideciduous, moist evergreen, and wet evergreen rainforest forms the natural vegetation, but these now occur in patches as the land is cleared for agriculture. Along the coast, mangroves line river mouths, especially in the southwestern half of the ecoregion (Hughes and Hughes 1992; Sayer et al. 1992).

In this tropical, wet ecoregion, the length of the rainy season varies along the coast. In the west, near Axim, the rainy season lasts for most of the year and only August is dry, whereas in the east it is bimodal. Precipitation averages about 2,125 mm per year in the west and decreases moving east, from 1,625 mm per year at Cape Three Points to less than 800 mm at the eastern edge of the coast. The primary rainy season east of Cape Three Points lasts from March to July, with rain continuing to a lesser degree from October to November (Hughes and Hughes 1992). Mean temperatures along the coast in the southeastern portion of the ecoregion average about 24°C in the coolest month and 27°C in the warmest month. Evaporation in the ecoregion can be extremely high, at 1,630 mm per year near Axim and increasing to 2,058 mm in the east (Hughes and Hughes 1992).

From west to east, the main rivers of the ecoregion are the Bia, Tano, Ankobra, and Pra (called the Ofin in its upper course). The lower course of the Tano River inundates 75 km^2 of forested swamps before flowing into the Aby lagoon complex (Hughes and Hughes 1992). The Bia River also feeds into this complex. Lake Bosumtwi (38.5 km^2), an endorheic crater lake created by a meteor impact approximately 1 million years ago, lies in the northeastern corner of the ecoregion (John 1986; Turner et al. 1996).

Outstanding or Distinctive Biodiversity Features

Several endemic fish and frogs live in or along the forested and mangrove-lined streams of this coastal ecoregion. About 10 percent of the 105 fish species in the ecoregion are endemic to it. The fish fauna consists of characids, cichlids, cyprinids, rivulines, and mormyrids; bagrid, mochokid, and clariid catfishes; and a few freshwater pipefish and clupeids. Fourteen percent of the fifty-five frog species are also endemic. Most endemic frog species are from the *Phrynobatrachus* genus. This ecoregion is particularly rich in mollusks, with twenty-one species present, although none are endemic.

There are several fish with highly restricted distributions in the ecoregion. Endemic cichlids such as *Chromidotilapia bosumtwensis* live only in Lake Bosumtwi (FishBase 2001), and *Limbochromis robertsi* is known only from the Pra River. The spiny eel, *Aethiomastacembelus praensis*, is restricted to the upper reaches of the Pra River (Teugels et al. 1988).

Along the coast and in the north of the ecoregion several wetlands serve as breeding and resting grounds for waterbirds. In the Owabi Wildlife Sanctuary in the north, duck species such as *Anas querquedula* can be found in large numbers, especially during times of drought in Sahelian wintering grounds (Wetlands International 2002). Other aquatic birds present in this sanctuary include jacanas (Jacanidae), pygmy geese (*Nettapus auritus*), Goliath herons (*Ardea goliath*), and several other heron species (Ardeidae). Approximately 23,000 waterbirds from twenty-seven species inhabit the Muni Lagoon in the southeast (Wetlands International 2002).

Status and Threats

Deforestation, mining and industry, agriculture, and impoundments have altered the freshwater systems of this ecoregion. Deforestation, particularly of wooded floodplains and mangroves, causes erosion and loss of habitat for freshwater species. Effluent discharges have increased in the region because of outputs from industry, mining, agriculture, and urban centers (Hens and Boon 1999).

More than 90 percent of the original closed-canopy forest of Ghana has been logged since 1940 (Hens and Boon 1999). Much of the remaining forest is found in forest reserves or national parks, and outside these reserves most of the forested land has been converted to agriculture, with remnant forest patches and secondary thicket. Land use activities in the ecoregion include logging, cacao, oil palm, rubber and cola farming, subsistence agriculture, and mineral production (Sayer et al. 1992).

Gold, diamonds, manganese, bauxite, sand, salt, and kaolin are mined. Mining has the potential to increase erosion, change hydrological patterns, and contaminate waters with heavy metals, arsenic, cyanide, or acid mine drainage. The Pra and Ankobra River basins drain an area northwest of Accra where gold and manganese mines are located (Hens and Boon 1999). As of 1998 mercury was still widely used in artisanal mining to amalgamate gold, leading to mercury contamination of rivers (Coakley 2001b). Near Obuasi in the center of the ecoregion, mercury concentrations are higher than normal in plants and soils, especially near mining areas (Amonoo-Neizer et al. 1996). The dredging technique used to mine alluvial diamonds in the Pra River also has the effect of increasing the suspended particle load in downstream waters.

To date there are few major dams in the Ashanti ecoregion, except for the hydroelectric Ayame I and II dams on the Bia River. Lake Ayame, formed from the lower Bia impoundment, covers 300 km^2 (Hughes and Hughes 1992). At least one species, *Citharinus eburneensis,* found only in this ecoregion and the adjacent Eburneo [20], has disappeared since dam construction (Gourene et al. 1999).

Two Ramsar sites are located in the Ashanti ecoregion. The Owabi Wildlife Sanctuary is located in the north of the ecoregion, about 10 km northwest of Kumasi, and the Muni Lagoon site covers about 94.61 km^2 in the southeast (Wetlands International 2002).

Justification for Ecoregion Delineation

This ecoregion is defined by the previously forested Ashanti region of Ghana, which probably, with the Eburneo [20] ecoregion, remained during dry climatic phases and acted as an aquatic refuge (Hugueny and Lévêque 1994). The fish fauna of this ecoregion is primarily Nilo-Sudanian, with a portion of the fauna of limited distribution to the Ashanti and Eburneo [20] ecoregions. Examples of such species include *Marcusenius furcidens, Citharinus eburneensis,* and *Synodontis bastiani.*

Data Quality: High

Ecoregion Number: **18**
Ecoregion Name: **Cuvette Centrale**
Major Habitat Type: **Moist Forest Rivers**
Final Biological Distinctiveness: **Bioregionally Outstanding**
Final Conservation Status: **Relatively Stable**
Priority Class: **V**
Authors: **Ashley Brown and Michele Thieme**
Reviewers: **Uli Schliewen and David Kaeuper**

Location and General Description

This ecoregion, rich in fish and other aquatic fauna, encompasses the largest tract of lowland rainforest in Africa and occupies most of the low, flat portion of the Congo Basin called the Cuvette Centrale (Bailey 1986). The rivers of the ecoregion possess a rich fish fauna that is still largely unstudied. Bordered on the north by a plateau that slopes toward the Congo Basin, this ecoregion lies entirely in the Democratic Republic of the Congo (DRC).

The ecoregion lies at about 300 m a.s.l. and is almost totally flat. The main channel of the Congo widens as it flows downstream and is surrounded by vast tracts of swamp that connect intermittently with the river. The river flows over its banks into the forest between November and January (Bailey 1986). The Congo accumulates about 70 percent of its water volume in the Cuvette Centrale as it receives input from many large tributaries (Banister 1986). Downstream, the river braids into a mosaic of islands, sandbanks, and floating *Eichhornia* masses (Bailey 1986). Permanent blackwater swamps with acidic, low-oxygen waters occur in the ecoregion, and many of the forested southern tributaries are blackwater streams. The waters of the main Congo channel often are a deep brown color (Roberts 1973).

Several large tributaries join the mainstem Congo as it flows through the Cuvette Centrale ecoregion. The Aruwimi, Itimbiri, and Mongala flow into the Congo from the northeast, and the Lulonga, Ikelemba, and Ruki join from the south. The lower valleys of the tributaries all contain swamp forests and *Raphia* palms. The Itimbiri Valley alone contains more than 1,500 km^2 of swamp (Hughes and Hughes 1992). Many northeastern tributaries, such as the Aruwimi, flow across rapids before joining the Congo (Bailey 1986).

The Cuvette Centrale ecoregion has an equatorial and wet climate, with rainfall fairly consistent year-round. Average annual rainfall in the Cuvette Centrale is 1,500–2,000 mm, and near the equator rain falls throughout the year (Bailey 1986; Lowe-McConnell 1987). Moving away from the equator, rainfall is more periodic, with distinct wet and dry seasons (Bailey 1986). Tributaries from the north flood from August to November and from the south flood from May to June (Lowe-McConnell 1987). Mean annual temperatures are about 24°C over most of the ecoregion, with mean daily maxima of 30°C and mean daily minima of about 20°C (Hughes and Hughes 1992).

The ecoregion contains large seasonally and permanently flooded areas. As the Congo flows through the Cuvette Centrale it ranges from 3 to 15 km in width and 3 to 10 m in depth (Hughes and Hughes 1992). The Congo flows over a broad alluvial plain with seasonally flooded forests that extend beyond its banks for up to 50 km. Seasonally inundated forests also cover islands in the river, and permanent swamp forest grows along the river. Rainfall occasionally inundates areas of forest not flooded by rivers, forming swamps in depressions.

The vegetation of the ecoregion is primarily tropical rainforest with a canopy that can exceed 45 m in height. Although there are a few dominant species, floodplain forests have a rich flora. The vegetation composition varies, largely depending on the soil type (Hughes and Hughes 1992). *Mitragyna stipulosa* usually dominates on muddy soils along slow-flowing water. *M. stipulosa* is associated with a variety of species, including *Alstonia congensis, Macaranga* sp., *Nauclea diderrichii, N. pobeguinii,* and many others. In the understory, *Cyrtosperma senegalense* and *Marantochloa congensis* are common. In areas with sandy soil that experience less flooding, *Guibourtia demeusei* dominates the forests. These forests have a low canopy of 30 m and a sparse understory. Tall forest vegetation on the islands of the Congo includes *Lannea welwitschii, Ficus mucuso, Oxystigma buchholzii,* and *Pseudospondias microcarpa* (Hughes and Hughes 1992).

Outstanding or Distinctive Biodiversity Features

The Cuvette Centrale has a rich fish fauna, with 300–400 fish species expected, about 240 species documented, and at least 12 known endemics (Poll and Gosse 1963). However, few biological studies have been conducted in this ecoregion, and further surveys should locate additional species and new endemics, particularly because rapids and waterfalls isolate the Cuvette Centrale fish fauna from the lower and upper sections of the river. The Cuvette Centrale contains a variety of habitats including open waters, creeks, coves, meadows of aquatic vegetation, permanent swamps, and floodplains (Hughes and Hughes 1992).

The base of the food chain in many of the nutrient-poor blackwater rainforest streams begins with allochthonous organic material that falls into the water. In addition to woody matter, terrestrial input in the form of insects, seeds, or fruits provides an important source of food for several fish species. Large characoids and cyprinids feed on terrestrial insects, seeds, or fruits that have fallen in the water, and several species may feed primarily on these items (Beadle 1981; Lowe-McConnell 1987). For example, in the streams flowing into the Congo near Yangambi, of 670 individuals in 16 families, 84 percent were feeding almost exclusively on terrestrial insects (Lowe-McConnell 1987).

The main river channel supports a wide variety of species adapted to pelagic, benthic, and shoreline habitats. Pelagic species include schooling plankton feeders such as *Microthrissa* and *Barbus* (Lowe-McConnell 1987). Schools of *Microthrissa* migrate upstream and downstream seasonally (Hughes and Hughes 1992). Catfish (bagrids and mochokids) are found in the slow-flowing portions of the Congo's main channel. Larger species of Mormyridae, Characidae, Citharinidae, and Cyprinidae often are found in the main channel, with smaller species of these families found in streams (Lowe-McConnell 1987). *Barbus* and *Hydrocynus vittatus* (which preys on *Barbus*) are also found in the open waters (Hughes and Hughes 1992). The benthic fish fauna of the main river channels includes many insectivores and some detritus feeders, omnivores, and piscivores (Lowe-McConnell 1987). Insect and detritus feeders include many *Barbus, Chrysichthys, Synodontis,* and *Tylochromis* species. Some piscivores such as *Chrysichthys, Mormyrops,* and *Polypterus* also feed near the bottom (Hughes and Hughes 1992). Large catfish (bagrids and mochokids) live primarily in the large river channels. Clariids live on the muddy bottoms of rivers, streams, and swamps (Lowe-McConnell 1987).

Because of the relative stability of flow levels in the central portion of the Congo River, different habitat types (such as swampy inundated areas) are present year-round. The relative stability of these habitats has allowed many fish species to become specialized to narrow niches. For example, *Eugnathichthys* spp. are fin eaters, attacking larger fish, and *Campylomornyrus* spp. use their long rostral probe to feed on bottom-dwelling insects among rocks or in the mud (Roberts 1973). Despite the specialization of some species, many Congo River species are able to survive a wide range of temperature conditions. The wide main river channel receives less shade from streamside forests and is more affected by wind than smaller tributary streams and as a result has higher water temperatures (22.5–33°C) than the smaller streams (Lowe-McConnell 1987).

Many fish species live in the swamps and marginal waters between the rivers and inundated forests (Lowe-McConnell 1987). Many clariids, anabantids (*Ctenopoma* and *Microctenopoma*), and lungfish (*Protopterus*) are able to withstand highly anoxic conditions and live in the swamps that border the middle Congo and adjacent shallow blackwater lakes (Beadle 1981). Several species are adapted to these conditions and either use accessory breathing organs to survive in the low-oxygen waters or live directly under the water surface, where oxygen is in greater supply (Banister 1986). Clariidae have arborescent accessory breathing organs in the gill chamber, and *Ctenopoma* (Anabantidae) have labyrinthine accessory breathing organs. *Protopterus dolloi* can build nests in mud and decaying vegetation, with a tunnel "chimney" to the atmosphere through which the fish is able to breathe. Other species such as *Hepsetus odoe* stay near the water surface. Most swamp fish feed on insects or other fish (Banister 1986, Lowe-McConnell 1987).

Many fish leave the rivers and spawn in the forest during high water. Juveniles often remain in the floodplains to feed and grow, despite the receding floodwaters (Hughes and Hughes 1992). The majority of species spawn when the floods originating from the northern tributaries occur in September and October, although another breeding season occurs between April and June. The fish hatch within several hours to 2 days and grow quickly in the inundated zone (Lowe-McConnell 1987). Parental care is present to a small extent. Lungfish construct a nest that the male guards and aerates. The air-breathers notopterids, osteoglossoids, and anabantids also guard their eggs and young (Roberts 1973). However, little is known about the reproductive biology of most of the species.

The ecoregion also hosts a rich avifauna, with several hundred species in the Congo Basin overall. Hartlaub's duck (*Pteronetta hartlaubii*), threatened elsewhere by habitat loss caused by deforestation, is found in this ecoregion (Hughes and Hughes 1992; Scott and Rose 1996).

In addition to the hippopotamus (*Hippopotamus amphibius*), several other aquatic mammals inhabit the Cuvette Centrale. The giant otter shrew (*Potamogale velox*) and the Congo clawless otter (*Aonyx congica*) depend on the ecoregion's rivers and streams. These two species feed on fish, crabs, frogs, and aquatic mollusks. Allen's swamp monkey (*Allenopithecus nigroviridis*) inhabits the ecoregion as well, feeding on fruits, leaves, and invertebrates, including crabs. Sitatunga (*Tragelaphus spekei*) and chevrotain (*Hyemoschus aquaticus*) also live along watercourses in the Cuvette Centrale. The chevrotain relies on the water as a refuge from predators (Kingdon 1997).

Eight mollusk species occur in the ecoregion. The distribution of snails in the ecoregion depends primarily on the acidity of the water. Although few mollusks can be found in brown-colored acidic waters, mollusks are common in the more nutrient-rich Congo mainstem and in northern tributaries that receive water from streams and rivers draining less impoverished upland soils (Brown 1994).

Status and Threats

Current Status

Civil strife in the DRC has been almost continuous since 1996 (East 1999). This strife has had devastating effects on the economy and on the welfare of human populations in the region. It has also limited nearly all conservation and management initiatives. International investment in logging has been deadlocked since the beginning of the conflicts, but mining and logging have continued, often illegally (Pan-African News Agency 2001).

Much of the inundated land of the Cuvette Centrale remains in pristine condition, although in areas along the river that flood seasonally there has been some land conversion to rice production (Hughes and Hughes 1992). Heavy pesticide use occurs at selected sites in this ecoregion, both for farming and for fishing.

As of the early 1990s, 86 percent of the forests of the DRC were still intact, but 37 percent of the total exploitable rainforest had been designated as timber concessions (Meditz and Merrill 1993). Deforestation is high in the northeastern corner of this ecoregion along its border with Uganda because of both the impact from an influx of refugees in the early 1990s from neighboring countries and the illegal extraction of logs from the DRC through Uganda. For example, Ugandan-Thai forest company DARA-Forest has exported about 48,000 m^3 of timber per year between September 1998 and 2000 from the Orientale Province of DRC (United Nations 2001). Most logging in the DRC is selective. Short-term effects of such logging include pulses of sediment delivered to rivers (D. Wilkie, pers. comm., 2000).

Sewage and industrial waste are dumped from the city of Kisangani (population approximately 500,000) directly into the Congo River. Agrochemical substances also make their way into the rivers and streams of the ecoregion (UNEP 1999).

Mining occurs largely on an artisanal scale, yet it can have intense impacts in localized areas (Hart and Hall 1996). Mining for gold, cassiterite (tin oxide), coltan, and diamonds occurs largely in the northeastern part of the ecoregion (United Nations 2001). Mining often disturbs riparian vegetation and causes an influx of sediments to waterways (D. Wilkie, pers. comm., 2000).

Fish and other freshwater organisms are a source of protein for many populations located along the river, and most rivers are fished (Bailey 1986; Hughes and Hughes 1992). Local wild herbs and roots have been used in the past as fish poisons, and increasingly pesticides such as DDT and Gamulinz are used to catch fish in large numbers (D. Wilkie, pers. comm., 2000). Increasing hunting pressures also threaten the dwarf crocodile, *Osteolaemus tetraspis,* although populations are currently robust (Riley and Huchzermeyer 1999).

As of 1986, water hyacinth, *Eichhornia crassipes,* had infested both the main channel of the Congo River as far upstream as Kisangani and the Aruwimi River. Water hyacinth impedes navigation and changes the quality of the water. It is intolerant of humic, acidic water and therefore is less of a problem in the tributaries of the central plain (Bailey 1986).

Civil war in this ecoregion has had a devastating effect on the human population and the structures of civil society, particularly in the northeastern portion. As a result, protected areas in the Cuvette Centrale have been extremely underfunded, if funded at all, or have been exploited for their natural resources during the ongoing conflict (Wilkie et al. 2001). For example, the Okapi Wildlife Reserve, which is located in the Ituri region and covers about 13,500 km^2 of forest, suffers from the effects of poaching and illegal mining of gold and coltan (columbotantalite). Coltan is an ore that provides tantalum for mobile phone components. In the summer of 2001, about 300–400 people were living and mining in the reserve (Reuters 2001). Mining has disturbed watercourses, and mining areas suffer from deforestation for fuelwood and poaching (Reuters 2001; Hart and Mwinyihali 2001). Other protected areas in the ecoregion include the Yangambi Reserve, situated near the Congo mainstem less than 100 km north of Kisangani, and the vast Salonga National Park, covering 36,000 km^2 of the southern portion of the central depression, including portions of several rivers (Sayer et al. 1992). There are thought to be fewer threats to the protected areas located in the central part of the ecoregion because population density is low and most of the pygmy inhabitants rely on sustainable practices for subsistence. However, the Salonga has had limited support from outside conservation organizations during the long conflict, and recent surveys have documented poachers and reported lower densities of large mammals (Hart and Mwinyihali 2001).

Future Threats

Continued political strife and deforestation pose the largest threats to the fresh waters of this ecoregion. Increased logging and mining are potential future threats if conservation management and enforcement remain inadequate. Coltan is increasingly mined in the DRC, especially in the east. Artisanal mining is easily undertaken and can destroy riparian vegetation and rivers (Redmond 2001). The possibility of political stability offers the best hope for sound environmental management in the future. However, increased mining, logging, and agricultural expansion in sensitive areas could be primary activities if careful planning and attention to environmental conservation do not accompany economic development.

Justification for Ecoregion Delineation

This ecoregion is defined by the flat central region of the Congo River from the Upper Congo Rapids [61] downstream to Lake Tumba [13]. The Cuvette Centrale is located in an ancient continental basin, invaded several times by the sea during the Mesozoic (Beadle 1981). Subsequently, an uplifting of land around the basin obstructed this drainage to the coast, forming a large endorheic lake system during the Pliocene. About 400,000 years ago, a coastal river, which is now called the Congo, cut back to capture the lake water (Beadle 1981). Most of the lake was drained, leaving the large swampy and seasonally flooded area still present today along the river in the flat basin, as well as the two major lakes, Tumba and Mai Ndombe, at the lowest part of the saucer-shaped Cuvette Centrale. Although dry phases during the Pleistocene altered the vegetation of the basin, the river system has remained stable throughout the Quaternary (Hughes and Hughes 1992). The environmental stability of the area over a long period, the range of available habitats in the bioregion, and the long isolation from other bioregions are thought to be conditions that favored the evolution of a rich biota, largely endemic to the Congo Basin (Beadle 1981).

Data Quality: Low

Ecoregion Number: **19**
Ecoregion Name: **Central West Coastal Equatorial**
Major Habitat Type: **Moist Forest Rivers**
Final Biological Distinctiveness: **Globally Outstanding**
Final Conservation Status: **Relatively Stable**
Priority Class: **III**
Author: **Victor Mamonekene**
Reviewer: **John Sullivan**

Location and General Description

In the Central West Coastal Equatorial ecoregion, coastal rivers flow through rainforests on their way to the Gulf of Guinea and the Atlantic Ocean. The main rivers are the lower Sanaga, Nyong, Ntem, Benito, and northern tributaries of the Ogooué (Abanga, Okano, and Ivindo). The ecoregion extends from the Sanaga River in Cameroon south to directly above the Gabon estuary. It extends about 400 km inland and includes parts of southern Cameroon, Equatorial Guinea, and northern Gabon.

The climate of this ecoregion is tropical and humid, with a mean annual rainfall between 1,500 and 2,700 mm. Mean annual temperature varies between 24 and 27°C. Flood regimes in the rivers of the ecoregion are bimodal, with peaks in April–June and October–December.

The topography of the ecoregion is dominated by a sedimentary coastal plain that gives way to an inland plateau. The dominant vegetation is littoral forest near the coast and a mixed mesophilous-evergreen forest inland. The estuaries support extensive mangrove forests.

The rivers of the ecoregion have a dense network of tributary streams. The Ivindo is a mostly rocky, deep river punctuated by rapids along its course through dense tropical forest. During peak flows, the rivers overspill their banks and flood the forest. This occurs especially at the head of the Ntem River basin, creating flooded swamps. The Nyong-Doume blackwater swamps are a rare habitat type located along the upper reaches of the Nyong, Doume, and Boumba rivers. Although these are separate drainages, the characteristics of these rivers are similar, and the most upstream portions connect with one another during flooding to create a vast expanse of flooded swamp. During the dry season, the flooded forest remains quite wet, with many freshwater pools. These forests have a low diversity of tree species, and swampy meadows of *Echinochloa* are found close to the rivers.

Outstanding or Distinctive Biodiversity Features

The forested rivers and streams of this ecoregion support a rich and endemic freshwater fauna. This is a very rich region with observed numbers of fish species exceeding predicted numbers for some river basins (Teugels and Guégan 1994). High numbers of fish, aquatic mollusks, amphibians, and aquatic reptiles depend on these freshwater systems for their survival.

More than 270 fish species are known from this ecoregion, 58 of which are endemic (G. Teugels, pers. comm., 2000). About one-third of the endemic fish are from the genus *Aphyosemion*. There are also several endemic genera within the Mormyridae (*Boulengeromyrus*, *Ivindomyrus*), Cichlidae (*Parananochromis*), and Clupeidae (*Thrattidion*). Characids, cichlids, rivulines, loach catfishes, claroteid catfishes, mochokid catfishes, cyprinids, gobies, and mormyrids are among the many fish families that are represented in the rich waters of this ecoregion. The Ivindo River appears to be the center of diversity of a riverine species flock of mormyrid fishes of the genus *Brienomyrus* in which most species are at present undescribed (Sullivan et al. 2002). Twelve distinct species in this genus are recorded in and near the rapids of Loa-Loa on the Ivindo River near Makokou, Gabon (C. D. Hopkins, pers. comm., 2003). Although the blackwater swamps are not well surveyed, it is known that several species of forest crocodiles (*Crocodylus cataphractus, Crocodylus niloticus,* and *Osteolaemus tetraspis*) and blackwater swamp fishes live in Nyong-Doume waters.

A stretch of approximately 40 km on the Dja River includes both the Nki Falls and the rapids of Chollet and separates coastal fish faunas from Congo fish faunas. Some waterfalls are more than 20 m high. Rapids and falls in this part of the Lower Guinean bioregion are rare. It is estimated that this area alone contains more than 150 fish species. Two of these have been found only in the region of Nki and Chutes de Chollet: an undescribed characin and an undescribed *Steatocranus* species. Intact biotas and unusual assemblages are already known to exist, although data from the most recent survey have yet to be analyzed completely.

The Monts de Cristal is a small coastal mountain range in northern Gabon that has high endemism for fish. The region is forested and at an approximate altitude of 800 m. The ecology of these mountains is similar to that of the Chaillu Massif, found further south. It is likely that the Monts de Cristal mountain streams support a small number of aquatic species, but several endemic cyprinodonts are known from them.

More than twenty aquatic mollusk species live in the ecoregion's waters. Members of the Neritidae, Ampullariidae, Planorbidae, and other families are present. There are also high numbers of frogs, with about 100 species known, including several endemic species.

Several large-bodied aquatic vertebrates, such as the West African black forest terrapin (*Pelusios niger*), West African mud turtle (*Pelusios castaneus*), and African slender-snouted crocodile (*Crocodylus cataphractus*) frequent the ecoregion's rivers. The dwarf crocodile (*Osteolaemus tetraspis*) also lives in the ecoregion, primarily in permanent pools in swamps and areas of slow-moving freshwater in the rainforests. The Atlantic hump-backed dolphin (*Sousa teuszii*) occasionally enters the lower stretches of these coastal rivers.

Status and Threats

Offshore oil exploitation and oil pollution of the estuaries are threats to the aquatic systems of this ecoregion. Logging is also a widespread threat in this forested area. Killifishes are particularly targeted for collection for the aquarium trade. A dam project on the Ntem River is awaiting financing. The Edea Dam on the Sanaga River has not generated a lake upstream, but no data are available to evaluate the effect of this dam on the ecosystem.

The ecological integrity of the Monts de Cristal is high. There is a small human population and few roads aside from the main route connecting Kougouleu with Médouneu.

The Minkébé Reserve in northeast Gabon has recently become a national park. It protects the headwaters of the Ntem River and part of the Ivindo River, a major tributary of the Ogooué. The Minkébé project is conducting research on fauna and flora. In this area of very low human population density, the managers will use the results of this research to develop a plan for rational conservation. Other new national parks in Gabon and in this ecoregion are the Ivindo, Mwagne, and Monts de Cristal.

Justification for Ecoregion Delineation

This ecoregion is defined by coastal rivers of the West Coast Equatorial bioregion (lower Sanaga, Nyong, Ntem, Benito, and northern tributaries of the Ogooué), which are characterized by similar fish species assemblages. The short stretch of falls on the Dja River separates the fish fauna characteristic of the Central West Coastal Equatorial [19] from that of the Sangha [27] ecoregion. The northern tributaries of the Ogooué River (Abanga, Okano, and Ivindo) are included in this ecoregion because their fauna is more similar to that of other rivers in the Central West Coastal Equatorial ecoregion than to the mainstem Ogooué. For example, possibly because of a 50-km stretch of rapids and falls that separates the Ivindo and Ogooué rivers and because of previous river captures, the fauna of the Ivindo is more similar to that of the Ntem than the Ogooué (Thys van den Audenaerde 1966; Sullivan et al. 2002). Questions remain concerning the evolutionary history of the Ivindo River fauna, including how it arose and why it is more similar to the Ntem River in Cameroon. During climatic fluctuations of the Pleistocene, this ecoregion may have been a refuge for ichthyofauna and other aquatic-dependent taxa of the West Coast Equatorial (or Lower Guinea) bioregion (Lévêque 1997).

Data Quality: Medium

Roman (1971) published a book on the fishes of Rio Muni (Equatorial Guinea), and Kamdem Toham (1998) and Kamdem Toham and Teugels (1998) studied the fish biodiversity of the Ntem River. Other taxa are less well studied.

Ecoregion Number:	**20**
Ecoregion Name:	**Eburneo**
Major Habitat Type:	**Moist Forest Rivers**
Biological Distinctiveness:	**Nationally Important**
Conservation Status:	**Endangered**
Priority Class:	**IV**
Authors:	**Michele Thieme and Robin Abell**
Reviewer:	**Christian Lévêque**

Location and General Description

The coastal rivers and streams of Eburneo are rich in aquatic organisms but have low levels of endemism among all taxa. Located principally in Côte d'Ivoire, these rivers traverse limited floodplains as they flow to the Atlantic Ocean.

The climate in the southern portion of this ecoregion is wet equatorial, receiving about 2,300 mm of rain per year, compared with 900 mm in the drier north (UNEP 1999). The climate in the south is characterized by two rainy seasons: a long one from May to July and a shorter one from August to September. In the north there is only one rainy season, from June to October.

The largest rivers in the ecoregion, from east to west, are the Comoé, Bandama, and Sassandra rivers. These rivers flow off of a plateau in the northern portion of the ecoregion and then continue south toward the coast. In general, the rivers have a gradual incline, and there are extensive lagoons along the coast, including the Ebrié, Tadio, and Aby lagoon complexes. The lagoons extend along 300 km of coastline and cover about 1,200 km^2 in area (Roth and Waitkuwait 1986). Floodplains are limited along this ecoregion's rivers, with the exception of the upper Bandama and upper Comoé, where seasonally flooded areas are extensive.

The original vegetation of the ecoregion reflects the level of rainfall. Sudanian savanna woodland predominates in the north and grades to a transition zone of forest-savanna mosaic and then to Guinean rainforest in the south. In the southwest and southeast of the ecoregion, inselbergs, rocky outcrops characterized by exposed ancient crystalline rock, dot the rolling landscape (Sayer et al. 1992). Much of the forest vegetation has been converted to other land uses.

Outstanding or Distinctive Biodiversity Features

Eburneo supports a high richness of aquatic mollusks. Thirty-three species are known from the ecoregion, the majority of which are snails. Four of the snails are endemic and several more are near-endemic, found also in neighboring Ashanti [17]. The fish fauna is also rich, with about 130 species, 10 of which are endemic to the ecoregion (Teugels et al. 1988).

The fresh and brackish water lagoons that dot the coastline are extremely productive and harbor abundant populations of fish and other aquatic taxa, including the vulnerable West African manatee (*Trichechus senegalensis*). Populations of manatee inhabit the Aby-Tendo-Her, the western Ebrié, and the Tadio-Nioumozou lagoon systems (Roth and Waitkuwait 1986). Rapids prevent the passage of manatees upstream in the ecoregion's rivers.

West of the Bandama River, a population of the vulnerable pygmy hippopotamus (*Hexaprotodon liberiensis*) lives along forested streams. Dwarf (*Osteolaemus tetraspis*), Nile (*Crocodylus niloticus*), and slender-snouted (*C. cataphractus*) crocodiles occur in Comoé National Park.

Status and Threats

Current Status

Cotton and sugar cane plantations occur in the northern part of the ecoregion, and coffee, cacao, palm oil, coconuts, rubber, bananas, and pineapples are grown around the coastal lagoons (Hughes and Hughes 1992). Cropland surrounds the Bandama River for almost its entire length, and sediment and pesticide runoff pollutes the river. Comoé National Park, in the northeast corner of the ecoregion, includes a portion of the headwaters of the Comoé River.

Pollution from agricultural sources is far less concentrated than the untreated industrial pollutants and human waste derived from Abidjan (Dufour et al. 1994). This city of more than 3 million inhabitants is on the edge of the Ebrié lagoon complex. A number of large companies are located in the city and have operations in food packaging, alcohol brewing, wood processing, cloth dyeing, and chemical production. One documented result of this pollution is elevated levels of heavy metals in fish inhabiting the Ebrié lagoon (Biney et al. 1994). The lagoons in this ecoregion have been severely degraded, experiencing eutrophication, anoxia, bacterial and microbiological contamination, and an overall lowering of water quality (UNEP 1999). Other threats to the coastal lagoons are increasing salinity from canals that have created new connections with the sea and extremely high fishing pressure. The Azagny National Park, which covers about 200 km^2 from the east bank of the Bandama River to the western portion of Ebrié lagoon, covers a small stretch of coastline (Hughes and Hughes 1992).

Dams on the Bandama Blanc and Sassandra rivers have created two major reservoirs, Lake Kossou and Lake Buyo, respectively. Schistosomiasis infections have increased in the vicinity of Lake Kossou since construction of the dam, presumably from the availability of more habitat for its snail host. These and several other dams along tributaries have compartmentalized the rivers by preventing fish movements up and downstream. The Bandama and Sassandra rivers are now considered reservoir rivers, whose flow regimes are tied to the energy demands of Côte d'Ivoire. The water hyacinth (*Eichhornia crassipes*) has invaded the human-made lakes of the Bandama River and several of the coastal lagoons.

Future Threats

Pollution from agricultural sources, changes in riverine hydrology and loss of connectivity caused by dams, and coastal development will continue to threaten the freshwater ecosystems of this ecoregion. Large portions of the waters of the ecoregion already face threats from agricultural and industrial pollution.

Justification for Ecoregion Delineation

This ecoregion is defined by the basins of the Comoé, Bandama, and Sassandra rivers. Although Roberts (1975) originally included this area in the Upper Guinea ichthyofaunal province, Hugueny and Lévêque (1994) determined that the fish species of this ecoregion are more similar to the Nilo-Sudanian fauna than to the Upper Guinean fauna (Roberts 1975; Hugueny and Lévêque 1994). Therefore, this ecoregion is included in the Nilo-Sudanian bioregion.

Data Quality: High

Ecoregion Number:	**21**
Ecoregion Name:	**Kasai**
Major Habitat Type:	**Moist Forest Rivers**
Final Biological Distinctiveness:	**Continentally Outstanding**
Final Conservation Status:	**Relatively Stable**
Priority Class:	**III**
Author:	**Ashley Brown**
Reviewers:	**Uli Schliewen and David Kaeuper**

Location and General Description

The occasionally torrential rivers of the Kasai Basin possess a rich aquatic fauna with high fish endemism. The Kasai River originates on the Lunda Plateau of Angola, flows through channels characterized by falls and rapids, and discharges an average of 1,200 m^3/s where it flows into the Congo River (Bailey 1986). The ecoregion largely follows the boundaries of the catchment of the Kasai River, which encompasses a vast area (900,000 km^2) in southwestern DRC and the northeast corner of Angola.

The Kasai ecoregion experiences a tropical climate, except in the more temperate uppermost Angolan highlands. Throughout the northern portion of the Kasai Basin there is one wet season and a mean annual temperature of 23°C. In the Bandundu, West Kasai, and East Kasai provinces, the daily temperature range is greater in the dry season than in the wet, with mean minima and maxima of 15°C and 32°C in July (dry season) and 19°C and 29°C in December (wet season) (Hughes and Hughes 1992). Where the headwaters of the Kasai originate in northern Angola, mean annual temperature is about 18°C. Mean annual rainfall in the northern half of the ecoregion is about 1,400–1,750 mm, whereas up to 2,000 mm/year falls over the Angolan tributaries.

The headwaters of the Kasai River begin in a series of highland massifs along the northern rim of the Lunda Plateau (with sources at elevations between 1,000 and 1,500 m a.s.l.) (Hughes and Hughes 1992). The land tends to slope gently toward the central Congo Basin, but as a result of numerous scarps located in the plateau, most of the rivers traverse falls and rapids before reaching the Cuvette Centrale (Marlier 1973; Hughes and Hughes 1992). The Cuvette Centrale is the low-lying portion of the Congo Basin that once made up part of an ancient inland lake formed during the Pliocene (Beadle 1981).

Tributaries to the Kasai originate at elevations up to 1,400 m a.s.l. in Angola and flow north into the DRC across a plateau of about 1,000 m a.s.l. Most of the Kasai River's rapids are located upstream from the city of Tshikapa at about 500 m a.s.l. (Kamdem Toham et al. 2003). Above this elevation, the upper reaches of the Kasai and its tributaries are interrupted by falls and rapids along the edge of the Lunda plateau (Marlier 1973). The rivers flow off the plateau and eventually flow into the mainstem Kasai. The Kasai descends to about 300 m a.s.l. at its lowest point in this ecoregion, and the mainstem river flows through the swamp forests of the central Congo Basin before entering the mainstem Congo River. The mainstem Kasai attains mean widths of 4 km for stretches of up to 700 km, and in its lower reaches it is studded with islands that become inundated in periods of high water. Major tributaries of the Kasai from east to west are the Kwango (Cuango), Wamba, Kwilu (Cuilo), Loange, Lulua, and Sankuru rivers.

The Kasai has a flooding regime that is unimodal in its upper reaches and becomes bimodal approaching its junction with the Congo River. The annual amplitude of flooding is about 3 m in its lower reaches (Marlier 1973). Seasonally and permanently inundated swamp forests grow in the lowest part of all river valleys in the ecoregion, in strips 100 m to 10 km wide. For example, along the Kwilu River and its tributaries there are eleven major swamps with a total area of about 1,550 km^2. Near Sandoa, there is a large wetland of 4,500 km^2 associated with the Lulua River (Hughes and Hughes 1992).

With the exception of swamp forests in the lowermost river reaches, the vegetation of the ecoregion is primarily savanna, with gallery forests lining the river valleys. A *Julbernardia-Brachystegia* complex dominates the vegetation of savannas in the south. Gallery forests are tall and dense and include species such as *Anthocleista vogelii*, *Mitragyna ciliata*, and *Phoenix reclinata* (Hughes and Hughes 1992). These forests become more extensive as the Kasai flows toward the central Congo Basin. In addition to gallery forests, extensive peat bogs exist in some valleys and shallow depressions adjacent to the rivers. Permanent swamp forests include the oil palm *Uapaca guineensis,* and in deep narrow valleys and ravines, *Mitragyna stipulosa* and *Sarcophrynium schweinfurthianum* dominate swamp forests directly adjacent to streams (Hughes and Hughes 1992).

Outstanding or Distinctive Biodiversity Features

This ecoregion has an incredibly rich fish fauna, with more than 200 known species, about one-quarter of which are endemic. However, few biological studies have been completed in recent years, and further surveys should reveal additional new species and endemics. For example, Angolan tributaries on the Lunda Plateau harbor an impressive array of endemic fishes, including the dwarf distichodontid *Dundocharax bidentatus* and one or more poeciliids, cyprinids, mormyrids, and cichlids (Poll 1967). The fish fauna is far from uniform throughout the ecoregion. Certain fish are characteristic of the savanna streams (e.g., *Aplocheilichthys* and *Hypsopanchax* species), and others are more common in the lower reaches that are surrounded by swamp forests (such as *Aphyosemion* and *Epiplatys* spp.) (Hughes and Hughes 1992).

Most rivers of this ecoregion contain rapids or waterfalls, with environments that are characterized by high levels of oxygen and habitats with low light intensity (Beadle 1981). In fish, specialization to life in rapids includes a reduction in eye size, flattened bodies, buccal suckers, paired fins, and depigmentation or blue coloration. In this ecoregion and elsewhere, long-snouted insectivorous *Campylomormyrus* species are most commonly found inhabiting rapids (Banister 1986). Other resident species include representatives of the genera *Chiloglanis* and *Euchilichthys*, which both possess ventral, sucker-like mouths with teeth aligned to rasp algae growing on rock surfaces (Banister 1986). In rapids of the lower Kasai River, specialized fish assemblages known only from a few historical specimens include the cichlids *Steatocranus rouxi* and *Teleogramma monogramma*. Interestingly, these taxa often have their closest relatives in Pool Malebo and the rapids of the Lower Congo but not in other parts of the Kasai drainage.

The Fwa River, a short (20-km) tributary of the Lubi located in the middle of the ecoregion, possesses five endemic cichlid species. Two of the endemic cichlids (*Cyclopharynx fwae* and *C. schwetzi*) belong to a genus found only in the Fwa (Roberts and Kullander 1994). The Fwa's stable environment might have facilitated speciation. Rains do not cause significant flooding in the river, and flow remains constant through the year. The river bottom, mainly fine white sand, alternates with cobblestones, boulders, and dense aquatic vegetation beds of *Vallisneria aethiopica*. The endemic cichlids dominate the fish fauna. Adult cichlids swim in open water in large numbers and do not have any predators in the river (Roberts and Kullander 1994). The five endemic cichlids all appear to be trophic specialists, with *Cyclopharynx* specializing on diatoms taken in with sand, *Thoracochromis* consuming mainly *V. aethiopica*, and *Schwetzochromis* eating filamentous algae. The young of *Thoracochromis callichromus* and *T. brauschi* are bright green and feed on and hide in *V. aethiopica*. This greenish coloration is unique among Congo cichlids and is presumably an adaptation for hiding among waterweeds (Roberts and Kullander 1994).

The Upper Sankuru and Lulua rivers in the southern and southeastern portion of the Kasai drainage support species such as *Tilapia rendalli* and *T. sparrmanii* that are otherwise found in the Zambezi drainage and southern Africa. This pattern indicates that the headwaters of the Upper Zambezi were once captured by the Kasai or that the swampy areas draining to both catchments permit species interchange.

Several lakes in the upland plateau of the Upper Kasai headwaters are at times isolated from nearby rivers and streams. These lakes include Lake Munkamba east of Kananga and Lake Gefu east of Kabinda. At least one lake contains a tilapiine cichlid (Thys van den Audenaerde 1964), which is currently synonymized with *Tilapia rendalli*. A closer inspection of additional specimens and material from other lakes may show that richness and endemism are higher than currently recognized.

Several species of large reptiles and mammals inhabit the Kasai ecoregion. These include *Crocodylus cataphractus* and *C. niloticus*, hippopotamus (*Hippopotamus amphibius*), African water rat (*Colomys goslingi*), giant otter shrew (*Potamogale velox*), and African clawless otter (*Aonyx capensis*). The clawless otter and the giant otter shrew, which feed on fish, crabs, frogs, and aquatic mollusks, both have distributions limited primarily to the Congo Basin (Kingdon 1997).

The ecoregion is also extremely rich in frogs, with about sixty species known from the Kasai. There are also about five known endemics, although both richness and endemism may be expected to increase with further research. Many frogs in the ecoregion are from the Hyperoliidae and Ranidae families.

Although little studied, the aquatic insect fauna is also thought to be rich, with many endemics. Historical collections from the rapids of the Kasai upstream from Tshikapa reveal a large number of species, including some endemics (Kamdem Toham et al. 2003).

Status and Threats

Current Status

The largest current threats to the ecoregion are civil strife and mining. In Angola, a civil war spanning 15–20 years devastated wildlife populations through intense hunting of large vertebrates (East 1999); the effects on aquatic species are undocumented. A fragile cease-fire now exists in Angola. Since 1996 civil unrest and war have continued practically unabated in the DRC. The effects of this strife have affected the economy of the DRC and the welfare of its people. Furthermore, nearly all conservation and management initiatives have been limited or halted. Although international investment in logging has been deadlocked since the beginning of the conflicts, artisanal and often illegal mining still poses a threat in the region (Pan-African News Agency 2001).

Much of the primary forest of this ecoregion has been disturbed and replaced by other vegetation. Large volumes of timber were exported during World War II, after which forest exploitation occurred along the railway from Illebo Port on the Kasai to Kananga, Kamina, Bkama, and Lubumbashi (Hughes and Hughes 1992). The Kasai Province is one of the areas particularly threatened by deforestation in the DRC (Sayer et al. 1992). According to recent reports, forest is disappearing along parts of the Kwilu and Kasai rivers (Shumway et al. 2003). In the southwest, swamp forest was undisturbed until recently, but as of 1992 clearance for rice agriculture had begun; the scale of this agriculture is unknown. *Loudetia phragmitoides* grasslands have replaced deforested areas that were once covered with periodically inundated swamp forests (Hughes and Hughes 1992). There is also heavy pressure on the forests for firewood and charcoal, which furnishes about 80 percent of energy consumption, including the majority of that used by the population in Kinshasa.

Farming is increasing, and there is rapid degradation of soils because arable lands are allowed to lie fallow only for short intervals. People are moving out of Kinshasa and back into more rural areas as the urban economy continues to falter and collapse.

The Kasai and tributaries including the Sankuru, Lubilanji, and Lulua are mined at an artisanal scale for gold and alluvial diamonds (Mino-Kahozi and Mbantshi 1997). Turbidity increases during mining as the gravel, sand, and other bottom sediments are disturbed by panning. Physical habitat alteration from mining also includes clearing of riparian vegetation in areas directly surrounding mining operations and the consequent destruction of fish spawning, nursery, and shelter areas (Mino-Kahozi and Mbantshi 1997). The cumulative impacts of these local operations on the freshwater fauna are unknown.

Most industrial-scale and international mining operations are currently on hold because of civil unrest. In the past these large-scale operations have had deleterious effects on the rivers of Kasai. Industrial mining of alluvial deposits for diamonds in the Kasai watershed has washed 6–7 million tons per year of silt, sand, and clay into the Mbuji-Mayi (Kanshi) River and its tributaries and the tributaries of the Lubilanji and Kasai rivers in the south of the ecoregion. Near the city of Mbuji-Mayi, in the middle of the ecoregion, mining processing plants discharged effluents directly into the Mbuji-Mayi River. Samples taken near the point of discharge from the Disele washery and the central plant showed a suspended matter load of 4 g/L (Mino-Kahozi and Mbantshi 1997).

Fishing in the ecoregion remains primarily artisanal. However, some fishers use toxic plant substances for fishing, a practice that can destroy fish populations, fingerlings, eggs, and other aquatic animals and plants (Mino-Kahozi and Mbantshi 1997). There was also a rise in the early 1990s in the use of chemical pesticides such as DDT and Gamulinz for fishing (D. Wilkie, pers. comm., 2000). Additionally, the widespread use of nets with small mesh sizes and fishing on spawning grounds raise concern about the potential for overfishing. Exploitation of large vertebrates, such as hippos and crocodiles, for bushmeat is causing serious declines in the populations of these animals (Shumway et al. 2003).

Industrial waste and sewage sometimes is dumped into this ecoregion's freshwater systems without treatment (UNEP 1999). In one example, caustic soda from a soft drink plant washed into the rivers adjacent to Mbuji-Mayi. Residues from sugar processing also caused localized asphyxiation of fish (Mino-Kahozi and Mbantshi 1997).

Future Threats

The southwest portion of the DRC is more densely populated than other regions of the country (East 1999). Therefore, future increases in population pose a threat, along with civil unrest, diamond and coltan mining, bushmeat hunting, and deforestation. Continuing political instability in the Kinshasa area probably would accelerate an increase in population in this ecoregion and result in increased pressure on resources. The possibility of oil exploitation in this region is also being explored (Shumway et al. 2003).

Justification for Ecoregion Delineation

The ecoregion largely follows the boundaries of the catchment of the Kasai River. The many falls and rapids along the course of the Kasai River and its tributaries restrict the movement of fish. This may explain the high level of endemism in this ecoregion (Beadle 1981). Rapids have prevented the dispersal of species to the lower reaches of rivers and headwater colonization by potentially competing species (Lévêque 1997). A few of the uppermost tributaries of the Kasai are included in the Zambezian Headwaters [76] ecoregion because of their historic connection with the rivers of that ecoregion and the Zambezian affinity of their fauna (Poll 1967).

Data Quality: Low

Ecoregion Number:	**22**
Ecoregion Name:	**Lower Congo**
Major Habitat Type:	**Moist Forest Rivers**
Final Biological Distinctiveness:	**Bioregionally Outstanding**
Final Conservation Status:	**Endangered**
Priority Class:	**IV**
Author:	**Victor Mamonekene**

Location and General Description

The Lower Congo River ecoregion extends from the mouth of the Congo River upstream to just below the Monts de Cristal rapids at Matadi and also includes left bank tributaries to the Congo River mainstem between Matadi and Kinshasa. The largest of these tributaries is the Inkisi River.

The Lower Congo River below the Monts de Cristal rapids has a discharge rate that varies between 30,000–50,000 m^3/s at low flow and 60,000–75,000 m^3/s at high flow. The river is 9.5 km wide at its mouth, and its freshwater flow continues through a deep channel into a submarine canyon offshore. The Congo River discharges an estimated 35,427,000 tons of dissolved substances into the ocean each year, and a plume of brown river water extends for more than 80 km offshore (Bailey 1986). The climate of the ecoregion is tropical and humid, with a rainy season from October to May and a dry season from June to September. The rainfall regime is bimodal, with a peak in November and another in March. Average annual rainfall is 1,888 mm, and average annual temperature is 23.8°C.

The main waterbody in this ecoregion is the Congo River, with a few large tributaries, such as the Mpozo and Lufu, entering at its left bank. The flood regime of the tributaries depends on local rains. However, when the Congo River is in full flood, it backs up the inflowing water from the tributaries and creates large pools in the tributary mouths (Mutambwe-Shango 1984). The flood regime of the Congo follows the bimodal pattern of the rains, with flood flows from October to December and from February to April.

The ecoregion extends from Banana to Matadi and is located in the southern part of the Mayombe Mountains, which run parallel to the coast all the way from northern Angola to Gabon. The vegetation is a mixture of dense Guinean-Congolian moist forest and grasslands. Downstream from Boma, there is a group of islands in the middle of the river, the largest of which is Mateba.

Outstanding or Distinctive Biodiversity Features

A brackish-water ichthyofauna occurs in the Congo River upstream to Matadi, where rapids stop brackish water species from moving further inland. The lower Congo River has a mangrove-lined estuary at its mouth, with many euryhaline fishes and freshwater representatives of marine families such as clupeids, gobies, and tetraodonts.

The ecoregion is moderately rich in fish, aquatic herpetofauna, and aquatic mollusks. About 200 fish are known from its waters, 11 of which are endemic. Mormyrids, cichlids, and cyprinids are particularly speciose. More than thirty frogs and reptiles are known from this ecoregion's fresh waters and seventeen mollusks. Five of the mollusks are endemic (*Lanistes congicus*, *L. intortus*, *Funduella incisa*, *Hydrobia luvilana*, and *Potadoma schoutendeni*).

Status and Threats

Current Status

Human population density in the Bas-Congo province of DRC is high, measured in 1990 at 42 people/km^2, with a growth rate of 2.5 percent from 1990 to 1995 (République Démocratique du Congo 1998). Clearing of the forest by burning to make way for agricultural activities has degraded the forest and created erosion. Gold mining in the rivers is common in the Mayombe Forest. Panning causes high turbidity, and sediments cover the bottom of the streams and the nests of some aquatic species. African bonytongue (*Heterotis niloticus*) has already been introduced to the waters of the Lower Congo ecoregion. The government of DRC has recognized overfishing as a problem in the Congo, resulting largely from nonselective techniques such as poisoning and the use of small-meshed seines (République Démocratique du Congo 1998).

Along the coast and estuary, mangroves have been cut, with resulting erosion and loss of nursery habitat for fish and other species. These areas are also disturbed by industrial pollution from tanker activity (République Démocratique du Congo 1998).

There is one reserve in the ecoregion and one bordering it on the coast. North of Matadi and in the ecoregion is the Luki Reserve, which is about 0.33 km^2 in extent and covers the entire basin of the Luki River, a tributary to the Congo (Lubini 1997). Near the mouth of the Congo River on the Atlantic Coast is the Parc National des Mangroves, a Ramsar site. This wetland park, which covers 660 km^2, was included on the Montreux Record of priority conservation sites in April 2000; impending threats at that time included development activity in the estuary and Moanda region, including the construction of transportation infrastructure and of a deep water harbor in Banana (Ramsar 2003).

Future Threats

Land conversion, gold mining, industrial pollution, and additional threats related to increased human populations are the main future threats in this ecoregion. The government of the DRC has recently announced plans to move forward quickly with development of the Grand Inga plant, located in the Lower Congo Rapids [60] ecoregion. Constructing this plant would involve blocking the river's flow and creating a large reservoir, which would affect the downstream ecology in this ecoregion (République Démocratique du Congo 2002).

Justification for Ecoregion Delineation

This ecoregion is defined by the lower course of the Congo River below the Monts de Cristal rapids and is characterized by a mix of brackish and freshwater species.

Data Quality: Low

Reasonably good collections exist for this ecoregion, but they need to be studied and catalogued. Mutambwe-Shango (1984) completed a hydrobiological study of the Luki Reserve, located north of Matadi.

Ecoregion Number:	**23**
Ecoregion Name:	**Madagascar Eastern Lowlands**
Major Habitat Type:	**Moist Forest Rivers**
Final Biological Distinctiveness:	**Continentally Outstanding**
Final Conservation Status:	**Critical**
Priority Class:	**II**
Author:	**John S. Sparks**

Location and General Description

The Eastern Lowlands comprise a narrow strip of land flanking Madagascar's eastern coast. Rivers flowing through the Eastern Lowlands have flat profiles compared with their steep upper reaches, have moderate to slow flows, and are often turbid (probably a result of extensive upstream deforestation). In this narrow coastal plain, rivers often meander and terminate in chains of brackish lagoons (Aldegheri 1972). Rainfall along the eastern coast of Madagascar reaches a mean high of approximately 350 mm per month on the Masoala Peninsula and in coastal areas immediately to the south, the wettest region of the island (Donque 1972). For the most part, precipitation in this ecoregion is evenly distributed throughout the year (Donque 1972). The substrate in most rivers ranges from coarse gravel to sand or mud, and the water is warm compared with middle and upper reaches of the same rivers.

Outstanding or Distinctive Biodiversity Features

Lowland *Pandanus* and palm swamps, as well as numerous estuarine habitats along the eastern coast, continue to support native fish faunas. A single *Pandanus* and palm swamp in Manombo Special Reserve, on the southeast coast, is the only remaining habitat of *Pantodon* sp., a small, endemic cyprinodontiform fish of uncertain familial affinity (Sparks 2003). Before the discovery of the southeastern Manombo population, which represents a distinct species (Reinthal and Sparks, unpublished data), the cogener *P. madagascariensis* had been collected only from a single coastal location to the north of Tamatave (a population and species now considered to be extinct).

The lower reaches of eastern drainages are more ichthyofaunally diverse than headwater regions. In terms of endemic taxa, these lower-elevation communities are rich primarily in *Bedotia*, but members of the endemic cichlid genera *Paretroplus*, *Paratilapia*, and *Ptychochromis* are also present.

Madagascar's eastern rivers and streams are also home to six crayfish species in the endemic genus *Astacoides* (Benstead et al. 2003). Recent studies have also revealed tens of new aquatic insect species, indicating that diversity and endemism of aquatic insects is high in this region (Benstead et al. 2000, 2003). For example, among the Trichoptera a relationship has been established between the eastern rainforest stream habitats and microendemism. The highest species diversity for mayflies on the island is also found in the eastern rainforests, with the number of described species inhabiting Madagascar increasing from 15 in the early 1990s to 100 in early 2002 (Benstead et al. 2003). Frogs are also abundant in this ecoregion, with about half of the seventy species being endemic. Several wetlands and forested streams along the coast are also important sites for waterbirds, including *Glareola ocularis*, *Phoenicopterus ruber*, the near-threatened *Lophotibis cristata*, the vulnerable *Ardeola idea*, and the vulnerable *Tachybaptus pelzelnii* (Project ZICOMA 2001; IUCN 2002).

Status and Threats

Rivers flowing through eastern coastal regions probably represent the most disturbed and degraded freshwater habitats on Madagascar because of extensive development and deforestation throughout the ecoregion and, to a lesser extent, in the upstream Madagascar Eastern Highlands [41]. Fishing pressure is high in many areas, as is habitat modification associated with agricultural practices. Erosion resulting from widespread deforestation has led to increased turbidity in the lower stretches of most eastern rivers. Very few coastal areas with intact forest remain, although exceptions include the Masoala Peninsula and a few small forest parcels in Manombo Special Reserve. Fish communities of lower eastern drainages are dominated by exotics, primarily Tilapiine cichlids. The major rivers flowing through this ecoregion (i.e., the Mangoro, Mananara, Rianila, Maningory, and Mananjary) have their origin in the Madagascar Eastern Highlands [41]. A number of these rivers remain pristine in their uppermost reaches, although some suffer from increased turbidity as a result of deforestation.

Justification for Ecoregion Delineation

The upper and lower eastern basins traditionally have been considered a single hydrographic region. Based on the results of recent ichthyofaunal surveys throughout eastern Madagascar, however, an elevational disjunct in species distributions has been documented (Reinthal and Stiassny 1991; Sparks and Stiassny 2003). Sparks and Stiassny (2003) recorded a total of sixty-nine native fish species from basins in the Eastern Lowlands and fifty-one species from the Eastern Highlands. In total, fifty-seven native species are restricted to the Madagascar Eastern Highlands [41] and the lowlands combined, whereas forty-eight native species are limited to western basins. Thus, eastern drainages are richer than western basins. In the eastern basins, the fish assemblages present in highland and lowland habitats differ considerably (Sparks and Stiassny 2003). Taking into account the substantial differences in their respective ichthyofaunal assemblages, the upper and lower eastern basins have been split at the edge of the coastal plain.

Data Quality: Medium

Of Madagascar's five major aquatic ecoregions, the Madagascar Eastern Highlands [41] and Madagascar Eastern Lowlands are certainly the best studied. This is mainly because a number of large towns and a network of regularly maintained roads along most of the eastern coast permit access to these basins.

Ecoregion Number:	**24**
Ecoregion Name:	**Malebo Pool**
Major Habitat Type:	**Moist Forest Rivers**
Final Biological Distinctiveness:	**Bioregionally Outstanding**
Final Conservation Status:	**Endangered**
Priority Class:	**IV**
Author:	**Victor Mamonekene**

Location and General Description

A rock-sill barrier in the channel of the Congo River causes it to expand into Malebo Pool directly above Kinshasa. A largely lentic-adapted fauna inhabits the pool that spans the border of the Republic of the Congo (ROC) and the DRC.

Malebo Pool is located between Douvre cliffs upstream and the Kintambo rapids downstream. It is about 35 km long and 25 km wide and seldom exceeds 10 m in depth. Water levels fluctuate by about 3 m within the year. Water moves quickly through the pool as it makes its way toward the ensuing rapids.

Many small islands are located in Malebo Pool; the largest is called Ile Mbamou, and it divides the mainstream into two channels. Extensive palm and papyrus swamps occur along its edges, and there is frequent passage of floating mats of *Eichhornia* (Bailey 1986).

The climate near Malebo Pool belongs to the Guineo-Equatorial Bas-Congolais type. It is characterized by a dry season from June to September and a rainy season from October to May, although there is a break in rainfall in January and February. High levels of precipitation occur from mid-November to mid-December and from March to April. These monsoonal rains originate from the Gulf of Guinea and Benguela current. Mean annual rainfall recorded at Brazzaville (1951–1960) is 1,370 mm. Mean monthly temperature ranges from minima of 17–22°C to maxima of 27–32°C.

Malebo Pool experiences two flooding events because of its location at the equator and the flooding of northern and southern tributaries of the Congo River at different times of the year. The first and smaller flood occurs from April to June and the second, larger one from October to January. The average annual flow of the Congo River at Pool Malebo is 30,000 m^3/s during the low water period and 60,000 m^3/s during the flooding season. In 1961 and 1962, when there were heavy rains in the basin, the flow reached 80,000 m^3/s.

According to the Cahen hypothesis, reported by Roberts and Stewart (1976), the Congo Basin is thought to have been an interior drainage throughout the Miocene and most of Pliocene, perhaps forming an interior lake. Then, in the late Pliocene or early Pleistocene, a Lower Guinean coastal river captured the Malebo Pool, and the Congo Basin drained to the coast. The freshwater fauna of Malebo Pool, particularly the fishes, is lentic adapted (*Citharinus* spp., *Distichodus* spp., *Tilapia* spp.), although there are some rheophilic (*Nannochromis nudiceps, Steatocranus gibbiceps, S. casuarius, S. tinanti*) forms from the downstream rapids (Poll 1939).

Outstanding or Distinctive Biodiversity Features

This ecoregion is rich in fish species and thought to be rich in aquatic insects but low in richness and endemism for other aquatic taxa. More than 200 fish species are documented from the ecoregion. Mormyrids (more than 40) are the most abundant and show the highest diversification, followed by the mochokids (about 25) and the citharinids (more than 20). Most endemic fishes are catfishes, with four endemic Amphiliidae, one Clariidae, and three Mochokidae, although several of these are known only from their type locality. Among endemic fish are the mountain catfish, *Leptoglanis mandevillei, L. brieni,* and *L. bouilloni* and an upside-down catfish, *Atopochilus chabanaudi*.

Only ten frogs and eight aquatic reptiles are known from the ecoregion. Among these are the African keeled mud turtle (*Pelusios carinatus*), Central African giant mud turtle (*Pelusios chapini*), and dwarf crocodile (*Osteolaemus tetraspis*).

Status and Threats

Current Status

Malebo Pool supports a large fishery that provides fish for local consumption in Kinshasa and Brazzaville. Pool Malebo is also a center of boat traffic because of its many wharfs, including Brazzaville and Kinshasa main harbors.

Future Threats

Increasing urbanization and agricultural activity on Mbamou Island and the mainland are causing an increase in erosion and runoff of sediments into Malebo Pool from the surrounding hillsides. Human activities in Brazzaville and Kinshasa contribute to the flow of organic and inorganic pollutants into the Congo River, polluting the river's waters. The demand for fish products in the markets of the two towns fringing Malebo Pool and the selective catch of aquarium fishes constitute the main threats.

Justification for Ecoregion Delineation

This ecoregion is defined by the extent of Malebo Pool directly above Kinshasa. A largely lentic-adapted freshwater fauna inhabits the pool.

Data Quality: Low

Available data on the aquatic fauna of this ecoregion are old. Poll (1939, 1959) has published on the ichthyological fauna of the Malebo Pool, and Sita (1980) has made a study of the vegetation. However, since those studies, there have been few studies on the aquatic flora and fauna reported from this ecoregion.

Ecoregion Number:	**25**
Ecoregion Name:	**Northern Upper Guinea**
Major Habitat Type:	**Moist Forest Rivers**
Biological Distinctiveness:	**Continentally Outstanding**
Conservation Status:	**Endangered**
Priority Class:	**I**
Authors:	Ashley Brown and Michele Thieme
Reviewers:	Christian Lévêque, Emmanuel Williams, and Samba Diallo

Location and General Description

The Northern Upper Guinea ecoregion lies on the western side of the Guinean range, extending from the foothills of the Fouta Djalon in Guinea southeast to Sierra Leone's southern border. The ecoregion also encompasses a small portion of Guinea-Bissau and Liberia. This ecoregion hosts a rich aquatic fauna with high endemism among fish, mollusks, amphibians, and crabs. Together with Southern Upper Guinea [28], Fouta Djalon [40], and Mount Nimba [42], this ecoregion forms the Upper Guinean bioregion, which has a distinct fish fauna (Lévêque 1997).

The short rivers of the ecoregion descend from the Guinean Dorsale and cross the coastal plain adjacent to the Atlantic Ocean. The rivers begin at elevations of around 500 m a.s.l. (and as high as 1,946 m a.s.l. at Mt. Bintumani in the Loma Mountains) (Hughes and Hughes 1992). Moving west, the gradient decreases and the landscape changes from undulating foothills to a coastal plain where riverine and floodplain lakes are common. Almost all of the coastal rivers have a torrential flow regime because of their steep downward slopes and rocky bottoms (DNE et al. 1999).

The main rivers are the Coliba (Tominé and Komba), Kogon, Tinguilinta, Fatala, Konkouré, Kolenté, Kaba, and Mongo rivers. The Tominé and Komba rivers join to form the Coliba River. The Coliba (basin size 17,807 km^2) drains the northwest side of the Fouta Djalon. It is 407 km long and flows into the Atlantic Ocean via an estuary shared with the Geba River in Guinea-Bissau. The Kogon River (basin size 7,288 km^2) flows northwest toward the border with Guinea-Bissau (379 km) and then flows along the border in a southwest direction until it empties into the estuary, Rio Komponi adjacent to the Atlantic Ocean. The Tinguilinta River (basin size 4,858 km^2) is 160 km long and flows through the Boké region, meeting the Atlantic Ocean at the Rio Nunez estuary in Kamsar. The Fatala River (basin size 692 km^2) is 205 km long, takes its source in the Fria region, flows through the Boffa region, and reaches the Atlantic ocean at Rio Pongo. The Konkouré River (basin size 17,046 km^2) receives the Kakrima River from the Labe Plateau and many other tributaries along its 139-km length. The Kolenté River (basin size 5,170 km^2) flows through the Kindia and Forecariah regions. It has a gentle slope throughout most of its 210-km length, and it meets the Atlantic Ocean in Sierra Leone. The Kaba (basin size 5,427 km^2) is 91 km long. It is formed by the Kaba and Mongo rivers coming from the Mamou region. It flows down to the Atlantic Ocean through Sierra Leone (Samoura et al. 1999).

The climate of the ecoregion is tropical and wet, with rainfall influenced by the moist southwest trade winds (Hughes and Hughes 1992). The ecoregion receives heavy but seasonal precipitation, with a mean annual rainfall of 3,000–5,000 mm at the coast and 2,000 mm inland (Sayer et al. 1992). The wet season lasts 5–6 months in the middle of the ecoregion near Conakry, with most of the yearly rainfall occurring between May and November. Similarly, in the Fouta Djalon highlands, the rainy season lasts for approximately 6 months. Even in the rainy season, precipitation is concentrated, with an average of 2,493 mm per year falling between August and September (Hughes and Hughes 1992). The dry season lasts at least 5 months.

According to Bazzo (2000), this ecoregion is formed by the raised edge of the Fouta Djalon plateau. The highest elevations have poor soils called "Bowe" with scrub and grass vegetation. This raised edge of the plateau has numerous rocky outcrops. Sandstone soils support relict mesophil forests. These include the species *Ceiba pentandra, Daniella oliveri, Parkia biglobosa, Bombax costatum, Khaya senegalensis, Erythrophleum guineense,*

Milicia excelsa, Terminalia ivorensis, and *Afzelia africana.* The lower elevations contain more iron and have deeper soils that affect the vegetation cover. Broad-leaved lowland forest covers the upland coastal plain, with seasonally inundated grasslands and riverine forests occurring along rivers, particularly in the southwestern portion of the ecoregion. Many secondary species occur here, such as *Hymenocardia acida, Daniella oliveri, Cordyla pinnata, Pterocarpus erinaceus, Prosopia africana, Imperata cylindrica, Andropogon gayanus,* and *Penisetum* sp. Floodplain lakes, which support beds of floating and submerged aquatic macrophytes, also occur in the coastal plain, and these lakes are surrounded by extensive tracts of swamp forest (Hughes and Hughes 1992).

Mangroves backed by freshwater swamp forests grow along most of the coast, especially along many of the riverine estuaries. Sediments brought downstream by the rivers constitute the soils of these swamps. Species such as *Pterocarpus santalinoides, Napoleonaea vogelii,* and *Mitragyna stipulosa* can be found in the swamp forests behind the mangroves (Sayer et al. 1992). In Guinea, mangrove forests cover almost the entire coastline for about 300 km. In 1965, the area of mangroves was estimated to be about 3,500 km^2, but today it is only 2,960 km^2, with a destruction rate of 4.5 km^2 or 4.2 percent per year. According to Rouanet (1957), cited in DNE et al. (1999), the Guinean mangroves covered 4,000 km^2 in 1957, so these mangroves have been reduced by more than 25 percent in 40 years. Seven mangrove species occur here: *Rhizophora mangle, Rhizophora racemosa, Avicenia africana, Drepanocarpus erectus, Banisteria leona, Conocarpus erectus,* and *Laguncularia racemosa* (Uschakov 1970; Matthes 1993; Diallo 1995). Other vegetation includes *Dalbergia, Sesuvium portulacastrum,* and *Paspalum vaginatum.*

Outstanding or Distinctive Biodiversity Features

The forested coastal streams and rivers of Upper Guinea support a diverse and endemic aquatic fauna (Lévêque et al. 1989). About 28 percent of the 160 fish species are endemic. Species from the Cyprinodontidae (*Aphyosemion* and *Epiplatys*) and Cyprinidae (*Barbus*) families dominate the endemics. There are also several endemic fish from the Mochokidae, Mormyridae, Claroteidae, and Cichlidae families. Ten endemic frogs, four endemic freshwater crabs, at least two endemic dragonflies (*Argiagrion leoninum* and *Allorhizucha campioni*) and five endemic mollusks also live in these waters.

The Konkoure River is one of the richest among the Atlantic basins, supporting eighty-eight fish species. About 20 percent of the Konkoure fish species are not shared with any other Upper Guinean basin. According to Daget (1962), even the faunistic barriers of the rapids that divide up the river could not entirely explain this level of endemism.

The endemic fishes are generally small-bodied and adapted to the swift currents and clear waters of the ecoregion (see ecoregion [28] for more detailed descriptions of adaptations). One-quarter of the endemic fish are rivulines, some of which are annuals. During the wet season, these annuals lay their eggs in the soil of temporary floodplain pools that desiccate in the dry season. These eggs hatch with the inundation of floodwaters in the rainy season (Lévêque et al. 1992).

Several nesting and overwintering birds can be found in the coastal floodplain, especially in the mangroves. Mangrove forests provide a variety of food sources for waterbirds such as fish, insects, and shellfish (Hughes and Hughes 1992). Species such as the great egret (*Ardea alba*), cattle egret (*Bubulcus ibis*), green-backed heron (*Butorides striatus*), purple heron (*Ardea purpurea*), pygmy goose (*Nettapus auritus*), and white-faced whistling duck (*Dendrocygna viduata*) nest in the freshwater and brackish swamps near the mouth of the Nunnez and Koumba rivers (Wetlands International 2002). At the mouth of Rio Pongo, the yellow-billed stork (*Mycteria ibis*), wooly-necked stork (*Ciconia episcopus*), and Goliath heron (*Ardea goliath*) breed. About 40 km north of Conakry on the coast, the Ramsar site Delta du Konkoure hosts large numbers of wintering Palearctic migrants (Wetlands International 2002).

The ecoregion also supports a variety of large aquatic reptiles and mammals. All three species of African crocodiles—the slender-snouted (*Crocodylus cataphractus*), Nile (*Crocodylus niloticus*), and dwarf (*Osteolaemus tetraspis*) crocodile—have historically inhabited the riverine floodplains and swamps of this ecoregion. The vulnerable West African manatee (*Trichechus senegalensis*), the common hippopotamus (*Hippopotamus amphibius*), and the vulnerable pygmy hippopotamus (*Hexaprotodon liberiensis*) also live in the ecoregion (IUCN 2002). Important areas for the pygmy hippopotamus include the Moa River around Tiwai Island and the Mahoi River in the Gola Forest (Sayer et al. 1992).

Status and Threats

Current Status

Deforestation and pollution from mining, agriculture, and industry threaten the rivers, small lakes, and wetlands of Northern Upper Guinea. Civil strife in this region has also affected the environment and complicated conservation efforts.

As population levels grow, human encroachment into forests, especially mangroves, also increases. The population of Guinea is growing at an estimated rate of 1.95 percent per year, and the population of Sierra Leone is growing at an estimated 3.67 percent per year (CIA 2000). People have cleared more than 97 percent of the terrestrial high forest in the southeastern portion of the ecoregion in Sierra Leone (Hughes and Hughes 1992). Much of the high forest has been replaced by secondary regenerating forest, called farmbush, which is cut periodically using a slash-and-burn farming system. Deforestation and overfishing of mangrove areas are serious concerns because they are important breeding and feeding habitats for aquatic fauna (Hughes and Hughes 1992). Mangroves are cleared for agriculture (often rice culture), fuelwood, timber, building poles, and fishtrap materials. Salt extraction involves clearing mangrove forests to make new extraction sites. As of 1980, 16 percent of the mangrove forests that once covered most of the coast had been cleared (Hughes and Hughes 1992). In coastal wetlands, rice farmers and fishers consider the West African manatee to be a pest and consequently hunt it heavily.

Pollution from industry, agriculture, and mining is becoming an increasing threat. Currently little processing of sewage and industrial waste occurs (Institute of Marine Sciences et al. 1998; UNEP 2000b). For example, the production of toxic effluents is about 8,000 m^3 per day in Conakry, of which only about 2,900 m^3 is processed before being discharged into the sea (Camara et al. 1999). Toxic effluents from industry and mining have polluted several rivers and bays near Conakry. Pollution is also concentrated near Kamsar mining sites (on the west coast) and Friguia-Kimbo, where red mud discharges into the Konkouré River (Institute of Marine Sciences et al. 1998). Additionally, Guinea is the world's second largest bauxite producer (CIA 2001). The dust coming from the processing of the bauxite in the Kamsar port covers all of the southern Boké areas and is washed into the rivers and coastal waters at the beginning of the rainy season. Agrochemical waste pollutes the rivers of the ecoregion, most heavily in irrigated areas such as the mouth of the Konkouré and the Fatala Delta. Sewage from domestic and industrial sources often goes through no processing facilities.

Years of civil war have destroyed the infrastructure and economy of Sierra Leone. Civil strife impedes food transport from relief agencies into affected areas and interrupts planting, putting pressure on natural resources (FAO 2000b). In Guinea the number of refugees is about 600,000 people. Civil strife also severely limited the conservation efforts that had begun before the conflict erupted (East 1999). In addition, the civil conflict caused most mining to be shut down in Sierra Leone, including bauxite and rutile mines. Diamond mining and smuggling is still ongoing, however (CIA 2001).

There are several Ramsar sites in Guinea and one Ramsar site, a number of forest reserves, and two proposed national parks in Sierra Leone, although enforcing laws often is difficult. Ramsar sites include Ile Alcatraz (along two offshore islands), Iles Tristao (at the mouth of the Kogon River), Rio Kapatchez, Rio Pongo, Ile Blance, and Delta du Konkouré. The largest of these sites, at 900 km^2, is the Delta du Konkouré, and the smallest is Ile Alcatraz at 0.01 km^2 (Wetlands International 2002). The Otamba-Kilimi proposed national park in the northwestern portion of the ecoregion protects several grassland floodplains, and the Loma Mountains proposed national park covers the mountain grasslands and forests of this mountain. The Mamunta-Mayoso Swamp Nature Reserve, near Makeni, covers swampland, but as of 1992 only local tribal chiefs had ratified this reserve (Hughes and Hughes 1992). The Gola Forest Reserves contain the largest areas of forest in the ecoregion and are linked to the tiny Tiwai Island Wildlife Sanctuary.

Future Threats

This ecoregion's highly endemic fauna faces a number of impending threats. Further deforestation of mangroves is of particular concern. Plans for future mines forebode increased direct pollution as well as deforestation of surrounding areas for fuel, construction materials, and agricultural conversion (Sayer et al. 1992). Continued political instability and associated displacement of human populations threaten the ecoregion.

Justification for Ecoregion Delineation

This ecoregion is defined by the basins of the Coliba (Tominé and Komba), Kogon, Tinguilinta, Fatala, Konkouré, Kolenté, Kaba, and Mongo rivers and is characterized by an endemic freshwater fauna. The high level of endemism among fishes in this ecoregion is postulated to be the result of isolation over time (Lévêque 1997). The Guinean range is an impassable barrier to the dispersal of fish from the Upper Guinean streams to the basins to the northeast, such as the Niger. Rapids and waterfalls in individual basins probably have served as additional barriers. Different habitats potentially contributed to the divergence as well; for example, forested streams characterize the Guinean region, whereas savanna streams predominate in the Nilo-Sudan ichthyofaunal province (Hugueny and Lévêque 1994). Another hypothesis for the high endemism in the bioregion is that the rivers and streams of this forested area acted as a refuge during dry climatic periods (Hugueny and Lévêque 1994).

Data Quality: High

The fish fauna of this ecoregion is fairly well investigated, although new species are discovered occasionally. The comprehensive volumes of *The Fresh and Brackish Water Fishes of West Africa* (Lévêque et al. 1990; Lévêque et al. 1992) cover this ecoregion. In contrast, the aquatic invertebrates are poorly known.

Ecoregion Number:	**26**
Ecoregion Name:	**Northern West Coastal Equatorial**
Major Habitat Type:	**Moist Forest Rivers**
Final Biological Distinctiveness:	**Globally Outstanding**
Final Conservation Status:	**Endangered**
Priority Class:	**I**
Author:	**Victor Mamonekene**
Reviewers:	**Angus Gascoigne and Robert C. Drewes**

Location and General Description

This ecoregion encompasses the coastal rivers and streams that feed the Gulf of Guinea, from the Cross River in Nigeria to the Bay of Cameroon in Cameroon and the fresh waters of the island of Bioko, which is part of Equatorial Guinea. The Cross, Ndian, and Meme rivers are the main waterways draining to the Rio del Rey estuary, and the Mungo, Nkam-Wouri, Doula, and Dibamba rivers drain to the Cameroon estuary. Most of these rivers originate in the Cameroonian highlands. Bioko has numerous fast-flowing rivers that radiate from the peaks of Pico de Basilé and the Southern Highlands. These rivers, streams, and estuaries support a rich aquatic fauna.

The ecoregion experiences a tropical, hot-humid seasonal climate. There is one rainy season, with the majority of rainfall occurring between July and September, and a dry season that typically lasts from November to February, peaking in December and January. However, some precipitation generally falls in the dry season. The annual temperature varies between 28°C and 30°C.

The larger rivers all experience a unimodal flood regime, coinciding with the rainy season. The mean annual discharge for the largest river, the Cross, is 173–913 m^3/s (Teugels et al. 1992).

Bioko, the largest island in the Gulf of Guinea, is part of the volcanic chain that includes Mt. Cameroon, São Tomé, Príncipe, and Annobón. It is dominated by three volcanic peaks: Pico de Basilé, which is the highest at 3,011 m, Gran Caldera de Luba (2,261 m), and Pico Biao (2,010 m). Pico de Basilé is the second highest mountain in western Central Africa, after Mt. Cameroon. Bioko is approximately 40 km offshore of Cameroon and has a total land area of 2,020 km^2 (Sunderland and Tako 1999). Rainfall is very high, with the heaviest in the southwest, reaching more than 10 m per year (Castelo 1994).

The ecoregion is located in an evergreen rainforest zone and includes parts of the Cameroonian highlands and mountain forests on Bioko. Reedbeds grow along the rivers, and mangroves occur in the estuaries; both of these vegetation types shade many of the smaller rivers and tributaries, limiting light exposure. The water of the Cross River has low conductivity (less than 25 μS/cm), low mineral content, and a low pH (less than 7.5). However, high conductivity values of 60–170 μS/cm have been documented in the Mungo River, and pH has been recorded there of 7.6–7.8.

The mangroves in this ecoregion's estuaries are extensive. For example, the Rio del Rey estuary contains about 1,500 km^2 of mangroves. These highly productive habitats are breeding grounds and nursery areas for crustaceans and fish (Gabche and Smith 2000).

Outstanding or Distinctive Biodiversity Features

Rivers in this ecoregion are particularly rich aquatically, with more freshwater fish species than those with similar drainage areas elsewhere in West Africa. More than 200 fish species inhabit these rivers, and about 130 frog species depend on the waters of this ecoregion for their survival. About 40 fish are considered near or strict endemics, and according to Teugels et al. (1992), 11 of the 132 freshwater fish species that occur in the Cross River basin probably are endemic. Many of the endemic fish are rivulines (genus *Aphyosemion*). These species are adapted for life in small, shaded streams in forests and swamps, and many feed on invertebrates that fall into the water from overhanging trees. About one-quarter (about thirty) of the frog species are estimated to be endemic. Many of them are not strict endemics but have small distributions that include parts of other ecoregions, including the Lower Niger-Benue [65] and Western Equatorial Crater Lakes [5].

Several marine or brackish water fish species occur in the lagoons and estuaries, including *Elops lacerta*, *Ethmalosa fimbriata*, *Sardinella* spp., *Pomadasys jubelini*, and *Caranx hyppos*.

On Bioko one species of the genus *Aphyosemion* (*A. oeseri*) is considered endemic to the northwest of the island and therefore must be considered severely threatened with extinction (Castelo 1994). *Barbus thysi* is also endemic to the island, and *A. volcanum* is near-endemic, occurring on Bioko and in streams on Mount Cameroon. A total of forty-eight dragonfly species (*Odonata*) have been recorded from Bioko (Dijkstra 2002). Twelve species are endemic to the region, of which four are found only on Bioko (Pinhey 1974).

The amphibian fauna of Bioko is poorly understood, in part because of poor sampling but also because of a lack of clarity and consensus on the relationships and taxonomy of African amphibians in general. A 1965 study by Robert Mertens listed one endemic caecilian and thirty species of five families of frogs, none endemic. A recent California Academy of Sciences (CAS) expedition (1998) confirmed the presence of at least twenty-seven frog species, including two from the vicinity of the Rio Iladyi in the Moca Valley (central highlands) that are undescribed (in the genera *Cardioglossa* and *Hyperolius*) and possibly endemic. Areas of greatest amphibian richness appear to be the well-watered areas of Moka Plateau (Valley), the southern highlands at elevations above 1,000 m (slopes of Gran Caldera de San Carlos and Lake Biao), and the forest slopes of Pico Basile below 500 m (the northern highlands; the higher reaches of Pico Basile are depauperate). Of the twenty-seven species confirmed by the recent CAS expedition, twenty-one need freshwater for reproduction, including the two undescribed, possible endemics; the rest reproduce by direct development.

Aquatic mammals inhabiting the ecoregion include the giant otter shrew (*Potamogale velox*), African water rat (*Colomys goslingi*), spot-necked otter (*Lutra maculicollis*), hippopotamus (*Hippopotamus amphibius*), and African clawless otter (*Aonyx capensis*). The vulnerable manatee (*Trichechus senegalensis*) also occurs in the ecoregion, sometimes ascending the rivers for considerable distances.

Status and Threats

Current Status

Logging activity around protected areas threatens the integrity of aquatic systems. The cities of Douala and Calabar are a source of pollution, especially in the Wouri and Mungo estuaries, respectively. On Bioko, the cities of Malabo and Luba also are sources of pollution. Pollution from large-scale agriculture affects the water quality of rivers such as the Wouri and Mungo, but no data exist to quantify the level of pollution (Republic of Cameroon 1999). In northern Bioko, cocoa agriculture dominates the landscape, and nearly all of the primary forest has been removed (Castelo 1994). Reid (1989) did not document overfishing in the Korup but states that data are limited. Poisons are commonly used to fish in Cameroon's interior waters, as are nets of small mesh sizes (Republic of Cameroon 1999). High fishing pressure and the use of toxic products such as lime and insecticides occur along the northern and western streams of Bioko (Castelo 1994). Declines in fish catches in the ecoregion's estuaries may be partially attributable to agricultural, industrial, and municipal pollution as well as to overfishing (Republic of Cameroon 1999; Gabche and Smith 2000).

Mangroves in this ecoregion have suffered from high levels of deforestation. In its national biodiversity action plan, the Republic of Cameroon identifies the Wouri River area as having undergone particularly substantial deforestation. Residents of fishing camps and coastal villages use the wood for fuel, fish smoking, home construction, tannin extraction, and traditional medicine. Unimpeded cutting of mangroves has occurred in Cameroon as a result of a lack of regulations (Republic of Cameroon 1999).

There are two protected areas in this ecoregion: Cross River National Park (Nigeria) and Korup National Park (Cameroon). The 4,000-km^2 Cross River National Park, created from Bashi-Okwango and Oban forest reserves, contains the largest remaining rainforest tract in Nigeria and is crossed by several tributaries to the Cross River. The World Conservation Monitoring Centre reported in 1985 that the Oban portion of the park was threatened by overgrazing on steep slopes, which contributed to erosion and the decline of water quality in rivers and streams (WCMC 1985; see Caldecott 1996 for a full review). The park has also been threatened by commercial logging (WWF International 1996). The 840-km^2 Korup National Park is crossed by the N'dian, Cross, and Akpa Korup rivers.

Future Threats

The continuation of logging activity and the intensification of agriculture are the major future threats. The burgeoning needs of a growing human population in the ecoregion will put pressure on the freshwater resources into the future.

Justification for Ecoregion Delineation

This ecoregion is defined by the coastal rivers and streams that feed the Gulf of Guinea, from the Cross River in Nigeria to the Bay of Cameroon in Cameroon and the fresh waters of the island of Bioko. This region acted as a refuge during the last ice age. Today, many species of trees and vertebrates occur only in this ecoregion, particularly in Korup and Cross River national parks. This ecoregion has part of its ichthyofauna in common with the Nilo-Sudan and Congo bioregions, in addition to possessing several of its own endemic species (Reid 1989). Bioko was connected to the continent some 10,000–12,000 years ago, and the island shares a majority of its fish species with those of the mainland (A. Kamdem Toham, pers. comm., 2000).

Data Quality: High

Data quality is considered high for most taxa, except for aquatic insects. Reid (1989) conducted a hydrobiological survey of Korup rainforest, and Teugels et al. (1992) published data on fish diversity of the Cross River Basin.

Ecoregion Number:	**27**
Ecoregion Name:	Sangha
Major Habitat Type:	Moist Forest Rivers
Final Biological Distinctiveness:	Nationally Important
Final Conservation Status:	Relatively Stable
Priority Class:	V
Authors:	Emily Peck and Michele Thieme
Reviewer:	Uli Schliewen

Location and General Description

The rivers and streams of the Sangha ecoregion run through some of the densest forests in the Afrotropical realm and support an impressive array of aquatic species. Extensive closed-canopy forests and large areas of natural savanna woodland occur in the north and central portion of the ecoregion. This landscape grades into the inundated floodplains, permanent swamp forests, and swamps of the Cuvette Centrale in the south (Hughes and Hughes 1992; Sayer et al. 1992).

The Sangha ecoregion spans the borders of three Central African countries: Cameroon, Central African Republic (CAR), and ROC. It encompasses the Sangha River basin, extending from the headwaters of the Mambéré River in the north on the Cameroon-CAR border, almost down to Malebo Pool in southern ROC. The watershed for the Congo Basin, roughly following the Congo-

Gabon border, defines the western boundary of the ecoregion, with the Massif du Chaillu marking the southwestern edge and the headwaters of the Boumba River marking the northwestern extent (Hughes and Hughes 1992; Sayer et al. 1992).

The Sangha River and its tributaries travel through four principal climatic zones. The northernmost climatic zone (6–8°N) receives a mean annual rainfall of 1,500 mm, with peak rainfall occurring in August and a dry season occurring from December to January. The second climatic zone is located 1–6°N and has an annual rainfall of approximately 1,700 mm. The wet season extends for several months, with most rainfall during August, which is also the coolest month. Here, too, a dry season exists during December and January. The mean monthly temperature for this region ranges from 13°C, recorded in January, to 40°C, recorded in March. The central portion of the ecoregion, between 1°N and 2°S, experiences a true equatorial climate. There is no dry season and little variation throughout the year from the mean annual temperature of 25°C. The mean annual precipitation ranges from 1,700 to 2,000 mm. Further south, the Baleke Plateau in south-central ROC has a subtropical climate with a mean annual rainfall of 1,700 mm, ranging from 1,300 mm/year along the Congo River to 2,200 mm/year in the hills along the border of Gabon. The mean annual temperature in this climatic region ranges from 18°C to 32°C. A distinct dry season exists from June to September and a wet season from October to May (Hughes and Hughes 1992; Sayer et al. 1992).

The headwaters of the Sangha include the Kadey River and its tributaries. The Kadey joins the Sangha in the southernmost part of CAR, along the border with Cameroon. Further south, the Ngoko River flows along the Cameroon-ROC border to converge with the Sangha at Ouesso, ROC. The Sangha River then flows south through the Cuvette Centrale, eventually draining into the mainstem Congo River. Several additional rivers, all of which eventually flow into the mainstem Congo River, drain the central and southern parts of the ecoregion. These include the Likouala River and its tributary the Kouyou, as well as the Alima, Nkeni, and Lefini rivers (Hughes and Hughes 1992).

Permanent swamps, periodically inundated swamp forest, large expanses of dense evergreen and deciduous rainforest, and savanna characterize the terrestrial landscape of the ecoregion. Savanna occurs primarily in the north of the ecoregion where the headwaters of the Sangha River begin. Evergreen rainforest grades into swamp forests as the eastward-flowing rivers of this ecoregion descend in elevation. The riparian swamp forests of the upper Sangha and its tributaries range in width from a few meters up to 7 km (Hughes and Hughes 1992). Permanent swamps increasingly dominate the landscape in the central and southern portions of the ecoregion. *Raphia* swamps proliferate along the main course of the Sangha as it passes south through the Cuvette Centrale and meets the Congo River. The shallow water of these swamps is deep brown to black, deoxygenated, and highly acidic (pH 3.5–5.2) (Hughes and Hughes 1992; Laraque et al. 1998). The Likouala and Kouyou rivers south of the mainstem Sangha also course through extensive swamp forest. Large areas of floating grasses often occur in the rivers. Further south, a mosaic of marshy savanna, boggy steppe, and temporarily inundated swamp forest exists along the Alima, Nkeni, and Lefini rivers (Hughes and Hughes 1992; Sayer et al. 1992; Wetlands International 2002).

Outstanding or Distinctive Biodiversity Features

Most of the rivers and streams of this ecoregion have been little explored from a biological standpoint, but recent surveys in Cameroon and CAR indicate very high fish species richness. These investigations have documented more than 200 species only in the Cameroonian portion of the Sangha drainage below the Nki falls and around 300 species in the entire drainage, including the Upper Dja and portions in the Cuvette Centrale (U. Schliewen, pers. comm., 2003; Hughes and Hughes 1992; Sayer et al. 1992).

Dominant fish families identified in the swamp forests of this ecoregion include Alestiidae, Aplocheilidae, Cichlidae, Claroteidae, Cyprinidae, Mochokidae, Mormyridae, and Schilbeidae. Schooling species of Clupeidae (*Microthrissa* and *Odaxothrissa*) and small species of *Barbus* live in open water habitat (Hughes and Hughes 1992). In the permanently inundated, low-oxygen, and acidic waters lives a specialized fauna adapted for this habitat. Adaptations include swimbladder lungs to assist respiration in *Polypterus* and *Protopterus* species and accessory respiratory organs above the gills in *Clarias* (Lowe-McConnell 1987; Kingdon 1989).

Maximum water levels in the Cuvette Centrale are reached during the rainy season between November and January. At this time, fish move into the flooded forests along the river channels to feed and breed.

The Sangha ecoregion also provides important habitat for freshwater crabs, amphibians, reptiles, and aquatic mammals. According to Cumberlidge and Bokyo (2000), eight species of freshwater crabs live along the upper Sangha. Their research discovered two new species, *Sudanonautes africanus* and *S. sangha*. It also confirmed two little-known species of *Potamonautes* and two species of *Sudanonautes* (Cumberlidge and Bokyo 2000). More than sixty-seven species of aquatic-dependent frogs live in the ecoregion, including six endemics. Three crocodile species occur in mainstem Congo River wetlands: *Crocodylus cataphractus*, *C. niloticus*, and *Osteolaemus tetraspis*. Among these, *O. tetraspis* is endemic to lowland equatorial rainforests and is also considered imperiled (Riley and Huchzermeyer 1999). Aquatic and semi-aquatic snakes in the ecoregion include *Boulengerina annulata, Natriciteres olivacea, Naja melanoleuca,* and *Python sebae*. Populations of Allen's swamp monkey (*Allenopithecus nigrovirdis*), endemic to the Cuvette Centrale, the giant otter shrew (*Potamogale velox*), and two otter species (*Aonyx congicus* and *Lutra maculicollis*) also depend on the rivers of the ecoregion (Hughes and Hughes 1992; Wetlands International 2002).

Status and Threats

Current Status

The Sangha is one of the most biologically diverse ecoregions in tropical Africa, and several national parks and reserves already exist in the ecoregion (Sayer et al. 1992). The Dzanga-Sangha Faunal Reserve is located in southernmost CAR. This faunal reserve borders the Nouabalé-Ndoki National Park in northwestern ROC and the Lobéké National Park in Cameroon. A trinational nature reserve was created when the CAR, ROC, and Cameroon signed an accord in December 2000 (Kamdem Toham et al. 2003). The swampy floodplains of Gualougo triangle have recently been gazetted and added to the southern border of the Nouabalé-Ndoki National Park.

Located in northwestern ROC are the Odzala-Koukoua National Park, Lekoli-Pandaka Faunal Reserve, and M'boko Hunting Reserve. In addition, the Lefini Faunal Reserve covers 6,300 km^2 of land in southeastern ROC, although industrial-scale cassava cultivation has been documented there (Wilkie et al. 2001). Part of the Boumba-Bek-Nki Reserve in Cameroon falls in the Sangha ecoregion (Sayer et al. 1992). Currently, many protected areas in the ecoregion suffer from insufficient funds and inadequate infrastructure (Wilkie et al. 2001). Additionally, environmental laws and regulations are not always strictly enforced.

At present, legal and illegal logging and hunting activities take place on protected and unprotected lands. Logging and associated activities (e.g., influx of human population, runoff of chemicals used to protect logs from wood-destroying insects, siltation from slash-and-burn agriculture that follows forestry activities) are the greatest threat to the forested rivers of this ecoregion.

Logging has greatly accelerated in recent years. For example, in northern ROC, an estimated 400 km^2 was logged each year before 1996, whereas 1,500 km^2 was exploited in 1996. Additionally, the total area of forest concessions before 1996 in the Likouala and Sangha regions of ROC was 10,000 km^2, whereas in 1996 alone an additional 10,400 km^2 of land was added to forestry concessions (Fay and Vedder 1997). More than twenty logging concessions have been allocated in northern ROC, covering all exploitable forest. Opening up the forests for logging will also encourage mineral prospecting and possibly mineral exploitation in this ecologically sensitive region. The government's capacity to monitor and regulate these activities effectively, at least in the near term, is unlikely (D. Kaeuper, pers. comm., 2002). In Cameroon, timber is an important sector of the export and domestic economy, and most logging occurs in nationally owned forest reserves. Tree species targeted for removal include walnut, mahogany, and red cedar. Shifting agriculture in conjunction with commercial logging increasingly threatens forest habitat near the more densely populated areas. Commercial crops cultivated in cleared forest around the perimeter of the Cuvette Centrale include oil palm, cocoa, and coffee. Cotton is also produced in the ecoregion (Wilkie et al. 1992).

Some of the approximately 100,000 refugees who have moved into ROC from DRC have settled and are exploiting ecologically sensitive zones along the Oubangui River and in the region where the Likouala, Likouala aux Herbes, and Sangha rivers flow into the Congo River below Mbandaka. Arable locations in this part of the ROC and northward along the Oubangui are few and sensitive to population pressures (agriculture, hunting, and fishing, all of which had been based, until now, on small, self-sufficient populations). Refugee groups are putting pressure on wildlife, fish, and farming resources in this area.

Industrial gold mining (near ROC's border with Cameroon) and alluvial diamond mining (CAR) in the upper reaches of the Sangha and its tributaries increase turbidity and industrial pollution. Long-term residents along both the Sangha and the Oubangui have noted high and increasing levels of silt, which are affecting the facility of river traffic (D. Kaeuper, pers. comm., 2002).

Future Threats

According to the 2000 Forest Resource Assessment conducted by the FAO, 300 km^2 of forest is lost every year in CAR, 174 km^2 in ROC, and 2,217 km^2 in Cameroon (FAO 2001c). Continued unsustainable logging is a serious threat to the biological integrity of the ecosystem. The dramatic increase and availability of small arms and military weapons in this region (ROC, CAR, and DRC) caused by continuing political instability and conflict, and better access through logging concessions, are an increasing threat to wildlife in the region, increasing poaching and bushmeat trade.

Justification for Ecoregion Delineation

This ecoregion largely follows the watershed of the Sangha River Basin. The river systems of the Sangha ecoregion are permanent and have existed since before the major earth movements of the Miocene. The flat central part of the ecoregion, the Cuvette Centrale, is part of the ancient continental Congo Basin, invaded by the sea during the Mesozoic. After this invasion by the sea, the peripheral land surrounding the Cuvette Centrale was uplifted, accentuating the basin (Beadle 1981). The Dja, an upper tributary of the Sangha, is completely isolated from the Ngoko-Sangha by a large waterfall, the Nki Falls in Cameroon. The fauna upstream of Nki Falls is completely different (more like that of the Nyong and other basins in the Central West Coastal Equatorial [19]) from that of the downstream portion of the river (pure Congo fauna). Therefore, the upper Dja is included in the Central West Coastal Equatorial [19] ecoregion instead of the Sangha.

Data Quality: Medium

Ecoregion Number:	**28**
Ecoregion Name:	Southern Upper Guinea
Major Habitat Type:	Moist Forest Rivers
Final Biological Distinctiveness:	Bioregionally Outstanding
Final Conservation Status:	Endangered
Priority Class:	IV
Authors:	Ashley Brown and Michele Thieme
Reviewer:	Christian Lévêque

Location and General Description

Major rivers of the Southern Upper Guinea ecoregion include the Mano, Lofa, St. Paul, St. John, Cestos, and Cavally. The ecoregion covers nearly all of Liberia, a portion of southern Guinea, and part of western Côte d'Ivoire. The short, partly torrential rivers and streams of this ecoregion support an endemic freshwater fish and crab fauna (Hugueny and Lévêque 1994).

Most rivers of the ecoregion originate in the Wologisi Range in the north or the Nimba Range in the southern highlands (500–1,000 m a.s.l.) and flow short distances to the ocean. The upper courses of the rivers, which are slowly eroding the plateau region, are rocky and torrential (Lévêque et al. 1990). The relief along the coast is moderately steep, and cataracts abound in the courses of the rivers. For example, the Mano River encounters more than fifteen waterfalls throughout its lower course (Hughes and Hughes 1992). There are few floodplains in this ecoregion, although near the mouths of the rivers swamps and mangrove forests occur. Historically, the upper and middle reaches of the rivers flowed through moist lowland forests that were evergreen toward the coast and semi-evergreen further inland.

The climate of Southern Upper Guinea is hot and humid. The ecoregion receives high levels of rainfall: more than 4,000 mm along the coast and about 2,000 mm inland. Rain falls an average of 180 days per year, mainly from April or May to October. River discharges peak in September and October, and low levels occur in February and March. Temperatures throughout the ecoregion average 24°C in the coolest month and 27°C in the warmest month (Hughes and Hughes 1992).

Outstanding or Distinctive Biodiversity Features

The rivers and streams of Southern Upper Guinea are rich in aquatic insects, fish, freshwater decapods (including seven endemics out of nine crab species), and frogs. About one-fifth of the 151 fish species are endemic, with high levels of endemism in the Cyprinodontidae, Cyprinidae, and Cichlidae families. Some genera recorded in the Upper Guinea bioregion are also represented in the Lower Guinea and Congo bioregions but have never been found in between: *Doumea, Paramphilius, Microsynodontis, Parailia, Ichthyborus,* and *Caecomastacembelus*. Some species are also common to Upper and Lower Guinea and Congo: *Mormyrus tapirus, Isichthys henryi, Papyrocranus afer,* and *Xeosmystus nigri* (Lévêque et al. 1990). These occurrences support the view that Upper Guinea was a refuge zone for fish during dry climatic periods.

Many fish of Southern Upper Guinea possess adaptations for life in swift rivers with rocky bottoms. Species of the Amphiliidae family, with elongated and humped forms, can resist currents and stay on the river bottom (Welcomme 1985). *Amphilius* spp. cling to the substrate using stiff pectoral spines and a sucker-like mouth, whereas *Chiloglanis* spp. and *Labeo* spp. possess buccal suckers (Lévêque et al. 1990, 1992).

Several rare mammal species inhabit Southern Upper Guinea. The vulnerable pygmy hippopotamus (*Hexaprotodon liberiensis*) survives in some of the remaining forested areas (East 1999). The hippopotamus (*Hippopotamus amphibius*) and the vulnerable West African manatee (*Trichechus senegalensis*) also live in the rivers of this ecoregion (IUCN 2002). Several smaller mammals depend on the short, swift rivers for feeding. For example, the African water rat (*Colomys goslingi*) feeds on aquatic insects, crustaceans, and other invertebrates, and the endangered, endemic Mt. Nimba otter shrew (*Micropotamogale lamottei*) feeds on crabs, fish, and insects in streams (Kingdon 1997; IUCN 2002).

Habitat partitioning has occurred among the three species of crocodiles that inhabit the ecoregion. Nile crocodiles (*Crocodylus niloticus*) live in the mangrove swamps and river mouths. Slender-snouted crocodiles (*C. cataphractus*) live in larger rivers that run through rainforest, and dwarf crocodiles (*Osteolaemus tetraspis*) live in forested small streams (Kofron 1992).

Several bird and frog endemics inhabit Southern Upper Guinea. The endangered rufous fishing owl (*Scotopelia ussheri*) is largely limited to the rainforests of this ecoregion, although small populations may also survive elsewhere in Côte d'Ivoire (Fry et al. 1988). Eleven of fifty-two frog species are endemics, with the majority of endemics in the *Phrynobatrachus* genus. Many *Phrynobatrachus* species migrate between different habitats to avoid desiccation (Rödel 2000).

Status and Threats

Current Status

Liberia was once almost completely forested, but now about two-thirds of forest cover has been removed through agricultural expansion, logging, and mining. Runoff from slash-and-burn agriculture and overgrazing exerts a continuing pressure on the rivers and streams of this ecoregion (Kofron 1992). Mangroves are cleared for fuelwood, home construction, and rice culture (Hughes and Hughes 1992). Road building, landfilling, and fuelwood collection have degraded a mangrove forest at the mouth of the Mensurado Creek in the western portion of the ecoregion (Sayer et al. 1992). However, the civil strife that lasted most of the 1990s disrupted civil society to such an extent that agriculture and woodcutting were limited (East 1999).

Before the start of civil war in 1990, Liberia was one of the world's largest iron ore producers, primarily from the Mt. Nimba area. During times of heavy rainfall, lateritic (rich in iron and aluminum oxides) material washes down rivers adjacent to mining operations (UNEP 1999). There is evidence that iron ore dust and other residues have polluted the Mano and St. John rivers (Hazelwood 1979), and the endangered Mt. Nimba otter shrew is threatened by mining in Liberia and proposed mining in Guinea (Kingdon 1997; IUCN 2002). Deforestation accompanying the construction of mining settlements increases erosion, exacerbating threats to aquatic systems.

Inadequate disposal of solid wastes and sewage continues to be a problem in the countries of Southern Upper Guinea. Treatment facilities often insufficiently treat sewage, allowing the release of raw sewage into waterways (UNEP 1999). As of 1979, Monrovia was the only Liberian city with a sewage treatment system (Hazelwood 1979). Whether this is still functional is unknown.

Currently there is little formal nature protection in this ecoregion. A national park had been proposed that would cover a large area between the Cestos and Sehnwehn rivers, in the middle of the ecoregion, but has not yet been created (Hughes and Hughes 1992). Liberia, which encompasses most of the ecoregion, is focused on surviving its continuing civil war, and conservation activities have been sporadic.

Future Threats

Continued deforestation, mining, and fishing threaten the freshwater systems of the ecoregion. Deforestation can cause erosion and seriously affect the rich mangrove communities in the coastal zone. Increased mining poses a threat, especially in iron ore regions near the Nimba Mountains and the Wologosi Range (Szczesniak 2001). Increased fishing pressure, already high in mangrove areas, threatens sustainable fisheries.

The ecoregion lacks strong conservation policies and management of natural resources. Proper sewage treatment and disposal is needed to mitigate the ecological and health hazards that result from raw sewage entry into waterways.

Justification for Ecoregion Delineation

This ecoregion is defined by the basins of the Mano, Lofa, St. Paul, St. John, Cestos, and Cavally rivers. Southern Upper Guinea is part of the Upper Guinea bioregion, which is characterized by a distinct ichthyofauna that includes many endemics. This high endemism probably is the result of long-term geographic isolation and stable and wet climatic regimes. The Guinean range is an impassable barrier to the dispersal of fish from the Upper Guinean streams to Nilo-Sudanian basins.

Rapids and waterfalls in individual basins probably have served as additional barriers (Lévêque et al. 1990; Lévêque 1997). Different ecological conditions (forested Upper Guinean streams and savanna Sudanian streams) also potentially contributed to the species divergence (Hugueny and Lévêque 1994). It is hypothesized that the high endemism may also result in part from the forested rivers of this ecoregion acting as a refuge for aquatic fauna during dry climatic periods (Lévêque 1997).

Data Quality: Medium

Ecoregion Number: **29**
Ecoregion Name: **Southern West Coastal Equatorial**
Major Habitat Type: **Moist Forest Rivers**
Final Biological Distinctiveness: **Globally Outstanding**
Final Conservation Status: **Vulnerable**
Priority Class: **I**
Author: **Victor Mamonekene**
Reviewer: **John Sullivan**

Location and General Description

The waters of this rainforest ecoregion are exceptionally rich in freshwater species. The major rivers, the Ogooué, Kouilou-Niari, and Nyanga, descend slowly to the west coast from inland plateaus and rolling hills. The ecoregion extends from northern Gabon through the eastern portion of Congo and Cabinda (Angola) and ends above the Congo River Basin in DRC.

This ecoregion is situated on and south of the equator. It experiences a dry season that peaks from June to September, a rainy season from October to May, and a short period of low precipitation in January and February. About 1,800 mm (annual mean) of rainfall is recorded at Mayumba on the coast (Gabon). Mist and fog at higher elevations (Mayombe mountain range and Chaillu Massif) play an important role in maintaining atmospheric humidity during the dry season. The temperature regime is unimodal, with maximum temperatures in March and April and minima in July and August. The mean annual temperature is 24°C.

The Ogooué, Nyanga, and Kouilou-Niari rivers experience a bimodal flood regime. The first flood period occurs in November, and the second in April. Mean annual discharge of the Ogooué River is about 4,758 m^3/s, and that of the Kouilou River is about 913 m^3/s.

Moving east, this ecoregion's coastal plain gives way to the

Mayombe Range, which is interrupted by the Tchibanga and Ndéndé-Mouila gorges in which savanna vegetation dominates. Beyond the Mayombe Range rises the Chaillu Massif, with its highest peak being Mt. Iboundji at 972 m a.s.l.

Phytogeographically, this ecoregion is in the Guineo-Congolian region, which is characterized by a dense moist forest. In many places where the native vegetation has been cleared, secondary forest or agricultural lands now occur. Dry grasslands predominate on sandy soils near the coast. Along the rivers, open marshes with *Papyrus* and *Vossia* are common. Mangrove forests thrive in the estuaries and lagoons that extend along the coast. More inland and in the north, around Koulamoutou and Lastourville, the moist forest extends interspersed with secondary forest (White 1983).

Outstanding or Distinctive Biodiversity Features

Diversity is high among freshwater fish, aquatic herpetofauna, and aquatic mollusks. Recorded species include more than 230 fish, more than 50 reptiles and amphibians, and about 15 mollusks. Among the fish, cyprinids, cyprinodonts, and mormyrids show particularly high species diversity. Additionally, about one-quarter of the fish species are endemic to this ecoregion (G. Teugels, pers. comm., 2000). There is an especially high diversity of snoutfishes (Mormyridae) and killifishes (Aplocheilidae) in the Ogooué Basin. The Ogooué (including its tributary, the Ivindo) probably is the center of speciation for the mormyrid genus *Brienomyrus* (Sullivan et al. 2002).

Additionally, the Kouillou-Niari region is a contact zone between the Ogooué and Congo rivers; though unstudied, it is suspected to be rich in freshwater species and endemics (Kamdem Toham et al. 2003). The Chaillu Massif in Gabon has steep gradients and dense forest and is also suspected to harbor many endemic small stream fishes. Several endemic cyprinodonts are known to live in these mountain streams, and surveys may reveal more endemic fish plus other taxa (Kamdem Toham et al. 2003).

Among large vertebrates present in the ecoregion are the African keeled mud turtle (*Pelusios carinatus*), spotted-neck otter (*Lutra maculicollis*), and swamp otter (*Aonyx congicus*). The vulnerable West African manatee (*Trichechus senegalensis*) sometimes visits the lower portions of these rivers. Hippopotamus (*Hippopotamus amphibius*) and elephants (*Loxodonta africana*) are also present.

Status and Threats

Current Status

Although the rivers, streams, and headwater forests of this ecoregion are mostly pristine, deforestation and subsequent erosion are increasing problems. Oil exploitation and logging are the main activities that pollute the freshwater systems of this ecoregion. River sediments are filtered for gold, causing degradation at a local scale (Lahm 2002).

There is an active fishery in the Ogooué and Kouilou rivers and all coastal lakes and lagoons of these basins. However, data are lacking on the level of fishing pressure and impacts on fish populations. *Chrysichthys* spp., *Tilapia* spp., and *Heterotis niloticus* (a recently introduced Nilo-Sudanian species) constitute the majority of the catch.

Introduced species, including African bonytongue (*H. niloticus*) and various nonnative tilapia, create a moderate threat for the ecoregion's native fish fauna. *Oreochromis niloticus* has been widely introduced via aquaculture operations, particularly in the Ivindo system. Investigation and promotion of responsible aquaculture with native species should be implemented.

The Moukoukoulou Dam, located on the Bouenza River halfway between Brazzaville and Pointe-Noire, is the largest dam in Congo. This dam has cut off the migration of shrimp and fish up the river (Kamdem Toham et al. 2003).

Gabon has recently created a series of national parks. New protected areas are the Mwagne, Pongara, Akanda, Plateaux Batéké, Birougou, and Waka national parks. Existing protected areas that have been upgraded to national park status are the Moukalaba-Doudou, Lopé, and Petit-Loango. In Congo-Brazzaville are the Dimonika and Conkouati reserves.

Future Threats

Future threats involve increases in the magnitude and extent of current threats. The Sounda Dam on the Kouilou-Niari (Congo), whose construction began in 1996, constitutes a major future threat for that river basin, especially because no preimpoundment impact studies were conducted. However, the project was temporarily halted in 1997 because of the war.

Justification for Ecoregion Delineation

This ecoregion is defined by the basins of the Kouilou-Niari and Nyanga rivers and the mainstem of the Ogooué River. The northern tributaries of the Ogooué River (Abanga, Okano, and Ivindo) are excluded from this ecoregion because their fauna is more similar to that of rivers in the Central West Coastal Equatorial [19] ecoregion than to that of the mainstem Ogooué. For example, possibly because of a 50-km stretch of rapids and falls that separates the Ivindo and Ogooué rivers and because of previous river captures, the fauna of the Ivindo is more similar to that of the Ntem than the Ogooué (Thys van den Audenaerde 1966; Sullivan et al. 2002). During climatic fluctuations of the Pleistocene period, the forested West Coast Equatorial bioregion (Lower Guinea) is considered to have been a refuge for many species. Various Amphiliidae, Mochokidae, and Cyprinidae and some anguilliform fish are limited to the Guinean rivers and adapted to the cooler waters and hydrological regimes of this forested region (Lévêque 1997).

Data Quality: Low

Few data are available to describe the biodiversity of this ecoregion's freshwater systems. Many rivers and streams have not yet been sampled, such as the Nyanga and upper Kouilou rivers and many small coastal basins (Daget and Stauch 1968; Teugels et al. 1991; Dowsett and Dowsett-Lemaire 1991; Mamonekene and Teugels 1993). Little information is available on the life history and ecology of freshwater species, and data on species distributions are scarce for aquatic taxa other than fish.

The Muséum National d'Histoire Naturelle's extensive fish collections cover only parts of the Ogooué Basin, and various sections of the Ogooué River are in need of specific studies. The middle Ogooué rapids between Njolé and Lastoursville are shallow gradient rapids for more than 200 km of the main channel. Rapids are particularly difficult to sample, but they probably have their own fauna. The lakes and swamps of northern Lambarene and Ngomo in the Ogooué River Basin may be important examples of large lowland equatorial swamp, and surveys of Odonata are needed. The large Ogooué Delta, between Port-Gentil and Lambarene, is a priority survey area for aquatic mammals and waterbirds.

Ecoregion Number:	**30**
Ecoregion Name:	Sudanic Congo (Oubangui)
Major Habitat Type:	Moist Forest Rivers
Final Biological Distinctiveness:	Bioregionally Outstanding
Final Conservation Status:	Relatively Intact
Priority Class:	V
Authors:	Emily Peck and Michele Thieme
Reviewers:	Uli Schliewen and David Kaeuper

Location and General Description

The Oubangui River, with a catchment of more than 777,000 km^2, is a major tributary of the Congo River. The river drains the savanna-covered, elevated plateaus (500–700 m) of the CAR and then flows into one of the most biologically diverse wetlands in all of Africa: the Cuvette Centrale. The northern border of this ecoregion marks the divide between the Congo River and Lake Chad basins, and the eastern border is the divide between the Congo and Nile basins. The ecoregion encompasses the entirety of the Oubangui Basin (with the exception of the Uélé Basin [74]) and a portion of the mainstem Congo River from its confluence with the Oubangui downstream to Malebo Pool. Located between major river basins, the Oubangui's freshwater fauna is considered to be transitional between the Nilo-Sudanian and Congolian faunas (Roberts 1973; Bailey 1986). The ecoregion lies in the southern portion of the CAR and includes the western boundary of the DRC along the Oubangui and Congo rivers and the western border of the ROC.

Four principal climatic zones exist in the Oubangui ecoregion. The southern area of the ecoregion (1°N–2°S latitude) experiences a Congolese equatorial climate. With no dry season and little variation in temperature, the climate is hot, humid, and stable. The Congolese equatorial climate has a mean annual rainfall of more than 1,700 mm and a mean annual temperature of 26°C. Slightly further north lies the subequatorial climatic zone, where temperatures vary from 13°C in January to 40°C in March. In this zone, the annual precipitation is more than 1,500 mm; however, a 2-month dry season occurs between December and January. From 4–8°N latitude, the Sudano-Guinean climate zone is characterized by an annual rainfall of approximately 1,400 mm and a 3- to 5-month dry season from June through September. Mean annual temperature is 24°C, with a range from 18°C to 32°C. In the far north, the tropical Sudanian climate zone has a 5-month dry season from November to March. Less than 1,200 mm of precipitation falls each year (Sayer et al. 1992).

The headwaters of the Oubangui River begin in the CAR. The Chinko River and its affluent the Vivado, the Ouarra and its affluent the Goangoa, the Kerre, and the Mbokou rivers all begin in the far southeastern corner of CAR and flow down gentle slopes to join the Mbomou River. Floodplains are abundant in this section of the basin and follow the riverbanks for a total of more than 1,320 km along these rivers, inundating more than 6,500 km^2 of land during the wet season. The Mbomou flows along the border of CAR and DRC for several hundred kilometers until it is joined by the Kotto River; below this confluence the river is called the Oubangui. Forested floodplains line the Oubangui as it flows west along the border. The floodplain contracts and the river valley narrows where the river flows over the Kouimba Rapids (4°37′N, 20°27′E). Rapids are also located upstream along the Mbomou, Uélé, and Kotto rivers. Several tributaries join the north bank of the Oubangui River, including the Kouma, Tomi, Ombella, M'Poko, M'Bali, and Lobaye, from east to west. Inundated swamp forests cover approximately 2,310 km^2 of the floodplains of these tributaries.

In western CAR the river turns south and flows along the border of the ROC and DRC through virtually unbroken primary rainforest into the Cuvette Centrale. The Oubangui River is slow-flowing and wide (4–15 km), with islands in its channel. Northeast of its confluence with the Congo, the Oubangui is joined by the Giri River and spreads out across a large floodplain, forming the Giri or Bangala swamp. These swamps have black, acidic waters (pH 3.5–5.2) derived from the surrounding floodplain forests (Roberts 1973; Bailey 1986; Hughes and

Hughes 1992). The Oubangui eventually joins the Congo, which, at this point, is a slow-flowing, braided maze of alluvial islands, sandbanks, and floating beds of grass. Right-bank tributaries of the Congo River below its junction with the Oubangui include the Likouala aux Herbes, Sangha, Likouala, Alima, Nkeni, and Lefini rivers. Swamp forests predominate on the floodplains in this section, covering more than 6,800 km² that includes parts of the Likouala aux Herbes swamps (Beadle 1981; Hughes and Hughes 1992). The Likouala aux Herbes drains Lac Telé, a shallow lake possibly formed by a meteorite crater. The Kasai River joins the left bank of the Congo about 150 km upstream of where this ecoregion ends at Malebo Pool.

A variety of vegetation types cover the terrestrial landscape of the Oubangui ecoregion. In the Sudanian zone in the north, savanna predominates, with some semihumid forest, cropland, and gallery forest interspersed. Primary tropical rainforest blankets the landscape in the southern equatorial latitudes; dense evergreen forest occurs in the west-central areas, and freshwater swamp forest occurs along the lower course of the Oubangui River (Sayer et al. 1992).

Outstanding or Distinctive Biodiversity Features

The tropical rainforests, floodplains, and swamps of the Oubangui ecoregion are some of the most undisturbed freshwater habitats in central Africa. Though largely unstudied, their biotas probably are among the most biologically diverse in central Africa. Of the nearly 700 fish species documented in the Congo Basin, more than 140 reside in the Oubangui ecoregion. Further study is expected to reveal many additional species. Major fish families represented in the ecoregion are Alestiidae (Characidae), Amphiliidae, Aplocheilidae, Channidae, Cichlidae, Citharinidae, Claroteidae, Clariidae, Clupeidae, Cyprinidae, Mastacembelidae, Mochokidae, Mormyridae, Polypteridae, Schilbeidae, and Tetraodontidae (Lévêque et al. 1988).

The lower course of the Oubangui River is characterized by lower levels of endemism than other rivers in the inner Congo Basin. However, upstream areas and rapids, which are often located in savanna regions, harbor some endemics, although these species are known only from small collections. Twelve endemic fishes are known from this ecoregion, including rivulines, cichlids, mochokids, and cyprinids. Recent collections in the Sangha River drainage indicate that some fish species previously considered to be endemic to the Oubangui ecoregion (e.g., *Haplochromis oligacanthus*) are in fact shared with the Sangha basin.

The Likouala aux Herbes and Lac Télé areas, which are connected to swamps of the Congo-Sangha-Oubangui confluence, are poorly investigated. However, preliminary surveys and historical collections indicate high species richness in these blackwater habitats, with a fauna that is similar to that of blackwater habitats in the Cuvette Centrale [18] ecoregion.

Other aquatic taxa also thrive in this ecoregion. Five turtle species and three crocodile species (*Crocodylus cataphractus, C. niloticus,* and *Osteolaemus tetraspis*) have been documented. Among these, *O. tetraspis* is endemic to the lowland equatorial rainforests of West and Central Africa. It inhabits shallow river basins bordered by swamps; thus, the Likouala swamps of the Oubangui ecoregion provide prime habitat for this increasingly hunted species (Riley and Huchzermeyer 1999). Of more than forty frog species, four are endemic: *Hyperolius brachiofasciatus, H. schoutedeni, Xenopus pygmaeus,* and *Ptychadena straeleni*. Allen's swamp monkey (*Allenopithecus nigroviridis*), a Congo Basin endemic, lives in Lac Télé (Wetlands International 2002).

The Oubangui also supports a rich avifauna. The wetlands of Lac Télé and Likouala aux Herbes are especially important for migratory bird species in the families Ciconiidae, Areidae, and Pelicanidae, with about 250 bird species known from the area (Wetlands International 2002). Recent surveys by the African Waterfowl Census revealed more than twenty waterbird species living in the Likouala aux Herbes swamps. Species present in notable congregations included African jacana (*Actophilornis africanus*), cattle egret (*Bubulcus ibis*), and great egret (*Ardea alba*) (Dodman et al. 1999). Additionally, the CAR, which covers much of the northern portion of the ecoregion, has particularly high bird richness, with approximately 700 species recorded, including the locally rare shoebill *Balaeniceps rex,* which inhabits riverine swamps of the Oubangui (Sayer et al. 1992).

Status and Threats

There are several large protected areas in this ecoregion, but many of these reserves suffer from insufficient funds and inadequate infrastructure (Wilkie et al. 2001). In the far northeast of the ecoregion and in CAR, the Zemongo Faunal Reserve includes 10,100 km² along the Vivado River, and the Yata-Ngaya Faunal Reserve covers the headwaters of the Kotto River. The status of these protected areas is unknown because of their remoteness, the state of insecurity in the northeast of CAR, and their proximity to the Sudanian border (Blom and Yamindou 2001). In southwestern CAR on the border with ROC, the Basse Lobaye Biosphere Reserve is situated along the Lobaye River, covering 182 km² of inundated, evergreen forest. Though small, this international reserve remains undisturbed. Further west, the organization for Forested Ecosystems of Central Africa works in the Ngotto Forest, a 3,250-km² tract of dense semideciduous forest including *Raphia* swamp forest along the Lobaye, M'Baéré, and Bodingué rivers. They have established a protected zone in the forest and areas for sustainable logging (ECOFAC 2001). Lac Télé/Likouala aux Herbes Community Reserve, a wetland of international importance under the Ramsar Convention, lies in the north of the ROC (Dodman et al. 1999). The reserve was created in 1998 to conserve the biodiversity and ecology of the wetland while permitting local people to take resources that they depend on for their survival. Since 2000, the Congolese government has been working with the Wildlife Conservation Society–Congo to manage the reserve (Wildlife Conservation Society–Congo 2001).

Industrial gold mining (near the ROC's border with Cameroon) and alluvial diamond mining (CAR) in the upper reaches of the Sangha and its tributaries could increase turbidity and industrial pollution. Long-term residents along the Sangha and Oubangui have noted high and increasing levels of silt, which are affecting the facility of river traffic (D. Kaeuper, pers. comm., 2002).

Logging and logging concessions have greatly accelerated in recent years. For example, in the northern ROC, before 1996, 400 km^2 was exploited each year, whereas in 1996, 1,500 km^2 was exploited. Additionally, the total area of forest concessions before 1996 in the Likouala and Sangha regions of ROC was 10,000 km^2, whereas in 1996 alone an additional 10,400 km^2 of land was attributed to forestry concessions (Fay and Vedder 1997). More than twenty logging concessions have been allocated in northern ROC covering all exploitable forested areas, and logging contracts require roads to be built through old-growth forests that will connect Bangui (CAR) and Ouesso (ROC), Makoua, and Owando. Other contract requirements include constructing wood-processing facilities and support facilities in each of the exploitation zones, which will draw population into each and cause attendant pressures on food resources. Opening up the forests for logging will also encourage mineral prospecting and possibly mineral exploitation in this ecologically sensitive region. The government's capacity to monitor and regulate these activities effectively, at least in the near term, is unlikely (D. Kaeuper, pers. comm., 2002).

The recent conflicts in CAR have limited industrial logging operations, but an influx of refugees from DRC in the south and from Sudan in the northeast and movements of internally displaced people probably are affecting the forest and other vegetative cover in these parts of the country (Blom and Yamindou 2001). Some of the approximately 100,000 refugees that have moved into the ROC from the DRC have settled and are exploiting ecologically sensitive zones along the Oubangui River and in the region where the Likouala, Likouala aux Herbes, and Sangha rivers flow into the Congo River below Mbandaka. Arable locations in this part of the ROC and northward along the Oubangui are few and sensitive to population pressures (agriculture, hunting, and fishing, all of which originally were based on small, self-sufficient populations). Refugee groups are putting pressure on wildlife, fish, and farming resources in this area.

An agreement has been signed between the presidents of central African states to divert waters from the Oubangui River to refill Lake Chad (D. Kaeuper, pers. comm., 2002). Such a diversion may cause the transfer of organisms between the Congo Basin and the Lake Chad Basin, with unknown consequences. The withdrawals could also affect the level of the Oubangui River, depending on how much water is withdrawn.

The dramatic increase and availability of small arms and military weapons in this region (ROC, CAR, and DRC) caused by continuing political instability and conflict increases the threat to wildlife in the region through poaching and the bushmeat trade.

Justification for Ecoregion Delineation

This ecoregion is defined by the Oubangui Basin, excluding the Uele River Basin [74] and extending downstream of the confluence of the Oubangui and Congo rivers to Malebo Pool [24]. High habitat diversity and hydrogeographic barriers have shaped the ichthyofauna of the Oubangui drainage. Numerous waterfalls separate upstream and downstream sections of the main river and its tributaries. Some of these rapids may have been formed after river capture of Nilo-Sudanic catchments by the Oubangui drainage. Evidence of river capture comes from several Nilo-Sudanic fish species or subspecies occurring in the Oubangui drainage that are otherwise absent in the Congo Basin, except in catchments further northeast with a similar history (e.g., the Lindi and Aruwimi). Nilo-Sudanic species include the cichlid fishes *Sarotherodon galilaeus galilaeus* and *Tilapia zillii*. It should be noted that the Uele River, a tributary to the Oubangui, defines a separate ecoregion [74] because isolation by rapids and waterfalls has created a distinct fauna there.

Data Quality: Medium

Ecoregion Number:	**31**
Ecoregion Name:	**Upper Congo**
Major Habitat Type:	**Moist Forest Rivers**
Final Biological Distinctiveness:	**Nationally Important**
Final Conservation Status:	**Relatively Intact**
Priority Class:	**V**
Author:	**Ashley Brown**
Reviewers:	**Uli Schliewen and David Kaeuper**

Location and General Description

The upper course of the Congo River, called the Lualaba, is a large, slow-flowing river for most of its course, except when broken by the Portes d'Enfer (Gates of Hell) rapids in its upper reaches. The Lualaba in this ecoregion starts at an elevation of about 800 m a.s.l. and flows northwards through savanna into the central equatorial plain, where it gradually descends to the Upper Congo Rapids [61] at about 500 m a.s.l. (Marlier 1973; Bailey 1986; Library of Congress 1993). Riverine forests and swamps border the slow-flowing reaches (Hughes and Hughes 1992).

This ecoregion's most upstream point occurs where the Luvua River, a tributary to the Lualaba, exits Lake Mweru [6] and flows northwest to meet the Lualaba. At the confluence of the Luvua and Lualaba, the Lualaba River doubles in width and many elongate islands occupy the middle of the channel. About 100 km downstream, the Lukuga River, which drains the overflow from Lake Tanganyika, enters the Lualaba (Bailey 1986). The Lukuga is more heavily mineralized than the Lualaba and Luvua rivers, but its small relative discharge has a minor effect on the Lualaba's water chemistry. Below the Lukuga confluence, the river begins its descent of 135 m over a distance of 880 km, beginning with the Portes d'Enfer at Kongolo, a 50-km series of rapids stretching from Kimbombo to Kindu (Bailey 1986). Here, the gorge narrows to 200–600 m wide as the river flows over the rapids. Below these rapids, the river broadens to 2–3 km and swells with inputs from the Elila, Ulindi, and Lowa tributaries, each descending from eastern highlands and together contributing 2,000–5,000 m^3/s of water (Bailey 1986; Hughes and Hughes 1992). The upper course of the Lomami River, which joins the mainstem Congo further downstream in the Cuvette Centrale [18], drains a substantial area in this ecoregion. Similarly, the Lindi River is largely contained in this ecoregion, but its confluence with the Congo is technically in the Cuvette Centrale [18] ecoregion. The Upper Congo ecoregion lies entirely in the DRC and extends from the base of the Shaba Plateau to the Upper Congo Rapids [61].

The climate of the ecoregion is tropical and very wet, especially near the equator. Rainfall increases along the length of the Lualaba as it flows north toward the equator. Rainfall is 1,400 mm per year near Kasonga, shortly downstream of where the Luama joins the Lualaba. It increases to 1,700 mm per year further north near Kindu and increases to more than 2,000 mm per year approaching Kisangani. There is no distinct dry season in this region. October is the wettest month, and December, January, and February are the driest. Temperatures vary little, and mean annual temperature for most of the DRC is 24°C (Hughes and Hughes 1992).

The Lualaba experiences a bimodal flood regime in its northern reaches near the equator. However, some tributaries remain unimodal, including the Elilia and the Ulindi rivers. The Lowa is irregular, tending toward bimodality, with maxima occurring from April to May and again from November to December (Marlier 1973).

The terrestrial landscape through which the Lualaba flows undergoes a transition from savanna in the south to equatorial forest in the north. *Acacia caffra* is common in riparian vegetation, and *Nymphaea* spp. lives in open waters. The Lualaba is lined by *Cyperus papyrus, Typha domingensis, Pycreus mundtii,* and *Paspalidium geminatum. Aeschynomene elaphroxylon* and *Hibiscus diversifolius* make up areas of swamp woodland (Hughes and Hughes 1992).

Outstanding or Distinctive Biodiversity Features

The alternating rapids and slow-flowing waters of the Upper Congo support a rich aquatic fauna, despite the low primary productivity of some areas. The ecoregion has an extremely rich mollusk fauna, with forty-two species recorded, twenty-two of which are endemic. Most of the endemic species are from the Thiaridae family. Members of the genus *Cleopatra* usually can be found in small, slow-flowing or stagnant waterbodies. *Lanistes nsendweensis* is found among rocks beneath falls and in forest streams (Brown 1994).

More than 150 fish species and 6 endemics are known from this ecoregion. It is likely that new surveys in this ecoregion will reveal new species and more endemics. Dominant families include Alestiidae, Cyprinidae, and Cichlidae. Three cyprinid barbs (*Barbus marmoratus, B. nigrifilis, B. papilio*), one claroteid catfish (*Amarginops platus*), kneriids (*Parakneria thysi*), one mochokid catfish (*Chiloglanis marlieri*), and a cichlid (*Tylochromis elongatus*) are considered endemic.

Bird species present in the ecoregion include the Goliath heron (*Ardea goliath*) and the pink-backed pelican (*Pelecanus rufescens*) (Hughes and Hughes 1992). The ecoregion also hosts a rich mammalian and herpetofauna. The aquatic genet, *Osbornictis piscivora,* is poorly known but is confined to the Congo Basin (Kingdon 1997). Other wetland mammals include *Hippopotamus amphibius* and several otter species (*Aonyx capensis, A. congicus,* and *Lutra maculicollis*). Two crocodile species, the Nile crocodile (*Crocodylus niloticus*) and the slender-snouted crocodile (*C. cataphractus*), are also present (Riley and Huchzermeyer 1999). More than thirty frog species are known from the ecoregion, with three of them being endemic or near-endemic (*Hyperolius diaphanus, H. frontalis,* and *Afrana amieti*).

Status and Threats

Current Status

Currently the threats to the ecoregion include deforestation, continued civil strife, mining, poor sewage disposal, and the invasive water hyacinth (*Eichhornia crassipes*). Civil strife in the DRC has been almost continuous since 1996 and has had devastating effects on the economy and the welfare of human populations in the region (East 1999). It has also limited nearly all conservation and management initiatives.

As of the early 1990s, 86 percent of the forests in the DRC were still intact. However, logging poses a serious threat. The DRC ranked among the top ten countries worldwide with the highest net loss of forest area between 1990 and 2000 (FAO 2001d). Deforestation is most severe in the Kivu Province, which lies in the Upper Congo.

Moving upstream beyond the Portes d'Enfer rapids, palm plantations and palm oil extraction factories are situated along the wider section of the Lualaba. Above 500 m, cotton is grown, and in the uplands tea and *Arabica* coffee are produced (Bailey 1986). The extent to which the war has disrupted agricultural operations is unknown.

The effects of mining threaten the freshwater systems of the Upper Congo in certain localities. The DRC's main gold centers are located on the eastern perimeter of the Congo Basin (Bailey 1986). Deforestation surrounding mining towns threatens regions of Shaba where dry evergreen forest naturally grows, with cleared land normally replaced by wooded savanna (Hughes and Hughes 1992). The Kivu region, which lies next to and partially in the Upper Congo ecoregion, is rich in tin resources (Mino-Kahozi and Mbantshi 1997). Historically, regulations on mining activity have not existed in the region (East 1999). Many of the rivers are being mined on a small scale for diamonds and gold. This activity has affected the Ulinzi (Ulindi) and Luvua rivers. These small mining operations usually divert or dam smaller watercourses as part of the mining activity (Mino-Kahozi and Mbantshi 1997).

Poor waste disposal and inadequate sewage treatment plants also threaten the ecoregion's waterways and their biota. Bad maintenance of sanitation facilities and the discharge of large amounts of soil, mineral pollutants (old batteries, aluminum cans, mercuric salts, antiseptic soaps), and organic pollutants are common (Mino-Kahozi and Mbantshi 1997). Sewage and industrial waste often are dumped untreated from cities directly into the rivers (UNEP 1999).

Currently no portion of the ecoregion is protected. The Maïko National Park is located along the Lowa River's upper reaches, although it lies just outside the ecoregion (Sayer et al. 1992).

Future Threats

The largest future threats include deforestation and civil strife, with mining as a secondary threat. Political stability would offer the potential of sound environmental management.

Justification for Ecoregion Delineation

This ecoregion extends from the outlet of Lake Mweru downstream to the Upper Congo Rapids [61] ecoregion and is distinguished by an endemic mollusk fauna and believed to harbor a largely undescribed endemic fish fauna. Because rapids isolate the upper Congo fish fauna from the rest of the Congo River system, many fish species of the Cuvette Centrale [18] are not found in the Upper Congo (Banister 1986; Lévêque 1997). Several fish of the Nilo-Sudanian bioregion occur in the upper Lualaba that are not found in the main Congo River channel because they are confined by the rapids at Portes d'Enfer. These species include *Polypterus bichir, Ctenopoma muriei, Oreochromis niloticus,* and *Ichthyborus besse* (Beadle 1981). Debate continues over the origin of these Sudanian species in the upper Lualaba (Lévêque 1997). The upper Lualaba also shares a substantial number of species with the Zambezi River (Poll 1976; Lévêque 1997).

Data Quality: Low

Surveys have not been conducted in this ecoregion in many years, so recent data are lacking. Much historic data on fish and aquatic invertebrates remain uncataloged.

Ecoregion Number:	**32**
Ecoregion Name:	Upper Niger
Major Habitat Type:	Moist Forest Rivers
Final Biological Distinctiveness:	Nationally Important
Final Conservation Status:	Vulnerable
Priority Class:	V
Author:	Ashley Brown
Reviewers:	Christian Lévêque and Emmanuel Williams

Location and General Description

High rainfall in the Fouta Djalon and central Guinean highlands feeds the steep, fast-flowing rivers and streams of the Upper Niger ecoregion (Welcomme 1986b). This ecoregion contains a fish fauna adapted to live in swift currents and rock-bottomed streams. The Upper Niger is also the crucial source of floodwaters that support the rich Inner Delta [7] downstream and provide important flow for the lower Niger. This ecoregion lies primarily in the countries of Guinea and Mali, with smaller portions in Burkina Faso and Côte d'Ivoire.

Two rivers, the Upper Niger and its tributary, the Bani, define this ecoregion. The Upper Niger and its tributaries flow through the southern and western portions of the ecoregion, whereas the Bani River flows through the eastern half. The two rivers meet in the Inner Delta at Mopti. The Niger's source is located 240 km from the Atlantic Ocean, on the landward side of the Fouta Djalon highlands (John 1986). At an altitude of 1,000 m a.s.l. the Niger (as the River Tembi) emerges from a deep ravine, from which it descends through rocky stretches with rapids and waterfalls (Lowe-McConnell 1985; John 1986). In Mali, the Niger cuts through the Manding Plateau (760 m a.s.l.), an extension of the Fouta Djalon Mountains (Hughes and Hughes 1992). In the northeastern corner of the ecoregion, it flows through 60 km of steeply descending rapids (Beadle 1981). As the slope of the land declines, the river slows and widens to the point where it enters the vast floodplains of the Inner Niger Delta [7]. From its headwaters to the inner delta, the Upper Niger flows for a total of about 750 km and drops approximately 300 m in altitude over its course (Welcomme 1986b). Tributaries in Guinea that join the Upper Niger on its right bank include the Mafou, Niandan, Milo, and Sankarani rivers. The Tinkisso River joins the Niger on its left bank near the Malian border with Guinea.

The climate of the Upper Niger ecoregion is tropical, with a wet season lasting roughly from May to November. Rainfall in the southern portion of the ecoregion is about 1,500–2,000 mm per year on the high plateau of eastern Guinea. As the river flows into Mali toward the floodplains, rainfall decreases to 600–1,275 mm per year (Hughes and Hughes 1992).

Headwater tributaries begin to flood in May, with a peak in September (Welcomme 1986b). Downstream, these waters flow into the Inner Niger Delta [7] and inundate the inner delta floodplain, with a total influx of water from the Upper Niger of 73 km^3 per year (Welcomme 1986b).

The Upper Niger and its tributaries flow through a mix of woodland and savanna. The headwaters in Guinea drain transitional savanna intermittent with large bands of deciduous and semideciduous gallery forest (Hughes and Hughes 1992). Further downstream in Mali, the Niger and Bani rivers flow through tropical lowland moist forest, which shifts to dry grassland as the rivers approach the Inner Delta (Welcomme 1986b). As it flows into Mali, the river becomes broad and studded with islands (Hughes and Hughes 1992). Because the Upper Niger flows over weathered and strongly leached pre-Cambrian rocks, its water carries a low sediment load (Rzóska 1985).

Outstanding or Distinctive Biodiversity Features

A rich fish fauna specialized to live in the steep, rapidly flowing streams of the Upper Niger is one of the distinguishing elements of this ecoregion's aquatic biodiversity. This ecoregion also supports several species of aquatic mammals and reptiles and provides a critical water supply to the Inner Delta and lower Niger.

The Upper Niger ecoregion is host to about 150 fish (8 endemics), many of which have physiological modifications for life in swift currents. The cyprinid *Garra waterloti*, found only in this ecoregion, the Fouta Djalon, and the headwaters of the Senegal-Gambia, possesses an adhesive buccal disc for clinging to the rocky bottom in fast riffles (Lowe-McConnell 1985; Welcomme and De Mérona 1988). The only endemic cichlid in the Niger River, *Gobiocichla wonderi*, has a small, elongated body that is well suited for taking refuge from the current in crevices between rocks along the bottom of the streams (Lowe-McConnell 1985; Lévêque et al. 1991). The majority of this ecoregion's fish species are shared with the other ecoregions in the Niger River Basin. Dominant fish families include Alestiidae, Aplocheilidae, Citharinidae, Cyprinidae, Mochokidae, and Alestiidae.

Several aquatic mammals, reptiles, and waterbirds also occur in the Upper Niger. These include the hippopotamus (*Hippopotamus amphibius*), marsh mongoose (*Atilax paludinosus*), African clawless otter (*Aonyx capensis*), and vulnerable West African manatee (*Trichechus senegalensis*). The manatee can travel up smaller tributaries when they are sufficiently deep to allow passage (Happold 1987). Birds that live along the riverbanks of this ecoregion include Egyptian plover (*Pluvianus aegyptius*), rock pratincole (*Glareola nuchalis*), Goliath heron (*Ardea goliath*), and spur-winged goose (*Plectropterus gambensis*).

Status and Threats

Current Status

The Upper Niger is generally less disturbed than downstream portions of the Niger River Basin (e.g., the Niger Delta [58]), but dams, deforestation, fishing, and pollution nonetheless threaten the ecoregion's aquatic biodiversity.

Droughts and dams have reduced flow and altered the hydrology of the Upper Niger and its tributaries, in turn leading to reduced flows reaching the Inner Delta and lower Niger. Droughts in 1973 and 1984 that lowered flows in the Upper Niger and Bani rivers resulted in a reduction in flood duration and the size of the inundated area in the Inner Delta (Laë 1995). Dams can exacerbate the effects of drought by reducing already low flows (Laë 1997). The Selingue Dam on the Sankarani River in the southeastern portion of the ecoregion irrigates about 20 km^2 of land and produces electricity. A decision to withhold floodwater at the Selingue Dam during a drought in the 1980s resulted in a severe reduction of bourgou grass (*Echinochloa stagnina*) growth in the Inner Delta (Oyebande and Balogun 1996).

Deforestation and habitat degradation from pastoralism, farming, and timber and fuel collection affect the health of the freshwater systems of this ecoregion. Deforestation as a result of firewood collection is an ongoing concern; Mali's National Energy Bureau recently announced that almost 100 percent of the country's domestic fuel needs are met by firewood (Integrated Regional Information Networks 2000). Rice is grown on riverine floodplains of the Niger and tributaries such as the Tinkisso and Lélé rivers (Hughes and Hughes 1992).

The Niger River passes through the Parc National du Haut Niger for 60 km in eastern central Guinea. Crespi (1998) reported that fish caught in this park were found to be large and mainly adults, signifying that the fish stocks of the area were not overexploited at that time. Additionally, Guinea recently designated six new large Ramsar sites in the Upper Niger totaling more than 45,000 km^2 (Niger-Mafou, Niger-Niandan-Milo, Niger Source, Niger-Tinkisso, Sankarani-Fié, and Tinkisso) (Ramsar 2002a).

Future Threats

Drought, deforestation, and diversion of water for irrigation are the largest threats to waterways of the Upper Niger ecoregion. Population increases of 2.98 percent in Mali and 1.95 percent in Guinea mean that encroachment pressures on the Upper Niger River systems probably will increase. Related increases in fishing and agriculture, especially without regulation, also pose a threat.

Justification for Ecoregion Delineation

This ecoregion is defined by the Upper Niger River basin above the Inner Niger Delta [7]. The Upper Niger probably flowed into the Senegal River during the Pliocene and early Pleistocene. This connection would account for similarities between the fish faunas of the two rivers (Lowe-McConnell 1985). The dry period after the early Pleistocene produced sand dunes that barred flow to the Senegal and diverted it to the basin west and northwest of Tombouctou (Beadle 1981). The Upper Niger has followed its current course since the Pleistocene, although flow has been interrupted during dry periods (Lowe-McConnell 1985).

Data Quality: High

Ecoregion Number:	**33**
Ecoregion Name:	**Cape Fold**
Major Habitat Type:	**Mediterranean Systems**
Final Biological Distinctiveness:	**Globally Outstanding**
Final Conservation Status:	**Critical**
Priority Class:	**I**
Author:	**Genevieve Jones**
Reviewer:	**Paul Skelton**

Location and General Description

The Cape Fold is a small ecoregion that encompasses a phenomenal diversity of landscape types and correspondingly high levels of biotic diversity and endemism. The diversity of aquatic systems is linked to the different climatic regimes in the area. Located at the southernmost part of the African continent, the Cape Fold ecoregion is bounded on the south and west by the cold Atlantic Ocean and in the southeast by the warm Indian Ocean. Thus, many of the inland ecosystems experience a strong marine influence. The ecoregion is bound on the northwestern side by the Orange River and in the interior by the arid Karoo.

In the Cape Fold ecoregion, Brown et al. (1996) describe five subregions, each dominated by different climates and thus resulting in different aquatic ecosystems. The fynbos (bordered by the Orange River in the north and the Atlantic Ocean in the west) and the southern coastal bioregions receive 600–2,000 mm/year of winter rainfall and support oligotrophic (nutrient-poor), peat-stained, acidic rivers. These rivers include the Olifants, Berg, and Breede rivers in the fynbos bioregion and the Bloukraans, Elands, Silwer, Kaaimans, Duiwe, Homtini, and Touws rivers in the southern coastal bioregion. The southern inland bioregion lies to the north of the southern coastal bioregion and includes the Couga, Baviaanskloof, and Olifants rivers. The rainfall is low (200–600 mm per annum), and the aquatic systems have pH levels in the neutral range, with low conductivities and clear water. The alkaline interior bioregion lies inland of the fynbos and surrounds the southern inland and coastal bioregions. It supports alkaline seasonal or ephemeral waters and includes parts of the Doring, Gamtoos, and Gouritz rivers. The drought corridor bioregion has erratic rainfall and thus supports seasonal rivers, including the Great Fish, Sundays, Kowie, and Bushmans (Van Nieuwenhuizen and Day 2000b).

Nonriparian wetlands of the ecoregion are also diverse and include acid sponges, restioid marshes, peat-stained systems, and other perennial and seasonal wetlands. A number of coastal lakes and alkaline saltpans of variable sizes and shapes and important estuarine areas (such as Langebaan Lagoon, Berg River Mouth, and Wilderness Lakes) also occur. Pans, some permanent in wetter areas, are associated with the low-lying Karoo and coastal plain (Silberbauer and King 1991).

The main geological type of the southwestern cape is Table Mountain Group sandstones, but Malmesbury shales and alluvium areas occur as well (Silberbauer and King 1991). Compounds such as tannins and phenols derived from fynbos vegetation run into the freshwater systems and result in black waters, which are characteristic of the southwestern Cape (Britton 1991). Some areas in the ecoregion rise to 1,500 m (Silberbauer and King 1991), but the coastal plain, which reaches an average altitude of 500 m, extends far inland and affects the ecological processes and types of wetlands.

The evolutionary history of the ecoregion is controversial, but the breaking up of the continents and climate change appear to have been the main factors affecting the evolution of present-day biotic communities in the area (Linder et al. 1992). The first flowering plants appeared on the continent after the breakup of Gondwana. This accounts for some of the distinctions between the Cape Floral Kingdom and other southern continents. Climatic changes probably eliminated many taxa, and the remaining families evolved into the flora seen today (Cowling and Richardson 1995).

Outstanding or Distinctive Biodiversity Features

Although the aquatic fauna of this ecoregion is not completely documented, diversity and endemism of several aquatic taxa, particularly invertebrates, amphibians, and fish, are remarkably high (Wishart and Day 2003). Furthermore, the temperate acidic waters of this ecoregion are a rare habitat type to which many aquatic organisms have adapted. This is particularly the case among fish (e.g., the redfin minnows *Pseudobarbus* spp.) and invertebrates (Harrison and Agnew 1962; Skelton 1988, 1994; Skelton et al. 1995).

There are about thirty indigenous fish in the freshwater systems of the Cape Fold ecoregion, sixteen of which (53 percent) are endemic. Eleven of these are critically endangered, three are vulnerable, and one is near-threatened (Skelton 1987). The fish fauna is dominated by cyprinids, and there are three anguillids, two claroteids, three gobiids, one anabantid, and one galaxiid.

There are also two endemic genera, *Austroglanis* and *Pseudobarbus,* and the near-endemic *Sandelia.* Current genetic work indicates that the galaxiid and *Sandelia* taxa both represent species complexes rather than single species.

Richness and endemism are also high among anurans. Of the thirty-eight described frogs in the ecoregion, about half are considered endemic. There is also one endemic monotypic genus, *Microbatrachella,* whose only representative, the endangered *M. capensis,* is among the smallest frog species in the world. It prefers to breed in undisturbed vleis (areas of shallow, temporary standing water) and shallow pans in the fynbos of the southern cape, and it is endangered by habitat loss.

Five areas have been designated as Ramsar sites because of their importance for aquatic species, particularly for waterbirds. The Wilderness Lakes are an unusual group of hydrologically connected coastal lakes that support juvenile fish nurseries and a number of waterbirds. At times, they have supported 5 percent of the population of Cape shoveller (*Anas smithii*), a southern African endemic. Rare plants (*Ferraria foliosa, F. densepunctulata, Cerycium venoum* [possibly extinct], and *Cullumia floccosa*) have been recorded at Verlorenvlei, one of the few freshwater coastal lakes. This large system on the west coast supports thousands of waders of at least eleven species and a number of locally rare species such as the African fish eagle (*Haliaeetus vocifer*), greater and lesser flamingos (*Phoenicopterus ruber* and *P. minor*), little bittern (*Ixobrychus minutus*), and Caspian terns (*Hydroprogne caspia*). De Hoop Vlei supports 15 percent of the Cape shoveller world population and 7 percent of the yellow-billed duck (*Anas undulata*) world population (Shmueli et al. 2000). Langebaan Lagoon is home to notable populations of mollusks and crustaceans and acts as a nursery for juvenile fish. The lagoon supports 12 percent of the locally rare African oystercatcher (*Haematopus moquini*) world population, hosts approximately 37,000 birds (mostly waders) in summer, and provides breeding grounds for seabirds that are endemic to the southern African and Namibian coastline. The Heunings Estuary provides habitats and breeding grounds for 15 percent of the locally rare Damara tern (*Sterna balaenarum*) population, a species that is endemic to southern Africa (Cowan and Marneweck 1996).

Status and Threats

Current Status

Degradation of freshwater habitats by water abstraction, runoff of pesticides and other pollutants, interbasin transfers, and dams related to agricultural and urban development is widespread throughout the ecoregion. However, the more arid, inland catchments generally are intact or stable because these lands are inhospitable and of little agricultural value. The majority of coastal catchments are moderately altered, with some parts significantly to critically altered (particularly the catchments connected with urban areas). One example of interbasin exchange is a tunnel connection transferring water from the Riviersonderend (Breede River system) to the upper Berg River, substantially lowering the winter base flow, raising the summer base flow, and delaying floods in the donor river. Additionally, the transfer of water to the Berg River has caused an increase in flow of 830–4,500 percent during low-flow months (Snaddon et al. 1998).

Much of the land in the ecoregion is cultivated. Plantations are associated mostly with the east coast and parts of the fynbos bioregion. Cultivation of alien plants is associated with the upper reaches of the river systems, and urbanization is associated with their lower reaches (Van Nieuwenhuizen 2000). Urbanization is extensive in areas surrounding Cape Town and Port Elizabeth. Degraded water quality occurs in both urban and cultivated lands from irrigation return flows, runoff of agricultural chemicals, and effluent from wastewater returns.

Alien plants and animals threaten the functioning of many freshwater systems in the Cape Fold ecoregion. Alien vegetation covers less than 10 percent of most catchments in the ecoregion; however, in certain catchments along the coast alien species cover 15 percent or more of the land. The most intrusive alien aquatic and riparian plants are water hyacinth (*Eichhornia crassipes*), salvinia (*Salvinia molesta*), water lettuce (*Pistia stratiotes*), parrot's feather (*Myriophyllum aquaticum*), red water fern (*Azolla filiculoides*), and black wattle (*Acacia mernsii*). Black wattle occurs in thickets that suppress native vegetation and impede water flow. Although parts of the ecoregion are free of alien fish, as many as eleven exotic species have been recorded in streams of the fynbos bioregion. In the Olifants River, bass (*Micropterus* spp.) and trout (*Oncorhynchus mykiss* and *Salmo trutta*) have eliminated several indigenous *Pseudobarbus* species from different reaches of the system (Impson et al. 2000).

Some wetlands and parts of rivers fall into formally conserved areas that are fairly pristine. National Water Act No. 36 of 1998 should promote the conservation of a broader range of wetland habitats, not only those that are breeding areas for fish or birds. The five designated Ramsar sites in the ecoregion are under the supervision of either the National Parks Board (Wilderness Lakes and Langebaan) or the Western Cape Nature Conservation (De Hoop Vlei, Heunings Estuary, and Verlorenvlei) (Cowan and Marneweck 1996). Limited financial resources mean that designating additional important wetland sites that are not in such protected areas probably would not be feasible.

Despite the fact that all but two of the fish species have been recorded in formally conserved areas, Skelton et al. (1995) believe that they are poorly conserved. Legislation such as the Impact Assessment Regulations of 1996 and the Water Act of 1998 may alleviate threats to indigenous fish (Impson et al. 2000).

Future Threats

The increasing human population places growing stresses on the freshwater systems of the ecoregion. Specifically, further introductions of alien species, agricultural expansion, and urbanization, along with their effects on water quality and quantity, are the largest future threats. For example, development of a steel mill at Saldanha Bay, near Langebaan Lagoon and the Berg River floodplain, will use large amounts of water in an area of low water resources (Davies and Day 1998). Global climate change and related sea level rise is also of great concern for the coastal freshwater systems of this ecoregion.

Justification for Ecoregion Delineation

The ecoregion is defined by the coastal rivers of the southern tip of the continent and is bound on the northwestern side by the Orange River and in the interior by the arid Karoo. Although the history of the area generally has been investigated in connection with the floral kingdom, the same processes would have affected aquatic fauna. For instance, the freshwater fish fauna is highly endemic and has ancient origins (Skelton 1994). One species complex (*Galaxias* spp.) has phylogenetic relatives in South America, Australia, and New Zealand and therefore is probably of Gondwanan age and derivation. The close phylogenetic affinities of other endemic species are uncertain, but it appears that the region is an isolated evolutionary arena.

Data Quality: High

Extensive systematic and ecological research has been carried out on particular river systems. A few of the larger permanent wetland systems have been well studied (such as the Wilderness Lakes, Langebaan Lagoon, and some urban systems), but there is little understanding of the ecological processes that occur in many others, particularly small and temporary wetland systems.

Ecoregion Number:	**34**
Ecoregion Name:	Permanent Maghreb
Major Habitat Type:	Mediterranean Systems
Final Biological Distinctiveness:	Globally Outstanding
Final Conservation Status:	Critical
Priority Class:	I
Authors:	Michele Thieme and Ashley Brown
Reviewer:	Ammar Boumbezeur

Location and General Description

This ecoregion, which extends along the northernmost part of Africa, contains a freshwater fauna with many European elements and high endemism for a dry region. Rivers and streams drain the Atlas Mountains and flow either into the Mediterranean Sea in the northeast, into the Atlantic Ocean in the west, or into the interior. Productive chotts (shallow, irregularly flooded depressions) or sebkhas (irregularly flooded depressions) also provide important habitat for waterbirds. Parts of Tunisia, Algeria, and Morocco are covered by this ecoregion.

This ecoregion experiences a Mediterranean climate with hot dry summers, cool wet winters, and extremely variable rainfall. Along the coastal strip, precipitation varies from 90 to 1,000 mm per year, with the lowest values in southern Morocco and the highest in northeast Algeria. Inland, rainfall varies substantially according to altitude and location. At high altitudes, rainfall ranges between 200 and 700 mm, but in rainshadow areas and at lower altitudes, precipitation is typically no more than 200 mm and can be completely absent in some years. Extremes in temperature are experienced throughout the ecoregion, with temperatures reaching 50°C at some inland locations and dropping to below freezing in the winter. Inland areas may experience huge diurnal fluctuations, but along the coast the daily temperature is more constant (Hughes and Hughes 1992).

Perennial and intermittent streams originate in the Atlas Mountains complex, comprising from west to east the Anti-Atlas, High Atlas, Moyen Atlas, Saharan Atlas, and Tell Atlas. These streams flow either to the coast or inland. The principal rivers flowing to the coast are the Sebou, Oumer, Rbia, Tensift, and Sous to the Atlantic and the Moulouya, Cheliff, and Medjerda to the Mediterranean. The Cheliff River is the largest river in Algeria and is unusual in that it originates in the Saharan portion of the Atlas and stretches about 700 km to the Mediterranean Sea (Roberts 1975). Many of the interior-flowing streams go underground and contribute water to the extensive aquifers that underlie the ecoregion. Others flow into chotts, flooding them during the winter and creating habitat for breeding and nonbreeding waterbirds. The largest chotts in the ecoregion are Chott Ech Chergui (Algeria), Chott Melrhir (Algeria), Chott Merouane (Algeria), Chott el Hodna (Algeria), and Chott el Jerid (Tunisia). The chotts typically are brackish to saline and support little aquatic animal life (Dumont 1987).

Reedbeds, including *Phragmites australis, Scirpus lacustris,* and *Typha* spp., dominate the fringes of coastal ponds, lakes, and intermittent waterways. Salt-tolerant species are widespread along the coast and inland. *Halocnemum strobilaceum* is the most common plant lining the inland chotts (Hughes and Hughes 1992).

Outstanding or Distinctive Biological Features

This ecoregion is notable for the Mediterranean versus Afrotropical nature of its freshwater fauna; its arid, high-elevation streams; and its productive wetlands. The four main fish families are Cyprinidae, Cichlidae, Cobitidae, and Salmonidae, with more than half of the forty fish species being endemic. There are twenty-six species of bivalves and snails in the ecoregion.

One species of this diverse group is the critically endangered *Margaritifera auricularia,* a large (up to 20 cm) freshwater mussel that has almost disappeared from the rivers and streams of Morocco (Altaba 1990; IUCN 2002).

Cold water springs, which support many endemic stenothermal species, are found throughout the highlands (Giudicelli and Dakki 1984). Four endemic blackflies (Diptera: Simuliidae) are associated with the torrential high-altitude streams of the High Atlas, with a total of twenty-four blackfly species occurring there (Giudicelli et al. 2000).

This ecoregion also hosts several herpetofauna species not found elsewhere in Africa. The endemic Moroccan green toad (*Bufo brongersmai*) and Moroccan spadefoot toad (*Pelobates varaldii*) inhabit the low-lying chotts along the Atlantic coast of Morocco. The only two salamanders in Africa live in this ecoregion. The endemic newt, *Pleurodeles poireti,* is limited to the temporary wetlands of northeastern Algeria and northern Tunisia, and *Pleurodeles walti* lives along the coast of Morocco in addition to Spain and Portugal (Schleich et al. 1996).

This ecoregion lies in a main migration route between Europe and Africa for tens of millions of Palearctic migrants, and many of the coastal wetlands and inland waters are important sites for waterbirds using the Atlantic Coastal Flyway (Magin 2001b). Merja Zerga, a coastal lagoon with extensive wetland habitats, hosts between 50,000 and 200,000 staging and wintering waterbirds from 100 or more species. These include the vulnerable marbled teal (*Marmaronetta angustirostris*) and the critically endangered slender-billed curlew (*Numenius tenuirostris*) (Wetlands International 2002; IUCN 2002). Merja Sidi Boughaba, a brackish to freshwater groundwater-fed coastal lagoon, is an important breeding ground for birds including *M. angustirostris* and the regionally rare coot *Fulica cristata.* Lac d'Afennourir and Baie de Khnifiss are particularly rich, with 287 and 179 bird species, respectively (Wetlands International 2002). Lac Ichkeul in northern Tunisia is one of the most important sites in the Mediterranean for wintering Palearctic waterbirds (Amari and Azafzaf 2001). Two of the most numerous species recorded at Lac Ichkeul are wigeon (*Anas penelope*) (50,000 in 1977) and pochard (*Aythya ferina*) (120,000 in 1986) (Scott and Rose 1996). Lac Tonga, Mekhada marshes, and Lac Fetzara, three Ramsar sites in northeastern Algeria, are the most important sites for breeding waterbirds (Tonga and Mekhada for *Oxyura leucocephala* and *Aythya nyroca*) and wintering Palearctic waterbirds (Mekhada, Fetzara, and Tonga) in that country (Boumezbeur 2001; Boumezbeur 2002).

Status and Threats

Current Status

The aquatic habitats and species of this ecoregion are highly sensitive to the effects of invasive species, water withdrawals, overgrazing, tourism, urbanization, and infrastructure development. Hunting pressure on avifauna is also high at selected wetlands (e.g., Chott Ech Chergui) (Hughes and Hughes 1992).

Nonnative fish populations are increasing in North Africa and threaten endemics and other natives. For example, the introduced North American mosquitofish (*Gambusia affinis*) competes with the native *Aphanius iberus.*

Water scarcity is a primary concern in this xeric region. To irrigate agricultural fields and supply water to cities and towns, Morocco, Algeria, and Tunisia have built numerous small dams along their few perennial rivers. Tunisia aims to develop 90 percent of its surface water and 100 percent of its groundwater resources by 2010 through the construction of 21 dams, 235 hillside reservoirs, and 610 deep tube wells (FAO 1997c). As of 1990, there were thirty-four operational dams in Morocco, and the majority of the 11 km^3 of water withdrawn annually in Morocco was used for agriculture. Morocco has placed more than 10,000 km^2 of land under irrigation, representing about 75 percent of cultivable land (FAO 1997b). To irrigate such a vast amount of land the state has built numerous hill reservoirs. Besides changing the natural flow regime and blocking movements of species, these irrigation projects are also causing soil salinization (Matoussi 1996).

Disposal of sewage from growing coastal cities and pollution from agricultural and grazing lands pose additional threats. For example, upstream of the city of Beja, Tunisia, the Beja Wadi has high water quality and a diverse freshwater fauna, but below the city, species diversity and water quality drop significantly (Mouelhi et al. 1998). In the Sebou River in Morocco, metallic and organic pollution are derived from industrial, domestic, and agricultural runoff. Pesticide and fertilizer runoff from the agricultural fields of the Gharb plain, along with industrial wastes from paper mills, sugar plants, tanneries, food industries, wool mills, and chemical plants and sewage from associated towns, are the major sources of these pollutants (Bennasser et al. 1997; Khamar et al. 2000). Lac d'Afennourir has experienced eutrophication, probably as a result of surrounding grazing activities. Other lakes also suffer from the effects of grazing, including desertification (Wetlands International 2002).

Several protected areas occur in this ecoregion. The Lac Tonga Ramsar site, located in Algeria near the border with Tunisia, comprises a complex of wetlands fed by the Oued-el-Hout and Oued-el-Eurg. The area is also protected as part of the Parc National d'El Kala (Wetlands International 2002). This park contains a freshwater lake and permanent freshwater marshes linked to the sea by an artificial channel (Wetlands International 2002). The Park National d'El Kala also includes three other lakes in addition to Tonga (Lakes Oubeira, Mellah, and Blue) and an extensive marsh, Mekhada, outside the limits of this park. Many of the coastal wetlands of the ecoregion are designated as Ramsar sites, including Merja Zerga, Merja Sidi Boughaba, Lac d'Afennourir, and Baie de Khnifiss in Morocco; Lac Oubeïra, Lac Tonga, Lac des Oiseaux, Marais Mekhada, and Lac Fetzara in Algeria; and Ichkeul in Tunisia (Wetlands International 2002).

However, many of these face the same threats as surrounding areas. For example, the salinity of Lake Ichkeul has increased because of reduced inflow from its tributaries, which have been dammed, and a resultant increase in inflow from the sea. The increase in salinity has caused the number of waterbirds at the lake to decline precipitously. There are plans for an increased allocation of freshwater to Lake Ichkeul to try to restore the system (Amari and Azafzaf 2001).

Future Threats

The highly fragile freshwater systems of this ecoregion face numerous threats, including runoff from industrial and agricultural sources, fragmentation from dams, overabstraction of water resources from the rivers and from dams, possible oil or gas spills, depletion of underground water sources, pressures from tourism, desertification, and salinization. The countries of the Maghreb are all considered water scarce and expect to face water shortages within the next decade or two.

Several oil and gas pipelines transport fuels from the interior to coastal ports. The density of pipelines and offshore oil and gas fields is highest in the eastern portion of the ecoregion, in Algeria and Tunisia (Fuller and Bush 1999). A spill of 11 million gallons of oil (equivalent to the size of the *Exxon Valdez* oil spill) occurred along the coast of Algeria in 1980, and the possibility of future accidents put coastal habitats at risk (Oil Spill Intelligence Report 1999).

Justification for Ecoregion Delineation

This ecoregion follows Roberts's (1975) delineation of the Maghreb ichthyofaunal province, delimited in the south by the Atlas Mountains. The freshwater fauna of this ecoregion shares a strong affinity with that of Europe's Mediterranean ecoregions. The origin of the fish fauna is Palearctic with some tropical elements, and bivalves and snails are mostly of Palearctic origin as well. This is the only African ecoregion with representatives of the fish genus *Pseudophoxinus,* a primarily European group. Similarly, the only Cobitidae representatives in Africa, *Cobitis marocancna* and *C. taenia,* occur in the highland streams of the Permanent and Temporary [92] Maghrebs. The only native Salmonidae species (*Salmo trutta*) in Africa also lives in this ecoregion (Roberts 1975; Lévêque 1990).

Data Quality: High

Ecoregion Number: **35**
Ecoregion Name: **Albertine Highlands**
Major Habitat Type: **Highland and Mountain Systems**
Final Biological Distinctiveness: **Bioregionally Outstanding**
Final Conservation Status: **Endangered**
Priority Class: **IV**
Authors: **Ashley Brown and Robin Abell**
Reviewers: **Marc Languy and Lauren Chapman**

Location and General Description

The Albertine Highlands ecoregion is defined by the basins of the high-elevation rivers and streams that drain to the Lualaba River in the far eastern part of the Congo Basin. On their westward course, the headwater streams flow from the summits of the Albertine Highlands through savanna plateaus, where swamps and wetlands surround them. The rivers then flow through broadleaf forest, cascading down waterfalls and rapids en route to the high-canopy rainforest of the Congo Basin. Wetlands are abundant along many of the rivers, and these provide habitat for important waterbird populations. Whereas the Albertine Highlands topographic feature extends across the Democratic Republic of the Congo (DRC), Uganda, Rwanda, Burundi, and Tanzania, this ecoregion is located entirely in the eastern DRC.

The Lualaba's eastern tributaries, which drain the western slopes of the Albertine Highlands, are the primary aquatic habitats of this ecoregion. Moving from south to north, these tributaries include the Luama, Elila, Ulinda, and Lowa rivers and the Lugula (tributary of the Ulindi) and the Oso (tributary of the Lowa). Maximum elevation in the ecoregion is above 3,000 m a.s.l. for some eastern summits (e.g., Mount Kahuzi and Mohi [Itombwe]), which give way in the west to elevated plateaus at about 1,500–2,000 m a.s.l. (Hughes and Hughes 1992). After passing through the numerous wetlands of these plateaus, the rivers descend sharply to about 800–1,000 m a.s.l., where the land levels off again in the middle and west of the ecoregion. A particularly extensive floodplain occurs along the banks of the Luama River for about 130 km, where the river flows through a low valley and drains a swamp about 600 km^2 in extent (Hughes and Hughes 1992).

The climate in this ecoregion is tropical and wet. Mean annual rainfall is greatest in the higher altitudes. Average precipitation grades from more than 2,200 mm per year in the center and west, at elevations of 1,000–1,300 m a.s.l., to more than 2,800 mm in eastern slopes above 1,800 m a.s.l. Precipitation decreases with distance from the equator. Mean annual temperature is about 24°C in the ecoregion. The mean daily maximum is about 30°C and the mean daily minimum is 12–15°C (Hughes and Hughes 1992).

Vegetation varies across the ecoregion with elevation. Wet rainforest and semideciduous forest grow below the plateaus (elevation 1,500–2,000 m a.s.l.), whereas the vegetation is primarily woody savanna at higher elevations on the plateaus (Hughes and Hughes 1992). In the large swamps that line streams on the savanna plateaus, *Typha domingensis* and *Cyperus latifolius* dominate. *C. latifolius* sometimes forms pure stands and covers large areas. *Miscanthidium violaceum*, *Scleria* sp., and *Nymphaea mildbraedii* grow further away from the plateau streams, and on the swamp edges a woody canopy about 3 m high is composed of *Hypericum lanceolatum*, *Maytenus acuminatus*, and *Myrica kandtiana*. At lower altitudes, *Uapaca guineensis*, *Ficus mucuso*, *Irvingia smithii*, *Klainedoxa* sp., and *Mitragyna stipulosa* dominate the riverside forest. Below 1,000 m, *Gilbertiodendron dewevrei* and *Staudtia stipitata* make up gallery forests, and *Mitragyna stipulosa* grows in semipermanent swamps. Monospecific stands of *Michelsonia microphylla* grow near the equator (Hughes and Hughes 1992).

Outstanding or Distinctive Biodiversity Features

The aquatic fauna of this ecoregion is incompletely known; data are particularly patchy for fishes, reptiles, and aquatic invertebrates. Among the known aquatic species are several fish adapted to torrential waters, a diverse frog fauna, important waterbird populations, and several aquatic mammals with limited distributions.

Fish richness appears to be low, with only sixteen described species (G. Teugels, pers. comm., 2000). However, the waters of this ecoregion are poorly sampled, and further investigation probably will reveal multiple new species. Interesting fish include *Chiloglanis* and *Amphilius* species possessing elaborate sucker-like mouths that allow them to cling to substrates and probably facilitate feeding on the vegetation growing on rocks (Welcomme and De Mérona 1988). Also of interest is *Kneria wittei*, which is found in streams up to 1,800 m in elevation.

Virunga National Park hosts several migratory bird species that use the area as a feeding and wintering ground, and large terrestrial mammals often congregate along rivers in the park (Wetlands International 2002). Two smaller, range-restricted mammals live in the waters of the Albertine Highlands ecoregion. The semi-aquatic and endangered Ruwenzori otter shrew (*Micropotamogale ruwenzorii*) inhabits the slopes of the Ruwenzoris in the northeast of this ecoregion, in addition to the Lake Victoria [57] ecoregion (Kamdem Toham et al. 2003). The aquatic genet (*Osbornictis piscivora*), endemic to the Congo Basin, is largely limited to the eastern part of the basin and is most commonly found in shallow headwater streams running through limbali forest.

The ecoregion hosts a highly endemic and rich aquatic frog fauna, in addition to several widespread reptiles. There are fifty-three known frog species, and twenty-two of them are endemic. Many of the endemics, such as the rare Ruwenzori river frog (*Afrana ruwenzorica*) and Albertine Rift reed frog (*Hyperolius alticola*), are from the Ranidae and Hyperoliidae families. *Crocodylus niloticus*, *Pelusios gabonensis*, and *Pelomedusa subrufa* occur here; researchers may discover other aquatic reptiles with further investigations.

Distributional data on aquatic macro-invertebrate assemblages in this ecoregion are largely unavailable, although some taxa have been well studied. For example, four freshwater mollusk species inhabit the ecoregion, including the endemic snail *Tomichia hendrickxi*. This species has been found in association with the macrophyte *Lemna* in thermal ponds with temperatures of about 35–40°C. The related *Tomichia kivuensis* is associated with habitats containing fast currents, gravel bottoms, and aquatic plants (Brown 1994).

Status and Threats

Current Status

Civil strife in the DRC has been almost continuous since 1996, and ethnic tensions in the region remain high. In addition, civil war in neighboring Rwanda and Burundi in the mid-1990s led to an influx of refugees into eastern DRC (UNESCO 2001). Lack of political stability and large refugee movements are the largest impediments to conservation in the ecoregion.

Cobalt, tin, tungsten, columbite (a niobium mineral), and tantalum deposits all exist in the ecoregion, as does gold. Because of civil strife, most industrial-scale mining operations have ceased, but artisanal mining continues, particularly for coltan (columbite-tantalite) and gold (Coakley 2001a). In areas adjacent to mining operations, riparian forests often are cleared, erosion increases, and fish spawning and nursery habitats are destroyed through siltation and sedimentation (Mino-Kahozi and Mbantshi 1997). Gold panning apparently is widespread in the northeast of the ecoregion, and it has been cited as a threat to populations of Ruwenzori otter shrew (Kingdon 1997; Kamdem Toham et al. 2003).

Deforestation presents a serious threat in the ecoregion. The estimate for deforestation in DRC (Zaire at the time) between 1981 and 1985 was 1,800 km^2 per year (Sayer et al. 1992). Exploitation of forests for fuelwood and conversion for grazing continues. For example, parts of the Kahuzi-Biega National Park have been encroached, and deforestation took place in parts of the Itombwe Massif that were formerly an intact forest-grassland mosaic (WWF-EARPO 2003). Near population centers, wetland herbaceous vegetation is heavily harvested for thatch used in basketry and fish traps (Hughes and Hughes 1992). The effects of deforestation on the ecoregion's aquatic biodiversity are undocumented, but adverse impacts are likely. For instance, the presence of the aquatic genet is highly correlated with the presence of limbali (*Gilbertiodendron* spp.) forest, suggesting that the genet's survival depends on the maintenance of these forests in an undisturbed state (Kingdon 1997). The Ruwenzori otter shrew, listed as endangered by IUCN, is threatened by human-induced habitat loss and ensnarement in fish traps (IUCN 2002).

The national parks of the ecoregion have been highly compromised by civil strife. The Kahuzi-Biega National Park encompasses wetlands of the plateau and mountainside areas (Hughes and Hughes 1992), but poaching has been very heavy in the eastern part of the park, a situation worsened by refugee influxes (East 1999). Agriculture, deforestation, firewood collection, and mining also occur within the boundaries of the park and threaten the integrity of adjacent freshwater streams (Demey and Louette 2001; WWF-EARPO 2003). Maiko National Park covers strips of swampy riverine forest in the northern section of the ecoregion along the Maiko River. The Maiko National Park has never received any significant support and lacks even basic infrastructure (WWF-EARPO 2003).

Future Threats

The largest future threats to the ecoregion are increasing population pressure, mining, and deforestation. Civil strife also poses a serious threat and could continue to undermine the management of mining activities and forests.

Justification for Ecoregion Delineation

This ecoregion is defined by the upper basins of the high-elevation rivers that drain to the Lualaba River in the eastern Congo Basin. Although the fauna is incompletely known, it is thought that this area may have served as an important refuge zone for aquatic species during dry periods (Lévêque 1997).

Data Quality: Low

Ecoregion Number:	**36**
Ecoregion Name:	**Amatolo-Winterberg Highlands**
Major Habitat Type:	**Highland and Mountain Systems**
Final Biological Distinctiveness:	**Bioregionally Outstanding**
Final Conservation Status:	**Vulnerable**
Priority Class:	**V**
Author:	**Belinda Day**
Reviewer:	**Paul Skelton**

Location and General Description

The headwater streams of the Amatolo-Winterberg Highlands ecoregion are poorly studied, yet there is enough information to confirm that they host a rich aquatic fauna and a number of relict and endemic species. This ecoregion covers a series of high-altitude Afromontane forests in the Eastern Cape Province of South Africa. It is situated on a relict arm of the Great Escarpment's southern rim. To the south and east of the Amatolo-Winterberg Highlands lies the low-altitude coastal plain (Hughes and Hughes 1992).

The climate of the ecoregion is essentially temperate because of its high altitude. Most rain falls in summer and autumn, with a mean annual range from 700 mm in low-lying areas to more than 2,000 mm on high-altitude peaks (Barnes 1998c). Although frontal systems produce some rainfall, most of the summer and autumn rain results from the onshore flow of warm, moist air into low-pressure systems situated near the coast or further inland. High-altitude areas are also subject to orographic rainfall as warm, moist air coming off the Indian Ocean is forced up onto the highlands (Preston-Whyte and Tyson 1988). Frost and snow may occur on the highest peaks in winter.

The headwaters of a large number of small streams drain this ecoregion. These include the headwaters of the Buffalo, the lower Great Fish River, the Keiskamma, and the Swart Kei. Streams are both perennial and ephemeral, but all are highly seasonal in their discharge. The streams typically are narrow, fast-flowing turbulent headwaters with rocky substrata, and many have sponges or bogs at their source and small areas of seasonally flooded marsh in flatter, low-altitude valleys (Hughes and Hughes 1992).

The area tends to be rugged, with steep cliffs and valleys and several peaks around 2,000 m a.s.l. that rise from the low-lying rolling grasslands and valley bushveld. Sandy loam soils predominate. A mixture of montane grasslands and fynbos heath is present at high altitudes, especially in rainshadow areas. Below the montane grasslands the highlands and peaks are covered with both wet and dry Afromontane forest. Scrub forests and rolling grasslands replace these montane forests at lower altitudes. Riparian valleys tend to be thickly forested, and shorter, dry forest species are found on the steeper slopes and gorges. Dominant trees in the forest canopy include *Rapanea, Podocarpus, Curtisia, Canthium, Celtis, Xymalos, Vepris, Ptaeroxylon,* and *Trichocladus*. In the south, toward the Mpofu Game Reserve, the topography becomes gentler and rolling grasslands are more prevalent (Barnes 1998c; Barnes et al. 2001).

The Amatolo-Winterberg Highlands extend south and east from the ranges of the Great Escarpment and are separated from the latter by the valleys of the Fish River in the west and the Great Kei to the north. The highlands represent a residual portion of the Great Escarpment that was bypassed by a more rapidly eroding Kei valley and retreating northern scarp (Wellington 1928 in Skelton 1986a), such that its relict Afromontane fauna and flora have survived.

Outstanding or Distinctive Biodiversity Features

Little is known about the aquatic fauna of this ecoregion. However, the taxa that are well studied, such as the Odonata, show some endemism. Three odonates (*Chloralesta apricans, C. tessellates,* and *Platycypha fitzimonsi*) are endemic to the ecoregion (I. DeMoor, pers. comm., 2001). Additionally, the forests of the Amatolo-Winterberg Highlands are a hotspot of terrestrial biodiversity and support a large number of endemic and threatened vertebrate and plant species (Barnes 1998c).

The rivers and streams of this highland ecoregion support a diverse herpetofauna and a number of aquatic species from other vertebrate taxa. There are nineteen known aquatic species of frogs and reptiles, including two endemic frog species. The endemic Amatola toad (*Bufo amatolicus*) is limited to the *Amatoles,* and the only populations of the endemic hogsback frog (*Anhydrophyrne rattrayi*) are found in this ecoregion (Barnes 1998c). Two riverine mammals (Cape clawless otter [*Aonyx capensis*] and marsh mongoose [*Atilax paludinosus*]) also occur.

Of the eight fish species known from this ecoregion, three are endemic. The Eastern Province rocky (*Sandelia bainsii*), the border barb (*Barbus trevelyani*), and *B. amatolicus* are limited to the tributaries of the Keiskamma and Buffalo rivers in the Amatole Forest Complex (Mayekiso and Hecht 1988). In the 2002 IUCN Red List, *Sandelia bainsii* is listed as endangered, and *B. trevelyani* is critically endangered (IUCN 2002). Both species are threatened by the presence of alien fish species and degradation and loss of habitat from excessive water abstraction.

Status and Threats

Current Status

The wetlands of this ecoregion face serious threats from alien invasions, growing human populations, and river impoundments. The lack of published research on the wetlands of the ecoregion hinders a comprehensive evaluation of conservation status.

A number of state forests support natural forest and a game reserve (Mpofu Reserve) in this ecoregion. In 1996, the natural forests were placed under the Directorate of Nature Conservation for the Eastern Cape Province, and plans are being prepared for their management. The Forest Conservation Act permits no grazing or hunting in these areas, but resource extraction (e.g., building materials, fuelwood) is allowed (Barnes 1998c).

There are also large tracts of commercial forest plantations of alien trees in the ecoregion and a number of smaller, private plantations. Wetland ecosystems adjacent to commercial alien forestry plantations face several threats. Erosion and siltation can occur, especially in pine plantations, where there is little undergrowth to bind the topsoil. Runoff of fertilizers or pesticides can lead to eutrophication and pollution of downstream wetlands. Furthermore, many of the alien plantations (mostly pines) are located in upper catchments above the natural forests. The high rates of evapotranspiration of the alien trees have caused water tables to drop, leaving less water downstream. Deprivation of water to indigenous forests may lead to a loss of ecosystem functioning, both in the forests themselves and in the wetlands adjacent to them (Jacot-Guillarmod 1988; Barnes 1998c; Barnes et al. 2001).

Sport fishing enthusiasts have introduced brown trout (*Salmo trutta*), rainbow trout (*Oncorhynchus mykiss*), and a number of bass species to the rivers and streams of the highlands. These alien fishes are voracious predators and are likely to have had serious effects on populations of both fishes and invertebrates.

Soil erosion is a problem throughout much of the ecoregion because of unsustainable logging of indigenous timber, especially *Podocarpus* spp., expansion of agricultural areas, and overgrazing of domestic livestock (Teague 1988; Barnes 1988, 1998c).

Impoundments are also a threat to the continued functioning of many riverine systems in the ecoregion. Many of the streams in Amatolo-Winterberg Highlands are dammed, even in their headwaters, and downstream dams outside the ecoregion may have upstream repercussions.

Future Threats

Rural population growth is a major threat in places in this ecoregion. The invasion of alien species, both terrestrial and aquatic, is likely to be an escalating problem, as are erosion, siltation, and eutrophication (Hart 1988; Jacot-Guillarmod 1988). Likewise, numerous dams (including small farm dams) are also causing changes in the natural functioning of riverine systems in this ecoregion.

Justification for Ecoregion Delineation

This ecoregion is defined by the Amatolo-Winterberg Highlands and is the southern extent of the montane escarpment aquatic region (Skelton 1993). The fishes of the Amatolo-Winterberg Highlands belong to the temperate fish fauna of southern Africa. These species are found only in southern Africa between the coastal rivers and streams of the Cape and the southern plateau tributaries of the Limpopo River. Elements of both the Cape and Karroid fishes live in the waters of this ecoregion. Typical Cape fauna are fish species found only in the Cape Fold Mountains and their outliers; the Karroid fishes are inland species whose distribution closely follows that of the geological strata of the Karoo Sequence (Skelton 1986a). As with the Drakensberg-Maloti Highlands [37], the presence here of a mixed fish fauna suggests that this ecoregion may prove to be a transition zone for other aquatic taxa as well.

Although the phylogenetics of the ecoregion's Cape fishes (*Sandelia bainsii* and *Barbus trevelyani*) are too poorly understood to reveal the species' biogeographic history, Skelton (1986a) suggests that the distribution of the Karroid fishes (examples from the Amatolo-Winterberg Highlands include *Labeo umbratus* and *Barbus anoplus*) is intimately linked to the evolution of the Orange-Vaal River Basin. When the supercontinent Gondwanaland broke up, the continental margins of southern Africa warped upward (Corbett 1979, as cited in Skelton 1986a). This resulted in the formation of two drainage systems: a huge system that drained the gently sloping land to the west of the escarpment (the Orange-Vaal River) and a much shorter, more powerful series of rivers that drained the eastern side. The Orange-Vaal system was more extensive during this time and drained much further north and south of its present catchment (Wellington 1958, as cited in Skelton 1986a). As they received higher rainfall and were eroding to a lower base level, the eastern systems gradually captured more and more of the peripheries of the inland drainage system. In the Amatolo-Winterberg Highlands, as in other places, fish species would have been captured from parts of the Orange-Vaal River and isolated in the new river systems. Here they underwent speciation, resulting in the formation of present-day endemics and range-restricted species (Skelton 1986a).

Data Quality: Medium

Surprisingly little information is available on the wetland ecosystems of the Amatolo-Winterberg Highlands. Future research on aquatic taxa may reveal a rich and varied relict Afromontane fauna, in keeping with the results from the studies of fishes in the ecoregion.

Ecoregion Number:	**37**
Ecoregion Name:	**Drakensberg-Maloti Highlands**
Major Habitat Type:	**Highland and Mountain Systems**
Final Biological Distinctiveness:	**Bioregionally Outstanding**
Final Conservation Status:	**Critical**
Priority Class:	**IV**
Author:	**Belinda Day**
Reviewer:	**Paul Skelton**

Location and General Description

The Drakensberg-Maloti Highlands are highly valued in southern Africa for their rivers' excellent water quality and high water yield. These rivers provide water to large areas of xeric South Africa. The ecoregion encompasses the whole of Lesotho, excluding the westernmost lowland areas. It also includes small parts of South Africa just south and north of Lesotho. The whole of the ecoregion lies above 1,850 m a.s.l. and slopes upward toward the east, where the border with the KwaZulu-Natal province of South Africa forms the rim of the southern African plateau. The steep formations along the uplifted edges of the plateau make up the Great Escarpment, of which the Drakensberg is the eastern portion. In this ecoregion, the highest part of the Drakensberg Escarpment forms a ridge at a height of more than 3,000 m, with spectacular drops of 500–700 m (Hughes and Hughes 1992).

The climate is temperate and strongly seasonal, with extreme temperatures of –20°C in winter and 31°C in summer recorded on the mountain peaks. The weather is also prone to sudden changes (Jacobsen 1999). The ecoregion receives summer rainfall (October–March), but precipitation is highly variable, both spatially and temporally. Mean annual rainfall averages between 1,300 and 1,900 mm per annum in the highlands, with figures of 2,000 mm recorded on the highest mountain peaks in the east. Winters are cold and extremely dry, although snow and frost are common in the eastern mountains from April to September. On most afternoons mists move westward down from the Drakensberg Escarpment onto the highland slopes (Hughes and Hughes 1992; Barnes 1998a).

The terrain of the highlands ranges from undulating hills to rugged peaks and headlands intersected by deep valleys and ravines. The highest peak in southern Africa, Thabana-Ntlenyana (3,482 m), is located in this ecoregion, and the whole of the eastern half of Lesotho lies above 2,440 m (Hughes and Hughes 1992; Barnes 1998a). The mountains and peaks of the Lesotho Highlands above 2,300 m consist of basalts, extruded approximately 180 million years ago during a series of tectonic events. Underlying the basalts are horizontal sedimentary strata topped by soft sandstones (Barnes 1998a).

The ecoregion is heavily dissected by many small streams, most of which make up the headwaters of the Senqu-Orange River, which is the largest catchment in the ecoregion. These rivers have eroded deep gorges through the overlying basalts and into the sandstones, and the resulting cliffs and valleys can be more than 1,000 m deep (Barnes 1998a). The Senqu rises in the northeast and is joined by a number of tributaries in the highlands (the Senqunyane, Malibamatso, Sinqa, Makhaleng, Matsoku, and Khubela rivers) before it flows southwest through the lowlands of Lesotho. Known as the Orange in South Africa, the river then flows westward through semi-arid and arid lands and empties into the Atlantic Ocean. The headwaters of a number of short, eastward-flowing rivers (the Tina, Keneka, Umzimvubu, Umzimkulu, and Umkomaas) also arise in the highlands (Lesotho Government 2000). The streams of this ecoregion tend to be steep (dropping up to 1,200 m in 60–70 km) and fast-flowing. Upon reaching the foothills and lowlands in the west, the rivers slow down and widen, often forming small floodplains and marshes (Hughes and Hughes 1992).

Pans (areas without external drainage, some of which seasonally dry out), marshes (often with reedbeds), tarns (shallow pools and pans that typically form on sandstones), and bogs and sponges (some of which have been reclassified as midslope and valleyhead fens) are common in the highlands (Lesotho Government 2000). They are especially prevalent in the northeast, where rainfall is highest, and decrease in both size and frequency to the south and west. Most occur in the soft sandstone of the alpine regions above 2,300 m. Bogs and sponges collectively cover thousands of hectares, but most are individually small (Hughes and Hughes 1992, Barnes 1998a). They are found at all altitudes, wherever there is a continuous or semicontinuous source of water. This distribution extends to extremely high altitudes, such as on the plateau along the rim of the Drakensberg Escarpment and on summits of peaks such as Thabana Ntlenyana. Pans are found in a few places in the highveld grasslands both at lower altitudes and on several of the plateaus, whereas marshes are less common because of reed harvesting. Marshes remain at Tebeteben, Mohlaka-oa-tuka, and Koro-Koro (Lesotho Government 2000). Tarns occur in depressions in soft sandstones, such as those found in the Sehlabathebe National Park (Hughes and Hughes 1992).

The Drakensberg-Maloti Highlands ecoregion falls in the Afromontane and Afroalpine bioregions (Stuart et al. 1990). The Lesotho Government (2000) classifies the vegetation into three types: Highveld grassland (1,400–1,800 m), Afromontane grassland (1,800–2,500 m), and Afroalpine grassland (above 2,500 m). Each of the various types of wetlands in the ecoregion is characterized by a different vegetation community. For example, montane sandstone pools are characterized by aquatic species such as *Limosella* spp., *Utricularia* spp., *Ilysanthes confertiflora*, and *Aponogeton ranunculiflorus*, whereas gravel-filled pools tend to support *Crassula galpinii*. *Crassula natans*, *Aponogeton spathaceum*, and *Limosella maior* are found in sandstone pools in the foothills.

Outstanding or Distinctive Biodiversity Features

The Drakensberg-Maloti Highlands contain rare examples of Afromontane and Afroalpine rivers and other wetlands (Lesotho Government 2000). The high-altitude wetlands found in this ecoregion have limited occurrences in the Eastern Cape and are found nowhere else in southern Africa (Wetlands International 2002).

The wetland fauna of this ecoregion is depauperate in comparison to other large river catchments in southern Africa, in part as a result of a long history of disturbance and human use (Jacot-Guillarmod 1963; Du Plessis 1969, as cited in Ferreira 1999). There are no records of the historical distribution of the aquatic fauna of Lesotho, so it is impossible to accurately estimate changes in biodiversity or abundance, although it is certain that abundance has decreased and extirpations may have occurred (Jacobsen 1999). Furthermore, very few studies have investigated population size, distribution, life history, or habitat needs of the wetland fauna, particularly the invertebrates. What is known about the invertebrates indicates that in contrast to the vertebrates, they may be very rich in species and include montane paleogenic representatives. For example, there are twenty-three odonate species in the ecoregion, including the endemic *Chloroestes draconicus* (M. Samways, pers. comm., 2001). Little has been published about the levels of endemism of wetland plant species, but the aquatic species *Aponogeton ranunculiflorus* is endemic to the ecoregion (Jacobsen 1999).

There are only a few large wetlands in the Drakensberg-Maloti Highlands that can support a diversity of wetland birds, with the result that there are only twelve water-dependent bird species. Two riverine mammals (the marsh mongoose [*Atilax paludinosus*] and the Cape clawless otter [*Aonyx capensis*]) (Ferreira 1999) live in the rivers of this ecoregion.

The Drakensberg-Maloti Highlands ecoregion is rich in amphibians at all altitudes. Jacobsen (1999) postulates that the high rainfall and diversity of aquatic habitats contribute to the high species count. Twenty-one aquatic-dependent frogs are known from this ecoregion, of which three are endemic and several are near endemic. These restricted-range species include *Rana vertebralis*, *R. dracomontana*, *Strongylopus hymenopus*, *Arthroleptella hewitti*, and *Heleophryne natalensis* (Stuart et al. 1990; Jacobsen 1999). According to the Lesotho Government (2000), the aquatic or umbraculate river frog (*Rana vertebralis*) and the Lesotho or Drakensberg river frog (*Rana dracomontana*) are imperiled. Four snake species are often found in wetland habitats in the Drakensberg-Maloti Highlands, but the only true wetland species is the common brown water snake (*Lycodonomorphus rufulus*).

There are ten fish species in the ecoregion, not including introduced rainbow and brown trout (*Oncorhynchus mykiss*, *Salmo trutta*), among other introduced fish species. The only endemic, the Maloti minnow (*Pseudobarbus quathlambae*), is known only from six high-altitude tributaries of the Orange River. It is critically endangered because of habitat degradation and alteration and competition with and predation by alien trout species (Stuart et al. 1990; Rall and Skelton 2001). Other fish species include *Barbus aeneus*, *B. kimberleyensis*, *B. anoplus*, *Labeo capensis*, *L. rubromaculatus*, *L. umbratus*, *Anguilla mossambica*, *Clarias gariepinus*, and *Austroglanis sclateri*. The rock catfish (*A. sclateri*) is considered to be a rare or intermediate species (Lesotho Government 2000).

Status and Threats

Current Status

Overgrazing, trampling, burning, clearing, and cultivation of wetlands have become progressively more intense in this ecoregion, especially in the foothills. Herds often are kept at densities far greater than the carrying capacity of the ecosystem (Kingdom of Lesotho 1989). Overgrazing and trampling, especially during winter, have resulted in the formation of active dongas (erosion gullies) along many of the headwaters and elevated sediment loads in many of the rivers. It is estimated that Lesotho loses about 2,000 tons of soil per square kilometer annually from erosion (Chakela 1981, cited in Lesotho Government 2000). Many of the swamps and marshes in the foothills and lowlands are silted up, and the extensive reedbeds have been reduced to thin strips along the riparian zone (Hughes and Hughes 1992). Where the topography permits it, subsistence agriculture (mainly maize and wheat) is intensive, and burning is used to promote new growth. In the low-lying areas of the ecoregion, the population is dense, maize and sorghum cultivation is intense, and overgrazing is severe. The ecological functioning of the entire highland grassland ecosystem is thus being impaired as grass cover is reduced and erosion accelerated (Stuart et al. 1990; Barnes 1998a; Lesotho Government 2000).

The introduction of alien fishes has had a major impact on the endemic Maloti minnow. All existing populations of the minnow are protected by barrier falls or inhabit marginal trout habitats. In addition to introduced brown and rainbow trout, there are introduced common carp (*Cyprinus carpio*), bluegill sunfish (*Lepomis macrochirus*), smallmouth bass (*Micropterus dolomieu*), and largemouth bass (*Micropterus salmoides*).

Sehlabathebe National Park covers a small (68 km^2) high-altitude area on the eastern border of Lesotho, along the rim of the Great Escarpment. It contains the headwaters of the Tsoelikana River and a number of representative montane wetlands and is one of the few refuges of the fish *P. quathlambae* and the aquatic plant *Aponogeton ranunculiflorus*. Until recently, it was the only national park in Lesotho, and Masitise Nature Reserve in the south was the only nature reserve. However, the country is establishing four additional reserves in the north at Bokong, Tsĕhlanyae, 'Muela, and Liphofung in the area of the Lesotho Highlands Water Project (LHWP) Phase 1A (Lesotho Government 2000). Natural sanctuaries for *P. quathlambae* (both within and outside its native range), uninhabited by invasive fish species, are also being maintained to protect this critically endangered fish (Lesotho Biodiversity Trust 2002).

The 2,428-km^2 Natal Drakensberg Park, also a Ramsar site, lies adjacent to the Sehlabathebe Park on the KwaZulu-Natal side of the border. This park contains representatives of a wide variety of wetland ecosystems. Although the park was created to protect water supplies, water is pumped from the park and transferred to the interior Gauteng Province. Additionally, there have been past difficulties associated with land claims, uncontrolled agriculture, grazing, and afforestation with exotic tree species (Wetlands International 2002).

In June 2001, the governments of South Africa and Lesotho signed a Memorandum of Understanding to establish the Maloti-Drakensberg Transfrontier Conservation and Development Project (Derwent 2001; South Africa Ministry of Environmental Affairs and Tourism 2001). The project is designed to cover 8,113 km^2 of mountainous area straddling the northeastern border of Lesotho and South Africa. This would include the existing Sehlabathebe and Natal Drakensberg parks and other protected areas. In July 2002 the World Bank's Global Environment Facility committed a large grant to fund the project, the completion of which was expected to take 2 years (Groenewald 2003).

Future Threats

The LHWP is currently under construction and when complete will transfer 2.2 billion m^3 of water per year from the headwaters of the Orange River to the tributaries of the Vaal River in South Africa. This is the largest interbasin transfer in Africa (Snaddon et al. 1998). Phase 1A of the project is complete, and Phase 1B is nearly finished. The closure of the LHWP dams will result in decreased discharge, which probably will cause changes in water chemistry, loss of the natural flow regime, and a lowered water table. Other potential effects include the introduction of alien species to both the donor and recipient rivers. The Katse Dam, one of a series of dams in the transfer, blocks the upstream spawning runs of indigenous fish species in the Malibamasto River. The possibility of earthquakes in the region has increased because of the weight of the volume of water being stored at high altitude; a number of small earthquakes have already occurred since the Katse Dam reservoir began filling (Snaddon et al. 1998).

Climate change is also of concern for the species adapted to the highland environs of this ecoregion. However, the recent designation of the transfrontier conservation area provides hope for the protection of representative wetland ecosystem types in the ecoregion.

Justification for Ecoregion Delineation

The delineation of this ecoregion follows the boundaries of the Drakensberg-Maloti Highlands and comprises the middle subregion of the Montane escarpment aquatic region (Skelton 1993). The Drakensberg-Maloti Highlands are a residual portion of the peripheral uplands of the Great Escarpment that has progressively been eroded and dissected northward and westward by the rivers on the southern and eastern southern African coasts (Truswell 1977; Corbett 1979 in Skelton 1986a). Relict fauna and flora of Afromontane species are present in the highlands, and a comprehensive treatment of the fish biogeography of the region is presented by Skelton (1994).

Data Quality: Medium

Little is known about the historical distribution, habitat needs, biology, or life history of most of the wetland taxa, particularly the invertebrates, in the Drakensberg-Maloti Highlands. However, it seems likely from the information available for the few taxa that have been studied that further investigation may reveal a diverse fauna that could include endemic or relict species. Because many taxa in Lesotho are seriously threatened, it is essential that research, monitoring, and conservation be made a priority before ecosystems are permanently degraded and destroyed.

Ecoregion Number:	**38**
Ecoregion Name:	**Eastern Zimbabwe Highlands**
Major Habitat Type:	**Highland and Mountain Systems**
Final Biological Distinctiveness:	**Nationally Important**
Final Conservation Status:	**Endangered**
Priority Class:	**IV**
Author:	**Belinda Day**
Reviewer:	**Paul Skelton**

Location and General Description

The Eastern Zimbabwe Highlands straddle the eastern border of Zimbabwe with Mozambique. The ecoregion forms a narrow, north-south belt, 450 km long, encompassing the high easternmost rim of the central southern African plateau.

Although the ecoregion falls in the tropics, the climate is essentially temperate because of its high altitude. Above 1,800 m temperatures are moderate, with summer maxima of 25°C and mean annual temperatures between 15°C and 17.5°C (Childes and Mundy 1998; Mtetwe 2000). At lower altitudes the climate is more subtropical, with temperatures often exceeding 35°C in summer (Childes and Mundy 1998). There is a warm dry season from August to October, a warm wet season from November to March, and a cool dry season between April and July. Precipitation is orographic: southeasterly monsoon winds, blowing across Mozambique from the Indian Ocean, are forced up onto the plateau and release their moisture as rain. Rainfall is high and highly predictable, with a prolonged rainy season from October to April (Mtetwe 2000). Most of the highlands receive a mean annual rainfall of more than 1,500 mm, and many of the highest peaks receive more than 3,000 mm (Hughes and Hughes 1992; Childes and Mundy 1998). Frost is common during winter (May–August), and mists or light rains are frequent during the dry season.

The topography of this ecoregion is diverse, with rolling hills lying below massive mountain peaks separated by dramatic river gorges, waterfalls, and steep-sided valleys. The land rises to the east, where the rim of the escarpment forms a high-altitude north-south ridge, broken in places by fertile lower-lying river valleys, including the Honde, Rusitu, and Burma valleys. In the south, altitudes are generally between 1,500 and 1,900 m. The highest peaks of the Chimanimani Mountains and Umkondo Highlands reach above 2,000 m (Childes and Mundy 1998). Further north, in the region to the southeast of Mutare, the highlands form a dissected plateau approximately 1,700 m above sea level. The highest region of the Eastern Highlands is the Nyanga Mountains in the north, with Mount Inyangi the highest point in Zimbabwe (2,592 m).

The Eastern Zimbabwe Highlands are deeply dissected by the headwaters of numerous rivers. Many streams flow east for a short distance before descending high waterfalls at the rim of the escarpment into Mozambique. The topography is spectacular, with deep ravines and gorges separated by steep valleys. The streams are high-altitude, narrow mountain torrents, with rapid flow and rocky substrata. Most streams are oligotrophic, apart from the Gairezi River, which is naturally eutrophic (Bell-Cross and Minshull 1988). Because the river valleys are narrow and steep, floodplains and swamps are very rare, even at lower altitudes (Hughes and Hughes 1992). However, springs and small lakes are numerous at high altitudes. Dambos, which are seasonally waterlogged grasslands found in the valley bottoms, are found on granites and granitic gneisses along the headwaters of many of the streams (Chabwela 1994a).

A number of rivers drain the Eastern Zimbabwe Highlands. The Nyanga Mountains form the watershed divide on which tributaries of the Pungwe, Save, and Lower Zambezi rivers arise. These tributaries include the Pungwe, Kairezi, Nyangombe, Nyazengu, and Ruenya rivers. The headwaters of the Odzani, Revue, and Nyamhwarara rivers drain the Nyanga Lowlands and Honde Valley. Tributaries of the Lower Save and Buzi (whose main tributaries are the Revue, Rusitu, and Haroni rivers) drain the Chimanimani mountains before flowing east through Mozambique. The Pungwe and Save basins, which cover the northeastern and southwestern highlands, respectively, are two of the major river systems in Zimbabwe (Sanyanga 1994).

Rainfall decreases from east to west, such that eastern slopes tend to be well forested but western slopes tend to support grasslands and woodlands adapted to the drier conditions. Soils in the Eastern Highlands are coarse-grained, old, and highly leached sandy loams. The dominant vegetation type above 1,800 m is Afromontane *Themeda* grassland, which contains a variety of herbaceous plants and includes some Afroalpine species. Between 1,650 and 1,800 m, *Syzygium* montane forest dominates. The wetter eastern mountain slopes and protected ravines and gullies are covered with Afromontane rainforest. Further west the grasslands are drier and flatter and are interspersed with scattered areas of dwarf *Brachystegia spiciformis* woodland. Below 1,650 m the climate is more subtropical, and medium-altitude rainforests dominate. Low-altitude forests occur further downslope, with *Brachystegia* woodlands predominating. The large commercial tea plantations in the Nyanga area are characterized by the presence of *Maranthes goetzeniana, Newtonia buchananii,* and *Xylopia aethiopica,* all lowland forest species. Streams in the Gurungwe Peak area to the southeast support *Breonadia microcephala* along their banks (Childes and Mundy 1998).

Soils in the Chimanimani Mountains in the south of the ecoregion are extremely nutrient poor and sandy, with poor water retention properties. These unusual soils have resulted in the evolution of floral endemics, which have more in common with the southern African Drakensberg species than with other Zimbabwean species. The dominant vegetation types on the plateau in the Chimanimani Mountains are montane grasslands, interspersed with *Erica, Protea,* and *Phillipia* scrub, with Afromontane forests in the wetter valleys. Dry montane forests are found below 1,500 m, and in drier areas miombo woodlands replace these forests. River valleys in the Save catchment are characterized by evergreen riparian forests (Campbell 1994). The forests include species such as *Acacia albida, Guibourtia conjugata, Cordyla africana, Salvadora angustifolia, Azima tetracantha, Terminalia gazensis, Croton megalobotrys, Ptelcopsis myrtifolia,* and *Xanthocercis zambesiaca* (Hughes and Hughes 1992).

Outstanding or Distinctive Biodiversity Features

Little work has been done on the freshwater fauna and flora of the ecoregion, but the area contains high rates of endemism in terrestrial flora and fauna. For example, the Chimanimani Mountains contain the highest plant biodiversity in Zimbabwe, with approximately sixty endemic species, although most are not wetland species (Childes and Mundy 1998).

The rivers and streams of the Eastern Highlands have at least thirty-two fish species and four endemics (*Labeo baldasseronii, Amarginops hildae, Varicorhinus pungweensis,* and *Parakneria mossambica*). Small barbs (*Barbus* spp.) are particularly common, and Bell-Cross and Minshull (1988) mention a minimum of twelve species. The Eastern Zimbabwe Highlands are also rich in amphibians. A number of rare and range-restricted amphibians occur in this ecoregion, three of which are endemic: *Bufo inyangae, Probreviceps rhodesianus,* and *Afrana inyangae.*

The region supports few rare or restricted-range wetland bird species, although the endangered blue swallow (*Hirundo atrocaerulea*) needs water for nest building and constructs nests only in the vicinity of wetlands (Childes and Mundy 1998). Breeding pairs of the vulnerable wattled crane (*Grus carunculatus*) also occur. The river valleys and riparian woodlands of the ecoregion are important for a number of bird species, and nine important bird areas occur in the highlands, mainly in forest-grassland mosaic habitats (Childes and Mundy 1998, 2001; Parker 2001).

Status and Threats

Current Status

Land use is varied in the ecoregion. Most of the lower-lying western areas are communal lands. These tend to have high population densities, usually are overgrazed, and are heavily deforested. Large commercial farms, including several very large tea plantations, resettlement farms, smallholdings, extensive pine and wattle forestry estates, and conservation areas are all found in the ecoregion. Freshwater systems in the commercial farming areas are subject to eutrophication caused by agricultural fertilizers, pollution from pesticides, widespread dam construction, and bankside irrigation (Matiza 1994b). Poor farming practices and overintensive cultivation of riverbanks and wetlands have resulted in the degradation and siltation of almost all of the middle- and low-altitude rivers and riparian wetlands (Bell-Cross and Minshull 1988; Matiza 1994a). The Save, for example, was once a large, fast-flowing river but is now heavily silted and is generally reduced to a series of pools during the dry season. As on all large river courses in the ecoregion, almost all the riparian forests in the Save catchment have been cleared for cultivation, especially in the communal areas (Campbell 1994). Areas of indigenous montane forest tend to be small and scattered, and enormous plantations of exotic softwoods have engulfed many of them (Childes and Mundy 1998, 2001).

Fortunately, the inaccessibility and presence of reserves has afforded many of the high-altitude Eastern Highlands headwater streams some degree of protection. Few, if any, of these systems are pristine, however, because of the presence of alien fishes and numerous impoundments. These dams halt the upstream movement of fish, such as *Labeo* and *Clarias* (Masundire 1994). Exotic brown trout (*Salmo trutta*) and rainbow trout (*Oncorhynchus mykiss*) were introduced into Nyanga rivers in 1929 for sport fishing. Many of the Eastern Highland streams, including those on the Chimanimani Mountains, are now stocked with trout and bass (*Micropterus* spp.), and there are many trout hatcheries and dams (Bell-Cross and Minshull 1988). These fish probably have already had serious impacts on the native fishes and invertebrates of the highlands.

Dambos, which are used for cultivation and as sources of water for grazing, have been subject to severe degradation. Over the past few years many have dried out and become badly eroded through overintensive agriculture, overgrazing, fires, and gully erosion (Matiza 1994b).

Another threat to the freshwater systems of the Eastern Zimbabwe Highlands is alien trees in riparian zones. Areas of remaining indigenous riparian montane forest tend to be small and scattered, and many have been engulfed by enormous plantations of exotic softwoods. There are extensive alien wattle (*Acacia dealbata* and *A. mearnsii*) and pine plantations in the highlands (Childes and Mundy 1998). In 1988 it was shown that 40 percent of the Nyanga National Park was already invaded by alien trees, and little has been done to halt or slow their spread (Childes and Mundy 1998, 2001).

There are no Ramsar sites in the Eastern Highlands, but there are a number of protected areas. Nyanga National Park (440 km²) covers a fairly large proportion of the Nyanga Mountains and a small area of the Nyanga lowlands. There are also large areas of privately owned nature reserves. Most of the Bunga mist forest in the middle highlands is protected by the Bunga Forest Botanical Reserve. Banti Forest Reserve lies in the central highlands on the border with Mozambique. Most of the Chimanimani Mountains on the Zimbabwean side of the border lies in the Chimanimani National Park, and Haroni-Makurupini forest, the largest and most pristine area of lowland forest in Zimbabwe, is located in the park. Much of the forest in the Haroni and Rusitu botanical reserves has been cleared for banana plantations, and the remaining forest is under severe threat from illegal clearing. The high plateau in Mozambique is unprotected. To the south, Chirinda Forest is under protection but is threatened by human encroachment (Childes and Mundy 2001).

Future Threats

The construction of dams, deforestation of catchment areas, overgrazing, pollution, and eutrophication are expected to continue to threaten the freshwater systems of this ecoregion (Matiza 1994a).

Justification for Ecoregion Delineation

This ecoregion is defined by the high-elevation easternmost rim of the central-southern African plateau located along the eastern border of Zimbabwe with Mozambique and encompasses the Nyanga and Chimanimani mountain ranges. Skelton (1993) considers this ecoregion to be part of the montane escarpment aquatic region in southern Africa, based on fish distributions. The montane escarpment aquatic region has generally high-gradient streams with cool temperatures and is fragmented into highland "islands" and includes this ecoregion, along with the Drakensberg-Maloti Highlands [37] and Amatolo-Winterberg Highlands [36]. The fish fauna is depauperate and includes several characteristic species such as mountain catlets (Amphiliidae) and chiselmouths (Skelton 1993; Tweddle and Skelton 1998).

Data Quality: Low

There are few publications on the freshwater ecosystems of the Eastern Zimbabwe Highlands, and many of the freshwater taxa, particularly the invertebrates, are all but unknown. For example, data on both the vegetation and invertebrates of Zimbabwean fresh waters are insufficient to allow the identification of freshwater bioregions (Mtetwe 2000). Studies of the aquatic invertebrate fauna of the highland streams would prove extremely useful in assessing the ecological distinctiveness of the Eastern Zimbabwe Highlands.

Ecoregion Number:	**39**
Ecoregion Name:	**Ethiopian Highlands**
Major Habitat Type:	**Highland and Mountain Systems**
Final Biological Distinctiveness:	**Bioregionally Outstanding**
Final Conservation Status:	**Endangered**
Priority Class:	**IV**
Author:	**Abebe Getahun**

Location and General Description

With a long history of isolation, the Ethiopian Highlands are known to harbor an endemic biota. The highlands extend from Eritrea in the north to Kenya in the south. According to Westphal (1975), uplift of the Ethiopian Highlands and Arabia occurred on an extensive scale after the regression of the Red Sea toward the southeast in the late Mesozoic to early Tertiary. The Great Rift Valley bisects the highlands into the eastern and western massifs, which are surrounded by escarpments. This ecoregion contains about 70 percent of Africa's highlands. The highest peak, Ras Dejen (Dashan), at 4,620 m, is in northern Ethiopia's Gondar region.

In the northwestern part of the highlands, the deep, steep-sided valleys of the major rivers separate blocks of mountains, and the upper courses of the big rivers such as the Tekezze and Abay (Blue Nile) plunge through deep gorges. This part of the Ethiopian Highlands is also the source of the headwaters for the Blue Nile. The Blue Nile watershed is the largest basin in Ethiopia. Rivers of this basin drain the great central plateau, and the Blue Nile descends about 1,450 m in a distance of only 350 km from its source to Khartoum. The Blue Nile flows to the north and forms a large floodplain in Egypt. The rich sediments carried by the Blue Nile are deposited downstream, greatly increasing the soil fertility of floodplains along the course of the Nile. Historically, nearly all of the sediments in the Nile Delta came from Ethiopia via the Blue Nile and Atbara rivers (Beadle 1981).

The southeastern portion of the Ethiopian Highlands includes the Sidamo, Bale, Arsi, and Harerge mountains. The highlands in this region are made up of volcanic rocks, and deep river cuts expose crystalline rocks (Ethiopian Mapping Authority 1988).

The Ethiopian Highlands receive about 950 mm or more of rainfall because of a double passage of the intertropical convergence zone. The high mountains east of Lake Tana and the southwestern mountains stand out as places of higher rainfall.

They receive 2,000 mm or more of rainfall each year (Westphal 1975). A rainfall regime that peaks in March–May and June–August is typical for the Ethiopian Highlands.

Rivers of the western highlands generally flow toward Sudan, whereas those of the eastern highlands tend to flow toward the Indian Ocean. The westward-flowing rivers (the Tekezze, Angereb, Atbara, Abay, Baro, and Akobo) form part of the Nile drainage basin. Three major highland lakes, Hayq, Ardebo, and Ashengie, lie near the edge of the western escarpment of the rift valley at altitudes between 2,000 and 2,500 m. Lake Hayq, located in northern Ethiopia's Wollo region, has an area of 5 km^2 and a maximum depth of 23 m and is noteworthy for its extremely clear water (Kebede et al. 1992). Lake Ardebo is located about 5 km southeast of Lake Hayq. This lake is smaller than Lake Hayq and flows into Hayq via the Anchercah River (Kebede et al. 1992). Lake Ashengie is located north of Lake Hayq in the Tigray region and sits at an altitude of 2,460 m. The lake covers an area of 25 km^2, with a maximum depth of 20 m and a mean depth of 14 m (Wood and Talling 1988). The lake is fed by a number of small streams from the surrounding areas, and there is no drainage out of the lake (Ethiopian Wildlife and Natural History Society 1996).

Outstanding or Distinctive Biodiversity Features

The fishes of the high mountain torrential streams are largely cyprinids (Harrison 1995; Getahun and Stiassny 1998) adapted to the swiftly flowing floodwaters that occur seasonally. Two genera of fishes (*Barbus* and *Garra*) dominate the fish fauna of these streams. *Clarias gariepinus, Varicorhinus beso,* and *Labeo* spp. are also found in high numbers. Endemism appears to be high among fish, but the fish fauna is not well known. Endemic fishes of the genus *Garra* (e.g., *G. dembecha, G. duobarbis,* and *G. ignestii*) have recently been described (Getahun 2000). Lake Hayq is believed to have no indigenous fish species, although the presence of *Clarias gariepinus* has been reported (Beckingham and Huntingford, in Kebede et al. 1992).

The Baro-Akobo Basin apparently is particularly rich in fish diversity (Golubtsov et al. 1995). The fauna is Nilo-Sudanic and is dominated by *Alestes, Bagrus, Barilius, Citharinus, Hydrocynus, Hyperopisus, Labeo, Malapterurus,* and *Mormyrus* genera. *Nemacheilus abyssinicus* is an endemic species found in this drainage basin, the Omo-Gibe drainage basin, and Lake Tana.

The invertebrate fauna is less well known than the fishes, and it is difficult to estimate endemism among the aquatic invertebrates. Harrison and Hynes (1988) indicated that *Dugesia* spp., *Baetis harrisoni, Pseudocloeon* sp., *Centroptilum sudafricanus, Afronorus peringueyi, Neoperla* spp., *Hydropsyche* sp., *Simulium* spp., nymphs of *Aeschna,* and chironomid larvae dominate the benthic communities in the stony runs and torrents of the Ethiopian Highlands. Compared with other highland ecoregions, the Ethiopian Highlands support a rich aquatic mollusk fauna, with more than twenty species described.

Status and Threats

Current Status

Because of past agriculture and timber-harvesting activities in this ecoregion, most of the rivers and streams have lost their riparian vegetation. This is particularly the case in the northern, northeastern, and central parts of the Ethiopian Highlands, inhabited by agriculturists. In these parts of the ecoregion, small patches of forest around churches are all that remain of the once densely forested highlands. In contrast, the southwestern and southeastern highlands are protected. The people in these regions depend on forest products and the presence of forest cover for their livelihood (e.g., shade for coffee plantations, hanging beehives, and shed and grazing ground for domestic animals). In degraded areas, overgrazing by high densities of cattle, sheep, and goats inhibits the regeneration of forest cover. The negative impacts of deforestation on aquatic systems are manifold and cause major perturbations to nutrient regimes, water temperature, turbidity, and pH. Sedimentation from erosion smothers benthic organisms, depresses oxygen levels, and reduces light penetration and photosynthesis. There are no major initiatives to reclaim these areas, nor are there adequate measures to stop further degradation.

Alteration of rivers and streams through damming and canalization are widespread. Changed flow regimes alter temperature, pH, nutrient, and sediment levels in downstream areas. In addition, unless carefully designed, dams affect the life cycles of migratory fishes and consequently hinder reproduction.

The human population in the Ethiopian Highlands is extremely high and continues to grow. Water quality problems caused by poor land use are exacerbated by the disposal of sewage and other wastes.

There are two national parks in this ecoregion: the Simien Mountains National Park in the northwest and the Bale Mountains National Park in the southeast. Rivers that lie in the Bale Mountains National Park, such as the Danka and Weyb rivers, are somewhat protected. However, these high-altitude rivers have a depauperate fish fauna, and only introduced rainbow trout (*Salmo gairdneri*) and brown trout (*S. trutta*) are found.

Several lakes in this ecoregion have been stocked. *Oreochromis niloticus* was stocked in 1978 in Lake Hayq (Kebede et al. 1992), and there is a commercial fishery for this species. Lake Ardebo and Lake Ashengie have also been stocked with *Oreochromis niloticus*.

Future Threats

Major threats to the aquatic systems in this ecoregion are loss of riparian forest, canal formation for irrigation purposes, and dam building. Several dams are proposed along the Abay (Blue Nile) and Tekezze rivers. There are also widespread irrigation projects that transport water via extensive systems of canals. These medium-sized projects are abundant in the northeastern highlands. Policies should be formulated for stream protection, including reforestation of riparian buffer zones.

Justification for Ecoregion Delineation

This ecoregion is defined by the two blocks of highlands in Ethiopia, separated by the rift valley and distinguished by a freshwater fauna adapted to the swift-flowing rivers of this high-altitude ecoregion. Despite the fact that the Ethiopian Highlands are separated from both the East African and the South Arabian mountains, the riverine fauna resembles that of eastern and southern Africa (Tudorancea et al. 1999), along with some elements of the Arabian Peninsula. Cyprinids are the dominant fish in the rivers of this ecoregion. For example, it is known that some small *Barbus* species (e.g., *Barbus paludinosus*, *B. trimaculatus*, and *B. radiatus*) have widespread distributions extending from South Africa to East Africa and into the highlands (Skelton et al. 1991). There are also fish groups (e.g., *Garra*) common to the Ethiopian Highlands and the Arabian Peninsula. These fish groups are estimated to have originated in the Lower Tertiary or late Cretaceous, before the separation of India and the Arabian Peninsula from continental Africa (Briggs 1987), whereas the Red Sea is believed to have separated the African continent from the Arabian Peninsula in the early Tertiary, between the Eocene and Oligocene epochs (Getahun 1998).

Data Quality: Medium

Historically, few scientific studies were made on the fauna of the river systems of Ethiopia; however, two recent studies have elevated the level of data available for the fish of this ecoregion (Getahun and Stiassny 1998; Golubstov et al. 2002). River systems in the Tekezze-Angereb Basin have not been studied at all because of past security problems. Preliminary reports indicate that the large river bodies of this basin support a rich fish fauna, and research is needed to confirm this. Some information on the benthic fauna of Ethiopian mountain streams and rivers is available in Harrison and Hynes (1988).

Ecoregion Number: **40**
Ecoregion Name: **Fouta-Djalon**
Major Habitat Type: **Highland and Mountain Systems**
Final Biological Distinctiveness: **Bioregionally Outstanding**
Final Conservation Status: **Vulnerable**
Priority Class: **V**
Author: **Michele Thieme**
Reviewers: **Christian Lévêque and Samba Diallo**

Location and General Description

Isolation of individual rivers, stability over geologic time, and numerous waterfalls and rapids that have restricted colonization by downstream species are believed to have encouraged the evolution of a large number of species limited to the rivers draining the Fouta-Djalon plateau. Headwaters for both north-flowing rivers, including the Senegal and Niger and the shorter, more torrential coastal rivers of Upper Guinea, emanate from the Fouta-Djalon. The area is characterized by sections of blocky, elevated plateau (about 600–1,500 m) separated by deep gorges through which rivers and streams descend to the coast or to feed larger rivers inland (Daget 1962). Most of the plateau and the lower plains of Gaoual and Koundara are in Guinea.

The Fouta-Djalon plateau is surrounded by a zone of forest-savanna transition, but the upland areas of the plateau are dominated by submontane vegetation, including the most prevalent species, Guinea plum tree (*Parinari excelsa*) and *Parkia biglobasa* (Martin 1991; Sayer et al. 1992). There is hardly any forest on the massif apart from some small, well-preserved forest reserves and a 4.5-km^2 pine plantation at Dalaba (DNE et al. 1999). The aquatic vegetation is dominated by Marantacea and Cyperacea. The rivers often are bordered by a forest gallery with *Pandanus*. Grasses bordering the rivers are dominated by *Hyparrhennia*, *Andropogon*, and *Pennisetum*. In many areas, the remaining forests constitute patches in the headwaters, galleries along the rivers, and scattered trees on the floodplains.

Several of the important mountainous areas in the Fouta-Djalon include Mont Loura (Mali), Fello Digui (Gaoual), Fello Sounga (Gaoual), and the raised edge of the piedmont (DNE et al. 1999; Bazzo 2000). Mont Lora is the highest peak in the region (1,538 m).

Rainfall in the region ranges between 1,600 and 2,200 mm/year, with 90 percent of it falling in the 6-month wet season (Hughes and Hughes 1992). The rivers descend rapidly from the plateau during the wet season, but because they drain areas dominated by sandstone, granite, and dolerite formations they have low particulate sediment loads (about 21 mg/L compared with their dissolved load (35 mg/L). The majority of the dissolved load comes from atmospheric inputs rather than rock weathering (Orange 1992).

The following tributaries and rivers originate in the Fouta-Djalon: the Tinkisso of the Niger Basin, the Bafing and Téné rivers of the Senegal Basin, the Gambia, Komba and Tominé rivers to the west, and the Kogon, Fatala, Konkouré, Kolenté, Kaba, and Mongo to the southwest. In addition to the Senegal, the Gambia, and the Konkouré, other major river systems beginning in the Fouta-Djalon include the Niger and the Kobal.

Floodplains are rare in the ecoregion, except in the Coliba Basin northwest of the Koundara prefecture. These cover an area of 300–400 km^2. Matthès (1993) has described many temporary and perennial lakes and ponds throughout the Fouta-

Djalon. Examples are the Mouké Djigué pond in the plain of Koloun in the Gaoual Prefecture, the pond of Brouwal in the Télimélé Prefecture, and lakes Kénè, Wèdou, and Kambouwol in the Lélouma Prefecture. These sites support traditional fishing activities and animal congregations (mammals, reptiles, birds, and insects).

Outstanding or Distinctive Biodiversity Features

The ecoregion supports a moderately endemic fish fauna with several relict species, and it is very rich in aquatic invertebrates. The fish fauna is dominated by fish from the Cyprinidae (minnows), Mormyridae (snoutfishes), and Alestiidae (African tetras) families. Nearly all of the endemic species are cyprinids, and most of these are barbs. About one-quarter of the sixty described fish species in the ecoregion are endemic.

Among aquatic plants, more than forty families have been identified along the rivers, lakes, ponds, and floodplains. This ecoregion is also home to an endemic Bromeliaceae (pineapple family), *Pitcairnia feliciana,* the only member of this family to live outside the Americas.

Flooding is intense during the wet season, when large amounts of rain fall in a short period of time. However, because of the great age of the plateau and its well-weathered nature, sediment load is low. The aquatic organisms of the plateau are adapted to periods of intense flooding followed by months with no rain and thus declining water levels.

Status and Threats

In the ecoregion many people use traditional agricultural practices: clearing forest by burning and using the open, fertile land to grow crops. These customs led to the loss of the majority of the forest cover of the Fouta-Djalon plateau early in the twentieth century, although remnants remain in forest reserves (Sayer et al. 1992). Mont Loura, Mont Kakoulima, Mont Gangan, Fello Digui, and Fello Sounga are in a state of total degradation following an overexploitation of their resources. Soil fertility has declined because of physical erosion and degradation of the soil structure.

Other threats in the Fouta-Djalon include road and dam construction, pollution, erosion, earthquakes, flooding, climate change, and sedimentation. Lack of forest cover, combined with increasing pressures on the systems from a growing population, is certain to have affected the freshwater systems of the region, although few data on the condition of rivers and streams of this ecoregion exist.

This ecoregion is ecologically important as a major source of water for a wide part of West Africa, such that the protection and management of this area are of international concern. Several projects aim at protecting and improving management and agricultural practices, and the Fouta-Djalon is under a regional integrated restoration and management program.

Justification for Ecoregion Delineation

This ecoregion is defined by the Fouta-Djalon plateau and is distinguished by an endemic aquatic fauna adapted to its headwater streams. Several factors facilitated evolution of the endemic fish fauna: individual rivers are extremely isolated from one another by the plateau; numerous waterfalls and rapids occur along their course, providing further isolation of headwater faunas from potential downstream colonists; and the cooler conditions afforded by the higher altitude of the plateau may have offered a refuge during periods of warming and desiccation at lower elevations. Initial uplifting in the late Jurassic created the Fouta-Djalon plateau. In the Miocene, tectonic movements caused further uplifting, and the plateau has been stable since that time. The plateau, as part of the Guinean range, has acted as a barrier to movements of aquatic species across it. For example, only a few fish species are known from both slopes of the Fouta-Djalon: *Amphilius rheophilus* and *A. platychir,* as well as *Barbus cadenati, B. guineensis,* and *B. dialonensis* (Lévêque 1997). In general, the fish fauna has affinities with the Upper Guinea bioregion.

Data Quality: Medium

Intensified data collection is needed.

Ecoregion Number:	**41**
Ecoregion Name:	Madagascar Eastern Highlands
Major Habitat Type:	Highland and Mountain Systems
Final Biological Distinctiveness:	Globally Outstanding
Final Conservation Status:	Critical
Priority Class:	I
Author:	John S. Sparks

Location and General Description

The Madagascar Eastern Highlands ecoregion encompasses a narrow strip of land extending from north to south, essentially the entire length of Madagascar, and covers the upper and middle reaches of the eastern coastal rivers. The ecoregion covers about 20 percent of the island, or about 80 percent of the eastern slope region (Aldegheri 1972). Terrain in the region is steep, with the summits of many eastern mountains located as little as 50 km from the coast (Aldegheri 1972). Dense evergreen forest is the principal vegetation type throughout this ecoregion, except at elevations above 1,800 m, where a shift to thicket and shrubland occurs (Lowry et al. 1997).

Rivers in this ecoregion are generally small to moderate in

size, rocky in substrate, and swift in current. Drainages are steep in their upper reaches, with numerous sections of rapids and cascades interspersed with flatter stretches of more moderate flows. These coastal rivers are short and terminate on a narrow coastal plain over a contracted continental shelf. In undisturbed areas, the water is clear and well oxygenated. Compared with rivers in western Madagascar, eastern rivers have lower pH (slightly acidic vs. neutral), higher conductivity, and much lower dissolved carbonate levels (Riseng 1997). Rainfall is high in the Madagascar Eastern Highlands, 2,500–3,000 mm annually according to Aldegheri (1972), and water levels vary widely with the occurrence of seasonal storms. In contrast to many river systems in western Madagascar, rivers in the east flow year-round.

From north to south, the major rivers of this region are the Bemarivo, Lokoho, Maningory, Rianila, Mangoro, Mananjary, and Mananara (Aldegheri 1972). Lake Alaotra, the largest lake on the island in surface area, is located in this ecoregion and empties into the Maningory River.

Outstanding or Distinctive Biodiversity Features

More than fifty species (some awaiting description) of freshwater fish, about half of which are endemic, are known to inhabit the rivers and streams of the Madagascar Eastern Highlands. Two fish families, Bedotiidae (rainbowfish) and Anchariidae (catfish), are endemic to Madagascar and are represented in this ecoregion (Sparks and Stiassny 2003). Endemic fish faunas are associated especially with *Pandanus* and palm swamps, as are found in the upper Mangoro and Mananara drainages. Because of land conversion, these habitats are now rare in Madagascar.

Both the Mangoro-Nosivolo and Mananara rivers harbor endemic cichlids, bedotiids, and anchariids and diverse other native species (Reinthal and Stiassny 1991; Sparks and Reinthal 2001). The upper sections of these two rivers are regions of noteworthy diversity in the ecoregion. A small stretch of rapids in the Nosivolo River near Marolambo is the only known habitat of the endemic cichlid *Oxylapia polli* (Reinthal and Stiassny 1991), and an area of only a few square kilometers in the Mananara drainage encompasses the distribution of the cichlid *Ptychochromoides vondrozo* (Sparks and Reinthal 2001).

In contrast, the headwaters of many other eastern rivers support markedly depauperate fish assemblages, and native fish species often are absent from the uppermost reaches. An example is the upper reaches of the Namarona River, with only two goby species having been collected there (J. Sparks, pers. obs., 1994, 1996). If native fishes occur at all in many of these headwaters, communities consist of at most a few gobioids (gobies and eleotrids), such as rock-climbing gobies of the genus *Sicyopterus*. If there are no significant barriers to upstream dispersal, eels (*Anguilla* spp.) often migrate into these areas. Fish communities in the middle reaches of these rivers are more diverse and are characterized by communities including endemic cichlids, bedotiids, gobioids, and anchariid catfish.

Rivers of the Madagascar Eastern Highlands (and Madagascar Eastern Lowlands [23]) are home to a number of closely related species of endemic rainbowfish of the genus *Bedotia*. The results of preliminary studies of this group suggest that each major eastern basin may contain a unique species (Sparks and Stiassny 2003; Loiselle and Stiassny 2003; Sparks and Smith 2004). Bedotiids of the genus *Rheocles* are also diverse in the middle and upper reaches of eastern rivers, although nearly all species (particularly *R. sikorae*, *R. wrightae*, and *R. pellegrini*) are quite rare and limited in distribution. Intact native fish communities can still be found in the middle reaches of many eastern rivers, including the Nosivolo, Ankavanana, and Mananara. Exotic species become more abundant as one moves downstream into the lower reaches of many eastern basins.

In addition to freshwater fishes, this ecoregion supports extremely high richness and endemism among aquatic frogs. About 130 aquatic frogs are described from the ecoregion, with nearly 65 percent being endemic. Many endemic frogs of the genus *Boophis*, which lay their eggs directly in water, are limited to midaltitude forests in eastern Madagascar (Glaw and Vences 1994). The middle and upper reaches of eastern rivers are also home to the nocturnal and rarely seen semi-aquatic tenrec, *Limnogale mergulus*. Tenrecs are an endemic family of insectivores of which only *Limnogale*, with webbed feet, is adapted for a semi-aquatic existence (Eisenberg and Gould 1970). Benstead et al. (2000) identified three additional mammal species, all endemic viverrids (*Fossa fossana*, *Galidea elegans*, and *Mungotictis decemlineata*), which depend at least partially on aquatic habitats. In the wetlands, both the critical Alaotra little grebe (*Tachybaptus rufolavatus*) and the critical Madagascar pochard (*Aythya innotata*) have not been observed recently and may be extinct (IUCN 2002). Other bird species in this ecoregion that are limited to marshland habitats on Madagascar include the endangered slender-billed flufftail (*Sarothrura watersi*), the Madagascar snipe (*Gallinago macrodactyla*), and the Madagascar rail (*Rallus madagascariensis*) (Langrand 1990). Forested regions in the ecoregion support a number of endemic and highly threatened crayfish species of the genus *Astacoides* (Hobbs 1987; Benstead et al. 2000). Preliminary studies indicate that diversity and endemism of aquatic insects also are high in this region (Benstead et al. 2000).

Status and Threats

Extensive deforestation throughout the region and the spread of exotic species are major threats to organisms inhabiting aquatic systems in the ecoregion (Benstead et al. 2000). Human population density is high throughout eastern Madagascar, and as a result, habitat conversion for agricultural use continues to pose a significant threat to freshwater habitats. Lake Alaotra, Madagascar's largest lake, is a prime example of a severely degraded lacustrine habitat. The Alaotra little grebe was known only from this lake and is believed extinct, as is the Madagascar pochard. Currently, the lake is dominated by exotic species, although an endemic bedotiid, *Rheocles alaotrensis,* remains extant. As a result of human encroachment, forest fragmentation is extreme in eastern Madagascar. Very few headwater regions are afforded any protection (including ichthyologically diverse systems such as the Mangoro-Nosivolo and upper Mananara systems), despite our knowledge that disturbances in these areas influence the stability of the entire basin and can have devastating consequences downstream. Unfortunately, freshwater systems have received little attention, and these habitats have not been conservation priorities in Madagascar until now.

Justification for Ecoregion Delineation

This ecoregion is defined by the upper reaches of the eastern coastal drainages above about 200 m elevation and distinguished by an endemic aquatic fauna.

Data Quality: Medium

Of the five major aquatic ecoregions of Madagascar, ichthyologists probably have explored the Madagascar Eastern Highlands and Madagascar Eastern Lowlands [23] most thoroughly. This is largely because a number of larger towns and a network of passable roads along the coast permit access to the upper and middle reaches of many eastern rivers. Yet many east coast drainages remain to be surveyed for fishes and other freshwater taxa. For example, a major gap exists in our knowledge of ichthyofaunal communities for the middle and upper reaches of rivers extending from the Masoala Peninsula south to Lake Alaotra (Sparks and Stiassny 2003). The middle and upper reaches of many of these rivers have yet to be inventoried for freshwater fishes and should be considered top research priorities.

Ecoregion Number:	**42**
Ecoregion Name:	**Mount Nimba**
Major Habitat Type:	**Highland and Mountain Systems**
Final Biological Distinctiveness:	**Bioregionally Outstanding**
Final Conservation Status:	**Endangered**
Priority Class:	**IV**
Author:	**Michele Thieme**
Reviewer:	**Christian Lévêque**

Location and General Description

Mount Nimba forms part of the southern extent of the "Guinean backbone," which stretches from northern Guinea to northern Côte d'Ivoire and rises 1,000 m above the surrounding lowland plains. This mountainous region is located at the intersection of Guinea, Ivory Coast, and Liberia.

Mount Nimba is part of an ancient mountain range, the Guinean range, which was upthrust between the end of the Jurassic and the end of the Eocene (Lévêque 1997). Since then, erosion has worn away the softer schists and granito-gneiss, exposing the underlying ore-containing quartzite (Lamotte 1983). This large ridge is about 40 km long and 8–12 km wide. Steep cliffs and quartzite peaks are prominent features of the landscape, along with deep valleys, high plateaus, rounded hilltops, and granite blocks. The highest peak in the ecoregion is Mont Richard Molard at 1,752 m (WWF and IUCN 1994).

Rainfall varies markedly with elevation and season. About 3,000 mm of rain falls at the highest altitudes, but much less, 700–1,200 mm, falls at the edges of the ridge (WWF and IUCN 1994). Most rainfall occurs between May and October.

Vegetation also changes with elevation. Grasslands cover the summits, dominated by *Loudetia kagerensis* and *Protea occidentalis* on the slopes. The plum tree (*Parinari excelsa*) dominates midaltitude (above 1,000 m) forests. A cloud or mist hangs over the mountain for months at a time at elevations over 850 m, promoting a lush growth of epiphytes in these forests (WWF and IUCN 1994). The forests give way to plains savanna that covers the piedmont below altitudes of about 500 m (Lamotte 1983).

Rivers descending the steep slopes of Mount Nimba run swiftly, often experiencing torrential floods during the rainy season. Rheophytes, plants that can live in running water, dominate the riparian vegetation (Hughes and Hughes 1992). The Cavally and Ya rivers, as well as tributaries of the Sassandra and Cess rivers, originate on Mount Nimba.

Outstanding or Distinctive Biodiversity Features

Moderate richness and high endemism of aquatic species, particularly among fish and amphibians, characterize the highland Mount Nimba ecoregion. The endemic aquatic fauna of Mount Nimba includes frogs, fish, a freshwater crab, and the endangered Mount Nimba otter shrew (*Micropotamogale lamottei*). Of the frogs, the endemic *Nectophrynoides occidentalis*, which occurs in montane grasslands, and *N. liberiensis*, also found on Mount Nimba, are totally viviparous. Species richness is high among aquatic invertebrates. For example, eighty-one dragonfly species (*Odonata*) have been identified from the Nimba region (Legrand 1985). The Cape clawless otter (*Aonyx capensis*) also lives in the mountain streams.

Status and Threats

Iron mining and slash-and-burn agriculture have severely degraded many rivers and streams in the ecoregion. Dumping of capping waste from mining activities has led to heavy metal pollution and sedimentation of rivers draining parts of the mountain, causing loss of riparian vegetation in some running waters and presumably decreased abundance of aquatic fauna (Collar and Stuart 1988). Population pressure is increasing in the region with an influx of refugees from Liberia, a growing local population, and workers coming to the mines (UNESCO 1999).

Mount Nimba has been a strict nature reserve since 1943 in Côte d'Ivoire and 1944 in Guinea. The Guinean section was recognized internationally as a biosphere reserve in 1980, and both reserves were added to the World Heritage list in the early 1980s. However, the Centre for Environmental Management of Mount Nimba (CEGEN) was established by the Guinean government after a large-scale mining project adjacent to the reserve was proposed. CEGEN would require any mining operator that worked in the region to follow guidelines provided by the World Heritage Committee and to take measures to protect the environment. Representatives of CEGEN and some conservationists have expressed hope that properly conducted mining activities could lead to economic benefits and funding for a sustained conservation effort in the reserve (WWF and IUCN 1994; UNESCO 1999). More recently, a trinational program for the integrated conservation of the Nimba Mountains has been initiated to outline strategies for transboundary communication, coordination, and planning for effective management of the zone (Toure 2002). Additionally, Guinea is undertaking a Global Environmental Facility–funded project, Conservation of the Biodiversity of the Nimba Mountains through Integrated and Participatory Management, which will address management of the Nimba Mountain Biosphere Reserve.

Justification for Ecoregion Delineation

This ecoregion is defined by the high-elevation Mount Nimba and distinguished by an endemic aquatic fauna. The Guinean Highlands, of which Mount Nimba forms a part, separate the coastal rivers and streams to the west from the north-flowing Niger River to the east. The highlands have formed a barrier to movement of aquatic species between these systems (Hugueny 1989). The fish fauna of the highlands has affinities with the Upper Guinea bioregion (Daget 1963). In the highlands, Mount Nimba's relative high elevation, the presence of rapids and waterfalls that has led to isolation, and the stability of the aquatic environment over time have promoted speciation.

Data Quality: Low

Although a biological research station exists on Mount Nimba, only a few studies have been completed on aquatic species.

Ecoregion Number:	**43**
Ecoregion Name:	Mulanje
Major Habitat Type:	Highland and Mountain Systems
Final Biological Distinctiveness:	Nationally Important
Final Conservation Status:	Endangered
Priority Class:	IV
Author:	Denis Tweddle
Reviewer:	Paul Skelton

Location and General Description

The Mulanje Massif is a huge, isolated block of mountains, more than 640 km^2 in extent, situated in the southeast corner of Malawi close to the Mozambique border (Eastwood 1979). The massif rises abruptly from the flat Phalombe plain (600 m above sea level), to a plateau at nearly 2,000 m, and then up to rocky peaks reaching 3,000 m.

The climate of the ecoregion is influenced largely by the equatorial area of low pressure and the position of the Inter-Tropical Convergence Zone (ITCZ). The ITCZ fluctuates in position over Malawi during the rainy season from October to March. The warm northeast monsoon is sometimes unstable, causing heavy storms to develop, especially over the north-facing slopes. Apart from the rainy season, southeast trade winds prevail. The airstream is cool and moist, and forced uplift causes heavy rainfall on the southeast-facing slopes and rainfall in all months, whereas northern Mulanje has a single rainfall maximum in December and January and a fairly long dry season. The southern slopes and plateau average 2,500 mm of rainfall annually, and the northern slopes receive about 1,000–1,500 mm. The high rainfall and extended rainy season result in many perennial streams flowing from the massif, with important implications for biodiversity.

The streams from the massif feed two systems. The northern streams flow into Lake Chilwa, a large (700 km^2), shallow lake at the headwaters of the Ruvuma River system. The southern streams coalesce into the Ruo River, a tributary of the Lower Shire River and ultimately the Lower Zambezi. All streams have torrential flow in the rainy season and during the dry months are much reduced in volume, with clear water protected from excessive sedimentation by the forest reserve of the massif.

The Phalombe plain is heavily populated and used extensively for agriculture. To the south of the mountain, the Lower Shire rift escarpment causes the Ruo River to drop rapidly to around sea level. This descent includes the 60-m-high Zoa Falls, which effectively isolate the Ruo River fish fauna from the Lower Zambezi fauna found in the lower reaches of the Ruo and in the Lower Shire. The area to the east of Mulanje in Mozambique is poorly known. Isolated smaller mountains in the area may have similar aquatic faunal characteristics to Mulanje.

Outstanding or Distinctive Biodiversity Features

The Mulanje Massif is an island of montane biodiversity. Its steep slopes and relative inaccessibility, together with the protection afforded by the forest reserves, have safeguarded its natural resources from large-scale exploitation and clearance for agriculture. The high relief results in much more rainfall than in surrounding areas and thus many more perennial streams.

On top of the plateau, the only fish species are introduced rainbow trout (*Oncorhynchus mykiss*) and an indigenous mountain catfish (*Amphilius hargeri*). The identity of the latter is in doubt because the species was described from one specimen collected on top of the mountain. The type specimen shows slight differences from the common and widespread *A. uranoscopus*, which is found, together with *A. natalensis*, at the base of the mountain. Further study is needed of specimens from the top of the mountain to determine the validity of *A. hargeri*.

In the Ruo River from the base of the plateau to Zoa Falls, at least ten fish species out of thirty-seven recorded are not found in neighboring river systems, including at least six probable endemics that are under investigation (Tweddle 1985; Tweddle unpublished data; Tweddle and Skelton 1998). Sampling in nearby Mozambique streams that drain to the East African coast is needed to verify the endemism of the Ruo River species.

Several relict fish species occur in the Ruo and in other Zambezi tributaries. These include *Hippopotamyrus ansorgii*, *Barbus eutaenia*, *B. lineomaculatus*, and *Opsaridium zambezense*. The perennial nature of the streams flowing from the southern slopes of the massif, and the barrier to upstream movement created by Zoa Falls, have allowed the long-term survival of a fish fauna that now shows major differences from those of neighboring river systems.

The northern streams of the massif do not contain any endemic fish species, although some have still to be described (Tweddle 1979). These perennial streams form a valuable refuge for the Lake Chilwa fauna during the periodic drying out of the lake (Kalk et al. 1979).

The biodiversity in other floral and faunal groups is also high but poorly known (Dudley 2000). There are two endemic amphibians, *Afrana johnstoni* and *Arthroleptis francei*, and two near-endemics, *Nothophryne broadleyi* and *Hyperolius spinigularis* (Broadley 2001), out of thirty known species (Poynton and Broadley 1985, 1986, 1987, 1988). There are at least twenty-two dragonflies, of which one, *Oreocnemis phoenix*, is an endemic species in an endemic genus (Wilson 1988). Predaceous diving beetles from the Dyticidae family are also well represented in the fauna (Dudley 2000).

Status and Threats

Current Status

The terrestrial and freshwater habitats of the massif remain healthy. The massif has until now been managed through the Mulanje Mountain Forest Reserve, which has prevented development or intensive agriculture. A Mulanje Mountain Conservation Trust Board has been established with support from the Global Environment Facility. The goal of the board is to establish an endowment trust with World Bank funding (Dudley 2000). Under the trust, zonation of the massif is proposed for forestry, tourism, research, biodiversity conservation, and natural resource management and sustainable use. Mulanje cedar (*Widdringtonia whytei*) is proposed as a key indicator species because of its high value and vulnerable status, with plans for management and regeneration of depleted cedar stands.

Exotic species do not appear to have had a major impact on the fish fauna. Reservoirs in the Ruo catchment at the base of the escarpment contain largemouth bass (*Micropterus salmoides*) and tsungwa (*Serranochromis robustus*), but no exotics have been found in the rivers. Spottail mosquitofish (*Phalloceros caudimaculatus*) are abundant in streams and dams on the Nswadzi tributary of the Ruo, well to the west of Mulanje, but have not spread to the rest of the system.

Exotic conifers (*Pinus patula*) on the plateau of the massif may have negative impacts on river flows, but this has not been quantified. Otherwise, the riverine habitats around the base of the massif are intact and natural, with no impoundments and only minor offtake through piped water schemes for villages around the base of the massif. The land around the base of the massif is covered extensively by tea estates, within which the steeper slopes of river and stream valleys are generally protected by natural vegetation. The tea plantations provide more effective ground cover than indigenous agricultural methods and thus help to protect against erosion and stream siltation. Outside the tea estates, extensive deforestation has affected tributaries, and consequently the lower Ruo is subject to more erratic flow and more frequent flood episodes than in the past.

Future Threats

Proposed mining of superficial bauxite deposits on the Lichenya Plateau in the southwest of the massif threatens water quality in several key tributaries of the Ruo River. Pollution from mine waste and heavy sedimentation can be expected, threatening the fish populations that are characteristic of clear, well-oxygenated mountain streams. Degradation of the streams below the massif is likely with increasing land clearance, and fishing with poisons by local inhabitants is known to occur in the streams, so public awareness campaigns about the values of the streams of this region are needed.

Justification for Ecoregion Delineation

This Mulanje Massif defines the boundaries of this ecoregion, whose headwater streams are isolated from downstream reaches by waterfalls, such that several endemic and relict species occur in and along these highland rivers. The massif is a quartz-syenite and granitic plutonic intrusion, uplifted, fractured, and eroded. Larger river valleys are truncated at the edge of the plateau, forming high waterfalls. For example, waterfalls on the Ruo River descend 200 m in a single fall. The plateau is surrounded on many sides by sheer rock walls up to 1,700 m in height (Chapman 1962; Dudley 2000). Floristically, Mulanje belongs to the Afromontane regional center of endemism (White 1983).

Data Quality: Medium

The fish fauna is generally well known, although further sampling is needed because some species are known only from few specimens (one each in the case of two unidentified *Barbus* species). Dudley (2000) summarizes information on all other flora and fauna and points out that many groups are poorly collected; further research on these taxa is essential, preferably under the umbrella of the Conservation Trust.

Ecoregion Number: **44**
Ecoregion Name: **Bijagos**
Major Habitat Type: **Island Rivers and Lakes**
Final Biological Distinctiveness: **Nationally Important**
Final Conservation Status: **Relatively Stable**
Priority Class: **V**
Authors: **Emily Peck and Michele Thieme**

Location and General Description

This ecoregion of forested, low islands supports a depauperate freshwater fauna. The Bijagos Archipelago, which is part of Guinea-Bissau, is located just off that country's coast opposite the mouth of the Rio Gêba on the mainland. The largest of these islands are Ilha Roxa, Ilha de Bubaque, Ilha Formosa, Ilha Caravela, Ilha de Uno, and the biggest, Ilha de Orango.

The archipelago includes eighty-eight islands covering an area of about 900 km². Falling tides daily expose an additional 1,000 km², about one-quarter of which is mangroves and three-quarters is mudflats (Robertson 2001). The climate of the ecoregion is tropical and humid, with an average annual rainfall of about 2,100 mm measured at 19 m a.s.l. (UNESCO 2001).

Extensive areas of oil palm, mangroves, mudflats, and climax woodlands cover the islands (Sayer et al. 1992). In places the coasts drop steeply to the ocean, and elsewhere extensive mangroves penetrate far inland (Sayer et al. 1992). Areas lacking mangroves are the northern shores of Caravella, Carache, and Bubaque islands; the western shores of Caravella, Uno, and Orango islands; and the rocky coasts of Unhocomo and Unhocomomozinho (Hughes and Hughes 1992).

Outstanding or Distinctive Biodiversity Features

The depauperate freshwater fauna of the Bijagos includes only one freshwater fish species, the benthopelagic *Barbus pobeguini*, and two freshwater mollusk species (*Neritina adansoniana* and *N. rubricata*). The Neritidae family has marine origins, and species in this family often are able to live in brackish or fresh water (Brown 1994).

Several aquatic mammal species inhabit this ecoregion. A population of hippopotamus (*Hippopotamus amphibius*) lives in the fresh, brackish, and marine waters of the islands (UNESCO 2001). The African clawless otter (*Aonyx capensis*) and the marsh mongoose (*Atilax paludinosus*) occur on the islands, and the largest population of the vulnerable West African manatee (*Trichechus senegalensis*) occurs here (Robertson 2001a).

The ecoregion hosts several species of reptiles and frogs, including two near-endemics. The near-endemic frogs are *Pseudhymenochirus merlini*, found also in Northern Upper Guinea [25], and *Phrynobatrachus fraterculus*, occurring also in Northern Upper Guinea [25] and Southern Upper Guinea [28]. Five marine turtle species breed in the archipelago (*Caretta caretta, Chelonia mydas, Lepidochelys olivacea, Eretmochelys imbricata,* and *Dermochelys coriacea*), and the vulnerable dwarf crocodile (*Osteolaemus tetraspis*) is found on the islands (Robertson 2001a).

The islands are of outstanding importance for wading birds. Hundreds of thousands of Palearctic waders winter on the mudflats that lie west of Bubaque Island (Dodman et al. 1999). About 710,000 were estimated to winter on the tidal flats of the archipelago in 1992–1993 (Salvig et al. 1994). The islands also include a number of heronries, in addition to breeding colonies of ibises, gulls, and terns (Robertson 2001a). A flock of greater flamingo (*Phoenicopterus ruber*) has also been observed at the mudflats west of Bubaque (Dodman et al. 1999).

Status and Threats

Current Status

Increasing human populations and tourism threaten the fragile ecosystems of the islands. About 27,000 people live on the islands, and populations are concentrated near the cities of Bubaque and Balmora (UNESCO 2001).

Forestry is one of the main economic activities in the ecoregion, with a large focus on the harvesting of natural palms. Rice agriculture occurs in the margins of flooded areas and in areas previously occupied by dry forest. Poorly managed forestry and agriculture operations could lead to increased erosion and sedimentation in the few waterbodies on the islands (United Nations Commission on Sustainable Development 1997).

The Bijagos archipelago lies in the Boloma-Bijagos United Nations Educational, Scientific and Cultural Organization (UNESCO) biosphere reserve. The Ilhas de Orango Islands National Park and part of the Poilão Marine National Park also occur within the boundaries of the biosphere reserve (UNESCO 2001).

Future Threats

The potential for inundation by rising sea levels is high. These islands are low lying and would be vulnerable to even small changes in sea level (Zinyowera et al. 1998).

Justification for Ecoregion Delineation

The Bijagos Archipelago defines the boundaries of this ecoregion, which is distinguished by its importance for wading birds. The ecoregion is close to mainland Africa, and its aquatic flora and fauna are very similar to those of the adjacent mainland. The islands are formed from areas of higher land that remained above water when the surrounding land was submerged.

Data Quality: Low

Ecoregion Number:	**45**
Ecoregion Name:	Canary Islands
Major Habitat Type:	Island Rivers and Lakes
Final Biological Distinctiveness:	Nationally Important
Final Conservation Status:	Endangered
Priority Class:	IV
Authors:	Ashley Brown and Michele Thieme
Reviewer:	Björn Malmqvist

Location and General Description

Ephemeral streams flowing from volcanic peaks down through deep gullies distinguish several of the subtropical Canary Islands. From west to east, the main islands are Hierro, La Palma, La Gomera, Tenerife, Gran Canaria, Fuerteventura, and Lanzarote. These islands and six related islets lie off the coast of Morocco in the Atlantic Ocean and form the Canary Islands autonomous region of Spain.

The islands can be divided into two groups based on their location and terrain. The eastern islands, including Fuerteventura, Lanzarote, and their surrounding islets, are of low elevation (maximum 800 m) and have a xeric landscape. The western islands—Gran Canaria, Tenerife, La Gomera, La Palma, and El Hierro—are younger than the eastern islands and more mountainous. The highest peak is El Teide on Tenerife, at 3,715 m (Vega 1998). The higher-elevation areas of these western islands receive moisture-laden winds from the northeast, and the precipitation supports rich vegetation.

The eastern islands are about 16–20 million years old, whereas the western islands of La Palma and El Hierro are only about 2–3 million years old (Clarke and Collins 1996). The islands Lanzarote and Fuerteventura in the east might once have been part of Africa, or they may have volcanic origins in common with the other islands. Volcanic eruptions have occurred in recent times on a few of the islands, with the most recent one on La Palma in 1971 (Clarke and Collins 1996).

Ravines and gullies (called barrancos) drain runoff, including snowmelt at times, from the high altitudes of the western islands. On Tenerife, there are now fewer than ten permanent streams, all of which are small and run for short distances (Malmqvist et al. 1993, 1995). The islands also have springs, ponds, aquifers, pools at ravine bottoms, and subterranean aquatic habitats. There are also artificial habitats such as channels and small reservoirs behind dams on many of the streams that radiate from the islands' peaks. Malmqvist et al. (1995) also identify madicolous habitats, areas with a constantly seeping film of water over bedrock. Freshwater habitats that the government of the Canary Islands considers of special interest are natural eutrophic lakes, petrifying (hard water) springs with tufa formations, and thermo-mediterranean riparian galleries and brush (Regional Department of Environment of the Canary Islands Autonomous Government 2001). Salt pans called the Salinas de Janubio also occur on Lanzarote Island (Clarke and Collins 1996).

The climate of the Canary Islands is subtropical. Low elevations are frost free in the winter, but summers are cool because of the influence of the ocean; mean monthly temperatures are from 17.5°C to 27°C between sea level and 350 m a.s.l. High-elevation areas can be cold, with snow on the highest peaks (El Teide) in winter and a mean monthly temperature below 7.5°C (Fernandopullé 1976).

Mean annual rainfall varies greatly by island and with elevation, with averages ranging from 135–325 mm/year for the eastern islands and 410–586 mm/year for the western islands, as measured between 1949 and 1967 (Fernandopullé 1976). These disparities occur because northeast trade winds pick up moisture from the cool Canary Current and deposit it at high elevations on the north-facing slopes of many of the western islands. The lower eastern islands receive little rain from these winds and therefore are much drier than the mountainous islands (Fernandopullé 1976). A bank of clouds often forms on the north slope at elevations of 600–1,700 m a.s.l., cooling these parts of the islands and creating a misting phenomenon called horizontal rain (Clarke and Collins 1996). The majority of rainfall occurs between October and March. Precipitation is between 100 and 350 mm per year in coastal zones, about 650 mm at elevations of 250–600 m; and about 400 mm at elevations above 600 m a.s.l.

Six vegetation types dominate the communities on the islands. Coastal halophytic vegetation, such as *Frankenia* sp. and *Astydamia latifolia,* grows at the lowest elevations. Saltwater marshes, saltwater lagoons, and dunes also occur along the coast. Between sea level and 400 m, coastal scrub, composed mainly of members of the spurge family (i.e., *Euphorbia balsamifera* and *E. canariensis*), grows up from the volcanic soil and between the crevices of cracked lava. Above the coastal scrub, the vegetation grades into thermophilic woodlands, including palm groves (e.g., *Phoenix canariensis*), juniper bushes (e.g., *Juniperus turbinata*), dragon trees (*Dracaena draco*), and mastic trees (*Pistacia atlantica*). On the moist, windward side of the islands, laurel forests with *Laurus azorica, Ocotea foetens,* and *Persea indica* and associated myrtle (*Myrica faya*) and heather (*Erica arborea*) are the next vegetative community. Moving further up slope, groves of pine (*Pinus canariensis*) grow to about 2,000 m a.s.l. On areas above 2,000 m, mountaintop scrub dominates, with species such as *Spartocytisus supranubius, Stemmacantha cynaroides,* and *Viola cheiranthifolia* on Tenerife and *Adenocarpus viscosus* and *Viola palmensis* on La Palma (Regional Department of Environment of the Canary Islands Autonomous Government 2001; UNESCO 2001). Hydrophilic vegetation, such as the canary willow (*Salix canariensis*), grows along the river beds of the gullies through which water runs almost permanently (Regional Department of Environment of the Canary Islands Autonomous Government 2001).

Outstanding or Distinctive Biodiversity Features

This ecoregion, though supporting a highly diverse and endemic terrestrial flora, is inhabited by few freshwater species because of a lack of permanent waterbodies, although there may be undiscovered freshwater species living in the little-studied subterranean waters of the islands. Kelly et al. (2001) argue that a high interisland dispersal of freshwater invertebrates has led to low single-island endemism in the Canary Island freshwater fauna compared with terrestrial invertebrates. In general, endemism in the Canaries appears to stem from the presence both of taxa that have originated through speciation after colonization and of relict populations that have survived here after disappearing elsewhere (Juan et al. 2000).

Only two fish species are known from the freshwater systems of the islands. The migratory *Anguilla anguilla* moves up into the more permanent sections of streams on the western islands after breeding in the ocean. The goby *Porogobius schlegelii* inhabits the brackish and fresh waters of the intertidal zone (Fish-Base 2001). The ecoregion does not host any endemic aquatic mammals, aquatic reptiles, frogs, or fishes.

Although it is species-poor in higher vertebrates, there is a rich aquatic invertebrate fauna inhabiting both epigean and hypogean habitats on these oceanic islands. Malmqvist et al. (1993) documented 127 taxa of stream macroinvertebrates in seven streams of Tenerife. Malmqvist et al. (1995) list more than thirty endemic freshwater macroinvertebrates from the islands, including one amphipod, two mayflies, one backswimmer, one water cricket, twelve aquatic beetles, five caddis flies, and more than ten Diptera. *Ancylus striatum* is endemic to the waters of the Canary Islands and Madeira (Nilsson et al. 1998). Additionally, twenty-two species of freshwater ostracods are known from the islands (Malmqvist et al. 1997). *Melita dulcicola,* one of three known freshwater amphipods on the islands, occurs in subterranean habitats on La Gomera (Stock and Vonk 1990).

The Salinas de Janubio hosts several waterbird species such as little ringed plovers (*Charadrius dubius*) and occasionally greater flamingos (*Phoenicopterus ruber*), purple herons (*Ardea purpurea*), and black-winged stilts (*Himantopus himantopus*) (Clarke and Collins 1996).

Status and Threats

Current Status

The largest threat to the freshwater systems of the Canary Islands is increasing water abstraction for agricultural and direct human uses (Malmqvist et al. 1995). On Tenerife, an extensive network of about 940 tunnels penetrates the island to abstract water. Water abstraction is causing many of the islands' surface waters to dry up. For example, on Gran Canaria, the number of streams dropped from 285 to 20 between 1933 and 1973, mainly because of water diversion and forestry (Alfonso Pérez 1980, in Crosskey 1988). As tourism increases in importance, it has stimulated an increased interest in conservation, but increasing numbers of tourists and tourism workers have also stressed the water resources of the islands. Already four of the five endemic stream-living insects and gastropods from Gran Canaria are considered extinct or on the edge of extinction (Nilsson et al. 1998).

Several nonnative species have been introduced into the fresh waters of the islands, including two frogs (*Hyla meridionalis* and *Rana perezi*), three fish (*Cyprinus carpio, Micropterus salmoides,* and *Gambusia* sp.), and one crayfish (*Procambrus clarkii*). These species probably compete with or prey on the native freshwater species, although no studies have been completed on their interactions and impacts.

The numerous small dams along the waterways of these islands block the migration of *Anguilla anguilla*. This species is considered vulnerable on the islands because of this threat, but population numbers remain strong in Tenerife's Barranco de Igueste de San Andrés and Barranco de Afur (Regional Department of Environment of the Canary Islands Autonomous Government 2001).

The Canaries have a surprising number of protected areas for such a small ecoregion. There are 145 protected nature areas that cover about 40 percent of the archipelago (Regional Department of Environment of the Canary Islands Autonomous Government 2001). However, water abstraction outside these areas still affects their freshwater systems.

Future Threats

The scarcity of water resources on the islands and the stresses on these resources from agriculture, the tourist industry, and growing human populations threaten the future of the freshwater ecosystems on the islands. The number of natural freshwater systems has already declined, and there are now only a few permanently flowing streams on Tenerife and Gran Canaria. There is a real risk of extinction for the endemic freshwater species that remain if water abstraction continues to outpace availability and leave less for the freshwater systems and the species that depend on them.

Justification for Ecoregion Delineation

The boundaries of this ecoregion are defined by the extent of the Canary Islands, which are distinguished by an endemic aquatic invertebrate fauna.

Data Quality: High

Ecoregion Number: **46**
Ecoregion Name: **Cape Verde**
Major Habitat Type: **Island Rivers and Lakes**
Final Biological Distinctiveness: **Nationally Important**
Final Conservation Status: **Endangered**
Priority Class: **IV**
Author: **Ashley Brown**
Reviewer: **Cornelius J. Hazevoet**

Location and General Description

The freshwater habitats of the volcanic Cape Verde Islands are composed primarily of temporary ribeiras (streams) that descend steep cliffs. Not unexpectedly, these streams possess an extremely depauperate fauna. Ten islands and five islets make up this ecoregion and are all part of the Republic of Cape Verde. Interior highlands of volcanic rock slope down to semidesert coastal plains with sparse vegetation and sandy beaches. The archipelago lies about 500 km off the west coast of Africa in the Atlantic Ocean.

The main islands in the north, also called the Barlavento Islands, are Santo Antão, São Vicente, Santa Luzia, São Nicolau, Sal, and Boavista. *Barlavento* means "windward," and the islands are so named because the northern Canary Current and the Northeast Trade Wind affect their climate. The northern islands are generally more rugged and mountainous, with peaks over 1,000 m a.s.l. (3,000 m a.s.l. on Fogo Island) (Hazevoet 1995). The Sotavento (Leeward) Islands in the south include Maio, Santiago, Fogo, and Brava. The southern islands have more gentle inclines and lower interior elevations. The largest island of Santiago (991 km^2) has two main mountain ranges and a rugged landscape that includes forested slopes and occasionally lush valleys. Santiago's coast in the south and southwest is stony, dry, and barren. Fogo Island has the only active volcano in the archipelago (Hazevoet 1995).

Temporary streams flow from the mountains of these islands, and a few brackish lagoons occur inland on Santiago and Boavista. Dry riverbeds temporarily flood during the rainy season to form ribieras. The ribieras can flow torrentially on the more mountainous islands for several weeks to months, while on the more level islands flooding ceases within several days. Ribeiras in the ecoregion include the Ribeira da Barca on northern Santiago and the Ribeiras da Vinha and de Calhau on São Vicente. The only permanent streams in the ecoregion are found on Santo Antão (Hazevoet 1995). Salt pans are found in the eastern islands of Sal, Boavista, and Maio.

The Cape Verde Islands lie in the dry belt of the Sahel and experience a tropical climate with low rainfall. The mean monthly temperature of the ecoregion as recorded at Praia on Santiago ranges from about 22°C in February to about 27°C in September. Estimated average annual precipitation for the islands is between 100 and 900 mm (Hazevoet 1995). A dry season lasts from December to July, and a wet season occurs from August to November, with the majority of precipitation falling between August and September. As a result of low and unpredictable rainfall, the islands are prone to extended periods of drought, the most recent lasting from 1968 to 1980, with no precipitation falling during that period (Sargeant 1997). Local variation in precipitation occurs; for example, higher altitudes of the northern islands, particularly those facing northeast, may receive precipitation from August to March (Hazevoet 1995).

Xeric-tolerant plant species dominate the vegetation of the islands. At the time of colonization, the vegetation communities of the islands probably were composed of semidesert and steppe vegetation and herbaceous savanna (Hazevoet 1995). Introduced species, overgrazing, and agriculture have caused substantial changes to the flora. Along the ribeiras and nearby irrigation canals, the reed *Arundo donax* and several trees and shrubs, including the indigenous tamarisk (*Tamarix senegalensis*), dominate. On higher-elevation windward slopes, *Euphor-*

bia dominates, with occasional endemic dragon trees (*Dracaena draco*) (dragon trees occur on São Nicolau only) and ironwood (*Sideroxylon marmulano*). On the northern islands in the west, at elevations below 600 m, herbaceous savanna and steppe vegetation is dotted with indigenous acacia (*Acacia albida*) and fig (*Ficus gnaphalocarpa*) (Hazevoet 1995). Whereas the coastal plains of the islands are described as semidesert, the mountains of the islands are sometimes sparsely forested (Sargeant 1997).

Outstanding or Distinctive Biodiversity Features

This ecoregion lacks native freshwater fish, aquatic frogs, and freshwater crabs. Santiago Island hosts mollusks from the families Hydrobiidae, Thiaridae, Lymnaeidae, Ancylidae, and Planorbidae (Rosa et al. 1999). Among these are both Palearctic species (*Hydrobia ventrosa, Lymnaea auricularia, Gyraulus laevis,* and an *Ancylus* sp.) and Afrotropical species (*Melanoides tuberculata, L. natalensis,* and *Bulinus forskalii*) (Brown 1994). A recently discovered endemic amphipod, *Melita cognata*, lives in groundwater-fed caves on Santiago (Stock and Vonk 1992).

Several endemic birds are associated with ribeiras and their surrounding vegetation. The endemic and endangered Cape Verde swamp-warbler (*Acrocephalus brevipennis*) inhabits reeds growing in ribeiras and adjacent to artificial ponds and reservoirs. The endemic and threatened Cape Verde purple heron (*Ardea bournei*) has breeding colonies in the ribeira valleys (Hazevoet 1995, 2001; Sargeant 1997).

Status and Threats

Current Status

Soil erosion, resulting from poor agricultural techniques and overgrazing, increases the turbidity and nutrient loads of streams. Overgrazing by goats substantially contributes to desertification and is particularly widespread on São Vicente and Boavista (Hazevoet 1995).

Agriculture forms a large part of the economy of the Cape Verde Islands and contributes sediment to the streams (Hazevoet 1995). On Santiago there are banana, manioc, and sugar cane plantations. Maize and beans are cultivated in the drier regions of the island of Santiago. On São Nicolau, sugar cane, banana, maize, and beans are cultivated. Agriculture also threatens water supplies; 5 percent of the cultivated land is irrigated (Hazevoet 1995). Unfortunately, improper land use such as the cultivation of crops on steep slopes has led to significant soil erosion.

In attempts to stop erosion, increase water infiltration, improve rangeland management, and increase vegetation for fuelwood, much of the land on Santiago and Santo Antão was afforested in the 1930s, 1940s, and 1950s (Van den Briel and Brouwer 1987; Hazevoet 1995). The land was planted mostly with nonnative trees, including oak (*Quercus* spp.) and pine (*Pinus* spp.). In the 1970s and 1980s, a second afforestation program began, and plants such as *Eucalyptus* spp. and *Lantana camara* were planted in a continued effort to add vegetation to the islands (Hazevoet 1995).

Future Threats

Drought and desertification are a constant threat to the freshwater systems of the ecoregion. Unabated grazing will only exacerbate the effects of drought. An increasing human population with increasing water needs also threatens the freshwater systems. Tourism on Sal is increasing, with many hotels recently built. Consequently, the south of this island is now heavily degraded (Hazevoet 1995).

Justification for Ecoregion Delineation

The Cape Verde Islands form the boundaries of this ecoregion, which hosts a few endemic aquatic invertebrates and ribeira-associated birds. The Cape Verde Islands are volcanic in origin and have never been connected with mainland Africa (Hazevoet 1995). This, combined with the inhospitable condition of the temporary watercourses, explains the paucity of freshwater fauna in the ecoregion. Both Afrotropical and Palearctic species of freshwater mollusks have been reported from the islands. The occurrence of Palearctic species so far south of their range in western Africa is of biogeographic interest (Brown 1994).

Data Quality: High

Ecoregion Number: **47**
Ecoregion Name: **Comoros**
Major Habitat Type: **Island Rivers and Lakes**
Final Biological Distinctiveness: **Bioregionally Outstanding**
Final Conservation Status: **Endangered**
Priority Class: **IV**
Author: **Ashley Brown**
Reviewer: **Roger Safford**

Location and General Description

The volcanic, mountainous Comoros Islands have a depauperate freshwater fauna that inhabits the lakes and streams of the ecoregion. Streams originate in forested highlands, dropping precipitously to flow through the coastal plain before reaching the ocean. There is also one crater lake, Dziani Boundouni, located in the southeast portion of Moheli Island, and two natural lakes on Mayotte: Dziani Karehani and Dziani Dzaha (crater) (Louette 1999). The Comoros Islands are located in the Indian Ocean northwest of Madagascar in the Mozambique Channel and cover about 2,171 km^2 (Mittermeier et al. 1999). From west to east, the islands include Grande Comore (Njazidja), Mohéli (Mwali), Anjouan (Ndzuani), and Mayotte (Maore). Mayotte is a French territory, and the other islands form the République Fédérale Islamique des Comores.

Whereas Mohéli and Mayotte both possess streams and other waterbodies, including crater lakes, the islands of Grande Comore and Anjouan possess few freshwater habitats. Mayotte, the oldest of the islands, has many meandering streams that flow from the highland rainforests of the island, in addition to two lakes, Dziani Karehani and Dziani Dzaha. On Mohéli, the freshwater but sulfurous Dziani Boundouni has frequent upwellings caused by subterranean volcanic activity (Wetlands International 2002). On Grande Comore, the youngest and largest island, the soil is thin and rocky, and there are no valleys or permanent watercourses. The 2,361-m-high Karthala Volcano on Grande Comore is still active, erupting every 10–20 years (Henkel and Schmidt 2000).

The ecoregion experiences a tropical climate, greatly influenced by its location in the Indian Ocean. Northwest monsoon winds begin in September, and the climate is hot (25–33°C) and humid from October to April. From May to September it is cooler and drier, with southerly winds and temperatures ranging between 16°C and 25°C (Battistini and Vérin 1984). Rainfall varies greatly throughout the islands and by elevation, ranging from a minimum of 900 mm/year to 6,000 mm/year (FAO Inland Water Resources and Aquaculture Service, Fishery Resources Division 1999), although few data exist for the wettest parts of the uplands. Periodically destructive cyclones hit the islands (Sayer et al. 1992).

Tropical rainforest remains in a few places at high altitudes (500–1,900 m a.s.l.), especially on Karthala, and dry forest, mangroves, baobabs, and Indo-Pacific scrub grow in parts of the lowlands (Stuart et al. 1990; Sayer et al. 1992). On lava flows, lichens, ferns, other herbaceous vegetation, and woody plants grow in succession (Henkel and Schmidt 2000). Thirty-three percent of the plant species in Comoros are endemic (Mittermeier et al. 1999). However, many plant species have been introduced to the islands. In 1979, out of 1,000 plant species identified from the Comoros, about 500 had been introduced (Henkel and Schmidt 2000).

Outstanding or Distinctive Biodiversity Features

Ten to fifteen fish species frequent the rivers and lakes of this ecoregion. However, most of the riverine fish are not limited to freshwater but also inhabit the brackish deltas and ocean. The freshwater fish fauna of the Comoros is dominated by freshwater gobies and catadromous species. These species inhabit all portions of the rivers, from the forested, headwater streams to the low-elevation marshes and estuaries. Species such as the rock flagtail (*Kuhlia rupestris*) prefer fast-flowing, clear rainforest streams such as those found on higher elevations of the islands. The catadromous and widespread Indonesian shortfin eel (*Anguilla bicolor*) and giant mottled eel (*Anguilla marmorata*) spawn at sea and move up the rivers to live as adults (Louette 1999). *Sicyopterus lagocephalus* is also catadromous but is restricted to Madagascar, Reunion, Mauritius, and the Comoros (FishBase 2001).

The ecoregion hosts sixteen freshwater snails, although all of these except those of the marine-derived Ellobiidae family are thought to have been brought to the islands by humans or birds. Neritids dominate the lower watercourses on Anjouan, whereas only *Lymnaea natalensis* and *Ceratophallus* sp. have been documented at elevations above 150 m (Brown 1994).

More than twenty dragonfly species (Odonata) are known from the ecoregion, especially representatives of the genera *Palpopleua* and *Anax* (Louette 1999). Several of these stream inhabitants are endemic (Samways 2003). The ecoregion also contains several caddis fly (Trichoptera) species, and the island of Anjouan has a trichoptera fauna that is similar to that of nearby continental Africa (Malicky 1989).

The freshwater Dziani Boundouni hosts a large population of little grebes (*Tachybaptus ruficollis*). Other birds seen at the lake are Madagascar squacco heron (*Ardeola idae*), striated heron (*Butorides striatus* of the Comoro-endemic race *rhizophorae*), cattle egret (*Bubulcus ibis*), great egret (*Casmerodius albus*), grey heron (*Ardea cinerea*), common moorhen (*Gallinula chloropus*), and greenshank (*Tringa nebularia*) (Wetlands International 2002). Several threatened wetland birds also occur on the islands: Madagascar heron (*Ardea humbloti*), *Ardeola idae,* and Madagascar harrier (*Circus maillardi*) (Safford 2001a). The latter species has been taxonomically rearranged, and this population is usually accepted as Madagascar serpent eagle (*Circus macrosceles*).

Two aquatic-dependent frogs occur on Mayotte but not the other islands: the near-endemic *Mantidactylus granulatus* and *Boophis tephraeomystax* (Louette 1999), which also occur on Madagascar and Nosy Bé islands (Glaw and Vences 1994). However, it is uncertain whether they are native and suspected that they might be accidental introductions.

Status and Threats

Current Status

The largest threat to the freshwater systems of the islands is the loss of forest cover and the resultant erosion and loss of perennial freshwater systems. Large tracts of forests have been cleared for timber or to open space for agriculture. In many places banana, palm, ylang-ylang (*Cananga odorata*), and manioc plantations have replaced the native vegetation (Henkel and Schmidt 2000). Forests have also been cut for fuelwood. Because the soils are of volcanic origin and are poor in organic material, deforested areas are especially susceptible to large-scale erosion, as is the case on Anjouan, where overcultivation has led to severe soil erosion. In the rainy season, the ocean around the island turns red from the runoff of lateritic soils. Freshwater and marine organisms are negatively affected by the resultant sedimentation and turbidity in aquatic habitats. Loss of forest has also caused perennial streams to become ephemeral on the three smaller islands, and it is estimated that up to 70 percent of the permanent watercourses on Anjouan have been lost to sedimentation (Sayer et al. 1992; Stattersfield et al. 1998; Safford 2001a).

On Mayotte the coastal areas and areas adjacent to streams are sustaining a much higher human population density than elsewhere on Mayotte. In particular, detergent from clothes washing in streams stains the rocks and even covers the wings of dragonflies. The lower elevations are occupied only by the widespread and eurytopic Odonata species, whereas the endemics are found only alongside the higher-elevation streams (Samways 2003).

Pollution from urban areas includes organic and liquid wastes, such as oil from vehicles and raw sewage, and also threatens the freshwater systems (Institute of Marine Sciences et al. 1998).

Few protected areas occur on the islands. Saziley National Park protects a small portion of Mayotte, but it is uncertain whether it encompasses any watercourses (Safford 2001c). Dziani Boundouni has been designated a Ramsar site, and as of 1998 a management plan was being prepared for the lake. Forest fires and nearby deforestation threaten the water quality of the lake (Wetlands International 2002).

Justification for Ecoregion Delineation

The Comoros Islands make up this ecoregion, which hosts a diversity of brackish and freshwater fish, dragonflies, caddis flies, waterbirds, and frogs. All of the freshwater fish of the islands are from secondary families. The Miocene volcanic origin of the islands and the fact that they have never been joined with a continent explain the lack of primary fish families (Louette 1999).

Data Quality: Low for Fish, Medium for Odonata

Ecoregion Number: **48**
Ecoregion Name: **Coralline Seychelles**
Major Habitat Type: **Island Rivers and Lakes**
Final Biological Distinctiveness: **Nationally Important**
Final Conservation Status: **Relatively Intact**
Priority Class: **V**
Authors: **Robin Abell and Ashley Brown**
Reviewer: **Justin Gerlach**

Location and General Description

With limited freshwater habitats, the almost completely flat Coralline Seychelles Islands host a depauperate freshwater fauna despite a highly endemic terrestrial fauna. The Coralline Seychelles comprise about 75 of the 115 or so islands of the Republic of Seychelles; the other islands are granitic in origin (Granitic Seychelles [49]). The archipelago is located in the western Indian Ocean about 1,600 km east of Mombassa, Kenya. The coralline islands average only a few meters above sea level and were formed by the buildup of coral reefs over an 82-million-year-old granitic basement that is found less than 1 km below the surface (Rocamora and Skerret 2001). Together, they cover an area of 214 km^2 (FAO Inland Water Resources and Aquaculture Service, Fishery Resources Division 1999). Some of the more prominent islands include the Aldabra group, the Farquhar group, the Amirantes group, and the geographically isolated Bird and Denis islands. Most available information for this ecoregion concerns Aldabra.

Aldabra Atoll, with an area of about 130 km^2, is the world's largest atoll (UNEP 1998; Stattersfield et al. 1998). The atoll, located in the extreme southwest of the Seychelles archipelago, consists of four main islands—Malabar (or Middle Island), Grand Terre (South Island), Polymnie, and Picard (West Island)—and a number of smaller islands. These islands enclose a large, shallow lagoon (about 300 km^2) that is bordered by mangroves (Fosberg and Renvoize 1980). Aldabra has a greater elevation than any of the other Coralline Seychelles, at about 4–8 m above mean low-tide level, with some dunes at 10–30 m (Skerrett 1999). The coralline islands have been submerged for several intervals in their history, the last interval being approximately 125,000 years ago (Rocamora and Skerrett 2001).

Freshwater habitats of the Aldabra group include freshwater ponds, abundant ephemeral pools, and crevices with freshwater and subterranean caverns with small reservoirs of water (Fosberg and Renvoize 1980). Surface freshwater habitats are most common during the rainy season. Lagoon systems are also present on Cosmoledo and Astove atolls, which are smaller than Aldabra but of regional significance for marine fauna and birds. These atolls lack freshwater systems. A single permanent brackish water pool is present on Assumption (as a result of mining activities).

The climate of the ecoregion is classified as wet tropical (FAO Inland Water Resources and Aquaculture Service, Fishery Resources Division 1999). Monthly mean temperatures in Aldabra are 22°C in August and 31°C in December (Aldabra Marine Program 2003). The more northerly coral islands, lying closer to the equator, have somewhat warmer temperatures. From November to April, the ecoregion experiences short, heavy rainshowers, high humidity, and higher temperatures; the remainder of the year constitutes the dry season, with lower temperatures and humidity and a steady breeze from the southeast (Calström 1995; CIA 2001). Average annual precipitation on the coralline islands is about 1,290 mm/year (FAO Inland Water Resources and Aquaculture Service, Fishery Resources Division 1999).

The vegetation of the ecoregion is xeric and includes dense thickets of the salt-tolerant *Pemphis acidula* on the rougher limestone and a mixed thicket of low trees, shrubs, herbs, and grasses on the higher, more consolidated rock (Stattersfield et al. 1998). Important families include Sapindaceae, Moraceae, and Tilliaceae. Mangroves line the lagoon on Aldabra Atoll and are an important habitat for various birds. About 20 percent of the flowering plants on the islands of the Aldabra group are endemic (Fosberg and Renvoize 1980; Skerrett 1999; Mittermeier et al. 1999).

Outstanding or Distinctive Biodiversity Features

This ecoregion hosts a recently evolved biota that includes a large number of endemic plants. However, the freshwater fauna is extremely depauperate because of the paucity of permanent freshwater habitats. The permanent freshwater fauna is limited to invertebrates. Five species of freshwater Heteroptera have been recorded (*Micronecta praetermissa, Anisops vitrea, Mesovelia vittigera, Limnogonus cereiventris,* and *Microvelia diluta diluta*) (Polhemus 1993). There are a small number of dragonflies but no caddis flies (J. Gerlach, pers. comm., 2003). Landlocked pools support algae and cyanobacteria (e.g., *Phormidium, Lyngbya, Pleurocapsa*), and other aquatic organisms that feed on these food sources (Braithwaite et al. 1989). The ecoregion lacks freshwater fish and frogs.

Numerous bird species inhabit or visit the islands, although only a portion is closely associated with aquatic habitats. Among Aldabra's endemic (one extinct) species and ten endemic subspecies are the Aldabra sacred ibis (*Threskiornis aethiopicus abbotti*), which inhabits tidal pools and coastal lagoons, and the Aldabran white-throated rail (*Dryolimnas cuvieri aldabranus*) (Seabrook 1990; Stattersfield et al. 1998). A small population of greater flamingo (*Phoenicopterus ruber*) breeds on the island. Several other waterbirds, including *Ardeola idea, Egretta dimorpha, Ardea cinerea,* and *Butorides striatus crawfordii,* are found in the Coralline Seychelles. The coralline islands also support seabird colonies of global and regional importance including several species of shearwaters, frigatebirds, terns, tropicbirds, and boobies (Rocamora and Skerrett 2001).

Status and Threats

Current Status

There are approximately ten human residents on Aldabra (a warden and rangers), and the untouched nature of this coralline atoll makes it unique. Although there is no large human habitation on the Aldabra Atoll, it has nonetheless been affected by introduced species including goats, rats, cats, and exotic vegetation (Fosberg and Renvoize 1980; Seabrook 1989, 1990; Coblentz et al. 1990; Stattersfield et al. 1998). It has been speculated that rats may have been the cause of the presumed recent (1986) extinction of the Aldabra warbler (*Nesillas aldabrana*). Campaigns to eradicate goats were undertaken in the late 1980s, destroying 75–85 percent of the total population, and efforts were made to contain the remaining population (WCMC 1984a; Coblentz et al. 1990). Further attempts at eradication in the 1990s reduced the population to fewer than twelve individuals, restricted to Grande Terre. There is some evidence that the goat population is increasing again.

Besides introduced species, the two main disturbances on the islands have been phosphate mining and planting of groves of introduced *Casuarina* and coconuts. Soil removal during mining on Assumption and loss of terrestrial vegetation could have affected the freshwater habitats of the islands (Fosberg and Renvoize 1980).

Aldabra mangroves and their turtles, fish, and tortoises were historically exploited, although populations appear to be recovering in areas that have been protected from overexploitation. Aldabra is a World Heritage Site and a Strict Nature Reserve (WCMC 1984a).

Future Threats

As tourism increases, islands that previously were infrequently visited become more susceptible to degradation. An airstrip was recently constructed on Assumption Island, permitting greater access to Aldabra Atoll and other islands. The continued protection of the protected areas of Aldabra will depend in large part on the ability of the Seychelles Island Foundation to raise adequate funding (WCMC 1984a).

Justification for Ecoregion Delineation

This ecoregion is defined by the Coralline Seychelles and separated from the granitic islands because the biogeographic affinities of the coralline islands lie much more with Madagascar and Africa than do those of the granitics (Rocamora and Skerrett 2001).

Data Quality: Medium

Ecoregion Number:	**49**
Ecoregion Name:	**Granitic Seychelles**
Major Habitat Type:	**Island Rivers and Lakes**
Final Biological Distinctiveness:	**Continentally Outstanding**
Final Conservation Status:	**Vulnerable**
Priority Class:	**II**
Authors:	**Robin Abell and Ashley Brown**
Reviewer:	**Justin Gerlach**

Location and General Description

About 1,500 km from the east coast of Africa lie the Granitic Seychelles, unique Indian Ocean islands with steep mountains and a primitive fauna that includes several endemic aquatic species. Forty-one of the 115 islands in the Seychelles archipelago are granitic, and the remainder is coralline. The granitic islands are the oldest oceanic islands in the world, a continental formation that was once part of Gondwanaland about 65 m.y.a. (Rocamora and Skerrett 2001). The islands are clustered in a small area, all situated within 90 miles of Mahé (Library of Congress 1994). The granitic islands have a total area of 241 km^2 (CIA 2001). The archipelago as a whole forms the Republic of Seychelles.

The islands are striking in their topography, with mountains rising steeply from the ocean to heights up to 905 m a.s.l. (Calström 1995). Some of the granitic islands have narrow coastal plains, and some are fringed by extensive coral reefs (Library of Congress 1994). Mahé, which is only 27 km long and about 11 km wide, is the highest of the islands, with a mountain ridge running along its length (Statistics and Database Administration Section MISD 2000). The granitic islands have many small, steep watercourses, but many are ephemeral.

The climate of the ecoregion is tropical. Average daily temperatures range from 24°C to 32°C (Georges 1998; Henkel and Schmidt 2000). Two wind systems, the southeast trade winds and the northwest monsoon, influence the seasonality of the islands (Rocamora and Skerrett 2001). From November to April, in conjunction with the northwest monsoon, the ecoregion experiences short, heavy rainshowers, high humidity, and higher temperatures. The remainder of the year constitutes the dry season, with lower temperatures and humidity and a steady breeze from the southeast (Calström 1995; CIA 2001). The highest rainfall typically occurs on the central highlands of Mahé, which receives up to 3,500 mm/year. Precipitation is markedly lower elsewhere, with an average annual value of 2,370 mm for the entire island of Mahé, 1,990 mm/year on Praslin, and 1,290 mm/year for most other islands (Calström 1995).

The native vegetation of these islands has many affinities with that of mainland Africa, Madagascar, and the Mascarencs, with palm forest the primary vegetation type. Common upland tree species include *Phoenicophorium borsigianum*, *Paraserianthes falcataria*, *Pterocarpus indicus* (a locally common introduction), *Adenanthera pavonina*, and native but cultivated coconut palms (*Cocos nucifera*). Six palm species are endemic to the Seychelles, including the imperiled, monotypic coco-de-mer palm (*Lodoicea maldivica*), which is limited to the islands of Praslin and Curieuse (and a small introduced population on Silhouette). At altitudes above 600 m, dense cloud forests occur (Henkel and Schmidt 2000). Specialist cloud forest plants are closely related to Southeast Asian species, reflecting the ancient biogeographic connections. In river valleys and marshes throughout the islands, various species of palms and screwpine (*Pandanus* spp.) were naturally abundant in historical times (Calström 1995); small areas of natural riverine vegetation can be found on some islands, most significantly on Silhouette.

Outstanding or Distinctive Biodiversity Features

This ecoregion hosts an aquatic fauna that is not particularly rich but includes several endemic species. Of the three native freshwater fish species, the cyprinodont *Pachypanchax playfairii* is endemic. The other two fish species are *Redigobius bikolanus* and *Anguilla bicolor*. The only endemic frog living on the islands is the vulnerable *Tachycnemis seychellensis* (IUCN 2002). The native Mascarene frog *Ptychadena mascareniensis* is common on the larger islands. Seven species of caecilian occur on several islands, with the greatest species diversity on Mahé, Praslin, and Silhouette islands; little is known of their conservation status. The vulnerable Seychelles mud turtle (*Pelusios seychellensis*) is also endemic to this ecoregion, although its taxonomic status is uncertain, and it may be extinct (Gerlach and Canning 2001; IUCN 2002). There are two endemic terrapin subspecies, *P. subniger parietalis* and *P. castanoides intergularis,* both recently proposed as critically endangered (Gerlach and Canning 2001; Gerlach 2002). There is also one endangered freshwater snail known only from small mountain streams above 250 m a.s.l. on Silhouette and Mahé islands and one endemic freshwater crab, *Seychellum alluaudi* (Gerlach 1997). Other crustacea include several widely distributed Indo-Pacific shrimps and crayfish. Ostracods and copepods have only recently been recorded in the islands; the only identified species to date is widespread in Europe, Africa, and Asia (Wouters 2002). Other more widely distributed freshwater fauna include four mollusk species from the family Nertididae, the cosmopolitan pond snail *Melanoides tuberculatus*, and, formerly, the estuarine crocodile (*Crocodylus porosus*) (extinct since the early 1800s) (Gerlach and Canning 1994).

The islands contain a rich dragonfly (Odonata) fauna, with twenty-two species and seven endemics. Endemics include *Allolestes machlachlani, Leptocnemis cyanops,* and *Teniobasis alluaudi.* Endemic species live in streams of higher-elevation forests and are capable of tolerating temporary drying of streams. More widespread species that live on the islands tend to inhabit pools more than streams, are not as dependent on forest cover, and are generally less tolerant of ephemeral aquatic habitats (Samways in prep.). Seven caddis fly species have been identified, all endemic and including endemic genera of biogeographic and ecological interest. Mayflies are represented by two endemic genera and a recently discovered unidentified species. There are also several species of diving beetles and aquatic bugs. Thirty percent of the aquatic insects are considered endemic, and most are limited to the least disturbed habitats (mountain streams and isolated pools). The aquatic beetle family Gyrinidae has only recently been collected in the islands and is represented by a widespread African species. Although not aquatic, several species of carabid ground beetle are closely associated with marshy habitats, as are unidentified pygmy mole crickets and some spiders (especially the lycosid *Trochosa urbana*).

There are two endemic subspecies of waterbirds in the Granitic Seychelles, *Butorides striatus degens* and *Bubulcus ibis sechellarum*. There are also eight endemic and threatened landbirds and globally and regionally important seabird congregations (Rocamora and Skerrett 2001). The only landbird associated with wetland habitats is the Seychelles black paradise flycatcher (*Terpsiphone corvina*), which currently breeds only on La Digue island. Although habitat loss has led to the decline in the range of this species, secure marsh areas are available on Silhouette and Curieuse to reverse the historic decline. The yellow bittern (*Ixobrychus sinensis*) breeds in the larger marshes; this highly threatened population is the only one of this Asian species in the African region (Gerlach and Skerrett 2002).

The critically endangered Seychelles sheath-tailed bat (*Coleura seychellensis*) is recorded from four islands, where it has been associated with marshy habitats (although not exclusively) (IUCN 2002). Only one surviving roost site is known, which contains up to thirty-two bats.

Status and Threats

Current Status

Two centuries of human presence on the islands has led to widespread habitat loss, the introduction of exotic species, and habitat fragmentation. The main threats to aquatic habitats have been drainage of coastal and upland marshes for agricultural and development purposes. Replacement of original forests by plantations and exotic species is especially severe in the lowlands (up to 300 m), and fragments of original forests remain mainly on high slopes on most islands (Calström 1995), although primary forest is retained on much of Silhouette Island. Endemic dragonflies are threatened by removal of cloud forest because they depend on streams with closed canopies (Samways in prep.). In the 1800s and early 1900s, forests were cut for plantations at middle elevations, and erosion became a problem. Measures were implemented to help control erosion; unfortunately, these measures included planting of the invasive tropical American shrub *Chrysobalanus icaco*. Lowland marshes have been drained and heavily colonized by introduced water lettuce (*Pistia stratiotes*); these habitat changes are noted for their effect on freshwater turtles, which probably have also suffered from predation by introduced cats and dogs and from pollution (Gerlach and Canning 2001).

Exotic species from a wide range of taxonomic groups have become established on the islands and are altering native habitats and biotas. Few are established in aquatic habitats; however, the aquatics that have had the largest effects are the tilapia (*Oreochromis mossambicus*) from southern Africa, water plants such as water lettuce (*Pistia stratiotes*), water hyacinth (*Eichhornia crassipes*), and Canadian pondweed (*Elodea canadensis*). The North American red-eared slider (*Trachemys scripta elegans*) has also been observed but has not yet been recorded breeding in the wild. These invasive species affect all significant-sized marshes in Seychelles with the exception of the Grande Barbe marsh on Silhouette.

Water for human consumption and irrigation is abundant in Seychelles, but there is little storage provision. Most water is taken from surface streams and is supplemented by desalination on Mahé and Praslin islands. Because aquifers on the main islands are unconfined and the water table is quite shallow, groundwater is vulnerable to contamination by human pollution and saltwater intrusion, particularly when it is overexploited or droughts occur. Government plans to increase agricultural production via improved irrigation threaten to increase pressure on the limited groundwater resources (FAO Inland Water Resources and Aquaculture Service, Fishery Resources Division 1999).

Tourism is a major part of the Seychelles economy, with an estimated 50 percent of the gross domestic product derived from tourism-related industries. Tourism began in earnest with the completion of an international airport in 1971. In 1993, a World Bank–funded project to improve tourism infrastructure, including roads and airports, was initiated. The government is attempting to build the tourism industry while simultaneously protecting the islands' environment, and caps have been placed on the number of beds on the largest islands (Library of Congress 1994).

National parks and preserves cover about 42 percent of the Seychelles Islands. The largest national park, the Morne Seychellois National Park, was designated to protect the main water catchment on Mahé. Similarly, a World Heritage Site, the Vallée de Mai, is situated in a national park on Praslin that contains the headwaters of two streams, Nouvelle De'Couverte River and Riviere Fond B'Offay, and also contains a portion of the Fond B'Offay River (WCMC 2001b). Otherwise, freshwater habitats receive little protection. The most significant unprotected freshwater habitats in the Granitic Seychelles are Police Bay (Mahé) and Grande Barbe (Silhouette) (Gerlach 2002).

The population of the Republic of Seychelles is about 80,000, the majority of whom live on the Granitic Seychelles (90 percent on Mahé, 7 percent on Praslin, and nearly 3 percent on La Digue) (Rocamora and Skerrett 2001; CIA 2001). On Silhouette the human population has dropped from 1,000 to 130, leading to decreased agricultural pressure and reforestation (Samways in prep.).

Future Threats

Increased consumption of water resources for human uses and irrigation, accompanied by development near coastal marshes, threatens the freshwater ecosystems of these islands. Tourism could lead to either increased environmental protection or degradation, depending on the ability of the government to manage the industry. Although eradication efforts have been successful on some islands, island habitats and their biotas will continue to be vulnerable to either the introduction of exotics or predation and competition by exotics where they have already been established (Rocamora and Skerrett 2001). Given the isolated nature of most Seychelles wetlands, it is possible to protect surviving habitat and ensure the preservation of the associated fauna. There is a need to maintain the largest wetlands as well as small areas currently in protected areas; the sites with the greatest potential for future conservation are Police Bay on Mahé and Grande Barbe on Silhouette (Gerlach 2002).

Justification for Ecoregion Delineation

The boundaries of the Granitic Seychelles Islands delineate this ecoregion, which is characterized by an endemic aquatic invertebrate fauna and several endemic species from other taxa.

Data Quality: High

Ecoregion Number:	**50**
Ecoregion Name:	São Tomé, Príncipe, and Annobón
Major Habitat Type:	Island Rivers and Lakes
Final Biological Distinctiveness:	Continentally Outstanding
Final Conservation Status:	Vulnerable
Priority Class:	II
Authors:	Ashley Brown and Robin Abell
Reviewer:	Angus Gascoigne

Location and General Description

On each of this ecoregion's islands, swift rivers marked by waterfalls and rapids descend from highland interiors and flow to the Gulf of Guinea. Located off the coast of Equatorial Guinea and Gabon, these volcanic islands include São Tomé (836 km^2), Príncipe (128 km^2), and Annobón (17 km^2), as well as several smaller islands. The islands are generally mountainous in the interior, sloping sharply down to the coasts. In São Tomé, one major mountain chain runs north to south, and a second runs northwest to southwest. Príncipe has two chains as well, both running east to west (Juste and Fa 1994). The highest peak on São Tomé is 2,024 m a.s.l., and on Príncipe it is 948 m a.s.l. On the much smaller island of Annobón the highest peak is 630 m a.s.l. (Juste and Fa 1994). Most of this ecoregion lies in the Democratic Republic of São Tomé and Príncipe. Annobón is part of Equatorial Guinea and lies 180 km southwest of São Tomé.

Rivers drain from the central highlands of São Tomé and the other islands, radiating toward the coasts. Most rivers are perennial and experience seasonal fluctuations. They flow through cuts in volcanic rock and are interrupted in places by waterfalls. The largest river of São Tomé is the Io Grande, which drains the southeastern portion of the island. Other large rivers on São Tomé include the Abade, the Manuel Jorge, and the Rio d'Ouro; the Rio Papagaio is the longest river on Príncipe (Juste and Fa 1994).

The islands have an oceanic equatorial climate with temperatures about 22–33°C near sea level. At higher altitudes, the temperature often drops to 9°C or lower (Sayer et al. 1992; Juste and Fa 1994). Annual rainfall varies dramatically depending on elevation. The maximum annual rainfall is 7,000 mm on São Tomé, 5,000 mm on Príncipe, and 3,000 mm on Annobón. The main dry season on São Tomé and Príncipe occurs from June to September and is called the *gravana*. A smaller dry season called the *gravanito* lasts from December to February. The seasons on São Tomé and Príncipe are more affected by the intertropical front than are the seasons on Annobón. On Annobón, oceanic winds cause a well-defined single dry season that lasts from May to October (Juste and Fa 1994).

Before conversion, the primary vegetation of the ecoregion was tropical rainforest (Juste and Fa 1994). Now much of the land is used for agrarian purposes. The remaining primary and secondary forests fall under the category of lowland forest, montane rainforest, mossy forest, and mangrove forest. The forests on São Tomé, Príncipe, and Annobón contain many endemic and imperiled species, such as *Rinorea thomensis*, *Afrocarpus mannii*, *Craterispermum montanum*, and *Pandanus thomensis* (WCMC 2001a).

Outstanding or Distinctive Biodiversity Features

Overall, freshwater faunal richness is extremely low, but there is high endemism among certain taxa. For instance, only five frog species live in this ecoregion, but all are endemic. *Leptopelis palmatus* lives in the lowland forests of only Príncipe and is distinct enough from the mainland *L. rufus* that it is considered a different species and a classic example of island gigantism. *Nesionixalus thomensis* and *Ptychadena newtoni* have been found only on São Tomé. *N. molleri* and *Phrynobatrachus dispar* live on both São Tomé and Príncipe (Schiøtz 1999; Christy 2001). The genus *Nesionixalus* is endemic to the two islands. There are only two species of freshwater fish: the frillfin goby (*Bathygobius soporator*) and the West African freshwater goby (*Awaous lateristriga*) (FishBase 2001). The widespread and brackish banded lampeye (*Aplocheilichthys spilauchena*) is found up to an altitude of 50 m in the Rio Papagaio on Príncipe (Rossignon 1999). All species are able to tolerate brackish water. Only three freshwater mollusk species live in the ecoregion, two from the Neritidae family and the vector of schistosomiasis, *Bulinus forskalii*. This species shows distinctive morphological characteristics in comparison with mainland forms (Brown 1991). Like gobiid fish, Neritidae species are primarily marine, but some lineages have adapted to live in brackish and fresh water. The freshwater crab *Potamonautes margaritarius* is endemic to São Tomé (Cumberlidge 1999). Four species of freshwater shrimp have been recorded from the islands: *Macrobrachium zariquieyi, M. chevalieri, Atya intermedia,* and *Atya sulcatipes* (Holthuis 1966).

The islands of São Tomé and Annobón also have volcanic crater lakes. On São Tomé, Lagoa Amelia, located in the center of the island at 1,400 m, is covered by a floating 2-m-thick layer of vegetation consisting of mosses, ferns, and orchids, including the endemic *Diaphananthe brevifolia*. The crater walls are home to the endemic giant begonia *Begonia crateris*. On Annobón, Lagoa A Pot is located in the north of the island and consists of open water populated with the shrimp species *A. sulcatipes* and the fish *Gambusia affinis,* introduced by the Spanish to control mosquito larvae in the colonial period. The depths of the two crater lakes are unknown.

Status and Threats

Current Status

Current threats to the islands' freshwater systems include deforestation for agriculture and timber, the introduction of exotic species, and pollution. The human population of São Tomé and Príncipe is estimated to be 165,000, with a growth rate of about 3 percent (CIA 2001). As the population increases, it exacerbates other problems such as pollution and deforestation.

São Tomé, Príncipe, and Annobón were uninhabited by people until the Portuguese colonized them in the fifteenth century. Land clearing for the production of sugar cane, then coffee and cocoa, decimated lowland forests. Many of these plantations were abandoned when the islands gained independence in 1975, and some secondary forest has since grown in these areas (UNEP 1998). Timber harvesting today largely occurs in these secondary forests, and much of the timber is used for fuelwood (Sayer et al. 1992) and house construction. Some plantations are being rehabilitated, and coastal mangrove forests are facing renewed threats (Sayer et al. 1992; UNEP 1998). Large quantities of sediments carried by rivers are derived at least in part from deforestation, and these sediment loads may threaten the freshwater fauna of the ecoregion (Institute of Marine Sciences et al. 1998).

Although few studies have been carried out in the islands to assess the status of the freshwater systems, there is some evidence that sewage, industrial pollution, and agrochemicals are affecting the water quality. Small-scale farmers use DDT, but its impact on nontarget species is not documented (Institute of Marine Sciences et al. 1998). The use of copper sulfate treatments in cocoa plantations has also caused freshwater pollution. Recent land privatization has led to an increase in market gardening and thus a loss of tree cover originally found in cocoa and coffee plantations.

Two terrestrial gastropod species, *Archachatina marginata* and *Opeas pumilum,* were recently introduced to the island of São Tomé (Gascoigne 1994), and further introduced species such as *Laevicaulis alte* have also been recorded since this date (A. Gascoigne, pers. comm., 2002). A number of alien mammals (monkeys, rats, mice, feral cats, and pigs) have also been introduced to the islands (UNEP 1998). It is not certain what the long-term effects of the presence of these mammals and gastropods have been on the freshwater fauna of this ecoregion.

There are currently no protected areas in São Tomé and Príncipe. The proposed Parque Naturais d'Obo on the two islands awaits legal ratification. The country is a member of the regional European Union–funded Conservation and Rational Use of Forest Ecosystems in Central Africa (ECOFAC) project. Annobón has been legally designated as a protected area but still lacks effective management.

Future Threats

Recent initiatives to encourage organic cacao production are expected to reduce pollution risks. However, land privatization on São Tomé and Príncipe is resulting in deforestation in many areas and probably will increase pressure on remaining primary forest.

Justification for Ecoregion Delineation

São Tomé, Príncipe, and Annobón form the boundaries of this ecoregion. These islands were formed by volcanic eruptions that occurred along a fracture in the earth's crust during the lower Tertiary and early Quaternary (Juste and Fa 1994). Because the islands have never been connected with each other or the mainland, which is about 220 km away from São Tomé and Príncipe and 340 km from Annobón, the flora and fauna are low in richness but high in endemism (Jones 1994).

Data Quality: Medium

The crater lakes on São Tomé and Annobón are unique habitats that offer opportunities for paleobotanical and paleoclimatological research. Gascoigne (1993, 1996) provides a bibliography on the fauna of the islands.

Ecoregion Number:	**51**
Ecoregion Name:	**Mascarenes**
Major Habitat Type:	**Island Rivers and Lakes**
Final Biological Distinctiveness:	**Continentally Outstanding**
Final Conservation Status:	**Critical**
Priority Class:	**II**
Authors:	**Robin Abell and Ashley Brown**
Reviewer:	**Roger Safford**

Location and General Description

The isolated islands of the Mascarenes ecoregion host a highly endemic aquatic biota, including several gobies that inhabit swift streams. The islands are of volcanic origin and are situated in a line along a midocean ridge, located east of Madagascar in the western Indian Ocean. The largest islands are the French Dependent Territory (Département) of Réunion (2,500 km^2) and the island of Mauritius (1,900 km^2). Mauritius, Rodrigues (110 km^2), and several smaller islands form the single independent nation of Mauritius. Réunion, about 690 km from Madagascar, is the most westerly of the islands, followed by Mauritius; Rodrigues lies some 570 km further to the east.

The islands have a rugged topography with many ravines and cliffs. On Réunion, Piton de la Fournaise (2,525 m) becomes active several times each year, and as a result the island's topography is particularly steep. The other major peak on Réunion is Piton des Neiges, a long-extinct volcano that reaches 3,069 m and is the highest peak in the Indian Ocean (Defos du Rau 1960). On the island of Mauritius, the highest point is Piton de la Rivière Noire, which reaches a height of only 828 m. Mauritius is considered to be the oldest of the Mascarene Islands and was formed by three periods of volcanic activity occurring between 10 million and 20,000 years ago (Montaggioni and Nativel 1988). The island consists of an irregular central plateau at about 300–600 m, surrounded by three mountain ranges and plains (Ramdin 1969). Rodrigues has even lower relief, with rolling hills rising only to 390 m. Although it consists primarily of basaltic lava and some volcanic dust, there are small areas in the south and the east of Rodrigues where wind-blown sand has accumulated to form limestone rocks. In these areas caves have formed (Sok Appadu and Nayamuth 1999). Many smaller islands also occur in this ecoregion; of particular interest is Round Island, a small island off the northern tip of Mauritius with a distinctive flora and fauna (especially reptiles) comprising many species found nowhere else.

At lower elevations, summer (December–April) temperatures average 30°C, whereas in winter (May–November) the average is closer to 25°C. However, at the highest elevations the temperature averages only 18°C, and ice sometimes forms. Southeasterly trade winds blow throughout the year. On Mauritius, rainfall varies from 890 mm in coastal areas on the leeward side of the island to 4,445 mm in parts of the central plateau, with large variations over short distances. Because of this rainfall pattern, both tropical moist and tropical dry forests grow on Mauritius. Abundant and heavy rains from cyclones cause pronounced erosion and landslides (WWF and IUCN 1994). The climate on Réunion is influenced by southeasterly trade winds and depressions. Mean annual temperatures are less than 16°C over large areas and up to 25°C in the drier, leeward lowlands (Le Corre and Safford 2001). Rainfall ranges from less than 2,000 mm on the leeward side of the island to 5,000 mm (and up to 9,000 mm locally) in the mountains. On Rodrigues, annual rainfall varies between 1,000 and 1,700 mm, and temperatures are roughly 24°C (Safford 2001b).

The main communities of native vegetation on the islands historically were probably lower montane wet evergreen forest, scrub and marsh vegetation in the uplands, evergreen or semideciduous forest in the rainshadow, and palm savanna at lower elevations (Le Corre and Safford 2001). Habitats ranged from coastal wetlands and swamp forests at low elevations, through lowland dry forest, rainforest, and palm savanna, to montane deciduous forests and finally (on Réunion) to heathland vegetation types on the highest mountains. Most of the original vegetation has been destroyed, and almost all remaining native plant communities have been badly degraded by introduced species (Stuart et al. 1990; WWF and IUCN 1994). Major plant families include Sapotaceae, Ebenaceae, Rubiaceae, Myrtaceae, Clusiaceae, Lauraceae, Burseraceae, Euphorbiaceae, Sterculiaceae, Pittosporaceae, and Celastraceae. There is a high diversity of palm species on the islands, including many endemic genera (WWF and IUCN 1994).

Fast-flowing rivers and streams are abundant on the Mascarene Islands. On the island of Mauritius, the Grand River South East (34 km) is the longest river, followed by Rivière du Poste (23 km), Grand River North West (22 km), Rivière La Chaux (22 km), and Rivière des Créoles (20 km). Several of these rivers are marked by waterfalls, the highest being Tamarin Falls or Sept Cascades (293 m). In addition to numerous artificial reservoirs, Mauritius has two natural crater lakes: Grand Bassin and Bassin Blanc (Government of Mauritius 2001). Rodrigues and Réunion are similarly endowed with high-gradient rivers. Mangroves occur along the coast near river mouths and estuaries. On Mauritius, small areas of mangrove are present at various locations including Terre Rouge, Riviere Noire, Baie du Cap, Riviere du Rempart, Trou d'Eau Douce, Poste Lafayette, Bras d'Eau, Roches Noires, and Poudre d'Or (Institute of Marine Sciences et al. 1998).

Outstanding or Distinctive Biodiversity Features

About thirty-five fish species use Mascarene freshwater habitats, and many move between fresh, brackish, and marine waters. About twenty fish species are tied to freshwater habitats for part of their life cycle and were included in our richness estimates. Of these, five are endemic to the Mascarenes (*Hypseleotris cyprinoides, Cotylopus acutipinnis, Glossogobius kokius, Gobius commersonii,* and *Oxyurichthys guibei*). Gobies dominate the freshwater fish fauna. *Cotylopus acutipinnis,* a goby endemic to Réunion, hatches in the sea and after its larval stage migrates to swift rainforest streams to live as an adult (FishBase 2001). *Sicyopterus lagocephalus,* a goby found predominantly in the Mascarenes, has a similar life history. Other fish families with freshwater species represented in this ecoregion include Anguillidae, Kuhliidae, and Eleotridae. According to FishBase (2001), Réunion has the highest number of native fish species (thirty), followed by Mauritius (twenty-one), and the two islands have many species in common. Rodrigues has a depauperate freshwater fish fauna (four species).

About twenty snail species are known from the fresh and brackish waters of the Mascarenes. The fauna is a mixture of those with Afrotropical (e.g., *Afrogyrus rodriguezensis* and *Bulinus cernicus*) and Asiatic (e.g., *Lymnaea mauritania* and *Gyraulus mauritianus*) affinities, and at least two are known to have been introduced. The endemic *Lantzia carinata* lives in or near a waterfall on the island of Réunion and apparently is adapted to cling to rocks (Brown 1994). This species was recently rediscovered after more than a century.

Two dragonfly species, *Argiocnemis solitaria* and *Platycnemis mauriciana*, are considered endemic to Mauritius and critically endangered (IUCN 2002). However, both species are known only from the type locality and therefore await taxonomic verification. The specific status of *P. mauriciana* is particularly in doubt (Samways 2002).

Until the arrival of humans, the islands supported several endemic (and often extraordinary) aquatic birds (Cheke 1987), but these are all extinct. The only surviving native freshwater bird species are green-backed heron (*Butorides striatus*) on all three main islands and common moorhen (*Gallinula chloropus*) on Mauritius and Réunion.

Status and Threats

Current Status

Despite their isolation, the Mascarene Islands are highly degraded, with vast losses of original habitat and resulting alteration of freshwater systems (Stuart et al. 1990). The natural habitats of these islands are heavily fragmented as a result of human activities, which began in earnest with French settlement of Réunion in 1642 and Dutch occupation of Mauritius in 1638 (UNEP 1998). Today, the country of Mauritius is distinguished by having one of the highest human population densities in the Afrotropics, at more than 580 people/km^2 on the island of Mauritius and about 330 people/km^2 on Rodrigues (Safford 2001b). On the island of Mauritius, most of the remaining forest is found in the southwest, around the Black River Gorge (WCMC 1993), and even here it is under great threat from ongoing invasion by exotic plants. Sugar cane, tea, and conifer plantations cover large tracts of the island, but forest clearance for these uses has ceased. On Rodrigues, most remaining forests are exotic, located mainly in valleys in the center of the island. There are some reserves, but even there introduced plants and animals are a major problem. On Réunion, most of the remaining forest blocks are found at higher elevations. In the lowlands, forest habitat is largely limited to the steep banks of rivers. Deforestation can lead to increased erosion and sedimentation of freshwater habitats, which poses a particular problem for species adapted to clear forested streams, as is the case for many Mascarene taxa.

Mascarene freshwater habitats, particularly in the lowlands, suffer substantially from organic and chemical pollution derived from industry, agriculture, and human settlements. The major economic activities on Mauritius are sugar cane production, textile manufacturing, and tourism. On Réunion sugar cane production is also predominant, with some subsistence agriculture (UNEP 1998). Sugar mills produce a range of wastes, including hot wastewater, oil waste, particulate matter, and solid and hazardous wastes; however, several mills have reportedly adopted technologies such as sedimentation ponds and wastewater recycling to reduce the pollution discharged to streams. The textile industry produces both liquid and airborne wastes, as do many of the other smaller industries, which include dyeing, ethanol distilling, battery manufacturing, soap and detergent manufacturing, galvanizing, food canning, and chemical manufacturing. Although these industries together use large quantities of water in their operations, the wastewater may bypass inland freshwater systems and be discharged directly to the sea after pretreatment at wastewater treatment facilities. Human wastes are also a source of pollution, with many settlements lacking access to sewage treatment or plumbing. Finally, tourism is placing increasing pressure on water supplies (Institute of Marine Sciences et al. 1998).

The widespread invasion of terrestrial plants and animals on the Mascarene Islands is a serious threat to the islands' biota, and exotics are equally dominant in the freshwater realm. Introduced fish species are nearly as great in number as natives, partly as a result of aquaculture. In Mauritius, the era of intentional introductions apparently began in the 1950s, when three tilapia species were introduced for the aquaculture industry. Then, in 1971–1972, Japanese (*Crassostrea gigas*), American (*Crassostrea virginica*), and European (*Ostrea edulis*) oysters were introduced for farming. In 1972, a hatchery with earthen freshwater ponds was built for the production of the giant freshwater prawn (*Macrobrachium rosenbergii*). In 1975 and 1976, six species of Indian and Chinese carp were introduced (*Labeo rohita*, *Catla catla*, *Cirrhinus cirrhosus*, *Ctenopharyngodon idellus*, *Hypophthalmichthys molitrix*, and *Cyprinus carpio*). Carp aquaculture was eventually discontinued because of a lack of consumer interest, yet all of these species have been reported in Mascarene fresh waters (Mauritius National Parks and Conservation Service 2000; FishBase 2001). In 1990, a hybrid of three tilapia species (*Oreochromis niloticus*, *O. mossambicus*, and *O. aureus*) was introduced for farming in freshwater ponds. Production of this fish still continues (Mauritius National Parks and Conservation Service 2000).

There are several protected areas on Réunion, including two nature reserves (Mare Longue and la Roche Ecrite), several protected biotopes (La Petite Ile, Piton des Neiges, and Grand Bénard), and a number of other reserves. Two of these, Etang de Saint Paul and Les Lagons, are wetland nature reserves that are in the planning stage (Le Corre and Safford 2001). On Mauritius, there are several protected areas, the largest being Black River National Park (66 km^2), which is being enlarged. In 2001, Mauritius also designated Rivulet Terre Rouge Estuary Bird

Sanctuary as its first Ramsar site (Wetlands International 2002). Some offshore islets are also reserved for conservation. On Rodrigues there are three protected areas (comprising only 0.58 km²), which include the only surviving remnants of natural vegetation. Two of these are small islands, and one is located on the mainland. Other areas have also been fenced to allow regeneration of woody habitats, which are dominated by exotic plant species; native vegetation restoration also takes place, necessitating more intensive management. The extent to which any of these areas afford protection to freshwater species and habitats is unknown.

Future Threats

The biggest future threats to Mascarene freshwater biodiversity are pressures from exotic species, pollution from further development, and excess water withdrawals. The native freshwater biota of the Mascarenes is poorly known, a fact that is remarkable considering the intensity of conservation efforts for its terrestrial biota, and there is the possibility that species may disappear before they are fully studied. On a positive note, in 2001 the Republic of Mauritius joined the Ramsar Convention, which establishes a framework for the protection of wetlands of high biodiversity importance.

Justification for Ecoregion Delineation

The Mascarene Islands define the boundaries of this ecoregion, which supports an endemic freshwater fauna that shares affinities with Madagascar, East Africa, and Asia.

Data Quality: Low

Ecoregion Number:	**52**
Ecoregion Name:	Socotra
Major Habitat Type:	Island Rivers and Lakes
Final Biological Distinctiveness:	Continentally Outstanding
Final Conservation Status:	Relatively Stable
Priority Class:	III
Author:	Wolfgang Wranik

Location and General Description

The islands of this ecoregion possess few permanent watercourses, but their temporary streams support an aquatic fauna that includes endemic crabs, mollusks, and insects. The ecoregion is named for its most prominent feature, Socotra Island.

The Socotra Archipelago is part of the Republic of Yemen and is situated in the northwestern part of the Indian Ocean. It is an eastern continuation of the Somali Peninsula and lies along the East-West Rift of the Gulf of Aden. The archipelago comprises the four islands Socotra (3,625 km²), Abd al Kuri (131 km²), Samha (41 km²), and Darsa (12 km²), and the rocks Sabuniyah, northwest of Socotra, and Ka'l Fir'awn, north of Abd al Kuri. The islands are separated from one another by shallow seas and from the mainland by a deep trench. Abd al Kuri lies about 100 km due east of Cape Guardafui in Somalia, and Socotra is about 360 km south of Ras Fartak on the mainland of Arabia (Mies and Beyhl 1998; Wranik 2003).

Socotra is composed of a basement complex of igneous and metamorphic rocks of Precambrian age overlain by sedimentary rocks, mainly limestone and sandstone. It is estimated that the archipelago has been isolated from the mainland since the end of the Cretaceous period (Mies and Beyhl 1998).

Topographically Socotra can be divided into three main zones: coastal plains, limestone plateau, and the Haghier Mountains. The coastal plains vary in width and length but are about 8 km at their widest. The largest unbroken stretch is the Noged, in the south, which is about 80 km long, whereas the north is much more irregular, less barren, and composed of a number of smaller plains separated from each other by rocky headlands. In some places the lowland plains are separated from the sea by ranges of hills and divided by watersheds into a series of small river basins reaching the sea through narrow gorges. Some of the north-flowing wadis terminate in fresh or brackish estuaries, separated from the sea by sandbars that are broken only by the heavy floods. However, no permanent streams cross the plains. The soils are alluvial, mainly of compacted gravel, stones, and coarse sands, but in some areas extensive sand dunes occur. The coastal plains are largely subdesertic. In general they are sparsely vegetated with herbaceous halophytic communities, open deciduous shrubland, and small local belts of *Avicennia marina* mangrove, and in some areas they are largely devoid of vegetation. However, after rain a good cover of grasses and herbs develops in many areas (Wranik 2003).

The limestone plateau extends over most of the island, averaging 300–700 m in altitude. It is dissected by a number of deep valleys and steep escarpments, often honeycombed with fissures, caves, and ledges and in places bounded directly by the sea. Soils are poorly developed, but in hollows and rock crevices fine gray clays may accumulate. The exposed summits are covered with sparse shrubland or low woody herb communities. The dominant plants are the shrubs *Croton socotranus* and *Jatropha unicostata* and various succulent and emergent trees (*Dendrosicyos, Adenium, Euphorbia, Sterculia, Boswellia, Commiphora*). In sheltered valleys and mountain areas the vegetation is more luxuriant (Wranik 2003).

The Haghier Mountains rise up in the center of Socotra in a series of bizarre pinnacles to a height of approximately 1,530 m. Deep and fertile red soils have accumulated in the valleys and on the gentler slopes by decomposition of the granites from the Haghier Massif. Semideciduous thicket develops in sheltered areas on the middle slopes of the granite mountains, and the higher slopes are covered in a mosaic of dense evergreen woodland and thicket, low shrubland, grassland, and open rock vegetation on the pinnacles. This includes the emergent *Dracaena cinnabari,* the bizarre-looking dragon's blood tree for which Socotra is famed. On the gentler slopes the vegetation has been cleared to form grassland pastures for cattle, and the open rock faces of the granite pinnacles are covered with lichens and low cushion plants (Wranik 2003).

The islands of the archipelago are characterized by a tropical, semi-arid climate where rainfall is scarce. The southern half of the main island probably is somewhat drier than the north. The SW-Monsoon blows from May to September. It is violent and brings hot, dry winds, often averaging 70 mph. On the exposed plateaus and coastal plains there is little if any rain, and at higher elevations the season brings drizzle and clouds. From November until March the area is under the influence of the NE-Monsoon. It is much gentler and cooler, and because it carries considerable moisture, this time of year is the rainy season. The transitional months between the monsoons are periods of uncertain light winds and sporadic rains. Tropical storms and cyclones bring torrential rain to the archipelago every few years (Wranik 2003).

No long-term climatic records are available, so there are no exact figures on annual rainfall. The only reliable data source is the short-term record measured at the Royal Air Force Station at Ras Karma in the north of the island during World War II (1943–1945). The data show a total annual rainfall for that particular area of 125–175 mm during the 3 years that records were kept, with the highest precipitation in November and December. The recorded measurements of a low, sporadic rainfall probably are typical for most of the plateau and plains areas (Wranik 2003). However, the highlands are likely to have a greater annual rainfall, possibly on the order of 380–620 mm and perhaps as much as 1,000 mm for the higher peaks of the Haghier. The peaks often are shrouded in clouds, and heavy mists and dew are common. These mists seem to be a main water source for vegetation in the higher altitudes and supply water for a number of animals (Wranik 2003).

As far as is known the mean annual temperatures on Socotra's plain areas vary from 17°C to 26°C (minimum) and 27°C to 37°C (maximum), but it is much cooler in the mountains, particularly at night. Frost is not reported. May is said to be the hottest month, when heat and humidity rise during the period of calm between the monsoons. September is also calm with high humidity (Popov 1957).

The Haghier Massif forms the most important watershed on Socotra Island, and numerous watercourses run both north and south from it. There are also a number of permanent springs. Particularly on the northern slopes the streams are permanent in their upper reaches, but on the plains they are mostly sporadic, carrying water only during the rains and just afterward. The streambeds usually are covered by stones and boulders and in some parts are lined by semi-aquatic plants such as sedges and grasses of the genera *Fimbristylus, Cyperus, Eragrostis, Cladium,* and *Paspalidium.* However, some of the river estuaries retain water for most of the year. On the southern coast water is scarce, limited to a few springs at the base of cliffs and several wells of brackish water. Although not well known, there is a large underground karstic system (Wranik 2003).

On all the smaller islands the conditions are more extreme than on Socotra. Large areas are fairly barren. Low woody herb communities dominate the vegetation. Woodland is absent, and shrubland is limited to a few sheltered places. On Abd al Kuri there is no surface water, but there are a few brackish wells. The human population is estimated at approximately 200–300 people, who engage mostly in fishing and trade. The islands of Samha and Darsa, also called The Brothers, are both flat-topped limestone plateaus with sheer sides. Samha has a limited freshwater supply and is inhabited by a small human population, whereas Darsa is uninhabited and seems to lack surface water (Wranik 2003).

Outstanding or Distinctive Biodiversity Features

The Socotra Archipelago is distinguished by a distinctive variety of plant and animal species. More than 850 species of flowering plants and ferns have been recorded, of which 230–270 (including some 15 species limited to Abd al Kuri) are considered to be endemic (Alexander and Miller 1996; Mies 2001). Among them are strange looking remnants of ancient floras, plants that long ago disappeared from the surrounding African-Arabian mainland. According to the World Conservation Monitoring Centre, the archipelago is the world's tenth richest island group in terms of endemic plant species (Davis et al. 1994).

The fauna of the archipelago also contains an exceptional number of endemics, although not as striking as the plants and poorly studied by comparison. The known number of species is only about 900. However, because most faunistic data date back to the major expeditions of approximately 100 years ago, a number of forms are known from very few or even single specimens. Many records need confirmation and a critical taxonomic revision. Large areas of the islands remain poorly explored (Forbes 1903; Evans 2000; Wranik 2003).

More than 175 bird species have been recorded on the islands. Forty-one species breed, whereas the others are migrants, winter visitors, and vagrants. Among the breeding birds, the most important are the endemic Socotra warbler (*Incana incana*), Socotra sunbird (*Nectarinia balfouri*), Socotra sparrow (*Passer insularis*), Socotra starling (*Onychognathus frater*), and Socotra cisticola (*Cisticola haesitata*) and the vulnerable Socotra bunting (*Emberiza socotrana*) (Kirwan et al. 1996; Evans 2000; BirdLife International 2000; IUCN 2002).

The freshwater communities are poorly known. Reconstructing the original freshwater fauna of Socotra Island is difficult because of inadequate surveys before the recent stocking (1990s) of *Aphanius dispar* as part of an antimalarial campaign (Wranik 1998). No freshwater fish were observed in rivers before this stocking, although a lack of data should not necessarily be equated with an absence of fish. Apparently stable populations of *A. dispar* persist in the streams, estuaries, and wells where they were stocked. A 1995 expedition to explore the freshwater fish fauna yielded only *A. dispar* (Al-Safadi 1998).

Five freshwater mollusk species are known from the islands, including two apparent endemics (*Ceratophallus socotrensis* and *Gyraulus cockburni*) (Brown 1994; Wranik 1998; Wranik 2003). There are three endemic freshwater crab species: *Socotrapotamon socotrensis*, *S. nojidensis*, and *Socotra pseudocardiosoma*. There are also records of phyllopods, ostracods, and copepods (*Ectocyclops, Halicyclops, Mesocyclops, Microcyclops, Paracyclops, Ropocyclops*) (Baribwegure and Dumont 2000). Amphipods include *Platorchestia platensis* (or *Indoweckelia* sp.), an interesting new endemic species that lives in water wells. Further studies promise exciting discoveries for other freshwater organisms including sponges, turbellarians, and nematodes. Aquatic habitats deep inside the cave systems have been isolated for a very long time and could be of special scientific interest and evolutionary importance (Wranik 2003).

In addition, the freshwater insect fauna probably will reveal new species. Currently known aquatic insects include eighteen odonates (with one endemic, *Enallagma granti*), six Heteroptera species, four Trichoptera, eighteen Coleoptera, and eleven Diptera (Leeson and Theodor 1948; Mattingly and Knight 1956; Kimmins 1960; Schneider 1996; Malicky 1999; Wranik 2003).

Status and Threats

Current Status

In 2000, the population of Socotra was estimated to be around 45,000 people. A large number live in and around the capital, Hadiboh, and in the coastal fishing villages around the island, and the rest live in highly dispersed settlements in the interior. Those living on the coastal plains are engaged mainly in fishing and trade, whereas the mountain people are subsistence farmers and pastoralists. They are seminomadic, and some inhabit caves during several months of the year. The most important elements of the local economy are livestock production and coastal fishing. A number of families practice subsistence farming with small-scale production of fruits and vegetables for local consumption. There are also handicrafts in the form of carpet weaving, basket making, and pottery. However, in terms of economic development Socotra has lagged seriously behind the mainland of Yemen.

Despite, or because of, the poor living conditions on Socotra, principles of cooperation, self-help, and community labor are well established, and there is also a whole range of relevant traditional rules and practices that are of ecological importance. These unwritten laws include regulations that control the cutting of live wood, forbid the use of other than dead wood as firewood, regulate grazing and the cutting of vegetation as fodder, and preserve important fruit-bearing trees. Traditionally the people practice rotational grazing, and the practice of transplanting and sowing certain plants and protecting them from livestock while they grow has been known and applied for a long time. A network of tribal councils strictly enforces these rules (Miller and Morris 2002).

No exact data are available on the numbers of livestock on Socotra. However, the actual numbers seem to be already at maximum sustainable levels. As yet there is no practicable way to provide supplementary fodder and water during the summer, so drought and diseases, sometimes at an epidemic level, continue to control livestock numbers (Wranik 2003). Pasture for goats, sheep, cattle, donkeys, and camels makes up about 80 percent of total land use (Mies and Beyhl 1998). Overgrazing occurs in parts of the ecoregion but is not uniform throughout. It is most concentrated in parts of the Haghier Mountains and the north-central limestone plateaus (Miller and Bazara'a 1998). Cattle grazing is also heavy in the central highlands of the island near Diksam, in the interior central highlands of the island (Miller and Bazara'a 1998).

Pollution is a problem in the ecoregion because there is no well-developed infrastructure to deal with sewage and other waste disposal. Sewage treatment is nonexistent in rural areas, and organic wastes are discharged directly into the environment, often to be eaten by goats and birds (Wranik 1998). Most of the rural population of the island depends on hand-dug wells and centrifugal pumps to draw water from the ground. In the Wadi Ayhaft and in the northern central section of the island near Hadiboh, the water supply comes from a cistern in the mountains and an extended pipe system (Wranik 1998).

Future Threats

The main threats to the island's environment are development of infrastructure, breakdown of traditional land management practices, introduction of exotic plants and animals, uncontrolled collecting of specimens by amateur and professional collectors, and climatic changes (Wranik 2003).

The opening of the island has stimulated plans to improve living conditions and infrastructure, such as a Socotra Sea Port, Socotra Airport, road development, electricity and water supply, medical care, and education, which will lead to an increasing influx of goods and technical facilities. This development will change the face of the island soon. In 1998, improvement of the airfield on Socotra and construction of a port close to Hadiboh began without an environmental impact assessment. As development on the island increases, so does the threat of the introduction of alien species. The care with which the development is planned and undertaken will determine the fate of the island's species (Wranik 2003).

Artificial water supplies and the importation of supplementary food could result in an increase in numbers of livestock. An increase in livestock numbers or even a disruption of the complicated patterns of seasonal livestock movements could be expected to quickly destroy the present fragile equilibrium between vegetation, people, and livestock. The vegetation plays a key role in holding the soil on the slopes and reducing surface runoff. Wood is used for various purposes on Socotra, including heating, cooking, fuel for lime manufacture, and building material. Any removal of the vegetation cover related to wood collection would result in accelerated soil erosion and the loss of surface water through increased runoff, creating a dangerous downward spiral for the island and its biota (Wranik 2003).

For a number of years IUCN, UNESCO, and other international organizations have identified the Socotra Archipelago as a very high-priority area for protection (Euroconsult 1994). In the Yemen National Environment Action Plan the establishment of a national protected area on Socotra was identified as a high-priority action. The High Committee for Development of Socotra has prepared a master plan for development of the archipelago in cooperation with the United Nations Development Programme, Global Environment Facility, and European Union, integrating biodiversity conservation, environmental management, and development objectives. In the Socotra Biodiversity Project, recommendations for critical areas and zoning for protection have been developed. In April 2000 the Yemen government formally approved a Conservation Zoning Plan for the Development of Socotra (Republic of Yemen 2000).

Justification for Ecoregion Delineation

This ecoregion follows the boundaries of Socotra Island and is distinguished by a depauperate but endemic freshwater invertebrate fauna.

Data Quality: Low

Further surveys of the aquatic insect fauna would be expected to reveal many more species (Schneider and Dumont 1998).

Ecoregion Number: **53**
Ecoregion Name: **Lake Malawi**
Major Habitat Type: **Large Lakes**
Final Biological Distinctiveness: **Globally Outstanding**
Final Conservation Status: **Vulnerable**
Priority Class: **I**
Authors: **Anthony J. Ribbink and Lucy Scott**
Reviewer: **Jos Snoeks**

Location and General Description

Lake Malawi/Niassa/Nyasa and its influents, Lake Malombe, and the Shire River in between the two lakes form this globally distinctive ecoregion (Tweddle et al. 1979; Ribbink 2001). The ecoregion hosts highly endemic species flocks of fishes, which make up one of the richest lake fish faunas in the world. Lake Malawi/Niassa/Nyasa (hereafter called Lake Malawi) is the southernmost lake of the Rift Valley and is bordered by the countries of Malawi, Mozambique, and Tanzania. Lake Malawi is the ninth largest lake in the world and the fourth deepest and has a surface area of about 28,000–31,000 km^2 (Bootsma and Hecky 1993; Ribbink 2001).

Three seasons are recognized. During the cool and dry period from May through August, temperatures near the Lake Malawi shore may drop to 15°C, with a daily average of 20–22°C. During the hot and dry phase between September and November, average air temperature is about 28°C but may reach 40°C. The average daily temperature during the wet season from late November through April is 25°C (Lopes 2001). Rainfall varies from about 600 to 2,200 mm per year depending primarily on location, with areas of higher altitude generally having higher rainfall (Hughes and Hughes 1992). Lake levels rise during the wet season, and annual fluctuations range between 0.4 m and 1.8 m (Lopes 2001). Flooding of the coastal plains occurs with increased rainfall in the wet season (Beadle 1981).

More than 200 rivers flow into Lake Malawi. Most are short, and many flow only in the rainy season. Major rivers include the Lufira, Songwe, Rukuru, Dwangwa, Bua, and Linthipe on the western coast and the Ruhuhu and Rio Lunho on the eastern coast. The Shire River, which links lakes Malawi and Malombe, shows seasonal and long-term variability in flow. Flows in this river diminished steadily from 1896 until they ceased in 1915. In 1935 lake levels rose sufficiently for flows to resume, and they have continued until the present day, varying in magnitude with rainfall (Beadle 1981; Lopes 2001). Lake Malombe is almost entirely dependent on the flow of the Shire River, and the lake dried out when flows ceased between 1915 and 1935 (Beadle 1981).

The tropical setting confers thermal stability to Lake Malawi, which is characterized by permanent stratification. This is maintained by temperature and density differences between the upper epilimnion, between 0 m and 125 m; the metalimnion, between 125 m and 230 m; and the anoxic hypolimnion, below 230 m (Gonfiantini et al. 1979). Full atmospheric exchange is limited to the surface mixed layer, which is well oxygenated but can be as little as 40 m deep. Limited vertical mixing does occur between strata, and oscillations of the thermocline caused by internal waves and wind-induced upwelling may cause advections of the metalimnion in the southeastern arm of the lake. The lengthwise orientation of the lake coincides with the southeast winds (*mwera*) that are channeled along the length of the lake and push surface waters north, to be replaced with deeper, cooler, nutrient-rich waters in the south. This upwelling maintains a thermal longitudinal gradient, with cooler surface waters in the southern part of the lake throughout the year. This upwelling system is the basis of the productive fisheries in the southern arms of the lake (Eccles 1974). Currents affect ecological processes such as nutrient cycling, plume dispersal, and fish productivity (Fryer and Iles 1972). Some indications suggest that there is a clockwise circular surface current in Lake Malawi, controlled by winds and radiant energy exchange and by inputs to the lake from rivers (Ribbink 2001). The flushing time of the lake is about 750 years (Bootsma and Hecky 1993).

Algae are the primary producers and the main source of organic carbon input into the lake, with phytoplankton in the pelagic zone and demersal algae on the lake floor. Nutrient cycles are driven by algae, which depend on the inflow of organic and inorganic nutrients from the catchments and on atmospheric deposition and the cycling of these nutrients in the lake. Certain forms of nitrogen and phosphorus have been shown to be contributed to the lake mainly through dry deposition and precipitation (Bootsma and Hecky 1999). Plant nutrient concentrations are found to increase with depth, with ammonia the dominant form of inorganic nitrogen in the deeper anoxic waters (Bootsma 1993; Patterson and Kachinjika 1995).

The topography in and around Lake Malawi is notable. The Livingstone Mountains in the northeast drop precipitously to cliffs at the lakeshore and continue their descent underwater (Beadle 1981; Hughes and Hughes 1992). The Lake Malawi shoreline tends to be steep and rocky in most places in the north, and the topography is less steep in the southern part of the ecoregion, where sandy bays and coastal plains are more common (Hughes and Hughes 1992). The lake itself contains numerous islands, islets, rocky outcrops, and reefs, which provide a rich and diverse underwater habitat for fish.

Outstanding or Distinctive Biodiversity Features

Endemism in the ecoregion is remarkably high. Among fish, 99 percent of the more than 800 cichlid species and more than 70 percent of the seventeen clariids are endemic (Snoeks 1999b; Ribbink 2001). Equally, the lacustrine invertebrates appear to have high levels of endemism, although exact numbers are unknown for these largely unstudied groups (Fryer 1959; Abdallah 2000).

Richness of taxa is exceedingly high in the ecoregion, with about 200 mammal, 650 bird, more than 30 freshwater mollusk, and more than 5,500 plant species (Brown 1994; Ribbink 2001). This richness is also reflected in the invertebrates and algae (Fryer 1959; Abdallah 2000). There are about 800 fish species, but there are also clearly recognizable geographic subpopulations, some having economic value, such as those sold as ornamental fishes. It has been suggested that as many as 3,000 recognizable fish taxa (species and populations) might be found in the lake. The latter number probably represents the largest number of fish taxa for any lake in the world (Ribbink 2001). There are about seventy noncichlid species (fourteen families) in the basin, of which a few, such as the lungfish and the trout, are introduced; about fifty, belonging to eleven families, are living in the lake itself (Snoeks 1999a). Worth mentioning are the endemic, mostly deep water–dwelling large catfishes of the genus *Bathyclarias*, which have been shown to have originated from a widespread, generalist species, *Clarias gariepinus*, which still occurs in the lake (Agnese and Teugels 2001). Several cyprinids are economically important, such as the sardine-like pelagic cyprinid *Engraulicypris sardella* (*usipa*), the salmon-like *Opsaridium microlepis* (*mpasa*), and the troutlike *O. microcephalum* (*sanjika*). It is difficult to summarize the large variety and complex nature of the more than 800 cichlid species. Just over 300 are scientifically described. The cichlids are represented mainly by lineages of mouth-brooding haplochromines. An important exception is the *chambo*, endemic *Oreochromis* spp., which are important food fish. The haplochromines can roughly be divided into the smaller, beautifully colored, mainly rock-dwelling *mbuna* and the non-*mbuna*. Malawi cichlids in general, but especially the *mbuna*, are very popular aquarium fishes. Several groups can be distinguished within the non-*mbuna*. There are two lineages of merely pelagic, predatory, and zooplanktivorous cichlids: *Rhamphochromis* (*ncheni*) and *Diplotaxodon* (*ndunduma*). *Utaka* are a group of zooplanktivorous and phytoplanktivorous schooling cichlids living over reefs and sandy habitats. Most of the remaining, very diverse demersal species are locally named *chisawasawa* or *kambuzi* (Turner 1995). All but four of the cichlid species are endemic to the lake (Ribbink 2001).

The evolutionary processes that have produced such incredibly high numbers of endemic cichlids have fascinated biologists for some time, leading to the use of metaphors such as "explosive speciation" to describe the rapid evolution of taxa (Fryer and Iles 1972; Liem 1980). The most widely accepted scenario of evolution of this highly diverse cichlid fauna is that riverine species with broad habitat ranges colonized the young lake and slowly became specialized to their new habitats in Lake Malawi. Changing lake levels during this time would have forced shallow-water stenotypes to live in habitats for which they were not adapted, placing them under severe selection pressure and causing incipient species flocks to become adapted to rocky, sandy, shallow, deep, vegetated, or mixed habitats (Fryer 1977; Owen et al. 1990). A prominent role for sexual selection has been suggested, based in part on the observation that many closely related species differ mainly in male color pattern (see Seehausen 2000 for a larger discussion).

The cichlids share several distinguishing characteristics. Many practice parental care, and most species in the lake basin are maternal mouthbreeders, breeding and releasing young in the same habitat. These cichlids also grow slowly and produce few young. Given these characteristics, most of the cichlids of this ecoregion are vulnerable to habitat degradation, sensitive to exploitation, and slow to recover from population declines (Ribbink 2001).

Several fish such as the *nchila* (*Labeo mesops*), *sanjika,* and *mpasa* are potamodromous, migrating annually up inflowing rivers to spawn. Large numbers of individuals congregate at river mouths and in the rivers before these spawning migrations, making them easy targets for fisheries. These potamodromous species are of special concern: most are endemic, they are prized as food fish, and they are subject to the twin threats of heavy exploitation and degraded spawning habitats in the rivers (Tweddle 1996).

Status and Threats

Current Status

Water quality is generally good; there are large areas of pristine forest, particularly in the catchments of Tanzania and Mozambique, and viable fish populations still occur in Lake Malawi and its tributaries. However, in many areas there are increasing rates of habitat degradation, erosion, sedimentation, and eutrophication. Additionally, fish communities have been modified through harvesting, with observed stock declines in parts of the lake (Turner 1994, 1996).

The sensitivity of the lake's limnology to anthropogenic activities in the catchment is striking, and it is clear that the entire ecosystem is fragile in both its short-term and longer-term responses to erosion and atmospheric input (Bootsma and Hecky 1993). Anthropogenic activities in the ecoregion, such as deforestation and inappropriate agricultural practices, are generating increased erosion and nutrient runoff in tributary catchments, with consequent sediment plumes and algal blooms in Lake Malawi. Increased sediment and nutrient loads can change the physical structure of rivers and alter water quality and algal composition in the lake. These changes often have cascading influences on the food chain. For example, sediment plumes can decrease the depth to which light penetrates in the water column, decreasing photosynthesis. In Lake Victoria, increased turbidity has been shown to decrease cichlid diversity by limiting the visual cues that allow proper mate choice (Seehausen et al. 1997a). The effects of an increased nutrient and sediment load may be delayed for many years as the cooler water of rivers sinks into the lake's hypolimnion, where the associated nutrients and sediments remain until waters mix with the epilimnion over time. The magnitude and extent of pollution in Lake Malawi is poorly known, but it is thought that increasing levels of nutrients, herbicides, pesticides, and toxic metals in the lake are caused both by air pollution and runoff from agricultural lands in the lake basin (Bootsma and Hecky 1999; Mkanda 2002). The Linthipe, Dwangwa, and Songwe basins, all in Malawi and entering the lake's western coast, have been identified as particularly degraded (Bootsma and Hecky 1999).

Along the sparsely populated eastern shores of Lake Malawi, ecological processes appear to function within their natural ranges of variation, and population sizes and structures of underwater communities are intact. Because the immediate catchments along the eastern shore are fairly undisturbed, any degradation of underwater habitats there probably is from lakewide, distant sources (Ribbink 2001).

Fish are estimated to provide 60–70 percent of the protein consumed by Malawians, with more than 70 percent of these fish coming from the Lake Malawi Basin (Turner 1995). Around the world, habitat degradation has been associated far more often with species extinction than harvesting (McAllister et al. 1997), but the behavior and stenotopy of the lake's cichlid fishes appear to increase their vulnerability to overexploitation (Reinthal 1993). Overfishing by small-meshed trawls and seines appears to have caused major changes in the fish communities in Lake Malombe and southern parts of Lake Malawi (Turner 1994). For example, the *chambo* (*Oreochromis* spp.) fishery in Lake Malombe declined because of the destruction of aquatic habitats by small-meshed nets (Banda and Hara 1997). Shallow-water cichlid species are particularly vulnerable to overharvesting by artisanal fishers because the fish are present as small, narrowly distributed populations (Ribbink 2001). The tendency of potamodromous species to aggregate during the spawning season also heightens their vulnerability (Tweddle et al. 1979; Tweddle 1996). Larger, more widespread populations occurring offshore are more resilient. Ideally, therefore, fishing activities should move into deeper water, but with more than 90 percent of the catch being taken by the artisanal fishery, the costs of moving offshore are prohibitive.

Lake Malawi National Park, encompassing part of the Nankumba Peninsula and thirteen islands in southern Malawi, was formed in 1980 and declared a United Nations Educational, Scientific and Cultural Organization World Heritage Site in 1984. It is the only truly aquatic park in the ecoregion, and it plays an important role in protecting populations of endemic cichlids. The Liwonde National Park encompasses a small southeastern section of Lake Malombe, and the Nkotakota National Park protects much of the Bua River catchment (Tweddle 1996).

Future Threats

Poverty, increasing human populations, and an almost total dependence on natural resources are the causes of deforestation, inappropriate agricultural practices, and heavy exploitation of wildlife and fishes in the lake basin. Habitat degradation in the catchments, changes to the river systems and flow regimes, and eutrophication of the waters are increasing and collectively result in changes to the ecological processes of the lake and declines in ecological resilience. Potamodromous fish species face the greatest extinction threat, followed by Lake Malawi's nearshore stenotopic cichlid fishes and the poorly known Amphibia. An increased commitment of time and resources to conserving this distinctive ecoregion will be necessary to secure its future and reverse current trends.

Ribbink (2001) recommends as priority actions the compilation of existing knowledge into a geographic information system decision-making framework that establishes priorities for conservation, research, and planning and the development of protected areas along the eastern catchments and nearshore regions of both lakes Malawi and Malombe. Conservation efforts focused on the unspoiled areas are likely to be more cost-effective than work in the heavily degraded regions.

Justification for Ecoregion Delineation

This ecoregion is based on boundaries of Lake Malawi and its basin. The fauna of Lake Malawi is separated from that of the Lower Zambezi by the Murchinson Falls in the lower Shire River, with the lowermost element of the cataracts, the Kapachira Falls, providing an absolute physical barrier to upstream movement of fish species (Tweddle et al. 1979). Lake Malombe is included in this ecoregion because the majority of its fishes and invertebrates are common to Lake Malawi and Lake Malombe, and as far as the fishes are concerned, most species are endemic to the two lakes (Turner 1996).

Lake Malawi developed as a consequence of tectonic activities in the Miocene, during the formation of the Great Rift Valley (Livingstone and Melack 1984). The rift formed when the land between two parallel faults subsided and the valley filled with water from the rivers that flowed into it. Although the rifting that formed the basin has a history of more than 20 million years, it is generally believed that water has been continuously present in the basin for about 2 million years (Fryer and Iles 1972; Crossley 1979; Johnson and Ng'ang'a 1990). The fishes that originally colonized the lake did so from the rivers that flowed into the Rift Valley (Lowe-McConnell 1987). Colonization was followed by explosive speciation resulting from narrow specialization as the cichlid fishes adapted to particular habitats. This evolutionary spectacle is an outstanding phenomenon, reflecting what is arguably the most rapid evolution and speciation of vertebrates anywhere in the world (Fryer and Iles 1972; Greenwood 1974; Ribbink 1994, 2001).

Data Quality: Medium

Overall, data on the systematics and ecology of the lake are poor, and extensive work is needed to improve the knowledge of the system. Sampling in the ecoregion has been concentrated only in certain areas, such that few lakewide datasets are available for informed decisions. However, the Southern African Development Community and Global Environment Facility Lake Malawi/Nyasa Biodiversity Conservation Project included a large fish systematics component (Snoeks 2004), as did an EU program that included studies on the trophic ecology of the demersal fish community (Irvine 2002). Data from these studies have improved the knowledge base. Additionally, data on commercially harvested fish and fisheries along the Malawi shores, particularly in the southeast arm of the lake, are good.

Ecoregion Number:	**54**
Ecoregion Name:	**Lake Rukwa**
Major Habitat Type:	**Large Lakes**
Final Biological Distinctiveness:	**Bioregionally Outstanding**
Final Conservation Status:	**Relatively Stable**
Priority Class:	**V**
Authors:	**Ashley Brown and Robin Abell**
Reviewer:	**Tim Davenport**

Location and General Description

Lake Rukwa is a saline lake of more than 85,000 km^2 located in the Rift Valley. The lake lies in the southwest corner of Tanzania, east of Lake Tanganyika's [55] southern tip. The Lake Rukwa ecoregion, comprising the lake and its catchment, is bordered in the southeast by the Mbeya Range, in the west by the slopes of the Ufipa escarpment and Mbizi (up to 2,664 m a.s.l.), and in the northeast by rocky cliffs and rolling hills that reach as high as 1,707 m at Mount Sange (Hughes and Hughes 1992; Baker and Baker 2001, 2002). Along the northern and western shores of the lake lie extensive wetlands that include both permanent swamps and temporary floodplains. The lake stretches lengthwise for about 165 km, with widths of 37 km in the north basin and a maximum width of 48 km near the middle of the lake (Seegers 1996).

There are several large tributaries to the lake. The Lupa, Chambua, and Songwe rivers drain the Mbeya Range and flow into the lake from the south, the Rungwa feeds the lake in the north, and the Momba River flows in from the west. In addition, several ephemeral rivers flow into the lake during the wet season. The lake is endorheic, with no external drainage (Hughes and Hughes 1992; Seegers 1996).

At high water, there is a single lake, but at lower levels the lake splits into two basins that differ in both size and depth. The water levels are high (Baker and Baker 2002), although there is evidence that the level is once again falling. The lakebed slopes downward to the east, resulting in greatest depths on the eastern shore. The south basin has maximum depths, reaching about 10–15 m. The north basin is shallower, and it occasionally dries completely. The water depth of the swamp barrier between the two basins usually is 1 m or less (Hughes and Hughes 1992; Seegers 1996).

The climate of the ecoregion is tropical and wet. There is one rainy season, with most precipitation falling from October to

April, although the Ufipa Highlands also experience rains in May and October (very rarely). Average annual rainfall ranges from about 650 mm in the south of the basin to about 900 mm in the north to about 2,500 mm in the Ufipa Highlands (Seegers 1996). In the southern portion the mean annual temperature is 21°C, with a mean maximum in the warmest month of about 28°C and a mean minimum in the coolest month of 12.7°C (Hughes and Hughes 1992). Surface water temperatures of the lake range from a minimum of about 20°C to a maximum of about 35°C (Seegers 1996).

Large areas, mostly in the northern and western portions of the basin, are inundated. The presence of dead or dying trees under water is evidence that these areas were flooded recently (Seegers 1996). At the deltas of the Luika, Songwe, Momba, and Chambue rivers (south lake) and Kavu and Rungwa rivers (north lake), there are also large swampy areas (Seegers 1996).

Floodplain vegetation surrounding Lake Rukwa is primarily grassland, with short grasses dominated by the salt-tolerant *Diplachne fusca, Sporobolus spicata,* and *S. robustus* (Hughes and Hughes 1992). The less saline swamps that surround the lake contain *Cyperus papyrus* and *Phragmites mauritianus. C. papyrus* can also be found along with *Oryza* sp., *Leersia* sp., *Vossia cuspidata,* and *Typha* sp. in deeper waters of the lake. Gallery forests of woodland, with many *Acacia* species, grow along the tributaries that feed the lake and along the lake's edges (Hughes and Hughes 1992).

Lake Rukwa lies in a rift in the Western Rift Valley, and the lake itself sits at 790 m a.s.l. The waters of the lake have high levels of sodium and high alkalinity (pH between 8.0 and 9.0) (Seegers 1996).

Outstanding or Distinctive Biodiversity Features

This ecoregion hosts an aquatic fauna that includes large aggregations of waterbirds and two endemic fish species flocks. A colony of about 40,000 great white pelicans (*Pelecanus onocrotalus*) occurs on Lake Rukwa, and about 20,000 white-winged terns (*Chlidonias leucopterus*) were observed at Lake Rukwa in January 1995 (Finlayson and Moser 1991; Wetlands International 2002). The fish flocks are of the cichlid genus *Haplochromis* and the catfish genus *Chiloglanis* (family Mochokidae). Both flocks have six known species (Seegers 1996). In total, there are about sixty fish species in the ecoregion, with nearly one-third endemic.

A variety of habitats are available to fish in the lake and its tributaries. Small fish, including species of the *Aplocheilichthys* and *Haplochromis* genera, young and semi-adult *Barbus,* and *Chelaethiops rukwaensis,* inhabit the aquatic macrophytes on the lake margins (Seegers 1996). Swampy deltas from the Luika, Songwe, and Chambue rivers provide habitat for another species assemblage that includes *Oreochromis* and *Tilapia* species. These two genera are also the most important commercially fished species of Lake Rukwa. The rocky habitat of the eastern shore appears to be inhabited by *Labeo cylindricus,* although in general these rocky shores host fewer fish species than other habitats in the lake. The Piti River contains typical riverine species including *Amphilius jacksonii, Leptoglanis rotundiceps,* and *Chiloglanis trilobatus* (Seegers 1996).

Large numbers of nonbreeding wetland birds are known to occur at the lake, although many of these records are from the 1950s (Vesey-FitzGerald and Beesley 1960). Species include *Pelecanus onocrotalus, Plegadis falcinellus, Plectropterus gambensis, Chlidonias leucopterus,* and *Rynchops flavirostris* (Baker and Baker 2001).

Other aquatic species of interest in the ecoregion include the marsh mongoose (*Atilax paludinosus*), spotted-necked otter (*Lutra maculicollis*), African clawless otter (*Aonyx capensis*), and hippopotamus (*Hippopotamus amphibius*). The Nile crocodile (*Crocodylus niloticus*) may occur in high densities, as evidenced by the continuing high annual rate of human mortality attributed to crocodiles.

Status and Threats

Current Status

Currently, the primary threat to Lake Rukwa's aquatic biodiversity is human population growth and associated pressures on natural resources, although the population in the catchment remains low. With expanding human settlements comes associated agriculture and livestock grazing (East 1999). For example, north of the Momba River, the Ufipa highlands are farmed for maize and wheat, leaving little natural vegetation (Seegers 1996).

Oreochromis esculentus and *Tilapia rendalli* have both been introduced to Lake Rukwa. These two species compete with the native *Oreochromis rukwaensis,* although it is unclear what the effects of the introduction will be on the *O. rukwaensis* populations (Seegers 1996). Artisanal fishing in dugout canoes has a long history in the area. Commercial fisheries (primarily for *O. rukwaensis*) have been established and abandoned several times (Hughes and Hughes 1992).

Between 2000 and 2002 the price of Rukwa-caught fish in Sumbawanga more than doubled. The reason given for this is that stocks had rapidly dwindled because of overfishing by immigrants, including Burundians and Rwandese. Certainly, fish numbers appear to have decreased in the past few years (T. Davenport, pers. comm., 2004).

The Uwanda Game Reserve covers much of the northern part of the lake (Sayer et al. 1992), although the reserve was almost completely flooded in the early 1990s. The Lukwati Game Reserve is also located along the northeastern shore (Baker and Baker 2001).

Future Threats

Changes to the hydrological regime from climate change or other unnatural causes could threaten Lake Rukwa because its depth already varies substantially with natural changes in rainfall (Seegers 1996). Increasing human populations probably will intensify pressures on the lake's fish populations, particularly if commercial fishing is permanently established; additionally, the possibility of mercury pollution from gold mines around the lake exists.

Justification for Ecoregion Delineation

This ecoregion is defined by the Lake Rukwa basin with its two endemic fish species flocks. Recent connections existed between the Rukwa Basin and both the Malagarasi system [11] and the Chambeshi (which drains to Lake Bangweulu [6]). There is also evidence of connections between the headwaters of the Manda and Kalambo rivers, which drain into Lake Tanganyika, and the Mfwizi River of the Rukwa drainage (Seegers 1996).

Data Quality: High

Important historical references for this ecoregion include Ricardo (1939), Verheyen (1939), Vesey-FitzGerald and Beesley (1960), and Vesey-FitzGerald (1964). The fish species of the lake are well known from Seeger's (1996) volume on the fish of the Lake Rukwa basin. Additionally, technical reports available at the Regional Natural Resources Office in Mbeya, such as Davenport (2000), review the status of the basin's ecosystems and provide comprehensive species lists of flora and fauna of the area.

Ecoregion Number: **55**
Ecoregion Name: **Lake Tanganyika**
Major Habitat Type: **Large Lakes**
Biological Distinctiveness: **Globally Outstanding**
Conservation Status: **Vulnerable**
Priority Class: **I**
Author: **Michele Thieme**
Reviewer: **Jos Snoeks**

Location and General Description

Lake Tanganyika has one of the richest lake faunas on earth (Worthington and Lowe-McConnell 1994). Its fauna exhibits extraordinarily high levels of endemism in several taxonomic groups at the species and genus levels. This lake is the second deepest in the world at 1,470 m deep, after Lake Baikal (Patterson and Makin 1998). The lake lies in a deep graben in eastern Africa's Great Rift Valley and is split between Burundi, Democratic Republic of the Congo (DRC), Tanzania, and Zambia. The ecoregion comprises the lake and its drainage basin.

The climate is tropical, with distinct wet and dry seasons. The dry season extends from May or June to August or September. During this time, steady southerly winds cause upwelling in the southern end of the lake and long internal waves that last for months after the winds have stopped. Air temperature in the ecoregion ranges from 18°C to 32°C (Spigel and Coulter 1996).

Lake Tanganyika is a meromictic lake, meaning that the non-circulating hypolimnion (bottom layer) does not mix with the circulating upper layer (epilimnion). The mixed, oxygenated layer extends to about 50–250 m (depending on season), limiting the distributions of aquatic life to this depth (Lowe-McConnell 1993). Beyond this depth, the environment is anaerobic, and accumulated particulate matter makes the water much denser than the above layer (Patterson and Makin 1998). Permanent stratification keeps much of the nutrients in the hypolimnion (Lowe-McConnell 1987).

The deep lake is narrow (about 650 km long by 50 km wide on average) with steep sides. The Kalemie shoal separates the lake into two main basins. The mean depth over the shoal is 500 m, and the depth on either side of the shoal is more than 1,000 m (Spigel and Coulter 1996). The volume of freshwater in the lake is equivalent to one-sixth of the earth's freshwater. The southern end of the lake has gentle sloping sides, and oxygenated waters extend down to a depth of about to 240 m during mixing of the circulating upper layer of water, allowing the benthic fish community to be more productive in this portion of the lake (Coulter and Mubamba 1993).

Rainfall and evaporation largely determine the water balance of the lake. Between 800 and 1,200 mm/year of rain falls in the vicinity of the lake (mean 1,050 mm/year), and evaporation averages 1,530 mm/year (Spigel and Coulter 1996; Patterson and Makin 1998). The many rivers and streams that enter the lake, including its major tributaries—the Rusizi, Malagarasi, and Lugufu rivers—play a small role in the lake's water balance. The only major outflow from the lake is the River Lukuga, which discharges to the Congo River via the Lualaba (Patterson and Makin 1998). Given the small influence of rivers, the flushing time of water in the lake (lake volume divided by river outflow) is an incredible 7,000 years (Spigel and Coulter 1996).

Outstanding or Distinctive Biodiversity Features

Lake Tanganyika is exceptional not only for its high level of species richness (animals and plants estimated at more than 1,400 species [Coulter 1991]) but also for high levels of endemism exhibited among several taxa. Fish, copepods, ostracods, shrimp, crabs, and mollusks are all represented by high numbers of endemic species. For example, 79 percent of 287 described fish species are endemic, with greater than 95 percent endemism in the Cichlidae family. More than 470 fish species (described and undescribed), including about 300 cichlids and more than 170 non-cichlids, are estimated to inhabit the lake basin (J. Snoeks, pers.

comm., 2002; De Vos and Snoeks 1994; Snoeks 2000). Lake Tanganyika is the only lake with species-rich lineages of substrate-brooding and mouthbreeding cichlids (Coulter 1991; Snoeks 2000). Seventy-four of eighty-five ostracod species (87 percent) and thirty-three of sixty-eight copepod species (49 percent) are also endemic (Patterson and Makin 1998).

A high level of endemism also exists at the genus level. Among mollusks, seventeen genera of Thalassoid prosobranchs are found exclusively in Lake Tanganyika, and the family Viviparidae has one endemic genus (*Neothauma*). Forty-nine of fifty-seven cichlid genera (86 percent) are endemic to the lake. The lake also supports species flocks in the Claroteidae and Mochokidae catfish families, the Centropomidae (*Lates*), and the Mastacembelidae spiny eels (Worthington and Lowe-McConnell 1994).

The pelagic fish community of Lake Tanganyika is unique among Africa's great lakes. Two species of endemic clupeids, *Limnothrissa miodon* and *Stolothrissa tanganicae*, feed on zooplankton in the pelagic zone and, in turn, provide food for the four predatory centropomids, *Lates angustifrons*, *L. mariae*, *L. microlepis*, and *L. stappersii*. These fish are the basis of the offshore fishery on the lake.

Only about one-third of the lake's shores have been sampled for fish, and an even larger area remains to be sampled for other taxa. Therefore, the numbers presented here probably represent only a fraction of the lake's actual diversity (Lowe-McConnell 1993). The Lake Tanganyika Biodiversity Project, funded by the Global Environment Facility, made significant progress in sampling unknown areas and contributing to knowledge of the lake's fauna (West 2001). Much taxonomic work for the lake's freshwater fauna is also incomplete, and numbers probably will change as a result of taxonomic revisions. Nonetheless, even with current limited knowledge, Lake Tanganyika undoubtedly has an exceptionally rich and endemic fauna.

Status and Threats

Current Status

Three national parks currently border the lake: Gombe Stream (Tanzania), Mahale Mountains (Tanzania), and Nsumbu (Zambia). Coulter and Mubamba (1993) recommended extending these land parks into the lake environment to create underwater reserves. The Lake Tanganyika Biodiversity Project sampled areas adjacent to the parks from 1997 and found these waters to contain 73 percent of the known fish species in the lake but only 52 percent of the known mollusks (Allison et al. 2000). Allison et al. (2000) recommend maintaining Mahale and Nsumbu's 1.6-km offshore zones, extending Gombe's boundary into the water to create an aquatic buffer zone and developing an integrated management plan for the fisheries of the Rusizi Delta. There are also a number nature reserves bordering the lake. Beyond these protected areas, the Biodiversity Special Study Team of the Lake Tanganyika Biodiversity Project recommends that an overall strategy of coastal zone management be implemented that would allow a gradation of activities to occur along the coast, depending on the biodiversity of an area and the level of threat, among other considerations. This comprehensive study has resulted in numerous specific recommendations that could help to maintain the biodiversity of Lake Tanganyika (Allison et al. 2000).

Coulter and Mbumba (1993) consider the majority of the lake to be under low levels of exploitation and unpolluted. However, they list three major threats to the integrity of the lake ecosystem: sedimentation from deforestation, water pollution near urban areas, and overfishing, which is changing species compositions and disrupting community interactions in some areas. Cohen et al. (1993) state that deforestation rates and related sedimentation generally are higher in the more heavily populated northern portion of the basin. Increasingly, sedimentation from watershed deforestation, road building, and other anthropogenic activities is causing the inundation of lacustrine habitats, particularly those of rock-dwelling fish species (Alin et al. 1999). As an example of the effects of sedimentation, the rate of outbuilding of the Rusizi River delta has increased by an order of magnitude because of deforestation in the basin (Patterson and Makin 1998). Examples of water pollution include nutrient loading (phosphorus and nitrogen) from sewage effluents, organic (oxygen-demanding) compounds in runoff, heavy metals from mining and leather tanning, pesticides from agricultural (coffee and cotton) runoff, and materials from oil exploration (Patterson and Makin 1998). The long residence time of water in the lake and the interconnectedness of fish and other animal communities make these threats potentially relevant to the entire lake, not only to isolated areas that are currently being affected.

Future Threats

Cohen (1994) recommends considering metapopulation dynamics when planning the placement of future underwater reserves because these processes are essential to maintaining the stenotopic, endemic species of the lake. He also recommends including a 5-km buffer in any freshwater reserve. The buffer zone around a nearshore reserve would extend out into the pelagic zone in one direction and upland to include land that drains into the lake in the other (Cohen 1992).

Global warming threatens the lake because it could increase the stratification between the warmer upper layer of water and the cooler lower layer, decreasing even further the limited amount of nutrient exchange between the nutrient-poor epilimnion and nutrient-rich hypolimnion. This exchange sustains the productive fishery of the lake (Watson et al. 1997), and a recent study shows that primary productivity of the lake has declined by about 20 percent, implying a decrease of about 30 percent in fish yields (O'Reilly et al. 2003; Verschuren 2003).

Since about the mid-eighties, exploration and scientific drilling, often sponsored by oil companies, have been ongoing in the East African region. Oil or gas drilling could threaten water quality; if an accidental spill or leak were to occur, both aquatic species and humans who depend on the lake for drinking water would be threatened (Lévêque 1997).

Justification for Ecoregion Delineation

Lake Tanganyika's great depth presumably has given its fauna an evolutionary advantage. Over its long history (the lake is probably 9–12 million years old), Lake Tanganyika served as a refuge for aquatic organisms during extremely dry periods when other waterbodies desiccated (Cohen et al. 1993). The split into two or three separate basins during low lake levels seems to have had important effects on the evolution and distribution of the ichthyodiversity and facilitated allopatric speciation (Coulter 1991; Snoeks 2000). The age of the lake has also facilitated a further differentiation of the fish fauna as compared with the faunas of lakes Malawi and Victoria. For example, the Lake Tanganyika cichlids are considered to have evolved from eight ancestral lineages, more lineages than for lakes Malawi or Victoria (Salzburger et al. 2002).

The fish fauna of Lake Tanganyika has strong affinities with the Congo Basin (Roberts 1975; De Vos and Snoeks 1994), and the two systems are still connected hydrologically. In the late Miocene and early Pliocene, the Congo Basin contained a large internal lake that covered the Cuvette Centrale (Coulter 1991). It is believed that Lake Tanganyika's fauna had its origins in this ancient environment, although the details and timeline are not complete. The twenty-three fish families of Lake Tanganyika are all present in the Congo Basin fauna.

Data Quality: Medium

Ecoregion Number:	**56**
Ecoregion Name:	**Lake Turkana**
Major Habitat Type:	**Large Lakes**
Final Biological Distinctiveness:	**Bioregionally Outstanding**
Final Conservation Status:	**Endangered**
Priority Class:	**IV**
Author:	**Emily Peck**
Reviewer:	**Jeppe Kolding**

Location and General Description

Lake Turkana is the largest lake in the eastern portion of the Rift Valley and the fourth largest lake by volume in Africa (Beadle 1981). Lying in a low closed basin at approximately 365 m a.s.l., the lake is situated primarily in northwestern Kenya, with only its northernmost end, the Omo Delta, inside Ethiopia. It is 260 km long, with an average width of 30 km, a mean depth of 31 m, and a maximum depth of 114 m. It has an area of approximately 7,560 km^2 and a volume of 237 km^2 (Coulter et al. 1986). Encompassing much more than the immediate Lake Turkana environs, the Lake Turkana ecoregion reaches north to include lakes Abaya and Chamo, and the headwaters of the Omo River in southwestern Ethiopia.

The climate of northwestern Kenya is hot, arid, and stable throughout the year. The mean monthly maximum air temperature ranges from 31°C to 33°C. Strong winds with a marked diurnal cycle blow from the south and southeast with little seasonality (Butzer 1971; Ferguson and Harbott 1982). The warmest and driest months are October through January. The period from April through August is somewhat cooler, with the highest likelihood of rain, although droughts occur on average every 6–7 years (Nicholson 1982; Kolding 1992). Less than 200 mm of rain falls in northwestern Kenya each year, and its occurrence is extremely unpredictable (Hughes and Hughes 1992).

The Ethiopian Rift Valley has a mean annual rainfall of 600 mm/year, receiving at least 50 percent of the precipitation between July and September. The western foothills of the Ethiopian Rift escarpment receive up to 1,000 mm of rainfall per year. This heavy rainfall causes the Omo River to flood (June–September), bringing nutrient rich waters into Lake Turkana (Beadle 1981). In the far north of the ecoregion, near Addis Ababa, Ethiopia, the mean annual precipitation is 1,302 mm/year (Hughes and Hughes 1992).

With no surface outlet, the water budget of the lake is a balance between river and groundwater inflow and evaporation. Evaporation rates are high, at around 2.3–2.8 m/year. An influx of about 19 km^3/year is needed to keep lake levels steady, and high interannual and intra-annual fluctuations in water level occur as a function of the rainfall in distant upland Ethiopia (Kolding 1992). Generally, the lake level fluctuates annually, with an amplitude of about 1–1.5 m, but it also undergoes long-term variations that exceed those of any other lake of natural origin (Butzer 1971). The mean retention time of water in the lake is a short 12.5 years (Kolding 1992).

The northwestern portion of the ecoregion is occupied by the highlands of the Ethiopian Massif (above 1,500 m a.s.l.). The northern portion of the eastern Rift Valley (40–60 km wide) lies between the steep slopes of the highlands and is occupied by a chain of lakes (FAO Inland Water Resources and Aquaculture Service, Fishery Resources Division 1999).

Of the twelve principal rivers that feed Lake Turkana, the River Omo is its only perennial tributary, supplying more than 90 percent of the lake's inflow (Beadle 1981). The Omo River drains the southwestern portion of the Ethiopian Massif and flows through the Rift Valley into Lake Turkana. Of the seasonal rivers that flow into the lake, the Turkwell and Kerio rivers are the largest contributors and enter the lake along its western edge and in its southern half (Hughes and Hughes 1992).

The salinity of Lake Turkana is higher than that of any other large African lake. This is because the lake has no outlet, and it has contracted in volume over the last 7,500 years. Very recent volcanic activity in the basin has also contributed to the high salinity of the lake (Beadle 1981). Water samples taken between 1931 and 1975 record salinity as ranging from 1.7 to 2.7 percent (Hughes and Hughes 1992; Spigel and Coulter 1996). The mean conductivity is about 3,500 µS/cm, with an estimated rise of about 0.45 µS/cm/year (Ferguson and Harbott 1982). Because of the volcanic origin of the catchment area, the water chemistry is dominated by sodium (more than 95 percent of the cations by weight) and bicarbonate, which generate a high alkalinity (pH about 9.3).The seasonal inflow of water and strong diurnal wind patterns keep the waters of Lake Turkana productive and well mixed. In fact, the oxygen content of Turkana's deepest water is never less than 70 percent (Beadle 1981).

Lake Abaya and Lake Chamo are located in the northeastern portion of the ecoregion. Five major rivers feed Lake Abaya, the most important of which is the Bilate. During the rainy season, overspill from Lake Abaya is carried to Lake Chamo via the Ualo River (Hughes and Hughes 1992).

The evergreen bush and woodland of the Ethiopian Massif grade into deciduous bush in the Rift Valley. Extensive seasonal floodplains exist along the Omo River Delta, at the northern tip of Lake Turkana. The most common emergent plants are the grasses *Paspalidium geminatum* and *Sporobolus spicatus*, with extensive beds of *Potamogeton* occurring in shallow bays (Hughes and Hughes 1992). Gallery forests of *Acacia elatior, Balanites aegyptiaca,* and *Hyphaena coriacea* grow along Lake Turkana's tributaries (Beadle 1981; Hughes and Hughes 1992). Swampy savanna forests of *Acacia* and *Ficus* species line the shores of Lake Abaya, and species of *Typha* and *Phragmites* are common along the banks of lakes Abaya and Chamo. The waters of Abaya and Chamo contain numerous submerged plants, such as *Ceratophyllum demersum, Hydrocotyle* sp., and *Potamogeton* spp., as well as floating plants such as *Lemna gibba, Nymphaea* spp., and *Ottelia ulvifolia* (Hughes and Hughes 1992).

Outstanding or Distinctive Biodiversity Features

Lake Turkana is unique among the larger lakes of the eastern Rift Valley in that its aquatic fauna is dominated by Nilotic riverine species rather than by species of the cichlid family (Lowe-McConnell 1993). Compared with other large African lakes, Turkana has low fish species richness, providing habitat for about fifty species, eleven of which are endemic. According to Hopson (1982), four fish communities live in the main lake: a littoral assemblage, an inshore assemblage, an offshore demersal assemblage, and a pelagic assemblage. Nearly all endemic species live in the offshore demersal or pelagic zone (Lowe-McConnell 1987). Endemic cichlids include three haplochromine species adapted for deep water: *Haplochromis macconneli, H. rudolfianus,* and *H. turkanae*. Other species endemic to Lake Turkana include *Barbus turkanae, Brycinus ferox, B. minutus, Labeo brunellii, Lates longispinis,* and *Neobola stellae*.

Lake Turkana is an important site for waterbirds, with up to 220,000 congregants having been recorded at one time and 84 waterbird species, including 34 Palearctic migrants, known from the lake (Bennun and Njoroge 1999). More than 100,000 *Calidris minuta* have been recorded at the lake, in addition to smaller congregations of other nonbreeding waterbirds (*Pelecanus rufescens, Phoenicopterus ruber, Vanellus spinosus, Charadrius hiaticula, C. asiaticus,* and *C. pecuarius*) (Bennun and Njoroge 1999). Bird species present near Lake Abaya include *Anhinga rufa, Bubulcus ibis, Casmerodius albus, Egretta garzetta, Haliaeetus vocifer,* and *Scotopelia peli* (Hughes and Hughes 1992).

Other aquatic animals in the ecoregion include *Hippopotamus amphibius, Crocodylus* spp., and an endemic freshwater turtle, the recently discovered and imperiled Turkana mud turtle (*Pelusios broadleyi*) (Hughes and Hughes 1992; Expert Center for Taxonomic Identification 2000). Lakes Abaya and Chamo support notably large populations of *Crocodylus niloticus* and *Hippopotamus amphibius* (Hughes and Hughes 1992). Three frog species are endemic to the ecoregion (*Bufo chappuisi, B. turkanae,* and *Phrynobatrachus zavattarii*).

Spawning migrations of fish are synchronized with the ecoregion's seasonal flooding, which occurs from June through September. During this time, various fish species migrate up the Omo River (*Hydrocynus forskalii, Alestes baremoze, Citharinus citharus, Distichodus niloticus,* and *Barbus bynni*) and other ephemeral affluents (*Brycinus nurse, Labeo horie, Clarias gariepinus,* and *Synodontis schall*) to breed, for periods of both long and short duration (Beadle 1981; Hopson 1982; Lévêque 1997).

Status and Threats

Current Status

Kenya has established several national parks and wildlife reserves in the southern portion of the ecoregion (East 1999). Created in 1974, the Sibiloi National Park covers an area of 2,500 km^2 along the northeastern shore of Lake Turkana. In 1984, the lake's three biggest islands, Northern, Central, and Southern Island, were also declared protected areas. These islands are breeding grounds for a population of 12,000 Nile crocodiles (*Crocodylus niloticus*). Mount Kulal Biosphere Reserve protects an area in the southeastern section of the ecoregion. Sections of several ephemeral tributaries to Lake Turkana are covered by the Nasolot National Reserve and the South Turkana National Reserve in Kenya (Hughes and Hughes 1992).

The northern portion of the ecoregion also contains several important protected areas. Along the banks of the Omo River, the Omo and Mago national parks and the Tama Wildlife Reserve protect habitat for the conservation of antelope communities (East 1999). The Stephanie Wildlife Reserve and Nechisar National Park also cover lands between lakes Abaya and Chamo. In total, the Lake Turkana ecoregion contains more than 8,100 km² of protected areas (MacKinnon and MacKinnon 1986).

Despite these parks and reserves, most of the ecoregion remains unprotected, with the greatest threat to the lake being a reduced water supply caused by drought. The rainfall pattern is very unpredictable, resulting in large natural fluctuations in water supply to the lake. Additionally, the Omo River delta has grown tremendously since the 1970s, presumably in response to increased sedimentation and decreased lake levels and river flows. Satellite photos show the rapid growth of the delta and the thick vegetation that covers it (Haack 1996; Evans 2001).

Since the start of commercial fishing at Lake Turkana in 1950, fish populations have fluctuated greatly, with an overall decline in landings occurring since 1984. However, the overall exploitation rate in Lake Turkana is still very low, and the general decline in the commercial landings can be attributed to the lowering of the lake level and in particular to the drying up of Ferguson's Gulf, which provided most of the tilapia. Correlation analyses (Kolding 1992) between the annual river discharges (represented by the mean annual lake level) and commercial catch rates indicated that discharge has a profound effect on the landings and that the rate of change in fish production correlates closely with variations in water inflow. The hydrological regime also has a profound impact on the fish community structure; the composition of pelagic fish stocks not subject to exploitation changed significantly between the 1970s and the 1980s (Kolding 1993).

Future Threats

Potential oil pollution threatens the ecoregion. Since 1982, oil exploration surveys have been carried out on several of the Rift Valley lakes, including Lake Turkana. In a closed system, an oil leak or spill could have catastrophic effects on the ecological integrity of the ecoregion's waters (Coulter et al. 1986).

Justification for Ecoregion Delineation

The Lake Turkana ecoregion is defined by the basins of lakes Turkana, Abaya, and Chamo and the Omo River Basin. Fish species in the ecoregion are mainly of Sudanian origin, providing evidence of a previous connection to the Sobat and Nile rivers (Beadle 1981). For example, the Nilotic species Nile tilapia (*Oreochromis niloticus*), *Bagrus domac*, and Nile perch (*Lates niloticus*) are abundant and common in lakes Turkana, Abaya, and Chamo (Hughes and Hughes 1992). The Omo River Basin also has several fish species in common with lakes Turkana, Abaya, and Chamo. The Turkana Basin was formed from tectonic movement during the early Miocene. Evidence suggests that Lake Turkana was once part of a larger body of water that included present-day Lake Baringo (south) and the Lotikipi Plains (west). In addition, a connection with the Nile and its tributary, the Sobat River, may have existed more than once during particularly wet periods of the Pleistocene, with the most recent connection occurring not more than 7,000 years ago (Beadle 1981; Dgebuadze et al. 1994).

Data Quality: High

Ecoregion Number:	**57**
Ecoregion Name:	**Lakes Kivu, Edward, George, and Victoria (including satellite lakes)**
Major Habitat Type:	**Large Lakes**
Final Biological Distinctiveness:	**Globally Outstanding**
Final Conservation Status:	**Critical**
Priority Class:	**I**
Author:	**Dalmas Oyugi**
Reviewer:	**Lauren Chapman**

Location and General Description

With an area of approximately 68,800 km², Lake Victoria is the largest tropical lake in the world and the second largest freshwater lake in the world (Spigel and Coulter 1996). In addition, it is the largest lake in Africa and contains one of the world's most important examples of rapid species radiations among its endemic haplochromine cichlid fauna.

Lake Victoria occupies a shallow depression 1,134 m a.s.l., between the West and East African rifts. Stretching 412 km from north to south and 355 km from west to east, Lake Victoria spans the borders of Tanzania, Uganda, and Kenya. Its catchment (more than 193,000 km²), reaches well into Rwanda and Burundi (Hughes and Hughes 1992). Also in the ecoregion, surrounding Lake Victoria, are lakes Kyoga, George, Edward, and the small but deep Lake Kivu.

The Lake Victoria ecoregion has an equatorial climate. In the north there are two rainy seasons—one during April and May and the other during October and November—whereas the south experiences one long rainy season from December to March (Burgis and Symoens 1987; Lowe-McConnell 1987). Average annual rainfall is 650–900 mm, with peak rains in April and a dry season between June and September (Beadle 1981). Temperature of the lake varies from 23°C to 27°C, with a mean temperature of 25°C throughout the year (Witte and Van Densen 1995). Wind speed is generally low during the wet season (0–3.5 m/s) because of the protection of the Rwenzori Mountains in the north and the Rift Escarpment to the east and south. Low wind speeds allow the buildup of a thermocline, which results in stratification in the lake (Talling 1966). Lake Victoria experiences annual stratification and overturn, with an associated upwelling of nutrients, as in temperate lakes (Lowe-McConnell 1987). Wind speeds increase during the dry season, between May and June, when strong southerly winds exceeding 15 m/s blow over the ecoregion (Spigel and Coulter 1996). These strong southerly winds cause high evaporation rates, water mixing, and a decrease in surface temperatures. Seasonal algal blooms, coupled with shallow water upwellings as a result of wind action, affect seasonal transparency patterns in the lake (Witte and Van Densen 1995).

The water balance of the lake is maintained primarily through rainfall and evaporation rather than inflows and outflows (Spigel and Coulter 1996). Because of this dependence on rainfall and evaporation, the residence time of water in Lake Victoria is 23 years (Cohen et al. 1996; Spigel and Coulter 1996). Lake level in Victoria has varied by about 2 m in the last century in response to changes in rainfall and evaporation. The lake basin itself is about 400,000 years old, but several recent studies suggest that the lake was completely dry for several thousand years and refilled only 15,000 years ago (Johnson et al. 1996, 2000; Talbot and Laerdal 2000).

Numerous rivers and streams drain into Lake Victoria. The principal affluent is the Kagera River, which enters the lake along its western shore, draining the highlands of Burundi and Rwanda. The Kagera River is about 360 miles long, and the Ruvuvu River is its principal tributary. A series of swamps (2–18 km wide) and small lakes occur along the course of the Kagera River, with several waterfalls in its upper reaches (De Vos et al. 2001b). The Nzoia River is also a perennial affluent of Lake Victoria. It drains the Elgon Massif, the Cherangani Hills, and Sergoit, entering the lake in the northeast. Inflows from rivers in the northwestern and southeastern portion of the ecoregion constitute the remainder of the riverine input. Rivers entering the lake from the northeast tend to be swift flowing, whereas rivers of the northwest tend to be sluggish and perennial (Hughes and Hughes 1992).

The only outlet from Lake Victoria is the Victoria Nile River, which flows through Lake Kyoga and then to Lake Albert, to the north and northwest, respectively (Hughes and Hughes 1992). Passing through extensive areas of swampland, the Victoria Nile enters Lake Kyoga from the south. Lake Kyoga is part of a permanently flooded series of shallow lakes and swamps called the Kyoga Lake–Kwania Swamp complex, which contains 3,416 km^2 of open water and shallow lakes and 2,184 km^2 of permanently flooded swamp (Hughes and Hughes 1992).

Located on the border of Rwanda and the DRC, Lake Kivu lies in the Western Rift Valley at an altitude of 1,463 m a.s.l. and is bordered by steep slopes that rise to elevations of more than 2,000 m. It is 100 km long and has a maximum width of 50 km. The lake is deep (about 480 m), and it is the most completely stratified lake in Africa. Lake Kivu formerly drained into Lake Edward. Volcanic activity severed this connection, and Lake Kivu now drains into Lake Tanganyika via the Ruzizi River (Worthington and Lowe-McConnell 1994).

Lakes Edward and George lie in the western portion of the ecoregion. Surrounded by extensive swamps, Lake George straddles the equator and is fed by numerous rivers, which drain the eastern slopes of the Rwenzori Mountains, the highlands of the Western Rift Valley, and the Virunga Massif. In general, Lake George is well mixed, although it does have a diurnal stratification cycle (Thompson 1976; Hughes and Hughes 1992). Lake George's main inflows are the rivers Nsong, Mbuku, and Bumlikwesi (Beadle 1981). The enormous swamp surrounding Lake George is dominated by papyrus (*Cyperus papyrus*), which typically makes up more than 95 percent of the plant biomass (Thompson 1976). Lake Edward is connected to Lake George by the Kazinga Channel, which is 40 km long and has a maximum width of less than 1 km (Beadle 1981). Just 5 km off its western shore in the DRC, Lake Edward plummets to a depth of 112 m and then slopes gradually up to its eastern shore in Uganda (Hughes and Hughes 1992). The major outflow from Lake Edward is the Semliki River, and its main inflows are the rivers Nyamugasani, Ishasha, Rutshuru, and Rwindi, with a smaller contribution by the Kazinga Channel (Beadle 1981).

The Lake Victoria ecoregion is also endowed with several small satellite lakes. A few of these include lakes Kanyaboli, Sare, and Namboyo in Kenya; lakes Nabugabo, Gigati, and Agu in Uganda; and lakes Ikimba and Burigi in Tanzania (Aloo 2003). These lakes may act as valuable sources of biodiversity because many of them are still undisturbed by human activities.

The ecoregion's lakes and shallow bays are home to many types of emergent and submerged vegetation (macrophytes). In the Lake Victoria Basin, the dominant macrophytes include *Cyperus papyrus*, *Miscanthidium violaceum*, *Phragmites mauritanius*, and *Typha domingensis*. The most extensive papyrus swamps (*C. papyrus*) in East Africa occur along the perimeter of Lake Victoria and along the perimeters of the other lakes in this ecoregion (Chapman et al. 2001). The valley swamps fringing the rivers flowing into Lake Victoria are dominated by *Miscanthidium violaceum*. These two swamp types also support a diversity of other plant species, with both types reported to contain more than thirty other species (Chapman et al. 2001). The interface between open water and permanent swamp sustains a distinctive plant and animal community. However, many fewer species are adapted to the often oxygen-poor environment of the permanent swamps and their dense stands of fibrous papyrus mats. Grasses and trees also grow on the seasonal floodplains, and stands of *Acacia* occur throughout the landscape adjacent to the lake and inflowing rivers (Hughes and Hughes 1992).

Lakes Victoria, Edward, and George are all part of the greater Nile catchment. They are separated from the lower Nile catchment by Murchison Falls, but evidence suggests that a connection between these lakes existed in the recent past, within the last several thousand years (Worthington and Lowe-McConnell 1994). During the uplift of the mid-Pleistocene, it is theorized that the rapid backponding of the Kagera and Katonga rivers overspilled the basin of Lake Victoria, draining into the Edward Basin well into the late Pleistocene (Beadle 1981). Even today, the Katonga River is able to flow in either direction, enabling species communication between all three lakes (Worthington and Lowe-McConnell 1994).

Lakes Edward and George have had a very dynamic history. In the early Pleistocene (2 m.y.a.), there may have been passable connections between Lake Edward and Albert basins, now separated by a section of rapids (300-m descent). Lake Albert is not included in this ecoregion because this separation probably has caused its fauna to remain distinctly riverine and Nilotic. The faunas of lakes Edward and George were very similar: fossil remains of *Lates, Hydrocynus,* and *Crocodylus niloticus* in the Edward Basin link the two faunas. Subsequently, the two lakes may have gone through a number of mass extinction and recolonization events (Beadle 1981). The last of these events probably occurred between 8,000 and 10,000 years ago and was associated with the eruption of volcanoes, which may have deposited enormous amounts of toxic ash on the lake, killing species such as *Lates,* which need well-oxygenated water (Beadle 1981; Schofield and Chapman 2000). Similarly, the Nile crocodiles (*Crocodylus niloticus*) were no longer present in the areas of lakes Edward and George after the eruption of neighboring volcanoes (Beadle 1981), presumably kept from colonizing the lake by difficulties in passing the Semliki rapids. Some species in lakes Edward and George are also found in Lake Albert (e.g., *Bagrus docmac, Oreochromis niloticus*), whereas other species (e.g., *Polypterus senegalus, Hydrocynus* spp.) and even families (such as Mastacembelidae, Characidae, Schilbeidae) that typify the Nilotic fauna from Lake Albert are absent (Greenwood 1966). Many of the cichlids from lakes Edward and George are closely related to species in Lake Victoria (Greenwood 1966, 1973; Beadle 1981; Kaufman et al. 1997); however, the number and distribution of endemic cichlids in the lakes are uncertain, and they hold the key to our understanding of the relationship between land form and evolution of cichlid faunas in the Lake Victoria Basin.

In contrast to the results of previous studies, a recent study on the phylogeny of the Lake Victoria–Edward cichlid species flock suggests that the flock is derived from the Congolese-Nilotic genus *Thoracochromis* and not from the East African riverine *Astatotilapia* (Seehausen et al. 2003). The rivers feeding Lake Victoria were tributaries of the Congo until uplifting of the region to its west about 400,000 years ago, and it is possible that ancestry with the Congolese *Thoracochromis* predates this event. Seehausen et al. (2003) also question the monophyly of the Lake Victoria–Edward flock and raise the possibility that the flock has arisen from hybrid swarms.

Outstanding or Distinctive Biodiversity Features

Lake Victoria's endemic haplochromine fauna is one of the world's most outstanding examples of explosive speciation and adaptive radiation (Greenwood 1981; Kaufman 1992; Kaufman et al. 1997). Recent studies have shown that sexual selection probably is the driving force behind reproductive isolation in the haplochromine cichlids of the Lake Victoria Basin and that the flexible and versatile pharyngeal jaw apparatus of these fishes has also played a role in the rapid diversification of these fish (Kaufman et al. 1997; Galis and Metz 1998; Seehausen and Van Alphen 1998, 1999; Seehausen et al. 1999). More than 600 endemic fish are known from Lake Victoria alone, although estimates of species numbers vary widely (Seehausen 1996; Kaufman et al. 1997; Turner et al. 2001). The total number of cichlid species in lakes Edward and George is about eighty, with nearly sixty of these being endemic (L. Chapman, pers. comm., 2003). Currently, twenty-eight fish species are known from Lake Kivu and its affluents. Nineteen of the twenty-eight are cichlids, and nine are noncichlids.

In addition to Cichlidae, the lakes of the Victoria ecoregion also host fish fauna from the families Alestiidae, Amphiliidae, Clariidae, Cyprinidae, Mochokidae, Mormyridae, Poeciliidae, and Protopteridae. About one-third of the approximately ninety noncichlid fish species are endemic to the ecoregion. In addition to the lacustrine species, there are also many riverine fish. For example, at least fifty-five fish species live in the Rwandan portion of the Kagera River. Several of these are anadromous fishes such as *Labeo* spp., *Clarias* spp., *Bagrus* spp., and *Barbus* spp., which use the river for spawning (Okedi et al. 1974).

From a biodiversity perspective, Lake Edward is very important. Preliminary surveys of the fish fauna of the lake (1996) have yielded many new cichlid species (L. Chapman, pers. comm., 2003). Additionally, Lake Edward is the least disturbed of the great lakes that contain endemic faunas because much of the lake lies in national parks in Uganda and the DRC. Finally, the riverine faunas in the tributaries of the Edward and George system are very rich and include the antecedents to the lacustrine radiations. However, many of the tributaries remain unexplored. The system is undergoing natural and anthropogenic turmoil. In addition to increasing pressures on environmental resources from fishing villages inside the protected areas, Nile crocodiles have become reestablished in the system after an 8,000-year absence, and hippopotamus populations are increasing in response to one decade of protection in the parks.

The ecoregion's swamps and wetlands also support numerous waterbirds. Among these, the vulnerable papyrus yellow warbler (*Chloropeta gracilirostris*), the vulnerable white-winged warbler (*Xenoligea montana*), the locally rare papyrus gonolek (*Laniarius mufumbiri*) and shoebill (*Balaeniceps rex*), and the more common Carruther's cisticola (*Cisticola carruthersi*), great egret (*Ardea alba*), and Baillon's crake (*Porzana pusilla*) inhabit wetlands bordering Lake Victoria (Bennun and Njoroge 1999, 2001). Congregations of *Chlidonias leucopterus, Egretta garzetta, Phalacrocorax africanus, P. carbo,* and *Larus cirrocephalus* also occur in marshes, bays, islands, and swamps along the margins of the lake (Baker and Baker 2001; Byaruhanga et al. 2001).

The ecoregion is also rich in other taxa, including aquatic-dependent reptiles, amphibians, and mammals, plankton, and freshwater mollusks. In particular, there is a high species richness of frogs, with more than sixty species known from this ecoregion, one-quarter of which are endemic (and confined mainly to forest habitats). Aquatic obligate vertebrate species in the Lake Victoria environs include five species of freshwater turtle, two aquatic snakes, a monitor lizard, the Nile crocodile (*Crocodylus niloticus*), three otter species (*Aonyx capensis, A. congicus*, and *Lutra maculicollis*), and hippopotamus (*Hippopotamus amphibius*) (Hughes and Hughes 1992). The ecoregion's waters also support an abundant mollusk fauna comprising fifty-four species, with about one-fifth endemic to the ecoregion (Brown 1994).

Status and Threats

Over the last century the Lake Victoria basin has undergone extreme ecological changes. The three major events that have caused these changes are an intensification of fishing activities, the introduction of nonindigenous fish, and increased human population density and associated increased agricultural and industrial activities (Balirwa et al. 2003).

The ecological balance of Lake Victoria began to deteriorate in the 1930s with a rise in commercial fishing. In the 1950s and early 1960s, when catches of indigenous tilapiine species declined, the Nile perch (*Lates niloticus*) and several exotic tilapiine species (*Oreochromis niloticus, O. leucostictus, Tilapia zillii*, and *T. rendalli*) were introduced into the Lake Victoria Basin (Ogutu-Ohwayo 1990; Witte et al. 1999). In the 1980s there was an explosive population increase of the Nile perch, in conjunction with a further decline of populations of several indigenous species (Ogutu-Ohwayo 1990; Kaufman et al. 1997; Witte et al. 1999). Commercial bottom trawling is thought to have contributed to the disappearance of large (15 cm standard length) haplochromine fishes between 1976 and 1982 (Lowe-McConnell 1993). Eutrophication, algal blooms, decreased water transparency, and lowered oxygen concentrations were also observed in the 1980s and could have also contributed to the declines in indigenous species (Hecky 1993; Hecky et al. 1994). Within a decade, Lake Victoria experienced "what is probably the largest mass extinction of contemporary vertebrates" (Seehausen et al. 1997b: 890), losing almost 200 endemic haplochromine cichlid species (Kaufman and Cohen 1993; Seehausen et al. 1997b; Witte et al. 1999). However, within the last several years the Nile perch fishery has showed signs of overfishing, and some of the indigenous fish have showed signs of resurgence in lakes Kyoga, Victoria, and Nabugabo (Chapman et al. 2003; Balirwa et al. 2003). Intense fishing pressure has led to an apparent decline in Nile perch biomass and a reduced size at maturity of Nile perch. These changes in the Nile perch population have coincided with the resurgence of some indigenous species. For example, in the Mwanza Gulf of Lake Victoria, Lake Victoria haplochromines have increased from 0.2 percent of the catch in 1987 to 21.3 percent of the catch in 1997 (Witte et al. 2000). It is expected that the resurging fauna in these lakes will differ from the original fauna because of the different selective pressures and environmental changes experienced. Ongoing morphological and genetic studies will give a clearer picture of the composition of the resurging species assemblages (Chapman et al. 2003). Among the haplochromine cichlids in sublittoral zones, only a few species in large quantities from a few trophic groups (mainly zooplanktivores and detritivores) appear to be recovering (Witte et al. 2000; Balirwa et al. 2003). Fishery models for the lake show that sufficient fishing pressure on the Nile perch should be maintained to ensure an abundance of haplochromine prey and allow continued resurgence of the indigenous species (Kaufman and Schwartz 2002). However, to prevent another serial collapse of stocks, pressure should not be so heavy that it threatens the sustainability of the Nile perch stock (Balirwa et al. 2003).

An increase in agriculture, fisheries, industrialization, and urbanization and a rapid rise in population have also occurred over the last 20 years, contributing to the degradation of the lake waters in this ecoregion. Population density around Lake Victoria and its satellite lakes is about 100 people/km^2, significantly higher than around the other African Great Lakes (Bootsma and Hecky 1993; Cohen et al. 1996). With this high density of rural peoples, pressure on natural resources is also high, resulting in intense agricultural activity and deforestation along the lake margins. Agriculture is a mainstay of the region's economy, employing approximately 85 percent of the population in Uganda and Tanzania. The main cash crops are rice, sugar cane, and coffee (FAO Inland Water Resources and Aquaculture Service, Fishery Resources Division 1999; CIA 2001). Changes in the fishery have also increased deforestation. Processing of Nile perch, a large, fatty fish, entails either frying or drying and smoking. Firewood is needed for both of these processing techniques. At Wichlum Beach (Kenya), the number of smoking kilns increased fivefold in 5 years (Witte et al. 1999). Additionally, sewage facilities are nearly nonexistent in the villages and towns lining the lake, such that raw sewage enters lake waters. Industrial activities in the ecoregion include textiles, leather tanning, paper mills, and breweries. Agriculture, industrialization, deforestation, and high population pressure have all led to increased erosion of the land, siltation, and eutrophication in the basin (Witte et al. 1999). One effect of eutrophication has been that the lake has lower oxygen levels below the hypolimnion, reducing or eliminating the deeper water habitats for many native species (Hecky et al. 1994).

The introduction and spread of water hyacinth (*Eichhornia crassipes*) have also drastically altered the functioning of the lakes in this ecoregion. Water hyacinth, a floating, invasive plant species, was first reported in Lake Kyoga, Uganda in 1988 and in Lake Victoria in 1989 (Twongo et al. 1995). By 1995 it had taken over large tracts of the lake and its shoreline, decreasing oxygen levels and light penetration. As a result, it hampered some fishery activities, reduced availability of water for agriculture and industry, degraded water quality, spread waterborne diseases, choked ports, and restricted access to key trade routes (Balirwa et al. 2003). As the negative economic and social repercussions of this invasive weed escalated, mechanical, chemical, manual, and biological control (the introduction of the weevils *Neochetina eichhornia* and *Neochetina bruchi*) measures were applied. By the late 1990s water hyacinth had disappeared almost completely. Concerns remain that the weed will have periodic outbreaks in the lake basin and will necessitate continued control measures. In fact, regrowth was witnessed in several areas of the basin in 2000 (National Agricultural Research Organization 2002).

Justification for Ecoregion Delineation

This ecoregion is defined by the basins of lakes Victoria, Edward, George, Kyoga, and Kivu and is characterized by a lacustrine fauna with cichlid species radiations typical of those in the Great Lakes bioregion. The mountains of the Karamoja district enclose the ecoregion in the northeast. In the east, the highlands of the Eastern Rift Valley comprise the Cherangani Hills, Elgeyo Escarpment, and Mau Escarpment. The Rwenzori Mountains and the Virunga Massif are the highlands that enclose the ecoregion in the west. The southwest is dominated by a chain of mountains in Rwanda, along the rim of the Western Rift Valley (Hughes and Hughes 1992).

Data Quality: High

The Lake Victoria Region has provided a model for relationships between land form and fish evolution. Our understanding of the evolution of these extraordinary faunas is increasing as faunal and paleolimnological surveys continue in the region. But there is a huge missing piece in the puzzle: the Lake Edward–Lake George region. Lake Edward is one of the less explored ichthyofaunas in Africa. Lake George is better known, although the extensive wetland around the swamp remains largely unexplored. The Lake Edward–Lake George system is very exciting biogeographically because it represents the confluence of the Albertine and Victoriine faunas.

Ecoregion Number:	**58**
Ecoregion Name:	Niger Delta
Major Habitat Type:	Large River Delta
Final Biological Distinctiveness:	Globally Outstanding
Final Conservation Status:	Critical
Priority Class:	I
Authors:	Ashley Brown and Michele Thieme
Reviewers:	Christian Lévêque, Emmanuel Obot, and Ojei Tunde

Location and General Description

The extremely productive Niger River Delta, one of the largest deltas in the world, supports an abundant aquatic fauna. As the Niger River flows into the delta it splits into multiple affluents and winds through lush swamp and mangrove forests before reaching the Gulf of Guinea. The delta wetlands lie entirely in Nigeria.

The climate in the delta is cooler than upstream, although it is still tropical. Temperatures near Port Harcourt in the east of the delta range from an average of 24°C during the coolest month to 26°C during the warmest month. The ecoregion is extremely humid, with the wet season lasting for 10–11 months, usually with only one dry month in December (Hughes and Hughes 1992). At Forcados, in the western part of the delta, rainfall averages 3,800 mm per year but decreases heading east to Port Harcourt, where the annual rainfall is 2,480 mm.

More than 80 percent of the delta floods seasonally, with swamps and pools remaining when floodwaters subside (Moffat and Linden 1995). In addition to precipitation, tidal movements and the Niger River flood determine the hydrological regime. The flood begins toward the end of the rainy season in August, peaks in October, and tapers off in December. The yearly rainfall determines some fluctuation in flow, but since 1968, after the completion of the Kainji Dam, the opening and closing of the dam sluices is also important. The delta experiences a strong tidal influence, with seawater penetrating as far inland as the riverine floodplain of the Nun River (Hughes and Hughes 1992).

The Niger Delta consists of three major sections: the upper riverine floodplain, the lower tidal floodplain, and an outer chain of coastal barrier islands (Hughes and Hughes 1992). The upper riverine floodplain stretches for 168 km from the head of the delta at Onitsha to the lower tidal floodplain. The Niger begins to separate into the Forcados and the Nun rivers in the upper riverine floodplain, where seasonal and permanent freshwater swamps occur. Seasonal swamp forest, dominated by *Anthocleista vogelii*, *Carapa procera*, and *Chrysobalanus orbicularis*, is inundated during the wet season. Permanent swamp forests, with forest floors that are inundated year-round, include members of the *Alstonia*, *Mitragyna*, and *Raphia* genera (Hughes and Hughes 1992).

Swamp forests of the upper riverine floodplain grade into mangrove forests on the lower tidal floodplain. In addition to the Nun, Forcados, and Orashi rivers, other main channels of the tidal floodplain are the Sombriero, Bonny, Brass, and New Calabar. Mangrove forest covers approximately 5,000 km^2 of the delta, and species include *Rhizophora racemosa* (up to 40 m height), *R. harrisonii*, and *R. mangle* (Hughes and Hughes 1992; Shumway 1999).

Sediments from the Niger River, sculpted by marine processes, have created the twenty major islands along the coast of the delta. In total, the islands have a surface area of 2,000 km², and several of them reach heights of 4 m a.s.l. (Hughes and Hughes 1992; Ajao 1994).

The Niger Delta is ancient and has a history of sedimentation spanning tens of millions of years, with three main depositional cycles: the middle Cretaceous, Paleocene, and Eocene to the Holocene. In the Pleistocene, during a period of low sea level, the Niger River cut a deep gulf into the ancient delta. The present Niger Delta was formed during the Holocene after the end of the last ice age. Over the last 6,000 years the delta floodplain has advanced across brackish water mangrove swamps, and the delta front beach ridges have been built up progressively further seaward. At present, two-thirds of the sediment delivered to the delta by the Niger is supplied by the Benue (Lowe-McConnell 1985).

Outstanding or Distinctive Biodiversity Features

Both blackwater and whitewater rivers flow into the highly productive Niger Delta, which supports an extremely rich freshwater fauna and includes the highest concentration of monotypic fish families in the world. Sixty percent of Nigeria's mangrove forests (the third largest in the world and the largest in Africa) are also found in the Niger Delta (Moffat and Linden 1995). The mangrove forest and freshwater swamp forests in the delta provide habitat for aquatic mammals, mollusks, and herpetofauna.

The freshwater fauna includes five monotypic fish families (Denticipidae, Pantodontidae, Phractolaemidae, Hepsetidae, and Gymnarchidae), the highest concentration of monotypic freshwater fish families of any ecoregion worldwide. Among these families, the endemic denticle herring (*Denticeps clupeoides,* family Denticipidae) and the hingemouth (*Phractolaemus ansorgii,* family Phractolaemidae) have the most limited distributions. The hingemouth possesses a completely alveolated swim bladder that functions as lungs and permits the species to survive in unoxygenated waters (Beadle 1981). *Pantodon buchholzi,* also from a monotypic family (Pantodontidae), is capable of aerial respiration with its swim bladder and also can leap out of the water for short distances and glide (FishBase 2001). These endemic families are found in a fauna comprising 150 freshwater fish, of which about 20 are endemic, including several freshwater representatives of marine families. Cichlids, citharinids, rivulines, and mormyrids dominate the fish fauna. The vulnerable, near-endemic freshwater stingray (*Dasyatis garouaensis*) and the endangered thorny freshwater stingray (*Urogymnus ukpam*) live in the delta (IUCN 2002).

Nutrient-rich, silt-laden whitewater rivers (e.g., the Niger and Orashi) and nutrient-poor blackwater rivers (e.g., the Sombreiro and New Calabar) converge at the delta. Blackwater rivers contain a majority (65 percent) of forest fish species, and whitewater rivers are dominated by savanna species (46.5 percent) but also contain a number of forest species (essay 3.5). Forest species typically are smaller and unlikely to undergo large migrations (Lowe-McConnell 1985). Savanna species are more likely to migrate to and use floodplains and to survive the more severe dry season of the savannas.

The delta provides habitat for various waterbirds and herpetofauna. The mangrove forests of the tidal floodplain are important for numerous waterbirds, including species of heron (Ardeidae), ibis (Threskiornithidae), and pelican (Pelecanidae). Eleven frog species, a variety of monitor lizards, and three turtle species also reside in the delta. Crocodiles include the slender-snouted (*Crocodylus cataphractus*), Nile (*C. niloticus*), and vulnerable dwarf crocodile (*Osteolaemus tetraspis*) (IUCN 2002).

Aquatic mammals present in the delta include the hippopotamus (*Hippopotamus amphibius*) and the vulnerable pygmy hippopotamus (*Hexaprotodon liberiensis*) (Moffat and Linden 1995). The vulnerable West African manatee (*Trichechus senegalensis*), semi-aquatic sitatunga (*Tragelaphus spekei*), marsh mongoose (*Atilax paludinosus*), and spotted-necked otter (*Lutra maculicollis*) also inhabit the ecoregion (Happold 1987; Sayer et al. 1992; Ajao 1994).

Status and Threats

Current Status

The Niger Delta is under intense pressure from urbanization and industrialization, oil exploration and exploitation, upstream impoundment, logging, agriculture, hunting, human population increase, invasive species, and fishing. The severity of these pressures threatens the health of this delicate ecosystem and the survival of the diverse aquatic fauna that depends on it.

Impoundments upstream have changed the flow regime and reduced sediment influx, with an estimated reduction of 70 percent of the original sediment input to the delta (Moffat and Linden 1995). Several other destructive practices in the delta exacerbate the effects of reduced sedimentation and lowered flows. Deforestation, particularly of mangroves with their complex root structure, accelerates erosion, as does the construction of harbor protection structures (e.g., jetties) (Ajao 1994). Lower flood flows since the construction of Kainji Dam have also reduced the inundation of swamp forest, subsequently reducing fish spawning areas (Moffat and Linden 1995). In studies at Oguta Lake, located at the inland edge of the delta, zooplankton and fish decreased after flow decreased from upstream dams (Moffat and Linden 1995).

Dredging of waterways and the practice of depositing dredged material on the beaches has led to changes in water circulation, sediment distribution, salinity, and freshwater inflows (Ajao 1994). In addition, the dredge spoils of acid sulfate soil can decrease forest regeneration productivity because of the high acidity of the soil when dry (Moffat and Linden 1995).

Introduced plants are widespread, threatening native plants and altering community dynamics in many waterways. Water hyacinth (*Eichhornia crassipes*) has completely covered many of the braided channels in the delta, blocking light penetration and causing a decrease in productivity (Sayer et al. 1992). For example, water hyacinth is estimated to cover an area of about 60 km² across a distance of 300 km of waterway in one portion of the delta (Ajao 1994). The alien Nypa palm (*Nypa fructicans*) was introduced between 1906 and 1912 and spread from the Cross River westward, thriving better in inland areas (Ajao 1994). The palm often replaces degraded mangrove forests because mangroves have a slow regeneration rate (Sayer et al. 1992; Moffat and Linden 1995).

Fishing pressure is increasing in the delta, by both artisanal and commercial fishers. In riverine areas overharvesting of juvenile fish is a major problem. Although there is a no-trawling zone up to 9.26 km from shore, trawling continues. Fishers use undersized mesh, traps, trawls, and gillnets to increase catch per unit effort. Of particular concern, the catch per unit effort decreased by 60 percent between 1980 and 1990, and the average body length of commercial fish species has also fallen between 1985 and 1995 (Moffat and Linden 1995).

Five to 10 percent of the mangrove forest in the delta has been lost, and the freshwater swamp is under encroachment pressure (Moffat and Linden 1995). Hippopotamus, manatee, monitor lizards, crocodiles, and even some heron species (*Ardea cinerea* and *A. goliath*) are now threatened in the delta (Ajao 1994). The freshwater swamp and barrier island forests are essential for preventing erosion, in addition to preserving biodiversity (Moffat and Linden 1995). Mangrove and freshwater swamp forests are harvested for fuel and lumber for houses and furniture. Agriculture, including rice and oil palm plantations, has also encroached on the seasonal swamp forests, with some oil palm plantations planned for development in primary forests (Hughes and Hughes 1992; Moffat and Linden 1995).

An increasing human population encroaches on the delta as farmers from degraded upland areas and people interested in the delta's oil development continue to move into the ecoregion (Moffat and Linden 1995). The most densely populated area is in the eastern portion of the delta, near Port Harcourt. To create more settlement areas for the growing population, sediments are pumped out of the delta and onto and adjacent to existing lands (Ajao 1994). Deforestation occurs along roads and paths cleared for pipelines, flowlines, and seismic lines. The infrastructure and human population brought into the delta by the oil industry are estimated by some to be even more severe in their impacts than oil pollution (Moffat and Linden 1995).

The concentration of oil and gas fields in the Niger Delta is one of the highest in Africa, with more than thirty oil fields in the delta and approximately the same number offshore (Hughes and Hughes 1992; Fuller and Bush 1999). Oil pipelines crisscross the delta and the seasonally flooded zones, but the western portion has fewer of them (Hughes and Hughes 1992). Many small spills have occurred in the region, releasing an estimated 2,300 m³ of crude oil into the delta annually (UNEP 1999). Oil pollution in the swamp and mangrove forests, shock waves from seismic surveys, pipeline routes, and gas flaring all have had a deleterious effect on the delta environment (Onu 2003). Nigeria flares more gas as a byproduct of oil production than any other country in the world (Moffat and Linden 1995).

Industrial effluents from steel refineries, thermal plants, and fertilizer plants sometimes flow into the rivers that feed the delta. Monitoring for compliance to effluent standards is negligible.

No significant wetland areas are protected in the delta (Ajao 1994). Three forest reserves exist: Upper Orashi, Nun River, and Lower Orashi, which comprise a total of 239 km². However, they are heavily exploited for their timber. Nine other forest reserves have been proposed. The only effective habitat protection is found in small sacred groves protected by communities. Community protection is also afforded to different animal species. In a number of lakes crocodiles receive protection, and in one area, Nembe, chimpanzees are protected (Bocian 1998).

Future Threats

A growing human population combined with a growing oil industry is the biggest threat to this ecoregion. The population of the delta grows at a rate of about 3 percent a year, and overall Nigeria's population already makes up one-fourth of the total population of Africa (Sayer et al. 1992).

The effects of upstream dams on river flows and sediment input to the delta are expected to continue to erode the delta. Several upstream reservoirs are now silting up, such that flooding has begun to increase again (Moffat and Linden 1995).

Potentially as harmful to aquatic species as changes in flow regimes is the threat of a large oil or gas spill (UNEP 1999). Moreover, as the oil industry continues to grow in the delta, problems can only be expected to increase. There exists increasing tension between villagers of the delta and multinational oil companies over environmental degradation caused by oil spills and lack of benefits of the oil industry for the local population. This tension has erupted in recent years, resulting in fatal violence, allegations of arson attacks on pipelines, and even disruption of oil spill cleanups (Reuters 2000). Onu (2003) proposes a policy framework for improving the regulatory process in the delta to reduce the environmental degradation and resultant conflicts.

Sea level rise from global warming poses an additional risk to the Niger Delta, especially because the delta occupies a large area at low elevation. If the sea level were to rise 1 m per 100 years (International Panel on Climate Change scenario), then it is projected that about 18,000 km² of the delta would be inundated with coastal waters in one century. If the sea level were to rise only 0.2 m, a conservative scenario, then more than 2,700 km² would still be at risk (Moffat and Linden 1995).

Justification for Ecoregion Delineation

This ecoregion is delineated based on the extent of the Niger Delta and is distinguished by a rich freshwater fauna and one endemic fish family (Denticipidae). The Benin River and Imo River mouths delimit the western and eastern boundaries of the ecoregion, and the city of Onitsha sits at its apex. Between the Benin and Imo rivers the delta stretches for about 500 km along the coast (Ajao 1994).

Data Quality: High

Ecoregion Number:	**59**
Ecoregion Name:	Nile Delta
Major Habitat Type:	Large River Delta
Final Biological Distinctiveness:	Nationally Important
Final Conservation Status:	Critical
Priority Class:	IV
Authors:	Emily Peck and Michele Thieme
Reviewer:	M. S. Farid and S. Baha El Din

Location and General Description

The Nile Delta is a fertile fluvial triangle wedged in the midst of one of the driest deserts in the world. Situated in northern Egypt, the Delta extends about 175 km from its apex at Cairo to the Mediterranean Sea and is about 260 km wide along the coast (Hughes and Hughes 1992).

The Nile Delta experiences a Mediterranean climate, with hot summers and cool, rainy winters. The summer season extends from April through October. The warmest month is generally August, with a mean monthly minimum temperature of 22°C and a mean monthly maximum of 30°C along the coast. The winter season extends from November through March. January is the coolest month, with mean monthly minimum and maximum temperatures of 9°C and 18°C along the coast. Average annual rainfall varies from 25 mm/year in Cairo to 200 mm/year along the coast (Hughes and Hughes 1992; FAO Inland Water Resources and Aquaculture Service, Fishery Resources Division 1999).

The Nile River previously braided into numerous channels as it flowed through the delta, moving and depositing unconsolidated, alluvial sediment from the upper reaches of the river to the complex of lagoons, marshes, lakes, temporary pools, agricultural lands, and shallow coastal areas along the Mediterranean Sea. Since the construction of the Aswan High Dam (completed in 1970), water flow through the delta has decreased dramatically, and its floodplains are no longer subject to annual flooding. As a result, the Nile River now occupies only two main channels: the Rosetta (western) and the Damietta (eastern). Today several small lakes occupy former river channels such as El Mannah, El Qatta, Faraonyat, Sinnéra, and San El Hagar. The main wetlands in the delta are the coastal lakes of Manzala, Burullus, Idku, and Maryut.

All deltas, including the Nile, experience phases of growth and shrinkage as a result of sediment input and redistribution by rivers and coastal processes. In the last 7,000 years the Nile Delta has generally been in the accretion phase. However, in the last 150 years the delta has entered an acute stage of subsidence, catalyzed by the construction of dams and barrages along the upper and lower Nile and the intensive regulation of the Nile's waters (Stanley and Warne 1998). The outer margins of the delta are eroding, and salinity levels of some of the coastal lands are rising as a result of seawater infiltration to the groundwater (Hughes and Hughes 1992; Baha El Din 1999). At the same time, the brackish lakes Manzala, Burullus, and Idku have decreasing salinity levels because of high year-round freshwater inflows and poor connections with the Mediterranean (Ramdani et al. 2001).

Outstanding or Distinctive Biodiversity Features

Changes in the aquatic vegetation of the delta reflect changes in the Nile's flow and sediment distribution. The *Cyperus papyrus* swamps that previously existed in the wettest areas of the delta disappeared with the closure of the Aswan High Dam. Reeds (*Phragmites australis* and *Typha* sp.) are now common throughout the delta wetlands, along with some sedge species (*Juncus* sp.) (Hughes and Hughes 1992). *Ceratophyllum* covers many of the delta lakes where water is fresh or slightly brackish. In lacustrine areas of higher salinity *Potamogeton pectinatus* and *P. crispus* predominate (Burgis and Symoens 1987).

Away from the lakes and swamps most of the Nile Delta is intensively cultivated. Only remnant patches of Mediterranean coastal vegetation remain, mainly on coastal sand dunes. Although the agricultural landscape is dominated by nonnative species, some native plants, including several locally rare species such as *Pistia stratiotes* and *Nymphaea lotus*, manage to persist along canals, drains, and roads.

Two endemic plant species are found: *Sonchus macrocarpus* (endemic to the Nile Delta) and *Zygophyllum aegyptium* (limited to Egypt and Libya). There are few globally threatened plant species.

The Nile Delta is part of one of the world's most important sites for migratory birds. More than 1 million ducks and shorebirds use the area on their annual migration from Europe to Africa, despite the fact that agriculture and dams have inextricably altered the floodplain delta ecosystem in the last 50 years (Denny 1991). Large numbers of passerines and nonpasserines also pass through the ecoregion in the spring and autumn.

The Delta lakes are also internationally important sites for wintering waterbirds, providing valuable habitat for several hundred thousand birds. Wetlands of special importance to migratory birds include lakes Manzala, Burullus, Idku, and Maryut (Baha El Din 1999, 2001; Wetlands International 2002). In addition, Lake Manzala hosts the largest population of wintering little gull (*Larus minutus*) and whiskered tern (*Childonias hybridus*) in Africa (Meininger and Sorensen 1993; Baha El Din 1999). The Nile Delta hosts one of the largest breeding populations of the purple gallinule (*Porphyrio porphyrio*) in the Mediterranean Basin. The Egyptian swallow (*Hirundo rustica savignii*) and Egyptian yellow wagtail (*Motacilla flava pygmaea*) are endemic to the Egyptian Nile Valley and Delta (S. Baha El Din, pers. comm., 2003). Ten globally threatened bird species are also known to occur in the delta.

The aquatic fauna of the Nile Delta ecoregion is a dynamic mix of freshwater and saline species. The freshwater fauna is typical of the Nilo-Sudan ichthyofaunal province, of which it is a part (Lowe-McConnell 1987; Lévêque 1997). The delta supports no endemic fish, but there is one endemic mollusk (*Biomphalaria alexandrina*). Of about thirty fish species present in the ecoregion, those from the families Cichlidae and Mormyridae are most common. Several marine species tolerant of freshwater also make their home in the delta, including *Anguilla* spp., *Mugil cephalus,* and *Solea vulgaris*.

The herpetofauna is a mixture of Afrotropical, Saharan, and Palearctic species. The nominate form of Bosc's lizard (*Acanthodactylus boskianus*) is endemic to the Nile Delta, where it is widespread and common on coastal dunes and near salt marsh habitats. On the other hand, the Nile soft-shelled turtle (*Trionyx triunguis*) and the Nile crocodile (*Crocodylus niloticus*) used to inhabit the region but have become extinct locally. The ecoregion provides habitat for six amphibian species, including the endemic Nile Valley toad (*Bufo kassasii*).

The mammal fauna of the delta is limited and characterized by species associated with fluvial habitats such as grass rat (*Arvicanthis niloticus*), Egyptian mongoose (*Herpestes ichneumon*), jackal (*Canis aureus*), red fox (*Vulpus vulpus*), and jungle cat (*Felis chaus*). Hippopotamus (*Hippopotamus amphibius*) and wild boar (*Sus scrofa*) were formerly a part of the Nile Delta mammalian fauna but became extinct during the past two centuries. Flower's shrew (*Crocidura floweri*) is endemic to the Nile Delta and Valley and may be extinct (S. Baha El Din, pers. comm., 2003).

Status and Threats

Current Status

Human settlement and agriculture occupy almost the entire delta landscape, except for a few small pockets of natural habitat, mostly in the extreme north of the delta. These areas are composed of wetlands, their surroundings, and coastal dunes and salt marshes. However, they are disappearing rapidly as the irrigation system of the delta is highly controlled and remaining wetland areas are converted to agricultural use (Wetlands International 2002). The wetlands of the delta have lost more than 50 percent of their original area in the past century to reclamation for agriculture, sedimentation, and erosion (in coastal lakes).

Three protected areas encompass parts of the Nile Delta ecoregion: Ashtom El Gamil Protected Area (160 km^2), Lake Burullus Protected Area (480 km^2), and the Nile Island Protected Area (<50 km^2). Perhaps most important of the three is Lake Burullus (which is also a Ramsar site) because of its substantial size and intact condition.

The completion of the first Aswan Dam reduced the annual flood pulse in the Nile Delta. The closure of the second Aswan High Dam (in 1970) stopped the flooding cycle entirely and the influx of sediments that occurred with it. The delta is now a subsiding and eroding coastal plain because of the lack of sediment input and retention (Stanley and Warne 1998). The Nile Delta has eroded by up to 2 km since the 1960s (Davies and Day 1998). Additionally, seawater intrusion of the delta threatens to increase salinity of the coastal lakes and rivers and threatens the quality of groundwater.

Without the seasonal floods, the delta is incapable of flushing out the waste and pollutants generated by an increasing human population (Stanley and Warne 1998). Farmers in the floodplain of the delta apply large amounts of fertilizers and pesticides to the land each year. These pollutants flow off the land and into the rivers and canals that feed the delta. Increasing amounts of polychlorinated biphenyls and organochlorides, such as lindane, malathion, and DDT, threaten both plant and animal populations, and this pollution may also threaten the groundwater reservoir (Abbassy et al. 1999).

Alien invasive species are a serious threat facing the native biodiversity. Water hyacinth (*Eichhornia crassipes*) is the most prevalent invasive species in the Nile Delta. Although it has some water purification properties and provides habitat for some aquatic species, it has displaced native floating aquatic vegetation such as *Pistia stratiotes* and *Nymphaea lotus,* which are now rare. It also causes the evaporation of large amounts of water; clogs canals and drains, hampering movement and fishing; and increases the sediment and nutrient load in infested waterbodies. This species is well established in Lake Manzalla, particularly in the south. The American crayfish (*Procambarus clarkii*) has infested most wetlands in the Nile Valley and Delta and is damaging the native aquatic fauna. The streaked weaver (*Ploceus manyar*) is an introduced bird species living in the delta. In addition, a recently introduced mollusk, *Biomphalaria glabrata,* is spreading and may be competing with the endemic *B. alexandrina* (Kristensen and Brown 1999).

Future Threats

Humans have dramatically changed the ecology of the Nile Delta, and further increases in human populations and related activities further threaten the remaining ecological functioning of the delta. Almost all of the 70 million people in Egypt live in the Nile Delta and Valley, with population densities averaging 1,000 people/km² (FAO Inland Water Resources and Aquaculture Service, Fishery Resources Division 1999). The disposal of resultant urban and industrial waste poses a serious threat to the integrity of the fresh waters of the Nile Delta (Abbassy et al. 1999). However, national and local laws governing environmental protection have recently become stricter and are actively being implemented. The potential for sea level rise to inundate the freshwater marshes and lagoons of the delta is a matter of serious concern. One recent study predicts that a 1.1-m rise within 1,000 years would cause the inundation of 72 percent of Rosetta City and the Rosetta branch of the Nile River (El-Raey et al. 1997).

Justification for Ecoregion Delineation

This ecoregion is delineated based on the extent of the Nile Delta and its marshes and is distinguished by a Nilo-Sudanian freshwater fauna with some brackish and marine elements.

Data Quality: High

Ecoregion Number: **60**
Ecoregion Name: **Lower Congo Rapids**
Major Habitat Type: **Large River Rapids**
Final Biological Distinctiveness: **Globally Outstanding**
Final Conservation Status: **Endangered**
Priority Class: **I**
Author: **Victor Mamonekene**

Location and General Description

In the Lower Congo Rapids ecoregion, the Congo River drops 280 m as it travels 350 km between the end of Malebo Pool and the city of Matadi (Beadle 1981). The river alternately passes through narrow, deep gorges (about 200 m wide) and wider stretches (about 2 km wide) but runs swiftly through both (Bailey 1986). This stretch of river encompasses thirty-two falls and rapids in the Mount Cristal Rapids. In addition to rapids with strong currents, there are also some calm sections of river between Brazzaville and Matadi. The ecoregion covers parts of the Republic of the Congo (ROC) and the Democratic Republic of the Congo (DRC). The largest tributaries to the Congo River in this stretch are the Djoué, Loufoulakari, Inkisi, Kwilu,* and Mpozo. Most of them flow from cliffs and over waterfalls as they descend to the mainstem river.

The climate of the ecoregion belongs to the Guineo-equatorial bas-Congolais type. Rains occur between March and April and between mid-November and mid-December. Precipitation originates offshore as monsoon rains under the influence of the Guinea and Benguela currents.

The fish fauna shows morphological and behavioral adaptations to life in fast-running waters. Gallery forests line the rivers, and tree-savanna vegetation covers the landscape. Over the last century the vegetation in this region has changed from lowland forest to wooded savanna (Shumway et al. 2003).

Outstanding or Distinctive Biodiversity Features

An endemic aquatic fauna occurs in the swift waters of the rapids of this ecoregion. Of the 129 fish species that have been recorded from the rapids, 34 are endemic (Roberts and Stewart 1976). A highly endemic, rheophyllic snail fauna also inhabits the rapids. This fauna includes four endemic monotypic genera (*Congodoma, Liminitesta, Septariellina,* and *Valvatorbis*), with sixteen endemics from a fauna of eighteen species (Brown 1994).

Most of the fishes and snails have morphological and behavior adaptations to fast-running water. Specializations among fish include reduction of eye size (micropthalism), a blue or bluish coloration, and modified body form (dorsoventrally depressed heads and bodies). Among the species adapted to fast-flowing water are cyprinids of the genera *Garra* and *Labeo;* catfishes of the genera *Atopochilus, Euchilichthys, Chiloglanis,* and *Gymnallabes;* cichlids of the genera *Steatocranus, Teleogramma, Lamprologus,* and *Leptotilapia;* and a group of endemic mastacembelids (Roberts and Stewart 1976). As examples, a mastacembelid eel (*Mastacembelus brachyrhinus*) and the endemic *Lamprologus lethops* are both cryptophthalmic, meaning their eyes are small and partially or completely covered by skin and other tissues (Roberts and Stewart 1976). Rheophyllic snails also exhibit adaptations, with the ability to adhere to rocks in the swift current and to tolerate large fluctuations in water level.

Status and Threats

Current Status

This ecoregion receives the full brunt of untreated sewage, sediment, and industrial chemical pollutants from the capitals of Brazzaville and Kinshasa, located on opposite sites of the Congo River just below Malebo Pool. Recent sampling in the river has found lead levels to be four times higher and cadmium levels to be three times higher than the U.S. Environmental Protection Agency's recommended levels for children (Shumway et al. 2003). Downstream from Brazzaville and Kinshasa, on both banks of the Congo River, there is extensive mining activity. Explosives are used to extract the sandstone that will be used to

---

* To be distinguished from the Kouilou River in Congo-Brazzaville and the Kwilu River that is a tributary of the Kasaï River.

make concrete for buildings and roads. The endemic fish of the rapids are also targets for the aquarium trade and are at risk of overexploitation.

This ecoregion contains the Inga hydropower station, which consists of two deteriorating plants, 351 and 1,424 megawatts in capacity, constructed in 1972 and 1982, respectively (Zhuwakinyu 2002). Apparently no environmental impact studies were conducted before or after construction of the dam; however, the effects on the biota may be low because the river is naturally divided into multiple channels at the Inga site, and the run-of-river dam blocks only one channel. The station supplies power to the DRC, the ROC, Zambia, Zimbabwe, and South Africa (Zhuwakinyu 2002).

Future Threats

The government of the DRC has recently announced plans to move forward quickly with development of the Grand Inga plant, with a projected capacity of 39,000 megawatts. Constructing this plant would involve blocking the river's flow and creating a large reservoir (République Démocratique du Congo 2002). The Inga station would serve as the hub of a pan-African electricity grid, with new interconnectors built to Egypt and Nigeria. Environmental impact statements could be undertaken as early as 2003 (Zhuwakinyu 2002).

Additional future threats relate to human population growth. Rates of population increase are high in both Brazzaville and Kinshasa, and pressures from these cities on the Congo River and its species are expected to increase.

Justification for Ecoregion Delineation

This ecoregion is delineated based on the 350-km stretch of rapids in the Lower Congo River and is distinguished by an endemic aquatic fauna adapted to the rapids. It is hypothesized that in the late Pliocene or early Pleistocene, a coastal Lower Guinean river captured Malebo Pool, connecting the previously interior Congo Basin to the ocean. The Congolian nature of the endemic species that occur in the rapids suggests that this fauna evolved before the river capture (Roberts and Stewart 1976). The rapids constitute a barrier to the movement of marine and brackish fishes from the delta into the interior of the Congo Basin.

Data Quality: Low

Little is known about the freshwater biota of this ecoregion. There is one paper on fishes of the Lower Congo Rapids (Roberts and Stewart 1976). In its National Biodiversity Action Plan, the government of the DRC identifies rapids and waterfalls as among the habitats most in need of biodiversity surveys (République Démocratique du Congo 1998).

Ecoregion Number:	**61**
Ecoregion Name:	Upper Congo Rapids
Major Habitat Type:	Large River Rapids
Final Biological Distinctiveness:	Bioregionally Outstanding
Final Conservation Status:	Vulnerable
Priority Class:	V
Authors:	Ashley Brown and Robin Abell
Reviewer:	Luc De Vos

Location and General Description

The Upper Congo Rapids ecoregion lies entirely in the Democratic Republic of the Congo (DRC) and extends downstream from near the city of Ubundu. Among the many rapids found in the ecoregion are the Tshungu rapids below Ubundu, the rapids near Wanie-Rukula, and the seven cataracts of Boyoma Falls near Kisangani, formerly called Stanley Falls. Several rapids also occur along the Lindi, Tshopo, and Maiko rivers before they meet the Congo (Hughes and Hughes 1992), and a number of prominent islands lie in the middle of the Congo.

In this ecoregion, the Lualaba River drops 60 m over a distance of just 100 km and emerges from the rapids with a new name: the Congo River. The Tshopo and the Lindi rivers join the Congo just below the rapids near Kisangani. Further downstream, the Lomami also joins the widening Congo (Hughes and Hughes 1992). Although the biota of the ecoregion has not been studied in detail in recent years, a rich fish fauna is known to inhabit the various rapids. The vegetation of this ecoregion, similar to that of the Central Congo, consists primarily of high tropical rainforest with a canopy that reaches up to 45 m (Hughes and Hughes 1992). Allochthonous inputs from riparian vegetation provide an important source of organic matter for aquatic food webs (Bailey 1986).

The climate of this ecoregion is humid and hot because it lies between 3°40'S and 1°N, just on the edge of the equatorial belt, where mean annual rainfall can exceed 1,800 mm (Hughes and Hughes 1992). Rain falls throughout the year, with October receiving the most precipitation at more than 200 mm. December, January, and February are the driest months (Bailey 1986). Mean annual temperature fluctuates little around 24°C year-round, and daily maxima and minima are about 30°C and 19°C, respectively (Bailey 1986).

Outstanding or Distinctive Biodiversity Features

A rich fish fauna (probably more than 150 species) and an endemic mollusk fauna inhabit the waters of this ecoregion. Predominant fish families are Mormyridae, Cyprinidae, Alestiidae, Citharinidae, Distichodontidae, Mochokidae, Schilbeidae,

Clariidae, Claroteidae, Amphiliidae, Aplocheilidae, and Cichlidae. Less species-rich but common families are Mastacembelidae, Tetraodontidae, Malapteruridae, Anabantidae, and Centropomidae.

Many of the fish that live in this ecoregion possess adaptations to fast-flowing waters (Beadle 1981). Aquatic animals generally encounter high oxygen levels but low light conditions under rocks or boulders, where they are likely to spend the most time (Beadle 1981). The Amphiliidae species of the genera *Amphilius, Phractura,* and *Doumea* possess stiff, enlarged pectoral fins to strengthen positioning among rocks in strongly flowing water (Welcomme and De Mérona 1988). *Chiloglanis* species (the so-called suckermouth catfishes), the rheophilous catfishes *Atopochilus* and *Euchilichthys* (family Mochokidae), and several Cyprinids of the genus *Labeo* have sucker-like mouths to cling to rocks in current and also to graze on algae growing on rocks. Very elongate *Phractura* species (family Amphilidae) show highly hydrodynamic body morphology. Other rheophilous (current-loving) fish include cichlids (e.g., genera *Lamprologus* and *Orthochromis*) and mastacembelids. The taxonomic status of many fish species in this ecoregion is ill-defined, and it is quite certain that further investigations in the rapids will reveal fish species new to science.

This ecoregion also hosts a rich aquatic invertebrate fauna. There are nine documented mollusk species, of which six are endemic. These endemics, all prosobranchs, are five species of *Potadoma* plus *Pseudogibbula cara*. The rapids are suspected also to be important for aquatic insects, but they have not been surveyed recently (Kamdem Toham et al. 2003).

The herpetofauna and aquatic mammal fauna in this ecoregion are similar to those in surrounding Congolese ecoregions. The slender-snouted crocodile (*Crocodylus cataphractus*), Nile crocodile (*Crocodylus niloticus*), and eleven frog species are known from this ecoregion, although the crocodile species have been under intense pressure from humans in recent decades (L. De Vos, pers. comm., 2002). The Congo clawless otter (*Aonyx congicus*) and giant otter shrew (*Potamogale velox*), with distributions limited primarily to the Congo Basin, are found here, as are the marsh mongoose (*Atilax paludinosus*) and the spotted-necked otter (*Lutra maculicollis*).

Status and Threats

Current Status

The largest threats to the Upper Congo Rapids ecoregion are continued civil strife and increased mining. Sound environmental management and protection will come only with political stability. Other threats include pollution from agriculture, industries, and human settlements and overfishing. Poaching of large vertebrates (including crocodiles, which have disappeared from many stretches in the river) has occurred over the last decades. Both industrial and artisanal diamond mining are prevalent near Kisangani. Artisanal mining has localized but intense effects (D. Wilkie, pers. comm., 2000).

Warring factions have struggled for control of Kisangani because of the rich natural resources in the surrounding area, and the DRC has been undergoing political upheaval and civil strife since 1996 (East 1999). This political situation has brought environmental protection to a standstill. Most international, industrial-scale mining and logging activities have also been suspended, although smaller-scale, often illegal mining continues (Pan-African News Agency 2001).

A United Nations Environment Programme report from 1999 describes the pollution in the region at that time (UNEP 1999). According to the report, 95 percent of industry discharge was dumped directly into the rivers and other freshwater systems of the DRC, and sewage was treated similarly. Waste from textile, chemical, plastic, and painting industries contained toxic metals such as chromium, mercury, lead, cadmium, and zinc. Food processing industries and construction industries also had a significant pollution impact. Agrochemicals such as DDT, DDD, and lindane were all used in the DRC and are all toxic compounds. It is unclear which of these industrial activities still operates in the ecoregion, although it is unlikely that waste treatment could have improved since the late 1990s.

This ecoregion lies in Kivu Province, where deforestation is severe (Sayer et al. 1992). Slash-and-burn agriculture in the Congo Basin has led to erosion, which increases turbidity in rivers and in turn decreases plankton production and fish survival. In aquatic systems where primary production is fueled by allochthonous organic matter, loss of streamside forests can lead to a decline in productivity (Marlier 1973).

Introductions of alien species should be prevented. So far the fishes in this ecoregion are nearly all native, but during the late 1980s the alien bonytongue (*Heterotis niloticus*) had begun to invade the Lualaba system (L. De Vos, pers. obs., 2002), probably after upstream colonization from lower parts of the Congo where it was introduced. The exotic water hyacinth (*Eichhornia crassipes*) has been found near Kisangani in the Congo River. Measures to control water hyacinth include chemical and mechanical removal attempts. The Boyoma Falls provide a natural and effective barrier to the spread of the water hyacinth upstream into the Lualaba River (Bailey 1986).

Future Threats

The largest threats to the ecoregion are continued civil unrest, population growth in Kisangani, and unregulated artisanal mining of diamonds in the region. The return of political stability might allow the resumption of industrial-scale mining, which comes with its own threats. This ecoregion currently has no protected areas. The upper Congo Rapids probably have high hydroelectric potential, but no large dams have been built in the area.

Justification for Ecoregion Delineation

The Congo Basin lies in an ancient continental lake basin (Beadle 1981). The steep sections of the Congo River, including the rapids of this ecoregion, are thought to have been formed by river capture of the upper Lualaba by the Congo (Marlier 1973). These rapids probably create a barrier to the movement of some fish species from the Congo River up into the Lualaba. This ecoregion is defined by the rapids of the upper Congo River between Ubundu and Kisangani and is distinguished by an aquatic fauna adapted to life in the rapids.

Data Quality: Low

Current surveys of the freshwater fauna are needed for this ecoregion.

Ecoregion Number:	**62**
Ecoregion Name:	**Bight Coastal**
Major Habitat Type:	**Savanna–Dry Forest Rivers**
Final Biological Distinctiveness:	**Continentally Outstanding**
Final Conservation Status:	**Critical**
Priority Class:	**II**
Authors:	**Ashley Brown and Robin Abell**
Reviewers:	**Christian Lévêque and Philippe Lalèyè**

Location and General Description

In this ecoregion, located along West Africa's Gulf of Guinea (or Bight of Benin), rivers flow down gently sloping plateaus to the flat coastal plain, where they form extensive swampy deltas and semicontinuous lakes and lagoons before reaching the ocean. The larger rivers of the ecoregion are the Mono (Togo and Benin), the Ouémé (Benin), and the Ogun-Oshun (Nigeria). The Mono River drains the southern and central parts of the Chaine du Togo mountain range and forms the border between Togo and Benin (Hughes and Hughes 1992). The Ouémé River (510 km long) drains most of Benin's low (200–300 m a.s.l.) southern plateau. The Ogun and Oshun rivers drain a low plateau in the southwest corner of Nigeria and flow into the Lagos and Lekki system of lagoons along the coast. Covering primarily the southern portions of Benin and Togo and southwest Nigeria, the ecoregion also extends slightly into the southeastern corner of Ghana.

The region experiences a distinctly bimodal rainfall pattern, with most annual precipitation falling between March and July and between September and November. Mean annual rainfall decreases from west to east, from 850 mm per year in the Mono River delta to 1,750 mm per year in the Ogun-Oshun Basin (Hughes and Hughes 1992). Temperatures are constant year-round, with little difference between daily maxima and minima (34°C and 22°C) (Sayer et al. 1992).

The ecoregion contains a mosaic of rivers, wetlands, coastal lagoons, and lakes with varying degrees of interconnectedness, both with each other and with the Gulf of Guinea. Along the coast, connecting channels between the lagoons often are transient, drying out in the dry season. Seasonal changes in precipitation and water inflow, and the resulting variable movement of water between coastal lakes and lagoons, produce fluctuating salinity. Many lagoons tend to be fresh to brackish during the wet season but have elevated salinity concentrations in the dry season. Major lagoons and coastal lakes are the Lagos (500–600 km^2) and Lekki (247 km^2) lagoons (Nigeria); lakes Nokoué (150 km^2) and Ahémé (85 km^2), Porto Novo lagoon (30 km^2), and the coastal lagoon (12 km^2) (Benin); and lakes Togo (46.6 km^2) and Vogan (8 km^2) (Togo). These waterbodies often are bordered by swampland that provides a link between individual lagoons and lakes, creating a vast wetland system (Hughes and Hughes 1992).

In the western portion of the ecoregion, floodplains occur along most rivers. In the more mountainous central area of Togo, floodplains occur as narrow strips, often only 25–50 m wide. There are numerous permanent swamps in the headwaters of the Okpara River, a tributary of the Ouémé, and strips of inundated forest occur along the Ouémé River (Hughes and Hughes 1992).

The vegetation of the ecoregion varies from the low, gently sloping plateau to the coastal plain. In the northern half of the ecoregion, open or fragmented deciduous forest dominates the vegetation, whereas it occurs in patches and bands in the south and in dense patches in the southwest (FAO Forest Resource Assessment Programme 1999). Dominant species in these medium-height deciduous and dry stands include *Antiaris africana*, *A. welwitschii*, and *Ceiba pentandra* (FAO Forest Resource Assessment Programme 1999). Moving south, the landscape is dominated by wooded savanna with deciduous woody trees (a mix of *Combretum*, *Terminalia*, and *Acacia* species), with an understory of tall grasses, shrubs, and herbs (White 1983). Further south, this gives way to secondary grassland and secondary wooded grassland, punctuated by riparian forest patches and wetlands. Tall grasslands (up to 3 m) include fire-resistant species of *Andropogon*, *Hyparrhenia*, and *Pennisetum* (Lawson 1986).

The sandy soils of the coastal plain support small stands of mangrove vegetation (dominated by *Rhizophora* and *Avicennia* species) around lagoons. The vegetation in the swampy deltas is a mixture of floodplain grasses, reeds, and cattails. Species of *Typha* and *Cyperus* dominate the reed swamps, and *Paspalum* and *Phragmites* species characterize grassy floodplains (Hughes and Hughes 1992).

Outstanding or Distinctive Biodiversity Features

The rivers and coastal lagoons of the Bight Coastal ecoregion host an aquatic fauna rich in fish and mollusks but low in endemism. The ecoregion contains only 6 endemic fish out of 153 fish species. Dominant fish families include Aplocheilidae, Cyprinidae, Cichlidae, and Mormyridae. About thirty mollusk species, with three endemics, are present in Bight Coastal. The ecoregion is also rich in aquatic frogs, with forty-nine species and six endemics, which are associated primarily with forests and wet savanna woodlands.

The system of coastal lagoons and lakes in this ecoregion is a unique habitat type. The salinity and volume of water in these lakes and lagoons vary throughout the year with the wet and dry seasons. The fish species composition of Lake Nokoue may vary with salinity, although the most common species (including *Ethmalosa fimbriata, Hemichromis fasciatus, Sarotherodon melanotheron,* and *Tilapia guineensis*) are present year-round (Laë 1992; Hughes and Hughes 1992).

In the Upper Ogun River, research suggests that the fish community exhibits seasonal variation in species composition (Adebisi 1988). At the onset of flooding, an increased relative abundance of piscivores has been observed, whereas omnivores were more dominant as floodwaters receded, and herbivores and insectivores dominated during periods of low water level (Adebisi 1988).

Based on studies in the Ouémé River, Welcomme (1979) suggested a general classification of fish species in the river and its floodplains; researchers have subsequently applied this classification to river systems around the world. The riverine fish communities are loosely categorized as white fish, black fish, and gray fish. White fish, such as Cyprinidae and Mormyridae, depend on main river channels for breeding and often migrate extensively to breed at the onset of floods (Lévêque 1997). Black fish, such as Anabantidae, Channidae, and Clariidae, live in the floodplain or marshy river fringes and often have adaptations to resist harsh environmental conditions such as anoxic waters. Black fish limit movements to lateral migrations. Gray fish, such as Cichlidae, Citharinidae, and Mochokidae, live in fringing vegetation, backwaters, and edges of floodplain lakes during the wet season and inhabit the main river channel during the dry season. With their lateral migrations into the floodplains to breed and feed, gray fish have more flexible behavior than either the black or white fish, and they adapt to changing hydrological conditions easily (Lévêque 1997).

Several large mammal and reptile species live in the coastal streams of this ecoregion. Mammals present include the vulnerable West African manatee (*Trichechus senegalensis*) and the hippopotamus (*Hippopotamus amphibius*) (IUCN 2002). The Nile crocodile (*Crocodylus niloticus*), the slender snouted crocodile (*C. cataphractus*), and the dwarf crocodile (*Osteolaemus tetraspis*) also live in the rivers of this ecoregion.

A rich and prolific avifauna lives on the Mono River (Hughes and Hughes 1992). Along the upper Mono, the Ramsar Reserve de Faune de Togodo, composed of ponds and swamps ideal for wading birds, is a stopover location for migratory birds (Wetlands International 2002). In Benin, ou Dendi in the northeast of the ecoregion supports large overwintering populations of comb duck (*Sarkidiornis melanotos*) and white-faced whistling duck (*Dendrocygna viduata*) (Scott and Rose 1996).

Status and Threats

Current Status

Much of Nigeria's forest and savanna has been destroyed or degraded by the expansion of agriculture, woodcutting, and overgrazing (East 1999). Logging in Benin (in the center of the ecoregion) is high, providing timber for the dense population of neighboring Nigeria (Sayer et al. 1992). Much of the mangrove forests in the southeast of the ecoregion has been cleared for fuelwood and timber, and these forests have been replaced by secondary vegetation such as *Cyperus articulatus* and *Paspalum vaginatum* (Hughes and Hughes 1992). Fuelwood collection and illegal hunting threaten mangroves in the central and western coast as well (Sayer et al. 1992). The swamp forests near Lagos Lagoon have been largely destroyed. Small-scale agriculture and cattle grazing occur on the floodplain margins in the east of the ecoregion (Hughes and Hughes 1992).

Fishing pressure is high in the coastal lagoons of this ecoregion. The highly productive lakes Nokoue and Porto Novo lagoon in Benin and the Lagos and Lekki lagoons along the densely populated coast of Nigeria are subject to intensive fishing pressure (Lalèyè et al. 1995a, 1995b; Lalèyè and Moreau in press).

Because most of the populations of Benin, Togo, and Nigeria are concentrated in coastal areas (Sayer et al. 1992), the coastal portion of the ecoregion is most significantly threatened. Benin is densely populated, with about 6 million inhabitants as of 2000, and most of the natural vegetation has been replaced by the expansion of plantations and agriculture (East 1999; WHO and UNICEF Joint Monitoring Programme for Water Supply and Sanitation 2000). Population in the coastal zone of Togo was estimated at 300 people/km^2 in 1992 (Sayer et al. 1992). Nigeria has a population of about 123 million, with a population growth rate of 2.67 (CIA 2001). Much of this population is concentrated in the Lagos and Lekki lagoons and further east along the coast in the Niger Delta [58]. Population density on the northern margins of Lagos lagoon is more than 250 people/km^2 (Hughes and Hughes 1992).

Water hyacinth (*Eichhornia crassipes*) has invaded the waters of Benin and Nigeria and proliferates in the upper reaches of the So and Ouémé rivers (Van Thielen et al. 1994; UNEP 1999). Although it is carried into the lagoons as well, the plant cannot survive in brackish water and dies during the dry season

(Van Thielen et al. 1994). Water hyacinth alters riverine habitats by creating barriers to movement and preventing insolation, both of which are associated with decreased fish abundance (Olaleye et al. 1993). Managers are using herbicides and biological control in an attempt to constrain the spread of the plant (Olaleye et al. 1993; Van Thielen et al. 1994).

Pollution from domestic sewage, industry, and mining also threatens the freshwater systems of Bight Coastal. Waste from the many industries in Lagos pollutes the waters of Lagos Lagoon (Hughes and Hughes 1992), and disposal of untreated sewage into waterways is a widespread problem throughout the ecoregion. For example, in Lagos, a city of 10.3 million as of 1995, there are no sewage treatment facilities (UNEP 1999). About 80 percent of Nigeria's manufacturing industries are located in and around Lagos. The lagoon suffers from anthropogenic eutrophication, which has decreased biodiversity (UNEP 1999). The situation is similar in other cities of the ecoregion. In the east, industrial waste, including runoff of clay and dissolved materials from mining, particularly threatens the coastal and riverine ecosystems. Untreated liquid industrial waste enters Lake Nokoue and Porto Novo lagoon in the center of the ecoregion.

Agriculture and associated pollution pose additional threats. Welcomme (1979) speculates that on floodplains where agriculture is practiced intensively, such as along the Ouémé River, much of the standing water and swamps will disappear through drainage and fill. Runoff from agrochemicals, fertilizers, and other biocides are of high concern in the west, where agriculture is increasing and agrochemicals are applied intensively. Pesticide use is on the rise in Benin as well (UNEP 1999).

About seventy small-scale irrigation dams currently exist in Togo, and a large-scale dam, the Nangbeto Dam, was completed in 1988 on the Mono River (Hughes and Hughes 1992). The hydroelectric dam was created as a joint project between Benin and Togo and partially funded by the World Bank. Coastal erosion is a problem at the mouth of the Mono River and all along the coast.

Currently, the Ramsar Reserve de Faune de Togodo protects about 310 km² along the Mono River (Wetlands International 2002). The reserve covers about 20 km of river that includes rapids, as well as many ponds and swamps that provide ideal habitat for waders and other waterbirds (Hughes and Hughes 1992; Wetlands International 2002). The Forestry Reserves of Ketou Forest, Dogo Forest, and the Forest of Ouémé all cover part of the Ouémé River (Hughes and Hughes 1992). In Nigeria, small nature reserves exist at Lekki and Omo (Sayer et al. 1992). The Olokemeji Forest Reserve protects one of the last areas of remaining virgin forest in western Nigeria.

Future Threats

As the ecoregion's population swells, continued deforestation, runoff from agricultural lands, and pollution pose future threats, along with potential new impoundments. Continued deforestation in the mangroves will also further destroy mangrove habitat and cause erosion. Sewage and waste treatment facilities must be improved throughout the ecoregion to prevent further pollution of the rivers and lagoons.

Justification for Ecoregion Delineation

This ecoregion is defined by the rivers that drain into the Bight of Benin, and it lies in the Dahomey Gap, an area of savanna interrupting the Guinean forest zones along the coast of West Africa, stretching from the Cavally River in Côte d'Ivoire to the Cross River in Ghana (Sayer et al. 1992; Hugueny and Lévêque 1994). The gap most recently formed around 4,000 years ago and has remained since then, but it was also present during previous dry phases associated with glacial maxima of ice ages, including at the height of the last glaciation around 18,000 years ago (Lévêque 1997). Given the low levels of endemism in the coastal rivers of this ecoregion and the Nilo-Sudanian nature of the fish fauna, it is hypothesized that many of these rivers desiccated during dry phases and were subsequently recolonized by fish from the Niger River (Lévêque 1997).

Data Quality: High

Ecoregion Number:	**63**
Ecoregion Name:	Cuanza
Major Habitat Type:	Savanna–Dry Forest Rivers
Biological Distinctiveness:	Continentally Outstanding
Conservation Status:	Vulnerable
Priority Class:	II
Author:	Lucy Scott
Reviewer:	Paul Skelton

Location and General Description

The coastal and escarpment rivers of western Angola contain a poorly known freshwater fauna that is suspected to be rich in species and to have a high number of endemics. This region covers a narrow coastal plain and a stepped escarpment rising to an altitude of more than 1,000 m (Hughes and Hughes 1992). All the westward-flowing rivers of Angola lie in this ecoregion. The principal drainages, from north to south, are the M'bridge, Loje, Dende, Bengo, Cuanza (Quanza), Cuvo, and Catumbela rivers.

Precipitation increases from an annual average of 500 mm along the coast in central Angola to about 1,000 mm in the central interior. The Benguela Current influences and reduces rainfall along the semi-arid coast. Temperatures remain stable year-round, with a mean annual temperature of about 20–22°C, although the average is higher in the coastal lowlands (Hughes and Hughes 1992).

Cloud forests on the escarpment appear to be in part a southern extension of the central African rainforests but have distinct Afromontane affinities. Inland, on the plateau, vegetation is predominantly miombo (*Brachystegia* and *Julbernardia*) woodland, containing an extensive network of grassy dambos. Dense gallery forests line the major watercourses. Rapids or falls occur in the mainstem of all large rivers throughout their descent down the western escarpment of Angola (Roberts 1975). Coastal savanna is the predominant vegetation on the coastal belt, although thicket is more common in the north (Stuart et al. 1990).

The Cuanza River, more than 970 km long, with its major tributaries the Lucala and the Luando, has the largest catchment (about 146,000 km^2) of all the rivers in this ecoregion. As the river descends below the 1,000-m contour, it begins to develop a series of extensive swamps. Near the junction of the Luando and Cuanza rivers is a large floodplain with tracts of permanent swamp, and floodplains occur sporadically downstream from here until Dondo. Wetland vegetation includes *Cyperus papyrus*, *Typha capensis*, *Phragmites mauritianus*, and *Echinochloa stagnina*. Lakes and lagoons then occur adjacent to the river until it meets the Atlantic Ocean, where there are extensive mangroves (Hughes and Hughes 1992; Dean 2001).

Outstanding or Distinctive Biological Features

Although the area is poorly known, richness and endemism among fish (and potentially other groups) are suspected to be high. No recent accounts are available, and knowledge is based mainly on collections made by early expeditions (Nichols and Boulton 1927; Fowler 1930; Trewavas 1936; Ladiges 1964). At least fourteen families of freshwater fish are represented in the ecoregion (Roberts 1975). Poll (1967) records about 109 freshwater fishes, of which 40 percent are cyprinids and 17 percent cichlids (Poll 1967; Roberts 1975). Possibly as many as thirty endemics (27 percent) are known from this ecoregion, but the real status is difficult to ascertain given the lack of comparative information from neighboring tributaries of the Congo and the Angolan reaches of the Cunene and the Okavango (e.g., see Greenwood 1984).

Aquatic frogs are abundant in this ecoregion's wetlands, with more than fifty species known, including the endemics Cuanza reed frog (*Hyperolius punctulatus*), *H. cinereus*, *Bufo grandisonae*, Congulu forest tree frog (*Leptopelis jordani*), Quissange forest tree frog (*L. marginatus*), and *Rana parkeriana*. Further studies probably would also reveal new species and endemics.

Several wetlands and floodplains in the ecoregion hold congregations of waterbirds that occur in numbers that are at least nationally significant. These include Quiçama, Mussulo, and the Luando Strict Nature Reserve. Among the waterbirds found in these important bird areas are the globally threatened *Grus carunculatus* and *Phoenicopterus minor* (Dean 2001).

About twenty aquatic mollusks are known from the waters of this ecoregion. About one-quarter of these species are shared with those of the Lower Congo [22] ecoregion.

Several aquatic-obligate mammals are known from the rivers and streams of this ecoregion, including the West African manatee (*Trichechus senegalensis*), giant otter shrew (*Potamogale velox*), and African water rat (*Colomys goslingi*) (Kingdon 1997).

Status and Threats

The status of the freshwater systems of the ecoregion is little known because civil conflict has continued there for several decades. The almost continuous civil war in Angola since 1974 has led to great instability, poor security, economic depression, displacement of the rural population, and a lack of infrastructure and basic services. Its effects on conservation, particularly on large mammals, have been devastating. Most of Angola's protected areas have been abandoned as their wardens were forced to leave for economic or security reasons, and they have become open areas for poachers and settlers (Huntley and Matos 1994). The Kissama (9,500 km^2) and Kangandala (630 km^2) national parks and the Luando (8,300 km^2) Nature Reserve cover parts of the Cuanza River catchment (WDPA Consortium 2003). Efforts are under way to rebuild and expand the infrastructure of the Kissama National Park for tourism purposes, to translocate and restore wildlife populations in the park, and to monitor and guide these changes through ecological research (Kissama Foundation 2001).

Only three of Angola's six existing dams (Cambambe, Biopo, and Matala) are operational; the other three were destroyed during the civil war. The construction of a 520-megawatt power station at Capanda on the Cuanza River is under way, although it is uncertain when it will be complete and operational (UNDP/World Bank Energy Sector Management Assistance Program 2001).

The potential for spills along the coast also exists because several oil and gas fields are currently in operation, particularly in the northern portion of this ecoregion, and Luanda is a major import and export terminal and refining center (Fuller and Bush 1999). As the economic and political situation in Angola improves, further development of the oil and gas industry is expected (UNDP/World Bank Energy Sector Management Assistance Program 2001).

Diamond mining has continued during the civil war at reduced levels and is expected to expand now that peace has returned. The major environmental damage resulting from diamond mining is the diversion of rivers to allow mining of alluvial diamond deposits. This practice causes soil erosion, and after the mining is completed, rivers often are not redirected to their original courses. Contamination of surface and ground waters also often occurs.

Justification for Ecoregion Delineation

The fish faunas of the rivers of this ecoregion are poorly known; however, it appears that they have greater affinity with the fauna of the Southern West Coastal Equatorial [29] bioregion than with the Congo or Zambezi bioregions. The Cuanza ecoregion is considered its own distinct bioregion because the Zambezian fauna is absent or poorly represented, and a number of endemic species have been described (Trewavas 1936, 1973; Roberts 1975). The Cichlidae and a number of species from other families in the upper Cuanza River are related to species from the Zambezi and Cunene river basins (Trewavas 1973). The earlier Vernay Angola Expedition (Nichols and Boulton 1927) and Gray African Expedition (Fowler 1930) also recorded a number of species from the upper reaches that are distinctly Zambezian, so the upper Cuanza Basin has been included in the Zambezian Headwaters [76] ecoregion.

Data Quality: Low

This is a critical ecoregion for major collection effort and study because it is very little known scientifically. The faunas and ecological functioning of the aquatic systems in the region are in need of investigation, and a significant amount of further study is needed before the degree of biological distinctiveness can be determined in this area.

Ecoregion Number:	**64**
Ecoregion Name:	Kenyan Coastal Rivers
Major Habitat Type:	Savanna–Dry Forest Rivers
Final Biological Distinctiveness:	Bioregionally Outstanding
Final Conservation Status:	Endangered
Priority Class:	IV
Authors:	Dalmas Oyugi and Michele Thieme

Location and General Description

Draining Mount Kenya and surrounding highlands, the coastal rivers of this ecoregion, with their associated swamps, floodplains, and lakes, host a depauperate fish fauna but a rich avifauna and herpetofauna and include some relict endemic forms of Odonata. This ecoregion encompasses the basins of the Tana and Athi/Galana rivers and those of several other coastal rivers that flow to the Indian Ocean, including the Goshi, Tiwi, Pemba, Umba, and Ramisi rivers. Many of the tributaries drain from Mount Kenya, the Aberdare Ranges, and the northern slopes of Mt. Kilimanjaro. Smaller rivers and streams also drain the coastal margins of eastern Africa, such as from the Shimba Hills and Arabuko-Sokoke forest. As its name suggests, the ecoregion covers Kenya's east coast inland to the walls of the Great Rift Valley, and it also extends marginally into northern Tanzania.

Rainfall is low in the ecoregion and occurs primarily during two rainy seasons. A long rainy season occurs between March and June, and a shorter one occurs between October and December. Much of the rest of the year is dry at lower altitudes, although rain still falls on higher-altitude inland mountains and occasional showers are found along the coast. At low elevations, rainfall varies from about 1,000 mm/year in the north to about 1,300 mm/year near the border with Tanzania (Hughes and Hughes 1992). Higher elevations receive far more precipitation, with a mean annual precipitation of about 2,000 mm on Mount Kenya. Mean air temperatures range from about 25°C to 27°C in the lowlands and decline with altitude; the temperature falls by about 0.2°C for every 1.6 km upstream (van Someren 1952; Hughes and Hughes 1992). The mean annual water temperature in streams varies from 12°C to 30°C.

The Tana River is the largest and longest river in Kenya (about 1,000 km), and the Tana River system drains the southern and eastern slopes of Mount Kenya and the eastern slopes of the Aberdare Ranges. Flooding of the river's lower reaches results from heavy rainfall on these mountains rather than from local rains in the arid lowlands (Wass 1995; Bennun and Njoroge 1999). The river flows through extensive wetlands, and its estuary has one of the best mangrove stands in Kenya (Kairo, Kenya Marine and Fisheries Research Institute, Mombasa, pers. comm., 2002). On its upper course the river flows north, then curves south just below the equator. The Tana's main tributary, the River Sagana, originates from a series of springs among open moorlands on Mount Kenya. As the Tana flows north, it receives the Galole, Hiraman, and Tula tributaries on its left bank. The last cataract on the river occurs at Kora. The river then flows toward Kipini, where it splits into several distributaries and forms a huge delta (Crafter et al. 1992).

The river passes through arid and semi-arid zones, meandering through alluvial floodplains and a thin belt of riverine forest. The forests support a diversity of species, including stands of *Garcinia livingstonei* and *Synsepalum brevipes* and characteristic species such as *Ficus* spp., *Phoenix reclinata, Acacia robusta,* and *Blighia unijugata* (Burgess and Clarke 2000). Oxbow lakes are prevalent along the lower course of the Tana and include lakes Vukoni, Pongi, Pacha, Lango La Shimba, Bilisa, Idsowe, Harakisa, Moa, Shakababo, Kongolola, Kitumbuini, and Dida Warede. The Tana flows into the delta through an artificial channel that was dug for navigation purposes between Belazoni and Ozi about a century ago. Some freshwater flowed through the old Tana River channel until recently, when farmers impounded it to provide water for irrigation. The delta is about 50 km long at its base where it opens into Formosa Bay.

The Athi River, called the Galana and Sabaki in its lower

course, is the second largest river in Kenya. The upper Athi River drains the eastern slopes of the Aberdare Range; the Tsavo River, its main tributary, drains the Kenyan slopes of Mt. Kilimanjaro. From Shakama (80 km upstream from the coast), the river descends about 91 m and becomes sluggish in flow. It is dotted with seasonal pools that dry out between rainy seasons. There are several permanent lakes high up on the floodplain, including Jilore, Merikano, Mekimba, and several smaller ones. These lakes are important breeding and nursery grounds for many fish species. The Sabaki River estuary is short, and the tidal portion is lined with fragmented mangrove stands and lagoons. As with the Tana, rains and snowmelt from upstream, rather than local rains, are the source of flooding. The Koroni swamp and marsh is located to the northeast of the river mouth. The riverine forest that surrounds the Athi River widens toward the sea and joins Arabuko Sokoke forest at Jilore. The bank is lined with macrophytes such as *Phragmites, Cyperus,* and *Saccarum* spp., whereas the small lakes are covered with dense *Nymphaea, Salvinia,* and *Pistia.*

Outstanding or Distinctive Biodiversity Features

The variety of habitats in the ecoregion includes rivers and their tributaries, mangrove forests, estuaries, small lakes, permanent swamps, and seasonal floodplains. Many fish use floodplains and swamps for breeding. *Pistia, Salvinia, Nymphaea,* and other aquatic plant species are common in flooded areas in the lower reaches of the rivers (D. Oyugi, pers. obs., 2001), and the endemic poplar tree (*Populus ilicifolia*) grows on sandy bars and banks in the Ewaso Ng'iro, Tana, and Athi rivers. These habitats and associated vegetative cover form the principal breeding and nursery grounds for many fish species (Whitehead and Greenwood 1959). Dominant fish families include Cyprinidae, Mormyridae, Clariidae, Claroteidae, and Mochokidae.

About fifteen of the fifty known freshwater fish species are endemic to this ecoregion. A few of the characteristic species are *Protopterus amphibius, P. annectens, Neobola fluviatilis, Synodontis serpentis, S. zanzibaricus, S. manni, Nothobranchius willerti,* and *Oreochromis spilurus.* The endemic *Parailia somalensis* is known only from the lower 200 miles of the Tana River, and the endemic subspecies *Petrocephalus catostoma tanensis* is limited to this river below Garissa (140 m a.s.l.) (Whitehead and Greenwood 1959; Whitehead 1962). The catadromous eel *Anguilla bengalensis* has been recorded far upstream (2,700–3,000 m) in the Ragati River in the bamboo zone during the onset of floods (van Someren 1952). There is also a species radiation in the Aplocheilid genus *Nothobranchius* (Lévêque 1997). The mountain headwater streams of Mt. Kenya, the Aberdares, and Taita Hills support a notable assemblage of amphibians, including the endemic species *Phrynobatrachus keniensis, Hyperolius montanus,* and *H. cystocandicans.* The lowland coastal forests support some notable aquatic fauna, particularly a Gondwana relict Odonata species (*Coryphagrion grandis*) that has its nearest relatives in Central and South America (Clausnitzer 2001). The amphibian *Mertensophryne micranotis* is also confined to the lowland forests of Kenya and Tanzania.

The freshwater prawns *Macrobrachium lepidactylus, M. rude, M. scabrinsculum, Caridina nilotica,* and *C. africana* live in the waters of this ecoregion and are caught in the trap fishery on the River Sabaki. The last two species also inhabit Lake Victoria.

About 429 bird species have been recorded at the Tana River Delta, with around 100 of these associated with water. Congregations of pelicans, egrets, storks, spoonbill, sandpipers, and terns occur, with an average of about 70,000 birds present (Bennun and Njoroge 1999). Many of these species breed in the delta (Crafter et al. 1992). The Tana River forests also host several near-threatened bird species, including the Malindi pipit (*Anthus melindae*) and the Basra reed warbler (*Acrocephalus griseldis*) (Bennun and Njoroge 1999). The Tana River cisticola (*Cisticola restrictus*) and the white-winged apalis (*Apalis chariessa*) are also known from the lower Tana Valley, although they have not been found for many years and may have been lost from this ecoregion (Stattersfield et al. 1998; Bennun and Njoroge 1999).

Mammals living in the ecoregion include marsh mongoose (*Atilax paludinosus*), African clawless otter (*Aonyx capensis*), and hippopotamus (*Hippopotamus amphibius*). Two endemic primates are also limited to the forests of the lower Tana River: Tana River mangabey (*Cercocebus galeritus*) (Kinnaird 1992) and Tana red colobus (*Procolobus rufomitratus*) (Kingdon 1997).

Status and Threats

Current Status

The main threats to the ecoregion are an increasing human population, river impoundments, overfishing, and deforestation. Humans have inhabited the coast of eastern Africa continuously since the late Pleistocene (Clarke and Karoma 2000), but populations now number in the tens of millions, with projected growth rates of more than 2 percent (CIA 2000).

Over the last 20 years, the Tana River has been the focus of dam construction in Kenya (Jumbe 1997). Reservoirs behind the Masinga, Kamburu, Gitaru, Kindaruma, and Kiambere dams support a fishing industry that has become increasingly profitable. In these reservoirs, riverine species from the genera *Barbus, Labeo,* and *Mormyrus* have declined drastically from loss of habitat, whereas tilapiine species have doubled in number since impoundment (Jumbe 1997). The dams increase the gross national product through electricity generation and provision of domestic and irrigation waters, but they also cause serious ecological disruptions. The downstream flood regime is altered; sediment transport to the lower Tana, its floodplains, and the delta is impeded; and anadromy and catadromy in some fishes are disrupted (Bennun and Njoroge 1999).

Fisheries in the wetlands, swamps, and rivers of the ecoregion provide an important source of protein and livelihood to local peoples. Additionally, many rivers are fished during the wet season, and then their floodplains are farmed or grazed during the dry season. In the Tana River, fisheries remain an artisanal activity, although many fishers use multifilament gillnets of small mesh sizes (De Vos et al. 2000). The largest fishes caught in the rivers Athi and Tana are *Protopterus annectens, Clarias gariepinus,* and *Clarotes laticeps,* although *M. tenuirostris, S. intermedius, L. gregorii, O. spilurus, Synodontis* spp., *B. zanzibaricus,* and *M. macrolepidotus* are also of economic importance (De Vos 2001).

High rates of deforestation in the ecoregion threaten the health of the ecoregion's rivers. Possible impacts of deforestation include a loss of water supply from forested regions, increased turbidity, greater light exposure to streams, increased flash flooding, and other disturbances. For example, the little Sagana River now carries a large amount of silt as a result of logging upstream on Mount Kenya (van Someren 1952). On the coast, the high pressure on forests comes from the need for wood for housing construction and for fuel. As of 1989, wood made up more than 90 percent of rural east African energy consumption (Clarke and Karoma 2000). Deforestation also occurs to open land for agriculture. Much of Kenya's original forest cover has been lost, but deforestation continues in the remaining patches of forest and may alter the ecosystem and reduce water supply.

A number of exotic aquatic species have been introduced to the eastern coast of Africa, including two snail species, *Physa acuta* and *Marisa cornuarietis,* from North and South America, respectively (Cooper 1996). The Louisiana red swamp crayfish (*Procambrus clarkii*) has also been introduced into East Africa to control schistosome-carrying snails (Cooper 1996). Higher-elevation rivers and streams have been stocked with rainbow trout (*Oncorhynchus mykiss*) or brown trout (*Salmo trutta*) since the beginning of this century. The original aquatic fauna of these systems is unknown.

Pollution from domestic sewage and aquacultural effluents is of concern along the coast of Kenya. Related habitat modification caused by urbanization and population pressures also occurs. Saline intrusion is occurring in some of the coastal rivers because of increased abstraction of groundwater (Institute of Marine Sciences et al. 1998).

Several rivers of this ecoregion are protected. In Kenya, the Tana River Primate National Reserve, Arawale National Reserve, Rahole National Reserve, Bisanadi National Reserve, North Kitui National Reserve, and Kora National Park all protect sections of the Tana River (Sayer et al. 1992). The Ngai Noethya National Reserve, Tsavo East National Reserve, Tsavo West National Reserve, and Chyulu National Park all border the Galana River and its tributaries (Hughes and Hughes 1992). The headwaters of the Tana and Athi rivers are covered by the Mt. Kenya and Aberdares national parks and contiguous forest reserves. The Taita Hills contain a number of forest reserves, and the coastal forests forming the catchment of the smaller coastal rivers are found in the Shimba Hills National Reserve, the small Arabuko-Sokoke National Park, and a larger number of forest reserves and Kaya forests protected as national monuments.

Future Threats

The major threats to this ecoregion are deforestation, impoundment of rivers for electricity generation, and diversion for irrigation schemes. Potential sites for dam development include Usueni, Adamson's Falls, Grand Falls, Kora Hills, and Karura (Tana and Athi Rivers Development Authority 1982). A series of dams constructed in the headwaters of the Tana River threaten the Tana floodplain forest, which needs a minimum frequency and duration of flooding (Bowell et al. 1995).

The excision, or degazettement, of protected forests on Mount Kenya was promoted by some Kenyan legislators in 2001 but was put on hold because of public opposition. The forests of this ecoregion play a crucial role in the functioning of the aquatic systems and in the maintenance of water supply for a large portion of the human population of Kenya (Bussmann 1999). Another protected area, the Tana River Primate Reserve, is threatened by the selective felling of *Ficus* and hunting (Kingdon 1997).

Another serious potential threat is the proposed mining of titanium in Kwale District along the coast. If implemented, the mining could have devastating impacts on the fauna and flora of the rivers and streams, especially the rivers Ramisi and Msambweni.

Justification for Ecoregion Delineation

This ecoregion encompasses the basins of the Tana and Athi/Galana rivers and those of several other coastal rivers that flow to the Indian Ocean. A distinctive fish fauna characterizes the basins of the Tana and Athi rivers, although recent studies show that this fauna may be more similar to that of the Shebelle-Juba catchments than previously thought. For example, the pancake catfish (*Pardiglanis tarabinii*) (Poll et al. 1972) was recently caught in the Tana River (De Vos 2001). Until March 2000, this species was known only from its holotype, which was from the Juba River in Somalia. Further studies should reveal whether these two ecoregions should be merged. The Athi and Tana rivers share only two species (*Clarotes laticeps* and *Mormyrus kannume*) with the Nile systems. However, these east-flowing rivers share affinities with the Lake Victoria fish fauna. There are about ten fish species (including *Petrocephalus catostoma* and *Chiloglanis* sp.) in common, suggesting that the headwaters of the east-flowing rivers may have originally connected with the Victoria Basin before faulting of the Rift Valley (Whitehead and Greenwood 1959; Whitehead 1962). This suggests that the east-west Pliocene hydrographic boundary was initially far to the west of present-day Lake Victoria.

Data Quality: Low

Whitehead (1959) noted that the Tana River is less studied than the Athi River because of its inaccessibility. Aside from a recent inventory undertaken in parts of both rivers by the Ichthyology Department of the National Museums of Kenya, there is little current information. The work of Whitehead (1959, 1960, and 1962) and Whitehead and Greenwood (1959) is several decades old. Lothar Seegers (Dinslaken, Germany) and Luc De Vos (National Museum of Kenya, Nairobi) are completing a checklist of Kenyan freshwater fishes. The Kenya Marine and Fisheries Research Institute has conducted biotic inventories of the creeks and bays of this ecoregion, with particular attention to Mida Creek and Gazi Bay.

Ecoregion Number:	**65**
Ecoregion Name:	**Lower Niger-Benue**
Major Habitat Type:	**Savanna–Dry Forest Rivers**
Biological Distinctiveness:	**Continentally Outstanding**
Conservation Status:	**Critical**
Priority Class:	**II**
Author:	**Ashley Brown**
Reviewers:	**Christian Lévêque and Emmanuel Obot**

Location and General Description

Biannual floods course through the two large rivers, the Niger and Benue, in this largely savanna ecoregion of West Africa. The Lower Niger-Benue ecoregion extends from below Tombouctou (at the northwest end of the Inner Niger Delta [7]) along the course of the Niger River and its tributaries downstream until it reaches the Niger Delta [58]. In this ecoregion, the Niger River stretches for more than 2,000 km through regions as varied as dry, rocky narrows in Mali to rainforest in southern Nigeria (Welcomme 1986b). The Benue River originates in the Adamaoua Massif in northern Cameroon and flows westward for 1,400 km until it meets the Niger River about 450 km above the delta, near the city of Lokoja, Nigeria. The Niger-Benue system flows through Niger, Mali, Burkina Faso, Guinea, Côte d'Ivoire, Benin, and Cameroon.

The climate of this ecoregion is tropical and warm year-round. Mean temperatures in the center of the ecoregion vary from 25°C in the coolest month to 34°C in the warmest month. The mean temperatures further east in Jos, Nigeria, are 20°C and 24°C for the coolest and warmest months. In the Benue headwaters of northern Cameroon, temperatures fluctuate more widely (Hughes and Hughes 1992).

As the lower Niger flows toward the delta, rainfall increases, influencing the surrounding vegetation and the hydrology of the river. Between Tombouctou and Gao, located about 500 km downstream, the Niger River runs through a xeric region with a mean annual rainfall of less than 400 mm/year. The mean annual rainfall increases to 800 mm/year in the middle of the ecoregion and about 700 km downriver of Gao and gradually becomes 1,400 mm per year at approximately 700 km above the delta (Grove 1985). As the river flows toward the southeast, the vegetation becomes lush and the river enters coastal rainforest near Onitsha, Nigeria (less than 100 km above the delta).

No floodplains exist along the 460-km stretch of the Niger between Ayourou and the Mekrou River confluence. Below this confluence and before the Kainji Dam, there are numerous floodplains, with a combined inundated area of about 5,000 km^2 (about 2,000 km^2 during the dry season) (Welcomme 1986b). The diminished extent of floodplains in this ecoregion compared with the Inner Delta [7] is the result of flood attenuation upstream. Swamps border the rivers in places and can be found in old river beds that capture water from groundwater sources and rainfall (John et al. 1993). Floodplain vegetation includes grasses such as *Echinochloa pyramidalis* and *Oryza barthii*. Gallery forest, including *Cola laurifolia* and *Khaya senegalensis*, lines the Niger as it flows through Benin. In the upper portion of the Benue River, about 1,000 km upstream of where the Benue joins the Niger, seasonal floods inundate the floodplains (Hughes and Hughes 1992). The upland vegetation along the Benue River consists of tropical savanna and moist savanna woodland.

The flood regime of the Niger downstream of Niamey, in the middle of the ecoregion, is bimodal. Floods originating in the headwaters of the Niger reach Nigeria in January and February and constitute the "black flood" (Hughes and Hughes 1992). Much of the silt and salt that the river carries out of the highlands is deposited in the inner delta floodplain or absorbed by aquatic vegetation, resulting in clearer water (John et al. 1993). "Black flood" refers to the dark water color derived from the influx of solute-rich water from tributaries downstream of the inner delta (Welcomme 1986b; John et al. 1993). Flooding from overland runoff of precipitation, mainly in the catchments of the Malendo and Sokoto rivers, constitutes the larger "white flood" in August–October (Hughes and Hughes 1992). These rivers are laden with kaolinitic colloids, giving the waters of the flood a milky color (John et al. 1993). The Benue flood, originating in the Adamaoua Mountains and reaching peak flow rates in August–September, reaches the Niger in October (Welcomme 1986b; Hughes and Hughes 1992).

The fish fauna of this ecoregion is typical of the Nilo-Sudan bioregion. Researchers believe that this large savanna river system may have acted as a refuge during periods of climatic drying and during subsequent wet periods was connected to other rivers in this ichthyological province. For example, the Chad Basin is still intermittently connected to the Niger via overflow of the Logone River into the Mayo Kebi, which joins the Benue River.

Outstanding or Distinctive Biodiversity Features

The Lower Niger-Benue contains a rich fish fauna with many species adapted to seasonal flooding. About 202 fish species, including 17 endemics, live in the rivers and streams of this ecoregion. Dominant fish families are Cyprinidae, Mormyridae, Mochokidae, Citharinidae, Aplocheilidae, and Alestiidae. Freshwater fish species known from the lower Niger River and coastal areas ascend into the lower Benue and middle Niger. Species such as *Pantodon buchholzi, Marcusenius abadii, Phago loricatus, Clarias jaensis, C. macromystax,* and *Arius gigas* move upstream (Lévêque et al. 1991).

Two endemic fish genera, *Dagetichthys* and *Dasyatis,* are present in the Niger-Benue ecoregion. The endemic *Dagetichthys lakdoensis,* a benthic, freshwater sole or flatfish, is the only freshwater representative of this family in Africa and lives primarily in the upper Benue (Welcomme 1986a). The freshwater stingray, *Dasyatis garouaensis,* inhabits only the waters of the Lower Niger-Benue and the Niger Delta (Welcomme 1986a; Lévêque et al. 1991).

Several fish species may undertake long-range migrations in this large savanna river. Most longitudinal migrations occur during the wet season, when fish travel to spawning grounds. Anecdotal evidence exists of fish traveling up to 640 km upstream toward the inner delta (Welcomme 1986a). One study found *Brycinus leuciscus* to travel at rates up to 9 km per day as floods receded, for a total distance of 400 km (Welcomme 1979). Several marine fish species, such as *Trachinotus goreensis, Mugil cephalus, Pomadasys jubelini,* and *Cynoglossus senegalensis,* migrate upstream as far as 500 km above the delta, and *Eleotris* spp. have been documented near Lokoja (300 km upstream from the delta) (Welcomme 1986a). The West African manatee, which lives in the Lower Niger, travels up tributaries in the wet season and occasionally becomes stranded as the waters recede (Happold 1987).

Many fish of the Niger River have special adaptations for surviving the anoxic conditions that often occur during the dry season. Lungfish (*Protopterus*) possess lungs for aerial respiration and can also estivate in a mucous cocoon in dry conditions, and other species burrow in the mud or produce drought-resistant eggs (Lowe-McConnell 1985). A representative of the *Polypterus* genus has lunglike modifications of the air bladder that allow it to breathe surface oxygen (Welcomme 1986a). Many species in the genera *Epiplatys, Aphyosemion,* and *Aplocheilichthys* have dorsally oriented mouths and a flattened head, which facilitates breathing from the surface film. Other species have developed arborescent (branching) respiratory organs (*Heterobranchus bidorsalis*), suprabranchial organs (*Ctenopoma kingsleyae*), or vascularized intestines (*Gymnarchus niloticus*) for capturing scarce oxygen (Lévêque 1997).

Among the eighty-eight frog species in the Lower Niger-Benue, sixteen may be endemic to the forests, savanna woodlands, and associated wetlands. A restricted-range savanna species (also found in the Gambia), *Phrynobatrachus francisci,* often is found on tufts of grass mounds in swamps during the wet season and hides from the heat in the damp crevices at the bottom of desiccated savanna pools during the dry season (Rödel 2000).

The Niger River hosts many Palearctic migrants, including high densities of ducks and geese (Anatidae), storks (Ciconiidae), herons (Ardeiidae), and other wading birds (Brouwer et al. 2001b; Wetlands International 2002). For example, the Parc National du "W" in Niger hosts important populations of *Ciconia nigra, Sarkidiornis melanotos, Dendrocygna viduata,* and *Plectropterus gambensis* (Brouwer et al. 2001a). The "W" du Benin National Park hosts *Ardea goliath, A. cinerea, Ciconia abdimii, Plegadis falcinellus,* and *Balearica pavonina* along the course of the Niger River (Cheke 2001). The lower Kaduna–middle Niger floodplain also hosts large colonies of *Merops malimbicus,* with more than 15,000 birds documented (Ezealor 2001).

Status and Threats

Current Status

Dam construction, drought, population growth, and agriculture pose the largest threats to the rivers and streams of the Lower Niger-Benue ecoregion.

The impacts to biodiversity of Kainji Dam, the largest dam on the Niger River, are numerous. Since impoundment in 1966, reduced flows have desiccated about 3,000 km^2 of floodplains, and fish catch has decreased significantly below the dam. Upstream, Kainji Lake has inundated 1,000 km^2 of floodplain (Welcomme 1986b). The dam has interrupted upstream movements of several fish species (*Chrysichthys nigrodigitatus* and *C. auratus*) and changed the species composition both upstream and downstream. Characins, mormyrids, and clariids have decreased downstream of the dam (Welcomme 1986a). Behind the impoundment, populations of clupeids and mochokids have increased, and migratory characins and cyprinids, mormyrids, and other bottom feeders have declined, presumably because of inundation of their benthic habitats by deoxygenated hypolimnetic waters (Welcomme 1986a; Lévêque 1997). Kainji Lake also submerged part of the Swashi River, which was a destination of spawning *Alestes, Brycinus,* and *Hydrocynus* species (Welcomme 1986a).

Numerous other dams erected for irrigation and hydroelectric power have altered the flow of the Niger and Benue rivers. Another major dam, the Lagdo Dam (40 m) in Cameroon, has reduced the flow contribution of the Benue to the Niger significantly; before the dam, the Benue contributed 60 percent of the total flow of the Niger River at the two rivers' confluence (FAO Workshop 2000). Lake Lagdo, formed by the dam, is 50 km long and occupies four Benue tributary valleys (Hughes and Hughes 1992). Water releases from the dam are minimal and occur only in times of heavy rainfall; the hydrology of the downstream floodplain has been significantly altered. Because the river level is so low below the dam, runoff erodes the steep riverbanks. Large-scale irrigation schemes near Lagdo have intensified human occupation of the floodplain, and deforestation on hills surrounding the river intensifies the erosion problem (Slootweg and van Schooten 1995).

The Sahelian droughts over the last several decades have further stressed the freshwater biota of the Lower Niger-Benue ecoregion and the human populations dependent on water and food from the rivers. During these droughts, water flow in the Niger at Niamey was extremely low, dropping to less than 1 m^3/s in 1974. There was no flow at all for several weeks in 1985 (Oyebande and Balogun 1996).

Increasing pressure on the river from grazing and agriculture is also of concern. The Niger and Benue rivers pass through some of the most densely populated parts of Africa, where the vegetation has been modified by centuries of agriculture. Pastoralism based on cattle, sheep, and goats has transformed the vegetation of the xeric area between Tombouctou and Gao. Floating rice cultures occupy large portions of the floodplain throughout the ecoregion (Welcomme 1986b). Timber from gallery forests is harvested along the river in Niger for construction and thatching (Hughes and Hughes 1992).

Fishing along the lower Niger may threaten fish populations in selected portions of the river. In Niger, the river is fished intensively, and in Nigeria illegal fishing practices, erosion, and siltation have depleted inland fisheries (Ita 1993). Frog populations also suffer from habitat loss and poaching (Happold 1987; IUCN 2002).

Several parts of the Niger and Benue rivers and their watersheds are protected. The National Park du "W" is a transboundary park between Benin, Burkina Faso, and Niger containing two Ramsar sites, one in Burkina Faso and one in Niger (Hughes and Hughes 1992; Wetlands International 2002). Three national parks protect Benue wetlands in Cameroon. The Faro National Park includes parts of the Faro and Deo rivers (tributaries to the Benue River), the Benue National Park encompasses a 50-km segment of the Benue River and sections of two of its upper tributaries, and the Bouba-Ndjidah National Park includes parts of the Lida River (120 km) and affluents of the Rei River (Hughes and Hughes 1992).

Future Threats

Water impoundment, population growth, and pollution from agriculture and industry threaten the future integrity of the rivers of this ecoregion. Niger plans to build a hydroelectric dam at Kandadji (100 km upstream of Niamey), which will also be used for irrigation (Pan-African News Agency 2000; FAO Workshop 2000). Mali, Niger, and Burkina Faso are planning a dam at Tossaye, which will irrigate 830 km^2 of land and generate 40 megawatts of energy (FAO Workshop 2000). These impoundment plans almost certainly would further alter the hydrology and ecology of the Niger. An increase in small dams in the headwaters of the Niger also poses a threat because most of these dams are for irrigation, and most irrigation water is never returned to the river channel (Oyebande and Balogun 1996).

The extremely rapid population growth in this ecoregion probably is the largest threat to the limited resources that are already under intense pressure. Nigeria alone has a population of 123 million and a growth rate of 2.67 percent (CIA 2000). Because the Niger and the Benue are both transboundary rivers, water rights are an issue of increasing importance. Currently there is a threat of a drinking water shortage in Niamey, which obtains most of its water from the Niger (Bechler-Carmaux et al. 1999). Continued cooperation efforts between countries in basin management (such as the Niger Basin Authority, formed by the nine riparian countries of the Niger) will be necessary to ensure sustainable use of the natural resources under such intense population pressures (Oyebande and Balogun 1996).

Justification for Ecoregion Delineation

This ecoregion comprises the Benue River Basin and the lower and middle portions of the Niger River Basin below the Inner Niger Delta [7] and above the Niger Delta [58]. In contrast to conditions in the two deltas, the river is characterized by broad channels with numerous sandbanks, and floodplains are limited in this ecoregion. The river flows largely through savanna vegetation, and the species assemblages are typical of the Nilo-Sudanian bioregion (Roberts 1975).

Data Quality: High

Ecoregion Number:	**66**
Ecoregion Name:	**Lower Zambezi**
Major Habitat Type:	**Savanna–Dry Forest Rivers**
Final Biological Distinctiveness:	**Nationally Important**
Final Conservation Status:	**Endangered**
Priority Class:	**IV**
Author:	**Denis Tweddle**
Reviewers:	**Paul Skelton and Brian Marshall**

Location and General Description

From the Cahora Bassa Dam, which forms the upper limit of the ecoregion, the Lower Zambezi River flows southeasterly for 593 km through Mozambique and to the Indian Ocean (Davies 1986). The largest tributary to the Lower Zambezi is the Shire River, which drains Lake Malawi. Below the Shire confluence, the river spreads out over a large floodplain-deltaic system. This ecoregion includes the Zambezi below Cahora Bassa and the Lower Shire River and falls largely in Mozambique, although it extends slightly into Malawi, Zambia, and Zimbabwe. Before the construction of the Kariba and Cahora Bassa Dams in the Middle Zambezi, floodwaters deposited nutrient-rich sediments on the floodplains along the Lower Zambezi River and its delta during each rainy season. Faunistically, the Lower Zambezi is a crossroads, containing elements of the Middle Zambezi, the eastern coastal rivers, and the Malawi region, in addition to brackish and marine species. This dynamic system provided breeding grounds for waterbirds, feeding and spawning areas for a diverse fish fauna, and habitat for mammals. It also helped to sustain major inshore sea fisheries, particularly for prawns.

About 85 percent of the rain in the Lower Shire Valley falls from mid-November to late March (Shire Valley Agricultural Development Project 1975), and the climate is dry from April or May to October. Annual rainfall increases from about 600 mm/year at Cahora Bassa to about 1,400 mm/year in the moderately wet and humid coastal zone. Temperatures are at their lowest in June (mean minimum 13°C and mean maximum 27°C in the Lower Shire Valley) and rise rapidly through September to a mean monthly maximum of 37°C in October before declining slightly with the onset of the rains.

From Cahora Bassa Dam down to Lupata Gorge, 70 km downstream of Tete, the Zambezi River generally follows a clearly defined channel. From below the gorge to the sea, the river is broad, often consisting of many anastomosing channels with shifting sandbanks. The Zambezi Delta starts at Mopeia, some 120 km from the coast, where the Rio Cuacula splits from the main channel and flows to the east toward Quelimane and the main Zambezi flows to the southeast.

Large amounts of sediment that the Zambezi River formerly distributed over its floodplains in the Lower Zambezi ecoregion now accumulate behind Kariba and Cahora Bassa dams, leading to changes in the morphology of the river downstream (Davies et al. 2000a). However, tributaries below the dams, such as the Shire, still bring in large quantities of sediment during the rains. The soils in the area are largely kaolinitic, and inland vegetation is undifferentiated, with *Colophospermum mopane* and *Adansonia digitata* being common (Davies 1986). The alluvial delta region is a mosaic of distinct vegetation types, with thirteen land cover categories recognized (Beilfuss et al. 2000), and extensive mangrove forests are present along the coast (Davies 1986).

Outstanding or Distinctive Biodiversity Features

The Zambezi Delta, with its associated woodlands, savanna, mangroves, and coastal dunes, is a regionally important area for conservation. Its wetlands form a mosaic of various woodland types on unstructured sandy substrates, and the southern portion around Marromeu is an area of major significance for wetland biodiversity, with an extensive area of papyrus, lagoons, aquatic grasslands, and mangroves (Timberlake et al. 2000; Timberlake 2000b). Müller et al. (2000) classified the vegetation in seven groups: forest/woodland, grassland, mangrove, savanna/palm savanna, wetland, dunes/beach, and fallow fields. A total of 445 taxa were sampled in the delta, of which 84 were classified as wetland plants.

Ninety-four fish species have been recorded in the Lower Zambezi (Bills 2000), of which fifty-five are primarily freshwater species, four are catadromous eels, and the remainder are estuarine. Most of the freshwater fishes are floodplain species. Several of these also occur in the Upper Zambezi but are absent from the Middle Zambezi because of a lack of suitable floodplain habitat (e.g., *Barbus haasianus* and two anabantids, *Microctenopoma intermedium* and *Ctenopoma multispine*). A smaller number of fish were recorded by Bills (2000) in the main channel, including rheophilic species such as *Labeo altivelis* and *L. congoro*, the two *Distichodus* spp., *Brycinus imberi*, and *Schilbe intermedius*. Estuarine species such as three gobies, the bull shark (*Carcharhinus leucas*), and the oxeye tarpon (*Megalops cyprinoides*) penetrate far into freshwater. The Lower Shire has sixty-three recorded species (Tweddle and Willoughby 1979), of which five are typical of the Lake Malawi system and have not been found in the rest of the Lower Zambezi. The Ruo River, a tributary of the Lower Shire, has a unique relict fauna above the 60-m-high Zoa Falls (ecoregion [43]). Apart from those in the Ruo, there are no known Lower Zambezi endemics.

The Zambezi Delta supports seventy-three waterbird species, including species of global concern, large breeding colonies of several species, and numerous Palearctic and intra-African migrants (Bento 2000). Species on the global Red List include *Grus carunculatus* and *Rynchops flavirostris* (Bento 2000; IUCN 2002). Thousands of pairs of white pelicans breed in the delta, in addition to large breeding colonies of storks and herons, including *Anastomus lamelligerus*, *Threskiornis aethiopicus*, *Ardea cinerea*, *Ardeola ralloides*, *Platalea alba*, and *Egretta* spp. Waterbird populations probably have decreased in the past 30 years because of the loss of the natural flooding cycle caused by the Kariba and Cahora Bassa dams. Overbank flooding of the extensive mosaic of habitats in the delta now rarely occurs, so the quality and quantity of waterbird habitat have declined (Bento 2000). The Zambezi Delta has been proposed as a wetland of international importance under the Ramsar Convention by the Mozambique government.

A recent survey found nineteen amphibian species in the delta, with another five probably occurring and four others possible (Branch 2000). Branch (2000) recommends further sampling in the delta to confirm the species present and further studies on the conservation status of the amphibians that occur here. Aquatic reptiles using the wetlands include the hinged terrapin (*Pelusios castanoides*), Nile monitor (*Varanus niloticus*), Nile crocodile (*Crocodylus niloticus*), and various frog-eating snakes (Branch 2000).

Dudley (2000) reported eighteen gastropod and three bivalve mollusk species from the delta, but other aquatic invertebrates in the Zambezi system have been inadequately studied (Dudley 2000; Marshall 2000a). The Odonata are better known than most (Pinhey 1981), with twenty-five species recorded from the Marromeu wetland complex in the Lower Zambezi Delta (Kinvig 2000). Marshall (2000a) noted the large gap in knowledge of the Zambezi's invertebrate species and recommended that conservation agencies focus on filling that gap.

Status and Threats

Current Status

Dam construction upstream, extensive clearing for sugar plantations, heavy logging in the forests, a long history of tree cutting along the main river, and regular extensive burning have markedly altered the hydrology of the Lower Zambezi River. The Cahora Bassa Dam has changed the flood regime of the Lower Zambezi such that the river no longer floods as much as it did before. Flood-dependent grasslands have been reduced, the floodplain vegetation zone is being populated with more upland species, and populations of waterfowl have declined. Sandbars used as nesting sites by African skimmers (*Rynchops flavirostris*) and other bird species are no longer formed, and erratic and poorly timed water releases from the dam inhibit nesting of the globally endangered wattled crane (*Bugeranus carunculatus*) (Bento 2000). The coastal zone is eroding because the floodplain is no longer replenished with sediment during seasonal flood events, causing the loss of coastal mangroves, which increases the susceptibility of white pelicans (*Pelecanus onocrotalus*) to disturbance.

River regulation has changed subsistence and commercial agricultural practices, reduced livestock grazing and carrying capacity, degraded floodplain fisheries and nearby inshore sea fisheries, changed natural resource use patterns, and increased the frequency and extent of fires (Beilfuss et al. 2000). The Shire is richer in dissolved ions than the Zambezi, causing its conductivity to rise markedly below the confluence (Hall et al. 1977); this effect probably is greater now that the flow of the Zambezi is reduced. The Lower Shire River is also severely affected by clearance of vegetation for farming: large areas of marsh vegetation are removed or burned each year to grow crops on land exposed by receding flood waters.

The nonnative fish kapenta (*Limnothrissa miodon*) has dispersed into the mainstream Zambezi below Cahora Bassa, and Nile tilapia (*Oreochromis niloticus*) is expected to invade from introductions in the headwaters, including the Middle Zambezi. Aquatic plant infestation is also a threat, although effective biological control agents are available for most species. Water hyacinth (*Eichhornia crassipes*) has invaded the Lower Zambezi ecoregion and caused problems in the Lower Shire by covering important fishing grounds such as the Bangula Lagoon. The South American weevils (*Neochetina bruchi* and *N. eichhorniae*) have been introduced to combat water hyacinth upstream in Malawi and Zimbabwe. An American water fern, *Azolla filiculoides*, covers most of the smaller channels in the delta (Müller et al. 2000) but can be controlled using a weevil, *Stenopelmus rufinasus*.

Beilfuss et al. (2000) recommend that flood releases from Cahora Bassa Dam be scheduled to simulate historic hydrological conditions and possibly rehabilitate the floodplain wetland ecosystem. Emergency flood releases from the dam in 1977 and 1978 led to dramatic improvements in the conditions for flora and fauna on the floodplain, and these were maintained for nearly 2 years. The emergency flood releases from the dam in March 2001 had the opposite effect: they caused major damage and disruption downstream because of an extremely sudden rise in water level and the magnitude of the flood peak. This result appears to illustrate the need to simulate historic conditions by releasing water early rather than waiting for the lake level to rise to maximum levels.

Future Threats

The Lower Zambezi ecoregion as a whole has a low human population density, but this is likely to change. Wildlife populations, already adversely affected by changes in the flood regime and by excessive hunting during the years of political instability and war in Mozambique, would be affected by increased human activities in the delta. The nomination and ratification of the Zambezi Delta as a wetland of international importance should help to counter these threats by creating new opportunities to increase international awareness and ecotourism. This could be particularly effective if linkages were made between the efforts to promote effective management and sustainable use of the Marromeu area with those in the Gorongosa National Park (Beilfuss et al. 2000).

Overfishing is also a threat. Although there is no firm evidence of declining catch rates, there is a distinct trend toward the use of smaller-meshed nets in the Lower Shire River (Tweddle et al. 1995). This is generally a sign of a fishery under stress, as excessive netting depletes larger fish species and fishers turn their attention to smaller species and juveniles.

Dam construction is a continuing threat, and proposals have been made for a number of dams and barrages on the Zambezi below Cahora Bassa. Preliminary feasibility studies are under way for the Mepanda Uncua Dam, about 60 km upstream of Tete, which would drown the last remaining section of the Cahora Bassa Gorge. Other dams have been proposed for Boroma and the Lupata Gorge, as have low-level barrages further downstream. These would change the nature of the river completely and aggravate the problems already being experienced in the lower floodplain and delta systems.

Justification for Ecoregion Delineation

The Lower Zambezi ecoregion begins at the base of the Cahora Bassa rapids and extends downstream to include the delta. From the Lupata Gorge downstream, the river is broad and often has anastomosing channels with shifting sandbanks (Timberlake 2000a). The origins of the Zambezi system are complex, but it is likely that the Upper Zambezi was once separate and probably joined the Limpopo (Skelton 1994). It was captured by the current Lower Zambezi during the mid-Pleistocene when the middle Zambezi cut back through the Batoka Gorge, a long (more than 100 km) east-west fissure through basalt, at the head of which is Victoria Falls, where erosion through similar limestone-filled fissures continues (Davies 1986). Victoria Falls is the western boundary of the Middle Zambezi system, separating it from the Upper Zambezi. Evidence from the distributions of fish (Jackson 1986) and other freshwater animals such as the Odonata (Pinhey 1978) support this hypothesis about the Zambezi system. A striking example is the occurrence of nine *Synodontis* species, seven of which are found only in the Upper Zambezi and the other two only in the middle and lower sections. The fish fauna of the Lower Zambezi has several Zambian and Congo elements not found in the Upper Zambezi, such as two *Distichodus* spp., *Mormyrops anguilloides*, *Mormyrus longirostris*, *Heterobranchus longifilis*, *Malapterurus shirensis*, and *Protopterus annectens*. Such species may have reached the Lower Zambezi partly through a northern link and partly by movement through the Middle Zambezi. The Lower Zambezi fauna contains all the elements of the Middle Zambezi, some east coast species typically found in floodplains, some species characteristic of the Malawi region, and some marine species (Tweddle and Willoughby 1979; Skelton 1994; Marshall 2000b). This ecoregion is also distinguished by the Zambezi Delta, with its convergence of freshwater and marine fauna, extensive mangroves, and floodplain habitats.

Data Quality: Low

Research in the Lower Zambezi has been very limited. After the major surveys of Peters in the 1840s (Peters 1868), few investigations were undertaken until recently. The Lower Shire River began to receive attention from the early 1970s because of its fishery importance (Tweddle et al. 1995). The 1999 survey of the Zambezi system, edited and compiled by Timberlake (2000a), represents a major step forward in our understanding of the Zambezi system. Further research is needed in all aspects of the ecoregion's biodiversity and conservation, and recommendations are made by the various authors in Timberlake (2000a).

Ecoregion Number: **67**
Ecoregion Name: Madagascar Northwestern Basins
Major Habitat Type: Savanna–Dry Forest Rivers
Final Biological Distinctiveness: Globally Outstanding
Final Conservation Status: Endangered
Priority Class: I
Author: John S. Sparks

Location and General Description

A variety of freshwater habitats occur in northwestern Madagascar, ranging from large rivers running through deciduous forest to steep and swift-flowing rivers draining the mountains and floodplain lakes. This ecoregion includes all west-flowing drainages from the northern tip of Madagascar (near Antsiranana) to and including the Mahavavy du Sud drainage basin, located to the southwest of Mahajunga. The satellite island of Nosy Be is also part of this ecoregion.

The northwestern basins of Madagascar, especially those draining the Tsaratanana Massif, receive more annual rainfall (more than 2,500 mm annually [Aldegheri 1972]) and are more diverse geomorphologically and biologically than the drier drainages of the Madagascar Western Basins [68] and Southern Basins [87] ecoregions. Compared with western rivers, those in the northwest are also less subject to large seasonal fluctuations in flow (Aldegheri 1972; Donque 1972). On average, northwestern and western basins have a higher pH (nearly neutral vs. mildly acidic), lower conductivity, higher temperatures, and significantly higher levels of dissolved carbonates (i.e., alkalinity) than eastern rivers (Riseng 1997). The oligotrophic floodplain lakes in this ecoregion exhibit higher conductivity than lakes in eastern Madagascar (Riseng 1997). From north to south, the major river basins in this ecoregion are the Mananjeba, North Mahavavy (Mahavavy du Nord), Sambirano, Ankofia, Sofia, Anjobony, Mahajamba, Betsiboka, and South Mahavavy (Mahavavy du Sud). Major lake basins are Lake Kinkony (Madagascar's second largest lake), Lake Andrapongy, and the lakes of the Sarodrano region. All of these oligotrophic floodplain lakes are extremely shallow and turbid.

Outstanding or Distinctive Biodiversity Features

The rivers and floodplain lakes of this ecoregion support rich and highly endemic freshwater fish faunas. A recent review found that seventy-one native freshwater fish have been recorded from northwestern rivers and lakes (Sparks and Stiassny 2003). Given these new data, the Northwestern Basins ecoregion is the most species-rich Malagasy ecoregion for freshwater fish and also contains the highest number of endemics (twenty-six species) (Sparks and Stiassny 2003). It is likely that fish diversity is still substantially underestimated, given that a majority of the basins draining the Tsaratanana Massif, as well as headwaters and upper-middle reaches of most of the ecoregion's major rivers, remain poorly surveyed.

Freshwater habitats include *tsingy* (karst) formations of the Ankarana reserve in the far north; steep, rocky clearwater rivers and streams draining the Tsaratanana Massif and Montagne d'Ambre; large, shallow oligotrophic floodplain lakes; large perennial rivers and their tributaries; and numerous crater lakes on the island of Nosy Be. A number of rivers and their tributaries in this ecoregion, most notably the Mahavavy du Nord, Sambirano, Ankofia-Anjingo, and Mangarahara-Amboaboa (Sofia tributary), have high habitat diversity for Madagascar and therefore contain rich and highly endemic freshwater fish faunas. For example, all but one species of *Paretroplus*, Madagascar's most diverse cichlid assemblage, have localized distributions limited to northwestern riverine and lacustrine habitats (Sparks 2002). These distributions often comprise only a single basin, few of which are currently protected. Thus, even localized disturbances could encompass the entire range of a species. In addition, northwestern Madagascar is home to a small species assemblage of killifish of the genus *Pachypanchax;* members of the endemic cichlid genus *Ptychochromis,* including at least two undescribed species; and distinct inland populations of an undescribed ariid catfish. The endemic atherinid genus *Teramulus* is known only from two extant populations, both located in isolated and undisturbed forested sections of this ecoregion.

Ichthyofaunal communities on the satellite island of Nosy Be are similar to those of the main island. This similarity makes sense, given the hypothesized recent connection of the island to the mainland. The small crater lakes on Nosy Be remain mostly undisturbed and are home to a number of Malagasy endemics, including members of the cichlid genera *Ptychochromis* and *Paratilapia*.

In addition to freshwater fishes, many of the larger lakes in this ecoregion support breeding pairs of the rare and highly endangered Madagascar fish eagle (*Haliaeetus vociferoides*) (Peregrine Fund data; pers. obs., 1994, 1996). The endemic freshwater turtle *Erymnochelys madagascariensis* is also found in the larger basins in the southern portion of this ecoregion (Glaw and Vences 1994). Healthy populations of the Nile crocodile, *Crocodylus niloticus* (considered by some to represent a distinct subspecies, *C. n. madagascariensis*), can still be found in northwestern Madagascar, including the crater lakes of Nosy Be and the subterranean rivers of Ankarana in the far north (Glaw and Vences 1994). The crocodiles in the Ankarana caves are considered to be the only cave-dwelling populations of crocodiles in the world. Additionally, the ecoregion probably supports high numbers of aquatic insects (Benstead et al. 2003).

Status and Threats

Major threats to freshwater communities in this ecoregion stem from deforestation, overexploitation, and the spread of numerous exotic species. For example, Lake Kinkony is subject to extreme fishing pressure, which has nearly extirpated the two species of endemic *Paretroplus* cichlids that inhabit the lake (Sparks 2002; P. Loiselle, pers. comm., 2002). In the Sarodrano lakes, an introduced predator, the Asian snakehead (*Channa* sp.), is considered to be the primary cause of the disappearance of another endemic cichlid, *Paretroplus menarambo* (Benstead et al. 2003). Unfortunately, this exotic taxon, which can tolerate extremely poor water conditions, is now widespread not only throughout this ecoregion but also in many other parts of Madagascar and threatens the continued survival of Madagascar's native fish species. Conversion of habitat for agriculture (rice and sugar cane) is a major problem in many lake basins. For example, a substantial portion of Lake Andrapongy is now devoted entirely to rice production.

Justification for Ecoregion Delineation

This ecoregion is defined by the basins in northwestern Madagascar, with an annual average of more than 1,400 mm of rain (FAO and International Institute for Applied Systems Analysis 2000). Compared with the dry western and southern basins, the wetter northwestern basins are substantially more diverse in habitat and are not as influenced by vast seasonal fluctuations in flow rate that could easily wipe out existing fish populations during periods of complete desiccation.

Data Quality: Medium

Following a number of recent ichthyofaunal surveys throughout northwestern Madagascar, focused primarily on endemic cichlids, the northwestern basins are now recognized as supporting diverse communities of native freshwater fishes. Our knowledge base in this ecoregion is adequate for isolated basins (e.g., many of the larger lakes, including Kinkony, Andrapongy, and those of the Sarodrano region, and portions of some rivers, including the Amboaboa, Ankofia, and Betsiboka) and extremely poor for others, especially the headwaters and upper to middle reaches of most rivers, including all those draining the Tsaratanana Massif. The middle to lower reaches of many larger rivers in the ecoregion are somewhat better studied. Overall, survey coverage in northwestern Madagascar is patchy, and with each new expedition a number of species new to science are encountered. Poorly surveyed and undisturbed habitats in this ecoregion should be a focus of future research efforts.

Ecoregion Number: **68**
Ecoregion Name: **Madagascar Western Basins**
Major Habitat Type: **Savanna–Dry Forest Rivers**
Final Biological Distinctiveness: **Continentally Outstanding**
Final Conservation Status: **Vulnerable**
Priority Class: **II**
Author: **John S. Sparks**

Location and General Description

This ecoregion includes the basins of the Manambolo, Tsiribihina, Mangoky, and Onilahy in western Madagascar. In contrast to the extremely wet eastern forests, western Madagascar is much drier, with an average annual rainfall of only about 20–25 percent that of the eastern slopes (Donque 1972). Deciduous forest is present in the wetter parts of this ecoregion, but drought-tolerant vegetation, such as Didiereaceae and Euphorbia thickets, dominates the landscape (Lowry et al. 1997). Western rivers are long, generally slow-flowing, and subject to seasonal fluctuations in water level and flow. Tidal influence is significant in the lower reaches of these watercourses (Kiener and Richard-Vindard 1972). Many of western Madagascar's smaller river basins are dry from April to November (Aldegheri 1972).

Fish communities contain numerous intrusive marine species, many of which migrate far inland (Kiener 1963). Brenon (1972) hypothesized that these incursions are largely the result of an expansive continental shelf and the presence of numerous coral reefs containing substantial fish populations. Marine taxa probably have colonized these streams subsequent to drought-induced extinctions of strictly freshwater taxa. The pattern of seasonal desiccation, then, may account for the depauperate assemblages of freshwater fishes inhabiting western basins and the relative dominance of marine species.

Outstanding or Distinctive Biodiversity Features

About fifty freshwater fish, including some endemics (three described), are known from this ecoregion. The larger and undisturbed rivers in western Madagascar (e.g., those adjacent to the Parc National de Isalo) historically supported extremely rare and localized freshwater fish assemblages, including members of the endemic genera *Ptychochromoides*, *Ptychochromis*, and *Ancharius* (undescribed species). Although isolated populations of *Ptychochromoides* and *Ancharius* survive in remote sections of this ecoregion where disturbance is limited, members of *Ptychochromis* have not been collected in western Madagascar for decades. Ichthyofaunal communities of the Isalo region (southern-central Madagascar) are remarkably similar to those encountered in the southeastern highlands (e.g., upper and middle reaches of the Mananara Basin). For example, closely related members of the catfish genus *Ancharius* and cichlid genus *Ptychochromoides* are present in both hydrological regions, which suggests that southwestern drainages of the Western Basins ecoregion and southeastern drainages of the Madagascar Eastern Highlands [41] were recently in contact (J. Sparks, unpublished data).

Distinctive freshwater habitats in this region include *tsingy* or karst formations characterized by fissures, subterranean streams, sinkholes, and caverns produced by erosion, located along the western coast, and isolated larger rivers in the central highlands with distinct assemblages of fish species. Lac Itasy, located to the east of the capital, Antananarivo, in the western central highlands, once supported a diverse assemblage of native fishes, although exotics have almost entirely replaced the native species.

In addition to freshwater fishes, the larger wetlands and lakes in this ecoregion provide habitat for the rare and highly endangered Madagascar fish eagle (*Haliaeetus vociferoides*) (Langrand 1990). The endemic big-headed turtle, *Erymnochelys madagascariensis*, the only member of the subfamily Podocneminae to occur in the Old World, is widespread but uncommon (because of exploitation as a food source) in the larger lakes and rivers of this ecoregion and in the Madagascar Northwestern Basins [67] ecoregion (Kuchling 1988; Glaw and Vences 1994).

Status and Threats

Many of the rivers of this ecoregion appear to be fairly intact. Population density is low in this part of Madagascar, and major threats are overfishing and localized habitat destruction. Because of the limited distributions of many of the endemic species, including the only extant populations of the cichlid *Ptychochromoides betsileanus*, an undescribed species of anchariid catfish, and the cave-dwelling eleotrid *Typhleotris pauliani*, even limited habitat destruction could cause species extinction.

Justification for Ecoregion Delineation

This ecoregion is defined by the basins in western Madagascar with an annual average of less than 1,400 mm rainfall (FAO and International Institute for Applied Systems Analysis 2000). Compared with the wetter northwestern basins, the dry western basins are substantially more influenced by vast seasonal fluctuations in flow rate that could easily wipe out existing fish populations during periods of complete desiccation.

Data Quality: Low

To date a comprehensive ichthyological survey of the western basins has not been completed, so comprehensive data are lacking. Many of the western basins are extremely difficult to access because roads are few and in poor condition.

Ecoregion Number: **69**
Ecoregion Name: Middle Zambezi Luangwa
Major Habitat Type: Savanna–Dry Forest Rivers
Final Biological Distinctiveness: Nationally Important
Final Conservation Status: Endangered
Priority Class: IV
Author: Helen Dallas
Reviewers: Paul Skelton and Brian Marshall

Location and General Description

The name of this ecoregion comes from the two main rivers that define it: the Middle Zambezi and the Luangwa. The Middle Zambezi flows through a series of narrow gorges and fault-defined valleys and has been extensively modified by two hydroelectric dams, which created two large reservoirs (lakes Kariba and Cahora Bassa) that have drowned about 60 percent of the length of the main river. Floodplains are limited, there are no extensive wetlands, and the ecology of the river is dominated by the regulating effects of the dams (Timberlake 1997). In contrast, the Luangwa River, a major tributary of the Zambezi, is unregulated and pristine. It runs in a southwesterly direction down the Luangwa Valley, an extension of the East African Rift Valley system, and drains much of eastern Zambia. In the dry season it is a slow-flowing, meandering river confined to the main channel and winding its way between shallow sandbanks. In the rainy season, the entire river bed, several kilometers wide in places, is inundated; water fills the oxbow lagoons and dambos (shallow, seasonally or permanently waterlogged, grass-covered depressions) and floods large areas of grassland. The ecoregion covers parts of Zambia, northern Zimbabwe, and western Mozambique; it extends from Victoria Falls to the Cahora Bassa Gorge on the Zambezi and from the headwaters of the Luangwa River to its confluence with the Zambezi.

This ecoregion experiences three distinct seasons: a cool dry season from May to August, a hot dry season from September to November, and a warm wet season from December to April. Most of the precipitation falls between November and March. Average rainfall ranges from 700 mm per year in the south along the Zambezi River to approximately 1,000 mm in the upper Luangwa Valley. In the hot season, daytime maximum temperatures average 27–38°C (Jachmann 2000).

The damming of the Zambezi River led to the creation of lakes Kariba (5,364 km^2) and Cahora Bassa (2,670 km^2) and changed the character of the Middle Zambezi. The regulation of river flow, a reduction in flood volume (Magadza 2000), and the deposition of sediments in the reservoirs have changed downstream habitats, whereas the creation of extensive new limnetic habitats has changed the composition of the fish populations (Timberlake 1997; Davies et al. 2000a). Some specialized rheophilic forms such as the cyprinids *Opsaridium zambezense* and *Barbus marequensis* have disappeared completely from the area flooded by Lake Kariba, whereas others, such as the labeos and distichodids that were formerly the dominant species in the river, have declined (Marshall 2000a). Aquatic weeds, particularly water hyacinth (*Eichhornia crassipes*), water fern (*Salvinia molesta*), and water lettuce (*Pistia stratiotes*), proliferated soon after the lakes were created, but their populations decreased as the lakes matured, and they are now only of nuisance value. A small area of alluvial terraces exists in the Mana Pools region between the two dams (Hughes and Hughes 1992) and apparently is subject to increased lateral erosion as a result of the lack of sediment being carried to it by the Zambezi River (Guy 1980–1981).

The Luangwa River, with a catchment area of 165,000 km^2, experiences flash flooding (Kasimona and Makwaya 1995), with the river rising rapidly to full flood during heavy rains. Important riverine habitats include oxbow lagoons, dambos, and the riparian fringe, the latter providing important habitat for terrestrial species. Seasonal flooding maintains this diversity of habitats. The river is very turbid during this period, and the clay content of its water apparently is the cause of the very high clay content of the water in Lake Cahora Bassa, into which it discharges.

Both the middle Zambezi and Luangwa valleys are fertile compared with the upper Zambezi (which consists mostly of nutrient-poor Kalahari sands) because they drain a geologically complex area with ancient volcanic rocks, metamorphosed sediments, and recent sedimentary systems as well as alluvial soils (Jachmann 2000). The natural vegetation gradually shifts from pure stands of the tree *Colophospermum mopane* that dominate the valley floor to miombo woodlands that stretch up the escarpments onto the plateaus to the north and south. Riparian forests line the rivers, whereas open savannas, with *Faidherbia albida* being an important and conspicuous species, dominate the vegetation on the alluvial soils (Wild and Barboza 1967; Jachmann 2000).

Outstanding or Distinctive Biodiversity Features

The Luangwa is one of the few remaining unregulated river systems in southern Africa. The ecoregion as a whole supports four aquatic mammals: the endangered marsh mongoose (*Herpestes palustris*), the African clawless otter (*Aonyx capensis*), the spotted-necked otter (*Lutra maculicollis*), and the hippopotamus (*Hippopotamus amphibius*). Sixty-one fish species have been recorded, one of which is endemic (*Oreochromis mortimeri*) (Marshall 2000a). Most of the fish of this ecoregion are widespread, although several are limited in distribution to the Zambezi and adjacent rivers (e.g., *Opsaridium zambezense*, *Synodontis zambezensis*, and *Pharyngochromis acuticeps*). Cichlids, cyprinids, and mormyrids dominate the largely riverine fish fauna. Several larger species such as the vundu (*Heterobranchus longifilis*), electric catfish (*Malapterurus shirensis*), nkupe (*Distichodus mossambicus*), and chessa (*Distichodus schenga*) are found in the Zambezi system only below Victoria Falls. Additionally, there are forty-nine amphibian species; four reptiles, including large populations of Nile crocodiles (*Crocodylus niloticus*); and twenty-seven molluscan species (Dudley 2000), one of which, *Gabbiella zambica*, is endemic.

Information on aquatic invertebrates is very limited, although a recent biodiversity study of the Odonata recorded eighteen zygopteran (damselfly) species and fifty-two anisopteran (dragonfly) species (FitzPatrick 2000). One species of primitive dragonfly, *Archaeophlebia victoriae*, has been recorded from the Victoria Falls region (Timberlake 1997), and three wetland butterfly species, *Catacroptera cloanthe*, *Leptosia alcesta*, and *Leptotes pulcher*, have been recorded from the Mana Pools area (Gardiner 2000).

Alluvial systems such as those around Mana Pools support a rich fauna. For example, the Mana Pools National Park hosts about 40 fish species, a healthy population of crocodiles, and more than 380 bird species (Hughes and Hughes 1992). These riparian habitats provide an important staging area for migratory waterbirds and are especially important for Lilian's lovebird (*Agapornis lilianae*), which is potentially threatened by the cage bird trade (Irwin 1981).

Extensive populations of crocodiles and hippopotamus occur in the rivers and lagoons of the Luangwa River Basin. Two mammal subspecies, Thornicroft's giraffe (*Giraffa camelopardalis thornicrofti*) and Cookson's wildebeest (*Connochaetes taurinus cooksoni*) (Berry 1973b; Jachmann 2000), occur in the geographically isolated Luangwa Valley, as does the puku (*Kobus vardonii*), a water-dependent antelope belonging to the tribe Reduncini, found only in Africa (Cotterill 2000).

About 700 bird species have been recorded in the Luangwa Valley (Scott 1993). Many waterbird species feed in the receding waters of the river and oxbow lagoons. These include yellow-billed storks (*Mycteria ibis*), saddle-billed storks (*Ephippiorhynchus senegalensis*), openbills (*Anastomus lamelligerus*), white pelicans (*Pelecanus onocrotalus*), great white egrets (*Egretta alba*), and goliath herons (*Ardea goliath*). Southern crowned cranes (*Balearica regulorum*) congregate in large flocks at the salt pans, and tens of thousands of knob-billed (*Sarkidiornis melanotos*) and white-faced ducks (*Dendrocygna viduata*) breed in the flooded mopane woodlands (Dowsett 1971). The region provides excellent feeding opportunities in the warm rainy season for Palearctic migrants and intra-African migrants such as the red-chested cuckoo (*Cuculus solitarius*), the European swallow (*Hirundo rustica*), swifts (*Apus* spp.), hobbies (*Falco cuvierii* and *F. subbuteo*), and bee-eaters (*Merops* spp.), as well as birds of prey such as the steppe eagle (*Aquila nipalensis*) and steppe buzzard (*Buteo buteo*). Southern carmine bee-eaters (*Merops nubicoides*) nest in the steep sandy banks of the river, and the valley is a major wintering ground for the eastern population of the white stork (*Ciconia ciconia*) (Stuart et al. 1990; Leonard 2001).

Status and Threats

The creation of lakes Kariba and Cahora Bassa transformed much of the Middle Zambezi River from a lotic (running water) environment into an extensive lentic (standing water) one. Littoral and benthic habitats, once limited in extent, are now widespread (Timberlake 1997). Water releases from Kariba differ from the natural hydrological regime, notably by reducing peak floods, and this has allowed aquatic plants to grow in areas where they could not have established themselves before. In particular, dense beds of *Phragmites* now grow on previously bare sandbanks. These were the principal sites of the African skimmer (*Rynchops flavirostris*), and this may have contributed to their decline in the area, although the loss of nesting habitat drowned by the two reservoirs probably is more important. The rocky gorges between the Victoria Falls and Lake Kariba support the largest population of rock pratincoles (*Glareola nuchalis*) in southern Africa, and proposed new dams on this section of the river would eliminate them. The Luangwa River is largely unmodified, and much of the Luangwa Valley is formally protected as national parks and game management areas.

Hydroelectric development, water abstraction, water pollution, and uncontrolled exploitation of lakeshore resources are the main threats to the Middle Zambezi River (du Toit 1994). Proposed dams at Batoka Gorge 50 km downstream of Victoria Falls, and at Devil's Gorge at the head of Lake Kariba, would have a major impact by drowning the last fast-flowing, rocky stretch of water in the system. Many animal species, not all of them fish, would be affected by this loss. Below Lake Kariba, the Zambezi is a regulated river, causing constant shifts in streambed morphology through erosion, with possibly detrimental effects on crocodile habitat (Magadza 2000). The discharges from the Kariba Dam, through the turbines and less often by spillage, do not correspond with what are thought to be the normal fish breeding seasons in the Mana Pools area (Kenmuir 1976, cited by Magadza 2000), although the extent to which fish breeding has been affected is not known. Intense fishing in the Zambezi since the end of the war also appears to have affected some populations in this ecoregion.

There are no immediate major threats to the Luangwa Valley except for the spread of the exotic Nile tilapia (*Oreochromis niloticus*), which is invading from the Middle Zambezi, where it is now widespread. This species poses a serious threat to the endemic *Oreochromis mortimeri*. An increase in illegal netting in the seasonally flooded lagoons seems to have reduced the numbers of certain bird species, such as white pelicans.

Justification for Ecoregion Delineation

The origins of the Zambezi system are complex, but it is likely that the Upper Zambezi was once separate and probably joined the Limpopo (Skelton 1994). It was captured during the mid-Pleistocene, when the Middle Zambezi cut back through the Batoka Gorge, a long (more than 100 km), east-west fissure through basalt, at the head of which is Victoria Falls, where erosion through similar limestone-filled fissures continues (Davies 1986). Victoria Falls is the western boundary of the Middle Zambezi system, separating it from the Upper Zambezi.

Evidence from the distributions of fish (Jackson 1986) and other freshwater animals such as the Odonata (Pinhey 1978) support this hypothesis about the Zambezi system. A striking example is the occurrence of nine *Synodontis* species, seven of which are found only in the Upper Zambezi and the other two only in the middle and lower sections. The fish fauna of the Middle Zambezi has several Zambian/Congo elements not found in the Upper Zambezi, such as two *Distichodus* species, *Mormyrops anguilloides*, *Mormyrus longirostris*, *Heterobranchus longifilis*, *Malapterurus shirensis*, and *Protopterus annectens*. In addition, the Middle Zambezi has some east coast species, suggesting that the Middle Zambezi formed part of an eastern drainage basin that included the Lower Zambezi, Lower Kafue, Luangwa, and Shire rivers (Bell-Cross and Minshull 1988).

Data Quality: Medium

Most scientific investigations carried out in this area have been related to lakes Kariba and Cahora Bassa; the former is particularly well studied (see bibliography in Timberlake 1997, 2000a). Almost nothing has been done on fish or other aquatic organisms in the riverine sections of the mid-Zambezi, either above or below Kariba, or in the Luangwa. Limited studies on the Luangwa River and its associated floodplains have focused largely on larger mammals (Ansell 1965; Berry 1973a, 1973b) and birds (Feely 1964; Dowsett 1971; Scott 1993). Dunham (1989, 1990, 1991a, 1991b, 1994) has worked on the ecology of the Mana Pools area.

Ecoregion Number:	**70**
Ecoregion Name:	**Pangani**
Major Habitat Type:	**Savanna–Dry Forest Rivers**
Final Biological Distinctiveness:	**Bioregionally Outstanding**
Final Conservation Status:	**Endangered**
Priority Class:	**IV**
Authors:	**Dalmas Oyugi and Michele Thieme**
Reviewer:	**Julius Sarmett**

Location and General Description

This ecoregion covers primarily the Pangani River Basin, which drains mounts Meru and Kilimanjaro, and the North and South Pare and West Usambara Mountains. Several small coastal basins flowing into the Indian Ocean are also found in this ecoregion, particularly the rivers draining the East Usambara Mountains to the sea at Tanga. The ecoregion extends from just north of the Tanzania-Kenya border to south of the Pangani River mouth.

Rainfall varies substantially within the ecoregion. The lowlands receive an average of only about 500 mm of precipitation a year. The slopes of the mountains receive 1,200–2,000 mm/year of rain, and at the highest elevations of Meru and Kilimanjaro precipitation declines to 200 mm/year (as snow) (Hughes and Hughes 1992). Rainfall is strongly seasonal, with a long rainy season from March to May, when the Intertropical Convergence Zone (ITCZ) passes north, and a short rainy season between November and December with the southward passage of the ITCZ. The dry season lasts from July to October, although there can be showers in coastal areas and on the mountains during this period, and mist precipitation forms from clouds in the mountain forests.

The catchment of the Pangani River covers about 42,000 km^2 and comprises most of the ecoregion. The river flows southeast from the slopes of mounts Meru and Kilimanjaro and from the North and South Pare and West Usambara Mountain blocks of the Eastern Arc Mountains before draining into the Indian Ocean at Pangani (Dadzie et al. 1988). Major tributaries of the Pangani include the Ruvu, Kiluletwa, and Mkomazi rivers. Lake Jipe is a shallow lake located on the Tanzania-Kenya border at 37°40′E, 3°40′S (to the east of Pare Mountains in Tanzania). It is 19 km long, 2 km wide, and only a few meters deep (Dadzie et al. 1988). It is fed by the River Lumi through a swamp at the north, with the water originating on Mount Kilimanjaro and the Pare Mountains. Water leaves the lake through the Ruvu River at its northwestern end. A second lake, Chala, lies in a crater 19 km north of Lake Jipe and on the flank of Mount Kilimanjaro. It has steep, rocky shores and deep waters. The Kikuletwa River begins on Mount Meru in the northwest and joins with streams from Mount Kilimanjaro before connecting with the Ruvu and becoming the Pangani (Baker and Baker 2001). The Pangani River is dotted with some small cataracts, the most outstanding of which is Soni Falls.

The Umba, Sigi (Tanga), Mkulumuzi, and Msangazi rivers also drain this ecoregion into the Indian Ocean, flowing primarily from the East and West Usambara mountains. These catchments are also included administratively in the Pangani Basin, such that its administrative area is 56,300 km^2. The Umba eventually reaches the Indian Ocean at the town of Vanga in Kenya, and the Sigi flows into the sea north of the town of Tanga in Tanzania. Swamps such as those above the Kalimawe Dam also occur in the ecoregion.

The terrestrial vegetation in this ecoregion is a mosaic of different types depending on altitude and availability of water. Vegetation types include alpine deserts and montane moorlands on the high mountains, montane and submontane forest, lowland coastal dry forest, coastal scrub, *Acacia* and *Brachystegia* woodland, riparian and swamp forests, floodplain vegetation, and mangrove forests (Hughes and Hughes 1992; Burgess and Clarke 2000). Although many plant species are endemic to the forests embedded in this ecoregion, the swamps and riparian forests generally are dominated by widespread species (e.g., *Cyperus papyrus, Milicia excelsa, Antiaris toxicaria, Phragmites* spp., and *Ficus* spp.) (Burgess and Clarke 2000). Grasses such as *Echinochloa pyramidalis, Cynodon dactylon,* and *Oryza* spp. dominate many of the floodplains of the ecoregion. Dry woodland composed of *Acacia* and *Combretum* also grows on the floodplain of the Pangani. Mangrove forests line river mouths throughout the ecoregion (Hughes and Hughes 1992; Kemp et al. 2000).

Outstanding or Distinctive Biodiversity Features

A depauperate yet distinctive freshwater fish fauna inhabits this ecoregion. About one-third of the approximately thirty-five known freshwater fish species are endemic. Endemic species include *Barbus venustus, Rhabdalestes tangensis, Ctenochromis pectoralis, Oreochromis hunteri, O. jipe,* and *O. pangani. O. hunteri* is limited to Lake Chala; the juveniles of this species feed on algae and debris between boulders and occur together with crabs such as *Potamon platycentron* (Lowe 1955). Although *Garra dembeensis* is not endemic to the ecoregion, the Pangani River is the southern limit of its distribution. Cyprinids are the best-represented element of the fish fauna, although cichlids, anguillids, rivulines, mochokids, and other families are also present. Crustaceans such as *Macrobrachium idyella* and *M. rude* have also been reported to occur in the Pangani River Basin.

About fifteen fish species occur in Lake Jipe, including the endemic *O. jipe*. Other species reported from Jipe include *Rhabdalestes tangensis, Barbus paludinosus, Aplocheilichthys* spp., *Haplochromis* spp., *Astatotilapia bloyeti*, and *Clarias gariepinus*. Cooler water temperatures limit the distribution of some species to higher altitudes in the ecoregion.

A rich aquatic-associated amphibian fauna is known from this ecoregion, with several species endemic to the montane regions. For example, *Hyperolius tanneri* is known only from a swamp at 1,410 m a.s.l. in the western Usambaras, *Phrynobatrachus keniensis* occurs in upland meadows of Kenya and Mount Meru, and *Parhoplophryne usambarica* and *Phrynobatrachus krefftii* are limited to the Usambara Mountains.

The coastal portion of the ecoregion contains parts of two endemic bird areas (Coastal Forests and Eastern Arc Mountains) (Stattersfield et al. 1998), but the endemic and restricted-range bird species are confined to forests or coastal woodland habitats, and none are associated with wetland habitats. However, a number of wetlands important for birds occur in the ecoregion, such as the Nyumba ya Mungu Reservoir, located on the Pangani River downstream of the confluence of the Ruvu and Kikuletwa rivers. This site hosts significant congregations of wetland birds, including the largest known colony of *Ardea cinerea* in East Africa and significant numbers of *Egretta ardesiaca, Dendrocygna bicolor, Charadrius pecuarius, Sterna nilotica, Chlidonias hybridus,* and *Rynchops flavirostris* (Baker and Baker 2001). Lake Jipe in Kenya also supports a diverse assemblage of wetland birds (Bennun and Njoroge 1999).

Status and Threats

Current Status

Humans have inhabited the coast of eastern Africa continuously since the late Pleistocene (Clarke and Karoma 2000). Human populations now number in the tens of millions, with projected growth rates of more than 2 percent (CIA 2000). Population densities are particularly high and increasing on the slopes of the ecoregion's mountains, where rainfall is most plentiful. Population levels in the West Usambara Mountains have increased more than twenty-three–fold since the turn of the twentieth century (Newmark 1998). Population pressure strains scant water resources and often causes degradation of the river systems. For example, sewage treatment facilities in Arusha and Moshi municipalities are ineffective, resulting in pollution of several rivers in the basin (Maganga et al. 2001).

Water sources are under pressure for domestic supply, irrigation, and hydropower generation. There are five dams in the basin, and Tanzania has generated hydroelectric power from the waters of the Pangani River since the 1930s (Maganga et al. 2001). The dam that creates the Nyumba ya Mungu reservoir provides hydroelectric power and water for irrigation downstream, and these uses often compete with one another (Sevaldsen 1997). There is also an active fishery in the reservoir, where both native fishes from River Pangani and introduced species live (Bailey 1969; Hughes and Hughes 1992). Another hydropower scheme is also operational on the lower Pangani River close to Pangani town.

Additionally, the foothills of the Usambara Mountains are dotted with dams varying in height from about 4.5 to 6 m. These were constructed largely for flood control and irrigation. Because of natural processes and an increase in agricultural land use, floodwaters carry high sediment loads off the upstream lands, causing sediment to accumulate behind the dams and significantly increasing turbidity in the ponds. Both large and small dams are limiting the transport of sediments downstream and probably blocking the movements of species in this ecoregion's rivers.

Although there is a ban on commercial forestry in much of the highlands, conversion of forest to agricultural lands or cutting of trees for fuelwood or construction material is still of concern (Rodgers 1993; Johansson et al. 1998). Mixed farming (the production of food and cash crops in addition to livestock) is common in the region. Coffee, tea, sisal, and timber are the main commercial agricultural products, with much subsistence agriculture as well.

Forests in this ecoregion are highly fragmented, and only

patches of forest remain in the Usambara and Pare Mountains (Newmark 1998). Loss of forest cover may lead to increased sedimentation, flash flooding, and resultant changes in aquatic species composition. Existing forest patches play a critical role in sustaining functioning aquatic systems and maintaining the recharge of groundwater resources (Rodgers 1993).

Fisheries in the wetlands, swamps, and rivers of the ecoregion provide an important source of protein and livelihood to local peoples. Additionally, many riverine floodplains are fished during the wet season and then farmed or grazed during the dry season.

Several fish species have been introduced into the waters of this ecoregion. The introduction of alien species into lakes Jipe and Chala has changed the species composition and community structure of the lakes. Some of the fish species introduced into Lake Jipe include *Tilapia rendalli, T. zillii, O. esculentus, O. variabilis, O. machrochir, O. pangani,* and *O. korogwe*. Species such as *O. esculentus* and *O. variabilis* were introduced from Lake Victoria, whereas *O. macrochir* was introduced from the Congo Basin. Lowe (1955) warned of the potential negative impact of these introductions on the native fauna. Hybridizations have occurred between native and introduced fish, rendering taxonomic identifications difficult. Brown trout (*Salmo trutta*) and rainbow trout (*Oncorhynchus mykiss*) have been introduced into the cooler, high-elevation streams. The effect of these introduced species probably will remain unknown because few early surveys of these high-elevation streams were conducted.

The headwaters of the Pangani Basin fall in a number of protected areas, including Arusha National Park on Mount Meru, Kilimanjaro National Park in Tanzania, and a small part of Tsavo West in Kenya. Most of the forests on the Pare Mountains and West and East Usambara Mountains are found in Forest Reserves managed by the Forest and Beekeeping Division of the Tanzanian government (Johansson et al. 1998), although some forests are also located in the Amani Nature Reserve.

Future Threats

The aquatic systems in this ecoregion are threatened by a range of anthropogenic impacts, especially agricultural activities on the mountains in the catchment. Considerable forest loss has already occurred in the catchment and is still proceeding (Newmark 1998).

Sedimentation of streams as a result of agriculture is a potential threat to the biological functioning of the systems. For example, sedimentation is a problem in the Lake Jipe and Chala catchments. Since 1951, Lake Jipe has been silting in, with the result that its banks are extending inward, reducing the capacity of the lake to absorb floodwaters (Lowe 1955). High silt loads also hamper penetration of light into the lakes and reduce primary productivity.

Justification for Ecoregion Delineation

This ecoregion encompasses the Pangani, Umba, Sigi (Tanga), Mkulumuzi, and Msangazi river basins and is characterized by a fish fauna with about 30 percent endemism. This ecoregion is part of the Eastern and Coastal bioregion, whose fish fauna probably arrived within the last 12,000 years, since the last interpluvial (Roberts 1975).

Data Quality: Low

Apart from the works of Lonnberg (1910), Lowe (1955), Bailey (1966, 1969), Dadzie et al. (1988), and Eccles (1992), the freshwater systems of this ecoregion are little studied. However, L. Seegers (pers. comm., 2001) is preparing a book on the fishes of the Pangani River. Despite Lowe's 1955 recommendation to carry out a comprehensive taxonomic review of the native fauna before the proliferation of alien species, little work has been done, and the taxonomy is now much more complicated. For instance, Lowe (1955) associated the four spined tilapiines with those of Lake Victoria, yet Bailey (1966) deemed them to be exceptionally distinct. There is therefore a need not only to unravel the tilapiine taxonomic disparities but also to document the aquatic diversity of the ecoregion by completing comprehensive taxonomic and ecological studies.

Ecoregion Number:	**71**
Ecoregion Name:	**Senegal-Gambia Catchments**
Major Habitat Type:	**Savanna–Dry Forest Rivers**
Final Biological Distinctiveness:	**Nationally Important**
Final Conservation Status:	**Vulnerable**
Priority Class:	**V**
Author:	**Michele Thieme**
Reviewers:	**Christian Lévêque and Papa Samba Diouf**

Location and General Description

The ecoregion spans parts of southern Mauritania, Senegal, southwestern Mali, Gambia, Guinea-Bissau, and Guinea and includes the Senegal, Saloum, Casamance, Gébe, and Gambia river basins. The Senegal and Gambia rivers are the ecoregion's largest, with drainage areas of about 441,000 km^2 and 77,000 km^2, respectively (Lévêque 1997).

Rainfall decreases from south to north and from the coast inland. At the southern end it averages about 1,500 mm/year, whereas the northern portion receives only about 400 mm/year.

During the rainy season (between June and October, duration depending on location), the ecoregion's major rivers often experience pronounced flooding. For example, historically, the Senegal River flooded 5,000 km² of land during its peak flood (Welcomme 1979). However, since the construction of two dams (Manantali and Diama dams) along the river, flooding has been significantly curtailed (Hamerlynck and Duvail 2003). In the dry season, saltwater moves into the deltas of the lowland coastal rivers (including the Saloum, Casamance, and Gébe), forcing strictly freshwater fish species to move inland. The penetration of seawater far inland permits the growth of mangrove forests 70–100 km inland in the southern portion of the ecoregion.

Terrestrial vegetation ranges from semidesert Sahelian grassland and shrubland in the north to progressively moister Guinea savanna in the south. In the wetter, southern portion of the ecoregion, seasonally inundated swamp forests line the rivers. Floodplain vegetation includes perennial grasses and sedges, in addition to reedmace (*Typha domingensis*) in the main channels.

Outstanding or Distinctive Biodiversity Features

This ecoregion is distinguished by rich coastal deltas that support large populations of migrant birds. The rivers of this ecoregion support a moderately rich aquatic fauna, but levels of endemism are low, with only three endemic frogs and one endemic fish species. The desiccation of the basins and recent recolonization by a Nilo-Sudanian fauna explains the low level of endemism.

The Senegal Delta wetlands, the Gambia Delta, and Geba-Corubal are prominent feeding grounds for migrant birds, with an estimated 3 million Palearctic and Afrotropical species using the Senegal Delta alone (Wetlands International 2002). Djoudj National Park hosts 366 bird species, including garganey (*Anas querquedula*), shoveler (*A. clypeata*), pintail (*A. acuta*), black-tailed godwit (*Limosa limosa*), greater and lesser flamingo (*Phoenicopterus ruber* and *P. minor*), great white pelican (*Pelecanus onocrotalus*), and avocet (*Recurvirostra avosetta*). Mangroves provide important habitat for some bird species, and this ecoregion contains the most northern mangroves on the African continent. This ecoregion also forms the northern limit for the West African dwarf crocodile (*Osteaoaemus tetraspis*) and West African manatee (*Trichechus senegalensis*).

Historically, the annual inundation of riverine floodplains created extensive feeding and breeding grounds for fish. Floods triggered migrations in a number of fish species, particularly in the Senegal River. Movements include both longitudinal migrations within rivers and lateral migrations onto the floodplain (Lévêque 1997).

Status and Threats

Dams, irrigated agriculture, and regional climate change (desertification) are the most prevalent threats to the aquatic biodiversity of this ecoregion's coastal rivers. Several Ramsar sites and national parks have been established in portions of the coastal deltas to protect feeding grounds for migrant birds. These include Djoudj and Diawling national parks and the Chat Tboul and Ndiael reserves in the Senegal Delta and the Saloum Delta National Park.

Dams impound the Bafing River at Manantali and the Senegal River near the delta at Diama, affecting the hydrology and ecology of these rivers. The dams have led to an expansion of irrigated agriculture into new areas and of aquatic vegetation in waterways after water levels stabilized. As a result, an intermediate schistosomiasis host, the snail *Biomphalaria pfeifferi*, has increased in abundance (Ernould et al. 1999). Above the Diama Dam, *Typha*, spp. *Vossia cuspidata*, and *Phragmites* spp., and the invasives *Pistia stratiotes* and *Salvinia molesta* grow in thick mats, preventing the passage of boats (Pieterse et al. 2003). The dams and dikes block the ingress of seawater, such that these plants flourish and alter the natural ecosystem. Additionally, the multiple dikes along the Senegal River prevent water from entering the floodplain along most of the river's course and block the movements of fish onto the floodplain. The dam at Diama has caused excessive downstream salinity and altered the fish species composition in the delta. Additionally, yields from floodplain fisheries have decreased with the changed flood regime and overfishing (Bousso 1997).

Dams in this ecoregion have also encouraged a shift from traditional flood recession agriculture to a more environmentally destructive irrigation agriculture, with 2,500 km² of land placed under irrigation since the dams were constructed (Thiam 1996; Lévêque 1997). Mainly in conjunction with rice agriculture, soil salinization is occurring on irrigated lands (Venema et al. 1997). This problem is further compounded by the decline in rainfall experienced in the region during the twentieth century. The most dramatic decrease began around 1970 and set in motion a process of increasing desertification (Debenay et al. 1994).

In recent years IUCN has undertaken a project to rehabilitate the Senegal Delta through managed flood releases (Hamerlynck and Duvail 2003). This restoration project has resulted in increased plant diversity, vegetation cover, pasture quality, waterbird populations, and fishery catch in the delta and partially restored estuarine habitats (Duvail and Hamerlynck 2003). For example, in Diawling National Park in Mauritania, fish catches have increased from less than 1,000 kg in 1992 and 1993 to more than 100,000 kg in 1999. Similarly, waterbird counts have increased from about 2,000 in 1992 to more than 38,000 in 1999. These dramatic changes are associated with managed flood releases that have increased the flooded surface area of three basins (Bell, Diawling, and Ntiallakh) in the lower delta since 1994 (Hamerlynck and Duvail 2003). In principle, managed flood releases from the upstream Manantali Dam on the Bafing River should also flood a minimum of 500 km² of floodplains each year, although in some years this target has not been reached (Hamerlynck and Duvail 2003).

Justification for Ecoregion Delineation

This ecoregion is defined by the Senegal and Gambia river basins and supports a Nilo-Sudanian freshwater fauna (Roberts 1975; Lévêque et al. 1990, 1992; Lévêque 1997). Uplifting of the earth's crust during the late Jurassic created the central Fouta Djallon Mountains, which became the source of the Gambia and Bafing (now a tributary to the Senegal) rivers. Further uplifting of these mountains occurred in the late Miocene (Lévêque 1997). In low-lying areas north of the mountain range, the Niger and Senegal rivers occasionally became connected during wet periods, and Nilo-Sudanian fishes are thought to have reached the Gambia from the Senegal by crossing low-lying country in between their lower courses. The Senegal and Gambia rivers are thought to have been nearly dry during the last interpluvial, from about 27,000 to 12,000 years ago (Roberts 1975). Thus, Nilo-Sudanian fish from the Niger River probably recolonized the Senegal-Gambia catchments during the last pluvial, about 12,000–8,000 years ago (Lévêque 1997).

Data Quality: High

Ecoregion Number:	**72**
Ecoregion Name:	**Eastern Coastal Basins**
Major Habitat Type:	**Savanna–Dry Forest Rivers**
Final Biological Distinctiveness:	**Continentally Outstanding**
Final Conservation Status:	**Vulnerable**
Priority Class:	**II**
Authors:	**Dalmas Oyugi, Michele Thieme, and Ashley Brown**
Reviewer:	**Lucy Kashaija**

Location and General Description

This ecoregion, situated along the eastern coasts of Tanzania and Mozambique, extends from the Wami River Basin south to the Luala River Basin, directly above the Lower Zambezi [66]. The ecoregion includes the Ruaha/Rufiji, Ruvuma, and Lúrio rivers and other smaller coastal basins.

The climate is dominated by three features: the wet southeast trade winds off of the Indian Ocean, the southwest monsoon system from the Congo Basin, and the northeast trade winds from Ethiopia and Somalia (Hughes and Hughes 1992). Rainfall is high (1,800–3,000 mm) in the highlands of this ecoregion, particularly in the Uluguru and Udzungwa mountains in Tanzania, which are under the direct climatic influence of the Indian Ocean (Lovett et al. 1997), and in the Kipengere Mountains, where the Ruaha has its headwaters. Along the coast, rainfall is lower, with an average of about 1,100 mm/year at Dar es Salaam and an average of 800–900 mm in Mozambique (Hughes and Hughes 1992). Rains occur between January and October, depending on location in the ecoregion, with rains beginning later in the year as one moves south. Rains occur in Tanzania between March and May and then begin again in October and November. Temperatures range from about 20–32°C along the coast in Tanzania to as high as 40°C in the south, although temperatures are significantly moderated by altitude, and frosts can occur in the headwater mountains during the austral winter (Hughes and Hughes 1992).

In the Tanzanian portion of the ecoregion, rivers descend from the escarpments and highlands of the interior to the narrow coastal plain (15–30 km wide). From north to south, the major river drainages are the Wami, Ruvu, Rufiji (whose tributaries include the Great Ruaha, Kilombero, Luwego, and Mbarangandeu), Matandu, and Mbemkuru. The Wami River drains the Nguu, Nguru, Ukaguru, and Rubeho Mountains of the Eastern Arc mountain range. The shorter Ruvu River flows from the Uluguru Mountains onto the coastal plain. The Rufiji River, the largest in Tanzania, has several major tributaries, including the Ruhudji, with floodplains along the Kilombero Valley (up to 6,265 km^2) and the Great Ruaha, Njombe, and Luwegu rivers. These rivers have their major headwaters in the Udzungwa and Mahenge mountains of the Eastern Arc and the more southern Kipengere Range (southern highlands). The Rufiji is lined by a number of small temporary and permanent lakes that include the Tangalala Lake Complex, a series of connected lakes (Manongi 1993). Larger rivers of northern Mozambique include the Ruvuma, Messalo, Lúrio, Mocubúri, Ligonha, and Licungo, all of which drain the highland interior (900–2,500 m a.s.l.) to the coastal plain. The Ruvuma River is the third largest in Mozambique and has a catchment area of 155,400 km^2 (Hughes and Hughes 1992). Swamps that line the lower reaches of the Ruvuma include the Nhica (75 km^2), the Quitemba (25 km^2), and the Miula (70 km^2). Lake Nangade and several other oxbow lakes also occur in the river's floodplain. South of the Ruvuma, the Messalo, Montepuez, Megaruma, Lúrio, Mocubúri, and Monapo rivers all have seasonal flows, and most are lined by swamps. The lower courses of several of these rivers expand out into narrow, long lakes, such as Lake Biribizi on the Montpuez River (Hughes and Hughes 1992).

Vegetation adjacent to and within the freshwater systems of this ecoregion consists primarily of a coastal mosaic including large areas of miombo woodland, coastal dry forest and coastal scrub, riparian and swamp forests, floodplain vegetation, and mangrove forests (Hughes and Hughes 1992; Hatton and Munguambe 1998; Burgess and Clarke 2000). Swamp and riparian forest trees include *Pandanus rabaiensis, Baikiaea insignis,*

Syzygium cordatum, Ficus verruculosa, F. trichopoda, Voacanga thouarsii, Raphia farinifera, and *Parkia filicoidea. Cyperus papyrus* also grows in permanent swamps lining the rivers, often in association with *Phragmites* spp. and *Nymphaea capensis* (Hughes and Hughes 1992). Grasses such as *Echinochloa pyramidalis, Cynodon dactylon,* and *Oryza* spp. dominate many of the floodplains of the ecoregion, whereas hygrophilous (living or growing in moist places) grass species such as *Andropogon schirensis, Digitaria milanjiana,* and *Loudetia phragmitoides* dominate dambos (seasonally waterlogged, predominantly grass-covered, shallow depressions in the headwater zone of rivers that are generally less than 5 km^2) in the ecoregion (Hatton and Munguambe 1998). Mangrove forests line river mouths throughout the ecoregion (Kemp et al. 2000). The montane headwaters of the rivers in this ecoregion are clothed with montane cloud forests, which are rich in endemic species of plants and animals (Lovett and Wasser 1993; Burgess et al. 1998).

Outstanding or Distinctive Biodiversity Features

The habitats in the ecoregion include forested headwater streams, medium-sized rivers and their tributaries, mangrove forests, estuaries, small lakes, permanent swamps, dambos, deltas, and seasonal floodplains. About 30 percent of the nearly 100 described fish species are endemic. Among these, there is a radiation of the Aplocheilidae genus *Nothobranchius,* with nine endemics known from this ecoregion (Lévêque 1997). Characins, anguillid eels, rivulins, cyprinids, gobies, and mochokids are the most speciose groups in the fresh waters of this ecoregion.

The mountain headwaters of this ecoregion are less well known. A rich frog fauna is dependent on the moist mountain conditions and includes a number of endemic species. More than sixty species of aquatic-dependent frogs are known from the mountainous region in southern Tanzania alone (includes the Kipengere, Livingstone, and Udzungwa mountain ranges), seven of which are endemic. Several of these species are in the Hyperoliidae family, including *Hyperolius kihangensis,* known from swamps in the dense evergreen forests of the Udzungwas (Schiøtz 1999). The recently described Kihansi spray toad, *Nectophrynoides asperginis,* lives only in the fine mist created by the cascading waters of the Kihansi Falls in the Southern Udzungwa Mountains. Limited investigation has also shown that some of the Eastern Arc Mountains of Tanzania contain an important assemblage of Odonata (dragonflies and damselflies), including endemic species of the genera *Umma* and *Chlorocnemis* that are more commonly found in the Congo Basin (Clausnitzer 2001). The lowland coastal forests also support important dragonfly assemblages, including *Coryphagrion grandis,* a Gondwana relict species whose nearest relatives are found in Central and South America (Clausnitzer 2001).

This ecoregion's riverine habitats provide good habitat for wetland birds. The productive swamps and floodplain lakes of the lower Ruvuma River provide extensive habitat for a rich avifauna, including large numbers of weavers. The Kilombero floodplain also provides habitat for the endemic Kilombero weaver (*Ploceus burnieri*), which breeds in extensive riverine swamps fringed with *Phragmites mauritianus* (Stattersfield et al. 1998; Baker and Baker 2001). Other important wetlands for birds in this ecoregion are Lake Tlawi, a high-altitude lake in the Mbulu Highlands that supports a nonbreeding congregation of *Fulica cristata;* the Rufiji Delta, which supports thousands of migrant waterbirds; and the Usangu Flats along the Ruaha River, which support at least 418 bird species and congregations of *Dendrocygna bicolor, Balearica regulorum,* and *Plectropterus gambensis* (Baker and Baker 2001, 2002). The Rufiji and several of its tributaries run through the large Selous Game Reserve in southeastern Tanzania. With its extensive rivers and streams, this reserve is expected to hold significant numbers of *Gorsachius leuconotus, Ephippiorhynchus senegalensis, Scotopelia peli,* and *Rynchops flavirostris* (Baker and Baker 2001).

Aquatic mammals that live in the ecoregion include marsh mongoose (*Atilax paludinosus*), African clawless otter (*Aonyx capensis*), and hippopotamus (*Hippopotamus amphibius*). Dugong (*Dugong dugon*), listed by IUCN as vulnerable, can be found in the lower brackish reaches of some rivers in Mozambique, in the southern Rufiji Delta, and perhaps also in southern Tanzania (Hughes and Hughes 1992; IUCN 2002).

Status and Threats

Current Status

The main threats to the ecoregion are an increasing human population, river impoundments, overfishing, pollution, poor water management for agricultural uses, and deforestation. Humans have inhabited the coast of eastern Africa continuously since the late Pleistocene (Clarke and Karoma 2000). However, these populations now number in the tens of millions, with projected growth rates of more than 2 percent (CIA 2000).

Construction of a dam on the Kihansi River (a tributary to the Great Ruaha) in the north of the ecoregion has had observed impacts. The dam is changing not only the hydrology of the river but also the moist forest and spray zone microclimate of the gorge below (Lovett et al. 1997). The Kihansi spray toad is threatened with extinction by the diversion of water for the hydroelectric plant, although efforts are under way to breed the toad in captivity and restore some parts of its natural habitat (World Bank 2001). Two dams, the Kidatu and Mtera, have already been built on the Ruaha and Rufiji rivers, respectively (Iddi 1998).

Fisheries in the wetlands, swamps, and rivers of the ecoregion provide an important source of protein and livelihood to local peoples. Additionally, many rivers are fished during the wet season, and then their floodplains are farmed or grazed during the dry season.

The high rates of deforestation in the ecoregion also threaten the health of the ecoregion's rivers. Possible impacts of deforestation include a reduced water supply from forested regions, increased turbidity, greater light exposure of riverine habitats, and increased flash flooding. The high pressure on coastal forests is generated by a need for wood for housing construction and for fuel. As of 1989, wood made up more than 90 percent of rural East African energy consumption (Clarke and Karoma 2000). Deforestation also occurs to open land for agriculture.

Pollution, including solid waste from cities, runoff from fertilizers and pesticides, and mining waste, is a problem in this ecoregion. Solid waste disposal has become an increasing problem because of rapid urbanization in Tanzania (Mato 1999). Heavy metals from phosphate fertilizers or copper fungicides have accumulated in the soils near farms and could have an impact on nearby rivers (Ikingura et al. 1997). Mkuula (1993) reported that chemicals from chromium-tanned leather were being dumped outside factories adjacent to the Msibazi River near Dar es Salaam and the Ngerengere River near Morogoro (Mkuula 1993). There is high mercury content in soils near gold mining and processing centers in Tanzania, although thus far studies show low mercury concentrations in fish and aquatic plants (Ikingura et al. 1997).

Many managed areas offer some protection to the freshwater systems of this ecoregion. Mikumi National Park in Tanzania covers parts of the Rufiji River and several of its tributaries (Sayer et al. 1992). The large Selous Game Reserve in the Rufiji Basin protects wetlands within its 55,000-km^2 area (Manongi 1993). The Ruaha National Park occurs on the left bank of the Ruaha River for about 143 km, although there are concerns about rainfall in the Kipengere Mountains, which provide the source waters for the Great Ruaha River (Hughes and Hughes 1992; Baker and Baker 2001). The large Niassa Game Reserve in Mozambique covers 15,000 km^2 and includes part of the Ruvuma River and many of its tributaries. The Gile Game Reserve in Mozambique protects a small portion of the southern coast of the ecoregion (Sayer et al. 1992). Many of the mountain forests covering the headwaters of this ecoregion are found in forest reserves. These catchment reserves are managed for water by the Tanzanian Forest and Beekeeping Division. In 1992 some forest reserves were transferred to the Udzungwa Mountains National Park.

Future Threats

The largest future threats are the effects of an increasing human population, new impoundments on the waterways of this ecoregion, and water diversion for irrigation schemes. The steep drops of some rivers in the ecoregion offer the possibility of construction of more hydropower dams.

Conflict in northern Mozambique has caused the loss of much of the wildlife inhabiting formerly protected areas. Stabilization of the political situation and rebuilding of the country offer hope for the future of environmental management (East 1999).

Ecotourism is a possible sustainable economic activity for this ecoregion. The Eastern Arc forests, particularly the Udzungwa Mountains, may be used for recreation and ecotourism, as can the transboundary peace park proposed between the Niassa and Selous game reserves (Iddi 1998).

Justification for Ecoregion Delineation

This ecoregion includes coastal basins from the Wami River in Tanzania to the Luala River in Mozambique and includes the Ruaha/Rufiji, Ruvuma, and Lúrio rivers and other smaller basins. The freshwater fish fauna of the rivers of this coastal ecoregion has its greatest affinities with the Zambezi fauna (Lévêque 1997; Bills 1997). Many of these rivers may have been dry as recently as the last interpluvial, which accounts for their low species counts (Roberts 1975). The Tanzanian Shield rivers, meaning the rivers between the Athi Basin in the Kenyan Coastal Rivers [64] ecoregion and the Zambezi in the Lower Zambezi [66] ecoregion, at times converged with fluctuations in sea level and may account for the similarity in fauna among many of the coastal rivers of the Eastern Coastal ecoregion (Lévêque 1997). As more data become available on their fauna, the delineations of these ecoregions probably will be updated.

Data Quality: Low

Ecoregion Number: **73**
Ecoregion Name: **Southern Temperate Highveld**
Major Habitat Type: **Savanna–Dry Forest Rivers**
Final Biological Distinctiveness: **Bioregionally Outstanding**
Final Conservation Status: **Endangered**
Priority Class: **IV**
Author: **Lucy Scott**
Reviewer: **Paul Skelton**

Location and General Description

The Southern Temperate Highveld ecoregion is situated in the interior of South Africa, with the western boundary formed by the Magaliesberg, Pilanesberg, and Waterberg mountain ranges, the northern boundary formed by the Soutpansberg, and the eastern boundary formed by the Drakensberg Mountains (O'Hagan 1989; Duggan 1990). The highveld forms a rolling, grassy plateau 900–1,900 m a.s.l. and gradually slopes down to the coast in the southeastern cape (Cooke 1964). The dominant limnological features are rivers and seasonal pans. The main drainages are those of the west-flowing Vaal River (the main tributary of the Orange River) and some stretches of the middle Caledon and Orange rivers. The headwaters of the Crocodile, Marico, Sabie, Komati, Usutu, Pongola, and Tugela rivers also drain from the highveld plateau to the east and northeast (Gabie 1965). About 70 percent of the highveld consists of fine sedimentary rocks, with significant exposures of dolomite along the northern and western boundaries of some catchments. There are a variety of soil types, and most streams originate over Karoo sediments, with some originating over the Witwatersrand system and still others as springs in dolomite (O'Keeffe et al. 1989).

The highveld has a temperate climate, with summer rainfall and the highest hailstorm frequency in southern Africa. Rainfall on the highveld varies from 1,400 mm in the east to 400 mm further west. The highveld also experiences a fairly dry period during the cooler months (Pritchard 1971; O'Hagan 1989). Moving southward, the northeastern cape has a warm temperate climate, with a slightly higher average rainfall of 600–1,200 mm. The climate of the southeastern cape is that of a subtropical coastal belt influenced by the warm Mozambique-Agulhas current; rainfall is 600–800 mm annually, with 80–85 percent occurring as brief summer thunderstorms from October to March (Agnew 1986; Skelton 1994). Frontal, relief, and convectional rain occur throughout the ecoregion (Pritchard 1971). Gross evaporation increases from 1,300 mm in the east to about 2,000 mm in the southwest at the rate of 1 mm/km. Summers are hot, with long, dry, often windless periods broken by thunderstorms that generate both strong winds and local flooding. Temperatures usually average 10–18°C with frost at night in the cool season to 15–40°C in summer. The extensive central plateau receives only 27 percent of the mean annual precipitation, most of which falls into the Orange and Vaal catchment (650,000 km^2 in size) (Midgley et al. 1994).

The Orange River, the largest river system in Africa south of the Zambezi, has two major tributaries, the Caledon and the Vaal, whose basins fall in this ecoregion (Cambray et al. 1986). The Vaal River rises to the north of the Drakensberg range and flows 900 km across the interior plateau to join the Orange River near Douglas, draining an area of 194,000 km^2. The catchment slopes gently from elevations on the order of 1,800 m a.s.l. in the east to about 1,200 m in the west, with some steep areas in the headwaters of the Wilge River, a Vaal tributary on the southeast border of the Upper Orange catchment. The Orange and Vaal rivers are typical of many South African rivers in that they carry very large sediment loads, especially during floods. Rapids and pools are common in the upper reaches before the gradient decreases and the river begins to flow over flat, sandy streambeds with reeds (Kleynhans 1983). Increased flow rates and flooding in all rivers of the highveld usually occur during the spring and summer (September–March), and all indigenous fish species breed during this period.

The ecoregion extends to the coast in the eastern cape, where the coastal rivers may be divided into three types. The largest systems extend well inland and have tributary headwaters originating on the escarpment (the Gamtoos, Sundays, Great Fish, and Kei river systems). The moderate-sized rivers extend inland as far as mountains such as the Winterberg or Amatola ranges (Swartkops, Bushmans, Keiskamma, Buffalo, and Kowie rivers). Finally, there are small coastal rivers such as the Coega, Baakens, and Kasuka rivers (Skelton 1980). Riparian zone wetlands (oxbow lakes, pans, and high-altitude bogs) are unique to the midlands of the northeastern cape (Forsyth et al. 1997).

Many ephemeral windblown pans, which are filled only during times of high summer rainfall (e.g., the Lake Chrissie pan complex), dot the landscape of the highveld plateau (Allanson et al. 1990). Pans and other enclosed drainage basins are particular features of the western part of the ecoregion (Midgley et al. 1994).

The major vegetation type consists of open, undulating, hygrophilous (living or growing in moist places) *Cymbopogon-Themeda* grassland, with Nama Karoo vegetation toward the west (Midgley et al. 1994; Carruthers 1997; Low and Rebelo 1998) and some bushveld to the north (Stuart et al. 1990; Low and Rebelo 1998). Frosts, fire, and grazing maintain the grass dominance and prevent the establishment of trees. The Nama Karoo Biome occurs on the central plateau of the western half of South Africa at altitudes between 500 m and 2,000 m, with the majority between 1,000 m and 1,400 m (Low and Rebelo 1998). Trees are found mainly in river valleys further east and in parts of the ecoregion with a high water table (Pritchard 1971). The Vaal River runs through a small part of Kalahari grassland–*Acacia* wooded steppe (Cooke 1964). The vegetation of mountain ranges such as the Pilanesberg is rich floristically, with wiry, sour grassveld in the less rocky parts and a dense, mixed bushveld in the rugged parts (Brett 1989).

Outstanding or Distinctive Biological Features

The Orange River drainage and the rivers of the southeastern cape are low-gradient highveld rivers, a system type that is rare in Africa. The aquatic fauna is depauperate. In particular, fish and mollusks are far fewer in number than those of the Zambezi Basin. The Orange River fish fauna is dominated by cyprinids. Two endemic freshwater mollusks (*Burnupia vulcanus* and *Pisidium harrisoni*) and thirteen endemic Trichoptera have been recorded in the Great Fish River (Laurenson and Hocutt 1984; Eekhout et al. 1997). Three endemic Odonata (*Aseudagrion raalensis, P. inopinatum,* and *Agriocnemis falcifera*) are also known from the ecoregion.

The Tugela headwaters, the Great Fish River, the Keiskamma River, and streams and waterbodies of the eastern Transvaal escarpment from the Pongola to the Sabie, the Blyde River, and the Witwatersrand-Magaliesberg areas have been identified as hotspots of aquatic species richness. The Great Fish River, the eastern Transvaal escarpment from the Pongola to the Sabie, and the Blyde River are also hotspots of threatened fish species in South Africa, a high proportion of which occur in protected areas (Skelton et al. 1995). The Nile crocodile (*Crocodylus niloticus*) and the water monitor (*Varanus niloticus*) are two vulnerable species that live in the waters of this ecoregion.

Globally threatened fish include the Treur River barb (*Barbus treurensis*, Upper Blyde River, very restricted distribution), the border barb (*Barbus trevelyani*, Buffalo and Keiskamma rivers), the rock catfish (*Austroglanis sclateri*, Orange-Vaal system, original distribution much reduced), and the Eastern Province rocky (*Sandelia bainsii*, found in four rivers in the Eastern Cape). *Kneria auriculata* is limited to altitudes between 1,100 m and 1,400 m a.s.l., and five relict populations are known from tributaries to the Crocodile River (Kleynhans 1986). *Varicorhinus nelspruitensis* is endemic to the upper Incomati and Pongola systems. *Amphilius natalensis* lives only in tributaries of the Incomati and Olifants rivers, between 900 and 1,300 m. The rare Incomati rock catlet (*Chiloglanis bifurcus*) is endemic to the Incomati system and is found only between 900 and 1,200 m. *Aplocheilichthys katangae* is found in tributaries of the Limpopo and is considered rare in the Transvaal (Kleynhans 1986).

The ecoregion also hosts several endemic and vulnerable aquatic invertebrates. The Yellowwoods River is a tributary of the Buffalo and has one endemic caddis fly, *Cheumatopsyche lateralis*. *Potamonautes warreni*, a freshwater crab, and *Leander capensis*, a freshwater prawn, are also endemic to the Orange Basin.

In terms of waterbirds, the highveld is important for conservation of the vulnerable wattled crane (*Bugeranus carunculatus*), the crowned crane (*Balearica regulorum*), and the white-winged flufftail (*Sarothrura ayresi*) (Huntley 1978).

Status and Threats

Current Status

The Southern Temperate Highveld is a seriously threatened ecoregion. Extensive regulation of its waters has taken place, and rapid urban and industrial development continues to present a variety of threats to ecosystem integrity.

The Orange River system probably is the most highly regulated in Africa, with many impoundments and weirs and a number of major interbasin transfers (IBTs). The largest dams on the Orange are the Gariep and the Vanderkloof dams (Midgley et al. 1994). The Gariep Dam is the main regulator of the system, providing flood control, hydroelectric power, and recreation. Floods from the rivers above the dams are absorbed by the Gariep and Vanderkloof dams, reducing the frequency of 3,000 m^3/s floods and cutting the maximum possible flood (31,200 m^3/s) of the Orange River by 65 percent. The Vanderkloof Dam modifies the flow of the river, virtually nullifying seasonal differences (Cambray et al. 1986).

The Vaal Dam on the Vaal River is the principal source of water for the largest metropolitan region in southern Africa, the densely settled Witwatersrand, and other towns and extensive irrigation projects. The Vaal River Development scheme, with the Vaal Dam as the main storage unit, was initiated in the 1930s to serve the Vaal-Harts irrigation scheme (involving diversion of water into the Harts Valley). To meet water demands in the Vaal Basin, a number of IBTs now import water into the system, including the Tugela-Vaal, Usutu-Vaal, and Lesotho Highlands schemes. The Vaal River is augmented by water from nine other basins (Midgley et al. 1994; Davies and Day 1998).

There are at least eight existing or proposed major IBTs in this ecoregion. The Orange–Sundays–Great Fish River water transfer (350 million m^3/year) probably is the largest, carrying water from the Orange River more than 100 km to the Great Fish River and then on to the Sundays River, where it supplies farmers with water for irrigation (Midgley et al. 1994). The Orange-Fish water transfer has changed the character of the Great Fish from a seasonal to a perennially flowing river (Laurenson and Hocutt 1984). Although the diversity of the invertebrate fauna appears to be the same, only 33 percent of naturally occurring species are still found in the river, and pest blackfly (*Simulium chutteri*) larvae dominate the lower reaches of the river (O'Keeffe et al. 1992). Ten fish species (38 percent) have been introduced to the Great Fish River, with five species (*Barbus aeneus, Labeo capensis, L. umbratus, Clarias gariepinus,* and *Austroglanis sclateri*) having negotiated the Orange-Fish tunnel since it was opened in 1975 (Skelton 1980; Laurenson and Hocutt 1984).

Intensive irrigation has also had other effects on the environment. The Vaal-Hartz irrigation scheme has tripled the total dissolved solids (TDS) in the Hartz River. Irrigation in the Olifants River has raised the TDS from 75 to 4,000 mg/L, and values from the Fish River of 400 mg/L are reported despite the high quality of water supplied by the Orange River (Fuggle and Rabie 1983). Domestic sewage and urban and industrial effluents also increase TDS (Fuggle and Rabie 1983). The Jukskei-Crocodile system has received treated sewage effluent from parts of Johannesburg and exhibits severe eutrophication. Large urban populations (Kingwilliamstown and Zwelitsha) on the middle reaches of the Buffalo River contributed to a tenfold increase in concentrations of phosphorus and nitrogen (O'Keeffe et al. 1992).

Industrial air pollution from coal-burning power stations located in the eastern highveld has caused acid rain to fall in many parts of this ecoregion. This has resulted in the serious acidification of some rivers (the lowest pH measured for rain was 2.9) (Davies and Day 1998).

The increased afforestation of mountain catchments with exotic trees, such as *Pinus radiata*, leads to substantially decreased water runoff. Afforestation in the eastern escarpment has already been shown to decrease runoff to 20–35 percent of pristine levels (Fuggle and Rabie 1983).

After habitat degradation, invasive animals and plants are the greatest threat to the freshwater biota in southern Africa (O'Keeffe et al. 1992). *Physa acuta* and *Lymnaea columella* are highly invasive alien freshwater snails. Introduced fish include rainbow trout (*Oncorhynchus mykiss*), brown trout (*Salmo trutta*), common carp (*Cyprinus carpio*), grass carp (*Ctenopharyngodon idella*), silver carp (*Hypophthalmichthys molitrix*), bluegill sunfish (*Lepomis macrochirus*), smallmouth bass (*Micropterus dolomieu*), largemouth bass (*Micropterus salmoides*), Mozambique tilapia (*Oreochromis mossambicus*), and southern redbreast tilapia (*Tilapia rendalli*). Water hyacinth (*Eichhornia crassipes*) and Kariba weed (*Salvinia molesta*) are two ubiquitous, invasive plant species that have been introduced to the highveld. All major rivers on the highveld are infested with invasive exotic woody species, the most problematic exotics being the black wattle (*Acacia mearnsii*), silver wattle (*Acacia dealbata*), grey poplar (*Populus canescens*), weeping willow (*Salix babylonica*), and *Robinia pseudoacacia* (Forsyth et al. 1997).

Important nature reserves in the ecoregion include the Blyde River Canyon Nature Reserve (226 km^2), the Tussen-die-Riviere Nature Reserve, the Pilanesberg National Park, the Golden Gate Highlands National Park, and the Barberspan Nature Reserve (a 18-km^2 pan, part of a 31-km^2 waterfowl sanctuary) (Duggan 1990).

Future Threats

With the ever-increasing number of dams on the rivers of this ecoregion, water supply to lower reaches of the rivers is a large concern. The Lesotho Highlands Water Project (see ecoregion 37 description) is expected to lower the discharge of the Orange River substantially, reducing the amount of water flowing into the Atlantic Ocean. IBTs are also of great concern in terms of both changes to water flows and transfer of species between drainages. As more IBTs are completed, there is certain to be an increase in transfer of species into receiving rivers.

Justification for Ecoregion Delineation

The boundaries of this ecoregion follow the "Transvaal-Orange Free State" subregion of Skelton's (1993) highveld aquatic ecoregion. The fauna has mixed tropical and temperate affinities and shares many species with the Limpopo and the Zambezi rivers (Skelton 1990a; Skelton et al. 1995). Considerable exchange of the interior drainage (Orange River system) with that of the coastal systems must have occurred along the retreating escarpment (Skelton 1980).

Data Quality: High

The level of biological and ecological investigation in this ecoregion is high, and the threats to ecosystem integrity are known.

Ecoregion Number: **74**
Ecoregion Name: Uele
Major Habitat Type: Savanna–Dry Forest Rivers
Final Biological Distinctiveness: Continentally Outstanding
Final Conservation Status: Relatively Stable
Priority Class: III
Author: Emily Peck
Reviewer: Luc De Vos

Location and General Description

The rivers and streams in the Uele ecoregion drain from a high plateau along the northeastern border of the Congo Basin in the Democratic Republic of the Congo. The basin of the Uele River, a major tributary to the Ubangui, determines the boundaries of this ecoregion.

The Uele ecoregion is situated between 3°N and 5°N. Rainfall varies from west to east and seasonally. The lower-elevation, more densely forested western part of the ecoregion receives an average annual rainfall of 1,600–1,700 mm. The higher, savanna-covered eastern plateau experiences an average annual rainfall of 1,200–1,500 mm. The wet season usually extends from March to November and the dry season from December to February. The mean annual temperature for the ecoregion is 24°C (WCMC 1984b; Orange 1998).

The high plateau in the northeast of the ecoregion slopes west and south toward the central Congo Basin. Numerous scarps and cliff faces break up the gently sloping landscape. As a result, waterfalls and rapids mark all rivers running off the plateau toward the central Congo Basin. The Uele River and its affluents, the Bili, Uere, and Bomokandi rivers, drain woodland savannas in the north and east and mixed evergreen forests in the south and west. The Uele River begins in the Blue Mountains at an elevation of 1,620 m and then traverses the high plateau for 1,170 km before joining the Ubangui River in Yakoma, below 500 m (Orange 1998). The Uele River's catchment covers 139,700 km^2. The Bili is a blackwater river with a catchment area of 27,400 km^2. The Bili runs through more extensive areas of evergreen forests and carries a large amount of dissolved organic matter (Orange 1998).

A mosaic of Afromontane forest, gallery forest, wooded savanna, and grassland blankets the northern highlands of the Uele ecoregion. Savannas range from dense woodland to almost treeless grassland. *Loudetia arundinacea* and *Hyparrhenia* spp. dominate the grasslands of the far northeast. Numerous small rivers and papyrus swamps dissect these grasslands. *Combretum* spp. and *Terminalia* spp. dominate the savanna woodland vegetation. Other predominant species include *Bauhinia thonningii*, *Dombeya quinqueseta*, *Hymenocardia acida*, and *Acacia*, *Grewia*, and *Bridelia* spp. Gallery forest contains *Irvingia smithii, Erythrophleum suaveolens, Chlorophora excelsa,* and *Klainedoxa* and *Ficus* spp. (WCMC 1984b).

The montane forests of the Blue Mountains exhibit an altitudinal zonation of plant species. Species such as *Podocarpus, Prunus,* and *Ocotea* occur at intermediate elevations. At higher altitudes these forests transition to elfin forests and communities of bamboo, tree ferns, and lobelias. At the highest altitudes, Afroalpine moorlands occur with a variety of shrubs and grasses (Sayer et al. 1992). Fire plays an important role in maintaining the vegetation cover and often sweeps across the savanna during the dry season. In the southern and western parts of the ecoregion, a closed canopy of broadleaf and needleleaf evergreen forest blankets the landscape. Evergreen forest covers 30 percent of the course of the Uele River and 44 percent of the Bili River (Orange 1998).

Outstanding or Distinctive Biodiversity Features

The headwaters of the Congo Basin are poorly known biologically and hydrologically, including the Uele River and its affluents. Current information suggests that the rivers of the Uele Basin are rich in fish and aquatic herpetofauna.

At least 136 fish species have been documented. Predominant fish families are Mormyridae (with more than 30 species), Cyprinidae (more than 20 species), Alestiidae (more than 12 species), Citharinidae (at least 10 species), Distichodontidae (at least 7 species), Mochokidae (more than 12 species), Schilbeidae (7 species), and Cichlidae. Less species-rich but common families in the Uele ecoregion are Clariidae, Amphiliidae, Polypteridae, Mastacembelidae, Tetraodontidae, Malapteruridae, Anabantidae, Centropomidae, Hepsetidae, Channidae, and Bagridae.

Based on collected data, the ecoregion exhibits a low level of endemism in fish, although this may result from low sampling effort. Nine fish species are considered endemic, although seven of them are reported only from their type locality in the Uele system: *Amphilius notatus, Distichodus langi, Clariallabes simeonsi, Chrysichthys uniformis, Barbus schoutedeni, Barbus urotaenia, Hippopotamyrus macroterops, Hippopotamyrus retrodorsalis,* and *Petrocephalus hutereaui.* Further investigations in the Upper Congo River Basin may reveal occurrences of these species outside the Uele system.

Many frogs, reptiles, and mammals also depend on the freshwater systems of this ecoregion. About seventy frogs are known, of which about a dozen are considered endemic. The Central African giant terrapin (*Pelusios chapini*), West African flapshell turtle (*Cycloderma aubryi*), and slender-snouted crocodile (*Crocodylus cataphractus*) are among the aquatic reptiles native to the ecoregion. Aquatic mammals include the giant otter shrew (*Potamogale velox*), marsh mongoose (*Atilax paludinosus*), swamp otter (*Aonyx congicus*), spotted-necked otter (*Lutra maculicollis*), aquatic genet (*Osbornictis piscivora*), hippopotamus (*Hippopotamus amphibius*), and African water rat (*Colomys goslingi*).

Status and Threats

Both national and international protected areas exist in the Uele ecoregion. Garamba National Park, a World Heritage site, is located in the northeast part of the ecoregion and covers 5,000 km^2. In the park, open savannas of the high plateau in the north grade into a mosaic of savanna, gallery forest, and marshland in the south. This park provides habitat for hippopotamus (*Hippopotamus amphibius*), two otter species (*Aonyx congicus* and *Lutra maculicollis*), marsh mongoose (*Atilax paludinosus*), and waterbuck (*Kobus ellipsiprymnus*) (WCMC 1984b). Civil unrest and conflict in the region have negatively affected the park, with poaching and civil unrest prevalent. Garamba is surrounded by three game reserves: Azande (west), Gangala-Na Bodio (south), and Mondo-Missa (east). Other areas include Bomu Nature Reserve and the Bili-Uere Hunting Reserve, both of which exist in the north-central portion of the ecoregion. Bomu Nature Reserve is situated along the Bomu River and the border of the Central African Republic and is covered by wooded savanna and deciduous forest. Further south in Bili-Uere, deciduous evergreen forest grades into needleleaf evergreen forest (WCMC 1984b). Bili-Uere is under threat from agricultural activity and mineral exploitation, and Bomu suffers from poaching (WCMC 1984b). Infiltrations of poaching expeditions from Sudan to the north result in warlike confrontations between park authorities and the poaching gangs (L. De Vos, pers. comm., 2002). Although poachers typically target terrestrial species, their activities and presence can put pressures on freshwater habitats and species. In general, inadequate funding, administration, and infrastructure threaten the biological integrity of protected areas in the ecoregion (MacKinnon and MacKinnon 1986).

Human population is concentrated along a north-south axis in the eastern part of the ecoregion. Cropland is also concentrated in the east, near the centers of population. Among the variety of crops are coffee, plantains, and root crops (Miracle 1973). In addition to agriculture, much of the economy is based on mining. The main minerals are gold, silver, and copper. A group of gold mines exists north of Watsa in the western reaches of the ecoregion. However, production stopped in the late 1990s because of political unrest.

Future Threats

Civil unrest, ongoing conflict, and refugee movements are the largest threats to the natural resources of this region. Increased population growth along with increased mining and agricultural activity are also threats.

Justification for Ecoregion Delineation

Although poorly known ichthyologically, this ecoregion occurs in a savanna-forest transition zone, so it is suspected that it may contain an aquatic fauna that is unique to this zone. However, currently nearly all fish species known from the Uele area also occur in the Ubangui system. The ecoregion is defined by the basin of the Uele River.

Data Quality: Low

Ecoregion Number:	**75**
Ecoregion Name:	Volta
Major Habitat Type:	Savanna–Dry Forest Rivers
Biological Distinctiveness:	Bioregionally Outstanding
Conservation Status:	Critical
Priority Class:	IV
Authors:	Ashley Brown and Michele Thieme
Reviewers:	Christian Lévêque and Andy Osei Okrah

Location and General Description

The Volta is one of West Africa's largest rivers, draining an area of 390,000 km^2 (Petr 1986). It is fed by the Black Volta, the White Volta, the Red Volta, and the Oti rivers, which together drain the plateau in the north, the Atakora Mountains in the east, and several highland areas in the west. The Volta Basin covers parts of six countries: small portions are in Mali, Côte d'Ivoire, Togo, and Benin, and the majority of the basin falls in Burkina Faso and Ghana.

The north of the ecoregion is dry, with intermittent streams. Rainfall increases from about 500 mm in northern Burkina Faso to about 1,200 mm in the region of Lake Volta, decreasing again in the lower Volta to about 1,000 mm near the delta. Peak rainfall is in August, although the south of the basin experiences a longer wet season (May–October) than the north (June–September). The average rainfall for the entire basin is about 800 mm/year, with the highest (1,500 mm/year) occurring in the southwestern uplands. The vegetative cover follows the rainfall pattern, with dry xerophytic savanna grasslands predominant in the north, savanna woodland covering the middle of the basin, and rainforest in the southwest (Hughes and Hughes 1992).

Four major tributaries contribute their seasonal floodwaters to the Volta River. The intermittent Red and White Volta rivers, which are dry from January until the onset of rains in May, originate in Burkina Faso, meandering across the low-elevation (250-m) plateau that slopes slightly south. These two rivers merge before joining the Volta. The Black Volta, a permanent river, arises in the highlands southwest of Koudougou, flowing north until its junction with the Sourou River. During flood season, the Black Volta's waters enter the Sourou and form a large swampy area called the Mare aux Hippos. Extensive floodplains are present along both the upper Black Volta and the Sourou. These two rivers then join and flow south into the mainstem Volta.

The Oti River, which flows from the east, is the Volta's largest tributary, providing 30–40 percent of its annual flow. A major tributary of the Oti, in turn, is the Pendjari River, which arises from the Atakora Mountains in Benin, forms the border of Benin and Burkina Faso, and eventually merges with the Oti River after entering Ghana.

As with rainfall, flows vary substantially throughout the basin. Many of the tributaries and headwater streams of the Volta are intermittent in their upper reaches but become permanent in their lower reaches. In the middle of the basin extensive floodplains occur along the savanna-lined rivers. The Oti and its tributaries, in particular, support large amounts of inundated low-lying plains (Petr 1986; Hughes and Hughes 1992; Lévêque 1995).

In 1964 the Volta River was dammed at Akosombo and Kpong, creating Lake Volta, the largest artificial lake in Africa with a surface area of approximately 8,300 km^2 (Biney 1990). Since 1966, the lake has inundated the confluences of all its major tributaries. Before the dam was built, the Black and White Voltas merged to become the Volta River. About 175 km downstream, the Oti joined the Volta, and 200 km downstream further the river flowed through the quartzite walls of Ajena-Akosombo Gorge. Below the gorge, the river spilled onto adjacent floodplains and into coastal lagoons and ponds during the flood season before it flowed into the sea (Petr 1986).

Outstanding or Distinctive Biodiversity Features

The Volta Basin has low endemism and moderate richness among all taxa. Only nine fish, one crab, and one frog are known endemics. About 145 fish species, 40 aquatic reptiles, and 25 aquatic mollusks inhabit the fresh waters of the Volta. Aquatic mammals present in the ecoregion are marsh mongoose (*Atilax paludinosus*), African clawless otter (*Aonyx capensis*), spot-necked otter (*Lutra maculicollis*), hippopotamus (*Hippopotamus amphibius*), and the vulnerable West African manatee (*Trichechus senegalensis*). The single endemic frog, *Phrynobatrachus francisci*, congregates along floodplains and ponds to spawn during the rainy season (Rödel 2000).

Cyprinids, mormyrids, mochokid catfishes, and characins dominate the fish fauna of the Volta. Most of these riverine species are insectivores, substrate feeders, or fish predators (Payne 1986). Many, including species of *Alestes, Citharinus, Distichodus,* and *Labeo,* are adapted to the seasonal floods, moving upstream and onto the floodplains to spawn and feed at the onset of flooding (Lowe-McConnell 1985).

Many species that have adaptations to riverine habitats have declined or been lost from the system as a result of damming.

The endemic fish *Steatocranus irvinei* is a rheophyllic species with ventral fins modified as suckers to attach itself to rocks in rapidly flowing waters. This riverine species was originally found in Senchi Rapids and rapids of the Bui Gorge on the Black Volta (Petr 1986). The freshwater prawn *Palaemon palucidens* was once abundant in the lower Volta and supported an active fishery but has declined since dam construction (Petr 1986).

Status and Threats

Current Status

Construction of dams on the Volta at Akosombo and Kpong has changed the ecology of the river and its floodplains. The reservoir flooded a vast area of wetland and displaced the resident large mammal fauna (Hughes and Hughes 1992). In 1981, the much smaller Kpong Dam closed, creating the Kpong Headpond only 80 km from the delta of the Volta River (Biney 1990). Although fish yield has increased behind Akosombo Dam, downstream fisheries have declined and the composition of the fish community has changed. Predominantly herbivorous fish replaced the insectivorous characins and cyprinids (Lowe-McConnell 1985). Additionally, Mormyridae species have declined in the waters behind the dam, presumably because their benthic habitats are now inundated with deoxygenated hypolimnetic waters (Lévêque 1997). The Akosombo Dam has drastically reduced the sediment supply to the sea, resulting in erosion of the coastlines of Togo and Benin at a rate of 10–15 m/year (World Commission on Dams 2000). The reduced nutrient supply to the floodplains of the delta has led to a decline in agricultural productivity and an increase in mangrove cutting (Rubin et al. 1998). Changes in the flow regime below the dam have also caused a decrease in fish abundance in the coastal Avu and Keta lagoons through lower water levels and higher salinities in the delta (Hughes and Hughes 1992).

The dams have compartmentalized the basin, preventing fish from migrating upstream to their spawning grounds with the onset of flooding. Populations of known migratory species, such as *Alestes baremoze* and *A. dentex*, presumably have declined, although there is a lack of studies to confirm this.

Forests originally dominated the southern area of Lake Volta, but most have been cut. Recently, runoff from deforested areas has been causing increased sedimentation in the lake (Kalitsi and Evans-Appiah 1999).

The White, Black, and Red Volta rivers originate in the northern portion of the ecoregion, where drought and soil erosion are serious problems (Fontes et al. 1999). A few of the headwater reaches lie in protected areas that provide some defense against poor land use. In the northwest part of the ecoregion, the floodplains of the Black Volta remain little disturbed, and the gallery forests along its banks are protected in the north for 100 km within the borders of the Deux Bales Classified Forest (Burkina Faso) (Hughes and Hughes 1992). Further south, the Black Volta runs through Bui National Park (Ghana) for about 60 km, and the Red Volta passes through Pô National Park for about 100 km in Burkina Faso. The upper floodplains of the Pendjari River are located in Pendjari National Park (Benin) in the northeastern portion of the ecoregion (Hughes and Hughes 1992). This park is part of a complex of parks that, because of their combined size and level of intactness, are considered critical for conservation in West Africa (East 1999).

Future Threats

Already largely disturbed, this ecoregion also faces various future threats. High-intensity fishing occurs on the floodplains of the rivers in this ecoregion, often with the use of pesticides such as lindane (Hughes and Hughes 1992; Lévêque 1995). Treatment of rivers with chlorophoxim to control the spread of onchocerciasis (river blindness) among human populations is ongoing. Despite long-term monitoring of these pesticides on nontarget species, the long-term effects on the aquatic fauna are still little known (Lévêque 1997). Cattle grazing and rice cultivation have increased below the dams in the former floodplains (Hughes and Hughes 1992). Particularly in coastal areas, untreated wastes from human, domestic, municipal, mining, and agricultural sources pollute many of the waterways of this ecoregion (UNEP 1999). Observed lead concentrations in *Macrobrachium* sp. are higher than World Health Organization standards (Biney et al. 1994).

With the highest growth rate in Ghana, the city of Accra poses a looming threat to both coastal rivers and lagoons downstream unless regulations for urban and industrial wastes can be properly enforced. Sewage and other urban pollutants are particularly threatening to the freshwater and brackish water lagoons along the coast.

Another dam, Bui Dam on the Black Volta, has been in the planning stages since the 1960s. If completed, the resulting reservoir would occupy 21 percent of the Bui National Park, filling 1,820 km^2 of the park surface (Kalitsi and Evans-Appiah 1999). There are concerns that the development would eliminate the populations of *Hippopotamus amphibius* that reside in the park.

Justification for Ecoregion Delineation

This ecoregion is defined by the boundaries of the Volta River Basin. The freshwater ichthyofauna of the Volta Basin is very similar to that of the Niger, with which faunal exchanges probably have occurred during the Holocene (Lévêque 1997). The Black Volta may have previously been a tributary of the Niger River between Gao and Niamey and only more recently been captured by the Volta (Lowe-McConnell 1985). Similarly, the Pendjari River probably was connected to the Niger but now feeds into the Volta (Lévêque 1997). Despite these recent connections, several fish species are endemic to the ecoregion, including *Irvineia voltae, Steatocranus irvinei, Synodontis arnoulti,* and *Barbus bawkuensis.*

Data Quality: Medium

Ecoregion Number:	**76**
Ecoregion Name:	Zambezian Headwaters
Major Habitat Type:	Savanna–Dry Forest Rivers
Final Biological Distinctiveness:	Bioregionally Outstanding
Final Conservation Status:	Relatively Intact
Priority Class:	V
Author:	Lucy Scott
Reviewer:	Paul Skelton

Location and General Description

The fast-flowing Zambezian headwater streams contribute to two major sub-Saharan rivers, the Zambezi and the Okavango. The ecoregion extends from the Okavango catchment in the west to the headwaters of the Kafue River in the east and includes the headwaters of the Upper Zambezi River. The majority of the rivers drain the interior of Angola, but parts of the upper Zambezi and the upper Kafue flow through northern Zambia. The headwaters of the Zambezi share a common watershed divide with the Congo River Basin to the northeast and with the headwaters of the Cuanza and other westward-draining rivers of Angola to the northwest.

Mean annual rainfall is varied across the region, declining sharply from east to west, from an annual average of 1,400 mm in the upper catchment of the Kafue River to 50 mm at the mouth of the Cunene (Hughes 1997). Seasonality is marked, with maximum rainfall falling throughout the region from October to March. Air temperatures average 20–25°C, with little variation in the humid east (Davies 1986; Barnard 1998).

The rivers of this ecoregion are permanent and characterized by steep gradients in places. However, these rivers are unlike many mountain headwaters in that high-gradient zones are discontinuous (Allanson et al. 1990). The major Upper Zambezi tributaries in the ecoregion are the Lungwebungu, Luanginga, Cuando, Luena, Dongwe, and Kabompo rivers (Bell-Cross 1972). The major tributary of the Kafue is the Lunga River. The Okavango tributaries, the Cubango and the Cuito, arise on the plateau of central Angola, and the Cunene River (1,200 km long) arises in west-central Angola. The Cunene's flow is disrupted by a series of rapids and by the 122-m-high Ruacana falls, located at the divide between the rim of the continental plateau and the Atlantic coastal slope (Roberts 1975).

The vegetation is predominantly miombo (*Brachystegia, Julbernardia* spp.) woodland containing an extensive network of grassy dambos along drainage lines, with dense gallery forests along major watercourses and a few dense patches of evergreen forest. Dambos are seasonally waterlogged regions set into the landscape through the weathering action of the lateral flow of groundwater (Desanker et al. 1997). Nutrients in water flowing into the dambos are adsorbed by clay particles or assimilated by organisms in the soil, so dambos are regions of intense biological activity. The woodland is interspersed with cultivated land and urban and industrial areas associated with mining. A region of mopane (*Colophospermum mopane*) and thorn scrub savanna (*Acacia* and *Sclerocarya*) is found to the southwest. It includes stretches of dense dry *Baikiaea plurijuga* forest, *Pterocarpus* woodland savanna, and semi-arid grassland (Stuart et al. 1990; Hughes 1997).

Shales, sandstones, dolomites, and quartzites of the Katanga system underlie the tributaries of the Upper Zambezi and Kafue. The associated soils are deep (1.8 m) and friable, varying from clays to sandy clay loams with clay content increasing with depth (Hughes 1997). The soil is ferralitic, and mineral reserves generally are low in the Zambezian headwaters (Davies 1986). The predominant upland soils in woodland areas are infertile, consisting of alfisols, oxisols, and altisols. Lithosols on ridge crests in drier areas are well drained but give way to poorly drained vertisols in the dambos (Desanker et al. 1997).

Outstanding or Distinctive Biodiversity Features

The upland tropical rivers of the Zambezian Headwaters ecoregion support a fauna that is well adapted to its clear, fast-flowing rivers. *Kneria polli, Amphilius uranoscopus,* and *Parakneria fortuita* are typical of these waters, where they live among rocky substrata and feed on larval Chironomidae, Trichoptera, and Odonata. *Schilbe yangambianus,* otherwise known from the western Congo Basin, further substantiates links between the Congo and Zambian systems (Jackson 1986). Five fish species, *Paramormyrops jacksoni, Parakneria fortuita, Hypsopanchax jubbi, Barbus bellcrossi,* and *Schilbe yangambianus,* are endemic to the headwaters of these rivers. Two rare dragonfly species, *Aciagrion rarum* and *Monardithemis flava,* are near-endemics to this ecoregion. The Nile crocodile (*Crocodylus niloticus*) occurs here, and populations are considered to be of reasonable size but nonetheless vulnerable (Stuart et al. 1990).

Status and Threats

Protected areas in Angola are afforded minimum protection because of years of civil unrest in the country. In Angola, the Kameia National Park (14,450 km^2) and the Luiana (8,400 km^2) and Mavinga (5,950 km^2) partial reserves are close to the border with Zambia (Stuart et al. 1990), although management and protection of these areas have been extremely difficult because of the civil war. In Zambia, the Liuwa Plain National Park (3,660 km^2) includes part of the Luanginga River, and the West Lunga National Park (1,680 km^2) includes the Lunga and Kabompo rivers. These rivers are mostly undisturbed, with low population levels throughout most of the ecoregion. However, as peace comes to this region, farming and logging intensity is expected to increase, and dams probably will be built on these headwater rivers.

Deforestation occurs in the eastern portion of this ecoregion. In northwest Zambia, near Mwinilunga, there are isolated patches of pristine lowland and gallery forest that are threatened by human encroachment. Their clearance would have serious negative impacts downstream (Stuart et al. 1990). The copper belt is located in the eastern portion of this ecoregion, and future mineral exploitation poses a further threat to the freshwater systems of the ecoregion (Curtis et al. 1998).

Justification for Ecoregion Delineation

This ecoregion is defined by the headwaters of the Upper Zambezi River and the headwaters of the Okavango, Kafue, and Cunene rivers. The Okavango, Upper Zambezi, and Kafue rivers have a very similar aquatic ichthyofauna, which is closely related to that of the upper Cunene. According to Skelton (1994), the early Tertiary drainage of the confluent Cunene, Okavango, Upper Zambezi, and Kafue rivers flowed in a southwesterly direction, draining to the ocean in the vicinity of the present-day Orange River mouth. This, together with drainage capture along the Zambezian-Congolian headwater divide, accounts for much of the faunal similarity of the Zambezian headwaters. The Zambezian headwaters share a number of species with the southern tributaries of the Congo River; therefore, some of the most upstream portions of these Congo headwaters have been included in the Upper Zambezian Headwaters ecoregion (Bell-Cross 1972; Skelton 1994). The capture of the Upper Zambezi and Kafue by the Middle Zambezi in the Plio-Pleistocene had little effect on the nature of the fauna of this ecoregion because Kafue Gorge and Victoria Falls presented impassable barriers to the upstream migration of fish (Skelton 1994).

Data Quality: Low

The level of biological investigation in this ecoregion is low. Angola is very poorly known biologically, and a substantial amount of further investigation is needed in this ecoregion before an adequate evaluation of biological distinctiveness can be made.

Ecoregion Number: **77**
Ecoregion Name: **Zambezian Lowveld**
Major Habitat Type: **Savanna–Dry Forest Rivers**
Final Biological Distinctiveness: **Bioregionally Outstanding**
Final Conservation Status: **Vulnerable**
Priority Class: **V**
Author: **Helen Dallas**
Reviewers: **Paul Skelton and Brian Marshall**

Location and General Description

Perennial and seasonal rivers and associated floodplains, swamp forests, swamps, seasonally inundated pans and grasslands, and coastal lakes of this coastal plains ecoregion support an extremely rich and diverse biota. The ecoregion extends from south of the Zambezi Delta in central Mozambique through the Tugela River system in South Africa. The major rivers are the Pungwe, Búzi, lower Save, Limpopo, lower Inkomati, Umbeluzi, Maputo (including Pongola and Usutu), Mkuze, Hluhluwe, and Tugela rivers, together with several important lakes, namely the lakes of Kosi Bay, Lake Sibaya, Lake St. Lucia, Lake Satine, Lake Piti, Poelela Lagoon, and numerous other smaller lakes and pans on the southern Mozambican coastal plain.

The ecoregion has a warm and humid climate, varying from moist subtropical in the east along the coast to moderately subtropical inland in the western portion of the coastal plain (Maud 1980). Annual rainfall averages 1,000 mm on the coast, decreasing to 600 mm at the foot of the mountain range that forms the western edge of the coastal plains. Along the Limpopo Valley there is an extensive area of very low rainfall (less than 400 mm/year, duration of rainy season less than 50 days) (Torrance 1981). The rainfall is extremely variable, and drought is the norm, although severe floods do occur from time to time. Rainfall occurs mainly during summer, from October to March.

Toward the coast perennial and seasonal rivers are fringed in places by seasonally inundated pans and lagoons. Swamps and mangroves are common in the lowland reaches, and coastal lakes, both estuarine and freshwater, are scattered on the coastal plateau. The northernmost Pungwe, Búzi, and Save rivers are seasonal rivers, with winter low flows confined to streams within the channel and summer flows overflowing the channel into adjacent floodplains along their lower reaches. At the coast these rivers form tidal mangrove swamps that are almost continuous from the Pungwe to the Save River, extending 50 km inland at the Pungwe mouth (Hughes and Hughes 1992).

The Limpopo, Inkomati, and Maputo rivers are all historically perennial, although water abstraction in their upper catchments has resulted in periodic cessation of flow in the lower reaches during the dry season. The Limpopo (390,000 km^2) has an extensive floodplain in Mozambique, with hundreds of seasonal pans and approximately 170 km² of semipermanent swamps (Hughes and Hughes 1992). The Inkomati River flows through the coastal plains of Mozambique, where it forms floodplain swamps. The Pongola River, which forms part of the Maputo-Usutu River System, floods seasonally, and its extensive floodplain pans capture and retain floodwater when the river overflows its banks. It forms part of the Ndumo Game Reserve.

Several coastal lakes and lagoons, ranging from freshwater to saline, are scattered throughout southern Mozambique and northern KwaZulu-Natal in South Africa. Of these, Lake St. Lucia is the largest (1,555 km²) and possibly the most important saline coastal lake in Africa because of its faunal composition that includes both marine and freshwater elements and because of its extent, biotic diversity, and pristine character (Cyrus 1989). The highly variable nature of this system, in terms of hydrology and salinity ranges, has influenced the biota of the lake, which is diverse and adapted to variability (Cyrus 1989). Other lakes include the estuarine-linked Kosi Bay (110 km²), which is one of the best-preserved large estuaries in South Africa, and Lake Sibaya (77.5 km²), which is the largest freshwater coastal lake in South Africa.

There is a diverse floral landscape that includes open woodland, coastal thicket, and palm veld, with coastal grassland and dune forest along the coast. Swamp forest, papyrus swamp, marsh and sedge communities, mangroves, and riverine woodland occur among the numerous wetlands on the coastal plain. The floral assemblage is diverse, and some 2,180 species of flowering plants have been recorded in the St. Lucia system alone (Cowan 1995b).

The geology of the ecoregion consists almost entirely of flat, low-level coastal plain, with pale sandy soils overlying Cretaceous beds. The coastal plain is separated from the Indian Ocean by an almost continual line of forested dunes composed of both Holocene and Pleistocene sand deposits. Alluvial terraces run along the rivers that flow eastward (Maud 1980). The soils of the eastern part of the plain are very sandy and infertile, whereas those of the western part are inherently fertile to very fertile, and the pans often have a high clay and peat content.

Outstanding or Distinctive Biodiversity Features

The extensive inland and coastal wetlands in this ecoregion support a moderately rich aquatic biota. In particular, the aquatic herpetofauna and several groups of aquatic invertebrates are very rich. Large numbers of wetland birds also congregate at the many wetlands that dot the coastline.

The ecoregion has a very rich flora and is globally recognized as a center of plant diversity (www.ccwr.ac.za/wetlands/st_lucia_ris.htm). Lake St. Lucia, for example, has some 2,180 species of flowering plants, many of which are rare or endangered species. Because of its high diversity and endemism, the lake is of special value for maintaining the genetic and ecological diversity of the flora of this ecoregion. The Kosi system has a number of rare and threatened plants and has five mangrove species.

Eight aquatic or semi-aquatic mammal species live in and alongside the watercourses in the ecoregion, including the hippopotamus (*Hippopotamus amphibius*), water mongoose (*Atilax paludinosus*), southern reedbuck (*Redunca arundinum*), vlei otomys or African swamp rat (*Otomys irroratus*), African marsh rat (*Dasymys incomtus*), clawless and spotted-necked otters (*Aonyx capensis* and *Lutra maculicollis*), and a marine mammal, the endangered dugong (*Dugong dugong*). Several savanna animals use the coastal lakes and swamps seasonally, migrating to them in the dry season when inland pans and rivers are dry.

Avifaunal richness is high; for example, Lake St. Lucia supports more than 350 bird species and is the most important breeding area for waterbirds in South Africa. The Kosi system has 296 bird species. Among those inhabiting the swamp forests are the African finfoot (*Podica senegalensis*), the white-backed night heron (*Gorsachius leuconotus*), and Pel's fishing owl (*Scotopelia peli*). More than 20,000 waterbirds have been recorded at Lake Sibaya, an important wetland that provides a link between the Kosi Bay and Lake St. Lucia systems (Barnes et al. 2001).

The ecoregion is an important destination or stopover for migratory birds and is one of the principal avifaunal breeding areas in southern Africa. Lake St. Lucia is of particular importance for congregations of *Pelecanus onocrotalus*, *Sterna caspia*, *Platalea alba*, *Phoenicopterus ruber*, *Anas undulata*, *A. smithii*, *Recurvirostra avosetta*, and *Larus cirrocephalus* (Barnes et al. 2001).

Cichlids, cyprinids, gobies, and mochokid catfishes dominate the fish of this ecoregion. Many species are found in fresh, brackish, and saline waters, and several catadromous species spend part of their life cycle in the freshwater coastal rivers and streams (e.g., several members of the Anguillidae family). About 120 freshwater fish species inhabit these waters, of which 22 are endemic. The endemics are cichlids, cyprinids, gobies, kneriids, eleotrids, and aplocheilids and mochokid, amphiliid, and claroteid catfish. Interesting endemics include several rock catlets (*Chiloglanis* spp.) that live in rocky riffles and rapids; the Sibayi goby (*Silhouettea sibayi*), whose largest known population occurs in Lake Sibaya; and the brightly colored turquoise killifish (*Nothobranchius furzeri*), which is limited in distribution to the ephemeral pans of the Gonarezhou National Park in Zimbabwe (Skelton 1994).

The richness of several groups of aquatic invertebrates, including Odonata, Ephemeroptera, and Trichoptera, is high. About 150 Odonata species are known from the ecoregion, and Lake St. Lucia has 52 dragonfly species. It also has 115 species of benthic amphipods, and Lake Sibaya has one endemic copepod (*Tropocyclops brevis*). A particular butterfly species, a skipper (*Parnara micans*), is regarded as endemic to Kosi Bay (Campbell 1969). Richness is also high for freshwater mollusks, with twenty-seven species occurring in the ecoregion, although only one (*Eussoia leptodonta*) is endemic, and this one is known only from the estuary of the Komati River.

This ecoregion is one of the richest areas for reptile and amphibian species in southern Africa. About seventy amphibians have been recorded in the ecoregion, of which three, *Afrixalus delicatus, Hyperolius pickersgilli,* and *Cacosternum striatum,* are endemic. Seventeen reptile species have been recorded, including the water monitor (*Varanus niloticus*), African rock python (*Python sebae*), brown water snake (*Lycodonomorphus rufulus*), olive marsh snake (*Natriciteres variegata sylvatica*), northern green water snake (*Philothamnus irregularis*), common green water snake (*P. hoplogaster*), forest cobra (*Naja melanoleuca*), and Nile crocodile (*Crocodylus niloticus*). Lake St. Lucia and its associated freshwater systems support one of the largest populations of Nile crocodiles on the subcontinent, and it is estimated that 1,500 individuals each measuring more than 2 m long are present.

Status and Threats

Many areas in this ecoregion are considered wetlands of international importance and are protected under the Ramsar convention; they include Lake St. Lucia, Lake Sibaya, the Kosi Bay system, and the Ndumo Reserve (which includes part of the Pongola and Usutu rivers) in South Africa (www.ccwr.ac.za/wetlands/ramsar_map.html). A few national parks in Mozambique protect coastal lakes, but these represent less than 1 percent of the total area of lake habitats, and the full spectrum of diversity of the lakes has not been protected (Hughes and Hughes 1992).

Most of the rivers, including the Save, Limpopo, and Inkomati, that flow into Mozambique from the neighboring countries such as South Africa, Zimbabwe, and Botswana have been impounded, and abstraction of water for irrigation is widespread (Vas and Pereira 2000). These activities change aquatic habitats downstream by altering the flow regime of a system (especially by reducing peak floods and increasing dry weather flows) and blocking the movements of fish and other species.

Mining is a threat in some areas, including the coastal dunes, which are important sources of minerals (Burgis and Symoens 1987; Kruger et al. 1997). Indigenous vegetation, including swamp forest, is under threat from slash-and-burn methods of cultivation and from afforestation schemes involving alien species such as pines (*Pinus elliottii*) and *Eucalyptus* spp. Plantations of alien species occur throughout the ecoregion and use larger quantities of water than native vegetation.

Proposals to extend irrigation farming along the Inkomati and Limpopo rivers are likely to lead to shrinkage of the seasonal floodplains. The coastal lakes in Mozambique are heavily used for irrigation, and canals have been dug back from many small lakes so that marginal vegetation can be drained, irrigated, and cultivated (Hughes and Hughes 1992).

Many of the endemic fish are classified as rare or vulnerable and are threatened by human disturbance, the deterioration of riverine and estuarine habitats through water pollution and abstraction, and antimalarial and tsetse fly spraying programs (Skelton 1993). Alien fish invasions are also a threat to native fish populations. For example, *Oreochromis niloticus* has recently entered the Limpopo River system, and it is feared that it will interbreed with the native *Oreochromis mossambicus*. Other invasive fish include *Oreochromis macrochir, Cyprinus carpio, Micropterus salmoides, M. dolomieu, Pharyngochromis acuticeps,* and *Hypophthalmichthys molitrix* (van der Waal and Bills 2000).

A transfrontier park has recently been created that spans the boundaries of South Africa, Mozambique, and Zimbabwe. The Great Limpopo Transfrontier Park includes South Africa's Kruger National Park, Mozambique's Limpopo National Park, and the Gonarezhou National Park in Zimbabwe, as well as a corridor of land that links Gonarezhou National Park with Kruger National Park. The total surface area of this park will be approximately 35,000 km^2. Another transfrontier conservation project is the Lubombo Transfrontier Conservation Area (LTFCA), which will link the Tembe Elephant Park in South Africa with the Maputo Special Reserve in Mozambique. This will create the first major elephant stronghold along Africa's eastern coastline and will also conserve a unique mosaic of terrestrial and aquatic habitats along the East African coast (Bills 2001). The overall area of the LTFCA will be 25,000 km^2. The creation of the Futi Corridor between the Tembe Elephant Park and the Maputo Special Reserve will increase the size of the protected area in Mozambique from 780 km^2 to 1,600 km^2.

As a result of increasing human populations, overfishing is potentially a problem in the ecoregion, but the intensity of fishing, particularly in coastal pans and lakes, is largely unknown at present. An assessment of the fisheries should be conducted, taking into account the importance of fishing to local communities (Bills 2001).

Justification for Ecoregion Delineation

This ecoregion is at the interface between tropical and subtropical African biota, and the tropical Zambezian and temperate fish faunas overlap in this ecoregion. The ecoregion is defined by the low-lying portions of the coastal rivers south of the Zambezi Delta to Lake St. Lucia. This ecoregion is included in the Zambezian bioregion because the rivers have historically been connected with the Zambezi River (Skelton 1994; Marshall 2000a). As one moves south, the fish fauna changes from predominantly Zambezian in origin to temperate, with few Zambezian fishes remaining south of Lake St. Lucia. The ecoregion is considered part of the tropical east coast region, defined by Skelton (1993), which is characterized by low-gradient mature systems with floodplain reaches.

Data Quality: Medium

Many of the aquatic systems in this ecoregion have been extensively studied, particularly the coastal lakes of South Africa (Whitfield and Cyrus 1978; Whitfield and Blaber 1978–1979a, 1978–1979b, 1978–1979c; Allanson 1979; Bruton and Cooper

1980). However, less work has been completed on the coastal rivers in Mozambique.

Ecoregion Number:	**78**
Ecoregion Name:	Zambezian (Plateau) Highveld
Major Habitat Type:	Savanna–Dry Forest Rivers
Final Biological Distinctiveness:	Nationally Important
Final Conservation Status:	Critical
Priority Class:	IV
Author:	Helen Dallas
Reviewer:	Paul Skelton

Location and General Description

The Zambezi Highveld, situated on the great southern African central plateau (IUCN 1992), includes the headwaters and highland streams of the Zambezi River Basin in the north, the Save River in the east, and the Limpopo River in the south. The aquatic habitats found on this plateau are large and small rivers, numerous dambos, a few artificial reservoirs, and isolated floodplains. The ecoregion is located in Zimbabwe.

Although the ecoregion is within tropical latitudes, it has a cool climate because of its altitude (more than 600 m). There is a warm rainy season (November to March) followed by a cool, dry season (April to mid-August) and then a hot, dry season (mid-August to October) (Gratwicke 1999). Rainfall varies from less than 400 mm per year in the Save and Limpopo catchments to 1,000 mm in some of the central areas (Hughes and Hughes 1992).

The headwater streams of the Highveld are small and clear but revert to swollen and turbid rivers after the rains (Gratwicke 1999). Rivers and streams of this plateau flow in two directions, with some feeding the Zambezi River system and others feeding the Save River system. Both rivers then flow through Mozambique and into the Indian Ocean. Perennially waterlogged dambos are widespread and cover approximately 12,000 km^2 (Owen 1994). Most dambos occur at an altitude above 1,200 m and are associated with a mean annual rainfall greater than 800 mm. Most streams depend on dambos for their dry season flow (Magadza 2000). Two prominent wetlands are the Chipinda Pools and the Save-Runde floodplain. Chipinda Pools is a cluster of large perennial pools in the Lundi River valley, located in Gonarezhou National Park. The 40-km^2 Save-Runde floodplain, which includes the Tamboharta Pan, occurs in the southeastern part of the ecoregion and is one of the few floodplain areas in it (Chabwela 1994a).

This ecoregion falls mostly in the terrestrial Zambezian biogeographic zone, and the vegetation is predominantly dry miombo woodland. Grassland occurs along the Great Dyke, a broad ridge in the center of the ecoregion (IUCN 1992). Soils, which are largely derived from gneissic granite, are sandy and well drained, with low fertility (Campbell 1994).

Outstanding or Distinctive Biodiversity Features

Although moderately rich in aquatic species, this ecoregion has no known endemics. Aquatic mammals include the marsh mongoose (*Atilax paludinosus*), African clawless otter (*Aonyx capensis*), and hippopotamus (*Hippopotamus amphibius*). About thirty-nine fish species, thirty-eight aquatic amphibians, eight aquatic reptiles, and seventeen freshwater mollusks live in the waters of the Zambezian Highveld. The wetland butterfly *Mashuna mashuna* has been recorded in high-level dambos (Gardiner 2000).

Several of the river systems, including the Pungwe and Save rivers, have an impoverished fish fauna (Bell-Cross and Minshull 1988). The families Alestiidae, Amphiliidae, Anguillidae, Cyprinidae, Cichlidae, Clariidae, Kneriidae, Mochokidae, Mormyridae, and Schilbeidae are represented.

Little information is available on the aquatic ecology of the numerous dambos in the region, although they are known to provide cover and food for indigenous terrestrial fauna and migratory birds (Katerere 1994). Dambos also provide a distinctive habitat for aquatic vascular plants; of the 109 dambo species recorded, 8 are found exclusively in this habitat (Magadza 2000). Both the Chipinda Pools and the Save-Runde floodplain are rich in bird life and provide watering areas for several large mammals (Fishpool and Evans 2001).

Status and Threats

Dam construction and abstraction of water for irrigation have negatively affected many of this ecoregion's rivers and dambos. In recent years many dambos have dried out and become badly eroded, and they are under continual threat from agriculture, overgrazing, gully erosion, and fires (du Toit 1994). Although information on the ecological consequences of dambo destruction is limited, it could cause the extirpation of the many wetland vascular plant species known from this ecoregion's highveld flora (Magadza 2000).

Impoundments on the Save and Lundi rivers have altered the flow regime, with attendant habitat changes (Hughes and Hughes 1992). The Save has been reduced to a heavily silted river characterized by flash floods and trickle flow (du Toit 1994). Irrigation schemes along the course of the Save River and agricultural inputs have contributed to the eutrophication of local wetlands (du Toit 1994). Four artificial reservoirs on the upper reaches of the Manyame River are highly eutrophic because of runoff from Harare (Marshall 1994). Several informal impoundments on the Mazowe River (part of the Zambezi Basin) have been created through dredging and excavation of the riverbank and river bed in the process of gold panning (Mkwanda 1994).

Grazing, farming, and destruction of vegetation have led to

soil erosion and consequent sedimentation in the lower reaches of several rivers (Chabwela 1994b). In the Save-Runde junction important bird area, the Save River carries a high sediment load from upstream land use (Fishpool and Evans 2001). Continued overexploitation of the wetlands in the region for agriculture will have further negative effects on freshwater systems.

Alien fishes also threaten the native fauna of these streams. Gratwicke and Marshall (2001) found that in streams in Zimbabwe, total fish abundance was lowered by 50 percent or more in the presence of the exotic predators *Micropterus salmoides* and *Serranochromis robustus*. These exotics appeared to affect most the abundance of the small fluvial barbs that are threatened throughout much of southern Africa.

Justification for Ecoregion Delineation

This ecoregion is delineated based on the northern subregion of the highveld (temperate) aquatic region of Skelton (1993) and follows the contours of the interior plateaus of Zimbabwe above about 600 m. The fauna is largely Zambezian. It is believed that the fish fauna originated from the more tropical equatorial region and that deteriorating ecological conditions during the Pleistocene have resulted in reduced or depauperate faunas in many rivers of this ecoregion (Bell-Cross and Minshull 1988).

Data Quality: Medium

With the exception of fish (Marshall in prep.; Bell-Cross and Minshull 1988), ecological information is limited, particularly for invertebrates (Mtetwe 2000). A substantial amount of research has been undertaken on the several artificial reservoirs in the region, but little work has focused on the riverine environments or the dambo areas (Marshall in prep.). Marshall and Gratwicke (1998–1999) urge further monitoring of the occurrence and distribution of fish species throughout Zimbabwe. In particular, they cite the lack of information available for the southern barred minnow (*Opsaridium peringueyi*); all four specimens of this species are more than 40 years old, and the species has not been collected in recent years, suggesting that it may be imperiled.

Ecoregion Number:	**79**
Ecoregion Name:	**Karstveld Sink Holes**
Major Habitat Type:	**Subterranean and Spring Systems**
Final Biological Distinctiveness:	**Bioregionally Outstanding**
Final Conservation Status:	**Endangered**
Priority Class:	**IV**
Author:	**Lucy Scott**
Reviewer:	**Paul Skelton**

Location and General Description

The Karstveld sinkhole lakes and caves are distinctive subterranean habitats supporting a substantial number of endemic aquatic species within several taxonomic groups. *Karstveld* is the name given to extensive dolomite and limestone formations located to the southeast and east of the Etosha Pan in Namibia. The area covers the Otavi Mountains and continues west through the northern parts of Outjo District as far as Otjovasandu (Irish 1992).

Sinkhole lakes and caves are karst features characteristic of dolomite that are formed as water penetrates cracks in the dolomite and leaches away the rock. This process forms water-filled underground caves, and sinkholes result when the roof of a cave collapses. The karstic waters are permanent and mainly underground (Irish 1992; Barnard 1998; Curtis et al. 1998).

The Karstveld ecoregion receives an average rainfall of 600–700 mm annually, with distinct wet and dry seasons including summer rains (Curtis et al. 1998). Mountainous savanna dominates the vegetation in the Karstveld ecoregion.

The Karstveld groundwater system consists of a number of distinct waterbodies including subterranean cave lakes, cenotes, and hemicenotes. Cave lakes occur in laterally developed horizontal cave systems. Cenotes are water-filled sinkholes, and hemicenotes are cenotes with openings that are small relative to the size of the water surface below ground. In the Karstveld ecoregion, the main waterbodies are the cave lake of Aigamas Cave (33 km northwest of Otavi); Dragon's Breath Cave, the largest underground lake in the world, with an approximated volume of 1,140,000 m^3; the hemicenotes Harasib (63 m deep) and Aikab (58 m deep); and the cenotes Lake Otjikoto (58 m deep) and Lake Guinas (130 m deep) (Irish 1992). The sinkhole lakes are moderately clear, and water temperatures average 19–27°C (Stuart et al. 1990).

Outstanding or Distinctive Biodiversity Features

The Karstveld systems have low species richness but a high level of endemism, even between waterbodies within the system. At least six stygobiotic amphipods of the family Ingolfiellidae, four isopods, and two endemic fish are known from the waters of this ecoregion. Among the amphipods, *Trogloleleupia dracospiritus* occurs only in Dragon's Breath Cave, and *T. gobabis* is found only in Arnhem Cave. The aquatic isopod *Namibianira aikabensis* occurs in the Aikab hemicenote and is limited in distribution to Namibian caves (Marais and Irish 1997). Many of the waterbodies of the Karstveld remain unexplored biologically, and future sampling probably would reveal new species (Irish 1992; Curtis et al. 1998).

There are three native fish species known to the Karstveld. *Clarias cavernicola*, endemic to Aigamas Cave, is southern Africa's only true cavefish (Skelton 1993). This depigmented fish has small eyes and feeds on bat droppings, animal carcasses, and insects that fall into the water from overhead (Skelton 1994). The species is endangered and may have a population of less than 200 individuals (Irish 1992; Barnard 1998). The endangered polychromatic Otjikoto tilapia (*Tilapia guinasana*) is endemic to Lake Otjikoto (Skelton 1987; Irish 1992). *Pseudocrenilabrus philander* is known from lakes Otjikoto and Guinas and is widespread elsewhere in southern Africa (Skelton 1993).

Dragon's Breath Cave contains the largest subterranean lake in the world. At least sixteen invertebrate species, including the endemic amphipod *Trogloleleupia dracospiritus* and one bat, *Hipposideros caffer,* inhabit the cave. This cave supports a detritus-based system, with all input to the food web coming from allochthonous sources, largely from dried bat guano (Irish 1992).

Status and Threats

Current Status

The unique habitats of the Karstveld sinkhole lakes and caves are threatened by the constant demand for water in this arid region. The Eastern National Water Carrier is the largest IBT in Namibia, and it draws a significant proportion of water from large natural aquifers in the Karstveld area, in the vicinity of Grootfontein. This has caused a drop in the water table and resulted in springs at ground level in the ecoregion drying up, having significant implications for the wetland biota (Irish 1992; Davies and Day 1998). Irrigation draws water from the waters of lakes Guinas and Otjikoto, and the Tsumeb mine also draws a significant volume from the latter (Barnard 1998). Pesticides used on the cultivated fields surrounding lakes Guinas and Otjikoto may make their way into the waters either through seepage or via insectivorous bats that defecate in the caves (Irish 1991).

The introduction of exotic animal species has also had an effect on the species composition of the lakes. *Pseudocrenilabrus philander* was the only recorded species in Lake Otjikoto before the introduction of *Tilapia guinasana* from Lake Guinas. The two species have different nesting habitats, but *P. philander* is threatened by the more recent introductions of *Oreochromis mossambicus, Poecilia reticulata,* and *Xiphophorus hellerii* (Skelton 1987). *Tilapia guinasana* is naturally rare and is currently found only in lakes Guinas and Otjikoto and a few artificial ponds, rendering it highly susceptible to the impacts of habitat destruction. The population in Lake Otjikoto has declined because of the introduction of *Oreochromis mossambicus,* although *Tilapia guinasana* is itself introduced into this lake (Skelton 1987).

Salinization and pollution also threaten the Karstveld habitats and their unique faunas (Barnard 1998). The pumps used to extract water from Lake Guinas contribute diesel pollution to the lake, and litter pollutes Lake Otjikoto (Irish 1991). Pesticide runoff from surrounding agricultural lands is a potential threat to both lakes (Skelton 1987). Lake Otjikoto is a national monument, but other waterbodies of the system are not officially protected (Davies and Day 1998).

Future Threats

The major future threat to the Karstveld system is likely to be the loss of aquatic habitat caused by water abstraction. Namibia is the most arid country in southern Africa, and water is in great demand. Ecosystems already threatened by the abstraction of water are likely to be more vulnerable to further introductions of aliens and even to single pollution events (Irish 1991, 1992; Barnard 1998).

Justification for Ecoregion Delineation

The boundaries of this ecoregion are defined by the extent of Karstveld, the extensive dolomite and limestone formations located to the southeast and east of the Etosha Pan in Namibia. The biogeographic affinities of the Karstveld fauna are Zambezian, and a number of taxa are shared with the Okavango, Cunene, and Zambezi rivers. The freshwater fauna is characterized by stygobiotic, endemic species. The lengthy period of isolation that the Karstveld fauna has experienced and the particularly stable conditions in the lakes have led to the evolution of an endemic fauna (Skelton 1987, 1990b).

Data Quality: Medium

Several of the caves and their lakes in this ecoregion have been mapped and explored (Irish 1991, 1992; Marais and Irish 1997), although further studies are warranted. These studies should determine the status of isolated populations and the effects that water abstraction has had on these systems. Irish (1991) recommends the collection of information on the cryptic fauna of springs, the stygobiotic fauna (aquatic species living permanently underground in the dark zone of caves) of unsampled subterranean waters, and more detailed geohydrological information.

Ecoregion Number: **80**
Ecoregion Name: **Thysville Caves**
Major Habitat Type: **Subterranean and Spring Systems**
Final Biological Distinctiveness: **Globally Outstanding**
Final Conservation Status: **Endangered**
Priority Class: **I**
Authors: **Ashley Brown and Robin Abell**
Reviewer: **Gordon McGregor Reid**

Location and General Description

A depauperate but distinctive aquatic fauna lives in the subterranean streams of the Thysville Caves. The caves are located just south of the mainstem Congo and downstream of Malebo Pool. The subterranean complex consists of two suites of caves (Banister 1986). They are located on two separate Congo tributary systems and do not appear to possess subterranean connections to each other, although the fauna inhabiting them is quite similar. This cave ecoregion lies at about 600–800 m a.s.l., and the surrounding land is even and flat and sits at about 500 m a.s.l. In the dark environment of the caves, photosynthesis is not possible; the energy supply providing the foundation for the cave's food chain comes from allochthonous inputs from the surrounding forest.

The climate of the Thysville Caves ecoregion is tropical and wet. The mean annual temperature in the ecoregion is approximately 24°C (Hughes and Hughes 1992). Mean annual rainfall in the nearby city of Mbanza-Ngungu is about 1,500 mm/year (Hughes and Hughes 1992).

Outstanding or Distinctive Biological Features

The Thysville Caves ecoregion hosts two fish species, one of which is subterranean, and several bird, aquatic mammal, and mollusk species. The most famous inhabitant is the endemic Congo blind barb (*Caecobarbus geertsii*). There is speculation that *Barbus holotaenia* is the closest surface water ancestor of this hypogean species (Banister 1986). However, because *B. holotaenia* lives outside the ecoregion, some researchers suggest that the true ancestor could actually be a more localized, as-yet undiscovered species from a nearby surface stream, most of which are poorly surveyed. Populations in the two separate caves appear to have the same characteristics, but it is unknown whether the populations are the same species and able to interbreed (Banister 1986). *C. geertsii* is the only species in its genus and one of only three species of blind cavefish known from tropical West Africa (Beadle 1981). The other fish in the ecoregion, the pinkfin alestes (*Alestes grandisquamis*), is a pelagic surface water fish with a widespread distribution in the Congo Basin.

Several aquatic mammals, mollusks, and waterbirds live above ground in this small ecoregion. Inhabitants include the marsh mongoose (*Atilax paludinosus*) and the mollusks *Etheria elliptica* and *Lentorbis benguelensis*. Birds include the Egyptian goose (*Alopochen aegyptiacus*), spur-winged goose (*Plectropterus gambensis*), and African pygmy-goose (*Nettapus auritus*).

Status and Threats

The current status of the ecoregion is unknown. The region is heavily populated, and the caves are located on the edge of a degraded forest (Kamdem Toham et al. 2003). The primary threats to the Thysville Caves include changes in the hydrology of the small rivers that feed these caves, increasing human population, and associated deforestation in the area.

Justification for Ecoregion Delineation

This ecoregion is defined by the extent of the Thysville Cave system and is characterized by a depauperate but endemic aquatic fauna.

Data Quality: Low

These caves are poorly known scientifically and should be surveyed for other stygobiotic taxa (Kamdem Toham et al. 2003).

Ecoregion Number: **81**
Ecoregion Name: Dry Sahel
Major Habitat Type: Xeric Systems
Final Biological Distinctiveness: Nationally Important
Final Conservation Status: Relatively Intact
Priority Class: V
Authors: Ashley Brown and Michele Thieme
Reviewer: Christian Lévêque

Location and General Description

Scattered oases throughout the Dry Sahel support a depauperate aquatic fauna that has evolved drought-resistant life stages. This vast ecoregion extends from the Atlantic coast in the west to the border of the Nile ecoregions in the east and is situated between the Lake Chad catchment in the south and the Maghreb in the north.

In lowland areas, ephemeral streams (also called oueds or ouadis), which experience sporadic flooding from rainfall upstream at higher elevations, occur across the landscape. Water often is just below the land surface, and oases dot the desert where water reaches the surface. Permanent or semipermanent pools (also called *gueltas*) occur along stream beds in mountainous areas.

Rainfall averages about 30–50 mm per year in the Sahara, although some places receive no rain at all, and mountainous areas may receive up to 100 mm. Rains come for a short period, normally during the summer months of July and August. However, rainfall is unreliable, and vast areas of desert may receive no rain for years at a time. The Sahara also experiences extremely strong winds that can help disperse the resting stages of crustaceans and aquatic insects (Rzóska 1984).

Outstanding or Distinctive Biodiversity Features

The Dry Sahel ecoregion has freshwater pockets that sometimes host surprisingly rich assemblages of species. Crustaceans, aquatic insects, fish, and amphibians inhabit permanent or seasonal pools. Given the arid, harsh conditions of the desert, nearly all species have hibernating or resting stages to avoid desiccation (Dumont 1987).

Fifteen fish species occur in the widely scattered and small isolated patches of water in the Sahara. Borkou-Ennedi-Tibesti area, in the east of the ecoregion, has the highest fish diversity, with eight fish species in Tibesti and seven in Ennedi (Lévêque 1990). Altogether, eleven fish species have been identified in Borkou-Ennedi-Tibesti waterbodies (Lévêque 1997). All of these are relictual fish populations from when these drainages were connected to the Niger or the Chad systems (Lévêque 1990). Many of the species are able to survive harsh environmental conditions. The widespread North African catfish (*Clarias gariepinus*) is highly omnivorous and is also capable of aerial breathing. In the Mauritanian portion of the ecoregion, a few fish species are known: *Barbus macrops, Barbus pobeguini, Clarias anguillaris,* and *Sarotherodon galilaeus* (Lévêque 1997). The gray monitor (*Varanus griseus*) is common, and the Nile monitor (*Varanus niloticus*) has been also recorded. There are also crocodiles in Tibesti and the Ennedi Massif (Wetlands International 2002).

The invertebrate fauna of this ecoregion is rich. Many invertebrates have been recorded in the Mauritanian waterbodies. These include crustaceans, mollusks, and the African freshwater medusa (*Limnocnida tanganyicae*). *Pseudagrion hamoni,* a relict damselfly species, and *Ischnura saharensis* and *I. senegalensis* also occur. In the Air Mountains, odonates, mollusks, and water beetles are all common. The odonate fauna of the ecoregion includes *Anax imperator, Orthetrum brachiale,* and *I. saharensis* (Dumont 1987).

The Dry Sahel ecoregion provides habitat for a surprisingly large number of migrating and native waterbirds, particularly along the coast. Banc D'Arguin National Park, located on the Atlantic coast in northern Mauritania, hosts more than 2 million waders and thousands of other migratory birds (Shine et al. 2001). Although this coastal wetland receives little freshwater input, it is highly productive because of upwellings in the Atlantic and is an extremely important feeding ground for migratory waterbirds, supporting the largest concentration of wintering waders in the world. Among the many species visiting the park are black tern (*Chlidonias niger*), greater flamingo (*Phoenicopterus ruber*), ringed plover (*Charadrius hiaticula*), grey plover (*Pluvialis squatarola*), red knot (*Calidris canutus*), redshank (*Tringa totanus*), bar-tailed godwit (*Limosa lapponica*), European spoonbill (*Platalea leucorodia*), and white pelican (*Pelecanus onocrotalus*).

Status and Threats

Current Status

Human population density throughout the ecoregion is generally low, but aquatic systems are under human pressure because of the scarcity of water. Livestock overgrazing has severely degraded some of the rangelands and polluted some of the waterways. The application of insecticides to combat desert locust (*Schistocerca gregaria*) may contaminate ephemeral ponds because of their small size and may also contaminate larger waterbodies (Lahr 1998).

Desertification and climate change are threatening the landscape. Since the 1950s, the Sahara Desert has been advancing southward. The causes of the desiccation appear to be a mix of natural and human-induced changes, including sea surface temperature variation in the oceans and loss of vegetation to aridity (Sinclair and Fryxell 1985; Zeng et al. 1999). If the spread continues, it will pose a significant threat to the wetlands found in the southwest portion of the ecoregion and their biota.

In areas of human habitation, much of the oasis vegetation has been replaced by planted date palms (*Phoenix dactylifera*). In the wet region of the Ounianga Kebir and Ounianga Serir, date palms have often replaced Hyphaene around lakes (Hughes and Hughes 1992). More than 25,000 date palms grow in the Ounianga Kebir oasis. The rock pools and gorges of the Ennedi Massif are practically uninhabited at present, and only nomads use the water for domestic needs. Several Ennedi Massif ponds are protected as part of the Fada Archei Faunal Reserve (Hughes and Hughes 1992). In oases south of the Emi Koussi and around the Tibesti, much of the natural vegetation that included *Acacia, Ficus, Hyphaene,* and *Tamarix* has also been replaced by date palms. Small-scale irrigation gardening occurs around the Tibesti, and goats, camels, and sheep graze there. Oasis pools also are fished occasionally.

In western Mauritania, small human settlements often are found near waterbodies, with associated fishing, date palm plantations, occasional pastoralism, wood gathering, and horticulture. Cattle grazing in some gueltas can contribute to eutrophication (Dumont 1987). The Saoura Valley is densely populated, with numerous villages, irrigation agriculture, and palm plantations. Lake le Bheyr is used to water livestock and for fishing, with livestock also grazing in the wet areas surrounding the lake. The lake is very eutrophic and turbid (Dumont 1987).

Future Threats

The largest threat to the waters of the Dry Sahel is a growing human population. The possibility of continued growth around water sources may lead to an increase in severity of human impacts on this limited resource and the fauna it supports. Dam building threatens downstream water supply, particularly in the Atlas Mountains. Nomadic use of the Banc d'Arguin has been increasing with growing desertification in the Sahel (Wetlands International 2002). Climate change and continued desertification also threaten the Dry Sahel. Currently a program called AGRHYMET (a special international institution of the Permanent Interstates Committee for Drought Control in the Sahel) is in effect to monitor and understand the climatic and land changes occurring in the Sahel (Diouf et al. 2000).

Justification for Ecoregion Delineation

This ecoregion is defined by the extent of the Saharan, southern Libyan, and Nubian deserts and the very northern portion of the Sahara-Sahelian transition zone. During the late Pleistocene three major wet periods occurred across the Sahara, resulting in a higher prevalence of watercourses. Rock drawings from later than 5,000 B.C. in the central Sahara even depict hippopotamus (Beadle 1981). In the eastern Sahara, freshwater wetlands probably occurred across an area of about 15,000 km^2 between 8,800 and 4,500 B.P., including the large paleo-lake called the West Nubian Lake (Pachur and Rottinger 1997). During the last humid period (within the last 10,000 years), connections probably existed between the Chad Basin and the drainages of the mountain ranges in the southern Sahara, including the Adrar des Iforas, Tassili N'Ajjer, Hoggar, Borkou-Ennedi-Tibesti, and Jebel Marra (Lévêque 1990). These connections presumably facilitated exchange of freshwater faunas between the basins. Remnants of the Sudanian fauna inhabit the few remaining watercourses and oases in the Sahel ecoregion. Because of the short time period (5,000–6,000 years) since their isolation, there has been little divergence of the fauna from their Sudanian stock (Lévêque 1990).

Data Quality: Medium

Ecoregion Number:	**82**
Ecoregion Name:	Etosha
Major Habitat Type:	Xeric Systems
Final Biological Distinctiveness:	Nationally Important
Final Conservation Status:	Vulnerable
Priority Class:	V
Author:	Lucy Scott
Reviewers:	Paul Skelton, Holger Kolberg, and Ben C. W. van der Waal

Location and General Description

The seasonally wet oshanas and pans of the Etosha ecoregion provide habitat for a surprisingly diverse freshwater fauna. Etosha's drainage system, which straddles the border of Angola and Namibia, includes the ephemeral shallow streams, interlinked watercourses locally known as oshanas (grass-covered temporary water channels), and pans (circular depressions temporarily holding water, often endorheic) of the Cuvelai River. The main source of water to Etosha Pan is from the Cuvelai drainage system. The Cuvelai River originates in Angola, and its catchment lies between the Cunene Basin in the west and the Okavango Basin in the east (van der Waal 1991; Barnard 1998).

The Etosha ecoregion has an arid climate. Rain falls mainly in summer, and precipitation increases slightly from west to east and decreases greatly from north to south. The average annual rainfall is 400–550 mm and falls mostly from December to March (Barnard 1998). A number of ephemeral rivers, some of them with origins as far north as the Angolan highlands nearly 400 km north of Etosha Pan, at altitudes of up to 1,450 m a.s.l., feed the oshanas and pans of this ecoregion. At the highest altitudes the source streams are perennial and well defined, fed by annual rainfall of more than 1,000 mm on average (van der Waal 1991). Moving south from the headwaters of this system, the land becomes flatter, and the oshanas meander toward the Namibian border. Rainfall declines to less than 500 mm per year, and the anastomosing channels eventually connect to form a large inland inverted delta in northern Namibia, with a southward gradient of more than 1:10,000. The oshanas receive local rain on average 2 out of 3 years and drain an area of 50,000 km^2 to flow via the westernmost Oshana Etaka, Cuvelai, and Oshigambo into the interconnected Omadhiya pans (including Lake Oponono). The Ekuma River flows south from this series of lakes to drain into the Etosha Pan, a flat, saline depression roughly 7,000 km^2 (Miller 1997). On average, large floods in the Cuvelai drainage system reach the Etosha Pan only once every 7–10 years (Barnard 1998; Mendelsohn et al. 2000). In some wet years, the Etosha Pan receives additional water from the Omuramba Owambo, an oshana entering the Etosha Pan from the east through Fischer's Pan (van der Waal 1991; Mendelsohn et al. 2000). A buildup of fish numbers and a subsequent mass migration of young fish take place down the rivers into the oshana region during good rain years, when the dense local human population in the oshana region harvests the abundant fish life.

The landscape in the Etosha ecoregion consists of saline desert with dwarf savanna fringe around the Etosha Pan. In the savanna fringe is a mosaic of grasslands, shrublands, and woodlands. The annual grass *Sporobolus salsus* grows on the pan itself after good rains or flooding, and blue-green algae cover the surface of the pan during the rainy season. The sedge *Cyperus marginatus* is also common on the pan margins, as are several perennial grasses (Mendelsohn et al. 2000). Halophytic vegetation lines the edge of the pan, consisting principally of *Sporobolus spicatus, S. ioclados, S. tenellus, Odyssea paucinervis,* and the small shrubs *Suaeda articulata* and *Salsola tuberculata* (Mendelsohn et al. 2000). *Atriplex vestita* and *Sporobolus tenellus* are also present, as are the occasional patches of annuals such as *Chloris virgata, Diplachne fusca, Dactyloctenium aegyptium,* and *Eragrostis porosa* (White 1983).

The Etosha Pan is a salt pan, lying on a basement of impermeable limestone that has been eroded away by wind to form a depression into which floodwater flows and then evaporates, leaving salt deposits. Clay pans in the area are much smaller depressions in sandy areas that are permeable to water, and the soils are rich in nutrients (Barnard 1998; Mendelsohn et al. 2000).

The Oshana region is an inverted delta more than 100 km wide north of Etosha and stretches into southern Angola. This region has higher-lying mopane savanna, but the watercourses are grass covered during the dry season, dominated by *Echinochloa, Elytrophorus,* and the sedge *Fuirea*. During the wet part of the annual cycle, aquatic plants including *Marsilea, Aponogeton, Utricularia, Nymphaea, Nymphoides, Ipomoeia,* and *Ottelia* are found. Pans generally are more saline, and *Sporobolus spicatus* and *Odyssea paucinervis* dominate (Clarke 1998).

Outstanding or Distinctive Biodiversity Features

When the ephemeral Etosha and Cuvelai pans receive floodwaters, they are transformed into important habitats for numerous wetland species. The ecoregion has several endemic aquatic species, including sixteen crustaceans (all Ostracods) and four insects (Curtis et al. 1998). About 150 plants, 40 crustaceans, and 19 fish have been recorded in the oshanas (van der Waal 1991; Barnard 1998). Etosha National Park supports at least 340 birds (Simmons et al. 2001), and about 40 percent of the wetland birds in the Cuvelai drainage are included in the Namibian Red Data Book (Kolberg et al. 1996).

The biodiversity of this ecoregion is entirely dependent on seasonal flooding events. Amphibians and aquatic invertebrates, such as fairy shrimp, that are adapted to long dry spells come to life when the floodwaters enter the ephemeral pans and oshanas (Mendelsohn et al. 2000). The Etosha Pan and oshanas are one of two regular breeding sites in southern Africa for lesser and greater flamingos (*Phoenicopterus minor* and *P. ruber,* respectively). Historically, the combined population of these birds was known to reach 1 million individuals in some years; in more recent times, numbers have reached only about 20,000 (Barnes 1998c). Breeding success has been limited, and the population here is no longer considered a viable breeding population (Simmons 1996; Simmons et al. 2001). There are also large breeding colonies of great white pelican (*Pelecanus onocrotalus*) and chestnut-banded plover (*Charadrius pallidus*), and the Etosha Pan supports the only breeding population (only about sixty individuals) of vulnerable blue cranes (*Grus paradisea*) outside South Africa (Simmons et al. 2001).

Status and Threats

Current Status

The Etosha Pan is located entirely in the boundaries of Namibia's 22,270-km² Etosha National Park, which is a large, well-protected conservation area. The Etosha Pan is included in the Etosha Pan, Lake Oponono, and Cuvelai drainage Ramsar site (Kolberg et al. 1996; Barnard 1998; Wetlands International 2002). The 6,600-km² Mupa National Park in Angola covers part of the northern portion of the ecoregion (Stuart et al. 1990).

Translocations of forty-six fish species from the Cunene to the Cuvelai occurred with the construction of a canal from the Cunene to the Cuvelai catchment and filling of the Olushandja Dam in the Oshana Etaka. Most of these species were unable to survive in the harsh conditions of the oshanas and pans (DeMoor and Bruton 1988). However, as of 1991, at least three cichlids had been recorded in permanent waterbodies away from the canal that drains Olushandja Dam, and by 1999 three new species were recorded in the Omadhiya Pans north of Etosha Pan (van der Waal 2000b). More nonnatives from the canal and reservoir are expected to colonize the waters of this ecoregion in the future (van der Waal 1991).

Future Threats

Namibia has a small human population of less than 2 million people but has one of the highest population growth rates in the world, and an increasing human population is the single greatest threat to the Etosha Pan ecoregion. The oshana region of the Cuvelai system supports about 45 percent of Namibia's population, who participate in subsistence fishing (particularly through the wet season) and farming on the floodplains and seasonal wetlands. Most of the people living in the area rely on seasonal ponds and shallow wells for their water supply (Wetlands International 2002). Wood is the main construction material, and this has led to deforestation, especially in the central Cuvelai. Pressure on the land is greatest in the areas surrounding the towns of Oshakati, Ombalantu Oshikango, and Ondangwa in northern Namibia, where human population levels and cattle densities are high (Kolberg et al. 1996; Mendelsohn et al. 2000). Lateral erosion and the filling of channels and pans is a major threat that affects sustainable production potential of the system.

Justification for Ecoregion Delineation

This ecoregion is defined by the Cuvelai drainage basin, which includes the ephemeral Etosha Pan. The biogeographic affinities of the Etosha ecoregion are Zambezian because of past river connections to the east (Skelton 1993, 1994). The Owambo Basin, in which the greater Etosha/Cuvelai drainage is situated, formed 70 million years ago. Since then it has been filling up with sand, silt, and clay washed in and blown from higher ground. Large rivers once drained into the basin during periods of higher rainfall to form large, shallow lakes that often dried up quickly, leaving behind concentrations of salts that are responsible for the brackish quality of much of the groundwater. The Cunene River flowed into the Owambo Basin until about 12 million years ago, when continental uplift changed the slope of the land, causing the Cunene ultimately to flow west from the Ruacana Falls to the Atlantic. The Oshana Etaka to the west of the Etosha system is a deep watercourse and may represent part of the original course of the Cunene River (Mendelsohn et al. 2000). The ecoregion is characterized by a freshwater fauna adapted to seasonal flooding and long dry spells.

Data Quality: Medium

The state of knowledge is moderate, and research could be directed toward the study of the ecological functioning of the system.

Ecoregion Number:	**83**
Ecoregion Name:	Horn
Major Habitat Type:	Xeric Systems
Final Biological Distinctiveness:	Nationally Important
Final Conservation Status:	Relatively Stable
Priority Class:	V
Authors:	Ashley Brown and Michele Thieme

Location and General Description

The temporary rivers, wadis, and sinkholes of the Horn ecoregion support a depauperate freshwater fauna adapted to fluctuating environmental conditions. The ecoregion corresponds to the xeric Horn of Africa, the most easterly portion of the continent, covering the northern portion of Somalia and small portions of Ethiopia and Djibouti. Most of the ecoregion is covered with barren desert or sparsely vegetated mixed scrub and grassland.

A mountain chain runs parallel to the Gulf of Aden, reaching the highest point in Somalia on Shimbiris Mountain at 2,416 m (WWF and IUCN 1994). The land slopes down from this northern ridge to an interior plateau at about 600–1,000 m a.s.l. in the central portion of the ecoregion. The low-lying coastal plain covers a narrow strip along the Gulf of Aden and the Indian Ocean.

Nearly all waterbodies in this arid ecoregion are ephemeral. The northern mountains drain primarily to the north toward the Gulf of Aden and to the southeast, where several wadis cut through the plateau and flow to the Indian Ocean (Hughes and Hughes 1992). Among the many small, ephemeral drainages flowing into the Gulf of Aden are the Durdur and Hodmo rivers. From north to south, the major drainages flowing into the Indian Ocean from the interior plateau include the ephemeral Jaceyl, Dhuudo, and Nugaal rivers. In the valley of the Nugaal River, there are several tugs (temporary watercourses that spreads across flat land) and bullehs (endorheic depressions) (Hughes and Hughes 1992). Along the tip of the Horn there are also pans, springs, and pools in many of the small northern wadis that discharge to the Gulf of Aden and the Indian Ocean. Sinkholes, salt pans, and subterranean water sources can be found inland.

The climate of the ecoregion is arid to semi-arid, with an annual rainfall of less than 250 mm. Rainfall can be particularly sparse at low elevations; for example, Berbera, on the Gulf of Aden, has a mean annual rainfall of 59 mm. Drought periods are common throughout the ecoregion. Rainfall is bimodal, with wet seasons from mid-April to June and then again from October to December (FAO Inland Water Resources and Aquaculture Service, Fishery Resources Division 1999). The climate is hot and humid on the coast and hot and arid inland. In the north, temperatures can reach 42°C on the gulf coast and can be as low as 0°C in the highlands (Hughes and Hughes 1992).

The vegetation of the ecoregion is primarily deciduous shrub or bushland inland, with grassy scrublands on the coast. The most common tree species belong to the deciduous genera *Acacia* and *Commiphora*. The understory consists of shrubby herbs less than 1 m high, such as *Acalypha, Barleria,* and *Aerva*. At lower elevations where rainfall is less consistent, vegetation becomes semidesert scrubland. Around sinkholes in the interior limestone country and near tugs and bullehs, *Acacia tortilis* grows in association with *Commiphora* spp. (Hughes and Hughes 1992). Thickets of dense vegetation often indicate surface or subsurface water. Evergreen and semi-evergreen scrub grows in the mountains, and Afromontane vegetation including juniper forest grows at the highest elevations (Hughes and Hughes 1992).

Outstanding or Distinctive Biodiversity Features

The aquatic fauna of this ecoregion is poorly known; however, it is suspected that the ecoregion hosts a depauperate freshwater fauna able to live in an environment with highly variable water chemistry and flows. Eight freshwater fish species are known to inhabit the Horn. The monotypic Somalian blind barb (*Barbopsis devecchii*), the only ecoregional endemic, lives in caves of the ecoregion. Many of the other fish species are adapted to life in challenging conditions. The lungfish (*Protopterus amphibius*) is capable of aerial breathing and estivation (burrowing in the mud) for survival during anoxic or dry periods, respectively. *Aphanius dispar* is found in oasis pools with hypersaline to fresh water (FishBase 2001). There are also several euryhaline fish. For example, the Jarbua terapon (*Terapon jarbua*) lives in river mouths and intertidal areas and also travels up rivers, and *Stenogobius gymnopomus* and *Syngnathus abaster* are primarily estuarine species (FishBase 2001).

Status and Threats

Current Threats

Because of civil strife in Somalia, information is lacking on the current status and threats to the ecoregion's freshwater systems. It can be assumed that the few existing water resources are under pressure from human populations, although these populations are thought to be low. The vegetation of most montane areas is still considered intact, even if small-scale logging of *Juniperus* is occurring in places. However, uncontrolled harvesting of *Acacia* woodlands (particularly *A. bussei*) for charcoal production is a problem on the adjacent plains. The major threats to the ecoregion are thought to be intensive grazing by goats and other livestock (including cattle in the mountains) and cutting of trees for timber, charcoal, and fuelwood.

Future Threats

Continued civil strife, possible effects of climate change, and overgrazing of riparian vegetation are the main future threats to fresh waters of this ecoregion. The prolonged period of political instability in the ecoregion has resulted in the breakdown of management authorities charged with the conservation of natural resources.

Justification for Ecoregion Delineation

This ecoregion is defined by the xeric Horn of Africa and is characterized by a depauperate and poorly known aquatic fauna.

Data Quality: Low

Ecoregion Number:	**84**
Ecoregion Name:	Kalahari
Major Habitat Type:	Xeric Systems
Final Biological Distinctiveness:	Nationally Important
Final Conservation Status:	Relatively Intact
Priority Class:	V
Author:	Liz Day
Reviewer:	Paul Skelton

Location and General Description

The semi-arid Kalahari ecoregion has no perennial natural surface water and includes some of the largest salt pans in the world (Hughes and Hughes 1992). The endorheic Ntwetwe and Sua pans together make up the Makgadikgadi system, a salt pan complex occupying a total of about 12,000 km^2 (Allan et al. 1995; Tyler and Bishop 1998). This system is one of four major pan systems in southern Africa; the others are Etosha, Hakskeenpan, and Grootvloer-Verneukpan (Lloyd and Le Roux 1985). The Makgadikgadi is fed by the seasonal Boteti River, which drains from the Okavango Delta, and the episodic Nata River, which rises in the Hwange National Park near the southwestern border of Zimbabwe (Hughes and Hughes 1992). North of Makgadikgadi, two other large pans, Nxai and Kgama Kgama, together cover about 230 km^2 (Hughes and Hughes 1992), and immediately south of Makgadikgadi is a pan named Lake Xau. In addition to these large pans in the north, the Kalahari also includes numerous smaller endorheic pans, including several in the Central Kalahari Game Reserve in central Botswana. Pans in this reserve are associated with the fossil river valleys of Deception and Okwa (Tyler and Bishop 1998). This ecoregion includes the northern portion of the Kalahari Desert and its endorheic systems, which sometimes flow into the numerous small pans in the south, the Makgadikgadi Depression, or the Okavango Delta [12].

Rainfall in the Kalahari ecoregion is infrequent, unreliable, and patchy (Van Rooyen 1984), with progressively less rain falling from north to south; mean annual rainfall in northern Kalahari is 400–550 mm, and in central Kalahari it is 250–450 mm (Harrison et al. 1997). Most of this rain falls in summer (October–March), often in the form of short-lived thunderstorms (Hughes and Hughes 1992). Summers are hot (reaching 45°C), and winters are cold (as low as –6°C). Potential evapotranspiration is high, exceeding precipitation rates in every month of the year (Ellery and McCarthy 1998). After heavy rainfall, ephemeral rivers with small catchments flow briefly, and the numerous pans, normally bare or covered with sparse grass and herbs, may retain water for a short time (Tyler and Bishop 1998). Many of these pans have calcrete floors, which aid in water retention (Harrison et al. 1997). In the north of the ecoregion, occasional limestone outcrops give rise to small freshwater pans and springs (Hughes and Hughes 1992).

The two main pans of the Makgadikgadi system retain water (at about depths of 15–25 cm) for somewhat longer periods than other pans in the ecoregion. Direct precipitation contributes substantially to the water level of these pans. In addition, water enters Sua Pan from the Nata River during the summer rainy season, and much later in the year, residual water from the Okavango system enters Ntwetwe via the Boteti River, the endpoint of the Okavango and Boteti system (Hughes and Hughes 1992). The pans are alkaline, and salinity usually is extremely high during periods of inundation. For most of the year, however, the pans are completely dry and salt-encrusted.

Nxai and Kgama Kgama Pans rely almost exclusively on rainfall for inundation. Further south, Lake Xau formerly received water from the Boteti River during conditions of high flow, but the Mopipi Dam now intercepts this water (Tyler and Bishop 1998).

The Makgadikgadi pans consist of salt-encrusted sands, with an aquatic biota dominated during the rainy season by blue-green algae. Along the margins are halophytic plant communities; *Sporobolus spicatus* and *Odyssea paucinervis* grasses dominate the saline fringes, and species such as *Portulaca oleracea*, *Sporobolus tenellus*, and *Suaeda fruticosa* occur in salt marshes along the slightly wetter fringe areas (Hughes and Hughes 1992). The surrounding vegetation community includes grasslands, low tree and bush *Acacia* savanna, and stunted mopane (*Colophospermum mopane*) woodland (Tyler and Bishop 1998). The few islands isolated in the large pans are densely vegetated with woodland species. A grassy peninsula dividing Sua and Ntwetwe pans contains a granite outcrop, Kubu Island, dotted with stunted baobab trees (*Adansonia digitata*) (Comley and Meyer 1994).

Like the greater Makgadikgadi pans, Nxai and Kgama Kgama pans are both set in savanna woodland or open forest. Nxai Pan, which occupies a fossil lakebed, usually is covered by short grass. The pan contains numerous small, treed islands with *Acacia erubescens, A. nigrescens, C. mopane, Syzygium cordata,* and *Adansonia digitata* (Hughes and Hughes 1992).

The channel of the Boteti River supports well-developed riparian woodland. In the river valley *Hyphaene* palms extend north to Nxai Pan. Local people use their fruit, vegetable ivory, to make necklaces and walking stick heads.

Outstanding or Distinctive Biodiversity Features

Barren, highly saline, subject to extremes of temperature and dramatic, unpredictable fluctuations in inundation, the Makgadikgadi pans are inhabited only by plants and animals with specializations that allow them to withstand these conditions. Invertebrates, mainly crustaceans (anostracans, conchostracans, and notostracans), are the major aquatic inhabitants of these pans (Lovegrove 1993). These organisms are opportunistic, many of them passing the dry months or years as heat- and desiccation-resistant eggs (Kok 1987). When standing water becomes available, the eggs quickly hatch and the organisms mature and begin reproducing, sometimes asexually, until conditions begin to deteriorate again, whereupon they mate sexually, produce fertile eggs, and die. The egg stages of these animals protect them from abrasion by windblown particles, from very intense sunlight, and from tiny invertebrate predators. In the eggs, the embryos withstand long periods of diapause by becoming ametabolic (Lovegrove 1993).

During the wet season, these crustaceans provide food for greater flamingos (*Phoenicopterus ruber*), while lesser flamingos (*P. minor*) feed on blue-green algae. The Makgadikgadi pan system provides globally important breeding sites for both species (Harrison et al. 1997). Along with Etosha in Namibia, Makgadikgadi is the only place in southern Africa where these species breed in large numbers on a fairly regular basis, with the largest known African breeding colony of *P. ruber* occurring at Sua Pan. When the waters dry out in this pan, the young chicks migrate across the pan to the Nata Delta. Many thousands of chicks die en route (Penry 1994). Among other birds that breed or feed in the pans during periods of inundation is the vulnerable wattled crane (*Grus carunculatus*), which in wet years flocks to the pans from its stronghold in the Okavango Delta (Harrison et al. 1997).

Across the ecoregion, the scarcity of surface water limits the distributions of many species, and few birds use the pans when they are dry. Amphibian species are limited to hardy opportunistic species, able to estivate for long periods and emerge to lay rapidly maturing eggs as soon as patches of standing water appear (Harrison et al. 2001). Fish enter the pans only with the floodwaters (Hughes and Hughes 1992; Skelton 1993). These immigrants die when the pans desiccate. The nutrient-enriched water created in the pans therefore is most important for the tens of thousands of migrant wading birds attracted to the pans' rich source of food.

The salt-encrusted sandy edges of the Makgadikgadi pans also provide habitat for an endemic lizard, the spiny agama (*Agama hispida makgadikgadiensis*) (Finlayson and Moser 1991). Historically, large mammalian herbivores moved between areas of standing water or good grazing associated with the mineral-enriched pans. These herbivores often traveled vast distances at a time, followed by their predators (Hughes and Hughes 1992). Today, veterinary cordon fences restrict the movement of these herds, whose numbers have also been decimated by past hunting activities (Tyler and Bishop 1998). Large herds of blue wildebeest (*Connochaetes taurinus*), for example, once occurred in the Makgadikgadi area, but their numbers have been drastically reduced, possibly by as much as 90 percent (Hughes and Hughes 1992).

Status and Threats

Few people occupy this ecoregion because of the erratic and very limited rainfall, coupled with infertile soils and a generally inhospitable climate. As a result, human impact is low in the Kalahari (Hall-Martin 1984). Many of the large pans are largely undisturbed (Hughes and Hughes 1992), as are isolated islands in the centers of the larger pans.

Localized impacts do occur along the margins of the pans. These include the extraction of soda ash from Sowa Pan, which is located between flamingo feeding and nesting grounds (Steyn 1990; Hughes and Hughes 1992); tourist vehicle traffic over the pan margins (resulting in disturbance to feeding and nesting birds); and the hunting of mammals and large birds. Overgrazing of cattle on pans and along the fringes of the Boteti River, as well as farming, are ongoing causes of habitat degradation in the ecoregion. The scarcity of water in the region also leads to overabstraction of what little water is available, from wells and from the rivers themselves. For example, Lake Xau no longer receives water from the Boteti River, which is now impounded. Future water diversion from any of the major rivers of the region would pose severe threats to the pan ecosystems. Diversion of water from the Nata River would result in the loss of breeding and feeding grounds for waterbirds, particularly flamingos.

National parks and other conservation areas afford some protection to parts of the ecoregion. The largest of these parks is the 51,000-km² Central Kalahari Game Reserve (Herremans 1998). Unfortunately, the largest pan system, Makgadikgadi, receives little formal protection, although the northern section of Sua Pan is protected as the community-run Nata Sanctuary (Hughes and Hughes 1992). The 2,500-km² Makgadikgadi Game Reserve in the west includes a small portion of the western section of Ntwetwe Pan.

The most serious future threats for the wetlands of the Kalahari ecoregion are those associated with continued water abstraction, overgrazing, and increased access to the pan systems by safari and other user groups.

Justification for Ecoregion Delineation

This xeric ecoregion is defined by the northern portion of the Kalahari Desert and its endorheic systems. The Makgadikgadi region is believed to be the relic of an ancient inland-draining sea, formed some 140 million years ago with the uplifting of the edge of southern Africa (Ellery and McCarthy 1998). During this time, the Cuando, Kafue, and Upper Zambezi rivers drained into the Makgadikgadi depression. Subsequent capture of these rivers by the Lower Zambezi diverted flow from these systems into the Indian Ocean, with the result that the Makgadikgadi lost most of its catchment (Thomas and Shaw 1991). The xeric ecoregion is characterized by a depauperate freshwater fauna specialized to survive in ephemeral waters.

Data Quality: Low

Ecoregion Number: **85**
Ecoregion Name: **Karoo**
Major Habitat Type: **Xeric Systems**
Final Biological Distinctiveness: **Nationally Important**
Final Conservation Status: **Vulnerable**
Priority Class: **V**
Author: **Liz Day**
Reviewer: **Paul Skelton**

Location and General Description

Tree-lined fossil river beds and ephemeral, endorheic pans dot the semi-arid landscape of the Karoo ecoregion. The pans include the Verneukpan-Grootvloer system (Barnes 1998b), one of a series of four major pan systems found in southern Africa (Baard et al. 1985). The ecoregion as a whole includes two distinct terrestrial habitat types (Lowe-McConnell 1996): the succulent Karoo, which contains the region known locally as Namaqualand, located on the west coast of South Africa, and the more easterly Nama Karoo (Branch and Braack 1989). The succulent Karoo is separated from the Nama Karoo by the Bokkeveldberg Mountains, and its rivers drain either directly into the Atlantic Ocean (e.g., Groen River) or into the Olifants River, a perennial system immediately south of the ecoregion. The ecoregion is located almost entirely in the Northern Cape province of South Africa.

The Nama Karoo is traversed by numerous intermittent rivers, including the Carnarvonleegte, Vis, and Sak. The Sak originates in the Nuweveldberge Mountains, is joined by the Vis River, and flows into the vast pans of Brandvlei. Beyond Brandvlei to the north, the Sak flows into Grootvloer Pan, a large, flat, alluvial floodplain that, during high summer rainfall periods, links the Sak River to the Orange River in the north. By contrast, Verneukpan, a system immediately east of Grootvloer, is an internal drainage basin (Lloyd and Le Roux 1985). Carnarvonleegte is one of many episodic rivers that flow into the Verneukpan system. Like Grootvloerpan, Verneukpan, when full, may also provide a passageway to the Orange River, via the Hartbeesrivier in the north. Measuring some 33.5 km long and 11 km wide, Verneukpan is the largest pan in the ecoregion and in all of South Africa (Lloyd and Le Roux 1985). Other major pans in this area include Brandvlei, Flaminkvlei, and Vanwyksvlei.

The succulent Karoo and Nama Karoo each occupy distinct climatic zones, the former falling in the winter rainfall region and the latter in the summer rainfall region (Harrison et al. 1997). Rainfall in both biomes is unpredictable, patchy, and low; the succulent Karoo averages 20–290 mm per year, and the Nama Karoo averages 100–520 mm (Lovegrove 1993). Temperatures also fluctuate dramatically, both daily and seasonally. Winter days are cool to mild, and nights are subject to frequent frost; summers are hot, with temperatures reaching 45°C (Barnes 1998b).

In the succulent Karoo, the winter rainfall regime prevents the widespread growth of trees and promotes an abundance of sclerophyllous shrubs (Harrison et al. 1997). By contrast, the vegetation of the Nama Karoo is characterized by dwarf shrubs and grasses (Lovegrove 1993). Vegetation on pans such as Verneukpan include species such as *Rhigozum trichotomum*, *Zygophyllum retrofractum* complex, *Lycium schizocalyx*, *L. oxycarpum*, *Pteronia mucronata*, *Stipagrostis* spp., and *Eriocephalus aspalathoides* (Lloyd and Le Roux 1985).

Dry for most of the year (Barnes 1998e), river beds in the Nama Karoo descend sharply from escarpments to meander across the flat plains of the Central Plateau. Lined by belts of riverine *Acacia karoo* thicket, the river beds create a network of riparian habitats that extends across the landscape (Barnes 1998e). Other riparian species include *Tamarix usneoides* and *Euclea, Ozoroa,* and *Acacia* shrubs (Barnes and Anderson 1998).

Pans such as Verneukpan and Grootvloerpan reach depths of up to 1.2 m during wet periods, although this happens rarely. For example, Verneukpan contained substantial water only five times during the period from 1885 to 1985 (Baard et al. 1985). In the succulent Karoo, surface flow in the coastal-draining rivers is also intermittent because most of the little runoff that the rivers receive is absorbed by the sandy river beds (Heydoorn and Grindley 1981).

Across the ecoregion, water in the rivers and many of the pans tends to brackish or even saline, particularly toward the end of the wet season (Channing 1987). Closed for years at a time, the estuaries of rivers such as Groenrivier are saline to hypersaline, becoming fresher only after being scoured open by occasional floods (Heydoorn and Grindley 1981).

Outstanding or Distinctive Biodiversity Features

The Grootvloer-Verneukpan system is the most outstanding wetland feature in the Karoo ecoregion (Baard et al. 1985). The only link between the Orange and the Sak River systems, Grootvloer plays an important role in fish migrations, allowing the free interchange of indigenous fish (e.g., smallmouth yellow fish [*Barbus aeneus*] and moggel [*Labeo umbratus*]) and other aquatic organisms between these two river systems during periods of high summer rainfall (Lloyd and Le Roux 1985).

The fish fauna is depauperate, with only four species known from its waters (*B. aeneus, B. anoplus, L. umbratus,* and *L. capensis*). Pools along the Sak River provide refugia for these species during dry periods, from which they may disperse under more favorable conditions (Hocutt and Skelton 1983).

Aquatic fauna are able to withstand long periods of drought in a state of diapause. Among invertebrates, diapause usually is endured by the egg. Once wet conditions occur, the eggs hatch quickly and the animals mature rapidly, often reproducing asexually several times during the short wet season and reproducing sexually only as conditions deteriorate and the wetlands dry up. The sexually reproductive stages of the life cycle produce fertile eggs, which are able to undergo diapause. Examples of invertebrates found in the pans include species of anostraca (e.g., *Streptocephalus* sp.), notostraca (e.g., *Triops namaquensis*), conchostraca (e.g., *Eocyzicus gigas*), cladocera (e.g., *Daphnia gibba*), and ostracoda (Lloyd and Le Roux 1985).

Standing water in the Karoo is a rare habitat, and the response of wetland- or riverine-associated animals in the ecoregion is to move to water wherever and whenever it occurs (Vernon 1986). Temporary water in the pans creates an attractive habitat for birds, and although there are no endemic waterbirds in the Karoo, this ecoregion provides an important area for the nationally rare black-necked grebe (*Podiceps nigricollis*) and black-winged stilt (*Himantopus himantopus*) (Vernon 1986). Perennial water at estuaries such as Groenrivier makes them extremely important wetland oases for water-associated regional and Palearctic migrant birds traversing the succulent Karoo coastline (Heydoorn and Grindley 1981). Frogs occurring in these areas are limited largely to opportunistic taxa, although several species are endemic to the succulent Karoo (e.g., namaqua caco [*Cacosternum namaquense*] and paradise toad [*Bufo robinsoni*] [Harrison et al. 2001]). These species have adaptations that allow them to survive long periods without water in the adult stage. Larval stages usually are very short, allowing the frogs to take advantage of short wet periods (e.g., among the cacos [Passmore and Carruthers 1995]). During hibernation, metabolic rates are drastically reduced. The ecoregion supports a low richness of Odonata, with about thirty-five species of mostly eurytopic, widespread species.

Status and Threats

Current Status

Numerous anthropogenic pressures have affected the integrity of river and pan systems in the Karoo ecoregion. These include the introduction of alien species such as carp (*Cyprinus carpio*) and goldfish (*Carassius auratus*) (Hocutt and Skelton 1983), pollution from pesticides, and high levels of livestock grazing along watercourses (Lloyd and Le Roux 1985; Barnes 1998b). Overgrazing and associated trampling have led to erosion, and other impacts affecting the structure of vegetation have resulted in the loss of riparian habitat (Barnes 1998e). It should be noted that one Karoo endemic has actually benefited from recent changes: the Namaqua prinia (*Phragmacia substriata*) has colonized the *Acacia karoo* trees that are encroaching along the watercourses (Vernon 1986). Other animals, especially waterfowl, exploit the availability of standing water in artificial reservoirs and sewage ponds (Barnes 1998e).

Despite the extent of impacts outlined here, certain areas in the Karoo ecoregion are protected in conservation areas. These include the 328-km^2 Karoo National Park in the south of the ecoregion and the 8-km^2 Carnarvon Nature Reserve east of Verneukpan (Lloyd and Le Roux 1985).

Future Threats

Future threats to the ecoregion lie in the continued erosion and degradation of the Karoo veld, disturbances from sand and diamond mining in the succulent Karoo region, and groundwater depletion for domestic and agricultural purposes.

Justification for Ecoregion Delineation

This semi-arid ecoregion is defined by the succulent Karoo and Nama Karoo regions and is characterized by a depauperate aquatic fauna with a southern temperate (cape) ichthyofauna (Skelton et al. 1995). The rivers of the Karoo ecoregion flow through an area interpreted as the relict channels and overbank deposits of large meandering mixed-load rivers. These rivers had their headwaters in the wetter Gondwanide mountain catchments in the south some 255 million years ago (Smith 1987). The floodplains supported a diverse fauna of mammal-like reptiles, many of which have been preserved in fossilized alluvial deposits (Smith 1993). The more recent paleohistory of the Karoo ecoregion can be interpreted from its fish fauna. The Orange River system forms a focus of distribution for several near-endemic lineages, with the distributions of *Barbus aeneus, B. anoplus,* and *Labeo umbratus* overlapping into the Karoo ecoregion (Skelton 1986a, 1994). Species such as *B. anoplus* also overlap into the Olifants catchment, and Skelton (1986a) suggests that the faunal association between the Orange and Olifants rivers probably relates to the period when the Upper Vaal–Orange drained via the Olifants mouth (from Paleogene to late Oligocene–early Miocene times) rather than via the present Orange mouth. In addition, the escarpment edge and its associated microclimate form an important biogeographic refugium, with relict populations of taxa that are associated with the moist montane grassland of the escarpment edge or regions of deep alluvial sands along the old river courses. Examples of these include amphibians such as *Cacosternum namaquense* (Branch and Braack 1989).

Data Quality: Medium

Ecoregion Number: **86**
Ecoregion Name: **Lower Nile**
Major Habitat Type: **Xeric Systems**
Final Biological Distinctiveness: **Nationally Important**
Final Conservation Status: **Vulnerable**
Priority Class: **V**
Author: **Emily Peck**
Reviewers: **Robert Collins and M. S. Farid**

Location and General Description

The lower Nile River provides a vital oasis for terrestrial and aquatic wildlife as it runs through the semi-arid Sahel and arid Sahara Desert of northern Sudan and Egypt. The ecoregion consists of the lower Nile River from Khartoum, where the White and Blue Nile rivers converge, downstream to the Nile Delta [59]; the areas draining to the lower Nile from the west; and areas draining from the east, including the Blue Nile and Atbara rivers up to but excluding their uppermost reaches. Both the Blue Nile and the Atbara originate in the Ethiopian Highlands [39] and carry large amounts of sediments, estimated annually at 1.4 billion metric tons (Waterbury 2002). Notable features of the ecoregion include four major cataracts over which the Nile flows before emptying into the Aswan High Dam reservoir, named Lake Nubia in Sudan and Lake Nasser in Egypt.

The arid desert climate in the central portion of the ecoregion becomes increasingly hostile toward the north. Two seasons occur in the ecoregion. A hot, dry summer exists from April though October, and a cooler winter exists from November through March. Northern Sudan experiences a high mean daily temperature of 35°C in summer and 20°C in winter and receives a mere 20 mm of annual rainfall (Hughes and Hughes 1992). At Aswan in Egypt, temperatures range from a mean monthly minimum of 8°C and maximum of 23°C in the coolest month (January) to a minimum of 25°C and maximum of 41°C in the hottest month (August). Little or no precipitation falls in the Egyptian deserts. The western desert in Egypt experiences periods of years without rainfall, and precipitation in the hills of the eastern desert is highly variable. Streams can flow violently for several days after a storm drops up to 100 mm of precipitation and then remain dry for several subsequent years (Hughes and Hughes 1992).

In the late spring, rain clouds from the South Atlantic reach the Ethiopian highlands and drop their precipitation, providing the annual Nile flood that reaches Egypt between mid-May and early July. Thereafter, the Nile surges with increasing volume for an average of 110 days, reaching its maximum height and volume in September. At the beginning of the flood in June, a parcel of water in the lower Nile takes 12 days to flow the 600 miles from the Aswan High Dam to Cairo, but the same journey takes only 6 days in full flood in September. By November, the flood returns to its lower discharge levels (Collins 2002).

The total mean average annual flow of the Nile is 88 billion m^3, of which the waters from Ethiopia (Blue Nile, Sobat, and Atbara) contribute 76 billion m^3, or 86 percent of the total Nile discharge (Kendie 2001; Collins 2002). The Blue Nile flood is so large that it blocks the flow of the White Nile at Khartoum, contributing an average 59 percent, or 52 billion m^3, to the Nile's total flow. Further north, the discharge from the Atbara contributes another 13 percent, or 11 billion m^3 (Collins 2002). Below the Atbara, the Nile flows through the harsh Nubian and Egyptian deserts for 2,415 km without receiving any more water before discharging into the Mediterranean Sea.

Lake Nubia or Nasser sits downstream of the Nile's great "S" bend that is located between the Atbara confluence and the third cataract. When the reservoir behind the Aswan High Dam (Sadd al-Aali) is full, it covers an area of 6,850 km^2 and holds

140 billion m³ of water. The reservoir can be divided into three sections. The southern portion is situated in Sudan and is mainly riverine, the middle section extends from Sudan north to Tushka in Egypt and has riverine characteristics only during seasonal floods, and the northern section, which lies entirely in Egypt, is fully lacustrine (Hughes and Hughes 1992).

The Aswan High Dam has significantly altered the Nile's hydrological regime (Beadle 1981). Previously, annual floods laden with nutrient-rich sediments sustained swamps between Khartoum and lower Egypt, and papyrus extended to the Nile Delta (Dumont 1986). The dam has caused the floodplains to virtually disappear (Welcomme 1979). When not in flood, the shallow (7–8 m average depth) lower Nile today has little energy and meanders through the wide, older channel, creating many in-channel islands (Abdelbary 1996; El-Sherbini et al. 1996).

A flat, featureless plain covers a majority of the ecoregion, although in the southeast the landscape rises sharply into the Ethiopian highlands, through which flows the Great Abbai (Blue Nile) from Lake Tana. Fed by powerful tributaries, the Abbai runs through a deep canyon before entering the plains of the Sudan. The Blue Nile gorge has its own unique habitat, with a narrow fringe of forest and scrub lining the river in its upper reaches (Rzóska 1978). From Khartoum downstream, the Nile valley is a broad, flat plain more than 300 km wide at its narrowest point and almost devoid of vegetation throughout the desert of northern Sudan. However, drought-tolerant plant species, such as *Polygonum* spp. and *Potamogeton* spp., occur in local stands and create narrow fringes along the mainstem Nile. Stands of *Phragmites* proliferate where the Nile flows into Lake Nubia (Dumont 1986; Hughes and Hughes 1992).

Below Lake Nasser, the Nile flows across a desert plateau, dividing it into two regions: the western desert, Sahara el Gharqiya, and the eastern desert, Sahara el Sharqiya (Hughes and Hughes 1992). The western desert is an arid sea of blown sand, becoming rocky toward the mainstem Nile. The eastern desert is bound by a discontinuous range of mountains that separates the Nile valley from the Red Sea. These mountains are extensively dissected by *wadis* (dry river beds that sometimes flow during flood events). The valley of the Nile River below Aswan Dam varies between 20 and 30 km in width and is confined by steep sides (Collins 2002).

At the northern edge of the ecoregion, west of the Nile mainstem, Lake Qârûn lies at the bottom of the Fayum depression, which is 71 km long and 20 km wide. Once fed by the Nile, Lake Qârûn now receives most of its flow as runoff from surrounding irrigated lands. As a result, its waters are becoming increasingly saline (Collins 2002).

Outstanding or Distinctive Biodiversity Features

The lower Nile River provides vital habitat in a desert environment for an array of fish and other wildlife. More than seventy fish species live in the ecoregion, many belonging to the families Alestiidae, Cichlidae, Citharinidae, Claroteidae, Cyprinidae, Mochokidae, and Mormyridae. No fish are endemic to the lower Nile ecoregion. In the northeastern corner of the ecoregion, Lake Qârûn supports large number of waterbirds; several grebes, as well as *Aythya fuligula*, *Fulica atra*, and *Anas crecca*, are abundant. In the winter of 1989–1990, more than 32,000 birds congregated at the lake (Baha El Din 2001). Three turtle species live in the lower Nile.

Of the nearly fifty indigenous mollusk taxa in the entire Nile Basin, nine species are endemic. In the Lower Nile ecoregion, fifteen gastropods and nine bivalves occur. Palearctic and Afrotropical gastropod species overlap in the Saharan portion of the Nile Basin and in parts of the headwaters of the Blue Nile. For example, Brown (1994) observed *Armiger crista* living together with *Ceratophallus natalensis* and *Segmentorbis angustus* in a pool alongside a tributary to the Blue Nile River.

Status and Threats

Current Status

All of the major rivers in the Lower Nile ecoregion are managed for flood control, irrigation, and electricity. Dams in the ecoregion include the Sennar and Roseires dams on the Blue Nile, the Jebel Aulia Dam just above the confluence of the White and Blue Nile, the Kashm le Girba Dam along the Atbara, and the Aswan High Dam along the mainstem Nile (Dumont 1986).

The Aswan High Dam was completed in 1971, creating Lake Nasser or Nubia. One of the largest dams ever built, it stands 4 km across, has a base width of 1 km (39 m at its crest), and is more than 100 m high (Davies and Day 1998). It has had a tremendous impact on the ecology of the Nile River and on the Nile Delta. The Nile historically provided nutrient-rich sediments to its floodplains during annual floods. These sediments are now trapped upstream of the dam in Lake Nasser. In total, since the formation of Lake Nasser, an estimated 1,000 km² of riverbank has been lost to the valley of the lower Nile and the delta, in addition to 6,000 tons of potassium and 17,000 tons of nitrogen that no longer are distributed along the course of the lower Nile (Davies and Day 1998). The Aswan High Dam impounds 50–80 percent of the flow of the Nile River. Peak flows, as measured 34 km downstream of the first dam at Aswan, completed in 1902, now reach a maximum of 2,500 m³/s, compared with 10,000 m³/s before dam closure. The earth's crust below the reservoir is laced with fractures, and the enormous weight of the reservoir's water has increased normal seismic activity and may have contributed to the 5.3 earthquake at Kalabsha on November 14, 1981. Since the closure of the Aswan High Dam the decline of nutrients in the flow of the Nile to the Mediterranean has dramatically reduced the sardine (*Sardinella aurita* and *Sardinella maderensis*) landings (96 and 36 percent, respectively) (Lévêque 1997).

Irrigated agriculture has a significant impact on the lower

Nile system. Numerous irrigation barrages exist along the lower Nile River and extract large amounts of water. In addition, two large reclamation projects extract water from the main river. The Sinai project pipes water through the Salaam [Peace] Canal to reclaimed land on the Sinai Peninsula, and as part of the Southern Egyptian Development Project, Toshka, the Sheikh Zayed Canal, has been built to deliver water from the Nile to a half-million acres of land in the western desert (Collins 2002). Because of the present lack of flooding in the Nile River valley, farmers must also apply fertilizers to the nutrient-poor soils that line the banks of the river. The Nile River is becoming progressively more saline because of the runoff of fertilizers and pesticides and saline return flows from the Nile valley (Hughes and Hughes 1992).

Future Threats

Pressures related to a growing human population, such as those from agriculture and infrastructure development, are expected to continue to increase and put heavy pressure on limited water resources. Several development plans exist that would lessen the already reduced flows in the lower Nile River. For example, the second phase of the Southern Egyptian Development Project, called the New Valley land reclamation project, will divert up to 5 billion m³ of water from Lake Nasser, run it through a network of canals, and irrigate more than 2,000 km² in the western desert of Egypt. This project will also necessitate building infrastructure and resettling 7 million people to the area (Swain 2002). Additionally, the Ethiopian government has recently begun building a dam on the Tekeze River, which becomes the Atbara River further downstream in Sudan. When complete, this dam will be 185 m high and equipped with a 300-megawatt power plant (BBC News 2002). The Sudan has begun building the initial infrastructure for the Merowe hydroelectric dam, located at the fourth cataract.

Other potential threats include changes in flow caused by climate change. Although the magnitude and direction of these changes are difficult to predict, they will certainly influence the ecology of the lower Nile, which is already seriously impaired (Strzepek et al. 1996).

Justification for Ecoregion Delineation

The boundaries of this ecoregion are defined by the lower Nile River from Khartoum, where the White and Blue Nile rivers converge, downstream to the Nile Delta [59]. The headwaters of the Blue Nile and Atbara rivers are separated into the Ethiopian Highlands [39] ecoregion because of their swift-flowing, steep nature and different aquatic fauna. The Nilo-Sudanian fauna of this ecoregion is adapted to the large, low-gradient, xeric rivers that experience seasonal flow. The valley of the Nile River was inundated by the Tethys Sea up through the Cretaceous period (approximately 65 m.y.a.) (Dumont 1986).

Five geologic phases of the Egyptian Nile can be distinguished, each separated by a dry period of no flow: the Eonile, Palaeonile, Protonile, Prenile, and the present Neonile (Rzóska 1978; Dumont 1986). The present Nile valley developed at the end of the Miocene. The rise of the high volcanic plateaus in Ethiopia, probably during the Oligocene, is responsible for the origin and direction of the Blue Nile and the Atbara River (Rzóska 1978). Tectonic movement and climatic changes have changed the Nile's course and flow many times. The Nile Basin has few endemic fish because of the frequent flow cessation that inhibited the evolution and persistence of aquatic species (Beadle 1981; Dumont 1986).

Data Quality: High

Ecoregion Number: **87**
Ecoregion Name: Madagascar Southern Basins
Major Habitat Type: Xeric Systems
Final Biological Distinctiveness: Bioregionally Outstanding
Final Conservation Status: Relatively Stable
Priority Class: V
Author: John S. Sparks

Location and General Description

The Madagascar Southern Basins ecoregion is characterized by its aridity. Encompassing the extreme southern part of Madagascar, this ecoregion receives only about 50 mm of rainfall per month (or 300–500 mm annually) (Donque 1972). As a result, many streams and rivers are dry from May to October (Aldegheri 1972; Benstead et al. 2000). Much of the ecoregion is essentially devoid of rivers or surface water (Aldegheri 1972), and overall the ecoregion has a depauperate aquatic fauna.

Karst formations characterized by fissures, subterranean streams, sinkholes, and caverns produced by erosion occur in the western portion of the ecoregion. Lake Tsimanampetsotsa, a soda lake, is located on the Mahfaly karst plateau along the western coast. Groundwater discharge that draws its water from a perched water table feeds the lake. The original natural stream course in the lake's catchment has disappeared, and only traces remain of an ancient network of river valleys that once existed under a more humid climate (Wetlands International 2002).

Rivers throughout the ecoregion are wide, shallow, and subject to extreme seasonality in flow. Vegetation is described as spiny desert, and members of the endemic family Didiereaceae dominate the landscape (Lowry et al. 1997). The major rivers in this ecoregion are the Mandrare, Manambovo, Menaranda, and Linta.

Outstanding or Distinctive Biodiversity Features

Distinctive habitats in this ecoregion include systems of limestone caves, located north and south of the Onilahy River and in the southwestern part of the Madagascar Western Basins [68] ecoregion. These caves are home to a small radiation of endemic blind cave eleotrids (*Typhleotris* spp.). Among these is the vulnerable blind fish (*Typhleotris madagascariensis*), found in the underground rivers and caves of the Mahafaly Plateau (IUCN 2002).

Lake Tsimanampetsotsa has hosted almost 100 *Phoenicopterus minor* in recent years and also harbors a population of 55 *Charadrius thoracicus*, a locally rare endemic plover that nests in grassy clumps around the west side of the lake (Wetlands International 2002; Project ZICOMA 2001).

Status and Threats

Fishing pressure is high in some parts of this ecoregion, and exotics are widespread. Human encroachment threatens the localized *Typhleotris* populations, which are all endemic to southwestern coastal Madagascar. No systematic studies of these fishes have been conducted to date, but data from preliminary investigations suggest that each cave system may contain a distinct species. Given such limited distributions, these fishes are extremely vulnerable to even minor disturbances. Fortunately, compared with many parts of the island, this region of Madagascar is sparsely populated.

Justification for Ecoregion Delineation

This ecoregion is defined by the arid southern portion of Madagascar, which receives less than 50 mm rain/month and is characterized by a depauperate freshwater fauna adapted to xeric conditions and underground rivers and caves.

Data Quality: Low

Few data are available for basins in this ecoregion or for the fishes that inhabit this remote part of the island. Southern Madagascar has not been the focus of ichthyofaunal collecting expeditions, given the desert-like climate, homogeneity of habitat, and few native fish species that would be expected to inhabit such areas. As in the Madagascar Western Basins [68], to date there has been no comprehensive ichthyological survey of the drainages in southern Madagascar. Ichthyologists have historically focused on other regions of Madagascar because few species would be expected to occur in xeric habitats, so data quality remains poor.

Ecoregion Number: **88**
Ecoregion Name: **Namib Coastal**
Major Habitat Type: **Xeric Systems**
Final Biological Distinctiveness: **Nationally Important**
Final Conservation Status: **Vulnerable**
Priority Class: **V**
Author: **Belinda Day**
Reviewers: **Paul Skelton and Peter Jacobson**

Location and General Description

The Namib Coastal ecoregion is one of the most arid areas on Earth but boasts diverse wetland ecosystems. It is situated along the Atlantic and includes the arid Skeleton Coast and the hyperarid Namib Desert. The ecoregion extends north of the Cunene River to include the Curoca River and extends south to but excludes the Orange River. The ecoregion contains a single perennial river, the Cunene; all others are ephemeral. The desert in the ecoregion consists of a dune sea, gravel plains, and granite inselbergs and is bordered in the east by an escarpment along most of its length. The ecoregion is primarily in Namibia and extends into southwestern Angola.

The climate of the ecoregion is extremely harsh and variable. As in most arid areas, rainfall is not only very low but also extremely unpredictable, and evaporation rates are high (Loutit 1991; Jacobson et al. 1995). There is a strong climatic gradient trending from the east, where the ephemeral rivers have their headwaters, to the west where they meet the sea. Rainfall decreases in these westward-flowing catchments from approximately 500 mm in the east to 0–25 mm/year at the coast (Jacobson et al. 1995). Rainfall tends to be intense and of short duration, with the result that soil infiltration rates are low. On average, only 2 percent of rainfall becomes surface runoff, 1 percent recharges aquifers, 14 percent is lost through evapotranspiration, and 83 percent evaporates (EEU: Environment Evaluation Unit 1990). Evaporation reaches a maximum of about 3,400 mm per year at the mouth of the Cunene River (Barnard 1998).

Despite the aridity, there are several types of wetlands in the Namib Desert (Logan 1960; Goudie 1972; Day 1990). Ephemeral systems include rain-fed endorheic pools or pans, rivers that flow for short periods, and the pools that remain in the river beds after flow has ceased (Day 1997). Endorheic pans range in size from large, shallow lakes like Sossusvlei, a huge clay-bottomed lake in the midst of the Namib dune field south of the Kuiseb River, to tiny depressions in granite inselbergs. Ephemeral rivers include a number of westward flowing "sand rivers" situated along the coast (Jacobson et al. 1995). These rivers are vital to the biological functioning of the entire ecoregion.

There are twelve major ephemeral rivers in the Namib coastal ecoregion, and their catchments vary in size from less than 2,000 km² (the Khumib River) to more than 30,000 km² (the Swakop River) (Loutit 1991). The ephemeral rivers depend on water from upstream to replenish their underground aquifers. Recharge of the aquifers tends to be slow because of the presence of clay and silt lenses in the river beds and is completed only after a few days of flooding (Bate and Walker 1993). Rainfall is highly seasonal, and river discharges may range within an order of magnitude from the dry to the wet season (Day 1997).

The perennial Cunene River flows over the rim of the continental plateau onto the Atlantic coastal slope; at the junction of these two formations is the 122-m-high Ruacana Falls. The Cunene below the falls is included in this ecoregion, and throughout this reach the river's flow is punctuated by a series of rapids (Roberts 1975). The lower 340 km of the Cunene has a steep gradient and flows swiftly in a narrow gorge with no floodplains or associated standing water and little vegetation (Curtis et al. 1998).

Linear oases of riparian vegetation line the river channels that cross the otherwise sparsely vegetated desert. Because of permanent underground flow, trees grow along the river courses where groundwater is sufficiently close to the surface (Jacobson et al. 1995). In the Kuiseb River, the four main species found in the river bed are *Acacia albida, A. erioloba, Euclea pseudebenus,* and *Tamarix usneoides*. These species grow in approximately that order away from the riverbank and decrease in size accordingly. They provide an extremely important forage supply for many domestic and wild animals during times of drought (Bate and Walker 1993; Jacobson et al. 1995).

A number of the ephemeral rivers have permanent or semipermanent wetlands at their mouths. These wetlands tend to be vegetated by species of sedge and reeds, such as *Typha, Scirpus, Phragmites, Cyperus,* and *Juncellus,* with large stands of *Suaeda* and *Tamarix* (Jacobson et al. 1995). Sandwich Harbor, about 50 km south of Walvis Bay, contains a freshwater marsh that is fed at least partially by fresh subsurface water. A number of estuaries, including the Uniab, have extensive reedbeds supported by semipermanent trickles and pools of fresh water (Day 1997). The estuaries of the other sand rivers normally are blocked by sand but are opened to the sea during occasional flooding events.

Outstanding or Distinctive Biodiversity Features

The linear oases formed by the ephemeral rivers are vital for supporting both aquatic and terrestrial biodiversity in the ecoregion. Flooding in these rivers brings enormous quantities of silt, nutrients, organic matter, and seeds downstream. The silt and organic debris are quickly colonized by algae, fungi, seedlings, and a host of invertebrates, which form the basis of a complex food chain that sustains the longitudinal ecosystems (Jacobson et al. 1995, 1996, 1999, 2000a, 2000b). These riverine corridors, as well as freshwater springs and pools, allow large herbivores (e.g., antelope, elephant, and rhinoceros) and carnivores (e.g., lions, black-backed jackals, and leopards) to move west through the desert to the coast during periods of prolonged rains (Jacobson et al. 1995).

The Cunene supports about 200 species of aquatic mollusks, crustaceans, insects, amphibians, and fish throughout its length. The lower Cunene supports six endemic fish, four endemic insects, and two endemic amphibians in this ecoregion. The endemic fish are *Chetia welwitschi, Kneria maydelli, Orthochromis machadoi, Thoracochromis albolabris, T. buysi,* and *Sargochromis coulteri* (Skelton 1993). Permanent seeps along the tributaries of the Cunene are critical for supporting certain of these species, such as *Kneria maydelli* (Barnard 1998). The lower Cunene is also the southernmost extent of the freshwater shrimp *Macrobrachium vollenhoveni* and the freshwater oyster *Etheria elliptica* (Curtis et al. 1998). More than 300 bird species depend on the riverine corridor of the Cunene River, and the highly localized *Cichladusa ruficauda* and *Estrilda thomensis* are associated with the riverine palms and thickets (Simmons et al. 2001).

Outside the Cunene, the ecoregion's waters are depauperate for most aquatic taxa and host few endemics. Only about 144 species of aquatic mollusks, crustaceans, insects, amphibians, and fish are known from the ephemeral rivers of Namibia, and the majority of them are insects (about 90 species) (Curtis et al. 1998). Although large populations of wetland birds congregate at the mouths of the ephemeral rivers, there are no endemic waterbirds. Similarly, there are no wetland mammals and only three freshwater fish species, none of which is endemic.

Although poorly known, the invertebrate fauna of the temporary ponds and hypersaline springs is extremely diverse and often highly specialized (Irish 1991). A number of aquatic invertebrate taxa, including the branchiopods and the ostracods, contain some endemics (Curtis et al. 1998). The microinvertebrate fauna in the permanent underground streams below the ephemeral sand rivers is completely unstudied but is likely to be distinctive. Further investigation of these ecosystems is likely to reveal new species and endemics.

Despite its aridity, the ecoregion is rich in frog species (Channing and Griffin 1993). About seventeen aquatic obligate frogs live along the waterbodies of this ecoregion. Most species live in the vleis (wetlands), pools, backwaters, and wetland vegetation that develop in the river beds after rain. Many frogs survive by burying themselves in the sand or mud during times of drought. Others are found in nonriverine temporary waterbodies, rock pools, vleis, and pans that develop after heavy rains, and still others are found only in permanent springs. One species, *Breviceps macrops,* appears to be endemic to the fog belt of the coastal dunes (Channing and Griffin 1993).

Sandwich Harbor is an extremely important bird habitat; counts have exceeded 300,000 birds in some years, and congregations of 50,000 wetland birds in summer and 20,000 in winter regularly occur (Simmons et al. 2001). The Walvis Bay wetlands, including the Kuiseb estuary, cover more than 35 km^2 and support a large proportion of southern Africa's flamingo populations and many migratory species (du Toit and Scuazzin 1995). The mouth of the Cunene River also hosts up to fifty-six wetland bird species and is a stopover on the migratory bird flyway to and from southern Africa (Simmons et al. 1993).

Status and Threats

Current Status

Water is the limiting resource in Namibia, especially in the arid Namib Coastal ecoregion. A rapidly expanding human population with an increasing demand for water has put pressure on the freshwater systems of the ecoregion. Because of the scarcity of water and strong conflicts between various users, almost all of the wetlands, apart from those in the most arid areas, are degraded to some extent by overexploitation, abstraction, or damming of their upper reaches (Day 1997).

The headwaters of this ecoregion's rivers are largely private rangelands, whereas the lower reaches are either state-owned conservation areas or communal rangelands (Jacobson et al. 1995). In the north, where 75 percent of the population lives, the land is farmed and grazed communally and is degraded (Day 1997). Historically, nomadic pastoralists grazed their animals on these lands and moved with the seasonal changes in vegetative cover. Permanent settlements occurred only along the perennial Cunene and around surface springs such as those near Windhoek (Day 1997). However, recent changes in land tenure have resulted in large sedentary human populations settling in the ecoregion. This, coupled with enormous population growth, has resulted in an ever-increasing demand for water (Jacobson et al. 1995).

Small dams on private farms are numerous in the eastern catchments of the ephemeral rivers, and they capture a significant proportion of river flow. Groundwater from the larger rivers, such as the Kuiseb, Omaruru, and Swakop, is abstracted for human consumption, industry (particularly mining), and livestock watering (Bate and Walker 1993). All of the coastal towns, including Walvis Bay and Swakopmund, depend on water pumped from alluvial aquifers and reservoirs (Jacobson et al. 1995). Except for those in the driest and most remote regions, all other freshwater wetlands, such as springs, temporary ponds, or riverine pools, are also used for the watering of livestock and human consumption.

Almost the entire Namib Coastal ecoregion lies in the Skeleton Coast Park and Namib-Naukluft Park or the restricted Diamond Area no. 1. Only the short stretch between Walvis Bay and Mile 108 is not located in a protected area. Apart from these areas, however, less than 1 percent of the nondesert areas are protected, resulting in the headwaters of these rivers receiving no protection. Two wetlands, Walvis Bay and Sandwich Harbor, have been proclaimed Ramsar sites, for which management plans are being developed (Day 1997).

One of the most serious threats to the ephemeral river ecosystems is the lowering of the groundwater table, through excessive abstraction and the damming of upstream river reaches (Day 1997). When the water table is lowered, roots of the trees growing in the river bed are no longer watered, and trees in the riverine forests die. Abstraction or mining of groundwater has already lowered the water table at many places along the coast. For example, McCullum (1994) found that the level of the Kuiseb aquifer had fallen from 2–3 m below the surface to a level of 13 m (only 4 m above bedrock).

A dam and water abstraction scheme at Caleque has also had negative downstream effects on the Cunene. The construction of a weir at Ruacana caused 1-m water level fluctuations, resulting in major mortality of freshwater fish (Barnard 1998).

Future Threats

A serious threat to the biodiversity and continued ecological functioning of the wetland ecosystems of the ecoregion is posed by the numerous introduced plants. It is highly possible that these plants may exclude indigenous species, resulting in a permanent change in wetland vegetation composition (Jacobson et al. 1995).

A planned dam and hydroelectric scheme on the Cunene River at Epupa Falls will inundate an estimated 90 km of river and riverine forest and cover about 295 km^2. In addition to the water abstraction scheme at Caleque, this project will have serious consequences for the downstream ecological functioning of the river. The proposed Epupa Dam could change the water quality and temperature of this coastal river and alter the dynamics of the coastal system. The warm Cunene flows into the cold Benguela Current, creating a warm water plume that provides favorable conditions for two rare turtles, the green turtle (*Chelonia mydas*) and soft-shelled turtle (*Trionyx triunguis*). The Epupa Dam would reduce the volume of water reaching the mouth and regulate the duration and intensity of floods, thereby decreasing essential nutrient inputs and reducing scouring (Simmons et al. 1993; Curtis et al. 1998).

The building of a dam on the Ugab River is also under consideration to provide water to Khorixas (Ashley 1995). Another dam is proposed at Donkersan on the Kuiseb River (Jacobson et al. 1995).

Safe pumping rates for the western ephemeral rivers are still not known (Bate and Walker 1993), and there is a strong likelihood that present pumping rates are lowering water tables along all of these rivers. Continued unregulated pumping, especially if coupled with a number of years of drought, is likely to result in the collapse and eventual loss of these unique longitudinal oases, with serious implications for the ecological functioning of the ecosystem (Bate and Walker 1993).

Without prompt and serious intervention, wetlands will continue to be lost in this ecoregion, possibly at an increasing rate. Water demand management, increased water recycling, and education on the finite nature of the water resources of the ecoregion should figure prominently in planning for this ecoregion. Although a minor problem at this stage, water pollution is worsening (Jacobson et al. 1995). All of the aforementioned threats necessitate the development of effective legislation for the use and conservation of water resources and wetlands.

Justification for Ecoregion Delineation

This ecoregion is defined by the coastal and largely ephemeral rivers and wetlands of the Skeleton Coast and Namib Desert, from above the Orange River north to the Curoca River. Paleohydrological studies of this ecoregion reveal that the climate has been highly variable over time and that it was substantially wetter (albeit temporarily) about 20,000 and again roughly 6,000 years B.P. (Jacobson et al. 1995). As rainfall decreased and evaporation increased at the beginning of each dry climatic period, many wetlands changed from permanent to temporary systems. Only the aquatic organisms that were adapted to periods of desiccation have survived in these systems. Similarly, there must have been many localized species extinctions and adaptations in permanent springs and pools that gradually became hypersaline as evaporation exceeded precipitation. Speciation (and thus endemism) was promoted by the island-like nature of the wetlands in the ecoregion, which are separated by huge arid "seas," thus minimizing the movement of genetic material between wetlands. The fauna of the upper Cunene River has affinities with the Zambezi River, so the headwaters of the Cunene are included in the Zambezian Headwaters ecoregion. The present-day Cunene is thought to have formed through a coastal drainage capturing the headwaters of the Cunene that previously connected to the Zambezi system (Skelton 1994).

Data Quality: Low

Wetlands in arid regions such as the Namib Coastal ecoregion are poorly researched. However, these wetlands could prove to be regionally significant and of immense conservation value (Breen 1991). Little is known about the ecology and life history of most of the wetland biota of the ecoregion (Jacobson et al. 1995), and even basic quantitative information and monitoring data on many taxa are inadequate or nonexistent because of a lack of staff and funds (Day 1997).

Ecoregion Number: **89**
Ecoregion Name: **Red Sea Coastal**
Major Habitat Type: **Xeric Systems**
Final Biological Distinctiveness: **Bioregionally Outstanding**
Final Conservation Status: **Vulnerable**
Priority Class: **V**
Author: **Abebe Getahun**

Location and General Description

The Red Sea Coastal ecoregion includes fresh, brackish, and saline waters. The ecoregion extends as a narrow strip along the shore of the Red Sea from Egypt to Djibouti for some 2,000 km. The plain is low and flat, but the topographic monotony is interrupted by a number of dry river beds and isolated hills.

Dry conditions prevail over the Red Sea coast, and rainfall is extremely variable both in quantity and in timing. Occasionally cyclonic storms of Mediterranean origin enter the Red Sea, and their effects are felt in the east of the ecoregion along the coast. These storms originate when colder air from the highlands meets warmer air over the Red Sea. On the coastal plains maximum rainfall occurs in December–February, and the annual rainfall is less than 300 mm (Westphal 1975). Mean temperatures for the hottest months are 30–35°C, and the humidity is high (Orme 1996).

The only perennial river in the ecoregion is the Awash River, which arises in central Ethiopia, crosses the rift valley, and drains into Lake Abhe. The lakes in this region include lakes Abhe, Afambo, Afdera, Asale, Bario, and Gamari. There are hot springs flowing into some of the lakes (e.g., Lake Afdera). Lakes Abhe, Afambo, Bario, and Gamari are adjacent to one another and interconnected because they are the ending for the Awash River, whereas lakes Afdera and Asale are located further north in the Dallol depression.

Lake Abhe, located at 240 m a.s.l. on the border between Ethiopia and Djibouti, is at the terminus of the Awash River. It is the largest of the lakes in the ecoregion, covering a surface area of 350 km^2. It is a saline lake (160 g/L) of moderate depth (maximum depth 37 m).

Lake Afdera, located 12.6°N and 41°E at an altitude of 80 m below sea level, sits in a depression that currently experiences volcanic and tectonic activities (Williams et al. 1977). Hot springs flow into the lake and are the only source of water other than scanty precipitation. Average annual rainfall is only about 100 mm (Wood and Lovett 1979). This saline lake has a surface area of 70 km^2 and a maximum depth of about 80 m (Wood and Talling 1988). The midday air, lake, and spring water temperatures in November have been recorded at 40°C, 33°C, and 50°C, respectively (Getahun and Lazara 2001). Unlike the other southern saline lakes in Ethiopia (lakes Abiata, Shala, Chitu, and Metahara), the pH of Lake Afdera is low and in the acidic range.

Lake Asale is found at an altitude of 155 m below sea level. The surface area is 55 km². The maximum depth is 40 m, and the salinity is high (276.5 g/L). Located only 10 km from the Red Sea and at an altitude of 155 m below sea level, the lake receives seawater through seepage. The predominance of chloride in Lake Asale creates an extremely high salinity. No data are available as to its biodiversity.

Outstanding or Distinctive Biodiversity Features

Because of the high salinity of the inland lakes in this ecoregion, the fauna that inhabit them are adapted to both fresh and salt waters (e.g., the cichlid fish *Danakilia franchettii*). Although little studied, these lakes are suspected to host a few endemic fishes. For example, in Lake Afdera there is one endemic monotypic fish genus, *Danakilia* (Trewavas 1983), and one endemic species, *Lebias stiassnyae* (Getahun and Lazara 2001). There are also hot springs flowing into Lake Afdera. Water from the hot springs dilutes the salinity of the lake, and most of the fish inhabit the area at the junction of the lake and the hot springs. The lowest depression on Earth (the Dallol depression) is found in this ecoregion. The depression extends from about 80 m to 160 m below sea level (Wood and Talling 1988). The ecoregion is xeric and experiences high air and water temperatures, to which the fauna are adapted.

Large numbers of Palearctic migrant birds visit the woody vegetation around Lake Abhe and adjacent lakes. The site is believed to hold more than 20,000 waterbirds annually. Some of these birds are white-faced tree duck (*Dendrocygna viduata*), white pelican (*Pelecanus onocrotalus*), squacco heron (*Ardeola ralloides*), cattle egret (*Bubulcus ibis*), little egret (*Egretta garzetta*), and marabou (*Leptoptilos crumeniferus*). Many Palearctic species also visit the site as an important staging point on their migration route to and from the Arabian Peninsula, both in autumn and in spring, including Basra reed warbler (*Acrocephalus griseldis*) (Ethiopian Wildlife and Natural History Society 1996, 2001). About 16,000 nonbreeding *Phoenicopterus ruber* and 600 *P. minor* have been observed on the Djibouti side of the lake (Magin 2001a).

Status and Threats

Wind and water erosion, overgrazing, and increasing soil salinity are the main threats to the biodiversity of this ecoregion. The ecosystem is likely to be converted to a true desert if either overgrazing continues unchecked or the climate becomes drier. The water level of Lake Abhe and connected lakes is gradually dropping because of droughts and removal of water upstream from the Awash River to irrigate farms (Ethiopian Wildlife and Natural History Society 2001).

Currently, edible salt is being harvested from Lake Afdera to be sold in landlocked Ethiopia (Getahun 2001). Salt production entails the building of shallow canals to evaporate surface water, so the water level of the lake is expected to drop. Additionally, spring waters, which are the main source of freshwater to the lake, suffer from overabstraction. Any activity that threatens the supply of freshwater to the lake could significantly harm the fish and other aquatic life of the lake.

Justification for Ecoregion Delineation

This ecoregion is delineated based on the basins of the coastal lakes that occur along the Red Sea from the Gulf of Tadjoura at the border with the Horn [83] ecoregion north to the Nile Delta [59]. These lakes are believed to have been formed as an extension of the Red Sea, so the fauna is more related to that of the Red Sea and the Mediterranean than of the southern lakes of the Rift Valley. The fish fauna is of marine origin. *Lebias dispar*, which is common in Lake Afdera, for example, is found in the Red Sea and the Mediterranean. *Danakilia franchetti* from Lake Afdera is also believed to be closely related to a species, *Iranocichla hormuzensis*, on the Arabian Peninsula in southwestern Iran (Trewavas 1983). This ecoregion is characterized by a depauperate fauna that is adapted to the high temperatures and fresh and saline waters of this xeric region.

Data Quality: Low

Very little is known about this ecoregion in general and about the saline lakes and their fauna in particular. Except for sporadic sampling, there have not been any thorough ecological or biological studies of the lakes in this ecoregion. The endemic fishes from Lake Afdera were collected from a small segment (about 100 m² area) at the shore where one of the hot springs joins the lake. It is not known whether these fishes live in the pelagic zone or along the shores where other springs join the lake. The exact number of springs joining the lake is not also known. Although Lake Afdera is reported to be biologically unproductive (Wood and Talling 1988), no mention has been made of the extent of its life forms (microbial, phytoplankton, zooplankton, and others).

Ecoregion Number:	**90**
Ecoregion Name:	**Shebelle-Juba Catchments**
Major Habitat Type:	**Xeric Systems**
Final Biological Distinctiveness:	**Continentally Outstanding**
Final Conservation Status:	**Vulnerable**
Priority Class:	**II**
Author:	**Abebe Getahun**

Location and General Description

The xeric systems of this ecoregion include the Wabi Shebelle and Juba basins, with the ecoregion extending from Kenya to Somalia along the coast of the Indian Ocean and inland to the Ethiopian Highlands [39]. All of the lower Wabi Shebelle and adjacent areas are very dry. The most reliable month for rain is April, followed by May, whereas no rain falls between October and September (Ethiopian Wildlife and Natural History Society 1996). The hottest months are February and March. During flooding, the rivers often spill over their banks and inundate adjacent floodplains. These floodplains often are covered in a tangled growth of small bushes and herbs, which include wild relatives of cotton. Large trees are not naturally found on the floodplains, but heat-tolerant species, including *Hyphaene thebaica,* have been planted in settlements (Ethiopian Wildlife and Natural History Society 1996).

Draining from the southeastern escarpment of the eastern Ethiopian highlands, the Wabi Shebelle and the Fafan rivers flow through the Somalian Desert, although they do not reach the Indian Ocean. The Wabi Shebelle is the major river of the central Somali region. Rising between the Arsi and Bale Mountains, it flows in a southeasterly direction to Somalia. In its lower section, the Wabi Shebelle and its main seasonal tributary from the east, the Fafan, cut through a series of wide, flat shelves of sedimentary rocks made of sandstone, limestone, and gypsum. Wabi Shebelle, with a catchment area of 205,407 km^2, winds for 1,340 km inside Ethiopia and a further 660 km in Somalia (Ethiopian Mapping Authority 1988). The Wabi Gestro, Ghenale River, and Dawa Parma River drain the southwestern escarpment of the eastern Ethiopian highlands. These rivers unite and become the Juba River, which eventually drains into the Indian Ocean (Westphal 1975). These Juba tributaries arise just east of Abaya and Chamo lakes but are separated from the lake drainages by a high mountainous divide. According to Roberts (1975), midway between the lower courses of the Wabi Shebelle and the Juba there is a low-lying limestone plateau with extensive underground waterways radiating out from it.

The rivers in this ecoregion are believed to host many Nilo-Sudanic fishes similar to those of the southern rift valley lakes (lakes Chamo and Abaya). It is believed that these lakes and the Shebelle-Ghenale river basins had former connections with the upper White Nile as recently as 7,500 years ago (Roberts 1975; McClanahan and Young 1996).

Outstanding or Distinctive Biodiversity Features

Endemic fish, including *Bagrus urostigma, Labeo boulengeri, Labeo bottegi,* and *Synodontis geledensis,* live in the rivers of this ecoregion. Most of the Nilotic species found in Lake Abaya, with the exception of *Hyperopisus bebe,* are also present in the Wabi Shebelle and Juba drainage (Roberts 1975). Another important feature of the area is the presence of subterranean waterways, which are inhabited by the endemic monotypic fish genera *Uegitglanis* and *Phreatichthys*. Both the clariid catfish (*U. zammaranoi*) and the cyprinid (*P. andruzzii*) lack visible eyes and are depigmented and scaleless.

Some of the rivers in this ecoregion (e.g., Ghenale River) support abundant fish populations. For example, Woldemariam (1972) in Tedla (1973) exaggerated that it is difficult to take a bath in the river because of the high numbers of fish. The vegetation alongside the middle section of the Ghenale River also supports populations of the vulnerable Prince Ruspoli's turaco (*Tauraco ruspolii*), white-winged collared-dove (*Streptopelia reichenowi*), and Jubaland weaver (*Ploceus dicrocephalus*) (Ethiopian Wildlife and Natural History Society 1996, 2001). Along the coast in Somalia are several areas of importance for waterbirds; for example, the Jasiira lagoon is known to support congregations of *Phoenicopterus ruber* and *Egretta gularis* (Robertson 2001b).

Status and Threats

Very little is known about the current status of the waters of this ecoregion because civil unrest has limited sampling. It is likely that riparian vegetation has come under increasing pressure as a result of agricultural development and from the pressures of a growing human population.

Exotic fish species (*Oncorhynchus mykiss* and *Salmo trutta*) have been introduced into the Danka and Weyb rivers, which are headwaters of the Juba River (Tedla and Hailemeskel 1981). There are no native species in these rivers or any record of their past existence. The cold temperature of the water might have prevented colonization of these waters by tropical fish species.

Justification for Ecoregion Delineation

A continuous escarpment, running in a wide curve from the Kenyan border to northern Somalia, forms the western and northern borders of this ecoregion, and the southwestern part of this escarpment forms the eastern wall of the Rift Valley. The escarpment rises northward and attains its maximum elevation of more than 3,000 m near the Chilalo Massif in Ethiopia. This ecoregion is characterized by a freshwater fauna adapted to xeric and seasonal rivers and by an endemic fauna adapted to subterranean waters.

Data Quality: Low

The native freshwater flora and fauna of this vast ecoregion have been poorly investigated. Although the riparian vegetation has been described to some extent, little is known about the upland vegetation. Without better information, identifying conservation priorities is difficult.

Ecoregion Number: **91**

Ecoregion Name: **Southern Kalahari**

Major Habitat Type: **Xeric Systems**

Final Biological Distinctiveness: **Nationally Important**

Final Conservation Status: **Endangered**

Priority Class: **IV**

Author: **Liz Day**

Reviewer: **Paul Skelton**

Location and General Description

Intermittent rivers and scattered seasonal pans occur in the arid landscape of this ecoregion, covered nearly entirely by a mantle of sand 3.5–35 m thick (Parris 1984; Van Wyk and Le Riche 1984; Skelton 1993). Spanning the northern border of the Northern Cape Province in South Africa and parts of southern Botswana and Namibia, the southern Kalahari ecoregion includes the Kalahari Gemsbok National Park (South Africa) and the adjacent Gemsbok National Park (Botswana).

In the north, an ancient watershed, the Bakalahari Schwelle, forms the divide between the Okwa River in the north and the Nossob and Molopo rivers in the south (Parris 1984). The Nossob is joined by the Auob River in its southerly reaches and flows from here in a southeasterly direction from Namibia. It joins the Molopo River on the South Africa–Botswana border near Ashkam. Historically, the Molopo flowed into the Orange River, upstream of Augrabies Falls; however, the link-up between Molopo and Orange River is not functional at present because of dune blockage (Moore 1999).

Two particularly important pans included in this ecoregion are the Hakskeenpan complex and Barberspan, both in the Northern Cape Province of South Africa. Unlike most of the pans, Barberspan is a permanent, shallow waterbody, fed by the natural diversion of the Harts River into Barberspan's fossil channel. The Hakskeenpan complex is one of four extensive pan systems in southern Africa: Etosha, Makgadikgadi, Hakskeenpan, and the Grootvloer-Verneukpan system (Lloyd and Le Roux 1985).

The climate in the Southern Kalahari ecoregion is semi-arid (Van Rooyen 1984). Temperatures fluctuate greatly, on both a seasonal and a daily basis, with mean maximum and minimum temperatures of 35.7°C and 19.5°C in January and 22.2°C and 1.2°C in July. Rainfall is unreliable and irregular, falling primarily during short-duration, high-intensity thunderstorms between November and April. Before the rainy season begins, strong northwesterly winds blow between September and November, producing dust storms. Mean annual precipitation varies from approximately 223 mm in the south to 250 mm in the north. Humidity is low and evaporation high (Parris 1984), the latter resulting in the characteristic large deficit in the annual water budget (Van Rooyen 1984). River water tends to be alkaline and turbid (Skelton 1993), but the rivers flow only briefly after rainfall, and rapid infiltration into the sandy Kalahari soil means that floods, although they do at times occur, are rare, and river flow ceases quickly (Parris 1984). Indeed, flood discharges from the Auob and Nossob rivers reach the Molopo system only after periods of exceptional rainfall (e.g., 1933, 1974) (Lancaster 1989). The pans in this area tend to be endorheic. Calc-pans, the most common form of pans, are underlain by lime with varying quantities of clay and retain standing water for short periods of time. Other kinds of pans include rarer rock pans, dune pans, and salt pans.

The southern Kalahari landscape is dominated by a 100- to 200-km-wide belt of linear dunes (Lancaster 1989). The terrestrial vegetation is largely Kalahari thornveld, comprising an extremely open shrub or tree savanna. Riparian plant communities are characterized by *Acacia erioloba* trees, whereas the pans support mainly a shrub veld of *Rhigozum trichotomum* and grasses (Van Rooyen 1984). During the dry season, vegetation in the pans tends to be short and sparse and the soils dry and heavily trampled.

The rivers of this ecoregion, called fossil rivers (Parris 1984), are relicts of a wetter epoch during the early Quaternary (Lancaster 1989). Although their vertical erosion was stopped by the onset of desert conditions, their present-day geomorphology reflects this wetter past. For example, the Nossob River once received enough water to flow as a mature river, with meanders and oxbow lake formations, and vertical erosion led to the formation of terraces (Malherbe 1984). Similarly, the proliferation of pans in this area also reflects a history of climatic change. As flow in rivers such as the Nossob decreased, the river was no longer able to erode wind-deposited sand fast enough, with the result that minor tributaries dammed against the higher base level of the Nossob and formed pans (Malherbe 1984). Other pans, further away from the river beds, were formed by deflation following periods of higher groundwater levels and deepened by wind action and the effect of trampling by animals (Lancaster 1989). The only permanently inundated pan, Barberspan, is a relict ephemeral pan in the fossil bed of the Palaeo-Harts River (Barnes 1998d). Earlier this century, the channel connecting this pan to the upper reaches of the Harts River was widened, changing the river into a perennial system (Noble and Hemens 1978). This has resulted in the wetland providing aquatic habitat to birds, amphibians, and invertebrate fauna throughout the year, and today Barberspan is noted as a refuge for waterbirds, including flamingos (Noble and Hemens 1978).

Outstanding or Distinctive Biodiversity Features

In addition to its ephemeral rivers, this ecoregion contains some of the largest concentrations of pans in southern Africa (Lancaster 1989). However, water availability in the Southern Kalahari is severely limited (Barnes and Anderson 1998) and species richness as a whole is low in the ecoregion, with no endemic obligate-aquatic fauna believed to occur here; for example, fish species tend to be hardy species or relicts (Skelton 1993). Similarly, the amphibian fauna is limited to hardy opportunistic species, able to breed at almost any time of year when water is available and to estivate, often over long periods of time (Harrison et al. 2001). For example, giant bullfrog species (*Pyxicephalus* spp.) estivate through the dry season in holes in the ground. Buried, they are protected from desiccation by a waxy cuticle, formed from mucus and layers of shed skin. In addition, the frogs store water in bladder-like outgrowths of their digestive tract, and their metabolic rate drops to less than one-quarter of its normal resting level (Lovegrove 1993). During the wet season, flooded ditches along the sides of roads also provide habitat for frogs.

Only a few widespread species of fish are known to frequent the intermittent waterways of the southern Kalahari. The southern mouthbrooder (*Pseudocrenilabrus philander*) lives in a few permanent and isolated springs and sinkholes in this ecoregion (e.g., Kuruman Eye, Wondergat sinkhole). During rare flood events, the sharptooth catfish (*Clarias gariepinus*) washes down the rivers from farm dams in their upper catchments and forms a briefly abundant food resource for human and animal populations (Haacke 1984). The presence of an accessory air-breathing organ enables this species to breathe air and survive in extreme conditions (Bruton 1979).

Both the pans and ephemeral rivers of the southern Kalahari form focal points for the large herbivores of the ecoregion, providing minerals to animals throughout the year and water during the rainy season (Barnes and Anderson 1998). The pans are also used by the Kalahari fauna variously for burrowing, grazing, saltlicks, and seasonal waterholes (Parris 1984). In addition, the trees associated with the river beds provide locally rare nesting and roosting habitat to birds. In the Kalahari Gemsbok National Park these birds include the globally threatened raptors Cape vulture (*Gyps coprotheres*) and whiteheaded vulture (*Trigonoceps occipitalis*) and the nationally threatened whitebacked vulture (*Gyps africanus*) and lappet-faced vulture (*Torgos tracheliotus*) (Barnes and Anderson 1998).

In a region where open waterbodies are scarce, Barberspan is vitally important for regionally nomadic waterbirds and as a stopover site for Palearctic migrant waders (Barnes 1998d). Waterbirds congregate here in numbers exceeding 20,000 individuals during the dry season (May–October), when the smaller wetlands in the surrounding area have dried up (Barnes et al. 2001). Most of these birds disperse with the rainy season, moving to the temporary pans to breed. Water depth in the pan is a critical determinant of the species using the pan at any one time. When water levels are low, waders such as ruff (*Philomachus pugnax*), lesser flamingos (*Phoenicopterus minor*), and greater flamingos (*P. ruber*) usually are present, with greatest abundance during dry years. In wet years, inundated grassland along the shoreline provides nesting habitat for herons, egrets, and great crested grebes (*Podiceps cristatus*) (Barnes 1998d). As well as being of international importance in terms of its bird habitats, Barberspan is also an example of a rare grass pan (Barnes 1998d).

Accorded national monument status in 1992, the Kuruman Eye is a dolomite spring around which the town of Kuruman in the Northern Cape of South Africa was created. The largest natural fountain in the Southern Hemisphere, this spring supplies water throughout even severe droughts and is used as a primary water supply for the town. The fish species *Pseudocrenilabrus philander* has been found to breed in the fountain area.

Status and Threats

Current Status

The erratic and very limited rainfall regime of this ecoregion, coupled with infertile soils and a generally inhospitable climate, have led to a small human population, with the result that the southern Kalahari is one of the few areas in the world where human impact is low (Hall-Martin 1984). Nevertheless, the ecoregion faces several threats. Fences block the migratory routes of large herbivores (Barnes and Anderson 1998). Several permanent villages have been established on or near the pans, as the result of a permanent water supply provided by the drilling of boreholes. Associated with these settlements is an increase in domestic dogs and livestock, resulting in disturbance, including disruption of nesting and breeding cycles, to birds and other animals using the pans and in the denudation of veld on and around the pans, increasing erosion of the pan edges (Parris 1988).

However, about 36,200 km^2 of the southern Kalahari is conserved within the boundaries of the Kalahari Gemsbok National Park and the adjacent Gemsbok National Park, which together make up the largest single conservation area in Africa (Hall-Martin 1984). Importantly, no fences or other barriers to migration exist between the boundaries of these two parks. Further south, Barberspan is situated in the Barberspan Nature Reserve. This pan was designated a Ramsar wetland in 1975 (Barnes 1998d), and the reserve has largely succeeded in minimizing human impact to the pan (Hughes and Hughes 1992).

Future Threats

Future threats to the ecoregion lie in the increases in permanent human populations around pans. Recent research suggests that Barberspan is in danger of slowly silting up, an impact that would significantly affect the ecological character of the site (Barnes 1998d). In the longer term, the location of the southern Kalahari in a semi-arid climatic zone makes the ecoregion very vulnerable to changes in climate, particularly further decreases in rainfall (Hall-Martin 1984). Natural water resources such as the Kuruman Eye are potentially threatened by overabstraction and by the close proximity of an urban environment, including introduction of alien fish and vegetation into the waterbody.

Justification for Ecoregion Delineation

This ecoregion is delineated based on the distribution of the relict river systems in the Kalahari that flowed into the Orange River Basin. The few aquatic species retain affinities with both the southern temperate and Zambezian faunas (Skelton et al. 1995) and are hardy and opportunistic species adapted to the harsh climate of this ecoregion.

Data Quality: Low

Hall-Martin (1984) recognized the dearth of existing knowledge on ecosystem processes in the southern Kalahari. The role of the ephemeral rivers in the maintenance and functioning of the broader ecosystem is largely unknown (Parris 1984, corroborated by personal comments (L. Day) with Kalahari Gemsbok Park's officials, 2001). Even basic species checklist information is limited for many taxonomic groups, with invertebrates being particularly poorly studied.

Ecoregion Number:	**92**
Ecoregion Name:	**Temporary Maghreb**
Major Habitat Type:	**Xeric Systems**
Final Biological Distinctiveness:	**Continentally Outstanding**
Final Conservation Status:	**Relatively Stable**
Priority Class:	**III**
Authors:	**Michele Thieme and Ashley Brown**
Reviewer:	**Ammar Boumbezeur**

Location and General Description

One perennial river, several temporary rivers, isolated oases, and saline lakes are among the few waterbodies in the xeric, depauperate Temporary Maghreb. Less than 100 mm of rain falls each year in much of this desert-covered ecoregion, which extends from the northwestern coast of Egypt, across Libya and Algeria, through the northern portions of Mali and Mauritania and ends at the Atlantic Coast in Western Sahara (Morocco) (Hughes and Hughes 1992). The Temporary Maghreb encompasses much of the Sahara Desert and includes the southern portion of the Atlas Mountains.

Oueds or wadis, defined as stream beds that are usually dry but flow periodically during the rainy season, drain the mountains of the Temporary Maghreb. This ecoregion contains the Oued Saoura, the largest temporary river of the Sahara and an extension of the Oued Guir system, which originates in the Moroccan Atlas and flows southeast. During large floods, it has been reported that the Oued Saoura flows 500–700 km into the desert (Hughes and Hughes 1992). In the extreme west of the ecoregion there are no permanent streams, and drainage courses are almost always dry (Dumont 1987). Temporary watercourses of note in the west include the Oued Draa, Oued Rheris, Oued Guier, and Oued Sassaf. Ephemeral streams also drain the highlands of the southern tip of Tunisia, flowing toward chotts (shallow irregularly flooded depressions) in the north. These watercourses include the Wadi al Farigh, Wadi al Hamin, and Wadi ash Shu'bah. Pools that are left behind when rivers stop flowing are called gueltas. The Gueltas of Ziza are located in the center of the ecoregion and include one inside a volcanic crater. These gueltas receive their water from the oueds of the Ziza Mountains (Hughes and Hughes 1992). Three gueltas, Issakarassene, Afilal, and Vallée d'Ihierir, are Ramsar sites in Algeria.

Numerous temporary pans, pools, sebkhets, and chotts are also scattered across the ecoregion and are especially prevalent near the coast, where rainfall is higher. A sebkhet (or sebkha or sabkhat) is a large, shallow depression that holds water seasonally, or sometimes for more than a year. Unlike sebkhets, chotts are irregularly flooded and may not be flooded every year (Davies and Gasse 1987). Large western sebkhets include Sebkhet Agsumal in Western Sahara (Morocco) and Sebkha Iquetti, Sebkha Oumm Ed Drous Telli, and Sebkha Oumm Ed Drouse Guebli in Mauritania. In the center of the ecoregion in Algeria lie the Tidihelt Depression, which includes Sebkha Azzel Matti, Sebkha Mekerrhane, and numerous chotts. Many chotts, oases, and semi-permanent lakes dot the landscape in the east. Near the coast in Libya lie two large sebkhets, Sabkhat Al Ghuzayytt and Sabkhat al Qunayy. In Egypt, the Siwa Depression contains about a dozen lakes on its floor. These lakes fluctuate seasonally and sometimes dry up. The nearby Qattara Depression also contains four semi-permanent saline lakes (Hughes and Hughes 1992).

Several subterranean basins occur beneath the Sahara Desert. Water from rainfall quickly seeps down through porous rock, filling the renewable underground aquifers (Beadle 1981). Artesian wells take advantage of these waters. The Nubian Sandstone Aquifer, a nonrenewable resource (i.e., fossil water or ancient reserve of underground water), covers an area greater than 2.5 million km^2 and underlies most of Egypt, eastern Libya, northern Sudan, and northern Chad. The rate of total extraction from the Nubian Sandstone Aquifer is about 1.03 billion m^3/year in Egypt, 850 million m^3/year in Libya, and 405 million m^3/year in Sudan (Centre for Environment and Development for the Arab Region and Europe 2001).

Although climate varies across the ecoregion, it is generally extremely arid, with unpredictable, low rainfall. Less than 100 mm of rain falls each year in the majority of the ecoregion, and parts receive no rainfall at all. The Siwa Depression receives no rain in most years, and mean annual precipitation is only about 8 mm. Coastal areas receive the most precipitation. For example, on the northeastern coast in Libya, mean annual rainfall ranges from about 110 mm in the east to about 370 mm in the west. Diurnal temperature ranges are greater in mountain and desert areas than on the coast. Temperatures can reach extremes: the absolute maxima and minima recorded at Adrar, Algeria are 50°C and below −4°C, respectively (Hughes and Hughes 1992).

Much of the aquatic vegetation in the ecoregion is halophytic. Plants from the halophytic Chenopodiaceae and Saliciornicaceae families line the saline sebkhets, and *Halocnemum strobilaceum* is the most common plant in the saline chotts (Hughes and Hughes 1992). In western Algeria, *Halopeplis amplexicaulis* is a pioneer plant often found on the salt banks of inland lakes (Tremblin 2000). Salt pans in the north of Libya support halophytic *Juncus, Limonium, Salicornia, Sarcocornia,* and *Suaeda. Phragmites australis, Typha capensis,* and *Cynodon dactylon* all grow along the Oued Saoura. In the central portion of Algeria, filamentous green algae and stands of *Typha capensis* and less often *Phragmites australis* grow in gueltas. Oases support species that grow directly in pools and swamps, such as *Juncus* spp., *Phragmites australis, Ceratophyllum demersum,* and *Potamogeton* spp. *Chara* sp., a stiff, attached macrophyte, grows in the Siwa and Quattra Depression lakes (Hughes and Hughes 1992). Two oases at Adrar, Algeria, Ouled Saïd and Tamentit, support more than 100 varieties of date palm (*Phoenix dactylifera*) (Boumezbeur 2002).

Outstanding or Distinctive Biodiversity Features

Overall, richness of the aquatic fauna of the Temporary Maghreb is low, although fish richness and endemism are both moderately high considering the scarcity of permanent waterbodies. About forty fish species are known from this ecoregion, about twenty of which are endemic.

The fish species that inhabit the waters of this ecoregion are adapted to the range of habitats and temperature extremes of its waters. The cooler waters of the mountains support Cobitidae and Salmonidae, families whose distributions generally are found in more northern latitudes. *Cobitis maroccana* is a short-lived species that occurs in watercourses with little current, and *Salmo trutta* is a coldwater species (FishBase 2001). At the other extreme, in the alkaline chotts that experience high temperature variability, species from the *Lebias* genus probably are the only permanent residents (Roberts 1975). The eggs of *Lebias fasciatus* are able to develop in the absence of water, in the sand or humid mud (LeBerre 1989).

The remaining freshwater fauna is depauperate, but invertebrates can be abundant. Only three aquatic frog species are known to inhabit the oases of the Temporary Maghreb, and the ecoregion hosts no aquatic mammals. A sponge, *Spongilla carteri*, has been found in gueltas in the southern portion of Algeria (Hughes and Hughes 1992). Largely Palearctic mollusks occur in the remnant waters of this ecoregion (Brown 1994). A dense invertebrate fauna inhabits the pools along the Oued Saoura and in the Qattara Depression (Hughes and Hughes 1992).

Wetlands of this ecoregion form important habitat for migrant and resident waterbirds. Birdlife is abundant in the Qattara Depression. The wetlands of northern Libya near the Gulf of Sirte and coastal lakes and inland chotts in Tunisia and Algeria support migrating waterbirds such as *Phoenicopterus ruber* and thousands of wintering ducks (Coulthard 2001; Amari and Azafzaf 2001; Robertson and Essghaier 2001).

Status and Threats

Current Status

The largest threat to this ecoregion is the exploitation of subterranean water for human use. Most North African countries tap into groundwater sources that are believed to have originated from both fossil water and rainwater (Beadle 1981). Libya is in the process of completing the Great Man-Made River, a system of more than 4,000 km of pipelines that transport groundwater from the Nubian sandstone aquifer in southern Libya to cities and farms along the Mediterranean coast. Similarly, Egypt extracts groundwater to supplement water that is being piped from the Nile River to farms of the New Valley project in Kharga and Daklha oases in northwestern Egypt (Murakami 1995). There has been much discussion about whether the aquifers are being recharged by runoff from mountainous areas in the southwest of the Sahara, but there is little consensus on this issue. For example, the estimated lifespan of the Man-Made River varies from 20 to 200 years, depending on calculated recharge. Concerns center on possible desiccation of the few freshwater oases caused by a drop in the water table and the salinization of groundwater sources caused by intrusion of coastal waters. In 1994, total estimated water withdrawal in Libya was about eight times the annual renewable water resources (FAO 1997a).

Agriculture and urban growth are putting pressure on the wetlands of the Temporary Maghreb. Subsistence agriculture in the Siwa Depression uses water from the freshwater springs and from wells. Date palm plantations also occur near many wetlands and oases, especially in the Oued Valley. In addition, human population and the tourism industry are increasing in the Saoura Valley and in urban centers along the coast and inland, such as at Ghat in Libya. An example of the effects of these growing urban centers is the discharge of raw sewage into the Oued Saoura from the town of Beni Abbès (Hughes and Hughes 1992).

Future Threats

Continued water abstraction schemes pose the largest future threat to the ecoregion if they are not well managed. An increasing human population means more pressure on the limited water resources. There are many gas and oil fields and extensive pipelines in Algeria and northern Libya (Fuller and Bush 1999). Potential oil spills or pipeline leaks pose a threat, especially in coastal zones.

Justification for Ecoregion Delineation

This ecoregion follows Doadrio (1994) in extending the Maghreb ichthyofaunal province southward to include the desert regions up to the Ahaggar Mountains. Like the Permanent Maghreb [34], the freshwater fauna of this ecoregion contains elements of both Palearctic (i.e., the fish families Cobitidae and Salmonidae) and Afrotropical origin, and the two ecoregions together form their own bioregion. However, the ephemeral nature of the waters of this ecoregion has led to the survival of fewer species than in the neighboring ecoregion. The freshwater fauna is characterized by species adapted to the ephemeral waters and harsh conditions of the Temporary Maghreb. The ecoregion is delimited by the Atlas Mountains in the north and the Ahaggar Mountains in the south.

Data Quality: High

Ecoregion Number:	**93**
Ecoregion Name:	**Western Orange**
Major Habitat Type:	**Xeric Systems**
Final Biological Distinctiveness:	**Bioregionally Outstanding**
Final Conservation Status:	**Endangered**
Priority Class:	**IV**
Author:	**Lucy Scott**
Reviewer:	**Paul Skelton**

Location and General Description

One of the five large rivers on continental Africa, the Orange River historically experienced large seasonal variations in flow, with summer floods bringing an influx of nutrients and sediments into the lower river. This ecoregion encompasses the western Orange River in its 1,200-km course downstream of its confluence with the Vaal River (Allanson et al. 1990). The Orange River flows westward, beginning in the wetter eastern half of South Africa and traversing increasingly arid country. After cascading over the Augrabies Falls (145 m high) and through a deep canyon, it flows through the rugged Richtersveld to emerge on a broad stretch of desert, across which it meanders for 480 km before reaching the Atlantic Ocean (Pritchard 1971; Midgley et al. 1994). This ecoregion lies in southern Namibia and northern South Africa.

In this ecoregion, the Orange River flows through an arid landscape where annual precipitation falls mostly in summer and averages less than 200 mm (less than 100 mm at the river mouth) (Pritchard 1971; Agnew 1986; O'Keeffe et al. 1992). The evaporation rate increases from west to east, from 1,600 mm at the coast to maximum levels of more than 2,700 mm a year in the interior (Midgley et al. 1994). Advective fogs are a common feature of the coastal strip, and many species of the Namib Desert have unique adaptations to harvest the moisture from these fogs (Cambray et al. 1986). Most of the western Orange River lies below 1,000 m, and the lower section forming the boundary between South Africa and Namibia lies below 350 m. This downstream section of the river has only one tributary, the Fish River, entering from the north, so there is little further addition to the flow (Alexander 1985; Agnew 1986; Allanson et al. 1990).

Like many South African rivers, the Orange River has a large sediment load, especially during floods. This is partly a natural feature of draining a semi-arid region and partly a consequence of poor land management and subsequent erosion. The foothills of Lesotho, central farmlands, and dunes of the Kalahari Desert all contribute sediment to the river (O'Keeffe et al. 1992). The water is turbid and alkaline, with moderate amounts of dissolved solids (Agnew 1986). The Orange is a highly regulated river, with a series of major impoundments along its upper course and one, Lake Boegoeberg, on its lower course (Allanson et al. 1990). The Augrabies Falls, at about 620 km from the mouth of the Orange River, is the only natural barrier to the movement of fish.

Along the upper part of the western Orange, from Buchuberg to Kakamas, there are rich strips of alluvial soil on both banks, deposited by summer floods of the river over many millennia. These floodplains are no longer inundated by seasonal floods but are irrigated by water abstracted from the Orange and transported via a network of canals and furrows (Midgley et al. 1994). Vegetation is classified as Karoo and Karoo shrub in the eastern portion of the ecoregion.

The western portion of the ecoregion is classified as desert and semidesert with succulent steppe vegetation (Stuart et al. 1990). Below Augrabies Falls, the Orange flows through increasingly arid areas with arenosols, lithosols, and weakly developed shallow soils (Cooke 1964; Cambray et al. 1986; Stuart et al. 1990). Below the falls, the 620-km reach has a low gradient with tree-lined banks and few submerged aquatic macrophytes (Curtis et al. 1998). Littoral sands cover the coast near Alexander Bay. The large discharge of water through the mouth limits tidal exchange and causes most mixing to occur at sea, thus limiting the extent of estuarine wetland areas (Cambray et al. 1986).

Outstanding or Distinctive Biodiversity Features

The western Orange River contains some endemic species, including three crustaceans and one fish, but it generally supports an impoverished freshwater fauna. However, compared with the upper Orange, the section of river included in this ecoregion has a more distinctive biota (Agnew 1986; Skelton et al. 1995; Curtis et al. 1998).

Thirteen indigenous fish species live in these waters (Skelton 1986b), and Baetidae, Simuliidae, Chironomidae, and Hemiptera dominate the macroinvertebrate fauna (Davies and Day 1998). Two endemic Trichoptera are found here, *Cheumatopsyche obtusa* and *Orthotrichia trilineata* (Eekhout et al. 1997). The Namaqua barb (*Barbus hospes*) is considered rare and is endemic to the lower Orange River below the Augrabies Falls, and an isolated population of *Mesobola brevianalis* (the southernmost neoboline cyprinid population in Africa) is also found in this stretch of river. The Orange-Vaal river endemic *Austroglanis sclateri* is rare and listed as data-deficient on the Red List (IUCN 2002). The channel-like nature of the gorge below the falls reduces habitat diversity to open flowing water interspersed with rocky rapids. Such conditions are likely to have been the major forces directing the anatomical adaptation of these two species to streamlined, active swimming forms (Skelton 1993).

The estuary at the mouth of the Orange consists of salt marsh, freshwater lagoons and marshes, sandbanks, and reedbeds, and it has been designated a Ramsar site (Cowan 1995a). This site is extremely important for breeding and migrant waterbirds, including Cape cormorant (*Phalacrocorax capensis*), Damara tern (*Sterna balaenarum*), common tern (*Sterna hirundo*), and Hartlaub's gull (*Larus hartlaubii*), with more than 25,000 individuals occurring at one time (Cambray et al. 1986; Stuart et al. 1990; Barnes et al. 2001).

Status and Threats

Current Status

The Orange River is one of the most regulated rivers in Africa, and water abstraction and the interruption of the natural flow regime are the greatest threats to the functioning of the system. Dry season water use by towns, crop irrigation (notably in spring before rivers swell naturally), hydroelectricity generation, and interbasin water transfers lead to substantial withdrawals (Seaman and Van As 1998). River regulation has drastically changed the natural hydrological regime from one of large flows in summer and negligible flows in winter to one of reduced summer and increased winter flows.

Historically, scouring flushing summer floods reached the mouth in most years, although flow levels varied greatly between years. In winter, the flow often was so low that the mouth closed, and severe summer floods were expected every 5–10 years. The flow regime is now determined completely by the release of water from impoundments, and regulation has stopped the mouth from closing in most years. When it did close in the winter of 1994, it was the first occasion in 16 years.

Parts of this ecoregion are under varying degrees of protection. The Orange River from its mouth upstream 10 km has been designated a Ramsar site and is protected to some extent by the Ais-Ais/Hunsberg Reserve Complex (Curtis et al. 1998). After the collapse of the salt marsh component of the estuary, the estuary was placed on the Montreux Record (which is a record of Ramsar sites where changes in ecological character have occurred, are occurring, or are likely to occur) in 1995. The rapid degradation of the estuary resulted from adjacent diamond mining activities at Lüderitz, to the north of the river, and upstream flow regulation of the river (Cambray et al. 1986; Stuart et al. 1990). *Barbus kimberleyensis* and *B. hospes* are protected in part by the Augrabies Falls National Park (94 km^2 protecting the falls and gorge) and the Richtersveld National Park. The Fish River Canyon Nature Reserve (461 km^2) covers the Fish River Canyon in Namibia, and the Spitzkop Nature Reserve (11 km^2) includes part of the Orange River catchment to the northeast of Augrabies Falls (Stuart et al. 1990). Apart from the introduction of two nonnative fish (DeMoor 1996) and two mollusks from North America (*Physa acuta* and *Lymnaea columella*), species composition of the western Orange appears to have changed little in the last three decades since major river regulation started. There was some concern over the status of *Barbus hospes,* but the species seems to have benefited to some degree from the increased and sustained low flows produced by river regulation (Skelton 1993).

Future Threats

Aridity precludes extensive land use in most of the lower Orange, and the human population density is as low as 1 person/km^2 (Cambray et al. 1986). Nevertheless, increasing exploitation of the Orange River is a growing threat. Water abstraction from the Orange and Vaal for industry, agriculture, and mining threatens the ecological functioning of the system. The expansion of industrial, agricultural, and mining interests in the catchment is likely to exacerbate existing problems (Davies and Day 1998). For example, water abstraction takes place at Prieska for the Prieska-Kenhardt copper fields, in the Richtersveld for the copper mines of Springbok and Okiep, and near the Orange River mouth for the diamond mines of Oranjemund and Alexander Bay (Midgley et al. 1994). Additionally, the Lesotho Highlands Water Scheme, in the headwater reaches of the system, supplies water to the Gauteng area of South Africa and has major environmental implications for the Orange River. It has been predicted that the scheme may reduce the amount of water at the mouth from 6.5 billion m^3 to 842 million m^3 on average (from an original flow of 11.5 billion m^3). This would result in a river mouth that is dry for years on end, apart from exceptional floods (Seaman and Van As 1998).

The ecosystem appears to be fairly robust, perhaps as a result of a long history of extreme conditions. The Orange River Re-planning Study being undertaken by the Southern African

Development Community will make recommendations on how to manage the lower Orange River into the future (SADC 2002). Among the proposed management plans for the western Orange is one to simulate a quasinatural flow pattern to manipulate mouth closure and opening. This would be a highly managed system but would give the ecosystem the best chance of functioning optimally.

Justification for Ecoregion Delineation

This ecoregion is defined by the mainstem of the Orange River and its tributaries below the confluence of the Orange and Vaal rivers. The Orange River is the northern limit of the southern temperate (Cape) ichthyofauna and the southern limit of the tropical (Zambezian) fauna of southern Africa (Skelton 1986b, 1994). Zambezian fish species in the Orange River include *Barbus trimaculatus, Clarias gariepinus, Barbus paludinosus, Mesobola brevianalis,* and *Tilapia sparrmanii.* The river is thought to have once flowed west to the Atlantic in the vicinity of the present-day Olifants River mouth before the Orange River was captured by the Koa tributary of the Lower Orange to isolate the Olifants River system (Skelton 1986b, 1994; Allanson et al. 1990).

Data Quality: Medium

Glossary

adaptation   Particular part of the anatomy, physiological process, or behavior pattern of a particular species that increases its chances of survival or ability to reproduce (Wilson 1992).

Afroalpine   African vegetation lacking trees, where frosts are common, typically above 3,000 m altitude.

Afromontane   African vegetation that can be with or without trees, where frosts are rare to absent, typically above 2,000 m and below 3,000 m altitude.

Afrotropical   The region of Africa south of the Sahara Desert.

allochthonous   Originating outside and transported into a given system or area (Lincoln et al. 1982).

alluvial   Relating to or consisting of any material that has been carried or deposited by running water.

amphibian   A member of the vertebrate class Amphibia (frogs, toads, and salamanders).

amphipod   Any of a large group of small crustaceans with a laterally compressed body, belonging to the order Amphipoda.

anadromous   Diadromous species that spawn in freshwater and migrate to marine habitats to mature (e.g., salmon).

anthropogenic   Caused or produced through human agency (Lincoln et al. 1982).

aquaculture   The cultivation of aquatic organisms (Lincoln and Boxshall 1987).

aquatic   Living in water.

aquifer   A formation, group of formations, or part of a formation that contains sufficient saturated permeable material to yield significant quantities of water to wells and springs (Maxwell et al. 1995).

arid   Referring to a climate or habitat with an annual rainfall of less than 250 mm, with evaporation exceeding precipitation and a sparse vegetation (Lincoln and Boxshall 1987).

artesian   Referring to underground water that moves under pressure and flows to the surface naturally.

assemblage   The combination of particular species that occur together in a specific location and have a reasonable opportunity to interact with one another (Matthews 1998).

avifauna   Bird fauna.

base flow   The part of stream discharge not attributable to direct runoff from precipitation, snowmelt, or a spring but instead coming from groundwater effluent.

basin   See **catchment**.

benthic   Living at or in or associated with structures on the bottom of a body of water (Brown and Gibson 1983).

biodiversity (also called biotic or biological diversity) The variety of organisms considered at all levels, from genetic variants belonging to the same species through arrays of species to arrays of genera, families, and still higher taxonomic levels; includes the variety of ecosystems, which comprise both communities of organisms in particular habitats and the physical conditions under which they live (Wilson 1992).

biodiversity conservation The goal of conservation biology, which is to retain indefinitely as much of Earth's biodiversity as possible, with emphasis on the biotic elements most vulnerable to human impacts (Angermeier and Schlosser 1995).

biodiversity vision Long-term goals for an ecoregion's biodiversity conservation and actions that identify key sites, populations, thresholds, and ecological processes.

biogeography The study of the geographic distribution of organisms, both past and present (Brown and Gibson 1983).

biological distinctiveness Scale-dependent assessment of the biological importance of an ecoregion based on species endemism, richness, relative scarcity of habitat type, and rarity of ecological or evolutionary phenomena. Biological distinctiveness classes are globally outstanding, continentally outstanding, bioregionally outstanding, and nationally important.

biomass (1) The total weight, at a given time, of living organisms of one or more species per unit area or of all the species of a community. (2) The weight of a taxon or taxa per unit of surface area or volume of water, expressed in units of living or dead weight, dry weight, ash-free weight, or nitrogen content (Armantrout 1998).

biome A classification of natural communities in a particular region based on dominant or major habitat, vegetation types, and climate. Also called **major habitat type**; see also **freshwater habitat type**.

bioregion A geographically related assemblage of ecoregion complexes that share a similar biogeographic history and thus have strong affinities at higher taxonomic levels (e.g., genera, families).

bioregionally outstanding Biological distinctiveness category.

biota The combined flora, fauna, and microorganisms of a given region (Wilson 1992).

biotic Biological, especially referring to the characteristics of faunas, floras, and ecosystems (Wilson 1992).

blackwater Water rich in humic acids and with low nutrient concentrations (Lincoln and Boxshall 1987).

bog Wet, spongy land that usually is poorly drained, highly acidic, and nutrient rich and is characterized by an accumulation of poorly to moderately decomposed peat and surface vegetation of mosses and shrubs (Armantrout 1998).

B.P. Before the present.

brackish Describing water with a salinity between that of normal freshwater and normal seawater.

caldera A large, basin-shaped volcanic depression, roughly circular in form (Bates and Jackson 1987).

canal An artificially created waterway (Armantrout 1998).

canopy Overhead cover of branches and foliage of adjacent vegetation (Armantrout 1998).

catadromous Diadromous species that spawn in marine habitats and migrate to freshwater to mature (e.g., eels).

catchment All lands enclosed by a continuous hydrological–surface drainage divide and lying upslope from a specified point on a stream (Maxwell et al. 1995) or, in the case of closed-basin systems, all lands draining to a lake.

channel A natural or artificial waterway that periodically or continuously contains moving water, has a definite bed, and has banks that confine water at low to moderate stream flows (Armantrout 1998).

channelization The process of straightening a stream, which usually involves lining it with concrete or rock. Channelization usually is undertaken to control flooding or divert water (Doppelt et al. 1993).

chotts Shallow, irregularly flooded depressions.

Cladocera Water fleas; order of diplostracan crustaceans containing about 450 mostly freshwater species (Lincoln and Boxshall 1987).

coltan A valuable black mineral combining niobite and tantalite; used in cell phones and computer chips.

community Collection of organisms of different species that co-occur in the same habitat or region and interact through trophic and spatial relationships (Fielder and Jain 1992).

connectivity Involving linkages of habitats, species, communities, and ecological processes across multiple scales; the opposite of fragmentation (Noss 1991, as cited in Federal Interagency Stream Restoration Working Group 1998).

conservation The planned management of natural resources; the retention of natural balance, diversity, and evolutionary change in the environment (Lincoln and Boxshall 1987).

conservation landscape A large area determined to be a priority for conservation, for which a detailed conservation strategy is designed in an ecoregional plan.

conservation targets Targets that guide priority setting and implementation of strategies. These include distinctive units of biodiversity (e.g., endemic species, areas of high richness); larger intact habitats; intact biotas; keystone habitats, species, and phenomena; large-scale ecological phenomena; and species of special concern.

continentally outstanding Biological distinctiveness category.

conversion The process of altering natural habitat for another use.

Copepoda Class of mostly small aquatic crustaceans exhibiting a great diversity of form and life history; includes many commensals, parasites of all major animal groups, and free-living forms, abundant in planktonic, benthic, and interstitial habitats (Lincoln and Boxshall 1987).

Cyprinidae Carps, minnows, barbs; very large family of omnivorous or herbivorous freshwater cypriniform teleost fishes with worldwide distribution except for Australia, New Zealand, and South America (Lincoln and Boxshall 1987).

Cyprinodontidae Killifishes, tooth carps, top minnows; family containing about 300 species of small (to 150 mm) mainly freshwater cyprinodontiform teleost fishes widespread in tropical and temperate parts of the New World (Lincoln and Boxshall 1987).

dam A barrier obstructing the flow of water that increases the water surface elevation upstream of the barrier (Armantrout 1998).

deforestation The permanent removal of forest and undergrowth (Lincoln and Boxshall 1987).

degradation The loss of native species and processes resulting from human activities such that only certain components of the original biodiversity still persist, often including significantly altered natural communities.

delta Flat plane of alluvial deposits between the branches at the mouth of a river, stream, or creek (Armantrout 1998).

depauperate Impoverished, moribund (Lincoln and Boxshall 1987).

desert An area with low rainfall and sparse to absent vegetation cover.

desertification The development of desert conditions as a result of human activity and climate changes (Lincoln and Boxshall 1987).

desiccation Process of dehydration or drying up (Armantrout 1998).

detritus A nondissolved product of disintegration or wearing away; pertains to small organic particles such as leaves and twigs. May also include material produced by erosion, such as soil, sand, clay, gravel, and rock, carried down a watercourse and deposited on an outwash fan or floodplain (Armantrout 1998).

diadromous Species that migrate between freshwater and marine habitats while spawning in one habitat and maturing in another (Nyman 1991).

diatom Microscopic algae with a siliceous skeleton that occurs as plankton or attaches to substrate (Armantrout 1998).

discharge (1) Rate at which a volume of water flows past a point per unit of time, usually expressed as cubic meters per second or cubic feet per second. (2) Intentional or unintentional release of substances into a waterway or waterbody that can result from spilling, leaking, pumping, pouring, or dumping. (3) Any addition of dredged or fill material into a waterway or waterbody (Armantrout 1998).

dispersal Multidirectional spread, at any time scale, of plants or animals from any point of origin to another, resulting in occupancy of other areas in their geographic range (Armantrout 1998).

distribution Occurrence, frequency of occurrence, position, or arrangement of animals and plants in an area (Armantrout 1998).

disturbance Any discrete event in time that disrupts ecosystem, community, or population structure and changes resources, substrate availability, or the physical environment (Fielder and Jain 1992).

diversion The removal of water from a waterbody.

diversity Variation that occurs in plant and animal taxa (i.e., species composition), habitats, or ecosystems in a given geographic location (Armantrout 1998).

drainage basin See **catchment**.

ecological integrity The likelihood that an area's populations, species, and assemblages will endure over time, given considerations of habitat intactness and population or species viability.

ecological processes Interactions between animals, plants, and their environment that ensure that an ecosystem's full range of biodiversity is adequately maintained. Examples include population and predator-prey dynamics, pollination and seed dispersal, nutrient cycling, migration, and animal dispersal.

ecoregion A large unit of land or water containing a geographically distinct assemblage of species, natural communities, and environmental conditions. The boundaries of an ecoregion encompass an area in which important ecological and evolutionary processes most strongly interact.

ecoregion complex A contiguous grouping of ecoregions that share biotic and ecological affinities.

ecosystem A community of organisms and their physical environment interacting as an ecological unit (Lincoln et al. 1982).

ecosystem service Benefit or service provided free by an ecosystem, or by the environment, such as clean water, flood mitigation, or groundwater recharge.

effluent (1) Discharge of liquid into a waterbody or emission of a gas into the environment; (2) a stream flowing out of a lake or reservoir (Armantrout 1998).

endangered species A species threatened with extinction (Lincoln and Boxshall 1987).

endemic A species or race native to a particular place and found only there (Wilson 1992).

endemic bird area A geographic region that contains at least two bird species as strict endemics in a region smaller than 50,000 km^2 (BirdLife International 2004b).

endemism Degree to which a geographically circumscribed area, such as an ecoregion or a country, contains species not naturally occurring elsewhere.

endorheic Closed basin (no exterior drainage).

ephemeral stream Stream where flows are short-lived or transitory and result from precipitation, snowmelt, or short-term water releases (Armantrout 1998).

epigean Pertaining to the biological domain at the ground surface or above it. Includes streams (Jennings 1998).

epilimnion Uppermost layer of water in a lake, characterized by an essentially uniform temperature where thorough mixing results from wind and wave action to produce a less dense but oxygen-rich layer of water (Armantrout 1998).

epiphytic Flora growing on the surface of macrophytes (Armantrout 1998).

erosion The wearing away of the land surface by wind, water, ice, or other geologic agents. Erosion results naturally from weather or runoff but often is intensified by human land use practices (Eckhardt 1998).

estuary A deepwater tidal habitat and its adjacent tidal wetlands, which are usually semi-enclosed by land but have open, partly obstructed, or sporadic access to the open ocean, in which ocean water is at least occasionally diluted from freshwater runoff from the land (Maxwell et al. 1995).

eutrophication Overenrichment of a waterbody with nutrients, resulting in excessive growth of organisms and depletion of oxygen concentration (Lincoln et al. 1982).

evolutionary phenomena In the context of WWF regional conservation assessments, evolutionary phenomena are patterns of community structure and taxonomic composition that are the result of extraordinary examples of evolutionary processes, such as pronounced adaptive radiations.

exotic species A species that is not native to an area and has been introduced intentionally or unintentionally by humans; not all exotics become established.

exploitation The act or instance of making productive use of a natural resource.

extinct Describes a species or population (or any lineage) with no surviving individuals.

extinction The termination of any lineage of organisms, from subspecies to species and higher taxonomic categories from genera to phyla. Extinction can be local, in which one or more populations of a species or other unit vanish but others survive elsewhere, or total (global), in which all the populations vanish (Wilson 1992).

extirpated Status of a species or population that has completely vanished from a given area but continues to exist in some other location.

extirpation Process by which an individual, population, or species is destroyed (Fielder and Jain 1992).

falls Free-falling water with vertical or nearly vertical drops as it falls over an obstruction (Armantrout 1998). Also known as **waterfalls**.

family In the hierarchical classification of organisms, a group of species of common descent higher than the genus and lower than the order; a group of genera (Wilson 1992).

fauna All the animals found in a particular place.

fen Low-lying peatland that is partially covered with fast moving, nutrient-rich, neutral to basic water that is rich in calcium. Fens often are dominated by sedges and rushes that form peat when they die and decay (Armantrout 1998).

fishery The industry of catching, processing, and selling fish or other aquatic animals; also the place where fish and other aquatic animals are caught, processed, or sold.

flood (1) Rising and overflowing of a waterbody onto normally dry land; (2) any flow that exceeds the bankfull capacity of a stream or channel and flows onto the floodplain (Armantrout 1998).

flood pulse Periodic increase in riverine productivity that occurs when rivers inundate floodplains (Armantrout 1998).

flooded grassland A grassland habitat that experiences regular inundation by water.

floodplains Areas that are periodically inundated by the lateral overflow of rivers or lakes or by direct precipitation or groundwater; the resulting physicochemical environment causes the biota to respond by morphological, anatomical, physiological, phenological, or ethological adaptations and produce characteristic community structures (Junk et al. 1989).

flora All the plants found in a particular place.

flow (1) Movement of water and other mobile substances from one location to another; (2) volume of water passing a given point per unit of time (Armantrout 1998).

fluvial Pertaining to or living in streams or rivers or produced by the action of flowing water (Armantrout 1998).

fragmentation Process by which habitats are increasingly subdivided into smaller units (Fielder and Jain 1992).

freshwater In the strictest sense, water that has less than 0.5 percent of salt concentration (Brown and Gibson 1983); in the context of this volume, refers to rivers, streams, creeks, springs, and lakes.

freshwater habitat type Aquatic systems that are fairly homogeneous with respect to size and thermal, chemical, and hydrological regimes.

freshwater marsh Area with continuously waterlogged soil that is dominated by emergent herbaceous plants but without a surface accumulation of peat (Armantrout 1998).

fynbos A type of endemic-rich shrubby vegetation found in the region around Cape Town in South Africa.

genera The plural of *genus*.

genus A group of similar species with common descent, ranked below the family (Wilson 1992).

geomorphology Geological study of the configuration, characteristics, origin, and evolution of land forms and earth features.

Global 200 Ecoregion A large region typically composed of several ecoregions that harbor outstanding biodiversity or representative assemblages and habitats.

globally outstanding Highest-level biological distinctiveness category.

Gondwanaland An ancient continent that included part of southern Africa and contained its own unique flora and fauna.

graben Portion of the earth's crust bounded on at least two sides by faults that have moved downward as a result of crustal activity to create an area that is lower than the adjoining land form (Armantrout 1998).

grassland A habitat type with landscapes dominated by grasses.

grazing The act of feeding by livestock on vegetation of pastures and ranges.

groundwater Water in the ground that is in the zone of saturation, from which wells and springs and groundwater runoff are supplied (Maxwell et al. 1995).

guild Group of organisms, not necessarily taxonomically related, that are ecologically similar in characteristics such as diet, behavior, or microhabitat preference or with respect to their ecological role in general.

habitat An environment of a particular kind, often used to describe the environmental needs of a certain species or community (Wilson 1992).

headwaters (1) Upper reaches of tributaries in a drainage basin; (2) the point on a nontidal stream above which the average annual flow is less than 5 cubic feet per second (Armantrout 1998).

herpetofauna All the species of amphibians and reptiles inhabiting a specified region.

Holocene An epoch of the Quaternary period, spanning the time from the end of the Pleistocene (8,000 years ago) to the present (U.S. Geological Survey 1999).

hydrographic The measurement or study of bodies of water and associated terrain.

hydrography The study, description, and mapping of oceans, lakes, and rivers, especially with reference to their navigational and commercial uses (Nevada Division of Water Planning 1998).

hydrological regime Water movement in a given area that is a function of the input from precipitation, surface water, and groundwater and the output from evaporation into the atmosphere or transpiration from plants (Armantrout 1998).

hydrology The science of waters of the earth; their occurrence, distribution, and circulation; their physical and chemical properties; and their reaction with the environment, including living beings (Nevada Division of Water Planning 1998).

hydrophyte Any plant adapted to live in water or very wet habitats (Lincoln et al. 1982).

hydropower Electrical energy produced by falling water.

hypogean Pertaining to below-surface environments.

hypolimnion Poorly oxygenated and illuminated lower layer or region in a stratified lake that extends from the metalimnion to the bottom and is removed from major surface influences (Armantrout 1998).

ichthyofauna All the fish species inhabiting a specified region (Brown and Gibson 1983).

ichthyogeographic Referring to the biogeography of fish species.

ichthyology The study of fishes (Lincoln and Boxshall 1987).

important bird area Key sites for bird conservation that either have significant numbers of one or more globally threatened species or are one of a set of sites that together hold a suite of restricted-range species or biome-restricted species or have exceptionally large numbers of migratory or congregatory species; sites are small enough to be conserved in their entirety (BirdLife International 2004a).

impoundment A body of water such as a pond, confined by a dam, dike, floodgate, or other barrier. It is used to collect and store water (Eckhardt 1998).

indigenous Native to an area.

intact habitat Undisturbed area characterized by the maintenance of most original ecological processes and by communities with most of their original native species still present.

interbasin water transfer A project that moves water between hydrologically unconnected catchments.

Intertropical Convergence Zone (ITCZ) An area where the northern and southern hemispheric trade winds converge, usually located between 10° north and south of the equator.

introduced species See **exotic species**.

invasive species Exotic species (i.e., alien or introduced) that rapidly establish themselves and spread through the natural communities into which they are introduced.

invertebrate Any animal lacking a backbone or bony segment that encloses the central nerve cord (Wilson 1992).

irrigation Movement of water through ditches, canals, pipes, sprinklers, or other devices from the surface or groundwater to provide water to vegetation (Armantrout 1998).

isopod A member of the crustacean order Isopoda, a diverse group of flattened and segmented invertebrates. Pill bugs are an example.

IUCN Acronym for the World Conservation Union.

karst Applies to areas underlain by gypsum, anhydrite, rock salt, dolomite, quartzite (in tropical moist areas), and limestone (Hobbs 1992).

lacustrine Pertaining to lakes, reservoirs, wetlands, or any large standing waterbody (Armantrout 1998).

lake A large inland body of freshwater or saltwater occupying a basin or hollow on the earth's surface, which may or may not have a current or single direction of flow (U.S. Department of Agriculture 1995).

landscape An aggregate of land forms, together with its biological communities (Lotspeich and Platts 1982).

lateral lakes A permanent zone of river inundation; lakes situated on the side of a river channel, often in the floodplain.

lentic Referring to standing freshwater habitats such as ponds and lakes (Brown and Gibson 1983).

levee A natural or artificial earthen obstruction along the edge of a stream, lake, or river. Usually used to restrain the flow of water out of a riverbank (Eckhardt 1998).

life cycle The entire lifespan of an organism from the moment it is conceived to the time it reproduces (Wilson 1992).

littoral Shallow shore area (less than 6 m deep) of a waterbody where light usually can penetrate to the bottom and that is often occupied by rooted macrophytes (Armantrout 1998).

lotic Referring to running freshwater habitats, such as springs and streams (Brown and Gibson 1983).

macroinvertebrates Invertebrates large enough to be seen with the naked eye (e.g., most aquatic insects, snails, and amphipods) (Maxwell et al. 1995).

macrophyte A plant that can be seen without the aid of optics (Armantrout 1998).

main channel Primary watercourse containing the major stream flow (Armantrout 1998).

major habitat type See **biome**.

mangrove A type of tree that can tolerate saltwater conditions.

marine Living in saltwater (Brown and Gibson 1983).

marsh Water-saturated, poorly drained wetland area that is periodically or permanently inundated to a depth of 2 m and supports an extensive cover of emergent, nonwoody vegetation, without peatlike accumulations (Armantrout 1998).

mediterranean climate Dry and hot summers with cool and moist winters, found at the northern and southern margins of Africa.

metalimnion Stratum between the epilimnion and hypolimnion that exhibits a marked thermal discontinuity, with a temperature gradient equal to or exceeding 1°C/m (Armantrout 1998).

metapopulation A group of two or more separated populations of the same species that regularly exchange genes.

Miocene A geological epoch in the Tertiary period (c. 26–5 million years B.P.) (Lincoln and Boxshall 1987).

miombo An eastern and southern African vegetation type dominated by tree species in the genera *Brachystegia*, *Julbernardia*, and *Isoberlinia*.

mollusk An animal belonging to the phylum Mollusca, such as a snail or clam (Wilson 1992).

montane Habitats typically above 1,500 m altitude in Africa.

moorland and heathland Vegetation types associated with Afromontane and Afroalpine regions of Africa.

mopane A southern African vegetation type dominated by the tree *Colophospermum mopane*.

Mormyridae Elephant fishes; family containing about 100 species of mostly small (to 0.5 m) primitive teleost fishes found in African freshwater habitats (Lincoln and Boxshall 1987).

morphology   The form and structure of an organism, with special emphasis on external features (Lincoln et al. 1982).

nationally important   A biological distinctiveness category.

nonindigenous species   See **exotic species**.

nonnative species   See **exotic species**.

non–point source pollution   A diffuse form of water quality degradation in which wastes are not released at one specific, identifiable point but from a number of points that are spread out and difficult to identify and control (Eckhardt 1998).

nutrient   Element or compound essential for growth, development, and life for living organisms such as oxygen, nitrogen, phosphorus, and potassium (Armantrout 1998).

nutrient cycling   Circulation of nutrient elements and compounds in and between the atmosphere, soil, parent rock, flora, and fauna in a given area such as a waterbody (Armantrout 1998).

oasis   Area of permanent water in a desert.

obligate species   A species that must have access to a particular habitat type to persist.

Odonata   Dragonflies and damselflies; order of large insects; larvae typically aquatic (Lincoln and Boxshall 1987).

oligotrophic   Having a low supply of plant nutrients (Eckhardt 1998).

overexploitation   Levels of collection, hunting, or fishing that are not ecologically sustainable. Also called **overharvest**.

overharvest   See **overexploitation**.

Palearctic realm   A biogeographic region that includes Africa north of the Sahara Desert.

Paleocene   A geological epoch in the Tertiary period (c. 65–54 million years B.P.) (Lincoln and Boxshall 1987).

pans   Dried lakes often encrusted by various salts.

pastoralists   People whose lifestyle involves the keeping of cattle and other livestock and regular movements across the landscape.

Pearson correlation   A measure of linear association between two variables. Values of the correlation coefficient range from –1 to 1. The absolute value of the correlation coefficient indicates the strength of the linear relationship between the variables, with larger absolute values indicating stronger relationships. The sign of the coefficient indicates the direction of the relationship.

peat bog   Bog with a dominant underlying material of peat (Armantrout 1998).

pelagic   (1) Open water areas of lakes, reservoirs, or seas away from the shore; (2) refers to organisms at or near the surface in water away from the shore (Armantrout 1998).

perennial   Stream, lake, or other waterbody with water present continuously during a normal water year (Armantrout 1998).

plankton   Small plants and animals, generally smaller than 2 mm and without strong locomotive ability, that are suspended in the water column and carried by currents or waves that may make daily or seasonal movements in the water column (Armantrout 1998).

Pleistocene   An epoch of the Quaternary period, spanning the time between 1.8 m.y.a. and the beginning of the Holocene at 8,000 years ago (U.S. Geological Survey 1999).

point source   Source of pollution that involves discharge of wastes from an identifiable point, such as a smokestack or sewage treatment plant (Eckhardt 1998).

polymorphism   The co-occurrence of several different forms (Lincoln et al. 1982).

population   In biology, any group of organisms belonging to the same species at the same time and place (Wilson 1992).

potamodromous   Life cycle strategy of a fish that includes migrations, spawning, and feeding entirely in freshwater (Armantrout 1998).

protected area   Any place receiving some form of conservation management, from strict protection to sustainable resource use.

protist An organism that belongs to the kingdom Protista, which includes forms with both plant and animal affinities (i.e., protozoans, bacteria, and some algae, fungi, and viruses) (U.S. Geological Survey 1999).

Quaternary A geological period from 2 m.y.a. to the present.

radiation The diversification of a group of organisms into multiple species because of intense isolating mechanisms or opportunities to exploit diverse resources.

Ramsar Convention An abbreviated name for the Convention on Wetlands of International Importance, also called the Convention on Wetlands (Ramsar, Iran, 1971).

Ramsar site Wetland designated by the contracting parties to the Ramsar Convention for inclusion in the *List of Wetlands of International Importance* because it meets one or more Ramsar criteria.

range (1) The limits of the geographic distribution of a species or group; (2) home range (Lincoln and Boxshall 1987).

rarity Seldom occurring either in absolute number of individuals or in space (Fielder and Jain 1992).

recharge Refers to water entering an underground aquifer through faults, fractures, or direct absorption (Eckhardt 1998).

refugia Habitats or environmental factors that convey spatial and temporal resistance or resilience to biotic communities affected by biophysical disturbances (Sedell et al. 1994).

rehabilitation (1) Action taken to return a land form, vegetation, or waterbody to as near its original condition as practical; (2) term implies making land and water resources useful again (primarily for humans) after natural or anthropogenic disturbances (Armantrout 1998).

relictual Referring to a species or group of organisms largely characteristic of a past environment or ancient biota.

representation The protection of the full range of biodiversity of a given biogeographic unit in a system of protected areas.

reservoir A pond, lake, tank, or basin (natural or human made) where water is collected and used for storage. Large bodies of groundwater are called reservoirs of water (Eckhardt 1998).

resident species Organisms normally found in a single habitat, ecosystem, or area (Armantrout 1998).

resilience The speed at which a habitat, population, or community is able to return to equilibrium after a perturbation (Pimm 1986).

restoration Management of a disturbed or degraded habitat that results in recovery of its original state (Wilson 1992).

rheophilous Thriving in running water; dwelling in flowing creeks.

rift valley A valley with steep sides; formed by a rift in the earth's crust.

riparian Referring to the interface between freshwater habitats (normally flowing waters) and the terrestrial landscape (Gregory et al. 1991).

riparian buffer zone A riparian area that is afforded some degree of protection (Wenger 1999).

river A natural stream of water, larger than a brook or creek.

riverine (1) Habitats that are formed by or associated with a river or stream; (2) wetlands and deeper water habitats in a channel that are influenced strongly by the energy of flowing water; (3) also applied to vegetation growing in a floodplain, in close proximity to watercourses with flowing water, or on islands in a river (Armantrout 1998).

runoff Surface water entering rivers, freshwater lakes, or reservoirs (Eckhardt 1998).

Sahel Geographic region in the southern margin of the Sahara Desert.

saline Salty.

salinity The amount of dissolved salts in a solution.

salinization The process by which soluble salts accumulate in soil.

salt pan An undrained natural depression in which water gathers and leaves a deposit of salt on evaporation (Vennie 2004).

savanna A habitat dominated by grasslands but with woodland and gallery forest elements.

sclerophyllous Relating to a type of vegetation characterized by hard, leathery, evergreen foliage that is specially adapted to prevent moisture loss; generally characteristic of regions with mediterranean climates.

sebkhas Irregularly flooded depressions.

secondary extinction The extinction of one species caused by the loss of another.

sedge Any of a group of perennial grasslike plants, usually with three-cornered, solid stems, common in low water or on wet and marshy ground (Hutchinson Dictionary of Plants 2000).

sediment Soil particles, sand, and minerals washed from the land into aquatic systems as a result of natural and human activities (Eckhardt 1998).

sedimentation (1) Action or process of forming and depositing sediments; (2) deposition of suspended matter by gravity when water velocity cannot transport the bed load (Armantrout 1998).

seine A large net that hangs vertically, with floats at the top and weights at the bottom, used to catch fish.

semi-aquatic Living partly in or adjacent to water (Brown and Gibson 1983).

semi-arid A term applied to regions or climates where moisture normally is greater than under arid conditions but still limits the growth of most crops (Vennie 2004).

silt (1) Fine soil that is 0.004–0.062 mm (0.00002–0.0003 in.) in diameter; (2) also applied to a soil or substrate containing a very high proportion of silt particles (Armantrout 1998).

siltation The deposition of finely divided soil and rock particles on the bottom of waterbodies (Eckhardt 1998).

sinkholes Depressions or cavities created by dissolution of limestone bedrock or collapse of caves. Typically found in karst landscapes.

site A localized natural habitat containing important biodiversity features.

speciation The process of species formation, in which one species evolves into two or more species (Quammen 1996).

species The basic unit of biological classification, consisting of a population or series of populations of closely related and similar organisms (Wilson 1992).

species assemblage The combination of particular species that occur together in a specific location and have a reasonable opportunity to interact with one another (Matthews 1998).

species flock A group of several ecologically diverse and closely related species that have evolved in a single macrohabitat, such as a particular lake basin (Lincoln and Boxshall 1987).

species radiation Refers to the evolution of a single species into several different species in the same geographic range (Wilson 1992). Also called adaptive radiation.

species richness A simple measure of species diversity calculated as the total number of species in a habitat or community (Fielder and Jain 1992).

sponge Any saclike simple invertebrate of the phylum Porifera.

spring A natural discharge of water as leakage or overflow from an aquifer through a natural opening in the soil or rock onto the land surface or into a waterbody (Hobbs 1992).

stakeholder Any person, group, or institution that affects or is affected by a particular issue or outcome.

stenotopic Able to exist only in a narrow range of habitats.

stock A race, stem, or lineage (Lincoln et al. 1982).

stream A general term for a body of flowing water (Maxwell et al. 1995); often used to describe a midsized tributary (as opposed to a river or creek).

sub-Saharan Africa Africa south of the Sahara Desert.

subspecies Subdivision of a species. Usually defined as a population or series of populations occupying a discrete range and differing genetically from other geographic races of the same species (Wilson 1992).

subterranean Underground.

subtropical An area where the mean annual temperature ranges from 13°C to 20°C (Brown and Gibson 1983).

surface water All waters whose surface is naturally exposed to the atmosphere, such as rivers, lakes, reservoirs, ponds, streams, impoundments, seas, and estuaries and all springs, wells, or other collectors directly influenced by surface water (Nevada Division of Water Planning 1998).

swamp   Tree- or tall shrub–dominated wetland characterized by periodic flooding and nearly permanent subsurface water flow through mixtures of mineral sediments and organic materials without peatlike accumulation (Armantrout 1998).

systematics   The classification of living organisms into hierarchical series of groups emphasizing their phylogenetic relationships; often used as equivalent to *taxonomy* (Lincoln et al. 1982).

taxon (pl. *taxa*)   A general term for any taxonomic category, such as a species, genus, family, or order (Brown and Gibson 1983).

temperate   An area in which the mean annual temperature ranges from 10°C to 13°C.

terrestrial   Living or occurring on land.

Tertiary   The first period of the Cenozoic era (after the Mesozoic era and before the Quaternary period), spanning the time between 65 and 1.8 m.y.a. (U.S. Geological Survey 1999).

thermocline   Layer of water between the warmer surface zone and the colder deep zone of a thermally stratified body of water.

tilapiine   A group of genera of tropical freshwater fishes in the family Cichlidae that are native to Africa and the southwestern Middle East; often called tilapias.

transboundary   A protected area that ranges across national borders.

transfrontier   See **transboundary**.

tributary   A stream or river that flows into a larger stream, river, or lake, feeding it water.

trona   A mineral occurring as a white crystalline fibrous deposit from certain soda brine springs and lakes.

trophic   Related to the processes of energy and nutrient transfer (i.e., productivity) from one level of organisms to another in an ecosystem (Armantrout 1998).

tropical   An area where the mean annual temperature is over 20°C.

turbidity   Refers to the extent to which light penetrates a body of water. Turbid waters are those that do not generally support net growth of photosynthetic organisms (Jeffries and Mills 1990).

typha   Reed maces; cattails.

upland   An area of land lying above the level where water flows or where flooding occurs.

Vertebrata   A subgroup of chordates, specifically those with backbones or spinal columns.

vision statement   A concise summary of the key goals of the biodiversity vision, intended to catalyze interest among target audiences outside WWF and focus conservation efforts on goals of primary importance.

wadi   A type of ravine or gully found in parts of Africa and the Arabian Peninsula through which water flows during the rainy season.

waterfalls   See **falls**.

watershed   See **catchment**.

water table   Depth below which the ground is saturated with water (Armantrout 1998).

wetlands   Lands transitional between terrestrial and aquatic systems where the water table usually is at or near the surface or the land is covered by shallow water. These areas are inundated or saturated by surface water or groundwater at a frequency and duration sufficient to support vegetation typically adapted for life in saturated soil conditions (Maxwell et al. 1995).

whitewater rivers   Waters that are less acidic, have a higher conductivity, and a low transparency compared with blackwater rivers. They show important changes in water level and support seasonal floodbank plain grasses; bottom vegetation is generally absent, and a bottom of sand, clay, and mud is present.

woody debris   Any woody material in an aquatic system, often providing habitat and nutrients.

xeric   Describes dryland or desert areas.

zoogeography   Study of the geographic distributions of animals and animal communities (Lincoln and Boxshall 1987).

Literature Cited

Abbassy, M. S., H. Z. Ibrahim, and M. M. Abu El-Amayem. 1999. Occurrences of pesticides and polychlorinated biphenyls in water of the Nile River at the estuaries of Rosetta and Damiatta branches, north of delta, Egypt. Journal of Environmental Science and Health Part B: Pesticides, Food Contaminants and Agricultural Wastes 34(2):255–267.

Abdallah, A. M. 2000. Nearshore benthic macroinvertebrates of Lake Malawi/Nyasa (East Africa). M.S. thesis. University of Waterloo, Ontario, Canada.

Abdelbary, M. R. 1996. Effects of the Aswan High Dam on Nile water and bed levels. Pages 444–463 in A. M. Shady, M. El-Moattassem, E. A. Abdel-Hafiz, and A. K. Biswas, editors. Management and development of major rivers. Oxford University Press, Calcutta, India.

Abell, R. A., D. M. Olson, E. Dinerstein, P. T. Hurley, J. T. Diggs, W. Eichbaum, S. Walters, W. Wettengel, T. Allnutt, C. J. Loucks, and P. Hedao. 2000. Freshwater ecoregions of North America: a conservation assessment. Island Press, Washington, DC, USA.

Abell, R., M. Thieme, E. Dinerstein, and D. Olson. 2002. A sourcebook for conducting biological assessments and developing biodiversity visions for ecoregion conservation. Volume II: freshwater ecoregions. World Wildlife Fund, Washington, DC, USA.

Abiya, I. O. 1996. Towards sustainable utilization of Lake Naivasha, Kenya. Lakes and Reservoirs: Research and Management 2(1996):231–242.

Acreman, M. C. and F. E. Hollis, editors. 1996. Water management and wetlands in Sub-Saharan Africa. IUCN, Gland, Switzerland.

Acropolis Kenya. 1994. An economic valuation of the costs and benefits in the lower Tana Catchment resulting from dam construction. Report prepared by Acropolis Kenya for Nippon Koei, Nairobi, Kenya.

Adebisi, A. A. 1988. Changes in the structural and functional components of the fish community of a seasonal river. Archiv für Hydrobiologie 113(3):457–464.

Admassu, D. 1994. Maturity, fecundity, brood size and sex ratio of Tilapia (*Oreochromis niloticus* L.) in Lake Awassa. SINET: Ethiopian Journal of Science 17(1):53–69.

Aggrey-Fynn, E. 2001. The contribution of the fisheries sector to Ghana's economy. Sustainable Fisheries Livelihood Programme, DFID/FAO, Cotonov, Benin.

Agnese, J. F. and G. G. Teugels. 2001. The *Bathyclarias-Clarias* species flock. A new model to understand rapid speciation in African Great lakes. Comptes Rendus de l'Academie des Sciences Serie III Sciences de la Vie 324(8):683–688.

Agnew, J. D. 1986. Invertebrates of the Orange-Vaal system, with emphasis on the Ephemoptera. Pages 123–134 in B. R. Davies and K. F. Walker, editors. The ecology of river systems. Dr. W. Junk Publishers, Dordrecht, The Netherlands.

Ajao, E. A. 1994. Coastal aquatic ecosystems, conservation and management strategies in Nigeria. Southern Africa Journal of Aquatic Sciences 20(1–2):3–22.

Alayo, P. D. 1973. Lista de peces fluviatiles de Cuba. Torreia, La Habana 29:1–55.

Alcamo, J., P. Döll, T. Henrichs, F. Kaspar, B. Lehner, T. Rosch, and S. Siebert. 2003. Development and testing of the WaterGAP 2 global model of water use and availability. Hydrological Sciences Journal 48(3):317–337.

Alcamo, J., R. Leemans, and E. Kreileman. 1998. Global change scenarios of the 21st century: results from the IMAGE 2.1 model. Elsevier, Oxford, UK.

Aldabra Marine Program. 2003. Atoll and other sites. Retrieved 2003 from the World Wide Web: http://www.aldabra.org.

Aldegheri, M. 1972. Rivers and streams on Madagascar. Pages 261–310 in R. Battistini and G. Richard-Vindard, editors. Biogeography and ecology in Madagascar. Dr. W. Junk, The Hague, The Netherlands.

Alexander, D. and T. Miller. 1996. Saving the spectacular flora of Socotra. Plant Talk 7:19–22.

Alexander, W. J. R. 1985. Hydrology of low latitude Southern Hemisphere land masses. Hydrobiologia 125:75–83.

Alimi, T. and O. A. Akinyemiju. 1990. An economic analysis of water hyacinth control methods in Nigeria. Journal of Aquatic Plant Management 28:105–107.

Alin, S. R., A. S. Cohen, R. Bills, M. M. Gashagaza, E. Michel, J. Tiercelin, K. Martens, P. Coveliers, S. K. Mboko, K. West, M. Soreghan, S. Kimbadi, and G. Ntakimazi. 1999. Effects of landscape disturbance on animal communities in Lake Tanganyika, East Africa. Conservation Biology 13(5):1017–1033.

Allan, D. G., M. T. Seaman, and B. Kaletja. 1995. The endorheic pans of South Africa. Pages 75–101 in G. I. Cowan, editor. Wetlands of South Africa. Department of Environmental Affairs and Tourism, Pretoria, South Africa.

Allan, J. D. and A. S. Flecker. 1993. Biodiversity conservation in running waters. BioScience 43(1):32–42.

Allanson, B. R., editor. 1979. Lake Sibaya. Monographiae Biologicae 36. Dr. W. Junk, The Hague, The Netherlands.

Allanson, B. R., R. C. Hart, J. H. O'Keefe, and R. D. Robarts. 1990. Inland waters of southern Africa: an ecological perspective. Monographiae Biologicae 64. Kluwer Academic Publishers, Dordrecht, The Netherlands.

Allen, G. R. 1989. Freshwater fishes of Australia. T. F. H. Publication, Neptune City, NJ, USA.

Allen, G. R. 1991. Field guide to the freshwater fishes of New Guinea, Publ. 9. Christensen Research Institute, Madang, Papua New Guinea.

Allison, E., V. Cowan, and R. Paley. 2000. Pollution control and other measures to protect biodiversity in Lake Tanganyika: biodiversity special study advice to the strategic action programme. RAF/92/G32. UNDP/GEF, Dar Es Salaam, Tanzania.

Alonso, L. E. and L. Nordin, editors. 2003. A rapid biological assessment of the aquatic ecosystems of the Okavango Delta, Botswana: high water survey. RAP Bulletin of Biological Assessment 27. Conservation International, Washington, DC, USA.

Aloo, P. A. 2003. Biological diversity of the Yala Swamp Lakes, with special emphasis on fish species composition, in relation to changes in the Lake Victoria Basin (Kenya): threats and conservation measures. Biodiversity and Conservation 12:905–920.

Al-Safadi, M. M. 1998. Freshwater fishes of Soqotra Island, Yemen. Pages 213–218 in H. J. Dumont, editor. Soqotra: proceedings of the first international symposium on Soqotra Island: present and future. United Nations Publications, New York, NY, USA.

Altaba, C. R. 1990. The last known population of the freshwater mussel *Margaritifera auricularia* (Bivalvia, Unionoida): a conservation priority. Biological Conservation 52:271–286.

Amari, M. and H. Azafzaf. 2001. Tunisia. Pages 953–973 in L. D. C. Fishpool and M. I. Evans, editors. Important bird areas in Africa and associated islands: priority sites for conservation. Pisces Publications and BirdLife International (BirdLife Conservation Series No. 11), Newbury and Cambridge, UK.

AMCEN. 2002. Report of the ministerial session, African Ministerial Conference on the Environment, ninth session, 4–5 July 2002, Kampala, Uganda. Retrieved 2003 from the World Wide Web: http://www.nemaug.org/amcen/index.htm.

Amonoo-Neizer, E. H., D. Nyamah, and S. B. Bakiamoh. 1996. Mercury and arsenic pollution in soil and biological samples around the mining town of Obuasi, Ghana. Water Air and Soil Pollution 91:363–373.

Angermeier, P. L. and I. J. Schlosser. 1995. Conserving aquatic biodiversity: beyond species and populations. American Fisheries Society Symposium 17:402–414.

Ansell, W. F. H. 1965. Hippo census on the Luangwa River. The Puku 3:15–28.

Armantrout, N. B., compiler. 1998. A glossary of aquatic habitat inventory terminology. American Fisheries Society, Bethesda, MD, USA.

Arnoult, J. 1959. Une nouvelle espèce de poisson aveugle de Madagascar: *Typhleotris pauliani* n.sp. Mémoires de l'Institut Scientifique de Madagascar 13:137–138.

Arthington, A. H., E. Baran, C. A. Brown, P. Dugan, A. S. Halls, J. M. King, C. V. Minte-Vera, R. E. Tharme, and R. L. Welcomme. 2004. Water requirements for floodplain rivers and fisheries: existing decision-support tools and pathways for development. WorldFish Center and Comprehensive Assessment of Water Management in Agriculture, Penang, Malaysia.

Ashley, C. 1995. Population dynamics, the environment and demand for water and energy in Namibia, Research Discussion Paper 7. Directorate of Environmental Affairs, Ministry of Environment and Tourism, Windhoek, Namibia.

Awaiss, A. and E. M. Saadou. 1998. Evaluation de la diversité biologique aquatique du Niger. Rapport de Consultation Projet NER/97/631/A/1G/99/MH-E/SE-CNEDD, Niamey, Niger.

Awaiss, A. and S. Seyni, editors. 1998. Utilisation durable de l'eau, des zones humides et de la diversité biologique dans les écosystèmes partagés (Bénin, Burkina Faso, Niger et Togo). Actes de Séminaire Atelier Sous-Régional, OMPO/RAMSAR/DFPP, Tapoa, 16–20 Novembre 1998.

AWF. 2000. African Wildlife Foundation (AWF) Heartland conservation planning report: participatory scoping meeting. Zambezi Heartland. Internal document. AWF, Kariba, Zimbabwe.

AWF. 2001. African Wildlife Foundation (AWF) Heartland conservation planning report: participatory scoping meeting. Four Corners Transboundary Natural Resources Management Area. Internal document. AWF, Kasane, Botswana.

AWF. 2003a. African Wildlife Foundation (AWF) African Heartlands program. Retrieved 2003 from the World Wide Web: http://www.awf.org/heartlands.

AWF. 2003b. Heartland Conservation Process (HCP): a framework for effective conservation in AWF's African Heartlands. African Wildlife Foundation (AWF), Washington, DC, USA.

Baard, E. H. W., A. L. De Villiers, and M. E. De Villiers. 1985. A contribution to the herpetofaunal distribution of the Verneukpan/Coppertom area, Northwestern Cape Province, South Africa. Journal of the Herpetological Association of Africa 35:35–32.

Baha El Din, S. M. 1999. Directory of important bird areas in Egypt. Palm Press, Cairo, Egypt.

Baha El Din, S. M. 2001. Egypt. Pages 241–264 in L. D. C. Fishpool and M. I. Evans, editors. Important bird areas in Africa and associated islands: priority sites for conservation. Pisces Publications and BirdLife International, Newbury and Cambridge, UK.

Bahuchet, S., editor. 1992. The situation of indigenous peoples in tropical forests, fishing populations. Retrieved 2001 from the World Wide Web: http://lucy.ukc.ac.uk/Sonja/RF/Ukpr/Report01.htm.

Bailey, R. G. 1966. The dam fisheries of Tanzania. East Africa Agricultural and Forestry Journal 32(1):1–15.

Bailey, R. G. 1969. The non-cichlid fishes of the eastward flowing rivers of Tanzania, East Africa. Revue de Zoologie Africaine 80:171–199.

Bailey, R. G. 1986. The Zaire River system. Pages 201–214 in B. R. Davies and K. F. Walker, editors. The ecology of river systems. Dr. W. Junk Publishers, Dordrecht, The Netherlands.

Baillie, J. and B. Groombridge. 1996. IUCN Red List of threatened animals. IUCN, Gland, Switzerland.

Bakarr, M., B. Bailey, D. Byler, R. Ham, S. Olivieri, and M. Omland, editors. 2001. From the forest to the sea: biodiversity connections from Guinea to Togo. Conservation International, Washington, DC, USA.

Baker, N. E. and E. M. Baker. 2001. Tanzania. Pages 897–945 in L. D. C. Fishpool and M. I. Evans, editors. Important bird areas in Africa and associated islands: priority sites for conservation. Pisces Publications and BirdLife International (BirdLife Conservation Series No. 11), Newbury and Cambridge, UK.

Baker, N. E. and E. M. Baker. 2002. Important bird areas in Tanzania: a first inventory. Wildlife Conservation Society of Tanzania, Dar es Salaam, Tanzania.

Balarin, J. D. 1986. National reviews for aquaculture development in Africa, Ethiopia. FAO Fisheries Circular 770(9):120.

Balirwa, J. S., C. A. Chapman, L. J. Chapman, I. G. Cowx, K. Geheb, L. Kaufman, R. H. Lowe-McConnell, O. Seehausen, J. H. Wanink, R. L. Welcomme, and F. Witte. 2003. Biodiversity and fisheries sustainability in the Lake Victoria Basin: an unexpected marriage? BioScience 53(8):703–715.

Balmford, A., A. Bruner, P. Cooper, R. Costanza, S. Farber, R. E. Green, M. Jenkins, P. Jefferiss, V. Jessamy, J. Madden, K. Munro, N. Myers, S. Naeem, J. Paavola, M. Rayment, S. Rosendo, J. Roughgarden, K. Trumper, and R. Kerry Turner. 2002. Economic reasons for conserving wild nature. Science 297:950–953.

Balon, E. K. and D. J. Stewart. 1983. Fish assemblage in a river with unusual gradient (Luongo, Africa Zaire system), reflections on river zonation, and description of another new species. Environmental Biology of Fishes 9:225–252.

Banarescu, P. M. 1995. Zoogeography of fresh waters. AULA-Verlag, Wiesbaden, Germany.

Banda, M. and M. M. Hara. 1997. Habitat degradation caused by seines on the fishery of Lake Malombe and upper Shire River and its effects. Pages 305–310 in K. Remane, editor. African inland fisheries, aquaculture and the environment. Fishing News Books, Malden, MA, USA.

Banister, K. E. 1986. Fish of the Zaire system. Pages 215–224 in B. R. Davies and K. F. Walker, editors. Ecology of river systems. Dr. W. Junk Publishers, Dordrecht, The Netherlands.

Banister, K. E. 1994. *Glossogobius ankaranensis,* a new species of blind cave goby from Madagascar (Pisces: Gobiodei: Gobiidae). Aqua 1(3):25–28.

Banister, K. E. and R. G. Bailey. 1979. Fishes collected by the Zaïre River Expedition 1974–75. Zoological Journal of the Linnean Society 66:205–249.

Barbier, E., M. Acreman, and D. Knowler. 1997. Economic valuation of wetlands: a guide for policy makers and planners. Ramsar Convention Bureau, Gland, Switzerland.

Baribwegure, D. and H. J. Dumont. 2000. Some freshwater cyclopoids (Crustacea: Copepoda) of the island of Soqotra (Indian Ocean), with the description of three new species. International Review of Hydrobiology 85(4):471–489.

Barker, J. A., R. J. Christensen, B. W. Neil, and J. R. Christensen. 1990. The power of vision [video]. Charthouse Learning Corporation, Burnsville, MN, USA.

Barnard, P., editor. 1998. Biological diversity in Namibia. Namibian National Biodiversity Task Force, Windhoek, Namibia.

Barnes, J. I. 1988. Environmental aspects of the Ciskeian Fish/Kat River Basins rural planning study. Pages 421–430 in M. N. Bruton and F. W. Gess, editors. Towards an environmental plan for the Eastern Cape. Rhodes University, Grahamstown, South Africa.

Barnes, K. N. 1998a. Important bird areas of Lesotho. Pages 281–294 in K. N. Barnes, editor. The important bird areas of southern Africa. BirdLife International, Johannesburg, South Africa.

Barnes, K. N. 1998b. Important bird areas of South Africa: an introduction. Pages 27–45 in K. N. Barnes, editor. The important bird areas of southern Africa. BirdLife International, Johannesburg, South Africa.

Barnes, K. N., editor. 1998c. The important bird areas of southern Africa. BirdLife International, Johannesburg, South Africa.

Barnes, K. N. 1998d. Important bird areas of the North-west Province. Pages 93–101 in K. N. Barnes, editor. The important bird areas of southern Africa. BirdLife International, Johannesburg, South Africa.

Barnes, K. N. 1998e. Important bird areas of the Western Cape. Pages 220–267 in K. N. Barnes, editor. The important bird areas of southern Africa. BirdLife International, Johannesburg, South Africa.

Barnes, K. N. and M. D. Anderson. 1998. Important bird areas of the Northern Cape. Pages 103–122 in K. N. Barnes, editor. The important bird areas of southern Africa. BirdLife International, Johannesburg, South Africa.

Barnes, K. N., D. J. Johnson, M. D. Anderson, and P. B. Taylor. 2001. South Africa. Pages 793–876 in L. D. C. Fishpool and M. I. Evans, editors. Important bird areas in Africa and associated islands: priority sites for conservation. Pisces Publications and BirdLife International (BirdLife Conservation Series No. 11), Newbury and Cambridge, UK.

Baron, J. S., N. L. Poff, P. L. Angermeier, C. N. Dahm, P. H. Gleick, N. G. Hairston Jr., R. B. Jackson, C. A. Johnston, B. D. Richter, and A. D. Steinman. 2002. Meeting ecological and societal needs for freshwater. Ecological Applications 12:1247–1260.

Bate, G. C. and B. H. Walker. 1993. Water relations of the vegetation along the Kuiseb River, Namibia. Madoqua 18(2):85–91.

Bates, R. L. and J. A. Jackson, editors. 1987. Glossary of geology, 3rd ed. American Geological Institute, Alexandria, VA, USA.

Battistini, R. and P. Vérin. 1984. Géographie des Comores. Agence de Co-opération Culturelle et Technique, Paris, France.

Bazzo, D. 2000. Atlas infogéographique de la Guinée maritime. Ministère de l'Agriculture et de l'Elevage et Ministère de la Pêche et l'Aquaculture. CNRS-IRD-CNSHB, Conakry, Guinea.

BBC News. 2002. Work starts on giant Ethiopian dam. Retrieved 2003 from the World Wide Web: http://news.bbc.co.uk/2/hi/business/2188785.stm.

Beadle, L. C. 1981. The inland waters of tropical Africa. Longman Group Limited, London, UK.

Bechler-Carmaux, N., M. Mietton, and M. Mathieu. 1999. The risk of a drinking water shortage in Niamey (Niger). Sécheresse 10(4):281–288.

Beilfuss, R., P. Dutton, and D. Moore. 2000. Land cover and land use change in the Zambezi Delta. Pages 31–105 in J. Timberlake, editor. Biodiversity of the Zambezi Basin wetlands. Biodiversity Foundation for Africa, Bulawayo/The Zambezi Society, Harare, Zimbabwe.

Bell-Cross, G. 1965. Physical barriers separating the fishes of the Kafue and Middle Zambezi River systems. Fisheries Research Bulletin of Zambia 4:97–101.

Bell-Cross, G. 1972. The fish fauna of the Zambezi River system. Arnoldia (Rhodesia) 5(29):1–19.

Bell-Cross, G. and J. L. Minshull. 1988. The fishes of Zimbabwe. National Museums and Monuments of Zimbabwe, Harare, Zimbabwe.

Béné, C., K. Mindjimba, E. Belal, T. Jolley, and A. Neiland. 2003a. Inland fisheries, tenure systems and livelihood diversification in Africa: the case of the Yaéré floodplains in Cameroon. African Studies 62(2):187–212.

Béné, C., A. Neiland, T. Jolley, B. Ladu, S. Ovie, O. Sule, O. Baba, Belal E., K. Mindjimba, F. Tiotsop, L. Dara, A. Zakara, and J. Quensiere. 2003b. Inland fisheries, poverty and rural livelihoods in the Lake Chad Basin. Journal of Asian and African Studies 38(1):17–51.

Béné, C. and A. E. Neiland. 2004. From participation to governance: a critical review of governance, co-management and participation in natural resources management with particular reference to inland fisheries in developing countries. WorldFish Center and Challenge for Food and Water, Penang, Malaysia.

Bénech, V. 1992. The northern Cameroon floodplain: influence of hydrology on fish production. Pages 155–164 in E. Maltby, P. Dugan, and J. C. LeFueve, editors. Conservation and development: the sustainable use of wetland resources. IUCN, Gland, Switzerland.

Bénech, V., J. R. Durand, and J. Quensière. 1983. Fish communities of Lake Chad and associated rivers and floodplains. Pages 293–356 in J. P. Carmouze, J. R. Durand, and C. Lévêque, editors. Lake Chad, Monographiae Biologicae 53. Dr. W. Junk Publishers, The Hague, The Netherlands.

Bennasser, L., M. Fekhaoui, J.-L. Benoit-Guyod, and G. Merlin. 1997. Influence of tide on water quality of lower Sebou polluted by Gharb Plain wastes (Morocco). Water Research 31:859–867.

Bennun, L. and P. Njoroge. 1999. Important bird areas in Kenya. Nature Kenya, the East Africa Natural History Society, Nairobi, Kenya.

Bennun, L. and P. Njoroge. 2001. Kenya. Pages 411–464 in L. D. C. Fishpool and M. I. Evans, editors. Important bird areas in Africa and associated islands: priority sites for conservation. Pisces Publications and BirdLife International, Newbury and Cambridge, UK.

Benstead, J. P., P. H. De Rham, J.-L. Gattolliat, F.-M. Gibon, P. V. Loiselle, M. Sartori, J. S. Sparks, and M. L. J. Stiassny. 2003. Conserving Madagascar's freshwater biodiversity. BioScience 53(11):1101–1111.

Benstead, J. P., M. L. J. Stiassny, P. V. Loiselle, K. J. Riseng, and N. Raminosoa. 2000. River conservation in Madagascar. Pages 205–231 in P. J. Boon, B. R. Davies, and G. E. Petts, editors. Global perspectives on river conservation: science, policy and practice. John Wiley & Sons, Chichester, UK.

Bento, C. 2000. Wetland bird survey of the Zambezi Delta. Pages 259–278 in J. Timberlake, editor. Biodiversity of the Zambezi Basin Wetlands. Biodiversity Foundation for Africa, Bulawayo/The Zambezi Society, Harare, Zimbabwe.

Berry, P. S. M. 1973a. A hippo count on the upper Luangwa River. The Puku 7:193–195.

Berry, P. S. M. 1973b. The Luangwa Valley giraffe. The Puku 7:71–92.

Bérubé, M. 1992. Une expérience de coopérative piscicole en République Centrafricaine. Pages 370–381 in G. M. Bernacsek and H. Powles, editors. Aquaculture systems research in Africa. IDRC-MR308e,f. International Development Research Centre, Ottawa, Canada.

Bibby, C. J., N. J. Collar, M. J. Crosby, M. F. Heath, C. Imboden, T. H. Johnson, A. J. Long, A. J. Stattersfield, and S. J. Thirgood. 1992. Putting biodiversity on the map: priority areas for global conservation. International Council for Bird Preservation, Cambridge, UK.

Bills, R. 1997. Freshwater of the Moebase region and their fish communities. Environmental impact assessment of the proposed TIGEN sands mineral mine, Zambezia Province, Mozambique. Coastal Environmental Services, Grahamstown, South Africa.

Bills, R. 2000. Freshwater fish survey of the Lower Zambezi River, Mozambique. Pages 461–485 in J. Timberlake, editor. Biodiversity of the Zambezi Basin wetlands. Biodiversity Foundation for Africa, Bulawayo/The Zambezi Society, Harare, Zimbabwe.

Bills, R. 2001. Inventory of fishes from the Maputo Special Reserve and Futi Corridor, Mozambique. A contract for the Transfrontier Conservation Areas Project, National Directorate of Conservation Areas, Ministry of Tourism, Mozambique. JLB Smith Institute of Ichthyology Investigational Report No. 63, Grahamstown, South Africa.

Biney, C. A. 1990. A review of some characteristics of freshwater and coastal ecosystems in Ghana. Hydrobiologia 208:45–53.

Biney, C., A. T. Amuzu, D. Calamari, N. Kaba, I. L. Mbome, H. Naeve, P. B. O. Ochumba, O. Osibanjo, V. Radegonde, and M. A. Saad. 1994. Review of heavy metals in the African aquatic environment. Ecotoxicology and Environmental Safety 28:134–159.

Binns, J. A., P. M. Illgner, and E. L. Nel. 2001. Water shortage, deforestation and development: South Africa's Working for Water Programme. Land Degradation & Development 12:341–355.

BirdLife International. 2000. Threatened birds of the world.

Lynx Edicions and BirdLife International, Barcelona, Spain and Cambridge, UK.

BirdLife International. 2004a. Important bird areas. Retrieved 2004 from the World Wide Web: http://www.birdlife.net/action/science/sites.

BirdLife International. 2004b. Endemic bird areas. Retrieved 2004 from the World Wide Web: http://www.birdlife.net/action/science/endemic_bird_areas/.

Bjørke, S. å. 2002. Vital climate graphics Africa: aridity zones. Retrieved 2003 from the World Wide Web: http://www.grida.no/climate/vitalafrica/english/25.htm.

Blom, A. and J. Yamindou. 2001. A brief history of armed conflict and its impact on biodiversity in the Central African Republic. Biodiversity Support Program, Washington, DC, USA.

Bocian, C. 1998. Preliminary observations on the status of primates in the Etiema Community Forest. Report for A. G. Leventis and the Nigerian Conservation Foundation, Lagos, Nigeria.

Boeseman, M. 1963. An annotated list of species from the Niger Delta. Zoologische Verhandelingen 61:3–48.

Bootsma, H. A. 1993. Spatio-temporal variation of phytoplankton biomass in Lake Malawi, Central Africa. Verhandlungen Internationale Vereinigung für Theoretische und Angewandte Limnologie (Proceedings of the 1992 SIL conference in Barcelona, Spain) 25(2):882–886.

Bootsma, H. A. and R. E. Hecky. 1993. Conservation of the African Great Lakes: a limnological perspective. Conservation Biology 7(3):644–656.

Bootsma, H. A. and R. E. Hecky. 1999. Water quality report: Lake Malawi/Nyasa Biodiversity Conservation Project. SADC/GEF Lake Malawi/Nyasa/Niassa Biodiversity Conservation Project, Senga Bay, Malawi.

Boston University. 2001. MODIS MOD12Q1 land cover product binary data from Boston University. Retrieved 2002 from the World Wide Web: http://duckwater.bu.edu/lc/mod12q1.html.

Boumezbeur, A. 2001. Atlas des zones humides Algériennes d'importance internationale. Direction Générale des Forêts, Ben Aknoun, Alger, Algeria.

Boumezbeur, A. 2002. Atlas des 26 zones humides Algériennes d'importance internationale. Direction Générale des Forêts, Ben Aknoun, Alger, Algeria.

Bousso, T. 1997. The estuary of the Senegal River: the impact of environmental changes and the Diama Dam on resource status and fishery conditions. Pages 45–65 in K. Remane, editor. African inland fisheries, aquaculture and the environment. Fishing News Books, Oxford, UK.

Bowell, R. J., A. Warren, H. A. Minjera, and N. Kimaro. 1995. Environmental impact of former gold mining on the Orangi River, Serengeti N.P., Tanzania. Biogeochemistry (Dordrecht) 28:131–160.

Bowmaker, A. P., P. B. N. Jackson, and R. A. Jubb. 1978. Freshwater fishes. Pages 1181–1231 in M. J. A. Werger, editor. Biogeography and ecology of southern Africa. Dr. W. Junk Publishers, The Hague, The Netherlands.

Braithwaite, C. J. R., J. Casanova, T. Frevert, and B. A. Whitton. 1989. Recent stromatolites in landlocked pools on Aldabra, western Indian Ocean. Palaeogeography Palaeoclimatology Palaeoecology 69:145–166.

Branch, B. 1998. Field guide to snakes and other reptiles of southern Africa. Struik Publishers, Cape Town, South Africa.

Branch, W. R. 2000. Survey of the reptiles and amphibians of the Zambezi Delta. Pages 377–392 in J. Timberlake, editor. Biodiversity of the Zambezi Basin wetlands. Biodiversity Foundation for Africa, Bulawayo/The Zambezi Society, Harare, Zimbabwe.

Branch, W. R. and H. H. Braack. 1989. Reptiles and amphibians in the Karoo National Park: a surprising diversity. Journal of the Herpetological Association of Africa 36:26–35.

Breen, C. M. 1991. Are intermittently flooded wetlands of arid environments important conservation sites? MADOQUA 17(2):61–65.

Brenon, P. 1972. The geology of Madagascar. Pages 27–86 in R. Battistini and G. Richard-Vindard, editors. Biogeography and ecology in Madagascar. Dr. W. Junk, The Hague, The Netherlands.

Brett, M. R. 1989. The Pilanesberg. Frandsen Publishers, Johannesburg, South Africa.

Brichard, P. J. 1980. Report for the government of Mozambique on organization of the ornamental fish trade. GCP/MOZ/006 (SWE) Field Document 1. FAO, Rome, Italy.

Briggs, J. C. 1987. Biogeography and plate tectonics. Elsevier, Amsterdam, The Netherlands.

Brismar, A. 2002. River systems as providers of goods and services: a basis for comparing desired and undesired effects of large dam projects. Environmental Management 29(5):598–609.

Britton, D. L. 1991. Fire and the chemistry of a South African mountain stream. Hydrobiologia 218:177–192.

Broadley, D. G. 2000. The herpetofauna of the Zambezi Basin wetlands. Pages 279–392 in J. Timberlake, editor. Biodiversity of the Zambezi Basin wetlands. Biodiversity Foundation for Africa, Bulawayo/The Zambezi Society, Harare, Zimbabwe.

Broadley, D. G. 2001. An annotated check list of the herpetofauna of Mulanje Mountain. Nyala 21:27–36.

Brooks, T., A. Balmford, N. Burgess, J. Fjeldsa, L. A. Hansen, J. Moore, C. Rahbek, and P. Williams. 2001. Toward a blueprint for conservation in Africa. BioScience 51:613–624.

Brouwer, J. 2003. Wetlands, biodiversity and poverty alleviation in semi-arid regions: a case study from Niger. Retrieved 2003 from the World Wide Web: http://www.iucn.org/themes/cem/cem/region/niger.htm.

Brouwer, J., S. F. Codjo, and W. C. Mullié. 2001a. Niger. Pages 661–672 in L. D. C. Fishpool and M. I. Evans, editors. Important bird areas in Africa and associated islands: priority sites for conservation. Pisces Publications and BirdLife International (BirdLife Conservation Series No. 11), Newbury and Cambridge, UK.

Brouwer, J., W. C. Mullié, and A. M. Issa. 2001b. A synopsis of bird biodiversity in Niger, with special emphasis on waterbirds and wetlands. In F. Pezold. Proceedings of the Conference on Biodiversity in the Niger River Basin, Bamako, Mali.

Brown, C. A., S. Eekhout, and J. M. King. 1996. National Biomonitoring Programme for Riverine Ecosystems: proceed-

ings of spatial frame workshop. Department of Environmental Affairs and Tourism, Pretoria, South Africa.
Brown, C. A. and J. M. King. 2000. Environmental flow assessments: concepts and methodologies. World Bank Water Resources and Environmental Management Guideline Series. Guideline 6. World Bank, Washington, DC, USA.
Brown, D. S. 1991. Freshwater snails of São Tomé, with special reference to *Bulinus forskalii* (Ehrenberg), host of *Schistosoma intercalatum*. Hydrobiologia 209:141–153.
Brown, D. S. 1994. Freshwater snails of Africa and their medical importance. Taylor & Francis, London, UK.
Brown, D. S. and T. K. Kristensen. 1998. Snail control and conservation: snails as friends as well as foes. In H. Madsen, C. C. Appleton, and M. Chimbari, editors. Proceedings of Workshop on Medical Malacology in Africa. Harare, Zimbabwe. Danish Bilharziasis Laboratory, Danida, Denmark.
Brown, J. H., and A. C. Gibson. 1983. Biogeography. C.V. Mosby, St. Louis, MO, USA.
Bruton, M. N. 1979. The survival of habitat desiccation by air breathing clariid catfishes. Environmental Biology of Fishes 4:273–280.
Bruton, M. N. and K. H. Cooper, editors. 1980. Studies on the ecology of Maputaland. Cape and Transvaal Printers, Cape Town, South Africa.
Bunn, S. E. and A. H. Arthington. 2002. Basic principles and ecological consequences of altered flow regimes for aquatic biodiversity. Environmental Management 30(4):492–507.
Burgess, N. D. and G. P. Clarke, editors. 2000. Coastal forests of eastern Africa, IUCN Forest Conservation Programme. IUCN, Gland, Switzerland.
Burgess, N., J. D'Amico Hales, E. Underwood, E. Dinerstein, D. Olson, I. Itoua, J. Schipper, T. Ricketts, and K. Newman. 2004. Terrestrial ecoregions of Africa and Madagascar: a conservation assessment. Island Press, Washington, DC, USA.
Burgess, N. D., M. Nummelin, J. Fjeldså, K. M. Howell, K. Lukumbyzya, L. Mhando, P. Phillipson, and E. Vanden Berghe, editors. 1998. Biodiversity and conservation of the Eastern Arc Mountains of Tanzania and Kenya. Special Issue of the Journal of the East African Natural History Society 87:1–367.
Burgis, M. J. and J. J. Symoens, editors. 1987. African wetlands and shallow water bodies. ORSTOM, Paris, France.
Bussmann, R. W. 1999. Growth rates of the important east African montane forest trees, with particular reference to those of Mount Kenya. Journal of East Africa Natural History Society 89:69–78.
Butzer, K. W. 1971. Recent history of an Ethiopian delta: the Omo River and the level of Lake Rudolf, Department of Geography, Research Paper No. 136. University of Chicago Press, Chicago, USA.
Byaruhanga, A., P. Kasoma, and D. Pomeroy. 2001. Uganda. Pages 975–1003 in L. D. C. Fishpool and M. I. Evans, editors. Important bird areas in Africa and associated islands: priority sites for conservation. Pisces Publications and BirdLife International, Newbury and Cambridge, UK.
Caldecott, J. 1996. Cross River, Nigeria. Pages 30–60 in J. Caldecott, editor. Designing conservation projects. Cambridge University Press, Cambridge, UK.
Calström, A. 1995. Seychelles: country report to the FAO International Technical Conference on Plant Resources. Retrieved 2001 from the World Wide Web: http://www.fao.org/WAICENT/FAOINFO/AGRICULT/AGP/AGPS/Pgrfa/pdf/seychell.pdf.
Camara, S., L. I. Bamy, M. Cisse, K. Bangoura, M. L. Keita, B. Kaba, and A. M. Balde. 1999. Analyse de la diversité des écosystemes marins et côtiers, identification des priorités pour sa conservation. Stratégie Plan d'Action Diversité Biologique. Direction Nationale de l'Environement, Conakry, Guinea.
Cambray, J. A., B. R. Davies, and P. J. Ashton. 1986. The Orange-Vaal system. Pages 89–122 in B. R. Davies and K. F. Walker, editors. The ecology of river systems. Dr. W. Junk Publishers, Dordrecht, The Netherlands.
Campbell, B. M. 1994. The environmental status of the Save Catchment. Pages 21–33 in T. Matiza and S. A. Crafter, editors. Wetlands ecology and priorities for conservation in Zimbabwe. Proceedings of a seminar on the wetlands of Zimbabwe. IUCN, Gland, Switzerland.
Campbell, G. G. 1969. A review of scientific investigations in the Tongaland area of Northern Natal. Transactions of the Royal Society of South Africa 38:305–316.
Cantrell, M. A. 1979. Invertebrate communities in the Lake Chilwa swamp in years of high level. Pages 161–173 in M. Kalk, A. J. McLachlan, and C. Howard-Williams, editors. Lake Chilwa: studies of change in a tropical ecosystem. Monographiae Biologicae, 35. Dr. W. Junk, The Hague, The Netherlands.
Carmouze, J. P., J. R. Durand, and C. Lévêque. 1983a. The lacustrine ecosystem during the "Normal Chad" period and the drying phase. Pages 527–560 in J. P. Carmouze, J. R. Durand, and C. Lévêque, editors. Lake Chad, Monographiae Biologicae 53. Dr. W. Junk Publishers, The Hague, The Netherlands.
Carmouze, J. P., J. R. Durand, and C. Lévêque, editors. 1983b. Lake Chad: ecology and productivity of a shallow tropical ecosystem. Monographiae Biologicae 53. W. Junk, The Hague, The Netherlands.
Carmouze, J. P. and J. Lemoalle. 1983. The lacustrine environment. Pages 27–63 in J. P. Carmouze, J. R. Durand, and C. Lévêque, editors. Lake Chad, Monographiae Biologicae 53. Dr. W. Junk Publishers, The Hague, The Netherlands.
Carpenter, S. R., S. G. Fisher, N. B. Grimm, and J. F. Kitchell. 1992. Global change and freshwater ecosystems. Annual Review of Ecology and Systematics 23:119–139.
Carruthers, V. 1997. The wildlife of southern Africa. Southern Book Publishers, Halfway House, South Africa.
Castelo, R. 1994. Biogeographical considerations of fish diversity in Bioko. Biodiversity and Conservation 3:808–827.
CAW. 2003. Centre for African Wetlands. Retrieved 2003 from the World Wide Web: http://www.afriwet.org/.
CBS News. 2004. Nigerian president says African Union ready to send up to 5,000 troops but needs hundreds of millions of dollars. Retrieved 2004 from the World Wide Web: http://cbsnewyork.com/international/un-sudan-ai/resources_news_html.
Center for International Earth Science Information Network, Columbia University, International Food Policy Research In-

stitute, and World Resources Institute. 2000. Gridded population of the world (GPW), Version 2. CIESIN, Columbia University, Palisades, NY, USA. Retrieved 2002 from the World Wide Web: http://sedac.ciesin.columbia.edu/plue/gpw.

Centre for Ecology and Hydrology. 2001. Managed flood releases from the Itezhi-Tezhi Reservoir, summary report. Retrieved 2002 from the World Wide Web: http://www.nwl.ac.uk/ih/www/research/images/ManFloodZambiasum4.pdf.

Centre for Environment and Development for the Arab Region and Europe. 2001. Nubian Sandstone Aquifer System Programme. Retrieved 2003 from the World Wide Web: http://isu2.cedare.org.eg/nubian/.

Chabwela, H. N. 1992. The ecology and resource use of the Bangweulu Basin and the Kafue Flats. Pages 11–25 in R. C. V. Jeffrey, H. N. Chabwela, G. Howard, and P. J. Dugan, editors. Managing the wetlands of Kafue Flats and Bangweulu Basin. Kafue National Park, Zambia. IUCN, Gland, Switzerland.

Chabwela, H. N. 1994a. Current threats to the wetlands of Zimbabwe. Pages 155–159 in T. Matiza and S. A. Crafter, editors. Wetlands ecology and priorities for conservation in Zimbabwe. Proceedings of a seminar on the wetlands of Zimbabwe. IUCN, Gland, Switzerland.

Chabwela, H. N. 1994b. The ecology and conservation status of the Save-Runde floodplain. Pages 43–46 in T. Matiza and S. A. Crafter, editors. Wetlands ecology and priorities for conservation in Zimbabwe. Proceedings of a seminar on the wetlands of Zimbabwe. IUCN, Gland, Switzerland.

Chabwela, H. N. 1994c. Status of wetlands of Zambia: management and conservation issues. ECZ, Lusaka, Zambia.

Chabwela, H. N. and W. Mumba. 1998. Integrating water conservation and population strategies on the Kafue Flats. In A. de Sherbinin and V. Dompka, editors. Water and population dynamics. American Association for the Advancement of Science (AAAS). Retrieved 2001 from the World Wide Web: http://www.aaas.org/international/psd/waterpop/Zambia.htm.

Chafota, J., N. Burgess, and S. Johnson, compilers. 2002. WWF ecoregion conservation programme: Lake Malawi/Niassa/Nyasa ecoregion. WWF-Southern Africa Programme Office, Harare, Zimbabwe.

Champion, P. D. and J. S. Clayton. 2001. A weed risk assessment model for aquatic weeds in New Zealand. Pages 194–202 in R. H. Groves, F. D. Panetta, and J. G. Virtue, editors. Weed risk assessment. CSIRO Publishing, Collingwood, Australia.

Channing, A. 1987. Opportunistic seasonal breeding by frogs in Namaqualand. Journal of the Herpetological Association of Africa 35:19–24.

Channing, A. and M. Griffin. 1993. An annotated checklist of the frogs of Namibia. MADOQUA 18(2):101–116.

Chao, N. L. and G. Prang. 2002. Decade of project PIABA: reflections and prospects. Ornamental Fisheries International Journal 39. Retrieved 2003 from the World Wide Web: http://www.ornamental-fish-int.org/piaba.htm.

Chapman, F. A., S. A. Fitz-Coy, E. M. Thunberg, and C. M. Adams. 1997. United States of America trade in ornamental fish. Journal of the World Aquaculture Society 28:1–10.

Chapman, J. D. 1962. The vegetation of the Mulanje Mountains, Nyasaland: a preliminary account with particular reference to the Widdringtonia forests. Government Printer, Zomba, Nyasaland (Malawi).

Chapman, L. J., J. Balirwa, F. W. B. Bugenyi, C. A. Chapman, and T. L. Crisman. 2001. Wetlands of East Africa: biodiversity, exploitation, and policy perspectives. Pages 101–132 in B. Gopal, editor. Wetlands biodiversity. Backhuys Publisher, Leiden, The Netherlands.

Chapman, L. J. and C. A. Chapman. 2003. Fishes of the African rainforests: emerging and potential threats to a little-known fauna. Pages 176–209 in T. L. Crisman, L. J. Chapman, C. A. Chapman, and L. S. Kaufman, editors. Conservation, ecology and management of African freshwaters. University of Florida Press, Gainesville, FL, USA.

Chapman, L. J., C. A. Chapman, P. J. Schofield, J. P. Olowo, L. Kaufman, O. Seehausen, and R. Ogutu-Ohwayo. 2003. Fish faunal resurgence in Lake Nabugabo, East Africa. Conservation Biology 17:500–511.

Cheke, A. S. 1987. An ecological history of the Mascarene Islands, with particular reference to extinctions and introductions of land vertebrates. Pages 5–89 in A. W. Diamond, editor. Studies of Mascarene Island birds. Cambridge University Press, Cambridge, UK.

Cheke, R. A. 2001. Benin. Pages 93–98 in L. D. C. Fishpool and M. I. Evans, editors. Important bird areas in Africa and associated islands: priority sites for conservation. Pisces Publications and BirdLife International, Newbury and Cambridge, UK.

Chenje, M., editor. 2000. State of the environment Zambezi Basin 2000. SADC, IUCN, ZRA, SARDC, Maseru, Lesotho; Lusaka, Zambia; and Harare, Zimbabwe.

Childes, S. L. and P. J. Mundy. 1998. Important bird areas of Zimbabwe. Pages 355–384 in K. N. Barnes, editor. The Important bird areas of southern Africa. BirdLife International, Johannesburg, South Africa.

Childes, S. L. and P. J. Mundy. 2001. Zimbabwe. Pages 1025–1042 in L. D. C. Fishpool and M. I. Evans, editors. Important bird areas in Africa and associated islands: priority sites for conservation. Pisces Publications and BirdLife International, Newbury and Cambridge, UK.

Christy, P. 2001. São Tomé and Príncipe. Pages 727–731 in L. D. C. Fishpool and M. I. Evans, editors. Important bird areas in Africa and associated islands: priority sites for conservation. Pisces Publications and BirdLife International (BirdLife Conservation Series No. 11), Newbury and Cambridge, UK.

Chu, K. Y., J. A. Vanderberg, and R. K. Klumpp. 1981. Transmission dynamics of miracidia of *Schistosoma haematobium* in the Volta Lake. Bulletin of the World Health Organisation 59:555–560.

CIA. 2000. CIA world factbook 2000. Retrieved 2001 from the World Wide Web: http://www.cia.gov/cia/publications/factbook/geos/by.html.

CIA. 2001. The world factbook 2001. Retrieved 2001 from the World Wide Web: http://www.cia.gov/cia/publications/factbook/index.html.

CIESIN, Columbia University, IFPRI, and WRI. 2000. Gridded population of the world (GPW), Version 2. Retrieved 2001 from the World Wide Web: http://sedac.ciesin.columbia.edu/plue/gpw.

Clarke, G. P. and N. J. Karoma. 2000. History of anthropic disturbance. Pages 251–262 in N. D. Burgess and G. P. Clarke, editors. Coastal forests of eastern Africa. IUCN, Gland, Switzerland.

Clarke, N. V. 1998. Guide to the common plants of the Cuvelai wetlands. Southern African Botanical Diversity Network, Windhoek, Namibia.

Clarke, T. and D. Collins. 1996. A birdwatcher's guide to the Canary Islands. Prion, Huntingdon, UK.

Clausnitzer, V. 2001. Notes on the species diversity of East African Odonata, with a checklist of species. Odonatologica 30(1):49–66.

Coakley, G. J. 2001a. The mineral industry of Congo (Kinshasa), USGS Minerals Information Database, Africa and the Middle East. Retrieved 2003 from the World Wide Web: http://minerals.usgs.gov/minerals/pubs/country/2001/cgmyb01.pdf.

Coakley, G. J. 2001b. The mineral industry of Ghana, USGS Minerals Information Database, Africa and the Middle East. Retrieved 2001 from the World Wide Web: http://minerals.usgs.gov/minerals/pubs/country/2001/ghmyb01.pdf.

Coblentz, B. E., D. Van Vuren, and M. B. Main. 1990. Control of feral goats on Aldabra Atoll. Atoll Research Bulletin 1–14.

Coenen, E. J. and G. G. Teugels. 1989. A new species of *Nannocharax* (Pisces; Distichodontidae) from South-East Nigeria and West Cameroon, with comments on the taxonomic status of *Hemigrammocharax polli* (Roman, 1966). Cybium 13:311–318.

Cohen, A. 1992. Criteria for developing viable underwater natural reserves in Lake Tanganyika. Mitteilungen Internationale Vereinigung für Theoretische und Angewandte Limnologie 23:109–116.

Cohen, A. S. 1994. Extinction in ancient lakes: biodiversity crises and conservation 40 years after J. L. Brooks. Ergebnisse der Limnologie 44:451–479.

Cohen, A., R. Bills, C. Z. Cocquyt, and A. G. Caljon. 1993. The impact of sediment pollution on biodiversity in Lake Tanganyika. Conservation Biology 7(3):667–677.

Cohen, A., L. S. Kaufman, and R. Ogutu-Ohwago. 1996. Anthropogenic threats, impacts, and conservation strategies in the African Great Lakes: a review. Pages 575–624 in T. C. Johnson and E. O. Odada, editors. The limnology, climatology and paleoclimatology of the East African lakes. Gordon and Breech Publishers, Amsterdam, The Netherlands.

Cole, G. A. 1994. Textbook of limnology, 4th ed. Waveland Press, Prospect Heights, IL, USA.

Collar, N. J. and S. N. Stuart. 1988. Key forests for threatened birds in Africa. Monograph No. 3. International Council for Bird Preservation, Cambridge, UK.

Collins, R. 2002. The Nile. Yale University Press, New Haven, CT, USA.

Comley, P. and S. Meyer. 1994. Traveller's guide to Botswana. New Holland Publishers, London, UK.

Compére, P. and A. Iltis. 1983. The phytoplankton. Pages 145–198 in J. P. Carmouze, J. R. Durand, and C. Lévêque, editors. Lake Chad, Monographiae Biologicae 53. Dr. W. Junk Publishers, The Hague, The Netherlands.

Conservation International. 2003. Critical Ecosystem Partnership Fund. Retrieved 2003 from the World Wide Web: http://www.cepf.net/xp/cepf/.

Cooke, H. B. S. 1964. The Pleistocene environment in southern Africa. In D. H. S. Davis, editor. Ecological studies in Southern Africa. Dr. W. Junk Publishers, The Hague, The Netherlands.

Cooper, S. D. 1996. Rivers and streams. Pages 133–170 in T. McClanahan and T. P. Young, editors. East African ecosystems and their conservation. Oxford University Press, New York, NY, USA.

Cornen, G., Y. Bandet, P. Giresse, and J. Maley. 1992. The nature and chronostratigraphy of Quaternary pyroplastic accumulations from Lake Barombi Mbo (West-Cameroon). Journal of Volcanology and Geothermal Research 61:367–374.

Costanza, R., R. d'Arge, R. de Groot, S. Farber, M. Grasso, B. Hannon, K. Limburg, S. Naeem, R. O'Neill, J. Paruelo, R. G. Raskin, P. Sutton, and M. van den Belt. 1997. The value of the world's ecosystem services and natural capital. Nature 387:253–260.

Cotterill, F. 2000. Reduncine antelope of the Zambezi Basin. Pages 145–199 in J. Timberlake, editor. Biodiversity of the Zambezi Basin wetlands. Biodiversity Foundation for Africa, Bulawayo/The Zambezi Society, Harare, Zimbabwe.

Cotterill, F. P. D. 2003. Insights into the taxonomy of tsessebe antelopes *Damaliscus lunatus* (Bovidae: Alcelaphini) in south-central Africa with the description of a new evolutionary species in south-central Africa. Durban Museum Novitates 28:11–30.

Cotterill, F. P. D. 2004. Drainage evolution in south-central Africa and vicariant speciation in swamp-dwelling weaver birds and flycatchers. The Honeyguide 50(1):7–25.

Coulter, G. W., editor. 1991. Lake Tanganyika and its life. Oxford University Press, Oxford, UK.

Coulter, G. W., B. R. Allanson, M. N. Bruton, P. H. Greenwood, R. C. Hart, P. B. N. Jackson, and A. J. Ribbink. 1986. Unique qualities and special problems of the African Great Lakes. Environmental Biology of Fishes 17(3):161–183.

Coulter, G. W. and R. Mubamba. 1993. Conservation in Lake Tanganyika, with special reference to underwater parks. Conservation Biology 7(3):678–685.

Coulthard, N. D. 2001. Algeria. Pages 51–70 in L. D. C. Fishpool and M. I. Evans, editors. Important bird areas in Africa and associated islands: priority sites for conservation. Pisces Publications and BirdLife International (BirdLife Conservation Series No. 11), Newbury and Cambridge, UK.

Couty, P. and P. Duran. 1968. Le commerce du poisson au Tchad. Office de la Recherche Scientifique et Technique Outre-Mer (ORSTOM), Paris, France.

Cowan, G. I. 1995a. South Africa and the Ramsar convention. Pages 1–20 in G. I. Cowan, editor. Wetlands of South Africa. Department of Environmental Affairs and Tourism, Pretoria, South Africa.

Cowan, G. I. 1995b. Wetland regions of South Africa. Pages 21–31 in G. I. Cowan, editor. Wetlands of South Africa. Department of Environmental Affairs and Tourism, Pretoria, South Africa.

Cowan, G. I. and G. C. Marneweck. 1996. South African National Report to the Ramsar Convention 1996. Department of Environmental Affairs and Tourism, Pretoria, South Africa.

Cowling, R. M. and D. M. Richardson. 1995. Fynbos: South Africa's unique floral kingdom. Fernwood Press, Cape Town, South Africa.

Crafter, S. A., S. G. Njuguna, and G. W. Howard, editors. 1992. Proceedings of a seminar on wetlands of Kenya, National Museums of Kenya, Nairobi, Kenya. IUCN, Gland, Switzerland.

Crespi, V. 1998. Preliminary study on the fishery resources of the River Niger in the Upper Niger National Park, Guinea. Fisheries Management and Ecology 5:201–208.

Cromie, W. J. 1982. Beneath the African lakes. Mosaic 12(1):2–6.

Crosskey, R. W. 1988. Taxonomy and geography of the blackflies of the Canary Islands (Diptera: Simuliidae). Journal of Natural History 22:321–355.

Crossley, R. 1979. Variations in the level of Lake Malawi. Malawian Geography 19:5–13.

Cumberlidge, N. 1999. The freshwater crabs of West Africa: Family Potamonautidae. IRD, Paris, France.

Cumberlidge, N. and C. B. Bokyo. 2000. Freshwater crabs (Brachyura: Potamoidea: Potamonautidae) from rainforest of the Central African Republic, Central Africa. Proceedings of the Biological Society of Washington 113(2):406–419.

Curtis, B., K. S. Roberts, M. Griffin, S. Bethune, C. J. Hay, and H. Kolberg. 1998. Species richness and conservation of Namibian freshwater macro-invertebrates, fish and amphibians. Biodiversity and Conservation 7:447–466.

Cyrus, D. P. 1989. The Lake St. Lucia system: a research assessment. Southern African Journal of Aquatic Sciences 15(1):3–25.

Dadzie, S., R. D. Haller, and E. Trewavas. 1988. A note on the fishes of Lake Jipe and Lake Chale on the Kenya-Tanzanian border. Journal of East Africa Natural History Society and National Museums of Kenya 192:46–52.

Daget, J. 1962. Les poissons du Fouta Djalon et de la basse Guinée. Mémoires Institut Francaise Afrique Noire 65:1–210.

Daget, J. 1963. La réserve intégrale du Mont Nimba. Poissons. (2ème note). Mémoires Institut Francaise Afrique Noire 66:573–600.

Daget, J., I. C. Gaigher, and G. W. Ssentongo. 1988. Conservation. Pages 481–491 in C. Lévêque, M. Bruton, and G. Ssentongo, editors. Biology and ecology of African freshwater fishes. ORSTOM, Paris, France.

Daget, J., J. P. Gosse, and D. F. E. Thys van den Audenaerde, editors. 1984. Check-list of the freshwater fishes of Africa, Vol. 1. ORSTOM-MRAC, Tervuren, Belgium.

Daget, J., J. P. Gosse, and D. F. E. Thys van den Audenaerde, editors. 1986a. Check-list of the freshwater fishes of Africa, Vol. 2. ORSTOM-MRAC, Tervuren, Belgium.

Daget, J., J. P. Gosse, and D. F. E. Thys van den Audenaerde, editors. 1986b. Check-list of the freshwater fishes of Africa, Vol. 3. ORSTOM-MRAC, Tervuren, Belgium.

Daget, J., J. P. Gosse, and D. F. E. Thys van den Audenaerde, editors. 1991. Check-list of the freshwater fishes of Africa, Vol. 4. ORSTOM-MRAC, Tervuren, Belgium.

Daget, J. and A. Stauch. 1968. Poissons d'eaux douces et saumâtres de la région côtière du Congo. Cahiers ORSTOM Série Hydrobiologie 2(2):1–49.

Davenport, T. R. B. 2000. Lake Rukwa Basin Integrated Project, Unpublished technical report for 1999–2000. WCS/CIC/Game Division, Mbeya, Tanzania.

Davies, B. R. 1986. The Zambezi River system. Pages 225–267 in B. R. Davies and K. F. Walker, editors. Ecology of river systems. Dr. W. Junk Publishers, Dordrecht, The Netherlands.

Davies, B. R., R. D. Beilfuss, and M. C. Thoms. 2000a. Cahora Bassa retrospective, 1974–1997: effects of flow regulation on the lower Zambezi River. Verhandlungen Internationale Vereinigung Limnologie 27:2149–2157.

Davies, B. and J. Day. 1998. Vanishing waters. University of Cape Town Press, Cape Town, South Africa.

Davies, B. and F. Gasse. 1987. Glossary. Pages 25–33 in M. J. Burgis and J. J. Symoens, editors. African wetlands and shallow water bodies. Editions de l'Orstom, Paris, France.

Davies, B. R., C. D. Snaddon, M. J. Wishart, M. C. Thoms, and M. Meador. 2000b. A biogeographical approach to interbasin water transfers: implications for river conservation. Pages 431–444 in P. J. Boon, B. R. Davies, and G. E. Potts, editors. Global perspectives on river conservation: science, policy and practice. John Wiley & Sons, Chichester, UK.

Davis, S. D., V. H. Heywood, and A. C. Hamilton. 1994. Socotra. Pages 293–316 in WWF and IUCN, editors. Centres of plant diversity: a guide and strategy for their conservation. 3 Volumes. IUCN Publications Unit, Cambridge, UK.

Day, J. A. 1990. Environmental correlates of aquatic faunal distribution in the Namib Desert. Pages 99–107 in M. K. Seely, editor. Namib ecology: 25 years of Namib research. Transvaal Museum Monograph 7. Transvaal Museum, Pretoria, South Africa.

Day, J. A. 1997. The status of freshwater resources in Namibia. Research Discussion Paper 22. Namibian Directorate of Environmental Affairs, Windhoek, Namibia.

Dean, W. R. J. 2001. Angola. Pages 71–91 in L. D. C. Fishpool and M. I. Evans, editors. Important bird areas in Africa and associated islands: priority sites for conservation. Pisces Publications and BirdLife International, Newbury and Cambridge, UK.

Debenay, J. P., J. G. Pages, and J. J. Guillou. 1994. Transformation of a subtropical river into a hyperhaline estuary: the Casamance River. Palaeogeography Palaeoclimatology Palaeoecology 107(1–2):103–119.

Defos du Rau, J. 1960. L'Ile de la Réunion. Institut de Géographie, Bordeaux, France.

De Graaf, M., E. Dejen, F. A. Sibbing, and J. W. M. Osse. 2000. *Barbus tanapelagius*, a new species from Lake Tana (Ethiopia): its morphology and ecology. Environmental Biology of Fishes 59(1):1–9.

Dejoux, C. 1983. The fauna associated with the aquatic vegetation. Pages 273–292 in J. P. Carmouze, J. R. Durand, and C. Lévêque, editors. Lake Chad, Monographiae Biologicae 53. Dr. W. Junk Publishers, The Hague, The Netherlands.

Delgado, C. L., J. Hopkins, and V. A. Kelly. 1998. Agricultural growth linkages in sub-Saharan Africa. Research Report 107. International Food Policy Research Institute, Washington, DC, USA.

Delgado, C., N. Wada, M. Rosegrant, S. Meijer, and M. Ahmed. 2003. Fish to 2020: supply and demand in changing global markets. International Food Policy Research Institute, Washington, DC, USA.

Demey, R. and M. Louette. 2001. Democratic Republic of Congo. Pages 199–218 in L. D. C. Fishpool and M. I. Evans, editors. Important bird areas in Africa and associated islands: priority sites for conservation. Pisces Publications and BirdLife International (BirdLife Conservation Series No. 11), Newbury and Cambridge, UK.

DeMoor, I. J. 1996. Case studies of the invasion by four alien fish species (*Cyprinus carpio*, *Micropterus salmoides*, *Oreochromis macrochir*, and *O. mossambicus*) of freshwater ecosystems in southern Africa. Transactions of the Royal Society of South Africa 51:233–255.

DeMoor, I. J. and M. N. Bruton. 1988. Atlas of alien and translocated indigenous aquatic animals in southern Africa. South African National Scientific Programmes Report, Report no. 144. Foundation for Research Development, Pretoria, South Africa.

Denny, P., editor. 1985. The ecology and management of African wetland vegetation. Dr. W. Junk, Dordrecht, The Netherlands.

Denny, P. 1991. Africa. Pages 115–148 in M. Finlayson and M. Moser, editors. Wetlands. International Waterfowl and Wetlands Research Bureau, London, UK.

Denny, P. 1993. Wetlands of Africa: introduction. Pages 1–31 in D. Whigham, D. Dykjova, and S. Hejny, editors. Wetlands of the world: inventory, ecology and management, Volume 1. Kluwer Academic Publishers, Dordrecht, The Netherlands.

Denny, P. 2001. Research, capacity-building and empowerment for sustainable management of African wetland ecosystems. Hydrobiologia 458:21–31.

Department of Water Affairs and Forestry, S. A. 2003. Working for Water Programme. Retrieved 2003 from the World Wide Web: http://www-dwaf.pwv.gov.za/wfw/default.asp.

de Rham, P. H. 1996. Poissons des eaux intérieures de Madagascar. Pages 423–440 in W. R. Lourenco, editor. Biogéographie de Madagascar. ORSTOM, Paris, France.

Derwent, S. 2001. Ukhahlamba! Africa Geographic 9(7):38–49.

Desanker, P. V. 2002. Impact of climate change on life in Africa. WWF, Washington, DC, USA.

Desanker, P. V., P. G. H. Frost, C. O. Frost, C. O. Justice, and R. J. Scholes. 1997. The Miombo network: framework for a terrestrial transect study of land-use and land-cover change in the Miombo ecosystems of Central Africa. IGBP Report 41. The International Geosphere-Biosphere Programme (IGBP), Stockholm, Sweden.

De Vos, L. 2001. Rediscovery of the giant catfish *Pardiglanis tarabini* (Siluriformes: Claroteidae). Ichthyological Exploration of Freshwaters 12:275.

De Vos, L., D. Oyugi, and J. K. Nguku. 2000. Fishes and fisheries of the lower Tana River. GEF Project Report. Ichthyology Department, National Museums of Kenya, Nairobi, Kenya.

De Vos, L. and L. Seegers. 1998. Seven new *Orthochromis* species (Teleostei: Cichlidae) from the Malagarasi, Luiche and Rugufu basins (Lake Tanganyika drainage), with notes on their reproductive biology. Ichthyological Exploration of Freshwaters 9:371–420.

De Vos, L., L. Seegers, L. Taverne, and D. Thys van den Audenaerde. 2001a. L'ichthyofaune du bassin de la Malagarasi (système du lac Tanganyika): une synthèse de la connaissance actuelle. Annales Musée Royal de l'Afrique Centrale, Sciences Zoologiques 285:117–152.

De Vos, L. and J. Snoeks. 1994. The non-cichlid fishes of the Lake Tanganyika basin. Archiv für Hydrobiologie Beihefte Ergebnisse Limnologie 44:391–405.

De Vos, L., J. Snoeks, and D. T. van den Audenaerde. 2001b. An annotated checklist of the fishes of Rwanda (East Central Africa), with historical data on introductions of commercially important species. Journal of East African Natural History 90(1–2):41–68.

Dgebuadze, Y. Y., A. S. Golubstov, V. N. Mikheev, and M. V. Mina. 1994. Four fish species new to the Omo-Turkana Basin, with comments on the distribution of *Nemacheilus abyssinicus* in Ethiopia. Hydrobiologia 286:125–128.

Diallo, A. 1995. Rapport sur les plantes flottantes de Guinée. Centre d'étude et de Recherche en Environement, Université de Conakry, Conakry, Guinea.

Dijkstra, K. D. B. 2002. Odonata of the Gulf of Guinea islands. Gulf of Guinea Islands' Biodiversity Network. Retrieved 2003 from the World Wide Web: http://www.ggcg.st/Species_Lists/odonata.htm.

Dinerstein, E., D. M. Olson, D. J. Graham, A. L. Webster, S. A. Primm, M. P. Bookbinder, and G. Ledec. 1995. A conservation assessment of the terrestrial ecoregions of Latin America and the Caribbean. The World Bank, Washington, DC, USA.

Dinerstein, E., G. V. N. Powell, D. M. Olson, E. D. Wikramanayake, R. A. Abell, C. J. Loucks, E. Underwood, T. F. Allnutt, W. W. Wettengel, T. H. Ricketts, H. E. Strand, S. O'Connor, N. Burgess, and M. Mobley. 2000. A workbook for conducting biological assessments and developing Biodiversity Visions for ecoregion conservation. Part I: terrestrial ecoregions. World Wildlife Fund, Washington, DC, USA.

Diouf, M., A. Nonguierma, A. Amani, A. Royer, and B. Some. 2000. Drought control in the Sahel: AGRHYMET results, state-of-the-knowledge and prospects. Secheresse (Montrouge) 11(4):257–266.

Directorate of Fisheries–Ghana. 2003. Ghana: post-harvest fisheries overview. DFID/Directorate of Fisheries, Accra, Ghana.

DNE, PNUD, and FEM. 1999. Evaluation de la diversité biologique de la Guinée: vision, buts, et objectifs de la strategie nationale pour sa conservation et son utilisation durable. PROJET/GUI/97/G32/A/1A/99 SNPA-DB. Ministère Mines Géologie et Environnement, Conakry, Guinea.

Doadrio, I. 1994. Freshwater fish fauna of North Africa and its biogeography. Annales du Musée Royal de l'Afrique Centrale, Zoologique 275:21–34.

Dobson, J. E., E. A. Bright, P. R. Coleman, R. C. Durfee, and B. A. Worley. 2000. A global population database for estimating populations at risk. Photogrammetric Engineering & Remote Sensing 66(7):849–857.

Dodman, T., H. Y. Béibro, E. Hubert, and E. Williams. 1999. African waterbird census 1998. Wetlands International, Dakar, Senegal.

Döll, P., F. Kaspar, and B. Lehner. 2003. A global hydrological model for deriving water availability indicators: model tuning and validation. Journal of Hydrology 270:105–134.

Dominey, W. J. 1987. Sponge-eating by *Pungu maclareni*, an endemic cichlid fish from Lake Barombi Mbo, Cameroon. National Geographic Research 3:389–393.

Dominey, W. J. and A. Snyder. 1988. Kleptoparasitism of freshwater crabs by cichlid fishes endemic to Lake Barombi Mbo, West Africa. Environmental Biology of Fish 22:155–160.

Donque, G. 1972a. The climatology of Madagascar. Pages 87–144 in R. Battistini and G. Richard-Vindard, editors. Biogeography and ecology in Madagascar. Dr. W. Junk, The Hague, The Netherlands.

Doppelt, B., M. Scurlock, C. Frissell, and J. Karr. 1993. Entering the watershed: a new approach to save America's river ecosystems. Island Press, Washington, DC, USA.

Dowsett, R. J. 1971. A preliminary survey of Zambia's wetlands and wildfowl. Unpublished report.

Dowsett, R. J. and F. Dowsett-Lemaire, editors. 1991. Flore et faune du Bassin du Kouilou (Congo) et leur exploitation, Tauraco Research Report No. 4. Tauraco Press in association with CONOCO, Jupille-Liege, Belgium.

Dowsett-Lemaire, F., R. J. Dowsett, and M. Dyer. 2001. Malawi. Pages 539–555 in L. D. C. Fishpool and M. I. Evans, editors. Important bird areas in Africa and associated islands: priority sites for conservation. Pisces Publications and BirdLife International (BirdLife Conservation Series No. 11), Newbury and Cambridge, UK.

Dudley, C. 2000. Freshwater molluscs of the Zambezi River Basin. Pages 487–526 in J. R. Timberlake, editor. Biodiversity of the Zambezi Basin wetlands. Biodiversity Foundation for Africa, Bulawayo/The Zambezi Society, Harare, Zimbabwe.

Dudley, C. O., D. E. Stead, and G. G. M. Schulten. 1979. Amphibians, reptiles, mammals, and birds of Chilwa. Pages 247–273 in M. Kalk, A. J. McLachlan, and C. Howard-Williams, editors. Lake Chilwa: studies of change in a tropical ecosystem. Monographiae Biologicae, 35. Dr. W. Junk, The Hague, The Netherlands.

Dudley, N. and S. Stolton. 2003. Running pure: the importance of forest protected areas to drinking water. World Bank/WWF Alliance for Forest Conservation and Sustainable Use, London, UK.

Dufour, P., A. M. Kouassi, and A. Lanusse. 1994. Les pollutions. Pages 309–333 in J. R. Durand, P. Dufour, D. Guiral, and S. G. F. Zabi, editors. Environnement et resources aquatiques de Côte d'Ivoire, Tome II: les milieux lagunaires. ORSTOM, Paris, France.

Dugan, P. J., E. Baran, R. Tharme, M. Prein, R. Ahmed, P. Amerasinghe, P. Bueno, C. Brown, M. Dey, G. Jayasinghe, M. Niasse, A. Nieland, V. Smakhtin, N. Tinh, K. Viswanathan, and R. Welcomme. 2002. The contribution of aquatic ecosystems and fisheries to food security and livelihoods: a research agenda. Pages 85–113 in Challenge Program on Water for Food: Background Papers. IWMI, Colombo, Sri Lanka.

Duggan, A. I. 1990. Illustrated guide to the game parks and nature reserves of southern Africa. The Reader's Digest Association, Cape Town, South Africa.

Dumont, H. J. 1986. The Nile River system. Pages 61–74 in B. R. Davies and K. F. Walker, editors. The ecology of river systems. Dr. W. Junk Publishers, Dordrecht, The Netherlands.

Dumont, H. J. 1987. Sahara. Pages 79–153 in M. J. Burgis and J. J. Symoens, editors. African wetlands and shallow water bodies. ORSTOM, Paris, France.

Dumont, H. J. 1992. The regulation of plant and animal species and communities in African shallow lakes and wetlands. Revue Hydrobiological Tropical 25(4):303–346.

Dunham, K. M. 1989. Long-term changes in Zambezi riparian woodlands, as revealed by photo panoramas. African Journal of Ecology 27:263–275.

Dunham, K. M. 1990. Biomass dynamics of herbaceous vegetation in Zambezi riverine woodlands. African Journal of Ecology 28:200–212.

Dunham, K. M. 1991a. Comparative effects of *Acacia albida* and *Kigelia africana* trees on soil characteristics in Zambezi riverine woodlands. Journal of Tropical Ecology 7:215–220.

Dunham, K. M. 1991b. Phenology of *Acacia albida* trees in Zambezi riverine woodlands. African Journal of Ecology 29:118–129.

Dunham, K. M. 1994. The effect of drought on the large mammal populations of Zambezi riverine woodlands. Journal of Zoology (London) 234:489–526.

Durand, J. R. 1980. Evolution des captures totales (1962–1977) et devenir des pêcheries de la région du lac Tchad. Cahiers ORSTOM, série Hydrobiologique 4:61–81.

Durand, J. R. 1983. The exploitation of fish stocks in the Lake Chad region. Pages 425–481 in J. P. Carmouze, J. R. Durand, and C. Lévêque, editors. Lake Chad: ecology and productivity of a shallow tropical ecosystem. Monographiae Biologicae 53. Dr. W. Junk Publishers, The Hague, The Netherlands.

du Toit, D. and T. Scuazzin. 1995. Sink or swim: water and the Namibian environment. Desert Research Foundation of Namibia, Windhoek, Namibia.

du Toit, R. F. 1994. Mid-Zambezi and Mana Pools: ecology and conservation status. Pages 35–42 in T. Matiza and S. A. Crafter, editors. Wetlands ecology and priority for conservation in Zimbabwe: proceedings of a seminar on wetlands ecology and priorities for conservation in Zimbabwe. IUCN, Gland, Switzerland.

Duvail, S. and O. Hamerlynck. 2003. Hydraulic modelling as a tool for the joint management of a restored wetland: sharing the benefits of managed flood releases from the Diama Dam reservoir. Hydrology and Earth System Sciences 17(1):133–146.

Dye, P., G. Moses, P. Vilakazi, R. Ndlela, and M. Royappen. 2001. Comparative water use of wattle thickets and indigenous plant communities at riparian sites in the Western Cape and KwaZulu-Natal. Water South Africa (Pretoria) 27:529–538.

Eadie, J. M., T. A. Hurly, R. D. Montgomerie, and K. L. Teather. 1986. Lakes and rivers as islands: species-area relationships in the fish faunas of Ontario [Canada]. Environmental Biology of Fishes 15:81–89.

Earth Observatory-NASA. 2001. News archive: Africa's Lake

Chad shrinks by 20 times due to irrigation demands, climate change. Retrieved 2001 from the World Wide Web: http://earthobservatory.nasa.gov/Newsroom/NewImages/images.php3?img_id=5199.

East, R., editor. 1999. Antelopes: global survey and regional action plans. Parts 1–3. IUCN/SSC Antelope Specialist Group, Gland, Switzerland and Cambridge, UK.

Eastwood, F. 1979. Guide to the Mulanje Massif. Lorton Publications, Johannesburg, South Africa.

Eccles, D. H. 1974. An outline of the physical limnology of Lake Malawi (Lake Nyasa). Limnology and Oceanography 19: 730–742.

Eccles, D. H. 1992. A field guide to the freshwater fishes of Tanzania. FAO, Rome, Italy.

Eckhardt, G. A. 1998. Glossary of water resource terms. Retrieved 2004 from the World Wide Web: http://www.edwardsaquifer.net/glossary.html.

ECOFAC. 2001. Central African Republic component: Ngotto Forest. Retrieved 2001 from the World Wide Web: http://www.ecofac.org/Composantes/CentrafriqueNgotto.htm.

Economic Commission for Africa. 2001. The state of demographic transition in Africa. ECA, Addis Ababa, Ethiopia.

Eekhout, S., J. M. King, and A. Wackernagel. 1997. Classification of South African Rivers, Volume 1. Department of Environmental Affairs and Tourism, Pretoria, South Africa.

EEU: Environment Evaluation Unit, University of Cape Town. 1990. Final environmental impact report for the proposed Omdel Dam on the Omaruru River, Unpublished report no. WR/90/7. Division of Water Quality, Department of Water Affairs, Windhoek, Namibia.

Eisenberg, J. F. and E. Gould. 1970. The tenrecs: a study in mammalian behavior and evolution. Smithsonian Contributions to Zoology 27:1–138.

Ellery, W. N. and T. S. McCarthy. 1998. Environmental change over the decades since dredging and excavation of the lower Boro River, Okavango Delta, Botswana. Journal of Biogeography 25(2):361–378.

El-Raey, M., Y. Fouda, and S. Nasr. 1997. GIS assessment of the vulnerability of the Rosetta area, Egypt to impacts of sea rise. Environmental Monitoring and Assessment 47(1):59–77.

El-Sherbini, A., M. El-Moattassem, and H. Sloterdijk. 1996. Water quality condition of the Nile River. Pages 162–175 in A. M. Shady, M. El-Moattassem, E. A. Abdel-Hafiz, and A. K. Biswas, editors. Management and development of major rivers. Oxford University Press, Calcutta, India.

Emerton, L. 1998. Economic tools for valuing wetlands in Africa. IUCN: The World Conservation Union, Eastern Africa Regional Office, Nairobi, Kenya.

Emerton, L. 1999. Balancing the opportunity costs of wildlife conservation for communities around Lake Mburo National Park, Uganda, Evaluating Eden Series Discussion Paper No. 5. International Institute for Environment and Development, London, UK.

Emerton, L., L. Iyango, P. Luwum, and A. Malinga. 1999. The economic value of Nakivubo Urban Wetland, Uganda. Uganda National Wetlands Programme and IUCN: The World Conservation Union, Eastern Africa Regional Office, Kampala, Uganda and Nairobi, Kenya.

Environmental Affairs. 2000. Lake Chilwa Wetland State of the Environment. Environmental Affairs Department, Government of Malawi, Lilongwe, Malawi.

ERGS Research Group, University of Kinshasa. 2003. Quantitative survey of Congo River biodiversity (March–April 2003). Innovative Resources Management, Congo River Environment and Development Project, Washington, DC, USA.

Ernould, J. C., K. Ba, and B. Sellin. 1999. Impact of the local water-development programme on the abundance of the intermediate hosts of schistosomiasis in three villages of the Senegal River Delta. Annals of Tropical Medicine & Parasitology 93(2):135–145.

Ernst, C. H. and R. W. Barbour. 1989. Turtles of the world. Smithsonian Institution Press, Washington, DC, USA.

Eswaran, H., P. Reich, and F. Beinroth. 2001. Global desertification tension zones. Pages 24–28 in D. E. Stott, R. H. Mohtar, and G. C. Steinhardt, editors. Sustaining the global farm: selected papers from the 10th International Soil Conservation Organization Meeting, May 24–29, 1999, West Lafayette, IN. International Soil Conservation Organization in cooperation with the USDA and Purdue University, West Lafayette, IN, USA.

Ethiopian Mapping Authority. 1988. National atlas of Ethiopia. Ethiopian Mapping Authority, Addis Ababa, Ethiopia.

Ethiopian Wildlife and Natural History Society. 1996. Important bird areas of Ethiopia: a first inventory. BirdLife International, Addis Ababa, Ethiopia.

Ethiopian Wildlife and Natural History Society. 2001. Ethiopia. Pages 291–336 in L. D. C. Fishpool and M. I. Evans, editors. Important bird areas in Africa and associated islands: priority sites for conservation. Pisces Publications and BirdLife International, Newbury and Cambridge, UK.

Euroconsult. 1994. Socotra Island Man and Biosphere Reserve: outline of a development programme. MAB Nomination Form. Prepared in co-operation with BMB and IHE. Assistance to the Technical Secretariat of the Environmental Protection Council of Yemen, Sana, Yemen.

Evans, C. A. 2001. Water features and water issues: Africa. Retrieved 2003 from the World Wide Web: http://eol.jsc.nasa.gov/newsletter/html_Mir/africa.htm.

Evans, M. I. 2000. Zoological results and conclusions of the multi-disciplinary expedition to the Soqotra Archipelago, February–March 1999. Report of the Socotra Biodiversity Project. Royal Botanic Garden, Edinburgh, UK.

Evans, T. E. 1996. The effects of changes in the world hydrological cycle on availability of water resources. Retrieved 2001 from the World Wide Web: http://www.fao.org/docrep/W5183E/w5183e04.htm.

Everard, M. and D. M. Harper. 2002. Towards the sustainability of the Lake Naivasha Ramsar site and its catchment. Hydrobiologia 488:191–203.

Everett, G. V. 1971. Sampling fish stocks in the Kafue River. Fisheries Research Bulletin of Zambia 5:297–304.

Expert Center for Taxonomic Identification. 2000. World Biodiversity Database v2.1. Retrieved 2002 from the World Wide Web: http://www.eti.uva.nl/Database/WBD.html.

Ezealor, A. U. 2001. Nigeria. Pages 673–692 in L. D. C. Fishpool and M. I. Evans, editors. Important bird areas in Africa and associated islands: priority sites for conservation. Pisces Publications and BirdLife International (BirdLife Conservation Series No. 11), Newbury and Cambridge, UK.

Ezealor, G. 2002. Critical sites for biodiversity conservation in Nigeria. Nigerian Conservation Foundation, Lagos, Nigeria.

FAO. 1995. AQUASTAT. Retrieved 2001 from the World Wide Web: http://www.fao.org/ag/agl/aglw/aquastat/main/index.stm.

FAO. 1997a. AQUASTAT: Libya. Retrieved 2003 from the World Wide Web: http://www.fao.org/waicent/faoinfo/agricult/agl/aglw/aquastat/countries/libya/index.stm.

FAO. 1997b. AQUASTAT: Morocco. Retrieved 2001 from the World Wide Web: http://www.fao.org/ag/agl/aglw/aquastat/countries/morocco/index.stm.

FAO. 1997c. AQUASTAT: Tunisia. Retrieved 2001 from the World Wide Web: http://www.fao.org/ag/agl/aglw/aquastat/countries/tunisia/index.stm.

FAO. 1997d. The Lake Chad Basin in Irrigation potential in Africa: a basin approach (FAO Land and Water Bulletin 4). Retrieved 2001 from the World Wide Web: http://www.fao.org/docrep/w4347e/w4347e00.htm.

FAO. 1997e. The Nile Basin in irrigation potential in Africa: a basin approach (FAO Land and Water Bulletin 4). Retrieved 2003 from the World Wide Web: http://www.fao.org/docrep/W4347E/w4347e0k.htm#the%20nile%20basin.

FAO. 1998. Database on Introductions of Aquatic Species (DIAS). Retrieved 2002 from the World Wide Web: http://www.fao.org/fi/statist/fisoft/dias/index.htm.

FAO. 1999. State of the world fisheries and aquaculture 1998. FAO, Rome, Italy.

FAO. 2000a. African dams. Retrieved 2002 from the World Wide Web: http://www.fao.org/ag/agl/aglw/aquastat/gis/index2.stm.

FAO. 2000b. FAO/GIEWS: foodcrops & shortages 11/00: Sierra Leone (6 November). Retrieved 2003 from the World Wide Web: http://www.fao.org/docrep/004/x8868e/pays/sil0011e.htm.

FAO. 2000c. The state of world fisheries and aquaculture. FAO, Rome, Italy.

FAO. 2001a. FAOSTAT agriculture data. Retrieved 2002 from the World Wide Web: http://www.fao.org/es/ESS/index.htm.

FAO. 2001b. Fishery country profile: Democratic Republic of Congo. Retrieved 2003 from the World Wide Web: http://www.fao.org/fi/fcp/en/COD/profile.htm.

FAO. 2001c. Global forest resources assessment 2000. FAO Forestry Paper 140. FAO, Rome, Italy.

FAO. 2001d. State of the world's forests 2001. FAO, Rome, Italy.

FAO. 2002. World fisheries statistics: country profile. Retrieved 2003 from the World Wide Web: http://www.fao.org/fi/statist/statist.asp.

FAO. 2003. FAOSTAT nutrition: food balance sheets. Retrieved 2003 from the World Wide Web: http://apps.fao.org/default.htm.

FAO Fisheries Department. 2002. The state of the world fisheries and aquaculture 2002. FAO, Rome, Italy.

FAO Forest Resource Assessment Programme. 1999. Global forest cover map. Working paper 19. Retrieved 2001 from the World Wide Web: http://www.fao.org/forestry/fo/fra/.

FAO Inland Water Resources and Aquaculture Service, Fishery Resources Division. 1999. Review of the state of the world fishery resources: inland fisheries, FAO Fisheries Circular, No. 942. FAO, Rome, Italy.

FAO and International Institute for Applied Systems Analysis. 2000. Global agro-ecological zones: 2000. Retrieved 2003 from the World Wide Web: http://www.fao.org/WAICENT/FAOINFO/AGRICULT/AGL/agll/gaez.

FAO Workshop. 2000. Electronic workshop: land-water linkages in rural watersheds. Retrieved 2001 from the World Wide Web: http://www.fao.org/ag/agl/watershed/archive.htm.

Fay, J. M. and A. Vedder. 1997. Fate of the forest: accelerated logging in the Central African Basin, Congo as a case study. Wildlife Conservation Society, New York, NY, USA.

Federal Interagency Stream Restoration Working Group. 1998. Stream corridor restoration: principles, processes, and practices. U.S. Government Printing Office, Washington, DC, USA.

Feely, J. M. 1964. Heron and stork breeding colonies in the Luangwa Valley. The Puku 2:76–77.

FEPA. 1991. Achieving sustainable development in Nigeria. National report for the 1992 United Nations Conference on Environment and Development. Federal Environment Protection Agency, Lagos, Nigeria.

Ferguson, A. J. D. and B. J. Harbott. 1982. Geographical, physical and chemical aspects of Lake Turkana. Pages 1–107 in A. J. Hopson, editor. Lake Turkana. A report of the findings of the Lake Turkana Project 1972–1975. Overseas Development Administration, London, UK.

Fernandopullé, D. 1976. Climatic characteristics of the Canary Islands. Pages 185–205 in G. Kunkel, editor. Biogeography and ecology in the Canary Islands. Dr. W. Junk, The Hague, The Netherlands.

Ferreira, S. 1999. Specialist report: wildlife and birds. Consultancy Report No: LHDA 648-F-19 (Vol. 1). Consulting Services for the Establishment and Monitoring of the Instream Flow Requirements for River Courses Downstream of LHWP Dams. Metsi Consultants and Lesotho Highlands Water Project for the Lesotho Highlands Development Authority, Maseru, Lesotho.

Fielder, P. L., and S. K. Jain, editors. 1992. Conservation biology: the theory and practice of nature conservation, preservation, and management. Chapman and Hall, New York, NY, USA.

Finlayson, M. and M. Moser, editors. 1991. Wetlands. Facts on File, Oxford, UK.

FishBase. 2001. Search FishBase. Retrieved 2001 from the World Wide Web: http://www.fishbase.org/search.cfm.

Fishpool, L. D. C. and M. I. Evans, editors. 2001. Important bird areas in Africa and associated islands: priority sites for conservation. Pisces Publications and BirdLife International (BirdLife Conservation Series No. 11), Newbury and Cambridge, UK.

Fitton, J. G. and H. Dunlop. 1985. The Cameroon line, West Africa, and its bearing on the origin of oceanic and continental alkali basalt. Earth Planetary Science Letters 72: 23–85.

FitzPatrick, M. 2000. Review of Odonata associated with the wetlands of the Zambezi Basin. Pages 527–564 in J. Timberlake, editor. Biodiversity of the Zambezi Basin wetlands. Biodiversity Foundation for Africa, Bulawayo/The Zambezi Society, Harare, Zimbabwe.

Fitzsimons, J. M. and R. T. Nishimoto. 1990. Territories and site tenacity in males of the Hawaiian stream fish goby *Lentipes concolor* (Pisces: Gobiidae). Ichthyological Exploration of Freshwaters 1:185–189.

Fjeldså, J., C. Rahbek, L. Hansen, and S. Galster. Ongoing. Zoological Museum of the University of Copenhagen database. Retrieved 2000 from the World Wide Web: http://www.zmuc.dk/commonweb/research/biodata.htm.

Fontes, J., M. Aizpuru, J.-L. Carayon, P. Larincq, S. Guinko, and M. Hien. 1999. Digital maps as an aid for vegetation resource characterization and surveys: a case study in a dry tropical environment in Burkina Faso. Secheresse (Montrouge) 10:19–25.

Forbes, H. O. 1903. Natural history of Sokotra and Abd-el-Kuri. Bulletin of the Liverpool Museums (Special Bulletin) 1–598.

Forsyth, G. G., D. B. Versveld, R. A. Chapman, and B. K. Fowles. 1997. The hydrological implications of afforestation in the North-Eastern Cape, Report 511/1/97. Water Research Commission, Stellenbosch, South Africa.

Fosberg, F. R. and S. A. Renvoize. 1980. The flora of Aldabra and neighboring islands. Her Majesty's Stationery Office, London, UK.

Fowler, H. W. 1930. The fresh-water fishes obtained by the Gray African Expedition 1929. With notes on other species in the Academy collection. Proceedings of the Academy of Natural Sciences of Philadelphia 82:27–83.

Frissell, C. A. and S. C. Ralph. 1998. Stream and watershed restoration. Pages 599–624 in R. J. Naiman and R. E. Bilby, editors. River ecology and management: lessons from the Pacific Coastal ecoregion. Springer-Verlag, New York, NY, USA.

Fry, C. H., S. Keith, and E. K. Urban. 1988. The birds of Africa, Volume III. Academic Press, London, UK.

Fryer, G. 1959. The trophic interrelationships and ecology of some littoral communities of Lake Nyasa with special reference to the fishes, and a discussion of the evolution of a group of rock-frequenting Cichlidae. Proceedings of the Zoological Society of London 132(2):153–281.

Fryer, G. 1977. Evolution of species flocks of cichlid fishes in African lakes. Zeitschrift für Zoologische Systematik und Evolutionsforschung 15:141–165.

Fryer, G. 1996. Endemism, speciation and adaptive radiation in great lakes. Environmental Biology of Fishes 45:109–131.

Fryer, G. H. and T. D. Iles. 1972. The cichlid fishes of the Great Lakes of Africa: their biology and evolution. TFH Publications, Neptune City, NJ, USA.

Fryxell, J. M. and A. R. E. Sinclair. 1988. Seasonal migration by white-eared kob in relation to resources. African Journal of Ecology 26:17–32.

Fuggle, R. F. and M. A. Rabie. 1983. Environmental concerns in South Africa. Juta & Co., Cape Town, South Africa.

Fuller, K. and P. Bush. 1999. Energy map of the world. The Petroleum Economist Limited, London, UK.

Gabche, C. E. and S. V. Smith. 2000. Cameroon estuarine systems. Land-Ocean Interactions in the Coastal Zone (LOICZ) Project of the International Geosphere-Biosphere Programme: a study of global change (IGBP) of the International Council of Scientific Unions (ICSU). Retrieved 2002 from the World Wide Web: http://data.ecology.su.se/mnode/Africa/Cameroon/cameroonintro.htm.

Gabie, V. 1965. Problems associated with the distribution of freshwater fishes in southern Africa. South African Journal of Science 61(11):383–391.

Galis, F. and J. A. J. Metz. 1998. Why are there so many cichlid species? Trends in Ecology and Evolution 13(1):1–2.

Gallais, J. 1967. Le Delta Intérieur du Niger, etudes de géographie régionale. Mém. IFAN 78. Larose, Paris, France.

Gardiner, A. 2000. Review of wetland Lepidoptera of the Zambezi basin. Pages 565–612 in J. Timberlake, editor. Biodiversity of the Zambezi Basin wetlands. Biodiversity Foundation for Africa, Bulawayo/The Zambezi Society, Harare, Zimbabwe.

Gascoigne, A. 1993. A bibliography of the fauna of the islands of São Tomé e Príncipe and the island of Annobon (Gulf of Guinea). Arquipelago Ciencias da Natureza 11:91–105.

Gascoigne, A. 1994. The dispersal of terrestrial gastropod species in the Gulf of Guinea. Journal of Conchology 35:1–7.

Gascoigne, A. 1996. Additions to a bibliography of the fauna of São Tomé e Príncipe and the island of Annobon, Gulf of Guinea. Addendum. Arquipelago Ciencias da Natureza 14A:95–103.

Gasse, F. and F. A. Street. 1978. Late quaternary lake-level fluctuation and environments of the northern rift valley and Afar regions (Ethiopia and Djibouti). Paleogeography, Paleoclimatology and Paleoecology 24:279–325.

Gawler, M., editor. 2002. Strategies for wise use of wetlands: best practices in participatory management. Proceedings of a workshop held at the 2nd annual conference on wetlands and development (November, 1998, Dakar, Senegal). Wetlands International, IUCN, WWF Publication No. 56, Wageningen, The Netherlands.

Georges, E. 1998. The Indian Ocean: Madagascar, Réunion, Mauritius, the Seychelles. Evergreen (Benedikt Taschen Verlag GmbH), First Edition Translations, Cambridge, UK.

Gerlach, J., editor. 1997. Seychelles Red Data Book: 1997. The Nature Protection Trust of Seychelles, Mahé, Seychelles.

Gerlach, J. 2002. Seychelles terrapin action plan. Phelsuma 10B:1–16.

Gerlach, J. and K. L. Canning. 1994. On the crocodiles of the western Indian Ocean. Phelsuma 2:54–58.

Gerlach, J. and L. Canning. 2001. Rapid declines in the critically endangered Seychelles terrapins (*Pelusios* spp.). Oryx 35(4):313–321.

Gerlach, J. and A. Skerrett. 2002. The distribution, ecology and status of the yellow bittern *Ixobrychus sinensis* in Seychelles. Journal of African Ecology 40:194–196.

Getahun, A. 1998. The Red Sea as an extension of the Indian Ocean. Pages 277–281 in K. Sherman, E. N. Okemwa, and M. J. Ntiba, editors. Large marine ecosystems of the Indian Ocean: assessment, sustainability, and management. Blackwell Science, Cambridge, MA, USA.

Getahun, A. 2000. Systematic studies of the African species of the genus *Garra* (Pisces: Cyprinidae). Unpublished Ph.D. thesis. City University of New York, NY, USA.

Getahun, A. 2001. Lake Afdera: a threatened saline lake in Ethiopia. SINET: Ethiopian Journal of Science 24(1):127–131.

Getahun, A. and K. J. Lazara. 2001. *Lebias stiassnyae:* A new species of killifish from Lake Afdera, Ethiopia (Teleostei: Cyprinodontidae). Copeia 1:150–153.

Getahun, A. and M. L. J. Stiassny. 1998. The freshwater biodiversity crisis: the case for conservation. Ethiopian Journal of Science 21(2):207–230.

Gichuki, N. N. 2003. Monitoring of birds in Lake Baringo and its catchment, Lake Baringo Land and Water Management Project report. GEF/UNOPS, Nairobi, Kenya.

Gilman, R. T., R. Abell, and C. E. Williams. 2004. Can conservation biology inform the practice of IRBM to promote sustainable water resource management? International Journal of River Basin Management 2(2):1–14.

Girard, O. and J. Thal. 2001. Mise en place d'un réseau de suivi de populations d'oiseaux d'eau en Afrique subsaharienne. Rapport de mission au Mali 9–23 Janvier 2001. ONCFS, Paris, France.

Githaiga, J. M. 1997. Utilization patterns and inter-lake movements of the lesser flamingo and their conservation in the saline lakes of Kenya. Pages 11–21 in G. Howard, editor. Conservation of the lesser flamingo in Eastern Africa and beyond. IUCN, Nairobi, Kenya.

Giudicelli, J., A. Bouzidi, and N. A. Abdelaali. 2000. Contribution to the faunistic and ecological study of the blackflies (Diptera: Simuliidae) of Morocco IV. The blackflies of the High Atlas mountain range. Description of a new species. Annales de Limnologie 36:57–80.

Giudicelli, J. and M. Dakki. 1984. The springs of the Middle Atlas and the Rif (Morocco): faunistics (description of 2 new species of Trichoptera), ecology and biogeography. Bijdragen Tot de Dierkunde 54:83–100.

Glaw, F. and M. Vences. 1994. A field guide to the amphibians and reptiles of Madagascar. Moos-Druck, Leverkusen, Germany.

Gleick, P. H., editor. 1993. Water in crisis: a guide to the world's freshwater resources. Oxford University Press, New York, NY, USA.

Godoy, R., D. Wilkie, H. Overman, A. Cubas, G. Cubas, J. Demmer, K. McSweeney, and N. Brokaw. 2000. Valuation of consumption and sale of forest goods from a Central American rain forest. Nature 406:62–63.

Goldschmidt, T. 1996. Darwin's dreampond: drama in Lake Victoria. MIT Press, Cambridge, MA, USA.

Golubtsov, A. S., A. A. Darkov, Y. Y. Dgebuadze, and M. V. Mina. 1995. An artificial key to fish species of the Gambela region (the White Nile Basin in the limits of Ethiopia). Joint Ethio-Russian Biological Expedition, Addis Ababa, Ethiopia.

Golubstov, A. S., Y. Y. Dgebuadze, and M. V. Mina. 2002. Fishes of the Ethiopian Rift Valley. Pages 167–258 in C. Todorancea and W. D. Taylor, editors. Ethiopian Rift Valley lakes. Backhuys Publishers, Leiden, The Netherlands.

Gonfiantini, R., G. M. Zuppi, D. H. Eccles, and W. Ferro. 1979. Isotope investigation of Lake Malawi. Pages 195–205 in Isotopes in lake studies. Proceedings on the application of nuclear techniques to study of lake dynamics organized by the International Atomic Energy Agency. IAEA, Vienna, Austria.

Goodman, S. M. and J. P. Benstead. 2003. The natural history of Madagascar. University of Chicago Press, Chicago, IL, USA.

Gordon, C., C. Cooper, C. A. Senior, H. Banks, J. M. Gregory, T. C. Johns, J. F. B. Mitchell, and R. A. Wood. 2000. The simulation of SST, sea ice extents and ocean heat transports in a version of the Hadley Centre coupled model without flux adjustments. Climate Dynamics 16:147–168.

Gordon, D. 2003. Technical change and economies of scale in the history of Mweru-Luapula's fishery (Zambia and Democratic Republic of Congo). Pages 164–178 in E. Jul Larsen, J. Kolding, R. Overa, J. R. Nielsen, and P. van Zwieten, editors. Management, co-management or no management? Major dilemmas in southern African freshwater fisheries, FAO Fisheries Technical Paper 426/2. FAO, Rome, Italy.

Gosse, J. P. 1963. Le milieu aquatique et l'écologie des poissons dans la région de Yangambi. Annales du Musée Royal de l'Afrique Centrale 116:113–271.

Gottfried, M. D. and D. W. Krause. 1998. First record of gars (Lepisosteidae, Actinopterygii) on Madagascar: late Cretaceous remains from the Mahajunga Basin. Journal of Vertebrate Paleontology 18:275–279.

Gottfried, M. D., L. L. Randriamiarimanana, J. A. Rabarison, and D. W. Krause. 1998. Late Cretaceous fish from Madagascar: implications for Gondwanan biogeography. Journal of African Earth Sciences, Special Abstracts Issue, Gondwana 10: Event Stratigraphy of Gondwana 27(1A).

Goudie, A. 1972. Climate, weathering, crust formation, dunes and fluvial features of the Central Namib Desert near Gobabeb, southwest Africa. Madoqua 1:15–31.

Gourene, G., G. G. Teugels, B. Hugueny, and F. E. T. Van Den Audenaerde. 1999. Evaluation of the ichthyological diversity of a West African river basin after the construction of a dam. Cybium 23:147–160.

Government of Mauritius. 2001. Official web portal: geography and climate. Retrieved 2001 from the World Wide Web: http://ncb.intnet.mu/govt/geograph.htm.

Gratwicke, B. 1999. The effect of season on a biotic water quality index: a case study of the Yellow Jacket and Mazowe Rivers, Zimbabwe. South African Journal of Aquatic Sciences 24(1–2):24–35.

Gratwicke, B. and B. E. Marshall. 2001. The relationship between the exotic predators *Micropterus salmoides* and *Serranochromis robustus* and native stream fishes in Zimbabwe. Journal of Fish Biology 58:68–75.

Green, J. and S. A. Corbet. 1973. Ecological studies on crater lakes in West Cameroon. The blood of endemic cichlids in relation to stratification and their feeding habits. Journal of Zoology, London 170:299–308.

Greenwood, P. H. 1966. The fishes of Uganda. The Uganda Society, Kampala, Uganda.

Greenwood, P. H. 1973. A revision of the "*Haplochromis*" and related species (Pisces: Cichlidae) from Lake George, Uganda. Bulletin of the British Museum (Natural History) Zoology 141–242.

Greenwood, P. H. 1974. The "*Haplochromis*" species (Pisces: Cichlidae) of Lake Rudolf, East Africa. Bulletin of the British Museum (Natural History) Zoology 141–165.

Greenwood, P. H. 1980. Towards a phyletic classification of the "genus" *Haplochromis* (Pisces, Cichlidae) and related taxa. Part II. The species from Lakes Victoria, Nabugabo, Edward, George, and Kivu. Bulletin of the British Museum (Natural History) Zoology 39:1–101.

Greenwood, P. H. 1981. The haplochromine fishes of the East African lakes. Kraus Int'l. Publishers, Munich, Germany.

Greenwood, P. H. 1983. The zoogeography of African freshwater fishes: bioaccountancy or biogeography? Pages 179–199 in R. W. Sims, J. H. Price, and P. E. S. Whalley, editors. Evolution, time and space: the emergence of the biosphere. Systematics Association special volume No. 23. Academic Press, London, UK and New York, NY, USA.

Greenwood P. H. 1984. The haplochromine species (Teleostei, Cichlidae) of the Cunene and certain other Angolan rivers. Bulletin of the British Museum (Natural History) 47(4):187–239.

Gregory, S. V., F. J. Swanson, W. A. McKee, and K. W. Cummins. 1991. An ecosystem perspective of riparian zones. BioScience 41:540–551.

Gren, I. and T. Söderqvist. 1994. Economic valuation of wetlands: a survey. Beijer Discussion Paper Series No. 54. Beijer International Institute of Ecological Economics, Royal Swedish Academy of Sciences, Stockholm, Sweden.

Griffin, J., D. Cumming, S. Metcalfe, M. t'Sas-Rolfes, E. Chonguica, M. Rowen, and J. Oglethorpe. 1999. Study on the development of transboundary natural resource management areas in southern Africa. Biodiversity Support Program, Washington, DC, USA.

Grimsdell, J. J. R. and R. H. V. Bell. 1975. Black lechwe research project, final report: ecology of the black lechwe in the Bangweulu Basin of Zambia. NCSR, Falcon Press, Ndola, Zambia.

Groenewald, G. 2003. Maloti-Drakensberg Transfrontier Conservation and Development Area. Retrieved 2003 from the World Wide Web: http://www.peaceparks.org.

Groombridge, B. and M. Jenkins. 1998. Freshwater biodiversity: a preliminary global assessment. WCMC Biodiversity Series 8. World Conservation Monitoring Centre, Cambridge, UK.

Grove, A. T. 1985. The environmental setting. Pages 9–13 in A. T. Grove, editor. The Niger and its neighbours: environmental history and hydrobiology, human use and health hazards of the major West African rivers. A.A. Balkema, Rotterdam, The Netherlands.

Grove, A. T., F. A. Street, and A. S. Goudie. 1975. Former lake levels and climatic change in the Rift Valley of southern Ethiopia. Geography Journal 141:177–202.

Groves, C. 2003. Drafting a conservation blueprint: a practitioner's guide to planning for biodiversity. Island Press, Washington, DC, USA.

Guy, P. R. 1980–1981. River bank erosion in the mid-Zambezi valley, downstream of Lake Kariba. Biological Conservation 19:99–112.

Haack, B. 1996. Monitoring wetland changes with remote sensing: an East African example. Environmental Management 20(3):411–419.

Haacke, W. 1984. The herpetology of the southern Kalahari domain. Supplement to Koedoe 1984:171–186.

Haggett, A. R. 1999. Is sport fishing in the panhandle of the Okavango in crisis? Tight Lines: The Angler's Friend 9–13.

Hall, A., I. M. C. B. S. Valente, and B. R. Davies. 1977. The Zambezi River in Mozambique: the physico-chemical status of the Middle and Lower Zambezi prior to the closure of the Cabora Bassa Dam. Freshwater Biology 7:187–206.

Hall-Martin, A. J. 1984. Symposium on the Kalahari ecosystem: summary and conclusions. Supplement to Koedoe 1984: 327–333.

Hamerlynck, O. and S. Duvail. 2003. The rehabilitation of the delta of the Senegal River in Mauritania. IUCN, Gland, Switzerland and Cambridge, UK.

Hammer, T. U. 1986. Saline lake ecosystems of the world. Dr. W. Junk, Dordrecht, The Netherlands.

Hansen, M., R. DeFries, J. R. G. Townshend, and R. Sohlberg. 2000. Global land cover classification at 1 km resolution using a decision tree classifier. International Journal of Remote Sensing 21:1331–1365.

Happold, D. C. D. 1987. The mammals of Nigeria. Clarendon Press, Oxford, UK.

Harper, D. M., G. Phillips, A. Chilvers, N. Kitaka, and K. Mavuti. 1994. Eutrophication prognosis for Lake Naivasha, Kenya. Verhandlungen-Internationale Vereinigung für Theoretische und Angewandte Limnologie 25(2):861–865.

Harrison, A. D. 1995. Northeastern Africa rivers and streams. Pages 507–517 in C. E. Cushing, K. W. Cummings, and G. W. Minshall, editors. River and stream ecosystems. Elsevier Science B. V., Amsterdam, The Netherlands.

Harrison, A. D. and J. D. Agnew. 1962. The distribution of invertebrates endemic to acid streams in the western and southern Cape Province. Annals of the Cape Provincial Museums 2:273–291.

Harrison, A. D. and H. B. N. Hynes. 1988. Benthic fauna of Ethiopian mountain streams and rivers. Archiv für Hydrobiologie Supplement 81:1–36.

Harrison, I. J. and G. J. Howes. 1991. The pharyngobranchial organ of mugilid fishes: its structure, variability, ontogeny, possible function and taxonomic utility. Bulletin of the British Museum (Natural History) Zoology 57:111–132.

Harrison, I. J. and M. J. Stiassny. 1999. The quiet crisis: a preliminary listing of the freshwater fishes of the world that are extinct or "missing in action." Pages 271–331 in R. MacPhee, editor. Extinctions in near time. Kluwer Academic/Plenum Publishers, New York, NY, USA.

Harrison, J. A., D. G. Allan, L. G. Underhill, M. Herremans, A. J. Tree, Parker V., and C. J. Brown, editors. 1997. The atlas of southern African birds, including Botswana, Lesotho, Namibia, South Africa, Swaziland and Zimbabwe, Volume 1. Non-passerines. BirdLife South Africa, Johannesburg, South Africa.

Harrison, J., M. Burger, and S. Ellis, editors. 2001. Conservation assessment and management plan for southern African frogs. First draft. January 2001. Southern African Frog Atlas Project. Avian Demographic Unit, University of Cape Town, Cape Town, South Africa.

Hart, J. A. and J. S. Hall. 1996. Status of eastern Zaire's forest parks and reserves. Conservation Biology 10(2):316–327.

Hart, R. C. 1988. Water quality and the environment: the Buffalo River situation. Pages 356–366 in M. N. Bruton and F. W. Gess, editors. Towards an environmental plan for the Eastern Cape. Rhodes University, Grahamstown, South Africa.

Hart, T. and R. Mwinyihali. 2001. Armed conflict and biodiversity in sub-Saharan Africa: the case study of the Democratic Republic of Congo (DRC). Biodiversity Support Program, Washington, DC, USA.

Hatton, J. and F. Munguambe, editors. 1998. The biological diversity of Mozambique. Ministry for the Coordination of Environmental Affairs, Maputo, Mozambique.

Hay, C. J., B. J. Zyl, and G. J. Steyn. 1996. A quantitative assessment of the biotic integrity of the Okavango River, Namibia, based on fish. Water SA 22(3):263–284.

Hay, W. W., R. M. DeConto, C. N. Wold, K. M. Wilson, S. Voight, M. Schulz, A. R. Wold, W.-C. Dullo, A. B. Ronov, A. N. Balukhovsky, and E. Söding. 1999. Alternative global Cretaceous paleogeography. Pages 1–47 in E. Barrera and C. C. Johnson, editors. Evolution of the Cretaceous ocean-climate system. Geological Society of America, Special Paper 332, Boulder, CO, USA.

Hazelwood, P. L. 1979. Liberia : draft environmental report on Liberia. Retrieved 2001 from the World Wide Web: http://www.wri.org/wdces/li79_99.html.

Hazevoet, C. J. 1995. The birds of the Cape Verde Islands. Dorset Press, Dorchester, UK.

Hazevoet, C. J. 2001. Cape Verde. Pages 161–168 in L. D. C. Fishpool and M. I. Evans, editors. Important bird areas in Africa and associated islands: priority sites for conservation. Pisces Publications and BirdLife International (BirdLife Conservation Series No. 11), Newbury and Cambridge, UK.

Hecky, R. E. 1993. The eutrophication of Lake Victoria. Verhandlungen Internationale Vereinigung Limnologie 25: 39–48.

Hecky, R. E., F. W. B. Bugenyi, P. Ochumba, J. F. Talling, R. Mugidde, M. Gophen, and L. Kaufman. 1994. Deoxygenation of the deep water of Lake Victoria, East Africa. Limnology and Oceanography 39(6):1476–1481.

Heino, J. 2002. Concordance of species richness patterns among multiple freshwater taxa: a regional perspective. Biodiversity and Conservation 11:137–147.

Henderson, L. and C. J. Cilliers. 2002. Invasive aquatic plants. A guide to the identification of the most important and potentially dangerous invasive aquatic and wetland plants in South Africa. Plant Protection Research Institute Handbook No. 16. ARC-Plant Protection Research Institute, Pretoria, South Africa.

Henkel, F.-W. and W. Schmidt. 2000. Amphibians & reptiles of Madagascar, the Mascarenes, the Seychelles & the Comoros Islands. Krieger Publishing Company, Malabar, FL, USA.

Henrichs, T., B. Lehner, and J. Alcamo. 2002. An integrated analysis of changes in water stress in Europe. Integrated Assessment 3(1):15–29.

Hens, L. and E. K. Boon. 1999. Institutional, legal, and economic instruments in Ghana's environmental policy. Environmental Management 24(3):337–351.

Heringa, A. C. 1990. Mali. Pages 8–14 in R. East, editor. Antelopes global survey and regional action plans. Part 3. West and Central Africa. IUCN/SSC Antelope specialist group. IUCN, Gland, Switzerland and Cambridge, UK.

Herremans, M. 1998. Conservation status of birds in Botswana in relation to land use. Biological Conservation 86:139–160.

Heydoorn, A. E. F. and J. R. Grindley. 1981. Estuaries of the Cape. Part II. Synopses of available information on individual systems. Report no. 3. Groen (CW7). CSIR Research Report 402. Creda Press, Cape Town, South Africa.

Hickley, P. and R. G. Bailey. 1987. Food and feeding relationships of fish in the Sudd swamps (River Nile, southern Sudan). Journal of Fish Biology 30(2):147–160.

Hiernaux, P. 1982. Les végétations et les fourragères dans les systèmes pastoraux. CIPEA, Mali.

Hoare, R. E. and J. T. Du Toit. 1999. Coexistence between people and elephants in African savannas. Conservation Biology 13(3):633–639.

Hobbs, H. H. 1987. A review of the crayfish genus *Astacoides* (Decapoda: Parastacidae). Smithsonian Contributions to Zoology 443:1–50.

Hobbs, H. H., III. 1992. Caves and springs. Pages 59–131 in C. T. Hackney, S. M. Adams, and W. H. Martin, editors. Biodiversity of the southeastern United States: aquatic communities. John Wiley, New York, NY, USA.

Hocutt, C. and P. H. Skelton. 1983. Fishes of the Sak River, South Africa with comments on the nomenclature a redescription of the smallmouth yellowfish *Barbus aeneus* (Burchell, 1822). Special Publication No. 32. JLB Smith Institute of Ichthyology, Grahamstown, South Africa.

Hoffman, L. C., J. J. Swart, and D. Brink. 2000. The 1998 production and status of aquaculture in South Africa. Water South Africa 26:133–135.

Hollis, G. E., W. M. Adams, and M. Aminu-Kano. 1993. The Hadejia Nguru wetlands. Environment, economy and sustainable development of a Sahelian floodplain wetland. IUCN, Gland, Switzerland.

Holthuis, L. B. 1966. The R/V Pillsbry deep-sea biological expedition to the Gulf of Guinea, 1964–65. 11. The freshwater shrimps of the island of Annobon, West Africa. Studies in Tropical Oceanography 4(1):224–239.

Hopson, A. J., editor. 1982. Lake Turkana. A report on the findings of the Lake Turkana project 1972–1975, Vols. 1–6. Overseas Development Administration, London, UK.

Horemans, B. 1998. The state of artisanal fisheries in West Africa in 1997. IDAF Technical Report No. 122. FAO, Cotonou, Benin.

Howard, G. W. and S. W. Matindi. 2003. Alien invasive species in Africa's wetlands. Some threats and solutions. IUCN, Eastern Africa, Nairobi, Kenya.

Howard-Williams, C. 1979. The distribution of aquatic macrophytes in Lake Chilwa: annual and long-term environmental fluctuations. Pages 105–122 in M. Kalk, A. J. McLachlan, and C. Howard-Williams, editors. Lake Chilwa: studies of change in a tropical ecosystem. Monographiae Biologicae, 35. Dr. W. Junk, The Netherlands.

Howell, P., M. Lock, and S. Cobb, editors. 1988. The Jonglei Canal: impact and opportunity. Cambridge University Press, Cambridge, UK.

Hughes, D. A. 1997. Southern African "FRIEND": the application of rainfall-runoff models in the SADC region. Report 3/97. Institute for Water Research, Grahamstown, South Africa.

Hughes, J. B., G. C. Daily, and P. R. Ehrlich. 1997. Population diversity: its extent and extinction. Science 278(5338): 689–692.

Hughes, R. H. and J. S. Hughes. 1992. A directory of African wetlands. IUCN, UNEP, and WCMC, Gland, Switzerland, Nairobi, Kenya, and Cambridge, UK.

Hugueny, B. 1989. West African rivers as biogeographic islands: species richness of fish communities. Oecologia 79:236–243.

Hugueny, B. and C. Lévêque. 1994. Freshwater fish zoogeography in West Africa: faunal similarities between river basins. Environmental Biology of Fishes 39:365–380.

Hulme, D. and M. Murphree, editors. 2001. African wildlife and livelihoods. James Curry, Oxford, UK.

Hulme, M., R. M. Doherty, T. Ngara, M. G. New, and D. Lister. 2001. African climate change: 1900–2100. Climate Research 17:145–168.

Huntley, B. J. 1978. Ecosystem conservation in southern Africa. Pages 1333–1385 in M. J. A. Werger, editor. Biogeography and ecology of southern Africa. Dr. W. Junk Publishers, The Hague, The Netherlands.

Huntley, B. J. and E. M. Matos. 1994. Botanical diversity and its conservation in Angola. Pages 53–74 in B. J. Huntley, editor. Botanical diversity in southern Africa. National Botanical Institute, Pretoria, South Africa.

Hutchinson, C. F., P. Warshall, E. J. Arnould, and J. Kindler. 1992. Development in arid lands. Environment 34(6):16–43.

Hutchinson Dictionary of Plants. 2000. Retrieved 2004 from the World Wide Web: http://www.tiscali.co.uk/reference/dictionaries/plants/.

ICLARM. 2002. Nairobi declaration. Conservation of aquatic biodiversity and use of genetically improved and alien species for aquaculture in Africa. ICLARM, CTA, FAO, IUCN, UNEP, CBD. World Fish Centre, Penang, Malaysia.

ICOLD. 1989. World Register of Dams: 1988 update. Central Office, International Commission on Large Dams, Paris, France.

Iddi, S. 1998. Eastern Arc Mountains and their national and global importance. Pages 18–26 in L. A. Depew, editor. Journal of East African natural history: biodiversity and conservation of the Eastern Arc Mountains of Tanzania and Kenya, 87. East Africa Natural History Society and National Museums of Kenya, Morogoro, Tanzania.

Ikingura, J. R., M. K. D. Mutakyahwa, and J. M. J. Kahatano. 1997. Mercury and mining in Africa with special reference to Tanzania. Water Air and Soil Pollution 97:223–232.

ILEC. 2001. World lake database. http://www.ilec.or.jp/database/database.html.

Iltis, A. and J. Lemoalle. 1983. The aquatic vegetation of Lake Chad. Pages 125–143 in J. P. Carmouze, J. R. Durand, and C. Lévêque, editors. Lake Chad, Monographiae Biologicae 53. Dr. W. Junk Publishers, The Hague, The Netherlands.

Impson, N. D., I. R. Bills, J. A. Cambray, and A. le Roux. 2000. Chapter 3: freshwater fishes. Pages 19–38 in G. D. P. Van Nieuwenhuizen and J. A. Day, editors. Cape Action Plan for the Environment: the conservation of freshwater ecosystems in the Cape Floral Kingdom. University of Cape Town, Cape Town, South Africa.

Inger, R. F. and C. P. Kong. 1962. Freshwater fishes of North Borneo. Fieldiana. Zoology 45. Chicago Natural History Museum, Chicago, IL, USA.

Institute of Marine Sciences, FAO, and SIDA. 1998. Overview of land-based sources and activities affecting the marine, coastal and associated freshwater environment in the eastern African region. UNEP Regional Seas Reports and Studies No. 167. UNEP, Dar es Salaam, Tanzania.

Integrated Regional Information Networks. 2000. Mali: firewood for fuel raises concerns about deforestation. Retrieved 2001 from the World Wide Web: http://www.reliefweb.int/IRIN/wa/countrystories/mali/20001003.phtml.

Invasive Species Specialist Group, I. 2000. Global Invasive Species Database. Retrieved 2002 from the World Wide Web: http://www.issg.org/database/welcome/.

Inventory of Conflict and Environment. 1997. ICE case studies: Nile River dispute. Retrieved 2001 from the World Wide Web: http://www.american.edu/ted/ice/nile.htm.

IPCC. 1992. Climate change 1992: the supplementary report to the IPCC scientific assessment. Cambridge University Press, Cambridge, UK.

IPCC. 2001. Climate Change 2001: impacts, adaptation and vulnerability. A report of Working Group II of the IPCC. Cambridge University Press, Cambridge, UK.

IRA. 2002. Baseline study of lakes Sagara and Nyamangoma wetlands and the surrounding environment in the Malagarasi/Muyovozi Ramsar Site, western Tanzania. Final report prepared by Institute of Resource Assessment (IRA), University of Dar es Salaam. Ministry of Natural Resources and Tourism: SIMMORS, Dar es Salaam, Tanzania.

Irish, J. 1991. Conservation aspects of karst waters in Namibia. Madoqua 17(2):141–146.

Irish, J. 1992. Cave investigations in Namibia: I. Biospeleology, ecology, and conservation of Dragon's Breath Cave. Cimbebasia 13:59–67.

Irvine, K. 2002. The trophic ecology of the demersal fish community of Lake Malawi/Niassa, Central Africa. Final report INCO-DC project ERBIC 18 CT 970195 (1994–1998).

Irwin, M. P. S. 1981. The birds of Zimbabwe. Quest Publishing, Harare, Zimbabwe.

Ita, E. O. 1993. Inland fishery resources of Nigeria, CIFA Occasional Paper no. 20. FAO, Rome, Italy.

IUCN. 1992. Protected areas of the world: a review of national systems. Volume 3: Afrotropical. World Conservation Monitoring Centre (WCMC) and International Union for Conservation of Nature (IUCN), Cambridge, UK and Gland, Switzerland.

IUCN. 1993. Environmental synopsis 1993: Nigeria. IUCN, Gland, Switzerland.

IUCN. 1994. Guidelines for protected area management categories. IUCN, Cambridge, UK and Gland, Switzerland.

IUCN. 1995–2003. IUCN wetlands and water resources projects. Retrieved 2003 from the World Wide Web: http://www.iucn.org/themes/wetlands/project.html.

IUCN. 2000. Vision for water and nature: a world strategy for conservation and sustainable management of water resources in the 21st century. IUCN: The World Conservation Union, Gland, Switzerland.

IUCN. 2001a. Economic value of reinundation of the Waza Logone Floodplain, Cameroon. Projet de Conservation et de Développement de la Région de Waza-Logone, Maroua, Cameroon.

IUCN. 2001b. One hundred of the world's worst invasive species: a selection from the global invasives species database. Invasive Species Specialist Group, Auckland, New Zealand.

IUCN. 2002. 2002 IUCN Red List of Threatened Species. Retrieved 2003 from the World Wide Web: http://www.redlist.org.

IUCN/SSC Freshwater Biodiversity Assessment Program. 2003. Safeguarding biodiversity in eastern Africa's inland waters: new IUCN/SSC project works to incorporate wildlife values into water resource development planning. Retrieved 2003 from the World Wide Web: http://www.iucn.org/themes/ssc/programs/freshwater/eastafrica.htm.

Jachmann, H. 2000. Zambia's wildlife resources: a brief ecology. Wildlife Resource Monitoring Unit, Environmental Council of Zambia, Lusaka, Zambia.

Jackson, P. B. N. 1961. Kariba studies. Ichthyology: the fish of the Middle Zambezi. Manchester University Press, UK.

Jackson, P. B. N. 1963. Ecological factors affecting the distribution of freshwater African fishes in tropical Africa. Annals of the Cape Provincial Museums (Natural History) 2:223–228.

Jackson, P. B. N. 1986. Fish of the Zambezi system. Pages 269–288 in B. R. Davies and K. F. Walker, editors. The ecology of river systems. Dr. W. Junk Publishers, Dordrecht, The Netherlands.

Jackson, R. B., S. R. Carpenter, C. N. Dahm, D. M. McKnight, R. J. Naiman, S. L. Postel, and S. W. Running. 2001. Water in a changing world. Ecological Applications 11:1027–1045.

Jacobsen, N. 1999. Specialist report: herpetofauna. Consultancy Report No: LHDA 648-F-19 (Vol. 2). Consulting services for the establishment and monitoring of the instream flow requirements for river courses downstream of LHWP dams. Metsi Consultants and Lesotho Highlands Water Project for the Lesotho Highlands Development Authority, Maseru, Lesotho.

Jacobson, P. J., P. L. Angermeier, and R. Loutit. 1996. The conservation significance of ephemeral rivers in northwestern Namibia. Ecological Society of America Abstracts 77(3):214.

Jacobson, P. J., K. M. Jacobson, P. L. Angermeier, and D. S. Cherry. 1999. Transport, retention, and ecological significance of woody debris within a large ephemeral river. Journal of the North American Benthological Society 18:429–444.

Jacobson, P. J., K. M. Jacobson, P. L. Angermeier, and D. S. Cherry. 2000a. Hydrologic influences on soil properties along ephemeral rivers in the Namib Desert. Journal of Arid Environments 45:21–34.

Jacobson, P. J., K. M. Jacobson, P. L. Angermeier, and D. S. Cherry. 2000b. Variation in material transport and water chemistry along a large ephemeral river in the Namib Desert. Freshwater Biology 44:481–492.

Jacobson, P. J., K. M. Jacobson, and M. K. Seely. 1995. Ephemeral rivers and their catchments: sustaining people and development in western Namibia. Desert Research Foundation of Namibia, Windhoek.

Jacot-Guillarmod, A. 1963. Further observations on the bogs of the Basutoland Mountains. South African Journal of Science 58(6):179–182.

Jacot-Guillarmod, A. 1988. Exotic plant invasion, man's activities, and indigenous vegetation. Pages 68–87 in M. N. Bruton and F. W. Gess, editors. Towards an environmental plan for the Eastern Cape. Rhodes University, Grahamstown, South Africa.

Jamu, D. M., J. Chimphamba, and R. E. Brummett. 2003. Land use and cover changes in the Likangala catchment of the Lake Chilwa Basin: implications for managing a tropical wetland in Malawi, southern Africa. African Journal of Aquatic Sciences 28(2):119–132.

Jansen, E. G. 1997. Rich fisheries–poor fisherfolk: some preliminary observations about the effects of trade and aid in the Lake Victoria fisheries. Socio-economics of the Lake Victoria Fisheries Report 1. IUCN Eastern Africa Regional Office, Nairobi, Kenya.

Jauro, A. B. 1998. Lake Chad basin Commission (LCBC) perspectives. Working paper presented at workshop for International Network of Basin Organizations, March 19–21 1998. Paris, France. Retrieved 2001 from the World Wide Web: http://www.oieau.fr/ciedd/contributions/atriob/contribution/cblt.htm.

Jeffries, M. and D. Mills. 1990. Freshwater ecology: principles and applications. Belhaven Press, London, UK.

Jenkins, M. D. 1987. Madagascar: an environmental profile. IUCN, Gland, Switzerland.

Jennings, J. N. 1998. Cave and karst terminology. Australian Speleological Federation Incorporated. Retrieved 2004 from the World Wide Web: http://werple.net.au/~gnb/caving/papers/jj-cakt.html.

Joffe, S. and S. Cooke. 1997. Management of water hyacinth and other invasive aquatic weeds: issues for the World Bank, Global IPM Facility. CABI Bioscience, Wallingford, UK.

Johansson, S. G., P. Cunneyworth, N. Doggart, and R. Botterweg. 1998. Biodiversity surveys in the East Usambara Mountains: preliminary findings and management implications. Journal of the East African Natural History Society 87(1–2):29–36.

John, D. M. 1986. The inland waters of tropical West Africa. E. Schweizerbart'sche Verlagsbuchhandlung, Stuttgart, Germany.

John, D. M., C. Lévêque, and L. E. Newton. 1993. Western Africa. Pages 47–78 in D. Whigham, D. Dykjova, and S. Hejny, editors. Wetlands of the world, 1. Kluwer Academic Publishers, Dordrecht, The Netherlands.

Johnson, N., C. Revenga, and J. Echeverria. 2001. Managing water for people and nature. Science 292(5519):1071–1075.

Johnson, T. C., K. Kelts, and E. Odada. 2000. The Holocene history of Lake Victoria. Ambio 29(1):2–11.

Johnson, T. C. and P. Ng'ang'a. 1990. Reflections on a Rift Lake. Pages 113–135 in B. J. Katz, editor. Lacustrine basin exploration: case studies and modern analogs. American Association of Petroleum Geologists Memoirs, Tulsa, OK, USA.

Johnson, T. C., C. A. Scholz, M. R. Talbot, K. Kelts, R. D. Ricketts, G. Ngobi, K. Beuning, I. Ssemmanda, and J. W. McGill. 1996. Late Pleistocene desiccation of Lake Victoria and rapid evolution of cichlid fishes. Science 273(5278):1091–1093.

Jones, P. J. 1994. Biodiversity in the Gulf of Guinea: an overview. Biodiversity and Conservation 3(9):772–784.

Jones, T., B. Phillips, C. E. Williams, and J. Pittock, editors. 2003. Managing rivers wisely: lessons from WWF's work for integrated river basin management. WWF International, Gland, Switzerland.

Juan, C., B. C. Emerson, P. Oromí, and G. M. Hewitt. 2000. Colonization and diversification: towards a phylogeographic synthesis for the Canary Islands. Trends in Ecology & Evolution 15:104–109.

Jubb, R. A. and I. G. Gaigher. 1971. Check list of the fishes of Botswana. Arnoldia Rhodesia 5(7):1–22.

Jul Larsen, E. 2003. Analysis of effort dynamics in the Zambian inshore fisheries of Lake Kariba. Pages 232–251 in E. Jul Larsen, J. Kolding, R. Overa, J. R. Nielsen, and P. van Zwieten, editors. Management, co-management or no management? Major dilemmas in southern African freshwater fisheries, FAO Fisheries Technical Paper 426/2. FAO, Rome, Italy.

Jumbe, J. J. 1997. The status of Tana River dam fisheries twenty years after dam construction. Pages 33–44 in K. Remane, editor. African inland fisheries, aquaculture and the environment. Fishing News Books, Malden, MA, USA.

Junk, W. J. 2002. Long-term environmental trends and the future of tropical wetlands. Environmental Conservation 29(4):414–435.

Junk, W. J., P. B. Bayley, and R. E. Sparks. 1989. The flood pulse concept in river-floodplain systems. Canadian Special Publication of Fisheries and Aquatic Sciences 106:110–127.

Juste, B. J. and J. E. Fa. 1994. Biodiversity conservation in the Gulf of Guinea Islands: taking stock and preparing action (Jersey Wildlife Preservation Trust, Jersey, June 4–6, 1993). Biodiversity and Conservation 3:759–771.

Kabii, T. 1997. An overview of African wetlands. In A. J. Hails, editor. Wetlands, biodiversity, and the Ramsar Convention. Ramsar Convention Bureau, Gland, Switzerland.

Kalitsi, E. A. K. and R. Evans-Appiah. 1999. Dams and ecosystems: assessing and managing environmental impacts (Ghana's experience). In presentation at WCD regional consultation, Cairo, Egypt.

Kalk, M. 1979. Zooplankton in Lake Chilwa: adaptations to changes. Pages 123–141 in M. Kalk, A. J. McLachlan, and C. Howard-Williams, editors. Lake Chilwa: studies of change in a tropical ecosystem. Monographiae Biologicae, 35. Dr. W. Junk, The Hague, The Netherlands.

Kalk, M., A. J. McLachlan, and C. Howard-Williams, editors. 1979. Lake Chilwa: studies of change in a tropical ecosystem. Monographiae Biologicae, Vol. 35. Dr. W. Junk, The Hague, The Netherlands.

Kamdem Toham, A. 1998. Fish biodiversity of the Ntem River Basin (Cameroon): taxonomy, ecology and conservation. Dissertation, Katholieke Universiteit Leuven, Belgium.

Kamdem Toham, A., J. D'Amico, D. Olson, A. Blom, L. Trowbridge, N. Burgess, M. Thieme, R. Abell, R. W. Carroll, S. Gartlan, O. Langrand, R. Mikala Mussavu, D. O'Hara, and H. Strand, editors. 2003. Biological priorities for conservation in the Guinean-Congolian forest and freshwater region. WWF-CARPO, Libreville, Gabon.

Kamdem Toham, A. and G. G. Teugels. 1998. Diversity patterns of fish assemblages in the Lower Ntem River Basin (Cameroon), with notes on potential effects of deforestation. Archiv für Hydrobiologie 141(4):421–446.

Karanja, F., L. Emerton, J. Mafumbo, and W. Kakuru. 2001. Assessment of the economic value of Pallisa District Wetlands, Uganda. Biodiversity Economics Programme for Eastern Africa, IUCN: The World Conservation Union and Uganda National Wetlands Programme, Kampala, Uganda.

Kasimona, V. N. and J. J. Makwaya. 1995. Present planning in Zambia for the future use of Zambezi River waters. Pages 49–56 in T. Matiza, S. Crafter, and P. Dale, editors. Water resource use in the Zambezi Basin. Proceedings of a workshop held at Kasane, Botswana, 28 April–2 May 1993. IUCN, Gland, Switzerland.

Kasulo, V. 1999. The impact of invasive species in African lakes. Pages 262–297 in C. Perrings, M. Williamson, and S. Dalmazzone, editors. The economics of biological invasions. Edward Elgar, Cheltenham, UK.

Katerere, D. 1994. Policy, institutional frameworks, and wetlands management in Zimbabwe. Pages 129–136 in T. Matiza and S. A. Crafter, editors. Wetlands ecology and priorities for conservation in Zimbabwe. Proceedings of a seminar on wetlands of Zimbabwe. IUCN, Gland, Switzerland.

Kaufman, L. S. 1992. Catastrophic change in species rich freshwater ecosystems. BioScience 42(11):864–868.

Kaufman, L. S., L. J. Chapman, and C. A. Chapman. 1997. Evolution in fast forward: haplochromine fishes of the Lake Victoria region. Endeavour 21(1):23–30.

Kaufman, L. and A. S. Cohen. 1993. The Great Lakes of Africa. Conservation Biology 7:632–633.

Kaufman, L. S. and J. Schwartz. 2002. Nile perch population dynamics in Lake Victoria: implications for management and conservation. Pages 257–313 in M. Ruth and J. Lindholm, editors. Dynamic modeling for marine conservation. Springer-Verlag, New York, NY, USA.

Kebede, E., G. Teferra, W. D. Taylor, and Z. Gebre Mariam. 1992. Eutrophication of Lake Hayq in the Ethiopian highlands. Journal of Plankton Research 14(10):1473–1482.

Kelly, L. C., D. T. Bilton, and S. D. Rundle. 2001. Population structure and dispersal in the Canary Island caddisfly *Mesophylax aspersus* (Trichoptera, Limnephilidae). Heredity 86:370–377.

Kemp, J., J. C. Hatton, and H. Sosovele. 2000. East African marine ecoregion: reconnaissance synthesis report. WWF, Dar Es Salaam, Tanzania.

Kendie, D. 2001. Egypt and the hydro-politics of the Blue Nile River. Northeast African Studies 6(1):141–169.

Kenmuir, D. H. S. 1976. Fish spawning under artificial flood conditions on the Mana flood-plain, Zambezi River. Kariba Studies 5:86–97.

Khamar, M., D. Bouya, and C. Ronneau. 2000. Metallic and organic pollution of water and sediments of a Moroccan aquatic system by urban wastewater. Water Quality Research Journal of Canada 35:147–161.

Khan, A. S., H. Mikkola, and R. E. Brummett. 2004. Feasibility of fisheries co-management in Africa. WorldFish Quarterly 27(1, 2):60–64.

Kiener, A. 1963. Poissons, pêche et pisciculture à Madagascar. Publication Centre Technique Forestier Tropical 24:1–224.

Kiener, A. and G. Richard-Vindard. 1972. Fishes of the continental waters of Madagascar. Pages 477–499 in R. Battistini and G. Richard-Vindard, editors. Biogeography and ecology in Madagascar. Dr. W. Junk, The Hague, The Netherlands.

Kifle, D. and A. Belay. 1990. Seasonal variations in phytoplankton primary production in relation to light and nutrients in Lake Awassa, Ethiopia. Hydrobiologia 196:217–227.

Kimmins, D. E. 1960. The Odonata and Neuroptera of the island of Socotra. Annals and Magazine of Natural History, 13th Series 3(7):385–392.

Kingdom of Lesotho. 1989. National Environmental Action Plan. Retrieved 2002 from the World Wide Web: http://www-wds.worldbank.org/servlet/WDSContentServer/WDSP/IB/2000/02/24/000009265_3970702135109/Rendered/PDF/multi_page.pdf.

Kingdon, J. 1989. Island Africa: the evolution of Africa's rare animals and plants. Princeton University Press, Princeton, NJ, USA.

Kingdon, J. 1997. The Kingdon field guide to African mammals. Academic Press, San Diego, CA, USA.

Kinnaird, M. F. 1992. Competition for a forest palm: use of *Phoenix reclinata* by human and nonhuman primates. Conservation Biology 6:101–107.

Kinvig, R. 2000. Odonata survey of the Zambezi Delta. Pages 559–564 in J. Timberlake, editor. Biodiversity of the Zambezi Basin wetlands. Biodiversity Foundation for Africa, Bulawayo/The Zambezi Society, Harare, Zimbabwe.

Kirwan, G. M., R. P. Martins, K. M. Morton, and D. A. Showler. 1996. The status of birds in Socotra and 'Abd Al-Kuri and the records of the OSME survey in spring 1993. Sandgrouse 17:83–101.

Kissama Foundation. 2001. Kissama Foundation. Retrieved 2003 from the World Wide Web: http://www.kissama.org/.

Kleynhans, C. J. 1983. A checklist of the fish species of the Mogol and Palala rivers (Limpopo system) of the Transvaal. Journal of the Limnological Society of Southern Africa 9(1):29–32.

Kleynhans, C. J. 1986. The distribution, status and conservation of some fish species of the Transvaal. South African Journal of Wildlife Research 16(4):135–144.

Kling, G. 1987. Comparative limnology of lakes in Cameroon, West Africa. Master's thesis. Duke University, Durham, NC, USA.

Kocher, T. D., J. A. Conroy, K. R. McKaye, J. R. Stauffer, and S. F. Lockwood. 1995. Evolution of NADH Dehydrogenase Subunit 2 in East African cichlid fish. Molecular Phylogenetics and Evolution 4(4):420–432.

Kofron, C. P. 1992. Status and habitats of the three African crocodiles in Liberia. Journal of Tropical Ecology 8(3):265–273.

Kok, D. J. 1987. Invertebrate inhabitants of temporary pans. African Wildlife 41(5):239.

Kolawole, A. 1987. Environmental change and the south Chad irrigation project, Nigeria. Journal of Arid Environments 13:169–176.

Kolberg, H., M. Griffin, and R. Simmons. 1996. The ephemeral wetlands of central northern Namibia. Pages 40–42 in A. J. Hails, editor. Wetlands, biodiversity and the Ramsar Convention. Ramsar Convention Bureau, Gland, Switzerland.

Kolding, J. 1992. A summary of Lake Turkana: an ever-changing mixed environment. Mitteilungen Internationale Vereinigung für Theoretische und Angewundte Limnologie 23:25–35.

Kolding, J. 1993. Trophic interrelationships and community structure at two different periods of Lake Turkana, Kenya: a comparison using the ECOPATH II box model. Pages 116–123 in V. Christensen and D. Pauly, editors. Trophic models of aquatic ecosystems, ICLARM Conference Proceedings No. 26. ICLARM, Penang, Malaysia.

Kolding, J., I. R. Bills, K. Mosepele, T. Mmopelwa, and S. Nengu. 2003. Fisheries and fish biodiversity in the Okavango Delta, Botswana: a rapid assessment. Pages 30–46 in K. Mosepele, editor. AquaRAP II: rapid assessment of the aquatic ecosystems of the Okavango Delta, Botswana: preliminary report. Conservation International, Okavango Program, Maun, Botswana.

Kone, B. 2002. Organisation socio-économique du Delta Intérieur du fleuve Niger. Pages 65–85 in E. Wymenga, B. Kone, and L. Zwarts, editors. Le Delta Intérieur du fleuve Niger: ecologie et gestion durable des resources naturelles. Wetlands International, RIZA, and Altenburg & Wymenga ecological consultants, Sévaré, Mali; Lelystad, The Netherlands; and Veenwouden, The Netherlands.

Konings, A. 1995. Malawi cichlids in their natural habitat, 2nd ed. Cichlid Press, El Paso, TX, USA.

Konings, A. 1998. Tanganyika cichlids in their natural habitat, 2nd ed. Cichlid Press, El Paso, TX, USA.

Kottelat, M. 1998. Fishes of the Nam Theun and Xe Banfai basins, Laos, with diagnoses of twenty-two new species (Teleostei: Cyprinidae, Balitoridae, Cobitidae, Coiidae and Odontobutidae). Ichthyological Exploration of Freshwaters 9:1–128.

Kottelat, M. and T. Whitten. 1996. Freshwater biodiversity in Asia with special reference to fish, World Bank Technical Paper No. 343. The World Bank, Washington, DC, USA.

Krause, D. W., J. H. Hartman, and N. A. Wells. 1997. Late cretaceous vertebrates from Madagascar: implications for biotic change in deep time. Pages 3–43 in S. M. Goodman and D. B. Patterson, editors. Natural change and human impact in Madagascar. Smithsonian Institution Press, Washington, DC, USA.

Kremen, C., V. Razafimahatratra, R. P. Guillery, J. Rakotomalala, A. Weiss, and J. S. Ratsisompatrarivo. 1999. Designing the Masoala National Park in Madagascar based on biological and socioeconomic data. Conservation Biology 13:1055–1068.

Kretschmar, M. 2003. Biodiversity patterns of multiple freshwater taxa in North America. Master's thesis. University of Lüneberg, Lüneberg, Germany.

Kristensen, T. K. and D. S. Brown. 1999. Control of intermediate host snails for parasitic diseases: a threat to biodiversity in African freshwaters? Malacologia 41(2):379–391.

Kruger, F. J., B. W. van Wilgen, A. V. B. Weaver, and T. Greyling.

1997. Sustainable development and the environment: lessons from the St. Lucia environmental impact. South African Journal of Science 93(1):23–33.

Kuchling, G. 1988. Population structure, reproductive potential, and increasing exploitation of the freshwater turtle *Erymnochelys madagascariensis*. Biological Conservation 43:107–113.

Kuwamura, T. 1997. The evolution of parental care and mating systems among Tanganyikan cichlids. Pages 57–86 in H. Kwanabe, M. Hori, and N. Makoto, editors. Fish communities in Lake Tanganyika. Kyoto University Press, Kyoto, Japan.

Ladiges, W. 1964. Beiträge zur zoogeographie und oekologie der süsswasserfische Angolas. Mitteilungen aus dem Hamburgischen Zoologischen Museum und Institut 61:221–272.

Laë, R. 1992. Les pêcheries artisanales lagunaires ouest-africaines: échantillonnage et dynamique de la ressource et de l'exploitation, Collection études et thèses. ORSTOM, Paris, France.

Laë, R. 1994. Effects of drought, dams and fishing pressure on the fisheries of the Central Delta of the Niger River. International Journal of Ecology and Environmental Sciences 20:119–128.

Laë, R. 1995. Climatic and anthropogenic effects on fish diversity and fish yields in the Central Delta of the Niger River. Aquatic Living Resources 8:43–58.

Laë, R. 1997. Effects of climatic changes and developments on continental fishing in West Africa: the examples of the Central Delta of the Niger in Mali and coastal lagoons. Pages 66–86 in Togo. K. Remane, editor. African inland fisheries, aquaculture and the environment. FAO and Fishing News Books, Oxford, UK.

Lahm, S. A. 2002. L'orpaillage au nord-est du Gabon historique et analyse socio-economique. CeNaReST/CARPE, Libreville, Gabon.

Lahr, J. 1998. An ecological assessment of the hazard of right insecticides used in desert locust control to invertebrates in temporary ponds in the Sahel. Aquatic Ecology 32:153–162.

Lalèyè, A. P., E. Baras, and J. C. Philippart. 1995a. Variations du régime alimentaire chez 2 espèces du genre *Chrysichthys* (*C. nigrodigitatus* et *C. auratus*) dans les lagunes du sud-Bénin. Aquatic Living Resources 8(4):365–372.

Lalèyè, A. P. and J. Moreau. In press. Resources and constraints of West African coastal waters for fish production. In Workshop on the biodiversity, management and utilization of West African fishes, organized in Ghana, 2–4 July 2002 by ICLARM (Philippines), Water Research Institute (WRI), Ghana and the Zoologisches Institut und Zoologisches Museum, Universität Hamburg (ZIM/UH), Germany.

Lalèyè, P., J. C. Philippart, and J. C. Heymans. 1995b. Cycle annuel de l'indice gonadosomatique et de la condition chez deux espèces de *Chrysichthys* (Siluriformes, Bagridae) du lac Nokoué et de la lagune de Porto-Novo au Bénin. Cybium 19:131–142.

Lamboj, A. 2002. *Chromidotilapia mrac*, a new species of Cichlidae (Teleostei: Perciformes) from Gabon. Ichthyological Explorations of Freshwater 13:251–256.

Lamboj, A. 2003. *Chromidotilapia melaniae* and *C. nana*, two new cichlid species (Perciformes, Cichlidae) from Gabon, Central Africa. Zootaxa 143:1–15.

Lamboj, A. and J. Snoeks. 2000. *Divandu albimarginatus*, a new genus and species of cichlid (Teleostei: Cichlidae) from Congo and Gabon, Central Africa. Ichthyological Explorations of Freshwater 11:355–360.

Lamboj, A. and M. L. J. Stiassny. 2003. Three new *Parananochromis* species (Teleostei, Cichlidae) from Gabon and Cameroon, Central Africa. Zootaxa 209:1–19.

Lamotte, M. 1983. The undermining of Mt. Nimba. Ambio 12(3–4):174–179.

Lancaster, N. 1979. The physical environment of Lake Chilwa. Pages 17–42 in M. Kalk, A. J. McLachlan, and C. Howard-Williams, editors. Lake Chilwa: studies of change in a tropical ecosystem. Monographiae Biologicae, 35. Dr. W. Junk, The Hague, The Netherlands.

Lancaster, N. 1989. Late quaternary palaeoenvironments in the southwestern Kalahari. Palaeogeography, Palaeoclimatology, Palaeoecology 70:367–376.

Langrand, O. 1990. Guide to the birds of Madagascar. Yale University Press, New Haven, CT, USA.

Laraque, A., M. Mietton, J. C. Olivry, and A. Pandi. 1998. Impact of lithological and vegetal covers on flow discharge and water quality of Congolese tributaries from the Congo River. Revue Des Sciences de L'eau 11(2):209–224.

Laurenson, L. B. J. and C. H. Hocutt. 1984. Colonisation theory and invasive biota: the Great Fish River, a case history. Environmental Monitoring and Assessment 6:71–90.

Lawson, G. W., editor. 1986. Plant ecology in West Africa: systems and processes. John Wiley & Sons, Chichester, UK.

LeBerre, M. 1989. Faune du Sahara. LeChavalier–R. Chabaud, Paris, France.

Le Corre, M. and R. J. Safford. 2001. La Réunion and Iles Eparses. Pages 693–702 in L. D. C. Fishpool and M. I. Evans, editors. Important bird areas in Africa and associated islands: priority sites for conservation. Pisces Publications and BirdLife International (BirdLife Conservation Series No. 11), Newbury and Cambridge, UK.

Leeson, H. S. and O. Theodor. 1948. Mosquitos of Socotra. Bulletin of Entomological Research 39:221–229.

Legrand, J. 1985. Additions to the Odonata fauna from Mount Nimba (Ivory Coast). Revue Francaise d'Entomologie (Nouvelle Serie) 7(1):37–38.

Lehner, B., T. Henrichs, P. Döll, and J. Alcamo. 2001. EuroWasser: model-based assessment of European water resources and hydrology in the face of global change. Kassel World Water Series 5. Center for Environmental Systems Research, University of Kassel, Kassel, Germany.

Leonard, P. M. 2001. Zambia. Pages 1005–1024 in L. D. C. Fishpool and M. I. Evans, editors. Important bird areas in Africa and associated islands: priority sites for conservation. Pisces Publications and BirdLife International, Newbury and Cambridge, UK.

Lesotho Biodiversity Trust. 2002. Maloti minnow policy and action plan. Retrieved 2003 from the World Wide Web: http://www.lhwp.org.ls/downloads/policies/Maloti%20Minnow%20Policy%20and%20Action%20Plan.doc.

Lesotho Government. 2000. Biological diversity in Lesotho: a country study. National Environment Secretariat, Maseru, Lesotho.

Lesotho Highlands Water Project. 2002. Overview of Lesotho Highlands Water Project. Retrieved 2003 from the World Wide Web: http://www.lhwp.org.ls/overview/default.htm.

Letouzey, R. 1985. Notice de la carte phytogéographique du Cameroun au 1: 500 000: region Afro-montagnarde et etage submontagnard. Institute de la Carte Internationale de la Végétation, Toulouse, France.

Lévêque, C. 1990. Relict tropical fish fauna in central Sahara. Ichthyological Explorations of Freshwaters 1(1):39–48.

Lévêque, C. 1995. River and stream ecosystems of northwestern Africa. Pages 519–536 in C. E. Cushing, K. W. Cummings, and G. W. Minshall, editors. River and stream ecosystems. Elsevier Science, Amsterdam, The Netherlands.

Lévêque, C. 1997. Biodiversity dynamics and conservation: the freshwater fish of tropical Africa. Cambridge University Press, Cambridge, UK.

Lévêque, C., M. Bruton, and G. Ssentongo, editors. 1988. Biology and ecology of African freshwater fishes. ORSTOM, Paris, France.

Lévêque, C. and D. Paugy, editors. 1999. Les poissons des eaux continentales africaines: diversité, écologie, utilisation par l'homme. IRD (ORSTOM) Editions, Paris, France.

Lévêque, C., D. Paugy, and G. G. Teugels, editors. 1990. The fresh and brackish water fishes of West Africa, Vol. 1. Coll. Faune Tropicale XXVIII. ORSTOM-MRAC, Paris, France.

Lévêque, C., D. Paugy, and G. G. Teugels. 1991. Annotated checklist of the freshwater fishes of the Nilo-Sudan river basins, in Africa. Revue d'Hydrobiologie Tropicale 24: 131–154.

Lévêque, C., D. Paugy, and G. G. Teugels, editors. 1992. The fresh and brackish water fishes of West Africa, Vol. 2. Coll. Faune Tropicale XXVIII. ORSTOM–MRAC, Paris, France.

Lévêque, C., D. Paugy, G. G. Teugels, and R. Romand. 1989. Inventaire taxonomique et distribution des poissons d'eau douce des bassins cotiers de Guinée et de Guinée Bissau. Revue d'Hydrobiologie Tropicale 22:107–27.

Library of Congress. 1993. Country studies: Zaire. Retrieved 2001 from the World Wide Web: http://lcweb2.loc.gov/frd/cs/kmtoc.html#km0006.

Library of Congress. 1994. Country studies: Seychelles. Retrieved 2001 from the World Wide Web: http://lcweb2.loc.gov/frd/cs/sctoc.html#sc0000.

Liem, K. F. 1980. Adaptive significance of intra- and interspecific differences in the feeding repertoires of cichlid fishes. American Zoologist 20:295–314.

Lincoln, R. J. and G. A. Boxshall. 1987. The Cambridge illustrated dictionary of natural history. Cambridge University Press, Cambridge, UK.

Lincoln, R. J., G. A. Boxshall, and P. F. Clark. 1982. A dictionary of ecology, evolution and systematics. Cambridge University Press, Cambridge, UK.

Linder, H. P., M. E. Meadows, and R. M. Cowling. 1992. History of the Cape flora. Pages 113–134 in R. M. Cowling, editor. The ecology of fynbos: nutrients, fire and diversity. Oxford University Press, Cape Town, South Africa.

Lippitsch, E. 1997. Phylogenetic investigations on the haplochromine Cichlidae of Lake Kivu (East Africa), based on lepidological characters. Journal of Fish Biology 51:284–299.

Livingstone, J. A. and D. M. Melack. 1984. Some lakes of sub-Saharan Africa. Pages 467–497 in F. B. Taub, editor. Lake and reservoir ecosystems. Ecosystems of the world, Vol. 23. Elsevier Science, New York, NY, USA.

Lloyd, J. W. and A. Le Roux. 1985. A conservation assessment of the Verneukpan-Copperton area. Department of Nature and Environmental Conservation, Cape Town, South Africa.

Logan, R. F. 1960. The central Namib Desert, southwest Africa, National Research Council Publication 758. National Academy of Sciences, Washington, DC, USA.

Loiselle, P. V. and M. L. J. S. Stiassny. 2003. *Bedotia*. Pages 867–868 in S. M. Goodman and J. P. Benstead, editors. The natural history of Madagascar. University of Chicago Press, Chicago, IL, USA.

Lonnberg, E. 1910. Fishes. Pages 1–8 in Y. Sjöstedt, editor. Wissenschaftliche Ergebnisse der schwedischen zoologischen Expedition nach dem Kilimanjaro, dem Meru und den umgebenden Massaisteppen Deutsch-Ostafrikas 1905–1906. P. Palmquist Aktiebolag, Stockholm, Sweden.

Lopes, S. 2001. Socio-economic aspects of the Mozambican coast. WWF Southern African Regional Programme Office, Harare, Zimbabwe.

Lopez, S. M. 2002. Papyrus conservation around Lake Naivasha: development of alternative management schemes in Kenya. Master's thesis. International Institute for Geo-information Science and Earth Observation, ITC, Enschede, The Netherlands.

Loth, P. E. and H. H. T. Prins. 1986. Spatial patterns of the landscape and vegetation of Lake Manyara National Park, Tanzania. ITC Journal 2:115–130.

Lotspeich, F. B., and W. S. Platts. 1982. An integrated land-aquatic classification system. North American Journal of Fisheries Management 2:138–149.

Louette, M., editor. 1999. La faune terrestre de Mayotte, Annales du Musée Royal d'Afrique Centrale (Sciences Zoologiques) 284. Musée Royal de l'Afrique Centrale, Tervuren, Belgium.

Lourenço, W. R. 1996. Biogeography of Madagascar. ORSTOM, Paris, France.

Loutit, R. 1991. Western flowing ephemeral rivers and their importance to wetlands in Namibia. Madoqua 17(2):135–140.

Lovegrove, B. 1993. The living deserts of southern Africa. Fernwood Press, Vlaeberg, South Africa.

Lovejoy, T. E. 1997. Biodiversity: what is it? Pages 7–14 in M. L. Reaka-Kudla, D. E. Wilson, and E. O. Wilson, editors. Biodiversity II: understanding and protecting our biological resources. National Academy of Sciences, Washington, DC, USA.

Lovett, J. C., J. Hatton, L. B. Mwasumbi, and J. H. Gerstle. 1997. Assessment of the impact of the lower Kihansi hydropower project on the forests of Kihansi Gorge, Tanzania. Biodiversity and Conservation 6:915–933.

Lovett, J. C. and S. K. Wasser, editors. 1993. Biogeography and ecology of the rain forests of eastern Africa. Cambridge University Press, Cambridge, UK.

Low, A. B. and A. G. Rebelo. 1998. Vegetation of South Africa,

Lesotho and Swaziland. Department of Environmental Affairs and Tourism, Cape Town, South Africa.

Lowe, R. H. 1955. New species of *Tilapia* (Pisces, Cichlidae) from Lake Jipe and the Pangani River, East Africa. The British Museums of Natural History 2(12):350–368.

Lowe, R. H. 1992. Nigeria. Pages 230–239 in J. A. Sayer, C. S. Harcourt, and N. M. Collins, editors. The conservation atlas of tropical forests: Africa. IUCN, London, UK.

Lowe-McConnell, R. H. 1985. The biology of the river systems with particular reference to the fishes. Pages 101–140 in A. T. Grove, editor. The Niger and its neighbors. A. A. Balkema, Rotterdam, The Netherlands.

Lowe-McConnell, R. H. 1987. Ecological studies in tropical fish communities. Cambridge University Press, Cambridge, UK.

Lowe-McConnell, R. H. 1993. Fish faunas of the African Great Lakes: origins, diversity and vulnerability. Conservation Biology 7:634–643.

Lowe-McConnell, R. H. 1996. Fish communities in the African Great Lakes. Environmental Biology of Fishes 45(3):219–235.

Lowry, P. P. I., G. E. Schatz, and P. B. Phillipson. 1997. The classification of natural and anthropogenic vegetation in Madagascar. Pages 92–123 in S. M. Goodman and B. D. Patterson, editors. Natural change and human impact in Madagascar. Smithsonian Institution Press, Washington, DC, USA.

Lubini, A. 1997. La végétation de la réserve de biosphère de Luki au Mayombe (Zaïre). Opera Botanica Belgica 10. Jardin Botanique National de Belgique, Meise, Belgium.

Lundberg, J. G., M. Kottelat, G. R. Smith, M. L. J. Stiassny, and A. C. Gill. 2000. So many fishes, so little time: an overview of recent ichthyological discovery in continental waters. Annals of the Missouri Botanical Garden 87(1):26–62.

MacKinnon, J. and K. MacKinnon. 1986. Review of the protected areas system in the Afrotropical realm. IUCN, Gland, Switzerland and Cambridge, UK.

Magadza, C. 2000. Human impacts on wetland biodiversity in the Zambezi Basin. Pages 107–122 in J. Timberlake, editor. Biodiversity of the Zambezi Basin wetlands. Biodiversity Foundation for Africa, Bulawayo/The Zambezi Society, Harare, Zimbabwe.

Magadza, C. H. D. 2003. Water resources management and water quality monitoring in an African setting. Readout 27(Guest Forum):1–13.

Maganga, F., J. Butterworth, and P. Moriarty. 2001. Domestic water supply, competition for water resources and IWRM in Tanzania: a review and discussion paper. Patrick Paper prepared for 2nd WARFA/Waternet Symposium: Integrated Water Resources Management: Theory, Practice, Cases; 30–31 October 2001, Cape Town, South Africa. Retrieved 2001 from the World Wide Web: http://www.nri.org/WSS-IWRM/Reports/Maganga.doc.

Magin, G. 2001a. Djibouti. Pages 233–239 in L. D. C. Fishpool and M. I. Evans, editors. Important bird areas in Africa and associated islands: priority sites for conservation. Pisces Publications and BirdLife International, Newbury and Cambridge, UK.

Magin, C. 2001b. Morocco. Pages 603–626 in L. D. C. Fishpool and M. I. Evans, editors. Important bird areas in Africa and associated islands: priority sites for conservation. Pisces Publications and BirdLife International (BirdLife Conservation Series No. 11), Newbury and Cambridge, UK.

Malaisse, F. and I. Kapinga. 1986. The influence of deforestation on the water balance of soils in the Lubumbashi region (Shaba, Zaire). Bulletin de la Société Royale de Botanique de Belgique 119:161–178.

Malherbe, S. 1984. The geology of the Kalahari Gemsbok National Park. Supplement to Koedoe 1984:33–44.

Malicky, H. 1989. Caddisflies (Insecta: Trichoptera) from the Seychelles, Comoro and Mascarene Islands. Annalen des Naturhistorischen Museums in Wien Serie B Botanik und Zoologie 93:143–160.

Malicky, H. 1999. Einige Köcherfliegen von der Insel Sokotra (Insecta, Trichoptera). Entomologische Zeitschrift mit Insektenbörse 109:12.

Malmqvist, B., C. Meisch, and A. N. Nilsson. 1997. Distribution patterns of freshwater Ostracoda (Crustacea) in the Canary Islands with regards to habitat use and biogeography. Hydrobiologia 347:159–170.

Malmqvist, B., A. N. Nilsson, and M. Báez. 1995. Tenerife's freshwater macroinvertebrates: status and threats (Canary Islands, Spain). Aquatic Conservation: Marine and Freshwater Ecosystems 5(139):1–24.

Malmqvist, B., A. N. Nilsson, M. Báez, P. D. Armitage, and J. Blackburn. 1993. Stream macroinvertebrate communities in the island of Tenerife. Archiv für Hydrobiologie 128(2):209–235.

Malmqvist, B. and S. Rundle. 2002. Threats to the running water ecosystems of the world. Environmental Conservation 29(2):134–153.

Mamonekene, V. and G. G. Teugels. 1993. Faune des poissons d'eaux douces de la Reserve de la Biosphere de Dimonika (Mayombe, Congo). Musée Royal de l'Afrique Centrale, Tervuren, Belgium.

Manconi, R., T. Cubeddu, and R. Pronzato. 1999. African freshwater sponges: *Makedia tanensis* gen. et sp. Nov. From Lake Tana, Ethiopia. Memoirs of the Queensland Museum 44:361–367.

Mandahl-Barth, G., C. Ripert, and C. Raccurt. 1974. Nature du sous-sol, repartition des mollusques dulcaquicoles et foyers de bilharzioses intestinal et urinaire au Bas-Zaire. Revue de Zoologie Africaine 88:553–584.

Manongi, F. J. 1993. River basin planning and management of wetlands. Pages 103–113 in G. L. Kamukala and S. A. Crafter, editors. Wetlands of Tanzania: proceedings of a seminar on wetlands of Tanzania, Morogoro, Tanzania, 27–29 November, 1991. IUCN, Gland, Switzerland.

Marais, E. and J. Irish. 1997. Cave investigations in Namibia IV. Aikab hemicenote, and other karst phenomena in the Etosha National Park. Madoqua 20(1):81–90.

Marlier, G. 1973. Limnology of the Congo and Amazon rivers. Pages 223–238 in B. J. Meggers, E. S. Ayensu, and W. D. Duckworth, editors. Tropical forest ecosystems in Africa and South America: a comparative review. Smithsonian Institution Press, Washington, DC, USA.

Marshall, B. E. 1994. Ecology and management of the Manyame lakes. Pages 55–68 in T. Matiza and S. A. Crafter, editors. Wetlands ecology and priorities for conservation in Zimbabwe.

Proceedings of a seminar on wetlands of Zimbabwe. IUCN, Gland, Switzerland.

Marshall, B. E. 2000a. Fishes of the Zambezi Basin. Pages 393–460 in J. Timberlake, editor. Biodiversity of the Zambezi Basin wetlands. Biodiversity Foundation for Africa, Bulawayo/The Zambezi Society, Harare, Zimbabwe.

Marshall, B. E. 2000b. Review of aquatic invertebrates of the Zambezi basin. Pages 613–652 in J. Timberlake, editor. Biodiversity of the Zambezi Basin wetlands. Biodiversity Foundation for Africa, Bulawayo/The Zambezi Society, Harare, Zimbabwe.

Marshall, B. E. In prep. Fish biology, fisheries and limnology in Zimbabwe: an introduction to the literature.

Marshall, B. E. and B. Gratwicke. 1998–1999. The barred minnows (Teleostei: Cyprinidae) of Zimbabwe: is there cause for concern? Southern African Journal of Aquatic Sciences 24:157–161.

Martin, C. 1991. The rainforests of West Africa. Birkhäuser Verlag, Basel, Switzerland.

Masundire, H. M. 1994. Effects of dam building on riverine wetlands. Pages 87–102 in T. Matiza and S. A. Crafter, editors. Wetlands ecology and priorities for conservation in Zimbabwe. Proceedings of a seminar on the wetlands of Zimbabwe. IUCN, Gland, Switzerland.

Matiza, T. 1994a. Foreword. Page 1 in T. Matiza and S. A. Crafter, editors. Wetlands ecology and priorities for conservation in Zimbabwe. Proceedings of a seminar on the wetlands of Zimbabwe. IUCN, Gland, Switzerland.

Matiza, T. 1994b. Wetlands in Zimbabwe: an overview. Pages 3–20 in T. Matiza and S. A. Crafter, editors. Wetlands ecology and priorities for conservation in Zimbabwe. Proceedings of a seminar on the wetlands of Zimbabwe. IUCN, Gland, Switzerland.

Mato, R. R. A. M. 1999. Environmental implications involving the establishment of sanitary landfills in five municipalities in Tanzania: the case of Tanga municipality. Resources Conservation and Recycling 25:1–16.

Matoussi, M. S. 1996. Sources of strain and alternatives for relief in the most stressed water systems of North Africa. In E. Rached, E. Rathgeber, and D. B. Brooks, editors. Water management in Africa and the Middle East: challenges and opportunities. International Development Research Centre, Ottawa, Canada.

Matthes, H. 1964. Les poissons du Lac Tumba et de la region d'Ikela. Annales du Musée Royal d'Afrique Centrale (Sciences Zoologiques) 126:1–204.

Matthes, H. 1993. Rapport préliminaire de la mission d'évaluation de la pêche continentale et de l'aquaculture en République de Guinée. FAO/SEP, Rome, Italy.

Matthews, W. J. 1998. Patterns in freshwater fish ecology. Chapman Hall, New York, NY, USA.

Matthews, W. J. and H. W. Robison. 1998. Influence of drainage connectivity, drainage area and regional species richness on fishes of the interior highlands in Arkansas. American Midland Naturalist 139:1–19.

Mattingly, P. F. and K. L. Knight. 1956. The mosquitos of Arabia. Bulletin of the British Museum (Natural History) Entomology 4(3):91–141.

Maud, R. R. 1980. The climate and geology of Maputaland. Pages 1–7 in M. N. Bruton and K. H. Cooper, editors. Studies on the ecology of Maputaland. Cape and Transvaal Printers, Cape Town, South Africa.

Mauritius National Parks and Conservation Service. 2000. Republic of Mauritius: first national report to the Convention on Biodiversity. Retrieved 2001 from the World Wide Web: http://www.biodiv.org/doc/world/mu/mu-nr-01-en.pdf.

Maxwell, J. R., C. J. Edwards, M. E. Jensen, S. J. Paustian, H. Parrott, and D. M. Hill. 1995. A hierarchical framework of aquatic ecological units in North America (Nearctic zone). U.S. Forest Service General Technical Report NC-176. St. Paul, MN, USA.

Mayekiso, M. and T. Hecht. 1988. Conservation status of the anabantid fish, *Sandelia bainsii*, in the Tyume River, South Africa. South African Journal of Wildlife Research 18(3):101–108.

McAllister, D. E., A. L. Hamilton, and B. Harvey. 1997. Global freshwater biodiversity: striving for the integrity of freshwater ecosystems. Sea Wind 11(3):1–140.

McClanahan, T. R. and T. P. Young, editors. 1996. East African ecosystems and their conservation. Oxford University Press, New York, NY, USA.

McCullum, J. 1994. Freshwater resources. Pages 181–205 in M. Chenje and P. Johnson, editors. State of the environment in southern Africa. Southern African Research and Documentation Centre, Harare, Zimbabwe.

McDowall, R. M. 1990. When galaxiid and salmonid fishes meet a family reunion in New Zealand. Journal of Fish Biology 37(Suppl. A):35–43.

McDowall, R. M. 1996. Introduction. Pages 9–14 in R. M. McDowall, editor. Freshwater fishes of southeastern Australia. A. H. & A. W. Reed, Sydney, Australia.

McLachlan, A. J. 1979. The aquatic environment: I. Chemical and physical characteristics of Lake Chilwa. Pages 59–78 in M. Kalk, A. J. McLachlan, and C. Howard-Williams, editors. Lake Chilwa: studies of change in a tropical ecosystem. Monographiae Biologicae, 35. Dr. W. Junk, The Hague, The Netherlands.

McNeely, J. A., H. A. Mooney, L. E. Neville, P. Schei, and J. K. Waage, editors. 2001. A global strategy on invasive alien species. IUCN in collaboration with GISP, Gland, Switzerland and Cambridge, UK.

Meditz, S. W. and T. Merrill, editors. 1993. Zaire: a country study. Retrieved 2001 from the World Wide Web: http://memory.loc.gov/frd/cs/zrtoc.html.

MedWet. 2004. Mediterranean Wetlands Initiative (MedWet). Retrieved 2004 from the World Wide Web: http://www.medwet.org/.

Mega-Cities. Mega-Cities: innovations for urban life, global network, Lagos, Nigeria. Retrieved 2001 from the World Wide Web: http://www.megacities.org/network/lagos.asp.

Meininger, P. L. and U. G. Sorensen. 1993. Egypt as a major wintering area of little gulls. British Birds 86:407–410.

Mendelsohn, J., S. el Obeid, and C. Roberts. 2000. A profile of north-central Namibia. Gamsberg Macmillan Publishers, Windhoek, Namibia.

Mengistou, S. and C. H. Fernando. 1991a. Biomass and production of the major dominant crustacean zooplankton in a tropical rift valley lake, Awassa, Ethiopia. Journal of Plankton Research 13:831–851.

Mengistou, S. and C. H. Fernando. 1991b. Seasonality and abundance of the dominant crustacean zooplankton in a tropical rift valley lake, Awassa, Ethiopia. Hydrobiologia 226:137–152.

Mengistou, S., J. Green, and C. H. Fernando. 1991. Species composition, distribution and seasonal dynamics of Rotifera in a rift valley lake in Ethiopia (Lake Awassa). Hydrobiologia 209:203–214.

Merron, G. S. and M. N. Bruton. 1995. Community ecology and conservation of the fishes of the Okavango Delta, Botswana. Environmental Biology of Fishes 43:109–119.

Meyer, A. 1993. Phylogenetic relationships and evolutionary processes in East African cichlid fishes. Trends in Ecology and Evolution 8(8):279–284.

Meyer, A., T. D. Kocher, P. Basasibwaki, and A. C. Wilson. 1990. Monophyletic origin of Lake Victoria cichlid fishes suggested by mitochondrial DNA sequences. Nature 347:550–553.

Midgley, D. C., W. V. Pitman, and B. J. Middleton. 1994. Surface water resources of South Africa. Report 1990.298/3.1/94. Water Research Commission, Pretoria, South Africa.

Mies, B. 2001. Flora und vegetationsökologie der insel Soqotra. Essener Ökologische Schriften 15.

Mies, B. A. and F. E. Beyhl. 1998. The vegetation ecology of Soqotra. Pages 35–81 in H. J. Dumont, editor. Proceedings of the first international symposium on Soqotra Island: present and future. United Nations Publications, New York, NY, USA.

Miller, A. G. and M. Bazara'a. 1998. Conservation status of the flora of the Soqotran Archipelago. Pages 15–34 in H. G. Dumont, editor. Proceedings of the first international symposium on Soqotra Island: present and future. United Nations Publications, New York, NY, USA.

Miller, A. G. and M. Morris. 2002. Final report on the conservation and sustainable use of the biodiversity of Soqotra archipelago. Report for the Global Environment Facility/UNDP Project. Royal Botanic Garden, Edinburgh, UK.

Miller, R. M. 1997. The Owambo Basin of northern Namibia. Pages 237–268 in R. C. Selley, editor. African basins. Sedimentary basins of the World. Elsevier, Amsterdam, The Netherlands.

Mina, M. and A. Golubtsov. 1995. Faunas of isolated regions as principal units in the conservation of freshwater fishes. American Fisheries Society Symposium 17(1995):145–148.

Mina, M. V., A. N. Mironovsky, and Y. Y. Dgebuadze. 1996. Lake Tana large barbs: phenetics, growth and diversification. Journal of Fish Biology 48(3):383–404.

Mino-Kahozi, K. and M. Mbantshi. 1997. Pollution and degradation of African aquatic environments and consequences for inland fisheries and aquaculture: the case of the Republic of Zaire. Pages 99–115 in K. Remane, editor. African inland fisheries, aquaculture and the environment. Fishing News Books, Osney Mead, Oxford, UK.

Miracle, M. P. 1973. The Congo Basin as a habitat for man. Pages 335–344 in B. J. Meggers, E. S. Ayensu, and W. D. Duckworth, editors. Tropical forest ecosystems in Africa and South America: a comparative review. Smithsonian Institution Press, Washington, DC, USA.

Mittermeier, R. A., N. Myers, P. R. Gil, and C. G. Mittermeier. 1999. Hotspots: Earth's biologically richest and most endangered terrestrial ecoregions. CEMEX, Monterrey, Mexico.

Mittermeier, R. A., N. Myers, J. B. Thomsen, G. A. B. Da Fonseca, and S. Olivieri. 1998. Biodiversity hotspots and major tropical wilderness areas: approaches to setting conservation priorities. Conservation Biology 12(3):516–520.

Mkanda, F. X. 2002. Contribution by farmers' survival strategies to soil erosion in the Linthipe River Catchment: implications for biodiversity conservation in Lake Malawi/Nyasa. Biodiversity and Conservation 11(8):1327–1359.

Mkuula, S. 1993. Pollution of wetlands in Tanzania. Pages 85–93 in G. L. Kamukala and S. A. Crafter, editors. Wetlands of Tanzania: proceedings of a seminar on wetlands of Tanzania. IUCN, Gland, Switzerland.

Mkwanda, R. 1994. The Mazowe River impoundments: ecology, utilisation and management status. Pages 69–72 in T. Matiza and S. A. Crafter, editors. Wetlands ecology and priorities for conservation in Zimbabwe. Proceedings of a seminar on wetlands of Zimbabwe. IUCN, Gland, Switzerland.

Moffat, D. and O. Linden. 1995. Perception and reality: assessing priorities for sustainable development in the Niger River Delta. Ambio 24:527–538.

Moguedet, G., P. Giresse, and Kinga-Mouzeo. 1990. Study of the Chad Basin water supply by catchment carried out in the Congo-Zaire Basin. Pages 523–539 in R. Paepe, R. W. Fairbridge, and S. Jelgersma, editors. Greenhouse effect, sea level, and drought. Kluwer Academic Publishers, Dordrecht, The Netherlands.

Mohr, P. A. 1961. The geology, structure, and origin of the Bishoftu explosion craters, Shoa, Ethiopia. Bulletin of the Geophysical Observatory, University College, Addis Ababa 2:65–101.

Montaggioni, L. and P. Nativel. 1988. La Réunion, Ile Maurice, géologie et aperçus biologiques. Masson, Paris, France.

Moore, A. E. 1999. A reappraisal of epeirogenic flexure axes in southern Africa. South African Journal of Geology 102:363–376.

Moore, A. E. and P. A. Larkin. 2001. Drainage evolution in south-central Africa since the breakup of Gondwana. South African Journal of Geology 104:47–68.

Moorehead, R. 1991. Structural chaos: community and state management of common property in Mali. Ph.D. thesis. University of Sussex, Sussex, UK.

Morjan, M. D., B. B. Nicholas, and L. I. Ojok. 2001. Report on the impact of conflict on the Boma National Park. New Sudan Wildlife Society, Nairobi, Kenya.

Mosepele, K. 2000. Preliminary length based stock assessment of the main exploited stocks of the Okavango delta fishery. M.Ph. thesis. Department of Fisheries and Marine Biology, University of Bergen, Norway.

Mosepele, K. and J. Kolding. 2002. Fish stock assessment in the Okavango Delta: preliminary results from a length based analysis. Pages 363–390 in T. Bernad, K. Mosepele, and L. Ramberg, editors. Proceedings of the "Environmental

Monitoring of Tropical and Subtropical Wetlands" conference. Maun, Botswana.

Moss, B. 1979. Algae in Lake Chilwa and the waters of its catchment area. Pages 93–103 in M. Kalk, A. J. McLachlan, and C. Howard-Williams, editors. Lake Chilwa: studies of change in a tropical ecosystem. Monographiae Biologicae, 35. Dr. W. Junk, The Hague, The Netherlands.

Moss, B. 1998. Ecology of fresh waters. Blackwell Science, Liverpool, UK.

Mouelhi, S., M. M. Kraiem, and J. Gagneur. 1998. Biological assessment of the water quality of the Beja Wadi (northern Tunisia). Bulletin de la Société d'Histoire Naturelle de Toulouse 134:33–45.

Moyle, P. B. and J. J. Cech Jr. 1996. Fishes: an introduction to ichthyology, 3rd ed. Prentice Hall, Englewood Cliffs, NJ, USA.

Mtetwe, S. 2000. Establishment of biomonitoring reference sites for Zimbabwe: a tool for effective integrated catchment management. Proceeding of the 1st WARFSA/WaterNet Symposium: Sustainable Use of Water Resources; Maputo, 1–2 November 2000, Zimbabwe.

Mugabe, J. and G. W. Tumushabe. 1999. Environmental governance: conceptual and emerging issues. H. W. O. Okoth-Ogendo and G. W. Tumushabe, editors. Governing the environment: political change and natural resources management in eastern and southern Africa. African Centre for Technology Studies, Nairobi, Kenya.

Müller, T., I. Mapaure, and R. Drummond. 2000. Zambezi Delta wetland plant survey. Pages 129–144 in J. Timberlake, editor. Biodiversity of the Zambezi Basin wetlands. Biodiversity Foundation for Africa, Bulawayo/The Zambezi Society, Harare.

Mundy, P. 2000. Wetland birds of the Zambezi Basin. Pages 213–278 in J. Timberlake, editor. Biodiversity of the Zambezi Basin wetlands. Biodiversity Foundation for Africa, Bulawayo/The Zambezi Society, Harare, Zimbabwe.

Munro, I. S. R. 1967. The fishes of New Guinea. Victor C. N. Blight, Sydney, Australia.

Murakami, M. 1995. Managing water for peace in the Middle East: alternative strategies. United Nations University Press, Tokyo, Japan.

Muruthi, P. 2004. African Heartlands: a science-based and pragmatic approach to landscape-level conservation in Africa. In N. Burgess, J. D'Amico Hales, E. Underwood, E. Dinerstein, D. Olson, I. Itoua, J. Schipper, T. Ricketts, and K. Newman, editors. Terrestrial ecoregions of Africa and Madagascar: a conservation assessment. Island Press, Washington DC, USA.

Mutambwe-Shango. 1984. Contribution à l'étude de l'écologie de la rivière Luki (sous-affluent du fleuve Zaïre): bassin versant-poissons. Université Paul Sabatier de Toulouse (Sciences), Toulouse, France.

Mutanga, J. G., W. Mwatha, O. Nasirwa, and N. Gichuru. 2000. Status and trends in the biodiversity of Eastern Rift Valley lakes in Kenya. Pages 6–34 in L. A. Bennun, H. Ndede, and N. N. Gichuki, editors. Conservation and sustainable use of biodiversity in Eastern Rift Valley lakes, Kenya. GEF/UNDP Consultancy Report. National Museums of Kenya, Nairobi, Kenya.

Mwalyosi, R. B. B. 1991. Vegetation survey for Lake Manyara. Institute of Resource Assessment, University of Dar es Salaam, Dar es Salaam, Tanzania.

Myers, G. S. 1938. Freshwater fishes and West Indian zoogeography. Annual Report of the Smithsonian Institution 1937: 339–364.

Myers, G. S. 1949. Salt-tolerance of fresh-water fish groups in relation to zoogeographical problems. Bijdragen tot de Dierkunde 28:315–322.

Myers, N. 1988. Threatened biotas: "hotspots" in tropical forests. Environmentalist 10:243–256.

Myers, N., R. A. Mittermeier, C. G. Mittermeier, G. A. B. da Fonseca, and J. Kent. 2000. Biodiversity hotspots for conservation priorities. Nature 403:853–858.

Nagelkerke, L. 1997. The barbs of Lake Tana, Ethiopia: morphological diversity and its implications for taxonomy, trophic resource partitioning, and fisheries. Special edition, Agricultural University, Wageningen, The Netherlands.

Nagelkerke, L. A., M. V. Mina, T. Wudneh, F. A. Sibbing, and J. W. M. Osse. 1995. In Lake Tana, a unique fish fauna needs protection. BioScience 45(11):772–775.

Nagelkerke, L. A. J. and F. A. Sibbing. 1996. Reproductive segregation among the *Barbus intermedius* complex of Lake Tana, Ethiopia. An example of intralacustrine speciation? Journal of Fish Biology 49:1244–1266.

Nagelkerke, L. A. J. and F. A. Sibbing. 1998. The *"Barbus" intermedius* species flock of Lake Tana (Ethiopia): I. the ecological and evolutionary significance of morphological diversity. Italian Journal of Zoology 65(suppl.):3–7.

Nagelkerke, L. A. J. and F. A. Sibbing. 2000. A revision of the large barbs (*Barbus* spp., Cyprinidae, Teleostei) of Lake Tana (Ethiopia), with a description of a new species, *Barbus osseensis*. Netherlands Journal of Zoology 50:179–214.

NASA. 2001. A shadow of a lake: Africa's disappearing Lake Chad. Retrieved 2003 from the World Wide Web: http://www.gsfc.nasa.gov/topstory/20010227lakechad.html.

Nasirwa, O. 2000. Conservation status of flamingos in Kenya. Waterbirds 23:47–51.

National Agricultural Research Organization. 2002. Experiences with managing water hyacinth infestation in Uganda. Fisheries Resources Research Institute, NARO, Jinja, Uganda.

Ndede, H., C. Situma, S. Gitau, P. Makenzi, and R. Ndetei. 2000. Impacts of human activities on landscape and natural resources of the Great Rift Valley Lakes: conservation and sustainable use of biodiversity in the Eastern Rift Valley lakes of Kenya. National Museums of Kenya, Nairobi, Kenya.

Neiland, A. E., C. Béné, E. Bennett, J. Turpie, C. K. Chong, A. Thorpe, M. Ahmed, R. A. Valmonte-Santos and H. Balasubramarian. 2004. River fisheries valuation: a global synthesis and critical review. WorldFish Center and Comprehensive Assessment of Water in Agriculture, Penang, Malaysia.

Neiland, A. E. and C. Béné. 2004. Review of river fisheries valuation in West and Central Africa. In A. Neiland,C. Béné, E. Bennett, J. Turpie, C. K. Chong, A. Thorpe, M. Ahmed, R. A. Valmonte-Santos and H. Balasubramarian. River fisheries valuation: a global synthesis and critical review. WorldFish Center and Comprehensive Assessment of Water in Agriculture, Penang, Malaysia.

Neiland, A. E., S. Jaffry, B. M. Ladu, M. T. Sarch, and S. P. Madakan. 2000. Inland fisheries of North East Nigeria including the Upper River Benue, Lake Chad and the Nguru-Gashua wetlands. Characterisation and analysis of planning suppositions. Fisheries Research 48:229–243.

NEMA. 1999. State of the environment report for Uganda. National Environment Management Authority, Kampala, Uganda.

NEPAD. 2002. The environment initiative. Retrieved 2003 from the World Wide Web: http://www.avmedia.at/nepad/indexgb.html.

NEPAD. 2003. Action plan of the New Partnership for Africa's Development (NEPAD). Retrieved 2003 from the World Wide Web: http://www.unep.org/gef/Documents/Brochures/radBC8B1.doc.

NEPAD and UNEP. 2003. Development of an action plan for the Environment Initiative of the New Partnership for Africa's Development (NEPAD). Programme Area 3. Prevention, control and management of invasive alien species. UNEP, AMCEN, GEF, NEPAD. UNEP, Nairobi, Kenya.

Nevada Division of Water Planning. 1998. Water words dictionary. Retrieved 2003 from the World Wide Web: http://www.state.nv.us/cnr/ndwp/dict-1/waterwds.htm.

Newmark, W. D. 1998. Forest area, fragmentation, and loss in the Eastern Arc Mountains: implications for the conservation of biodiversity. Journal of East African Natural History 87:29–36.

Newton, S. F., B. Haddane, E. H. Ali, G. Atta, M. Y. Al Salam, L. I. Ojok, and E. Yohannes. 1996. North and northeast African crane and wetland action plan. Pages 619–622 in R. D. Beifuss, W. R. Tarboton, and N. N. Gichuki, editors. Proceedings of the African crane and wetland training workshop. Wildlife Training Institute and International Crane Foundation, Botswana and Baraboo, WI, USA.

Nichols, J. T. and R. Boulton. 1927. Three new minnows of the genus *Barbus,* and a new characin from the Vernay Angola Expedition. American Museum Novitates 264:1–8.

Nicholson, S. E. 1982. Sub-Saharan rainfall in the years 1976–1980: evidence of continued drought. Unpublished master's thesis. Clark University, Worcester, MA, USA.

Nile Basin Initiative. 2003. Nile Basin Initiative Shared Vision Program. Retrieved 2003 from the World Wide Web: http://www.nilebasin.org.

Nilsson, A. N., B. Malmqvist, M. Báez, J. H. Blackburn, and P. D. Armitage. 1998. Stream insects and gastropods in the island of Gran Canaria (Spain). Annales de Limnologie 34(4):413–435.

Nishida, M. 1997. Phylogenetic relationships and evolution of Tanganyikan cichlids: a molecular perspective. Pages 1–23 in H. Kwanabe, M. Hori, and N. Makoto, editors. Fish communities in Lake Tanganyika. Kyoto University Press, Kyoto, Japan.

Njuguna, S. G. 1984. Aquatic plant life in Kenya: status and perspectives for future research. Endangered resources for development: proceedings of a workshop on the status and options for management of plant communities in Kenya. National Museums of Kenya, Nairobi, Kenya.

Nkotagu, H. H. Unpublished project proposal. The Malagarasi wetland environmental project (MWEP): human impacts on the Malagarasi wetland. University of Dar Es-Salaam, Tanzania.

Noble, R. G. and J. Hemens. 1978. Inland water ecosystems in South Africa: a review of research needs. South African National Scientific Programmes Report No. 34. CSIR, Pretoria, South Africa.

Norrgren, L., U. Pettersson, S. Orn, and P.-A. Bergqvist. 2000. Environmental monitoring of the Kafue River, located in the copperbelt, Zambia. Archives of Environmental Contamination and Toxicology 38:334–341.

Norris, S. M. and G. G. Teugels. 1990. A new species of *Ctenopoma* (Pisces, Anabantidae) from southeastern Nigeria. Copeia 2:492–499.

Noss, R. F. 1992. The Wildlands Project: land conservation strategy. Wild Earth Special Issue: The Wildlands Project 10–25.

Noss, R. F. and A. Y. Cooperrider. 1994. Saving nature's legacy: protecting and restoring biodiversity. Island Press, Washington, DC, USA.

Nyman, L. 1991. Conservation of freshwater fish: protection of biodiversity and genetic variability in aquatic ecosystems. SWEDMAR, Göteborg, Sweden.

Oak Ridge National Laboratory. 2002. LandScan Global Population 2002 Database. Retrieved 2003 from the World Wide Web: http://www.ornl.gov/sci/gist/landscan/.

O'Connor, C., M. Marvier, and P. Kareiva. 2003. Biological versus social, economic, and political priority setting in conservation. Ecology Letters 6:706–711.

Odusile, W. 2001. Battling the weeds of destruction. Retrieved 2001 from the World Wide Web: http://www.thisdayonline.com/archive/2001/03/26/20010326fea01.html.

Ogutu-Ohwayo, R. 1990. The decline of the native fishes of Lakes Victoria and Kyoga (East Africa) and the impact of introduced species, especially the Nile perch, *Lates niloticus,* and the Nile tilapia, *Oreochromis niloticus.* Environmental Biology of Fishes 27:81–96.

O'Hagan, T. 1989. Southern African wildlife. The Reader's Digest Association, Cape Town, South Africa.

Oil Spill Intelligence Report. 1999. Largest tanker spills: international oil spill statistics. Retrieved 2003 from the World Wide Web: http://www.coltoncompany.com/shipping/statistics/spillstanker.htm.

Ojany, F. F. and R. B. Ogendo. 1973. Kenya: a study in physical and human geography. Longman Kenya, Nairobi, Kenya.

Okedi, J., F. Chale, and P. Basasibwaki. 1974. The Kagera Rivers: preliminary observation units on fishery and limnology in EAFFRO annual report. East Africa Freshwater Fisheries Research Organization (EAFFRO), Jinja, Uganda.

O'Keeffe, J. H., D. B. Danilewitz, and J. A. Bradshaw. 1987. An "expert system" approach to the assessment of the conservation status of rivers. Biological Conservation 40:69–84.

O'Keeffe, J. H., B. R. Davies, J. M. King, and P. H. Skelton. 1989. The conservation status of southern African rivers. Pages 226–289 in B. J. Huntley, editor. Biotic diversity in southern Africa. Oxford University Press, Cape Town, South Africa.

O'Keeffe, J. H., M. Uys, and M. N. Bruton. 1992. Freshwater systems. Pages 277–315 in R. F. Fuggle and M. A. Rabie, editors. Environmental management in South Africa. Juta & Co., Cape Town, South Africa.

Olaleye, V. F., E. A. Akintunde, and O. A. Akinyemiju. 1993. Effect of a herbicidal control of water hyacinth (*Eichhornia crassipes*) on fish composition and abundance in the Kofawei Creek, Ondo State, Nigeria. Journal of Environmental Management 38(2):85–97.

Olivier, K. 2001. The ornamental fish market. FAO/GLOBEFISH Research Program 67. FAO, Rome, Italy.

Olivry, J. C., A. Chouret, G. Vuillaume, J. Lemoalle, and J. P. Bricquet. 1996. Hydrologie du Lac Tchad, Monographie Hydrologique, 12. ORSTOM, Paris, France.

Olson, D. M. and E. Dinerstein. 1998. The Global 200: a representation approach to conserving the earth's most biologically valuable ecoregions. Conservation Biology 12(3):502–515.

Olson, D. M., E. Dinerstein, P. Canevari, I. Davidson, G. Castro, V. Morisset, R. Abell, and E. Toledo, editors. 1998. Freshwater biodiversity of Latin America and the Caribbean: a conservation assessment. Biodiversity Support Program, Washington, DC, USA.

Omer-Cooper, J. 1930. Dr. Hugh Scott's expedition to Abyssinia: a preliminary investigation of the freshwater fauna of Abyssinia. Proceedings of the Zoological Society of London Part I:195–207.

Ondieki, C. M., editor. 2000. Impact of climatic and hydrological changes on biodiversity of the Kenyan Rift Valley Lakes, GEF Rift Valley Lakes project. National Museums of Kenya, Nairobi, Kenya.

Onu, N. C. H. 2003. The oil rich Niger Delta region: a framework for improved performance of the Nigerian regulatory process. Ambio 32(4):325–326.

Orange, D. 1992. Hydroclimatology of the Fouta Djalon massif and present dynamics of an old lateritic landscape. Science Geologiques 93.

Orange, D. 1998. Matter transports in humid tropical watershed in a forested area: the Uele River in Zaïre. Retrieved 2002 from the World Wide Web: http://www.mpl.ird.fr/hydrologie/document/monogras/oubangui/index104.htm.

O'Reilly, C. M., S. R. Alin, P.-D. Plisnier, A. S. Cohen, and B. A. McKee. 2003. Climate change decreases aquatic ecosystem productivity of Lake Tanganyika, Africa. Nature 424:766–768.

Orme, A. R. 1996. Coastal environments. Pages 238–266 in W. M. Adams. A. S. Goudie, and A. R. Orme, editors. The physical geography of Africa. Oxford University Press, Oxford, UK.

Owen, R. B., R. Crossley, T. C. Johnson, D. Tweddle, I. Kornfield, S. Davison, D. H. Eccles, and D. E. Engstrom. 1990. Major low levels of Lake Malawi and their implications for speciation rates in cichlid fishes. Proceedings of the Royal Society of London 240:519–553.

Owen, R. J. 1994. Irrigation and cultivation in dambo wetlands in Zimbabwe. Pages 103–112 in T. Matiza and S. A. Crafter, editors. Wetlands ecology and priorities for conservation in Zimbabwe. Proceedings of a seminar on the wetlands of Zimbabwe. IUCN, Gland, Switzerland.

Oyebande, L. and I. Balogun. 1996. Environmentally sound management of the Niger River system. Pages 200–218 in A. M. Shady, M. El-Moattassem, E. A. Abdel-Hafiz, and A. K. Biswas, editors. Management and development of major rivers. Oxford University Press, Calcutta, India.

Pachur, H.-J. and F. Rottinger. 1997. Evidence for a large extended paleolake in the eastern Sahara as revealed by spaceborne radar lab images. Remote Sensing of Environment 61:437–440.

Palfreman, A. 2001. Poverty and fisheries development: lessons from West Africa. Pages 39–48 in S. Goddard, H. Al-Oufi, J. McIlwain, and M. Claereboudt, editors. Proceedings on Fisheries, Aquaculture and Environment in the NW Indian Ocean. Sultan Qaboos University, Muscat, Sultanate of Oman.

Palmer, C. G. 1999. The application of ecological research in the development of a new water law in South Africa. Journal of the North American Benthological Society 18:132–142.

Palmer, C. G., R. Berold, W. J. Muller, and P.-A. Scherman. 2002. Some for all forever: water ecosystems and people. Report No. TT176/02. Water Research Commission, Gezina, South Africa.

Pan-African News Agency. 2000. Niger wants to build a power dam on River Niger. Retrieved 2001 from the World Wide Web: http://allafrica.com/stories/printable/20010110368.html.

Pan-African News Agency. 2001. U.N. report blames allies of exploiting resources. Retrieved 2001 from the World Wide Web: http://allafrica.com/stories/printable/200104180178.html.

Parker, V. 2001. Mozambique. Pages 627–638 in L. D. C. Fishpool and M. I. Evans, editors. Important bird areas in Africa and associated islands: priority sites for conservation. Pisces Publications and BirdLife International (BirdLife Conservation Series No. 11), Newbury and Cambridge, UK.

Parris, R. 1984. Pans, rivers and artificial waterholes in the protected areas of the south-western Kalahari. Supplement to Koedoe 1984:63–82.

Parris, R. 1988. Important role of the Kalahari Pans. African Wildlife 24:234–237.

Passmore, N. I. and V. C. Carruthers. 1995. South African frogs. Southern Book Publishers and Witwatersrand University Press, Johannesburg, South Africa.

Patterson, G. and O. Kachinjika. 1995. Limnology and phytoplankton ecology. Pages 1–67 in A. Menz, editor. The fishery potential and productivity of the pelagic zone of Lake Malawi/Niassa: scientific report of the UK/SADC pelagic fish resource assessment project. Natural Resources Institute, Kent, UK.

Patterson, G. and J. Makin, editors. 1998. The state of biodiversity in Lake Tanganyika: a literature review. Natural Resources Institute, Chatham, UK.

Payne, A. I. 1986. The ecology of tropical lakes and rivers. John Wiley & Sons, Chichester, UK.

Pellegrin, J. 1933. Les poissons des eaux douces de Madagascar et des îles voisines. Mémoires de l'Academie Malgache, Vol. 14, Tananarive, Madagascar.

Pellegrin, J. 1934. La faune ichtyologique des eaux douces de Madagascar. Annales de Sciences Naturelles 10(17):425–432.

Penry, H. 1994. Bird atlas of Botswana. University of Natal Press, Pietermaritzburg, South Africa.

Peters, C. M., A. H. Gentry, and R. O. Mendelshon. 1989. Valuation of an Amazonian rainforest. Nature 339:655–656.

Peters, H. M. and S. Berns. 1982. Die Maulbrutpflege der Cichliden. Untersuchungen zur Evolution eines Verhaltensmusters. Journal of Zoological Systematics & Evolutionary Research 20:18–62.

Peters, W. 1868. Reise nach Mossambique. IV Flussfische. Druck & Verlag, Berlin, Germany.

Pethiyagoda, R. 1991. Freshwater fishes of Sri Lanka. Wildlife Heritage Trust, Colombo, Sri Lanka.

Petr, T. 1986. The Volta River system. Pages 163–183 in B. R. Davies and K. F. Walker, editors. The ecology of river systems. Dr. W. Junk, Dordrecht, The Netherlands.

Phiri, G. and L. A. Navarro. 2000. Water hyacinth in Africa and the Middle East: a survey of problems and solutions. International Development Research Centre, Ottawa, Canada.

Pieterse, A. H., M. Kettunen, S. Diouf, I. Ndao, K. Sarr, A. Tarvainen, S. Kloff, and S. Hellsten. 2003. Effective biological control of *Salvinia molesta* in the Senegal River by means of the weevil *Cyrtobagous salviniae*. Ambio 32(7):458–462.

Pimentel, D., S. McNair, J. Janecka, J. Wightman, C. Simmonds, C. O'Connell, E. Wong, L. Russel, J. Zern, T. Aquino, and T. Tsomondo. 2001. Economic and environmental threats of alien plant, animal, and microbe invasions. Agriculture Ecosystems & Environment 84:1–20.

Pimm, S. L. 1986. Community stability and structure. Pages 309–329 in M. E. Soule, editor. Conservation biology: the science of scarcity and diversity. Sinauer Associates, Sunderland, MA, USA.

Pinhey, E. 1974. Odonata of the northwest Cameroons and particularly of the islands stretching southwards from the Guinea Gulf. Bonner Zoologische Beiträge 25:179–212.

Pinhey, E. C. G. 1978. Odonata. Pages 1049–1112 in M. J. A. Werger and A. C. van Bruggen, editors. The biogeography and ecology of southern Africa. Dr. W. Junk, The Hague, The Netherlands.

Pinhey, E. 1981. Checklist of the Odonata of Mozambique. Occasional Papers of the National Museums and Monuments 6:557–632.

Poff, N. L. 1997. Landscape filters and species traits: towards mechanistic understanding and prediction in stream ecology. Journal of the North American Benthological Society 16:391–409.

Polhemus, D. A. 1993. The Heteroptera of Aldabra Atoll and nearby islands, western Indian Ocean, Part 2: freshwater Heteroptera (Insecta): Corixidae, Notonectidae, Veliidae, Gerridae and Mesoveliidae. Atoll Research Bulletin 381:1–9.

Poll, M. 1939. Les poissons du Stanley Pool. Annales du Musée Royal du Congo Belge Tervuren I IV(1).

Poll, M. 1959. Recherches sur la faune ichtyologique de la region du Stanley Pool. Annales du Musée Royal du Congo Belge 71:75–174.

Poll, M. 1967. Contribution à la faune ichtyologique de l'Angola. Diamang, Lisbon, Portugal.

Poll, M. 1976. Poissons: exploration du Parc National de l'Upemba. Fondation pour Favoriser les Recherches Scientifiques en Afrique 73:1–127.

Poll, M. and J. P. Gosse. 1963. Contribution à l'étude systématique de la faune ichtyologique du Congo Central. Annales du Musée Royal d'Afrique Centrale (Sciences Zoologiques) 116:43–111.

Poll, M., B. Lanza, and A. Sassi. 1972. Genvre hoveau extraordinaire der Bagridae du fleauve Juba: *Pardiglanis tarabini* gen.n. sp.n. (Pisces Siluroformes). Monitore Zoologico Italiano, Firenze (N.S.) (Suppl.) 4(15):327–345.

Poll, M. and R. H. Renson. 1948. Les poissons, leur milieu et leur peche au bief superieur du Lualaba. Bulletin Agricole du Congo Belge 39:427–446.

Pollard, D. A., B. A. Ingram, J. H. Harris, and L. F. Reynolds. 1990. Threatened fishes in Australia: an overview. Journal of Fish Biology 37 (Suppl. A):67–78.

Popov, G. 1957. The vegetation of Socotra. Journal of the Linnean Society (Botany), London 55(362):706–720.

Porter, J. W. and I. L. Muzila. 1989. Aspects of swamp hydrology in the Okavango. Botswana Notes and Records 21:73–91.

Postel, S. and S. Carpenter. 1997. Freshwater ecosystem services. Pages 195–214 in G. C. Daily, editor. Nature's services: societal dependence on natural ecosystems. Island Press, Washington, DC, USA.

Poynton, J. C. and D. G. Broadley. 1985. Amphibia Zambesiaca 1. Scolecomorphidae, Pipidae, Microhylidae, Hemisidae, Arthroleptidae. Annals of the Natal Museum (Pietermaritzburg) 26:503–553.

Poynton, J. C. and D. G. Broadley. 1986. Amphibia Zambesiaca 2. Ranidae. Annals of the Natal Museum (Pietermaritzburg) 27:115–181.

Poynton, J. C. and D. G. Broadley. 1987. Amphibia Zambesiaca 3. Rhacophoridae and Hyperoliidae. Annals of the Natal Museum (Pietermaritzburg) 28:161–229.

Poynton, J. C. and D. G. Broadley. 1988. Amphibia Zambesiaca 4. Bufonidae. Annals of the Natal Museum (Pietermaritzburg) 29:447–490.

Preston-Whyte, R. A. and P. D. Tyson. 1988. The atmosphere and weather of southern Africa. Oxford University Press, Cape Town, South Africa.

Pritchard, J. M. 1971. Africa: the geography of a changing continent. Africana Publishing Corporation, New York, NY, USA.

Project ZICOMA. 2001. Madagascar. Pages 489–537 in L. D. C. Fishpool and M. I. Evans, editors. Important bird areas in Africa and associated islands: priority sites for conservation. Pisces Publications and BirdLife International (BirdLife Conservation Series No. 11), Newbury and Cambridge, UK.

Quammen, D. 1996. The song of the dodo: island biogeography in an age of extinction. Touchstone, New York, NY, USA.

Quensière, J. I., editor. 1994. La pêche dans le Delta Central du Niger. Karthala, Paris, France.

Rabinowitz, P. D., M. F. Coffin, and D. Falvey. 1983. The separation of Madagascar and Africa. Science 220:304–324.

Rall, J. and P. Skelton. 2001. Conservation of the Maloti minnow (Phase 1): distribution and conservation status. Final report, contract 1041. Lesotho Highlands Development Authority, Maseru, Lesotho.

Ramdani, M., R. J. Flower, N. Elkhiati, M. M. Kraiem, A. A. Fathi, H. H. Birks, and S. T. Patrick. 2001. North African wetland lakes: characterization of nine sites included in the CASSARINA project. Aquatic Ecology 35(3–4):281–302.

Ramdin, T. 1969. Mauritius, a geographical survey. University Tutorial Press, London, UK.

Ramsar. 1999. Resolutions and recommendations from the 7th Conference of the Contracting Parties of the Wetlands Convention (Ramsar, Iran, 1971), San Jose, Costa Rica, May 1999. Retrieved 2003 from the World Wide Web: http://www.ramsar.org/key_res_vii_index.htm.

Ramsar. 2000. United Republic of Tanzania joins the Ramsar Convention. Retrieved 2001 from the World Wide Web: http://www.ramsar.org/w.n.tanzania_joins.htm.

Ramsar. 2002a. Guinea designates six large Ramsar sites in the upper Niger River Basin. Retrieved 2003 from the World Wide Web: http://www.ramsar.org/w.n.guinea_names6.htm.

Ramsar. 2002b. Lake Chad Basin Commission 49th ministerial meeting reaches Ramsar-related decision, 15 January 2002. Retrieved 2003 from the World Wide Web: http://www.ramsar.org/w.n.chad_lcbc49_e.htm.

Ramsar. 2003. The Montreux record. Retrieved 2003 from the World Wide Web: http://www.ramsar.org/key_montreux_record.htm.

Rand McNally. 1987. Universal world atlas. Rand McNally, Chicago, IL, USA.

Redmond, I. 2001. Coltan boom, gorilla bust: the impact of coltan mining on gorillas and other wildlife in eastern DR Congo. Retrieved 2001 from the World Wide Web: http://www.bornfree.org.uk/coltan/coltan.pdf.

Regional Department of Environment of the Canary Islands Autonomous Government. 2001. Biodiversidad. Retrieved 2001 from the World Wide Web: http://www.gobcan.es/medioambiente/eng/biodiversidad/.

Reid, G. M. 1989. The living waters of Korup rainforest: a hydrobiological survey report and recommendations with emphasis on fish and fisheries. World Wide Fund for Nature Publication, Godalming, UK.

Reid, G. M. 1990. Threatened fishes of Barombi Mbo: a crater lake in Cameroon. Journal of Fish Biology 37(Suppl. A):209–211.

Reid, G. M. 1996. Ichthyogeography of the Guinea-Congo rain forest, West Africa. Proceedings of the Royal Society of Edinburgh, Section B, Biological Sciences 104:285–312.

Reinthal, P. 1993. Evaluating biodiversity and conserving Lake Malawi's cichlid fish fauna. Conservation Biology 7(3):712–718.

Reinthal, P. and M. L. J. Stiassny. 1991. The freshwater fishes of Madagascar: a study of an endangered fauna with recommendations for a conservation strategy. Conservation Biology 5:231–243.

Remane, K. 1997. African inland fisheries, aquaculture, and environment. Fishing News Books, Oxford, UK.

Republic of Cameroon. 1999. Biodiversity status strategy and action plan. Yaounde, Cameroon: United Nations Environment Program. Retrieved 2002 from the World Wide Web: http://www.biodiv.org/doc/world/.

Republic of Yemen. 2000. Presidential Decree Number 275 of year 2000, regarding the conservation zoning plan for Socotra Islands (Socotra, Samha, Darsa, Abd Al-Kuri and the associated small islands, rocks and rock outcrops) into areas for conservation and development. Sana'a, Yemen.

République Démocratique du Congo. 1998. Stratégie nationale et plan d'action de la biodiversité. Project ZAI/96/G31/C/1G/99. Retrieved 2002 from the World Wide Web: http://bch-cbd.naturalsciences.be/congodr/cdr-fra/contribution/strataction/strategie/conserv.htm.

République Démocratique du Congo. 2002. RD Congo's Inga Dam could power-up much of southern Africa. Retrieved 2002 from the World Wide Web: http://www.rdcongogov.info/BusinessIngaPwrUp.html.

Reuters. 2000. Planet ark: Shell says fears fire outbreak from Nigeria oil spill. Retrieved 2001 from the World Wide Web.

Reuters. 2001. Planet ark: Congo Okapi park battles miners, poachers, and war. Retrieved 2001 from the World Wide Web: http://www.planetark.org/dailynewsstory/.

Revenga, C., J. Brunner, N. Henninger, K. Kassem, and R. Payne. 2000. Pilot analysis of global ecosystems: freshwater systems. World Resources Institute, Washington, DC.

Revenga, C., S. Murray, J. Abramovitz, and A. Hammond. 1998. Watersheds of the world: ecological value and vulnerability. World Resources Institute and Worldwatch Institute, Washington, DC.

Ribbink, A. J. 1994. Lake Malawi. Pages 27–33 in K. Martens, B. Goddeeris, and G. Coulter, editors. Speciation in ancient lakes. Archiv für Hydrobiologie Beiheft Ergebnisse der Limnologie 44. E. Schweizerbart'sche Verlagbuchhandlung, Stuttgart, Germany.

Ribbink, A. J. 2001. Lake Malawi/Niassa/Nyasa ecoregion-based conservation programme: biophysical reconnaissance. WWF Southern African Regional Programme Office, Harare, Zimbabwe.

Ricardo, C. K. 1939. The fishes of Lake Rukwa. Journal of the Linnean Society of London 40(275):625–657.

Ricciardi, A. and J. B. Rasmussen. 1999. Extinction rates of North American freshwater fauna. Conservation Biology 13(5):1220–1222.

Richter, B. D., R. Mathews, D. L. Harrison, and R. Wigington. 2003. Ecologically sustainable water management: managing river flows for ecological integrity. Ecological Applications 13:206–224.

Ricketts, T. H., E. Dinerstein, D. M. Olson, and C. J. Loucks. 1999a. Terrestrial ecoregions of North America: a conservation assessment. World Wildlife Fund, Washington, DC.

Ricketts, T. H., E. Dinerstein, D. M. Olson, and C. Loucks. 1999b. Who's where in North America? Patterns of species richness and the utility of indicator taxa for conservation. BioScience 49(5):369–381.

Riley, J. and F. W. Huchzermeyer. 1999. African dwarf crocodiles in the Likouala swamp forests of the Congo Basin: habitat, density, and nesting. Copeia 199(2):313–320.

Riseng, K. J. 1997. The distribution of fishes and the conservation of aquatic resources in Madagascar. M.S. thesis. University of Michigan, Ann Arbor, MI, USA.

Roberts, T. R. 1973. Ecology of fishes in the Amazon and Congo Basins. Pages 239–254 in B. J. Meggers, E. S. Ayensu, and W. D. Duckworth, editors. Tropical forest ecosystems in Africa and South America: a comparative review. Smithsonian Institution Press, Washington, DC.

Roberts, T. R. 1975. Geographical distribution of African freshwater fishes. Zoological Journal of the Linnean Society 57:249–319.

Roberts, T. R. and S. O. Kullander. 1994. Endemic cichlid fishes of the Fwa River, Zaire: systematics and ecology. Ichthyological Explorations of Freshwaters 5(2):97–154.

Roberts, T. R. and D. J. Stewart. 1976. An ecological and systematic survey of fishes in the rapids of the lower Zaire or Congo River. Bulletin of the Museum of Comparative Zoology, Harvard 147:239–317.

Robertson, P. 2001a. Guinea-Bissau. Pages 403–409 in L. D. C. Fishpool and M. I. Evans, editors. Important bird areas in Africa and associated islands: priority sites for conservation. Pisces Publications and BirdLife International (BirdLife Conservation Series No. 11), Newbury and Cambridge, UK.

Robertson, P. 2001b. Somalia. Pages 779–792 in L. D. C. Fishpool and M. I. Evans, editors. Important bird areas in Africa and associated islands: priority sites for conservation. Pisces Publications and BirdLife International (BirdLife Conservation Series No. 11), Newbury and Cambridge, UK.

Robertson, P. 2001c. Sudan. Pages 877–890 in L. D. C. Fishpool and M. I. Evans, editors. Important bird areas in Africa and associated islands: priority sites for conservation. Pisces Publications and BirdLife International (BirdLife Conservation Series No. 11), Newbury and Cambridge, UK.

Robertson, P. and M. Essghaier. 2001. Socialist People's Libyan Arab Jamahiriya. Pages 481–487 in L. D. C. Fishpool and M. I. Evans, editors. Important bird areas in Africa and associated islands: priority sites for conservation. Pisces Publications and BirdLife International (BirdLife Conservation Series No. 11), Newbury and Cambridge, UK.

Robinson, C. T., K. Tockner, and J. V. Ward. 2002. The fauna of dynamic riverine landscapes. Freshwater Biology 47:661–677.

Rocamora, G. and A. Skerrett. 2001. Seychelles. Pages 751–768 in L. D. C. Fishpool and M. I. Evans, editors. Important bird areas in Africa and associated islands: priority sites for conservation. Pisces Publications and BirdLife International (BirdLife Conservation Series No. 11), Newbury and Cambridge, UK.

Röckner, E., K. Arpe, L. Bengtsson, M. Christoph, M. Claussen, L. Dümenil, M. Esch, M. Giorgetta, U. Schlese, and U. Schulzweida. 1996. The atmospheric general circulation model ECHAM-4: model description and simulation of present day climate, MPI-Report No. 218. MPI für Meteorologie, Hamburg, Germany.

Röckner, E., L. Bengstsson, J. Feichter, J. Lelieveld, and H. Rohde. 1999. Transient changes with a coupled atmosphere-ocean GCM including the tropospheric sulfur cycle. Climate Change 38:307–343.

Rödel, M. 2000. Herpetofauna of West Africa, Vol. 1 Amphibians of the West African savanna. Chimaira, Frankfurt, Germany.

Rodgers, W. A. 1993. The conservation of the forest resources of eastern Africa: past influences, present practices and future needs. Pages 283–327 in J. C. Lovett and S. K. Wasser, editors. Biogeography and ecology of the rain forests of Eastern Africa. Cambridge University Press, New York, NY, USA.

Roggeri, H., editor. 1995. Tropical freshwater wetlands: a guide to current knowledge and sustainable management. Kluwer Academic Publishers, Dordrecht, The Netherlands.

Roman, B. 1971. Peces de Rio Muni, Guinea Ecuatorial (Aguas dulces y salobres). Fundacion la Salle de Ciencias Naturales, Barcelona, Spain.

Rosa, F., M. Simoes, and F. Lagos Costa. 1999. Geographic distribution of the freshwater snails on the island of Santiago (Cape Verde–Cabo Verde): preliminary data. Garcia de Orta Serie de Zoologia 23:193–201.

Rosegrant, M. W., X. Cai, and S. A. Cline. 2002. World water and food to 2025: dealing with water scarcity. IFPRI, Washington, DC, USA.

Rosenzweig, M. 1995. Species diversity in space and time. Cambridge University Press, Cambridge, UK.

Ross, J. P., editor. 1998. Crocodiles: status survey and conservation action plan, 2nd ed. IUCN/SSC Crocodile Specialist Group, IUCN, Gland, Switzerland and Cambridge, UK.

Ross, K. 2003. Okavango: jewel of the Kalahari. Struik Publishers, Cape Town, South Africa.

Rossignon, O. 1999. Contribution à écologie des crevettes dulcaquicoles de São Tomé: du cadre limnologique à l'elevage. Faculté Universitaire des Sciences Agronomiques de Gembloux, Gembloux, Belgium.

Roth, H. H. and E. Waitkuwait. 1986. Repartition et statut des grandes espèces de mammifères en Côte d'Ivoire. III Lamantins. Mammalia 50:227–242.

Roux, F. and G. Jarry. 1984. Numbers, composition and distribution of populations of Anatidae wintering in West Africa. Wildfowl 35:48–60.

Rubin, J. A., C. Gordon, and J. K. Amatekpor. 1998. Causes and consequences of mangrove deforestation in the Volta Estuary, Ghana: some recommendations for ecosystem rehabilitation. Marine Pollution Bulletin 37(8–12):441–449.

Rzóska, J. 1974. The upper Nile swamps, a tropical wetland study. Freshwater Biology 4:1–30.

Rzóska, J. 1978. On the nature of rivers with case stories of Nile, Zaire and Amazon. Dr. W. Junk, The Hague, The Netherlands.

Rzóska, J. 1984. Temporary and other waters. Pages 105–114 in J. L. Cloudsley-Thompson, editor. Sahara Desert. Pergamon Press, Oxford, UK.

Rzóska, J. 1985. The water quality and hydrobiology of the Niger. Pages 77–98 in A. T. Grove, editor. The Niger and its neighbors. A.A. Balkema, Rotterdam, The Netherlands.

SADC. 2002. Pre-feasibility study of future developments and management options on the lower Orange River. Retrieved 2002 from the World Wide Web: http://www.sadcwscu.org.ls/programme/rsap/prog_regionalstrag_pcn30-22.htm.

Safford, R. J. 2001a. The Comoros. Pages 185–190 in L. D. C. Fishpool and M. I. Evans, editors. Important bird areas in Africa and associated islands: priority sites for conservation. Pisces Publications and BirdLife International (BirdLife Conservation Series No. 11), Newbury and Cambridge, UK.

Safford, R. J. 2001b. Mauritius. Pages 583–596 in L. D. C. Fishpool and M. I. Evans, editors. Important bird areas in Africa and associated islands: priority sites for conservation. Pisces Publications and BirdLife International (BirdLife Conservation Series No. 11), Newbury and Cambridge, UK.

Safford, R. J. 2001c. Mayotte. Pages 597–601 in L. D. C. Fishpool and M. I. Evans, editors. Important bird areas in Africa and associated islands: priority sites for conservation. Pisces Publications and BirdLife International (BirdLife Conservation Series No. 11), Newbury and Cambridge, UK.

Salvig, J. C., S. Asbirk, J. P. Kjeldsen, and P. A. F. Rasmussen. 1994.

Wintering waders in the Bijagos Archipelago, Guinea-Bissau 1992–1993. Ardea 82(1):137–141.

Salzburger, W., A. Meyer, S. Baric, E. Verheyen, and C. Sturmbauer. 2002. Phylogeny of the Lake Tanganyika cichlid species flock and its relationship to the Central and East African haplochromine cichlid fish faunas. Systematic Biology 51(1):113–135.

Samoura, A. B., S. T. Diallo, F. L. Keita, S. Sidibé, I. Crissé, and P. Koivogui. 1999. Analyse de la biodiversité des écosystemes des eaux continentales, DNE/projet Gui/97/G32/A/1G/99. Stratégie Plan d'Action Diversité Biologique. Ministère Mines, Géologie et Environnement, Guinea.

Samways, M. J. 2002. Threatened Odonata species of Africa. Odonataologica 31(2):151–170.

Samways, M. J. 2003. Threats to the tropical island dragonfly fauna (Odonata) of Mayotte, Comoro archipelago. Biodiversity and Conservation 12:1785–1792.

Samways, M. J. In prep. Conservation of a fugitive dragonfly fauna in the ancient tropical Seychelles Archipelago.

Sanderson, E. W., M. Jaiteh, M. A. Levy, K. H. Redford, A. V. Wannebo, and G. Woolmer. 2002. The human footprint and the last of the wild. BioScience 52(10):891–904.

Sanyanga, R. A. 1994. Tourism and wetlands management in Zimbabwe, with special reference to the Zambezi River system. In T. Matiza and S. A. Crafter, editors. Wetlands ecology and priorities for conservation in Zimbabwe. Proceedings of a seminar on the wetlands of Zimbabwe. IUCN, Gland, Switzerland.

Sargeant, D. 1997. Cape Verde: a birder's guide to the Cape Verde Islands. D. E. Sargeant, Norfolk, UK.

Sarunday, W. N. 1999. Conservation and sustainable use of biodiversity in Eastern Rift Valley lakes in Tanzania. GEF/UNDP/UNEP consultancy report. National Environment Management Council, Dar es Salaam, Tanzania.

Saunders, D. L., J. J. Meeuwig, and A. C. J. Vincent. 2002. Freshwater protected areas: strategies for conservation. Conservation Biology 16(1):30–41.

Sayeed, A. 2001. Economy versus environment: how a system with RD and GIS can assist in decisions for water resource management. A case study in Lake Naivasha, Central Rift Valley Province, Kenya. Master's thesis. ITC, Enschede, The Netherlands.

Sayer, J. A., C. S. Harcourt, and N. M. Collins, editors. 1992. The conservation atlas of tropical forests: Africa. IUCN, London, UK.

Schiøtz, A. 1999. Tree frogs of Africa. Edition Chimaira, Frankfurt, Germany.

Schleich, H. H., W. Kästle, and K. Kabisch. 1996. Amphibians and reptiles of North Africa: biology, systematics, field guide. Koeltz Scientific Publishers, Koenigstein, Germany.

Schliewen, U. K. 1996a. Barombi Mbo: summary of the background knowledge of the natural history of the lake and recommendations for conservation. Gesellschaft für Technische Zusammenarbeit (GTZ), Eschborn, Germany.

Schliewen, U. K. 1996b. Ichthyological survey of the Rumpi Hill waters. Gesellschaft für Technische Zusammenarbeit (GTZ), Eschborn, Germany.

Schliewen, U. K., K. Rassmann, M. Markmann, J. Markert, T. D. Kocher, and D. Tautz. 2001. Genetic and ecological divergence of a monophyletic cichlid species pair under fully sympatric conditions in Lake Ejagham, Cameroon. Molecular Ecology 10:1471–1488.

Schliewen, U. K. and M. L. J. Stiassny. 2003. *Etia nguti*, a new genus and species of cichlid fish from the River Mamfue, Upper Cross River Basin, Cameroon, Central Africa. Ichthyological Exploration of Freshwaters 14(1):61–72.

Schliewen, U. K., D. Tautz, and S. Paabo. 1994. Sympatric speciation suggested by monophyly of crater lake cichlids. Nature 368(6472):629–632.

Schluter, D. 2000. The ecology of adaptive radiation. Oxford University Press, Oxford, UK.

Schminke, H. K. 1987. The genus *Thermobathynella* (Capart, 1951) (Bathynellacea, Malacostraca) and its phylogenetic relationships. Revue d'Hydrobiologie Tropicale 20:107–112.

Schneider, W. 1996. The dragonfly fauna of Socotra Island. First International Scientific Symposium on Socotra Island: present and future. University of Aden and Sana'a, Sana'a, Yemen.

Schneider, W. and H. J. Dumont. 1998. Checklist of the dragonflies and damselflies of Soqotra Island (Insecta: Odonata). Pages 219–232 in H. J. Dumont, editor. Proceedings of the first international symposium on Soquotra Island: present and future. United Nations Publications, New York, NY, USA.

Schofield, P. J., and L. J. Chapman. 2000. Hypoxia tolerance of introduced Nile perch: implications for survival of indigenous fishes in the Lake Victoria Basin. African Zoology 35:35–42.

Schwanck, E. J. 1995. The introduced *Oreochromis niloticus* is spreading on the Kafue Floodplain, Zambia. Hydrobiologia 315:143–147.

Scott, A. J. 1993. A revised and annotated check-list of the birds of the Luangwa Valley national parks and adjacent areas. Occasional paper no 3. Zambian Ornithological Society, Lusaka, Zambia.

Scott, D. A. and P. M. Rose. 1996. Atlas of Anatidae populations in Africa and western Eurasia. Wetlands International Publication 41. Wetlands International, Wageningen, The Netherlands.

Seabrook, W. 1989. Feral cats (*Felis catus*) as predators of hatchling green turtles (*Chelonia mydas*). Journal of Zoology (London) 219:83–88.

Seabrook, W. 1990. The impact of the feral cat (*Felis catus*) on the native fauna of Aldabra Atoll, Seychelles (Indian Ocean). Revue d'Ecologie la Terre et la Vie 45:135–146.

Seaman, M. T. and J. G. Van As. 1998. The environmental status of the Orange River mouth as reflected by the fish community, Report 505/1/98. Water Research Commission, Pretoria, South Africa.

Sedell, J. R., G. H. Reeves, and K. M. Burnett. 1994. Development and evaluation of aquatic conservation strategies. Journal of Forestry 92:28–31.

Seegers, L. 1996. The fishes of the Lake Rukwa drainage. Annales du Musée Royal de l'Afrique Central, Sciences Zoologiques 287. Musée Royal de l'Afrique Central, Tervuren, Belgium.

Seehausen, O. 1996. Lake Victoria rock cichlids. Verduijn Cichlids, Zevenhuizen, The Netherlands.

Seehausen, O. 2000. Explosive speciation rates and unusual species richness in haplochromine cichlid fishes: effects of sexual selection. Advanced Ecological Research 31:237–274.

Seehausen, O., E. Koetsier, M. V. Schneider, L. J. Chapman, C. A. Chapman, M. E. Knight, G. F. Turner, J. J. M. van Alphen, and R. Bills. 2003. Nuclear markers reveal unexpected genetic variation and a Congolese-Nilotic origin of the Lake Victoria cichlid species flock. Proceedings of the Royal Society of London B 270:129–137.

Seehausen, O. and J. J. M. Van Alphen. 1998. The effect of male coloration on female mate choice in closely related Lake Victoria cichlids (*Haplochromis nyererei* complex). Behavioral Ecology and Sociobiology 42:1–8.

Seehausen, O. and J. J. M. Van Alphen. 1999. Can sympatric speciation by disruptive sexual selection explain rapid evolution of cichlid diversity in Lake Victoria? Ecology Letters 2:262–271.

Seehausen, O., J. J. M. Van Alphen, and F. Witte. 1997a. Cichlid fish diversity threatened by eutrophication that curbs sexual selection. Science 277:1808–1811.

Seehausen, O., J. J. M. Van Alphen, and F. Witte. 1999. Can ancient colour polymorphisms explain why some cichlid lineages speciate rapidly under disruptive sexual selection? Belgium Journal of Zoology 129(1):43–60.

Seehausen, O., F. Witte, E. F. Katunzi, J. Smits, and N. Bouton. 1997b. Patterns of the remnant cichlid fauna in southern Lake Victoria. Conservation Biology 11(4):890–904.

Sevaldsen, K. 1997. Competing interests and growing demand for water: irrigation in the lowlands of Kilimanjaro, the case of Makuyuni Ward. Department of Geography, Norwegian University of Science and Technology, Trondheim, Norway. Retrieved 2001 from the World Wide Web: http://www.svt.ntnu.no/geo/Forskning/Pangani/text/Abstr_Kirsti_Sevaldsen.html.

Seyoum, S. and I. Kornfield. 1992. Identification of the subspecies of *Oreochromis niloticus* (Pisces: Cichlidae) using restriction endonuclease analysis of mitochondrial DNA. Aquaculture 102(1–2):29–42.

Shambaugh, J., J. Oglethorpe, and R. Ham, editors. 2001. The trampled grass: mitigating the impacts of armed conflict on the environment. Biodiversity Support Program, Washington, DC.

Sheil, D. and S. Wunder. 2002. The value of tropical forest to local communities: complications, caveats and cautions. Conservation Ecology 6:9 Online: http://www.consecol.org/vol6/iss2/art9/.

Shine, T., P. Robertson, and B. Lamarche. 2001. Mauritania. Pages 567–581 in L. D. C. Fishpool and M. I. Evans, editors. Important bird areas in Africa and associated islands: priority sites for conservation. Pisces Publications and BirdLife International (BirdLife Conservation Series No. 11), Newbury and Cambridge, UK.

Shire Valley Agricultural Development Project. 1975. An atlas of the Lower Shire Valley, Malawi. Department of Surveys, Blantyre, Malawi.

Shmueli, M., I. Izhaki, A. Arieli, and Z. Arad. 2000. Energy requirements of migrating great white pelicans, *Pelecanus onocrotalus*. Ibis 142(2):208–216.

Shodjay, F. 1985. Entwicklung der Binnenfischerei in der Region Maj-Ndombe, Zaire. Unpublished report to German Technical Cooperation GTZ.

Shumway, C. A. 1999. Forgotten waters: freshwater and marine ecosystems in Africa. Strategies for biodiversity conservation and sustainable development. Global Printing, Alexandria, VA, USA.

Shumway, C. A., D. Musibono, S. Ifuta, J. Sullivan, R. Schelly, J. Punga, J.-C. Palata, and V. Puema. 2003. Congo River Environment and Development Project (CREDP) biodiversity survey: systematics, ecology and conservation along the Congo River, September–October 2002. New England Aquarium Press, Boston, MA, USA.

Sibbing, F. A., L. A. J. Nagelkerke, R. J. M. Stet, and J. W. M. Osse. 1998. Speciation of endemic Lake Tana barbs (Cyprinidae, Ethiopia) driven by trophic resource partitioning: a molecular and ecomorphological approach. Aquatic Ecology 32:217–227.

Silberbauer, M. J. and J. M. King. 1991. The distribution of wetlands in the south-western Cape Province, South Africa. South African Journal of Aquatic Science 17(1–2):65–81.

Simmons, R. E. 1996. Population declines, viable breeding areas, and management options for flamingos in southern Africa. Conservation Biology 10(2):504–514.

Simmons, R. E. 2000. Declines and movements of lesser flamingos in Africa. Waterbirds 23 (Special Publication 1):40–46.

Simmons, R. E., C. Boix-Hinzen, K. Barnes, A. M. Jarvis, and A. Robertson. 2001. Namibia. Pages 639–660 in L. D. C. Fishpool and M. I. Evans, editors. Important bird areas in Africa and associated islands: priority sites for conservation. Pisces Publications and BirdLife International (BirdLife Conservation Series No. 11), Newbury and Cambridge, UK.

Simmons, R. E., R. Braby, and S. J. Braby. 1993. Ecological studies of the Cunene River mouth: avifauna, herpetofauna, water quality, flow rates, geomorphology and implications of the Epupa Dam. Madoqua 18(2):163–180.

Sinclair, A. R. E. and J. M. Fryxell. 1985. The Sahel of Africa: ecology of a disaster. Canadian Journal of Zoology 63(5):987–994.

Sita, P. 1980. La végétation du Stanley Pool en relation avec celle des plateaux voisins. Université de Bordeaux III, Bordeaux, France.

Skelton, P. H. 1980. Aspects of freshwater fish biogeography in the Eastern Cape. The Eastern Cape Naturalist 24(3):16–22.

Skelton, P. H. 1986a. Distribution patterns and biogeography of non-tropical southern African freshwater fishes. Pages 211–230 in E. M. van Zinderen Bakker Sr., J. A. Coetzee, and L. Scott, editors. Palaeoecology of Africa and the surrounding islands, Southern African Society for Quaternary research, Vol. 17, Proceedings of the VIIth biennial conference. A.A. Balkema, Rotterdam, The Netherlands.

Skelton, P. H. 1986b. Fishes of the Orange-Vaal system. Pages 143–161 in B. R. Davies and K. F. Walker, editors. The ecology of river systems. Dr. W. Junk Publishers, Dordrecht, The Netherlands.

Skelton, P. H. 1987. South African Red Data Book: fishes. South African National Scientific Programs Report 40:199.

Skelton, P. H. 1988. A taxonomic revision of the redfin minnows

(Pisces, Cyprinidae) from southern Africa. Annals of the Cape Provincial Museums 16:201–307.

Skelton, P. H. 1990a. The conservation and status of threatened fishes in southern Africa. Journal of Fish Biology 37(Suppl. A):87–95.

Skelton, P. H. 1990b. The status of fishes from sinkholes and caves in Namibia. Journal of the Namibia Scientific Society 42:75–83.

Skelton, P. H. 1993. A complete guide to the freshwater fishes of southern Africa. Southern Book Publishers, Halfway House, South Africa.

Skelton, P. H. 1994. Diversity and distribution of freshwater fishes in East and southern Africa. Pages 95–131 in G. G. Teugels, J. F. Guégan, and J. J. Albaret, editors. Biological diversity in African fresh and brackish water fishes, Symposium PARADI. Annals of the Royal Central Africa Museum (Zoology) 275.

Skelton, P. H. 2001. A complete guide to the freshwater fishes of southern Africa, 2nd ed. Struik Publishers, Cape Town, South Africa.

Skelton, P. H., J. A. Cambray, A. Lombard, and G. A. Benn. 1995. Patterns of distribution and conservation status of freshwater fishes in South Africa. South African Journal of Zoology 30(3):71–81.

Skelton, P. H., D. Tweddle, and P. B. N. Jackson. 1991. Cyprinids of Africa. Pages 211–239 in I. J. Winfield and J. S. Nelson, editors. Cyprinid fishes: systematics, biology and exploitation. Chapman & Hall, London, UK.

Skerrett, A. 1999. Birds of Aldabra: 1. Bulletin of the African Bird Club 6.1. Retrieved 2001 from the World Wide Web: http://www.africanbirdclub.org/feature/aldabra.html.

Skinner, J. R., P. J. Wallace, W. Altenburg, and B. Fofana. 1987. The status of heron colonies in the Niger Delta, Mali. Malimbus 9:65–82.

Slootweg, R. and M. L. F. van Schooten. 1995. Partial restoration of floodplain functions at the village level: the experience of Gounougou, Benue Valley, Cameroon. Pages 159–166 in H. Roggeri, editor. Tropical freshwater wetlands: a guide to current knowledge and sustainable management. Kluwer Academic Publishers, Dordrecht, The Netherlands.

SMEC. 1986. Southern Okavango integrated water development Phase 1. Report to the Ministry of Mineral Resources, Water and Energy. SMEC, Gaborone, Botswana.

Smith, A. 1988. The great rift: Africa's changing valley. Sterling Publishing, New York, NY, USA.

Smith, R. D. and E. Maltby. 2003. Using the ecosystem approach to implement the Convention on Biological Diversity: key issues and case studies. IUCN, Gland, Switzerland and Cambridge, UK.

Smith, R. J., R. D. J. Muir, M. J. Walpole, A. Balmford, and N. Leader-Williams. 2003. Governance and the loss of biodiversity. Nature 426:67–70.

Smith, R. M. H. 1987. Morphology and depositional history of exhumed Perninan pointy-bars in the southwestern Karoo, South Africa. South African Journal of Sedimentary Petrology 57:19–29.

Smith, R. M. H. 1993. Sedimentology and ichnology of flood plain paleosurfaces in the Beaufort Group (late Permian, Karoo sequence, South Africa). Palaios 8:339–357.

Snaddon, C. D., M. J. Wishart, and B. R. Davies. 1998. Some implications of inter-basin water transfers for river ecosystem functioning and water resources management in southern Africa. Aquatic Ecosystem Health and Management 1:159–182.

Snoeks, J. 1994. The haplochromine fishes (Teleostei, Cichlidae) of Lake Kivu, East Africa: a taxonomic revision with notes on their ecology. Annales Musée Royale Afrique Centrale Science Zoologique 270:1–221.

Snoeks, J. 1999a. The non-cichlid fishes of the Lake Malawi/Nyasa system. Pages 49–55 in J. Snoeks, editor. Report on the systematics and taxonomy. SADC/GEF Lake Malawi/Nyasa Biodiversity Conservation project, Senega Bay, Malawi.

Snoeks, J., editor. 1999b. Report on the systematics and taxonomy. SADC/GEF Lake Malawi/Nyasa Biodiversity Conservation project, Senega Bay, Malawi.

Snoeks, J. 2000. How well known is the ichthyodiversity of the large East African lakes? Advances in Ecological Research 31:17–38.

Snoeks, J., editor. 2004. The cichlid diversity of Lake Malawi/Nyasa: identification, distribution and taxonomy. Cichlid Press, El Paso, TX, USA.

Société Nationale d'Electricité (SNEL), République Démocratique de Congo. 2003. Inga: the highest available capacity in Africa for Africa. SNEL, Kinshasa, DRC.

Sodsuk, P., B. A. MacAndrew, and G. F. Turner. 1995. Allozyme variation in *Oreochromis* species from Lake Malawi: implications for evolutionary relationships. Journal of Fish Biology 47:321–333.

Soils Incorporated (Pty) Ltd. and Chalo Environmental and Sustainable Development Consultants. 2000. Kariba Dam case study, prepared as an input to the World Commission on Dams, Cape Town. Retrieved 2003 from the World Wide Web: http://www.dams.org.

Sok Appadu, S. N. and A. R. Nayamuth, editors. 1999. Initial National Communication of the Republic of Mauritius under the United Nations Framework Convention on Climate Change. Retrieved 2001 from the World Wide Web: http://www.unfccc.de/resource/docs/natc/maunc1/.

Somé, L., D. Wilkie, and J. Oglethorpe, editors. 2001. Congo Basin information series. CARPE: phase 1 results and lessons learned. Biodiversity Support Program, Washington, DC, USA.

South Africa Ministry of Environmental Affairs and Tourism. 2001. Maloti-Drakensberg Transfrontier Conservation Area established. Retrieved 2002 from the World Wide Web: http://www.environment.gov.za/NewsMedia/MedStat/2001june11/MalutiSigning_11062001.htm.

South African Government. 1998. National Water Act, No. 36 of 1998. South African Government Gazette 398(19182).

Sparks, J. S. 2002. *Paretroplus dambabe*, a new cichlid fish (Teleostei: Cichlidae) from northwestern Madagascar, with a discussion on the status of *P. petiti*. Proceedings of the Biological Society of Washington 3:62–79.

Sparks, J. S. 2003. Pantanodon. Pages 871–872 in S. M. Goodman and J. P. Benstead, editors. The natural history of Madagascar. The University of Chicago Press, Chicago, IL, USA.

Sparks, J. S. and P. N. Reinthal. 2001. A new species of *Pty-*

chochromoides from southeastern Madagascar (Teleostei: Cichlidae), with comments on monophyly and relationships of the ptychochromine cichlids. Ichthyological Explorations of Freshwaters 12:115–132.

Sparks, J. S. and W. L. Smith. 2004. Phylogeny and biogeography of the Malagasy and Australasian rainbowfishes (Teloste: Melanotaenioidei): Gondwanan vicariance and evolution in freshwater. Molecular Phylogenetics and Evolution 33(3).

Sparks, J. S. and M. L. J. Stiassny. 2003. Introduction to Madagascar's freshwater fishes. Pages 849–863 in S. M. Goodman and J. P. Benstead, editors. The natural history of Madagascar. University of Chicago Press, Chicago, IL, USA.

Spigel, R. H. and G. W. Coulter. 1996. Comparison of hydrology and physical limnology of the East African Great Lakes: Tanganyika, Malawi, Victoria, Kivu and Turkana (with references to some North American Great Lakes). Pages 103–135 in T. C. Johnson and E. O. Odada, editors. The limnology, climatology, and paleoclimatology of the East African lakes. Gordon and Breach Publishers, Amsterdam, The Netherlands.

St. Louis, V. L., C. A. Kelly, E. Duchemin, J. W. M. Rudd, and D. M. Rosenberg. 2000. Reservoir surfaces as sources of greenhouse gases to the atmosphere: a global estimate. BioScience 50(9):766–775.

Stanley, D. J. and A. G. Warne. 1998. Nile Delta in its destruction phase. Journal of Coastal Research 14(3):794–825.

Statistics and Database Administration Section MISD. 2000. Seychelles in figures. Retrieved 2001 from the World Wide Web: http://www.seychelles.net/misdstat/_Geography_Climate_History_An/_geography_climate_history_an.htm.

Stattersfield, A. J., M. J. Crosby, A. J. Long, and D. C. Wege. 1998. Endemic bird areas of the world: priorities for biodiversity conservation. BirdLife International, Cambridge, UK.

Stewart, D. J. and T. R. Roberts. 1984. A new species of dwarf cichlid fish with reversed sexual dichromatism from Lac Mai-Ndombe, Zaire. Copeia 1984(1):82–86.

Steyn, P. 1990. Soda ash Botswana, the paradox of Sua. African Wildlife 44(4):244–247.

Stiassny, M. L. J. 1996. An overview of freshwater biodiversity: with some lessons from African fishes. Fisheries 21(9):7–13.

Stiassny, M. L. J. 1997. A phylogenetic overview of the lamprologine cichlids of Africa (Teleostei, Cichlidae): a morphological perspective. South African Journal of Science 93:513–523.

Stiassny, M. L. J. 2002a. Conservation of freshwater fish biodiversity: the knowledge impediment. Verhandlungen der Gesellschaft für Ichthyologie 3:7–18.

Stiassny, M. L. J. 2002b. Revision of *Sauvagella* Bertin (Clupeidae; Pellonulinae; Ehiravini) with a description of a new species from the freshwaters of Madagascar and diagnosis of the Ehiravini. Copeia 2002(1):67–76.

Stiassny, M. L. J., P. Chakrabarty, and P. V. Loiselle. 2001. Relationships of the Madagascan cichlid genus *Paretroplus*, with description of a new species from the Betsiboka River drainage of northwestern Madagascar. Ichthyological Exploration of Freshwaters 12(1):29–40.

Stiassny, M. L. J. and M. C. C. de Pinna. 1994. Basal taxa and the role of cladistic patterns in the evaluation of conservation priorities: a view from freshwater. Pages 235–249 in P. L. Forey, C. J. Humphries, and R. I. VaneWright, editors. Systematics and conservation evaluation. Clarendon, Oxford, UK.

Stiassny, M. L. J. and A. Meyer. 1999. Cichlids of the rift lakes. Scientific American 280(2):64–69.

Stiassny, M. and N. Raminosoa. 1994. The fishes of the inland waters of Madagascar. Pages 133–149 in G. G. Teugels, J. F. Guégan, and J. J. Albaret, editors. Biological diversity in African fresh and brackish water fishes, Symposium PARADI. Annals of the Royal Central Africa Museum (Zoology) 275.

Stiassny, M. L. J. and D. M. Rodriguez. 2001. *Rheocles derhami*, a new species of freshwater rainbowfish (Atherinomorpha: Bedotiidae) from the Ambomboa River in northeastern Madagascar. Ichthyological Exploration of Freshwaters 12(2):97–104.

Stiassny, M. L. J., U. K. Schliewen, and W. J. Dominey. 1992. A new species flock of cichlid fishes from Lake Bermin, Cameroon with a description of eight new species of *Tilapia* (Labroidei: Cichlidae). Ichthyological Exploration of Freshwaters 3:311–346.

Stock, J. H. and R. Vonk. 1990. Stygofauna of the Canary Islands: a freshwater amphipod from La Gomera (Canary Islands), *Melita dulcicola* new species. Annales de Limnologie 26(1):29–38.

Stock, J. H. and R. Vonk. 1992. The first freshwater amphipod (Crustacea) from the Cape Verde islands: *Melita cognata*, new species with notes on its evolutionary scenario. Journal of African Zoology 106:273–280.

Storey, B. C. 1995. The role of mantle plumes in continental breakup: case histories from Gondwanaland. Nature 377:301–308.

Storey, M., J. J. Mahoney, A. D. Saunders, R. A. Duncan, S. P. Kelly, and M. F. Coffin. 1995. Timing of hot spot-related volcanism and the breakup of Madagascar and India. Science 267:852–855.

Streelman, J. T., R. Zardoya, A. Meyer, and S. A. Karl. 1998. Multilocus phylogeny of cichlid fishes (Pisces: Perciformes): evolutionary comparison of microsatellite and single-copy nuclear loci. Molecular Biology and Evolution 15:798–808.

Strzepek, K. M., D. N. Yates, and D. E. D. El Quosy. 1996. Vulnerability assessment of water resources in Egypt to climatic change in the Nile Basin. Climate Research 6:89–95.

Stuart, S. N., R. J. Adams, and M. D. Jenkins. 1990. Biodiversity in Sub-Saharan Africa and its islands: conservation, management and sustainable use, Occasional Papers of the IUCN Species Survival Commission No. 6. IUCN, Gland, Switzerland.

Sturmbauer, C. and A. Meyer. 1992. Genetic divergence, speciation and morphological stasis in a lineage of African cichlid fishes. Nature 358:578–581.

Subramaniam, S. P. 1992. A brief review of the status of the fisheries of the Bangweulu Basin and Kafue Flats. Pages 45–55 in R. C. V. Jeffrey, H. N. Chabwela, G. Howard, and P. J. Dugan, editors. Managing the wetlands of Kafue Flats and Bangweulu Basin. Kafue National Park, Zambia. IUCN, Gland, Switzerland.

Sullivan, J. P., S. Lavoué, and C. D. Hopkins. 2002. Discovery and phylogenetic analysis of a riverine species flock of African electric fishes (Mormyridae: Teleostei). Evolution 56:597–616.

Sunderland, T. C. H. and C. T. Tako. 1999. The exploitation of *Prunus africana* on the island of Bioko, Equatorial Guinea. A report for the People and Plants Initiative. WWF-Germany/IUCN-SSC Medicinal Plant Specialist Group, Frankfurt, Germany.

Swain, A. 2002. The Nile River Basin Initiative: too many cooks, too little broth. SAIS Review 22(2):293–308.

Szczesniak, P. A. 2001. The mineral industries of Côte d'Ivoire, Guinea, Liberia, and Sierra Leone, USGS Minerals Information Database, Africa and the Middle East. Retrieved 2001 from the World Wide Web: http://minerals.usgs.gov/minerals/pubs/country/2001/ivgvlislmyb01.pdf.

Tadesse, Z. 1988. Studies on some aspects of the biology of *Oreochromis niloticus* L. (Pisces: Cichlidae) in Lake Zwai, Ethiopia. Unpublished MS. thesis. School of Graduate studies, Addis Ababa University, Addis Ababa, Ethiopia.

Takahashi, K., Y. Terai, M. Nishida, and N. Okada. 1998. A novel family of short interspersed repetitive elements (SINEs) from cichlids: the patterns of insertion of SINEs at orthologous loci support the proposed monophyly of four major groups of cichlid fishes in Lake Tanganyika. Molecular Biology and Evolution 15:391–407.

Talbot, M. R. and T. Laerdal. 2000. The late Pleistocene–Holocene palaeolimnology of Lake Victoria, East Africa, based upon elemental and isotopic analyses of sedimentary organic matter. Journal of Palaeolimnology 23:141–164.

Talling, T. F. 1966. The annual cycle of stratification and phytoplankton growth in Lake Victoria (East Africa). Internationale Revue der Gesamten Hydrobiologie 51:545–621.

Talwar, P. K. and A. G. Jhingran. 1992. Inland fishes of India and adjacent countries. Vol. I and II. A.A. Balkema, Rotterdam, The Netherlands.

Tana and Athi Rivers Development Authority. 1982. Tana Delta Irrigation Project: feasibility study. Haskkoning B. V., The Netherlands and Mwenge International Associates, Kenya.

Teague, W. R. 1988. Conservation planning in the Eastern Cape. Pages 402–419 in M. N. Bruton and F. W. Gess, editors. Towards an environmental plan for the Eastern Cape. Rhodes University, Grahamstown, South Africa.

Tedla, S. 1973. Freshwater fishes of Ethiopia. Dissertation. Haile Selassie I University, Addis Ababa, Ethiopia.

Tedla, S. and F. Hailemeskel. 1981. Introduction and transplantation of freshwater fish species in Ethiopia. SINET: Ethiopian Journal of Science 4(2):69–72.

Teferra, G. 1987. A study on a herbivorous fish (*Oreochromis niloticus* L.) diet and its quality in two Ethiopian Rift Valley Lakes, Awassa and Zwai. Journal of Fish Biology 30:439–449.

Teferra, G. 1988. Digestive efficiency and nutrient composition gradient in the gut of *Oreochromis niloticus* L. in Lake Awassa, Ethiopia. Journal of Fish Biology 33:501–509.

Teferra, G. 1989. Stomach pH, feeding rhythm and ingestion rate in *Oreochromis niloticus* L. (Pisces: Cichlidae) in Lake Awassa, Ethiopia. Hydrobiologia 174:43–48.

Teferra, G. and C. H. Fernando. 1989. The food habits of a herbivorous fish (*Oreochromis niloticus* L.) in Lake Awassa, Ethiopia. Hydrobiologia 174:195–200.

Teugels, G. G. and J. F. Guégan. 1994. Diversité biologique des poissons d'eaux douces de la Basse-Guinée et de l'Afrique Centrale. Pages 67–85 in G. G. Teugels, J. F. Guégan, and J. J. Albaret, editors. Biological diversity in African fresh and brackish water fishes, Symposium PARADI. Annals of the Royal Central Africa Museum (Zoology) 275.

Teugels, G. G., C. Lévêque, D. Paugy, and K. Traore. 1988. Etat des connaissances sur la faune ichtyologique des bassins côtieres de Côte d'Ivoire et de l'Ouest du Ghana. Revue d'Hydrobiologie Tropicale 21:221–237.

Teugels, G. G., G. M. Reid, and R. P. King. 1992. Fishes of the Cross River Basin (Cameroon-Nigeria): taxonomy, zoogeography, ecology and conservation. Annales du Musée Royal d'Afrique Centrale (Sciences Zoologiques) 266:1–132.

Teugels, G. G. and T. R. Roberts. 1990. Description of a small distinctively coloured new species of the characoid genus *Neolebias* from the Niger Delta, West Africa (Pisces; Distichodontidae). Revue de Zoologie Africaine 104:61–67.

Teugels, G. G., J. Snoeks, L. De Vos, and J. C. Diakanou-Matongo. 1991. Les poissons du bassin inferieur du Kouilou (Congo). Pages 109–139 in R. J. Dowsett and F. Dowsett-Lemaire, editors. Flore et faune du bassin du Kouilou (Congo) et leur exploitation, 4. Tauraco Press, Liege, Belgium.

Teugels, G. G. and D. F. E. Thys van den Audenaerde. 1990. Description of a new species of *Bryconaethiops* (Pisces, Characidae) from Nigeria and Cameroon. Ichthyological Exploration of Freshwaters 1(3):207–212.

Tharme, R. E. 2000. An overview of environmental flow methodologies, with particular reference to South Africa. Pages 15–40 in J. King, R. Tharme, and M. De Villiers, editors. Environmental flow assessments for rivers: manual for the building block methodology. Water Research Commission Technology Transfer Report No. TT131/00. Water Research Commission, Pretoria, South Africa.

Thiam, A. 1996. Pesticides in the Senegal River delta. Cahiers Agriculture 5(2):112–117.

Thomas, D. S. G. and P. A. Shaw. 1991. The Kalahari environment. Cambridge University Press, Cambridge, UK.

Thomas, J. D. 1998. The Achilles heel of the snail hosts of schistosomiasis in relation to control. In H. Madsen, C. C. Appleton, and M. Chimbari, editors. Proceedings of "Workshop on Medical Malacology in Africa. Harare, Zimbabwe." Danish Bilharziasis Laboratory, Danida, Denmark.

Thompson, K. 1976. Swamp development in the headwaters of the White Nile. Pages 177–196 in J. Rzóska, editor. The Nile: biology of an ancient river, Monographiae Biologicae 19. Dr. W. Junk Publishers, The Hague, The Netherlands.

Thys van den Audenaerde, D. F. E. 1964. Revision systematique des espèces congolaises du genre *Tilapia* (Pisces, Cichlidae), Annales du Musée Royal de l'Afrique Central, 80: Sciences Zoologiques, No. 124. Musée Royal de l'Afrique Centrale, Tervuren, Belgium.

Thys van den Audenaerde, D. F. E. 1966. Les *Tilapia* (Pisces, Cichlidae) de Sud-Cameroun et du Gabon étude systematique, Annales du Musée Royal de l'Afrique Central, 80: Sciences

Zoologiques, 153. Musée Royal de l'Afrique Central, Tervuren, Belgium.

Thys van den Audenaerde, D. F. E. 1968. Addendum to the freshwater fishes of Fernando Po. Revue de Zoologie et de Botanique Africaines 78(1–2):123–128.

Thys van den Audenaerde, D. F. E. 1972. Description of a small *Tilapia* (Pisces, Cichlidae) from West Cameroon. Revue de Zoologie et de Botanique Africaines 85:93–98.

Ticheler, H. 2000. Fish biodiversity in West African wetlands. Wetlands International, Wageningen, The Netherlands.

Tiéga, A. 1998. La Convention sur les Zones Humides (Ramsar, Iran, 1971): mise en oeuvre en Afrique et perspective d'avenir. Pages 147–157 in A. Awaiss and S. Seyni, editors. Utilisation durable de l'eau, des zones humides et de la diversité biologique dans les écosystèmes partagés (Bénin, Burkina Faso, Niger et Togo). Actes de Séminaire Atelier Sous-Régional, OMPO/RAMSAR/DFPP, Tapoa, 16–20 Novembre 1998.

Tiéga, A. 2001. Priorities for wetland biodiversity conservation in Africa. Pages 101–105 in G. Bergkamp, J.-Y. Pirot, and S. Hostettler, editors. Integrated wetlands and water resources management. Proceedings of a workshop held at the 2nd annual conference on wetlands and development (November, 1998, Dakar, Senegal). IUCN, Gland, Switzerland.

Tilman, D., J. Fargione, B. Wolff, C. d'Antonio, A. Dobson, R. Howarth, D. Schindler, W. H. Schlesinger, D. Simberloff, and D. Swackhamer. 2001. Forecasting agriculturally driven global environmental change. Science 292:281–284.

Timberlake, J. 1997. Biodiversity of the Zambezi Basin wetlands: a review of available information, Phase 1. Draft report for the IUCN, Harare, Zimbabwe.

Timberlake, J., editor. 2000a. Biodiversity of the Zambezi Basin wetlands. Consultancy report for IUCN ROSA. Biodiversity Foundation for Africa, Bulawayo/The Zambezi Society, Harare, Zimbabwe.

Timberlake, J. 2000b. Vegetation types of various wetland areas in the Zambezi Basin. Pages 1–30 in J. Timberlake, editor. Biodiversity of the Zambezi Basin wetlands. Biodiversity Foundation for Africa, Bulawayo/The Zambezi Society, Harare, Zimbabwe.

Timberlake, J., R. Drummond, P. Smith, and M. Bingham. 2000. Wetland plants of the Zambezi Basin. Pages 31–81 in J. Timberlake, editor. Biodiversity of the Zambezi Basin wetlands. Biodiversity Foundation for Africa, Bulawayo/The Zambezi Society, Harare, Zimbabwe.

Tockner, K. and J. A. Stanford. 2002. Riverine flood plains: present state and future trends. Environmental Conservation 29:308–330.

Torrance, J. D. 1981. Climate handbook of Zimbabwe. Department of Meteorological Services, Harare, Zimbabwe.

Toure, M. 2002. Initiating a tri-national programme for the integrated conservation of the Nimba Mountains. Second tri-national meeting, Côte d'Ivoire, Guinea, Liberia, February 12–15, 2002, NZérékoré, Guinea. Fauna and Flora International, Conservation International, and BirdLife International, Abidjan, Côte d'Ivoire.

Tremblin, G. 2000. Autecological behaviour of *Halopeplis amplexicaulis*, a pioneer plant of inland salt lakes in western Algeria. Secheresse (Montrouge) 11:109–116.

Trewavas, E. 1936. Dr. Karl Jordan's expedition to South-West Africa and Angola: the freshwater fishes. Novitates Zoologicae 40:63–74.

Trewavas, E. 1962. Fishes of the crater lakes of the northwestern Cameroons. Bonner Zoologische Beitraege 13:146–190.

Trewavas, E. 1973. A new species of cichlid fish of rivers Quanza and Bengo, Angola, with a list of the known Cichlidae of these rivers and a note on *Pseudocrenilabrus natalensis*. Fowler. Bulletin of the British Museum of Natural History 25:27–38.

Trewavas, E. 1974. The freshwater fishes of Rivers Mungo, Meme and Lakes Kotto, Mboandong and Soden, West Cameroon. Bulletin of the British Museum of Natural History (Zoology) 26:329–419.

Trewavas, E. 1983. Tilapiine fishes of the genera *Sarotherodon*, *Oreochromis* and *Danakilia*. British Museum (Natural History), London.

Trewavas, E., J. Green, and S. A. Cobet. 1972. Ecological studies on crater lakes in West Cameroon, fishes of Barombi Mbo. Journal of the Zoological Society of London 167:41–95.

Truswell, J. F. 1977. The geological evolution of South Africa. Purnell, Cape Town, South Africa.

Tudorancea, C., R. M. Baxter, and C. H. Fernando. 1989. A comparative limnological study of zoobenthic associations in lakes of the Ethiopian rift valley. Archiv für Hydrobiologie Supplement 83:121–174.

Tudorancea, C., C. H. Fernando, and J. C. Paggi. 1988. Food and feeding ecology of *Oreochromis niloticus* (Linnaeus, 1758) juveniles in Lake Awassa (Ethiopia). Archiv für Hydrobiologie Supplement 79:267–289.

Tudorancea, C., Z. GebreMariam, and E. Dadebo. 1999. Limnology in Ethiopia. Pages 63–118 in R. G. Wetzel and B. Gopal, editors. Limnology in developing countries, 2. International Association for Limnology, New Delhi, India.

Tudorancea, C. and A. D. Harrison. 1988. The benthic communities of the saline lakes Abijata and Shala (Ethiopia). Hydrobiologia 158:117–123.

Tudorancea, C. and A. Zullini. 1989. Associations and distribution of benthic nematodes in the Ethiopian Rift Valley lakes. Hydrobiologia 179:81–96.

Tuite, E. H. 1979. Population size, distribution and biomass density of lesser flamingo in the Eastern African Rift Valley. Journal of Applied Ecology 16:765–775.

Turner, B. F., L. R. Gardner, W. E. Sharp, and E. R. Blood. 1996. The geochemistry of Lake Bosumtwi, a hydrologically closed basin in the humid zone of tropical Ghana. Limnology and Oceanography 41(7):1415–1424.

Turner, G. F. 1994. Fishing and the conservation of the endemic fishes of Lake Malawi. Ergebnisse der Limnologie 44:481–494.

Turner, G. F. 1995. Management, conservation and species changes of exploited fish stocks in Lake Malawi. Pages 365–395 in T. J. Pitcher and P. J. B. Hart, editors. The impact of species changes in African lakes. Chapman and Hall, London, UK.

Turner, G. F. 1996. Offshore cichlids of Lake Malawi. Cichlid Press, Lauenau, Germany.

Turner, G. F. 1999. What is a fish species? Reviews in Fish Biology and Fisheries 9:281–297.

Turner, G. F., O. Seehausen, M. E. Knight, C. J. Allender, and R. L. Robinson. 2001. How many species of cichlid fishes are there in African lakes? Molecular Ecology 10:793–806.

Turner, J. L. 1981. Changes in multi-species fisheries when many species are caught at the same time. CIFA Tech. Pap./Doc. Tech. Comité des Pêches Continentales pour l'Afrique 8:201–211.

Turpie, J., B. Smith, L. Emerton, and B. Barnes. 1999. Economic value of the Zambezi Basin wetlands. IUCN Regional Office in Southern Africa, Cape Town, South Africa.

Tweddle, D. 1979. The zoogeography of the fish fauna of the Lake Chilwa Basin. Pages 177–182, 440–441 in M. Kalk, A. J. McLachlan, and C. Howard-Williams, editors. Lake Chilwa: studies of change in a tropical ecosystem. Monographiae Biologicae, 35. Dr. W. Junk, The Hague, The Netherlands.

Tweddle, D. 1983. The fish and fisheries of Lake Chiuta. Luso: J. Sci. Tech. (Malawi) 4(2):55–81.

Tweddle, D. 1985. The importance of the national parks, game reserves and forest reserves of Malawi to fish conservation and fisheries management. Nyala 11:5–11.

Tweddle, D. 1996. Fish survey of Nkhotakota Wildlife Reserve. JLB Smith Institute of Ichthyology Investigational Report 53:1–79.

Tweddle, D., D. S. C. Lewis, and N. G. Willoughby. 1979. The nature of the barrier separating the Lake Malawi and Lower Zambezi fish faunas. Ichthyological Bulletin of the J. L. B. Smith Institute of Ichthyology 39:1–10.

Tweddle, D., R. D. Makwinja, and G. Sodzapanja. 1995. Catch and effort data for the fisheries of the Lower Shire River and associated marshes. Malawi Fisheries Bulletin 31:1–49.

Tweddle, D. and P. H. Skelton. 1998. Two new species of Varicorhinus (Teleostei: Cyprinidae) from the Ruo River, Malawi, Africa, with a review of other southern African *Varicorhinus* species. Ichthyological Exploration of Freshwaters 8:369–384.

Tweddle, D. and N. G. Willoughby. 1979. An annotated checklist of the fish fauna of the River Shire south of Kapachira Falls, Malawi. Ichthyological Bulletin of Rhodes University 39:11–22.

Twongo, T., F. W. B. Bugenyi, and F. Wanda. 1995. The potential for further proliferation of water hyacinth in lakes Victoria, Kyoga, and Kwania and some urgent aspects of research. Africa Journal of Tropical Hydrobiology and Fisheries 6:1–10.

Tyler, S. J. and D. R. Bishop. 1998. Important bird areas of Botswana. Pages 103–122 in K. N. Barnes, editor. The important bird areas of southern Africa. BirdLife South Africa, Johannesburg, South Africa.

Tyler, S. J. and D. R. Bishop. 2001. Botswana. Pages 99–112 in L. D. C. Fishpool and M. I. Evans, editors. Important bird areas in Africa and associated islands: priority sites for conservation. Pisces Publications and BirdLife International (BirdLife Conservation Series No. 11), Newbury and Cambridge, UK.

UNDP/World Bank Energy Sector Management Assistance Program. 2001. Africa Gas Initiative: Angola, Vol. II. World Bank, Washington, DC, USA.

UNEP. 1996. Groundwater: a threatened resource. UNEP Environment Library No. 15. UNEP, Nairobi, Kenya.

UNEP. 1998. Island directory. Retrieved 2002 from the World Wide Web: http://www.unep.ch/islands/CMP.htm.

UNEP. 1999. Regional overview of land-based sources and activities affecting the coastal and associated freshwater environment in the West and Central African region. Regional Seas Reports and Studies No. 171. UNEP/GPA Co-ordination Office and West and Central Africa Action Plan, Regional Coordinating Unit, The Hague, The Netherlands, and Abidjan, Côte d'Ivoire.

UNEP. 2000a. Environmental impact of refugees in Guinea. UNEP, Regional Office for Africa, Nairobi, Kenya.

UNEP. 2000b. Global environment outlook 2000: GEO-2000. UNEP's Millennium Report on the Environment. Earthscan, London, UK.

UNESCO. 1999. Report of the 23rd session of the convention concerning the protection of the world cultural and natural heritage, World Heritage Committee, 29 November–4 December 1999 in Marrakesh, Morocco. World Heritage, Paris, France.

UNESCO. 2001. UNESCO-MAB biosphere reserve directory. Retrieved 2001 from the World Wide Web: http://www2.unesco.org/mab/br/brdir/directory/database.asp.

United Nations. 2001. Report of the panel of experts on the illegal exploitation of natural resources and other forms of wealth of the Democratic Republic of Congo. Retrieved 2002 from the World Wide Web: http://www.un.org/Docs/sc/letters/2001/357e.pdf.

United Nations. 2002. Report of the World Summit on Sustainable Development: Johannesburg, South Africa 26 August–4 September 2002. United Nations, New York, NY, USA.

United Nations Commission on Sustainable Development. 1997. Natural resource aspects of sustainable development in Guinea-Bissau: desertification and drought. Retrieved 2001 from the World Wide Web: http://www.un.org/esa/agenda21/natlinfo/countr/guineab/natur.htm#desert.

United Nations Population Division. 1999. World population prospects: the 1998 revision, Vol. 1. United Nations, New York, NY, USA.

United Nations Population Division. 2001. World population prospects: population database. Retrieved 2002 from the World Wide Web: http://esa.un.org/unpp/.

Uschakov, P. V. 1970. Observation sur la repartition de la faune benthique du littoral guinéen. Cahier de Biologie Marine 11:435–457.

U.S. Committee for Refugees. 2002. Worldwide refugee information—country report: Tanzania. Retrieved 2003 from the World Wide Web: http://www.refugees.org/world/countryrpt/africa/tanzania.htm.

U.S. Department of Agriculture. 1995. National resources inventory glossary. Retrieved 2004 from the World Wide Web: http://www.ftc.nrcs.usda.gov/doc/nri/all_toc.html.

U.S. Geological Survey. 1999. Paleontology glossary of terms. Retrieved 2004 from the World Wide Web: http://geology.er.usgs.gov/paleo/glossary.shtml.

U.S. Geological Survey. 2001a. Earthshots: Lake Chad, West Africa. Retrieved 2001 from the World Wide Web: http://edc.usgs.gov/earthshots/slow/LakeChad/LakeChadtext.

U.S. Geological Survey. 2001b. USGS Minerals Information

Database, Africa and the Middle East. Retrieved 2001 from the World Wide Web: http://minerals.usgs.gov/minerals/pubs/country/africa.html#cg.

Van den Briel, J. and R. Brouwer. 1987. Fuelwood scarcity and forest policy on Santo Antao, Cape Verde (Atlantic Ocean). Netherlands Journal of Agricultural Science 35:81–83.

van der Kamp, J. and M. Diallo. 1999. Suivi écologique du Delta Intérieur du Niger: les oiseaux d'eau comme bio-indicateurs. Recensements crue 1998–1999. Malipin Publication 99-02. Wetlands International and Altenburg & Wymenga, Sevaré, Mali and Veenwouden, The Netherlands.

van der Kamp, J., M. Diallo, W. Altenburg, and E. Wymenga. 2002a. Colonies nicheuses d'oiseaux d'eau. Pages 163–186 in E. Wymenga, B. Kone, and L. Zwarts, editors. Le delta intérieur du fleuve Niger: ecologie et gestion durable des resources naturelles. Wetlands International, RIZA, and Altenburg & Wymenga Ecological Consultants, Sévaré, Mali; Lelystad, The Netherlands; and Veenwouden, The Netherlands.

van der Kamp, J., M. Diallo, and B. Fofana. 2002b. Dynamique des populations des oiseaux d'eau. Pages 87–139 in E. Wymenga, B. Kone, and L. Zwarts, editors. Le delta intérieur du fleuve Niger: ecologie et gestion durable des resources naturelles. Wetlands International, RIZA, and Altenburg & Wymenga Ecological Consultants, Sévaré, Mali; Lelystad, The Netherlands; and Veenwouden, The Netherlands.

van der Kamp, J., M. Diallo, B. Fofana, and B. Kone. 2001. Bio-indicateurs dans le delta intérieur du fleuve Niger: suivi de populations d'oiseaux d'eau par dénombrements aériens 1999–2001. Mali-PIN Publication 01-03. Wetlands International and Altenburg & Wymenga, Sévaré, Mali and Veenwouden, The Netherlands.

van der Linde, H., J. Oglethorpe, T. Sandwith, D. Snelson, and Y. Tessema. 2001. Beyond boundaries: transboundary natural resource management in sub-Saharan Africa. Biodiversity Support Program, Washington, DC, USA.

van der Waal, B. C. W. 1991. Fish life of the Oshana Delta in Owambo, Namibia, and the translocation of Cunene species. Madoqua 17(2):201–209.

van der Waal, B. C. W. 1996. Some observations on fish migrations in Caprivi, Namibia. Southern African Journal of Aquatic Sciences 22:62–80.

van der Waal, B. C. W. 2000a. Fish as a resource in a rural river catchment in the Northern Province, South Africa. African Journal of Aquatic Science 25:56–70.

van der Waal, B. C. W. 2000b. Fish resource management study of the Omadhyia Wetland Complex, Oshana Region. Northern Namibia Environmental Project, Ministry of Environment and Tourism, Windhoek, Namibia.

van der Waal, B. C. W. and R. Bills. 2000. *Oreochromis niloticus* (Teleostei: Cichlidae) now in the Limpopo River system. South African Journal of Science 96(1):47–48.

van der Waal, B. C. W. and P. H. Skelton. 1984. Check list of the fishes of Caprivi. Madoqua 13:303–320.

Van Nieuwenhuizen, G. D. P. 2000. Chapter 11: conservation status of freshwater ecosystems. Pages 117–154 in G. D. P. Van Nieuwenhuizen and J. A. Day, editors. Cape Action Plan for the Environment: the conservation of freshwater ecosystems in the Cape Floral Kingdom. University of Cape Town, Cape Town, South Africa.

Van Nieuwenhuizen, G. D. P. and J. A. Day, editors. 2000a. Cape Action Plan for the Environment: the conservation of freshwater ecosystems in the Cape Floral Kingdom. Freshwater Research Unit, University of Cape Town, Cape Town, South Africa.

Van Nieuwenhuizen, G. D. P. and J. A. Day. 2000b. Chapter 2: freshwater ecosystems in the Cape Floral Kingdom. Pages 5–17 in G. D. P. Van Nieuwenhuizen and J. A. Day, editors. Cape Action Plan for the Environment: the conservation of freshwater ecosystems in the Cape Floral Kingdom. University of Cape Town, Cape Town, South Africa.

Van Rooyen, T. 1984. The soils of the Kalahari Gemsbok National Park. Supplement to Koedoe 1984:45–61.

van Someren, V. D. 1952. The biology of the trout in Kenya Colony. Government Printer, Nairobi, Kenya.

Van Thielen, R., O. Ajuonu, V. Schade, P. Neuenschwander, A. Adite, and C. J. Lomer. 1994. Importation, releases, and establishment of *Neochetina* spp. (Col.: Curculionidae) for the biological control of water hyacinth, *Eichhornia crassipes* (Lil.: Pontederiaceae), in Bénin, West Africa. Entomophaga 39(2):179–188.

van Wilgen, B. W., R. M. Cowling, and C. J. Burgers. 1996. Valuation of ecosystem services. BioScience 46(3):184–189.

Van Wyk, P. and E. Le Riche. 1984. The Kalahari Gemsbok National Park: 1931–1981. Supplement to Koedoe 1984:21–31.

Van Zalinge, N., T. S. Nao Thuok, and D. L. Tana. 2000. Where there is water, there is fish? Cambodian fisheries issues in a Mekong River Basin perspective. Pages 37–48 in M. Ahmed and P. Hirsch, editors. Common property in the Mekong: issues of sustainability and subsistence. ICLARM Studies and Reviews 26. ICLARM, Penang, Malaysia.

Vareschi, E. and J. Jacob. 1985. The ecology of Lake Nakuru: synopsis of production and energy flow. Oecologia 65:412–424.

Vari, R. P. and L. R. Malabarba. 1998. Neotropical ichthyology: an overview. Pages 1–11 in L. R. Malabarba, R. E. Reis, R. P. Vari, Z. M. S. Lucena, and C. A. S. Lucena, editors. Phylogeny and classification of neotropical fishes. EDIPURCRS, Porto Alegre, Brazil.

Vas, A. C. and A. L. Pereira. 2000. The Inkomati and Limpopo international river basins: a view from downstream. Water Policy 2:99–112.

Vega, J. B. I. 1998. Naturaleza de España: un tesoro para el año 2000. WWF/Adena, Madrid, Spain.

Venema, H. D., E. J. Schiller, K. Adamowski, and J. Thizy. 1997. A water resources planning response to climate change in the Senegal River Basin. Journal of Environmental Management 49:125–155.

Vennie, J., compiler. 2004. The North American Lake Management Society lake and water word glossary. Retrieved 2004 from the World Wide Web: http://www.nalms.org/glossary/glossary.htm.

Vergara, R. R. 1980. Principales caracteristicas de la ictiofauna dulceacuicola cubana. Ciencias Biologicas 5:95–106.

Verheyen, E., L. Rüber, J. Snoeks, and A. Meyer. 1996. Mitochondrial phylogeography of rock-dwelling cichlid fishes reflect historical lake level fluctuations in Lake Tanganyika. Philosophical Transactions of the Royal Society London 351(1341):797–805.

Verheyen, R. 1939. Notes sur la faune ornithologique de l'Afrique Central. Bulletin du Musée Royal d'Histoire Naturelle XV:61.

Vernon, C. J. 1986. A preliminary account of the avifauna of the Karoo Biome. Bontebok 5:52–64.

Verschuren, D. 2003. The heat on Lake Tanganyika. Nature 424:731–732.

Vesey-FitzGerald, D. 1964. Mammals of the Rukwa Valley. Tanganyika Notes and Records 62:1–12.

Vesey-FitzGerald, D. and J. S. S. Beesley. 1960. An annotated list of the birds of the Rukwa Valley. Tanganyika Notes and Records 54:91–110.

Vörösmarty, C. J., A. Askew, W. Grabs, R. G. Barry, C. Birkett, P. Döll, W. Grabs, A. Hall, R. Jenne, L. Kitaev, J. Landwehr, M. Keeler, G. Leavesley, J. Schaake, K. Strzepek, S.-S. Sundarvel, K. Takeuchi, and F. Webster. 2001. Global water data: a newly endangered species. Eos, Transactions, American Geophysical Union 82(5):54–58.

Wanzie, C. S. 1990. Water resources management and wildlife conservation in the Lake Chad Basin. Mammalia 54(4):579–585.

Wass, P. 1995. Kenya's indigenous forests: status, management and conservation. IUCN Forest Conservation Programme, Gland, Switzerland and Cambridge, UK.

Waterbury, J. 2002. The Nile Basin: national determinants of collective action. Yale University Press, New Haven, CT, USA.

Watson, R. T., C. Z. Marufu, and R. H. Moss. 1997. IPCC special report on the regional impacts of climate change: an assessment of vulnerability. Retrieved 2002 from the World Wide Web: http://grida.no/climate/ipcc/regional/019.htm#box2-5.

WCMC. 1984a. World Heritage Sites: Aldabra Atoll. Retrieved 2001 from the World Wide Web: http://www.wcmc.org.uk/protected_areas/data/wh/aldabra.html.

WCMC. 1984b. World Heritage Sites: Garamba. Retrieved 2001 from the World Wide Web: http://www.wcmc.org.uk/protected_areas/data/wh/garamba.html.

WCMC. 1985. Protected areas database: Cross River National Park. Retrieved 2002 from the World Wide Web: http://www.unep-wcmc.org/protected_areas/data/sample/0405p.htm.

WCMC. 1993. Ecologically sensitive sites in Africa. Volume III: South-Central Africa and Indian Ocean. The World Bank, Washington, DC.

WCMC. 2001a. Tree conservation database: São Tomé e Príncipe. Retrieved 2002 from the World Wide Web: http://www.ggcg.st/botany/trees_stp.htm.

WCMC. 2001b. Vallée de Mai Nature Reserve. Retrieved 2001 from the World Wide Web: http://www.wcmc.org.uk/protected_areas/data/wh/mai.html.

WDPA Consortium. 2003. 2003 World Database on Protected Areas. IUCN/UNEP, Washington, DC, USA.

Welcomme, R. L. 1972. The inland waters of Africa. CIFA Technical Paper No. 1. FAO, Rome, Italy.

Welcomme, R. L. 1979. Fisheries ecology of floodplain rivers. Longman, London, UK.

Welcomme, R. L. 1985. River fisheries, FAO Fisheries Technical Paper No. 262. FAO, Rome, Italy.

Welcomme, R. L. 1986a. Fish of the Niger system. Pages 25–48 in B. R. Davies and K. F. Walker, editors. The ecology of river systems. Dr. W. Junk Publishers, Dordrecht, The Netherlands.

Welcomme, R. L. 1986b. The Niger River system. Pages 9–23 in B. R. Davies and K. F. Walker, editors. The ecology of river systems. Dr. W. Junk Publishers, Dordrecht, The Netherlands.

Welcomme, R. L. 1988. International introductions of inland fish species, FAO Fisheries Technical Paper No. 294. FAO, Rome, Italy.

Welcomme, R. L. 1999. A review of a model for qualitative evaluation of exploitation levels in multi-species fisheries. Fisheries Management and Ecology 6:1–19.

Welcomme, R. L. 2001. Inland fisheries: ecology and management. Blackwell Science, Oxford, UK.

Welcomme, R. L. and B. De Mérona. 1988. Fish communities of rivers. Pages 251–276 in C. Lévêque, M. Bruton, and G. Ssentongo, editors. Biology and ecology of African freshwater fishes. ORSTOM, Paris, France.

Welcomme, R. L. and A. Halls. 2001. Some considerations of the effects of differences in flood patterns on fish populations. Ecohydrology and Hydrobiology 1:313–323.

Welcomme, R. L. and A. Halls. 2004. Dependence of tropical river fisheries on flow. In R. L. Welcomme and T. Petr, editors. Proceedings of the Second International Symposium on the Management of Large Rivers for Fisheries, Volume. Mekong River Commission, Phnom Penh, Cambodia.

Wenger, S. 1999. A review of the scientific literature on riparian buffer width, extent and vegetation. Office of Public Service and Outreach, Institute of Ecology, University of Georgia, Athens, GA, USA.

West, K. 2001. Lake Tanganyika: results and experiences of the UNDP/GEF conservation initiative (RAF/92/G32) in Burundi, D.R. Congo, Tanzania, and Zambia. Lake Tanganyika Biodiversity Project.

Westphal, E. 1975. Agriculture systems in Ethiopia. Joint publication of the College of Agriculture, Haile Selassie I University, Ethiopia and the Agricultural University, Wageningen, The Netherlands.

Wetlands International. 2002. Ramsar Sites Database: a directory of wetlands of international importance. Retrieved 2003 from the World Wide Web: http://www.wetlands.org/RDB/Ramsar_Dir/_COUNTRIES.htm.

Wetlands International. 2004. Ramsar Sites Database Service: maps and graphs: Africa. Retrieved 2004 from the World Wide Web: http://www.wetlands.org/RSDB/default.htm.

White, F. 1983. The vegetation of Africa, a descriptive memoir to accompany the UNESCO/AETFAT/UNSO vegetation map of Africa. Natural Resources Research 20:1–356. UNESCO, Paris, France.

Whitehead, P. J. P. 1959. Notes on collection of fishes from the Tana River below Garissa, Kenya. Journal of East Africa Natural History Society 23(101):167–171.

Whitehead, P. J. P. 1960. The river fishes of Kenya, Part II: the

lower Athi (Sabaki) River. East African Agricultural Journal 25(4):259–265.
Whitehead, P. J. P. 1962. Two new fishes from eastern Kenya. British Museum of Natural History, London, UK.
Whitehead, P. J. P. and P. H. Greenwood. 1959. Mormyrid fishes of the genus *Petrocephalus* in eastern Africa, with a description of *Petrocephalus gliroides* (Vinc.). Revue de Zoologie et de Botanique Africaines 60:1–3.
Whitfield, A. K. and S. J. M. Blaber. 1978–1979a. Feeding ecology of piscivorous birds at Lake St. Lucia. Part 1: diving birds. Ostrich 49:185–198.
Whitfield, A. K. and S. J. M. Blaber. 1978–1979b. Feeding ecology of piscivorous birds at Lake St. Lucia. Part 2: wading birds. Ostrich 50:1–9.
Whitfield, A. K. and S. J. M. Blaber. 1978–1979c. Feeding ecology of piscivorous birds at Lake St. Lucia. Part 3: swimming birds. Ostrich 50:10–20.
Whitfield, A. K. and D. P. Cyrus. 1978. Feeding succession and zonation of aquatic birds at False Bay, Lake St. Lucia. Ostrich 49:8–15.
WHO and UNICEF Joint Monitoring Programme for Water Supply and Sanitation. 2000. Global water supply & sanitation sector assessment 2000, part II: Bénin. Retrieved 2002 from the World Wide Web: http://www.whoafro.org/wsh/countryprofiles/coted'ivoire.pdf.
Wikramanayake, E., E. Dinerstein, C. J. Loucks, D. M. Olson, J. Morrison, J. L. Lamoreux, M. McKnight, and P. Hedao. 2002. Terrestrial ecoregions of the Indo-Pacific: a conservation assessment. Island Press, Washington, DC, USA.
Wild, H. and L. A. G. Barboza. 1967. Vegetation map of the Flora Zambesiaca area. Supplement to Flora Zambesiaca. M.O. Collins, Salisbury, Rhodesia.
Wildlife Conservation Society–Congo. 2001. Lac Télé Reserve. Retrieved 2002 from the World Wide Web: http://www.wcs-congo.org/lac.htm.
Wilkie, D. S., J. F. Carpenter, and Q. Zhang. 2001. The under-financing of protected areas in the Congo Basin: so many parks and so little willingness-to-pay. Biodiversity and Conservation 10:691–709.
Wilkie, D. S., J. G. Sidle, and G. C. Boundzanga. 1992. Mechanized logging, market hunting and a bank loan in Congo. Conservation Biology 6(4):570–580.
Williams, J. E., J. E. Johnson, D. A. Hendrickson, S. Contreras-Balderas, J. D. Williams, M. Navarro-Mendoza, D. E. McAllister, and J. E. Deacon. 1989. Fishes of North America endangered, threatened, or of special concern: 1989. Fisheries 14(6):2–20.
Williams, J. E. and R. R. Miller. 1990. Conservation status of the North American fish fauna in fresh water. Journal of Fish Biology 37(Suppl. A):79–85.
Williams, M. A. J., P. M. Bishop, and F. M. Dakin. 1977. Late quaternary lake levels in southern Afar and the adjacent Ethiopian Rift. Nature 267:690–693.
Williams, R. 1971. Fish ecology of the Kafue River and flood plain environment. Fisheries Research Bulletin of Zambia 5:305–330.
Wilson, E. O. 1992. The diversity of life. Harvard University Press, Cambridge, MA, USA.
Wilson, J. G. M. 1988. Check-list of dragonflies of Mulanje. Unpublished manuscript.
Winemiller, K. O. 1991. Comparative ecology of *Serranochromis* species (Teleostei: Cichlidae) in the upper Zambezi River floodplain. Journal of Fish Biology 39:617–639.
Winemiller, K. O. and L. C. Kelso-Winemiller. 1994. Comparative ecology of the African pike, *Hepsetus odoe,* and tigerfish, *Hydrocynus forskahlii,* in the Zambezi River floodplain. Journal of Fish Biology 45:211–225.
Winemiller, K. O. and L. C. Kelso-Winemiller. 1996. Comparative ecology of catfishes of the upper Zambezi River floodplain. Journal of Fish Biology 49:1043–1061.
Winkelmann, D. L. 1998. CGIAR Activities and goals: tracing the connections. Issues in Agriculture. The Consultative Group for International Agricultural Research, World Bank, Washington, DC, USA.
Wishart, M. J. and B. R. Davies. 1998. The increasing divide between first and third worlds: science, collaboration and conservation of third world aquatic ecosystems. Freshwater Biology 39:557–567.
Wishart, M. J. and J. A. Day. 2003. Endemism in the freshwater fauna of the South-Western Cape, South Africa. Verhandlungen Internationale Vereinigung Limnologie 28:1762–1766.
Wishart, M. J., J. Gagneur, and H. T. El-Zanfaly. 2000. River conservation in North Africa and the Middle East. Pages 127–154 in P. J. Boon, B. R. Davies, and G. E. Petts, editors. Global perspectives on river conservation: science, policy, and practice. John Wiley & Sons, Chichester, UK.
Wissmar, R. C. and R. L. Beschta. 1998. Restoration and management of riparian ecosystems: a catchment perspective. Freshwater Biology 40:571–585.
Witte, F., T. Goldschmidt, J. Wanink, M. van Oijen, K. Goudswaard, E. Witte-Maas, and N. Bouton. 1992. The destruction of an endemic species flock: quantitative data on the decline of the haplochromine cichlids of Lake Victoria. Environmental Biology of Fishes 34:1–28.
Witte, F., P. C. Goudswaard, E. F. B. Katunzi, O. C. Mkumbo, O. Seehausen, and J. H. Wanink. 1999. Lake Victoria's ecological changes and their relationships with the riparian societies. Pages 189–202 in H. Kawnabe, G. W. Coulter, and A. C. Roosevelt, editors. Ancient lakes: their cultural and biological diversity. Kenobi Productions, Ghent, Belgium.
Witte, F., B. S. Msuku, J. H. Wanink, O. Seehausen, E. F. B. Katunzi, P. C. Goudswaard, and T. Goldschmidt. 2000. Recovery of cichlid species in Lake Victoria: an examination of factors leading to differential extinction. Reviews in Fish Biology and Fisheries 10:233–241.
Witte, F. and W. L. T. Van Densen, editors. 1995. Fish stock and fisheries of Lake Victoria: a hand book for field observations. Samara Publishing, Cardigan, UK.
Wittenberg, R. and M. J. W. Cock, editors. 2001. Invasive alien species: a toolkit of best prevention and management practices. CAB International, in collaboration with GISP, Wallingford, UK.
Woldegabriel, G., J. L. Aronson, and R. C. Walter. 1990. Geology, geochronology, and rift basin development in the central sector of the main Ethiopian rift. Geological Society of America Bulletin 102:439–458.

Wood, C. A. and R. Lovett. 1979. Rainfall reliability in Ethiopia: tables and maps. SINET: Ethiopian Journal of Science 2:111–120.

Wood, R. B., M. V. Prosser, and R. M. Baxter. 1978. Optical characteristics of the rift valley lakes, Ethiopia. SINET: Ethiopian Journal of Science 1:73–85.

Wood, R. B. and J. F. Talling. 1988. Chemical and algal relationships in a salinity series of Ethiopian inland waters. Hydrobiologia 158:29–67.

World Bank. 2000. Chad-Cameroon petroleum development and pipeline project. Retrieved 2001 from the World Wide Web: http://www.worldbank.org/pics/ifcspi/3as4338.txt.

World Bank. 2001. Biodiversity at the World Bank: supporting the web of life. Retrieved 2002 from the World Wide Web: http://lnweb18.worldbank.org/essd/essd.nsf/GlobalView/BioMagazine%208-01FOR%20WF.pdf/$File/BioMagazine%208-01FOR%20WF.pdf.

World Commission on Dams. 2000. Dams and development. Earthscan Publications, London, UK.

World Resources Institute. 2003. Watersheds of the world CD. IUCN–The World Conservation Union, the International Water Management Institute (IWMI), the Ramsar Convention Bureau, and the World Resources Institute (WRI), Washington, DC, USA.

World Water Council. 2003. The 3rd World Water Forum: final report. Secretariat of the 3rd World Water Forum, Kyoto, Japan.

Worthington, E. B. and R. H. Lowe-McConnell. 1994. African lakes reviewed: creation and destruction of biodiversity. Environmental Conservation 21(3):199–213.

Wouters, K. 2002. On the distribution of *Cyprideis torosa* (Jones) (Crustacea, Ostracoda) in Africa, with the discussion of a new record from the Seychelles. Bulletin de l'Institut Royal des Sciences Naturelles de Belgique 72:131–140.

Wranik, W. 1998. Faunistic notes on Soqotra Island. Pages 135–198 in H. J. Dumont, editor. Soqotra: proceedings of the first international symposium on Soqotra Island: present & future. United Nations Publications, New York, NY, USA.

Wranik, W. 2003. Fauna of the Socotra Archipelago: field guide. Universität Rostock, Rostock, Germany.

Wright, P. C. 1997. The future of biodiversity in Madagascar: a view from Ranomafana National Park. Pages 381–405 in S. M. Goodman and B. D. Patterson, editors. Natural change and human impact in Madagascar. Smithsonian Institution Press, Washington, DC, USA.

Wudneh, T., M. A. M. Machiels, and W. L. T. van Densen. 1999. The selective impact of a developing fishery on the fish community of Lake Tana, Ethiopia. Pages 259–280 in W. L. T. van Densen and M. J. Morris, editors. Fish and fisheries of lakes and reservoirs in Southeast Asia and Africa. Westbury, West Yorkshire, UK.

WWF-EARPO. 2003. The Albertine Rift Montane Forests ecoregion: WWF strategic framework 2004–2014. WWF East Africa Regional Programme Office, Nairobi, Kenya.

WWF International. 1996. Forests for life. Retrieved 2002 from the World Wide Web: http://www.panda.org/resources/publications/forest/report/index.htm.

WWF and IUCN. 1994. Centres of plant diversity: a guide and strategy for their conservation. 3 vols. IUCN Publications Unit, Cambridge, UK.

WWF Living Waters. 2000. The great lake that nearly vanished. Retrieved 2001 from the World Wide Web: http://www.panda.org/livingwaters/chad.html.

Wymenga, E., B. Kone, and L. Zwarts, editors. 2002. Le delta intérieur du fleuve Niger: ecologie et gestion durable des resources naturelles. Wetlands International, RIZA, and Altenburg & Wymenga Ecological Consultants, Sévaré, Mali, Lelystad, The Netherlands, and Veenwouden, The Netherlands.

Yap, K. 2002. World trade of ornamental fish. PROMPEX (Peru) Comision para la Promocion de Exportaciones. Retrieved 2003 from the World Wide Web: http://www.prompex.gob.pe/prompex/Inf_Sectorial/Pesca/WorldTra.pdf.

Zakaria-Ismail, M. 1994. Zoogeography and biodiversity of the freshwater fishes of Southeast Asia. Hydrobiologia 285:41–48.

Zambezi River Authority. 2003. Zambezi River Authority. Retrieved 2003 from the World Wide Web: http://www.zaraho.org.zm.

Zeng, N., J. D. Neelin, K. M. Lau, and C. J. Tucker. 1999. Enhancement of interdecadal climate variability in the Sahel by vegetation interaction. Science 286(5444):1537–1540.

Zhuwakinyu, M. 2002. Grand power plan for Congo River prioritised. Engineering News, July 19.

Zimmerman, D. A., D. A. Turner, and D. J. Parson. 1996. Birds of Kenya and northern Tanzania. Princeton University Press, Princeton, NJ, USA.

Zinyowera, M. C., B. P. Jallow, R. Shakespeare Maya, and H. W. O. Okoth-Ogendo. 1998. Africa. Pages 29–84 in R. T. Watson, M. C. Zinyowera, and R. H. Moss, editors. IPCC special report on the regional effects of climate change: evaluation of vulnerability. Intergovernmental Expert Group on Climate Change.

Zwarts, L. and M. Diallo. 2002. Eco-hydrologie du Delta. Pages 45–63 in E. Wymenga, B. Kone, and L. Zwarts, editors. Le delta intérieur du fleuve Niger: ecologie et gestion durable des resources naturelles. Wetlands International, RIZA, and Altenburg & Wymenga ecological consultants, Sévaré, Mali, Lelystad and Veenwouden, The Netherlands.

List of Authors

Michele L. Thieme, M.S.
Freshwater Conservation Biologist
Conservation Science Program
World Wildlife Fund–US

Robin Abell, M.S.
Freshwater Conservation Biologist
Conservation Science Program
World Wildlife Fund–US

Melanie L. J. Stiassny, Ph.D.
Axelrod Research Curator of Ichthyology
Division of Vertebrate Zoology
American Museum of Natural History
and Adjunct Professor
Department of Ecology, Evolution and
 Environmental Biology
Columbia University

Paul Skelton, Ph.D.
Managing Director
South African Institute for Aquatic Biodiversity
and Professor
Rhodes University

Bernhard Lehner, M.S.
Freshwater GIS Specialist
Conservation Science Program
World Wildlife Fund–US

Guy G. Teugels, Ph.D.
Curator of Fishes
Vertebrate Section, Laboratory of Ichthyology
Africa Museum

Eric Dinerstein, Ph.D.
Chief Scientist and Vice President for Science
Conservation Science Program
World Wildlife Fund–US

Andre Kamdem Toham, Ph.D.
Congo Basin Forest Partnership Technical Director and
 Senior Ecoregional Conservation Coordinator
WWF–Democratic Republic of the Congo Program Office
Kinshasa, DRC

Neil Burgess, Ph.D.
Senior Conservation Scientist
Conservation Science Program
World Wildlife Fund–US

David Olson, Ph.D.
Director
Conservation Science Program
World Wildlife Fund–US
Current address:
South Pacific Program
Wildlife Conservation Society

List of Contributors*

E. K. Abbam
Water Research Institute
Accra, Ghana

Aboubacar Awaiss
WWF
Western Africa Regional Programme Office (WARPO)
Niamey, Niger

S. Baha El Din
Cairo, Egypt

Christophe Béné
WorldFish Center
Africa and West Asia Programme
Cairo, Egypt

Ammar Boumbezeur
Direction Générale des Forêts
Ben Aknoun, Alger, Algérie

Ashley Brown
Conservation Science Program
WWF-US
Washington, DC, USA

David S. Brown
London Natural History Museum
London, UK

Michael Brown
Innovative Resources Management
Washington, DC, USA

Randall E. Brummett
WorldFish
c/o Humid Forest Ecoregional Center
Yaoundé, Cameroon

Harry Chabwela
University of Zambia
Lusaka, Zambia

Jonas Chafota
WWF–Southern Africa Regional Programme Office
Harare, Zimbabwe

Lauren Chapman
University of Florida
Gainesville, FL, USA

Monica Chundama
WWF Zambia Coordination Office
Lusaka, Zambia

Robert Collins
University of California
Santa Barbara, CA, USA

Helen Dallas
Freshwater Research Unit
University of Cape Town
Cape Town, South Africa

*Includes essay and ecoregion description authors attendees at initial workshop as well as ecoregion description and species list contributors and reviewers.

Tim Davenport
Wildlife Conservation Society
Mbeya, Tanzania

Belinda Day
Freshwater Research Institute
University of Cape Town
Cape Town, South Africa

Liz Day
The Freshwater Consulting Group
Cape Town, South Africa

Luc De Vos
National Museum of Kenya
Nairobi, Kenya

Samba Diallo
Centre National des Sciences Halieutiques de Boussoura
 (CNSHB)
Conakry, Guinea

Papa Samba Diouf
WWF-International
West Africa Regional Office
Dakar, Senegal

Ignacio Doadrio
Museo Nacional de Ciencias Naturales (CSIC)
Madrid, Spain

Tim Dodman
Wetlands International
Orkney, UK

Robert C. Drewes
California Academy of Sciences
San Francisco, CA, USA

Patrick Dugan
WorldFish Center
Africa and West Asia Programme
Cairo, Egypt

Jean Marc Elouard
ORSTOM
Madagascar

Lucy Emerton
IUCN: The World Conservation Union
Colombo, Sri Lanka

M. S. Farid
Water Resources Research Institute
Cairo, Egypt

Lincoln Fishpool
BirdLife International
Cambridge, UK

Angus Gascoigne
Environmental Information and Technical Services
São Tomé et Príncipe

Meg Gawler
ARTEMIS Services
Prévessin-Moëns, France

Justin Gerlach
The Nature Protection Trust of Seychelles
Mahé, Seychelles

Abebe Getahun
Department of Biology
Addis Ababa University
Addis Ababa, Ethiopia

Nathan Gichuki
University of Nairobi
Nairobi, Kenya

Cornelius J. Hazevoet
Museu e Laboratório Zoológico e Antropológico
 (Museu Bocage)
Lisboa, Portugal

Geoffrey Howard
IUCN: Eastern Africa Regional Programme, Nairobi, Kenya

IUCN Eastern Africa Regional Programme
Nairobi, Kenya

Peter Jacobson
Department of Biology
Grinnell College
Grinnell, IA, USA

Genevieve Jones
Freshwater Research Unit
University of Cape Town
South Africa

David Kaeuper
Former U.S. Ambassador to the Republic of the Congo

Lucy Kashaija
WWF-Tanzania Programme Office
Dar es Salaam, Tanzania

Holger Kolberg
Directorate Scientific Services
Ministry of Environment and Tourism
Windhoek, Namibia

Jeppe Kolding
Department of Fisheries and Marine Biology
University of Bergen
Bergen, Norway

Bakary Kone
Wetlands International
Projet Mali PIN
Mopti, Mali

Ad Konings
Cichlid Press
El Paso, TX, USA

Thomas Kristensen
Danish Bilharziasis Laboratory
Charlottenlund, Denmark

Philippe Lalèyè
Université Nationale du Bénin
Faculté des Sciences Agronomiques
Cotonou, Bénin

Denis Landenbergue
WWF International
Living Waters Programme
Gland, Switzerland

Marc Languy
WWF Eastern Africa Regional Programme Office
Nairobi, Kenya

Christian Lévêque
Centre National de la Recherche Scientifique (CNRS)
Paris, France

Björn Malmqvist
Department of Ecology and Environmental Science
Umea University
Umea, Sweden

Christopher Magadza
University of Zimbabwe
Kariba, Zimbabwe

Victor Mamonekene
Institut de Développement Rural
Université Marien Ngouabi-Brazzaville
Brazzaville, Congo

Jimmiel Mandima
Project Officer
African Wildlife Foundation Zambezi Heartland
Kariba, Zimbabwe

Brian Marshall
Biology Department
University of Zimbabwe
Harare, Zimbabwe

Samuel Matagi
WWF
Eastern Africa Regional Programme Office
Nairobi, Kenya

Miranda Mockrin
Conservation Science Program
WWF-US
Washington, DC, USA

Ketlhatlogile Mosepele
University of Botswana–HOORC
Maun, Botswana

Musonda Mumba
University College
London, UK

Henry Mwima
Heartland Coordinator
African Wildlife Foundation Four Corners Heartland
Kasane, Botswana

Leo Nagelkerke
Wageningen University
Wageningen, The Netherlands

Emmanuel Obot
Nigerian Conservation Foundation
Lagos, Nigeria

Andy Osei Okrah
Friends of Rivers and Water Bodies
Kumasi, Ghana

Mike K. Oliver
USA

Dalmas Oyugi
Kenya National Museum
Nairobi, Kenya

Carolyn (Tally) G. Palmer
Institute for Water Research
Rhodes University
Grahamstown, South Africa

Emily Peck
Conservation Science Program
WWF-US
Washington, DC, USA

C. Bruce Powell
Rivers State University of Science and Technology
Port Harcourt, Nigeria

Gordon McGregor Reid
North of England Zoological Society
Zoological Gardens
Chester, UK

Anthony. J. Ribbink
South African Institute for Aquatic Biodiversity
Grahamstown, South Africa

Roger Safford
BirdLife International
Cambridge, UK

Michael J. Samways
University of Stellenbosch
Stellenbosch, South Africa

Julius Sarmett
Pangani Basin Water Office
Ministry of Water and Livestock Development
Hale-Tanga, Tanzania

Robert Schelly
Cornell University and American Museum
 of Natural History
New York, NY, USA

Uli Schliewen
Zoologische Staatssammlung München
Munich, Germany

Lucy Scott
South African Institute for Aquatic Biodiversity
Grahamstown, South Africa

Lothar Seegers
Dinslaken, Germany

Ole Seehausen
Department of Biological Sciences
University of Hull
Hull, UK

Jos Snoeks
Africa Museum
Tervuren, Belgium

John S. Sparks
Department of Ichthyology
American Museum of Natural History
New York, NY, USA

John Sullivan
Cornell University
Ithaca, NY, USA

Jean-Jacques Symoens
retired, University of Brussels
Brussels

D. Thys van den Audenaerde
retired, Africa Museum
Tervuren, Belgium

Louis Tsague
Garova Wildlife College
Garova, Cameroon

Ojei Tunde
Nigerian Conservation Foundation
Lagos, Nigeria

Denis Tweddle
South African Institute for Aquatic Biodiversity
Grahamstown, South Africa

Ben C. W. van der Waal
University of Venda for Science and Technology
Thohoyandou, South Africa

Fiesta Warinwa
Africa Wildlife Foundation
Nairobi, Kenya

Robin Welcomme
Imperial College
Suffolk, UK

David Wilkie
Boston University
Boston, MA, USA

Christopher E. Williams
WWF International
Living Waters Programme
Washington, DC, USA

Emmanuel Williams
Wetlands International
Dakar, Senegal

Wolfgang Wranik
University of Rostock
Rostock, Germany

Eddy Wymenga
Altenburg & Wymenga
Ecological Consultants
The Netherlands
Zoological Museum of the University of Copenhagen,
  Denmark

Index

Abstractions, 120
African Wildlife Foundation (AWF), 53–55, 57, 123, 124, 128
Agriculture, 5, 6, 13, 113, 131, 132, 136
Albertine Highlands, 244–46
Alien species, *see* Invasive alien species
Amatolo-Winterberg Highlands, 246–48
Anabantoids, 134
Annobón, 271–72
Aquatic habitat threats, *see* Threats
Aquatic resource ecological monitoring methods, 55
Aquatic Resources Working Group (ARWG), 127
Arum lily (reed), 38
Ashanti, 211–12

Bangweulu-Mweru, 94, 185–86
Barombi Mbo, 58–60
Barotse Floodplain, Zambia:
    modeling alternative wetland conservation and development scenarios for, 14, 16
Basin projects with ecosystem focus, 104
Basins, 2
    closed, 25, 26, 94
    conservation planning in, 103–4
    visionary work in, 112
    in West and Central Africa, contributions of fisheries to, 7, 8
    see also specific basins
Bermin, Lake, 58, 59
Bight Coastal, 299–301
Bijagos, 261–62
Biodiversity:
    patterns of, by major habitat type, 165
    see also Biological distinctiveness
Biodiversity index scores and latitude, patterns of:
    by major habitat type, 171
Biodiversity visions, 105
Biological distinctiveness:
    data, 157–63
        synthesis of, 44, 46
    methods for assessing, 139–43
    statistical analyses of, 165–71
    steps for evaluating, 140
Biological Distinctiveness Index (BDI), 29–30, 43, 44
    overlap of freshwater and terrestrial, 100
    preliminary, 41, 42
Biological distinctiveness overlap, 99
Biological (species and nonspecies) distinctiveness categorization, 142–43
Biological values:
    nonspecies, 30
    species, 29–30
Bioregions, 23
    defined, 23
    freshwater, 28
Biota threats, 31, 72, 74, 146
Bird areas, important, 45

Cameroon, 58–60
    see also Waza Logone Floodplain
Campylomormyrus, 37
Canary Islands, 262–64
Catch assessment surveys (CASs), 55
Catchment degradation, 120
Catchment management, *see* Integrated river basin management
Catchments, *see specific catchments*
Catfishes, 134
Central West Coastal Equatorial, 98, 215–16
Chad, Lake, 117
    catchment, 192–95
Chilwa, Lake, 173–75
Chiuta, Lake, 173–75
Chlorolestes apricans, 21
Cichlid fish, diversity of, 4, 36, 39, 53, 97, 108, 134
    in East African lakes, 35, 48
Cichlid species flocks, 94, 96, 132
    in small Cameroonian lakes, 58–60
    understanding evolutionary radiations and, 48–51

Climate (and global) change, 86, 115, 147
    effects on freshwater ecoregions, 86–89
    limitations in simulating the future, 86–87
Community-based natural resource management, 132–33
Comoros, 265–67
Congo:
    Lower, 220–22
    Upper, 236–38
Congo Basin, fish biodiversity in, 51–53
Congo Rapids:
    Lower, 46, 52, 97, 296–97
    Upper, 297–99
Congo River, 51–52
Conservation:
    sparked by international treaty, 116–18
    and sustainable development, challenges to, 112–16
    understanding evolution can help, 51
Conservation planning in ecoregions and river basins, 103–4
Conservation status:
    data on, 157–63
        statistical analyses of, 165–71
    final (threat-modified), 33, 75–77, 91, 92, 147
    methods for assessing, 145–47
    see also Snapshot conservation status
Conservation status index (CSI), 30–33, 91, 93
    integrating biological distinctiveness and, 33–34
    integrating biological importance and, 91–93
        representation, 94
Convention on Wetlands (Ramsar), 116–18
Coralline Seychelles, 267–68
Cuanza, 301–3
Cuvette Centrale, 52, 212–15

Dams, 114
Deltas:
    inland, 47
    large river, 25, 27, 97
Dragonflies (Odonata):
    conservation action, 21
    indicator qualities, 19
    species richness of Africa fauna, 19
    threatened species, 20, 21
    threats to, 19–21
Drakensberg-Maloti Highlands, 248–51
Dry Sahel, 336–38

Eastern Coastal Basins, 320–22
Eburneo, 216–18
Ecological changes caused by species introduction, 120–21
Ecological monitoring methods, 55
Ecological phenomena, 30, 41, 42, 44, 45, 142–43
Ecoregions, 3, 22–23
    area and latitude, 170
    classification, 33–34, 94–99
    data quality, 40
    defined, 22
    globally distinctive, 46
    visionary work in, 112

Ecosystem goods and services, valuation of, 130
Ecosystems:
    conserving biodiversity and maintaining their role in livelihoods, 117–18
    degradation, counting the economic costs of, 13, 15
    restoration, providing economic justification for, 14, 17
    see also Freshwater systems
Edward, Lake, 97, 287–91
Ejagham, Lake, 58–59
Elevation zones, 4
Endemism, see under Species biological values
Environmental governance, 112–13
Ethiopian Highlands, 253–55
Etosha, 338–40
Evolution, understanding of:
    can help conservation, 51
Evolutionary phenomena, 30, 41, 42, 44, 142–43
Evolutionary radiations, 48, 50, 142
Exotic species as biota threat, 31, 74, 146
Exploitation, 31, 74, 132, 146

Faunal groups, relationships of diversity between, 168
Fish:
    and food security, 9
    as source of income and employment
        direct income and employment, 6–8
        multiplier employment effect, 7, 8
    species endemism, 39
    species richness, 36–37
    see also under Species biological values
Fish assemblages:
    length of species in, 82
    resilience, 85
Fish biodiversity investigations, highlights from, 54
Fish biodiversity surveys, 53–57
    collection methods, 54
Fish consumption, 9
Fish trade, ornamental, 133–34
    role of children in, 133
Fisheries:
    as central element of household livelihoods, 9–10
    inland
        contributions to rural livelihoods and food security, 6–11
        and overexploitation, 115
    and the poor, 10
    river, neglect of, 6
    and rural development, 9–10
    sustaining, in face of growing demand for water, 129, 130
        governance and institutions, 129–30
    in Upper and Middle Zambezi sites, 55–57
Fishers, 7
    not necessarily full-time professionals, 6–8
Fishery resources in southern African landscapes, AWF experience in management of, 123–25
    conservation targets, 125
        native fishes, 126–27
        river systems, 125–26
        wetlands, 126

Heartland conservation planning, 125
    intervention strategies, 127
        multi-institutional partnerships, 127–28
Fishes:
    Madagascar's freshwater, 62–63, 68–70
        number of, 68
        regions of occurrence, 64–67
    ornamental, as rural livelihood option, 132–34
        lessons for Africa, 134–35
    see also Cichlid fish; Cichlid species flocks
Fishing:
    cast net, 7
    impacts on inland fish populations, 81–83
        associated problems, 85
        conservation measures, 85
        indicators, 83–84
        number of species, 82
Fishing pressure, 5
Floodplains, 25, 26, 94–95, 98
    see also specific floodplains
Fold, Cape, 96, 240–42
Food security, 6–9, 79
    see also Fisheries
Forest rivers:
    dry, 98–99, 325–27
    moist, 25, 26, 95–96, 98
Four Corners TBNRM area, 123, 125, 127
Fouta-Djalon, 255–56
Freshwater ecoregions, see Ecoregions
Freshwater systems, 1, 3, 35
    evolution and diversity, 2, 4
    see also Ecosystems

General circulation models (GCMs), 86–88
George, Lake, 97, 287–91
Global Invasive Species Programme (GISP), 80
Granitic Seychelles, 268–71
Great Lakes of eastern Africa, 48–50
Guinea:
    Northern Upper, 95, 224–26
    Southern Upper, 231–32
Guinean-Congolian Forest, 105, 108
Guinean-Congolian freshwater region, 105, 107, 108

Habitat types:
    rare, 30, 44, 47, 143
    see also Major habitat types
Heartland conservation planning, 125
Heartland Conservation Policy (HCP), 122
Heartlands, 123, 124
    see also specific heartlands
Herpetofauna, aquatic, 38, 40
Highland systems, 25, 26, 96
Hippopotamus (hippopotamus amphibious), 39
Horn, 340–41
Hyperolius horstocki (frog), 38

Icthyofauna, 62–63, 69

Icthyophages, 83
Important bird areas (IBAs), 45
Infrastructure, planned, 147
Inner Niger Delta, 186–90
Integrated river basin management (IRBM), 110, 111
Integrated water resource management, 111
Interbasin transfers (IBTs), 74, 114–15
Invasive alien species (IASs), 78, 113–14
    economic impact, 79–80
    information gaps, 80
    monitoring and risk assessment, 81
    as possible threat to food security, 79
    problem of, 78
    recommendations regarding, 81
    regional and local initiatives, 80
    water resources in danger, 78–79
Island rivers and lakes, 25, 26, 98
IUCN, see World Conservation Union

Kafue, 190–92
Kafue Flats, 110–11
Kafue lechwe, 110–11
Kalahari, 341–43
    southern, 354–56
Kampala, 14
Karoo, 343–45
Karstveld Sink Holes, 334–35
Kasai, 98, 218–20
Kenyan Coastal Rivers, 303–6
Killfish (Aplocheilidae), 134
Kilum Mountain Forest, 59
Kivu, Lake, 97, 287–91

Lake Kariba Research Station (ULKRS), 127, 128
Lake Mburo National Park, Uganda, 14, 18
Lake Naivasha Management Implementation Committee (LNMIC), 121, 122
Lake Naivasha Management Plan, 121
Lake Naivasha Riparian Association (LNRA), 121, 122
Lakes, 25, 26, 94–95, 98
    diversity of cichlid fish in East African, 48
    island, 25, 26
    large, 25, 27, 96–97
    small, 25, 26, 94
    in West and Central Africa, contributions of fisheries to, 7, 8
    see also specific lakes
Land-based threats (land degradation), 31, 71, 72, 74, 145–46
Land use change and habitat loss, 113
Lanistes carinatus, 37
Latitude, 170, 171
Lesotho Highlands Water Project (LHWP), 114
Livebearers (Poeciliidae), 134
Logging, 113
Lualaba, Upper, 98, 203–5

Madagascar:
    ecoregions, 68, 69
    see also specific habitats

Index   427

Madagascar Eastern Highlands, 96, 256–58
Madagascar Eastern Lowlands, 222–23
Madagascar Northwestern Basins, 97, 311–12
Madagascar Southern Basins, 347–48
Madagascar Western Basins, 313
Maghreb:
   Permanent, 96, 242–44
   Temporary, 99, 356–58
Mai Ndombe, 195–97
Major habitat types (MHTs), 23–28, 41
   area of, 28
   conservation status, 71, 72, 74, 75
   patterns of biodiversity and, 165–67
   threats to, 72, 74
   see also Habitat types
Malagarasi-Moyowosi, 98, 197–99
Malawi/Niassa/Nyasa, Lake, 108, 109, 278–81
Malebo Pool, 223–24
Mammals, aquatic, 38–39
Management, adaptive vs. dynamic, 122
Management forums, 122
Mascarenes, 272–75
Mediterranean systems, 25, 26, 96
MedWet Initiative, 118
Metacnemis valida, 21
Mining, 113
Minnows (Cyprinidae), 134
Miocene, 2
Mollusciciding, 101–2
Mollusks, aquatic, 37–38
   species endemism, 39–40
Montagne d'Ambre National Park, 33
Mountain systems, 25, 26, 96
Mulanje, 259–61
Mutonga-Grand Falls Dam, 15

Nairobi, Kenya, see Naivasha, Lake
Naivasha, Lake, 119
   biodiversity, 119
   economy, 119–20
   forests and other vegetation, 119
   management, 121, 122
      components in management process, 121–22
   threats, 120–21
Nakivubo Swamp, Uganda, 13, 14
Namib Coastal, 348–51
Namib-Naukluft National Park, Namibia, 29
Namibia, 29, 114
National Water Act (NWA), 136
   resource protection and use, 136–38
   sustainability and governance, 138
Natural resource-based economies, resource extraction in, 113
Natural resource management, community-based, 132–33
New Partnership for Africa's Development (NEPAD), xix, 80, 117
Niassa, Lake, 108, 109
Niger, Upper, 238–40
Niger Basin Initiative (NBI), 105, 106

Niger-Benue, Lower, 306–8
Niger Delta, 60–62, 97, 291–94
   Inner, 186–90
Nigeria:
   cost of inaction in, 80
   Niger Delta in, 60–62
   see also Niger Delta
Nile, Upper, 95, 205–8
Nile Basin Initiative (NBI), 110
Nile Delta, 294–96
Nile perch (*Lates niloticus*), 74
Nile River, 114
   lower, 345–47
Nimba, Mount, 258–59
North America, comparison with, 91, 93
Northern Eastern Rift, 181–83
Northern West Coast Equatorial, 95–96, 226–28
Nyasa, Lake, 108, 109

Odonata, see Dragonflies
Oil, 113
Okavango Floodplains, 94–95, 199–201
Okavango River, 114–15
Okavango wilderness, islands and waterways of, 23
Oku, Lake, 59
Orange, western, 358–60
Oxylapia polli, 69

Pallisa District, Uganda, 13
Pangani, 316–18
Pantanodon n. sp. "manombo," 69
Phosphorus levels, total, 57
Political instability and civil unrest, 112
Pollution, 120
Population density, 71, 73
Population increase, 147
Predator-prey relationships, 84
Priority class of ecoregions, data on, 157–63
Priority class overlap, 99–101
Priority ecoregions, 94–99
Priority-setting, 90
   data quality and, 91, 93
   exercises, 99–101
   integration matrix for, 33–34, 90–91, 93
   where to act first, 91

Quaternary, 2

Rainbow fishes, 134
Ramsar Convention, 116–18
Ramsar List/Ramsar sites, 117, 118, 121
   see also Naivasha, Lake
Rapids:
   large river, 25, 27, 97
   lower Congo, 52, 97, 296–97
   upper Congo, 297–99
Rare ecological phenomena, see Ecological phenomena
Rare evolutionary phenomena, see Evolutionary phenomena

Red Sea Coastal, 351–52
River deltas, see Deltas
River rapids, see Rapids
Rivers:
   dry forest, 25, 27
   island, 25, 26
   moist forest, 25, 26
   see also Basins
Rukwa, Lake, 281–83

Sangha, 228–30
São Tomé, Príncipe, 271–72
Sauvagella n. sp. "*ambomboa*," 69
Savanna, 25, 27, 98–99
   dry forest rivers, 325–27
Schistosomiasis, 101–2
Senegal-Gambia catchments, 318–20
Shebelle-Juba catchments, 352–53
Snail-borne diseases, 101
Snail diversity, aquatic:
   safeguarding human health vs., 101–2
Snapshot conservation criteria, synthesis of, 74–75
Snapshot conservation status, 31–33, 71–72, 74–75
   calculation of the value of, 146
   determination of, 145–46
Socotra, 98, 275–78
Sossusvlei, 29
South African Institute for Aquatic Biodiversity (SAIAB), 54
South African water law, new, 136–38
Southern African Development Community (SADC), 55, 127
Southern Eastern Rift, 94, 176–80
Southern Temperate Highveld, 322–25
Southern West Coast Equatorial, 95–96, 232–34
Species, introduced (aquatic), 5
Species assemblages, 4
Species biological values:
   data quality, 40
   endemism, 29–30, 36, 39–41, 59, 139–41
      data on, 149–55, 165–69
      percentage, 40, 165–69
   relationship of ecoregion area to, 168–69
   richness, 29–30, 35–39, 41, 141
      adjusting by area, 169, 171
      data on, 149–55, 165–69, 171
Species distinctiveness categorization, 142
Species distribution and endemism data, 139–42
Species radiations, see Evolutionary radiations
Spring systems, 25, 27, 97–98
Subterranean systems, 25, 27, 97–98
Sudanic Congo, 234–36
Swamps, 25, 26, 94–95, 98
   see also specific swamps

Tana, Lake, 94, 180–81
Tana River hydroelectric scheme, Kenya, 13, 15
Tanganyika, Lake, 48–50, 96–97, 283–85
Teramulus waterloti, 69
Terrestrial and freshwater Biological Distinctiveness Index, 100

Tetras (Alestiidae), 134
Threat assessment, future, 75–76
Threat levels, future, 72
Threats, 5
   aquatic habitat, 31, 72, 74, 146
   categories of, 31–32
   likelihood of future, 147
Thysville Caves, 97–98, 335–36
Tilapias, 113, 114
Total phosphorus (TP) levels, 57
Transboundary Natural Resources Management (TBNRM), 53, 56
   see also Four Corners TBNRM area
Transboundary programs, 110
Tumba, 201–3
Turkana, Lake, 285–87
Typhleotris n. sp., 69

Uele, 98–99
Uganda Wildlife Authority, 18

Verde, Cape, 264–65
Victoria, Lake, 48–50, 97, 287–91
Volta, 327–29

Water:
   for Africa's freshwater ecosystems, competing for, 128–31
      building partnerships, 131
   democracy and, 136
   managing, for aquatic ecosystems, 130–31
Water abstractions, 120
Water hyacinth, 113, 114
   economic costs and control expenditures associated with, 79–80
Water management, 114–15
Water use, effects on freshwater ecoregions, 87–89
Waterbirds, 45
Waterbodies, infested, 78–79
Waterfalls, 33
WaterGAP 2 model, 87–88
Watershed management, see Integrated river basin management
Waza Logone Floodplain, Cameroon, 14, 17
Western Equatorial Crater Lakes, 94, 183–85
Wetland conservation:
   generating funding and incentives for, 14
   identifying needs for financial and economic incentives for, 18
Wetland economic value, broadening the definition of, 11–12
Wetland ecosystems, economic values of, 11–12
Wetland management, using valuation for, 12–13
Wetland services, economic value for urban population, 13, 14
Wetland values, integrated into development planning, 12–13
Wetlands:
   future challenges in valuing, 14–15
   role in rural economy, 13
Wetlands of International Importance, 117, 118
World Conservation Monitoring Centre (WCMC), 100

World Conservation Union (IUCN), 80, 121

Xeric systems, 25, 27, 98

Zambezi:
- Lower, 308–11
- Upper and Middle
  - fish biodiversity surveys in, 54
  - fisheries in, 55–57

Zambezi Basin, contribution of fisheries to households' cash income in, 7, 8

Zambezi Department of Fisheries (DOF), 127, 128
Zambezi Floodplains, Upper, 208–10
Zambezi Heartland, 123, 125–27
Zambezi Luangwa, Middle, 314–16
Zambezi River, 53–57, 126–27
- Middle, 56–57
- Upper, 55–56

Zambezian Headwaters, 329–30
Zambezian Lowveld, 330–33
Zambezian (Plateau) Highveld, 333–34
Zimbabwe Highlands, Eastern, 251–53

Island Press Board of Directors

VICTOR M. SHER, ESQ. *(Chair)*
Sher & Leff
San Francisco, CA

DANE A. NICHOLS *(Vice-Chair)*
Washington, DC

CAROLYN PEACHEY *(Secretary)*
Campbell, Peachey & Associates
Washington, DC

DRUMMOND PIKE *(Treasurer)*
President
The Tides Foundation
San Francisco, CA

ROBERT E. BAENSCH
Director, Center for Publishing
New York University
New York, NY

DAVID C. COLE
President
Aquaterra, Inc.
Washington, VA

CATHERINE M. CONOVER
Quercus LLC
Washington, DC

WILLIAM H. MEADOWS
President
The Wilderness Society
Washington, DC

MERLOYD LUDINGTON
Merloyd Lawrence Inc.
Boston, MA

HENRY REATH
Princeton, NJ

WILL ROGERS
President
The Trust for Public Land
San Francisco, CA

ALEXIS G. SANT
Trustee and Treasurer
Summit Foundation
Washington, DC

CHARLES C. SAVITT
President
Island Press
Washington, DC

SUSAN E. SECHLER
Senior Advisor
The German Marshall Fund
Washington, DC

PETER R. STEIN
General Partner
LTC Conservation Advisory Services
The Lyme Timber Company
Hanover, NH

DIANA WALL, PH.D.
Director and Professor
Natural Resource Ecology Laboratory
Colorado State University
Fort Collins, CO

WREN WIRTH
Washington, DC